ORGANIC SYNTHESES

ORGANIC SYNTHESES

AN ANNUAL PUBLICATION OF SATISFACTORY
METHODS FOR THE PREPARATION OF
ORGANIC CHEMICALS
VOLUME 89
2013

MARK LAUTENS
VOLUME EDITOR

The procedures in this text are intended for use only by persons with prior training in the field of organic chemistry. In the checking and editing of these procedures, every effort has been made to identify potentially hazardous steps and to eliminate as much as possible the handling of potentially dangerous materials; safety precautions have been inserted where appropriate. If performed with the materials and equipment specified, in careful accordance with the instructions and methods in this text, the Editors believe the procedures to be very useful tools. However, these procedures must be conducted at one's own risk. Organic Syntheses, Inc., its Editors, who act as checkers, and its Board of Directors do not warrant or guarantee the safety of individuals using these procedures and hereby disclaim any liability for any injuries or damages claimed to have resulted from or related in any way to the procedures herein.

For general information on our other products and services or for technical support, please contact our Customer Care Department within the United States at (800) 762-2974, outside the United States at (317) 572-3993 or fax(317) 572-4002.

Wiley also publishes its books in a variety of electronic formats. Some content that appears in print may not be available in electronic formats. For more information about Wiley products, visit our web site at www.wiley.com.

Library of Congress Catalog Card Number: 21-17747
ISBN 978-1-118-74319-5

Printed in the United States of America
10 9 8 7 6 5 4 3 2

ORGANIC SYNTHESES

*Out of print.
†Deceased.

v

*Out of print.
†Deceased.

Out of print.
†*Deceased.*

*Out of print.
†Deceased.

NOTICE

Beginning with Volume 84, the Editors of *Organic Syntheses* initiated a new publication protocol, which is intended to shorten the time between submission of a procedure and its appearance as a publication. Immediately upon completion of the successful checking process, procedures are assigned volume and page numbers and are then posted on the Organic Syntheses website (www.orgsyn.org). The accumulated procedures from a single volume are assembled once a year and submitted for publication. The annual volume is published by John Wiley and Sons, Inc., and includes an index. The hard cover edition is available for purchase through the publisher. Incorporation of graphical abstracts into the Table of Contents began with Volume 77. Annual volumes 70–74, 75–79 and 80–84 have been incorporated into five-year versions of the collective volumes of *Organic Syntheses*. Collective Volumes IX, X and XI are available for purchase in the traditional hard cover format from the publishers. Beginning with Volume 88, a new type of article, referred to as Discussion Addenda, appeared. In these articles submitters are provided the opportunity to include updated discussion sections in which new understanding, further development, and additional application of the original method are described. Organic Syntheses intends for Discussion Addenda to become a regular feature of future volumes. The Editors welcome comments and suggestions from users concerning the new editions.

Organic Syntheses, Inc., joined the age of electronic publication in 2001 with the release of its free web site (www.orgsyn.org). Organic Syntheses, Inc., fully funded the creation of the free website at www.orgsyn.org in a partnership with CambridgeSoft Corporation and Data-Trace Publishing Company. The site is accessible to most internet browsers using Macintosh and Windows operating systems and may be used with or without a ChemDraw plugin. Because of continually evolving system requirements, users should review software compatibility at the website prior to use. John Wiley & Sons, Inc., and Accelrys, Inc., partnered with Organic Syntheses, Inc., to develop the new database (www.mrw.interscience.wiley.com/osdb) that is available for license with internet solutions from John Wiley & Sons, Inc. and intranet solutions from Accelrys, Inc.

Both the commercial database and the free website contain all annual and collective volumes and indices of *Organic Syntheses*. Chemists can draw structural queries and combine structural or reaction transformation queries with full-text and bibliographic search terms, such as chemical name, reagents, molecular formula, apparatus, or even hazard warnings or phrases. The preparations are categorized into reaction types, allowing search by category. The contents of individual or collective volumes can be browsed by lists of titles, submitters' names, and volume and page references, with or without reaction equations.

The commercial database at www.mrw.interscience.wiley.com/osdb also enables the user to choose his/her preferred chemical drawing package, or to utilize several freely available plug-ins for entering queries. The user is also able to cut and paste existing structures and reactions directly into the structure search query or their preferred chemistry editor, streamlining workflow. Additionally, this database contains links to the full text of primary literature references via CrossRef, ChemPort, Medline, and ISI Web of Science. Links to local holdings for institutions using open url technology can also be enabled. The database user can limit his/her search to, or order the search results by, such factors as reaction type, percentage yield, temperature, and publication date, and can create a customized table of reactions for comparison. Connections to other Wiley references are currently made via text search, with cross-product structure and reaction searching to be added in the near future. Incorporations of new preparations will occur as new material becomes available.

INFORMATION FOR AUTHORS OF PROCEDURES

Organic Syntheses welcomes and encourages submissions of experimental procedures that lead to compounds of wide interest or that illustrate important new developments in methodology. Proposals for *Organic Syntheses* procedures will be considered by the Editorial Board upon receipt of an outline proposal as described below. A full procedure will then be invited for those proposals determined to be of sufficient interest. These full procedures will be evaluated by the Editorial Board, and if approved, assigned to a member of the Board for checking. In order for a procedure to be accepted for publication, each reaction must be successfully repeated in the laboratory of a member of the Editorial Board at least twice, with similar yields (generally $\pm 5\%$) and selectivity to that reported by the submitters.

Organic Syntheses Proposals

A cover sheet should be included providing full contact information for the principal author and including a scheme outlining the proposed reactions (an *Organic Syntheses* Proposal Cover Sheet can be downloaded at orgsyn.org). Attach an outline proposal describing the utility of the methodology and/or the usefulness of the product. Identify and reference the best current alternatives. For each step, indicate the proposed scale, yield, method of isolation and purification, and how the purity of the product is determined. Describe any unusual apparatus or techniques required, and any special hazards associated with the procedure. Identify the source of starting materials. Enclose copies of relevant publications (attach pdf files if an electronic submission is used).

Submit proposals by mail or as e-mail attachments to:

Professor Charles K. Zercher
Associate Editor, *Organic Syntheses*
Department of Chemistry
University of New Hampshire
23 College Road, Parsons Hall
Durham, NH 03824

For electronic submissions: *org.syn@unh.edu*

Submission of Procedures

Authors invited by the Editorial Board to submit full procedures should prepare their manuscripts in accord with the Instructions to Authors which are described below or may be downloaded at orgsyn.org. Submitters are also encouraged to consult this volume of *Organic Syntheses* for models with regard to style, format, and the level of experimental detail expected in *Organic Syntheses* procedures. Manuscripts should be submitted to the Associate Editor. Electronic submissions are encouraged; procedures will be accepted as e-mail attachments in the form of Microsoft Word files with all schemes and graphics also sent separately as ChemDraw files.

Procedures that do not conform to the Instructions to Authors with regard to experimental style and detail will be returned to authors for correction. Authors will be notified when their manuscript is approved for checking by the Editorial Board, and it is the goal of the Board to complete the checking of procedures within a period of no more than six months.

Additions, corrections, and improvements to the preparations previously published are welcomed; these should be directed to the Associate Editor. However, checking of such improvements will only be undertaken when new methodology is involved.

NOMENCLATURE

Both common and systematic names of compounds are used throughout this volume, depending on which the Volume Editor felt was more appropriate. The Chemical Abstracts indexing name for each title compound, if it differs from the title name, is given as a subtitle. Systematic

Chemical Abstracts nomenclature, used in the Collective Indexes for the title compound and a selection of other compounds mentioned in the procedure, is provided in an appendix at the end of each preparation. Chemical Abstracts Registry numbers, which are useful in computer searching and identification, are also provided in these appendices.

ACKNOWLEDGMENT

Organic Syntheses wishes to acknowledge the contributions of Amgen, Inc. and Merck & Co. to the success of this enterprise through their support, in the form of time and expenses, of members of the Board of Editors.

INSTRUCTIONS TO AUTHORS

All organic chemists have experienced frustration at one time or another when attempting to repeat reactions based on experimental procedures found in journal articles. To ensure reproducibility, *Organic Syntheses* requires experimental procedures written with considerably more detail as compared to the typical procedures found in other journals and in the "Supporting Information" sections of papers. In addition, each *Organic Syntheses* procedure is carefully "checked" for reproducibility in the laboratory of a member of the Board of Editors.

Even with these more detailed procedures, the experience of *Organic Syntheses* editors is that difficulties often arise in obtaining the results and yields reported by the submitters of procedures. To expedite the checking process and ensure success, we have prepared the following "Instructions for Authors" as well as a **Checklist for Authors** and **Characterization Checklist** to assist you in confirming that your procedure conforms to these requirements. Please include a completed Checklist together with your procedure at the time of submission. Procedures submitted to *Organic Syntheses* will be carefully reviewed upon receipt and procedures lacking any of the required information will be returned to the submitters for revision.

Scale and Optimization

The appropriate scale for procedures will vary widely depending on the nature of the chemistry and the compounds synthesized in the procedure. However, some general guidelines are possible. For procedures in which the principal goal is to illustrate a synthetic method or strategy, it is expected, in general, that the procedure should result in at least 5 g and no more than 50 g of the final product. In cases where the point of the procedure is to provide an efficient method for the preparation of a useful reagent or synthetic building block, the appropriate scale may be larger, but in general should not exceed 100 g of final product. Exceptions to these guidelines may be granted in special circumstances. For example, procedures describing the preparation of reagents employed as catalysts will often be acceptable on a scale of less than 5 g.

In considering the scale for an *Organic Syntheses* procedure, authors should also take into account the cost of reagents and starting materials. In general, the Editors will not accept procedures for checking in which the cost of any one of the reactants exceeds $500 for a single full-scale run. Authors are requested to identify the most expensive reagent or starting material on the procedure submission checklist and to estimate its cost per run of the procedure.

It is expected that all aspects of the procedure will have been optimized by the authors prior to submission, and that each reaction will have been carried out at least twice on exactly the scale described in the procedure. It is appropriate to report the weight, yield, and purity of the product of each step in the procedure as a range. In any case where a reagent is employed in significant excess, a Note should be included explaining why an excess of that reagent is necessary. If possible, the Note should indicate the effect of using amounts of reagent less than that specified in the procedure.

Reaction Apparatus

Describe the size and type of flask (number of necks) and indicate how *every* neck is equipped.

"A 500-mL, three-necked, round-bottomed flask equipped with an 3-cm Teflon-coated magnetic stirbar, a 250-mL pressure-equalizing addition funnel fitted with an argon inlet, and a rubber septum is charged with . . . ."

Indicate how the reaction apparatus is dried and whether the reaction is conducted under an inert atmosphere. This can be incorporated in the text of the procedure or included in a Note.

"The apparatus is flame-dried and maintained under an atmosphere of argon during the course of the reaction."

In the case of procedures involving unusual glassware or especially complicated reaction setups, authors are encouraged to include a photograph or drawing of the apparatus in the text or in a Note (for examples, see *Org. Syn.,* Vol. 82, 99 and Coll. Vol. X, pp 2, 3, 136, 201, 208, and 669).

Reagents and Starting Materials

All chemicals employed in the procedure must be commercially available or described in an earlier *Organic Syntheses* or *Inorganic Syntheses* procedure. For other compounds, a procedure should be included either as one or more steps in the text or, in the case of relatively straightforward preparations of reagents, as a Note. In the latter case, all requirements with regard to characterization, style, and detail also apply.

In one or more Notes, indicate the purity or grade of each reagent, solvent, etc. It is desirable to also indicate the source (company the chemical was purchased from), particularly in the case of chemicals where it is suspected that the composition (trace impurities, etc.) may vary from one supplier to another. In cases where reagents are purified, dried, "activated" (e.g., Zn dust), etc., a detailed description of the procedure used should be included in a Note. In other cases, indicate that the chemical was "used as received".

> "Diisopropylamine (99.5%) was obtained from Aldrich Chemical Co., Inc. and distilled under argon from calcium hydride before use. THF (99+%) was obtained from Mallinckrodt, Inc. and distilled from sodium benzophenone ketyl. Diethyl ether (99.9%) was purchased from Aldrich Chemical Co., Inc. and purified by pressure filtration under argon through activated alumina. Methyl iodide (99%) was obtained from Aldrich Chemical Co., Inc. and used as received."

The amount of each reactant should be provided in parentheses in the order mL, g, mmol, and equivalents with careful consideration to the correct number of significant figures.

The concentration of solutions should be expressed in terms of molarity or normality, and not percent (e.g., 1 N HCl, 6 M NaOH, not "10% HCl").

Reaction Procedure

Describe every aspect of the procedure clearly and explicitly. Indicate the order of addition and time for addition of all reagents and how each is added (via syringe, addition funnel, etc.).

Indicate the temperature of the reaction mixture (preferably internal temperature). Describe the type of cooling (e.g., "dry ice-acetone bath") and heating (e.g., oil bath, heating mantle) methods employed. Be careful to describe clearly all cooling and warming cycles, including initial and final temperatures and the time interval involved.

xvii

Describe the appearance of the reaction mixture (color, homogeneous or not, etc.) and describe all significant changes in appearance during the course of the reaction (color changes, gas evolution, appearance of solids, exotherms, etc.).

Indicate how the reaction can be monitored to determine the extent of conversion of reactants to products. In the case of reactions monitored by TLC, provide details in a Note, including eluent, R_f values, and method of visualization. For reactions followed by GC, HPLC, or NMR analysis, provide details on analysis conditions and relevant diagnostic peaks.

"The progress of the reaction was followed by TLC analysis on silica gel with 20% EtOAc-hexane as eluent and visualization with *p*-anisaldehyde. The ketone starting material has $R_f = 0.40$ (green) and the alcohol product has $R_f = 0.25$ (blue)."

Reaction Workup

Details should be provided for reactions in which a "quenching" process is involved. Describe the composition and volume of quenching agent, and time and temperature for addition. In cases where reaction mixtures are added to a quenching solution, be sure to also describe the setup employed.

"The resulting mixture was stirred at room temperature for 15 h, and then carefully poured over 10 min into a rapidly stirred, ice-cold aqueous solution of 1 N HCl in a 500-mL Erlenmeyer flask equipped with a magnetic stirbar."

For extractions, the number of washes and the volume of each should be indicated as well as the size of the separatory funnel.

For concentration of solutions after workup, indicate the method and pressure and temperature used.

"The reaction mixture is diluted with 200 mL of water and transferred to a 500-mL separatory funnel, and the aqueous phase is separated and extracted with three 100-mL portions of ether. The combined organic layers are washed with 75 mL of water and 75 mL of saturated NaCl solution, dried over 25 g of MgSO$_4$, filtered through a 250-mL medium porosity sintered glass funnel, and concentrated by rotary evaporation (25 °C, 20 mmHg) to afford 3.25 g of a yellow oil."

"The solution is transferred to a 250-mL, round-bottomed flask equipped with a magnetic stirbar and a 15-cm Vigreux column fitted with a short path distillation head, and then concentrated by careful distillation at 50 mmHg (bath temperature gradually increased from 25 to 75 °C)."

In cases where solid products are filtered, describe the type of filter funnel used and the amount and composition of solvents used for washes.

"... and the resulting pale yellow solid is collected by filtration on a Büchner funnel and washed with 100 mL of cold (0 °C) hexane."

When solid or liquid compounds are dried under vacuum, indicate the pressure employed (rather than stating "reduced pressure" or "dried *in vacuo*").

".... and concentrated at room temperature by rotary evaporation (20 mmHg) and then at 0.01 mmHg to provide ...."

"The resulting colorless crystals are transferred to a 50-mL, round-bottomed flask and dried overnight in a 100 °C oil bath at 0.01 mmHg."

Purification: Distillation

Describe distillation apparatus including the size and type of distillation column. Indicate temperature (and pressure) at which all significant fractions are collected.

".... and transferred to a 100-mL, round-bottomed flask equipped with a magnetic stirbar. The product is distilled under vacuum through a 12-cm, vacuum-jacketed column of glass helices (Note 16) topped with a Perkin triangle. A forerun (ca. 2 mL) is collected and discarded, and the desired product is then obtained, distilling at 50–55 °C (0.04–0.07 mmHg) ...."

Purification: Column Chromatography

Provide information on TLC analysis in a Note, including eluent, R_f values, and method of visualization.

Provide dimensions of column and amount of silica gel used; in a Note indicate source and type of silica gel.

Provide details on eluents used, and number and size of fractions.

"The product is charged on a column (5 × 10 cm) of 200 g of silica gel (Note 15) and eluted with 250 mL of hexane. At that point, fraction collection (25-mL fractions) is begun, and elution is continued with 300 mL of 2% EtOAc-hexane (49:1 hexanes:EtOAc) and then 500 mL of 5% EtOAc-hexane (19:1 hexanes:EtOAc). The desired product is obtained in fractions 24–30, which are concentrated by rotary evaporation (25 °C, 15 mmHg) ...."

Purification: Recrystallization

Describe procedure in detail. Indicate solvents used (and ratio of mixed solvent systems), amount of recrystallization solvents, and temperature protocol. Describe how crystals are isolated and what they are washed with.

"The solid is dissolved in 100 mL of hot diethyl ether (30 °C) and filtered through a Büchner funnel. The filtrate is allowed to cool to room temperature, and 20 mL of hexane is added. The solution is cooled at −20 °C overnight and the resulting crystals are collected by suction filtration on a Büchner funnel, washed with 50 mL of ice-cold hexane, and then transferred to a 50-mL, round-bottomed flask and dried overnight at 0.01 mmHg to provide ...."

Characterization

Physical properties of the product such as color, appearance, crystal forms, melting point, etc. should be included in the text of the procedure. Comments on the stability of the product to storage, etc. should be provided in a Note.

In a Note, provide data establishing the identity of the product. This will generally include IR, MS, ^{1}H-NMR, and ^{13}C-NMR data, and in some cases UV data. Copies of the proton and carbon NMR spectra for the products of each step in the procedure should be submitted showing integration for all resonances. Submission of copies of the NMR spectra for other nuclei are encouraged as appropriate.

In the same Note, provide quantitative analytical data establishing the purity of the product. Elemental analysis for carbon and hydrogen (and nitrogen if present) agreeing with calculated values within 0.4% is preferred. However, GC data (for distilled or vacuum-transferred samples) and/or HPLC data (for material isolated by column chromatography) may be acceptable in some cases. Provide details on equipment and conditions for GC and HPLC analyses.

In procedures involving non-racemic, enantiomerically enriched products, optical rotations should generally be provided, but enantiomeric purity must be determined by another method such as chiral HPLC or GC analysis.

In cases where the product of one step is used without purification in the next step, a Note should be included describing how a sample of the product can be purified and providing characterization data for the pure material. Copies of the proton NMR spectra of both the product both *before* and *after* purification should be submitted.

Hazard Warnings

Any significant hazards should be indicated in a statement at the beginning of the procedure in italicized type. Efforts should be made to avoid the use of toxic and hazardous solvents and reagents when less hazardous alternatives are available.

Discussion Section

The style and content of the discussion section will depend on the nature of the procedure.

For procedures that provide an improved method for the preparation of an important reagent or synthetic building block, the discussion should focus on the advantages of the new approach and should describe and reference all of the earlier methods used to prepare the title compound.

In the case of procedures that illustrate an important synthetic method or strategy, the discussion section should provide a mini-review on the new methodology. The scope and limitations of the method should be discussed, and it is generally desirable to include a table of examples. Please be sure each table is numbered and has a title. Competing methods for accomplishing the same overall transformation should be described and referenced. A brief discussion of mechanism may be included if this is useful for understanding the scope and limitations of the method.

Titles of Articles

In cases where the main thrust of the article is the illustration of a synthetic method of general utility, the title of the article should incorporate reference to that method. Inclusion of the name of the final product is acceptable but not required. In the case of articles where the objective is the preparation of a specific compound of importance (such as a chiral ligand), then the name of that compound should be part of the title.

Examples

Title without name of product: "Stereoselective Synthesis of 3-Arylacrylates by Copper-Catalyzed Syn Hydroarylation" (*Org. Synth.* **2010**, *87*, 53).

Title including name of final product (not required): "Catalytic Enantioselective Borane Reduction of Benzyl Oximes: Preparation of (S)-1-Pyridin-3-yl-ethylamine Bis Hydrochloride" (*Org. Synth.* **2010**, *87*, 36).

Title where preparation of specific compound is the subject: "Preparation of (S)-3,3'-Bis Morpholinomethyl-5,5',6,6',7,7',8,8'-octahydro-1,1'-bi-2-naphthol" (*Org. Synth.* **2010**, *87*, 59).

Style and Format

Articles should follow the style guidelines used for organic chemistry articles published in the ACS journals such as *J. Am. Chem. Soc.*, *J. Org. Chem.*, *Org. Lett.*, etc. as described in the the ACS Style Guide (3rd Ed.). The text of the procedure should be constructed using a standard word processing program, like MS Word, and using the *Organic Syntheses* procedure template available at orgsyn.org. Chemical structures and schemes should be drawn using the standard ACS drawing parameters (in ChemDraw, the parameters are found in the "ACS Document 1996" option) with a maximum width of 15 centimeters. The graphics files should be inserted into the document at the correct location and the graphics files should also be submitted separately. All Tables that include structures should be entirely prepared in the graphics (ChemDraw) program and inserted into the word processing file at the appropriate location. Tables that include multiple, separate graphics files prepared in the word processing program will require modification.

Acknowledgments and Author's Contact Information

Contact information (institution where the work was carried out and mailing address for the principal author) should be included as footnote 1. This footnote should also include the email address for the principal author. Acknowledgment of financial support should be included in footnote 1.

Biographies and Photographs of Authors

Photographs and 100-word biographies of all authors should be submitted as separate files at the time of the submission of the procedure. The format of the biographies should be similar to those in the Volume 84–88 procedures found at the orgsyn.org website. Photographs can be accepted in a number of electronic formats, including tiff and jpeg formats.

ARNOLD R. BROSSI
1923–2011

The death of Arnold Robert Brossi on July 16, 2011 at age 87 in Bethesda, MD, is a loss to the chemistry community of a most charismatic, exuberant, engaging, and accomplished person. Born in Winterthur, Switzerland on December 19, 1923, Arnold received his doctorate in chemistry from the Swiss Federal Institute of Technology (ETH), Zurich, in 1952 under the mentorship of Professor Oskar Jeger for research on terpenes. On completion of his doctorate, he joined the chemistry research department of Hoffmann-La Roche in Basel, Switzerland, remaining there until 1963 when he was appointed director, subsequently vice-president, of chemistry research at the company's Nutley (New Jersey) facilities. In 1973, Arnold replaced Dr. Otto Isler as director of chemistry research at Roche's headquarters in Basel on Dr. Isler's retirement, and in 1976 returned to the USA as Chief of Natural Products at the National Institute of Diabetes & Digestive & Kidney Diseases (NIDDK) of the National Institutes of Health in Bethesda, MD. On his retirement from the NIH in 1991, Arnold was appointed an NIH emeritus scientist and continued to work at the NIH as a visiting scientist.

Arnold's research interests were broad with an emphasis on the synthesis and study of biologically active natural products, particularly alkaloids, in therapeutic areas such as malaria, neurodegenerative disorders, and cancer. In addition, he placed great emphasis on the importance of enantiomerically pure drug substances (fondly known at Roche as Brossi's rule.) He was an author on over 400 publications and patents.

In Nutley, Arnold directed not only chemistry but also at various times animal health research and microbiology, thus taking on a heavy burden of administration and management. However, despite this, his passion for chemistry did not wane and he was able to supervise a personal laboratory concerned with the chemistry of alkaloids. As an administrator and manager with charisma, Arnold was nonpareil. He fostered an atmosphere that was highly conducive to doing outstanding research and took a keen interest in the personal lives and work of all his scientists irrespective of their position, always ignoring the hierarchy and bureaucracy typical of large corporations. Under his leadership, organic chemistry flourished at Roche (Nutley) with, to give but a few examples, the chemistry of benzodiazepines, the discovery of proline-catalyzed enantioselective reactions, the synthesis of quinine, steroids and vitamins, the chemistry of indoles and polyether antibiotics.

During his tenure at the NIH, Arnold continued his research on alkaloids with an emphasis on neurodegenerative diseases, and was able to attract many eminent scientists to work in association with him, including the late Professor Nelson Leonard who joined him as a Fogarty-Scholar-in-Residence (1989–1990). In keeping with his life-long interest in international cooperation, Arnold actively directed the research of over 40 post-doctoral scientists from 17 countries.

Arnold held a number of academic positions, among them were research professor at the School of Pharmacy, University of North Carolina, adjunct professor of chemistry Georgetown University, faculty member Residential School of Medicinal Chemistry, Drew University (Madison, NJ), and adjunct professor at the National Institutes of Pharmaceutical Research and Development, Beijing.

Arnold was a member of the Advisory Board of Organic Syntheses, Inc. and was the Editor-in-Chief of Volume 53. He initiated a close, continuing relationship between Roche and Organic Syntheses, Inc in which many procedures were checked at the Nutley facilities. He was Editor-in-Chief for volumes 21–40 and co-editor for volumes 41–45 of *The Alkaloids*, and was on the editorial board of *Heterocycles* for which a special issue (volume 39, issue 2) was published in his honor.

He was a member of the Steering Committee on Malaria for the World Health Organization in Geneva (Switzerland), and was a founding member in 1966 of the biennial Mona Symposia on Natural Products, University of the West Indies (Jamaica), for which the 1994 symposium was held in his honor. In addition, he was an advisor for Plantaceutica, Inc, Research Triangle Park (North Carolina) and Advance Biofractures, Inc., Lynbrook, NY.

In recognition of his achievements, Arnold received many awards and honors, including the Alfred Burger Award for Medicinal Chemistry of the American Chemical Society (1991), the Charles Mentzer Prize of the French Society of Medicinal Chemistry (1989), the Hamis Medal of the Czechoslovak Chemical Society (1989), and the Medal of Merit from Palacky University, Czech Republic (1994). He was the recipient of honorary doctorates from Bowdoin College (Maine, 1984) and Adam Mikiewicz University, Poznan, Poland (1993). He was an honorary member of the Swiss Chemical Society, the Pharmaceutical Society of Japan, the American Society of Pharmacognosy, the Polish Chemical Society, and a Fellow of the American Association for the Advancement of Science.

Arnold was a true ambassador for chemistry, and although he prized individuality, he had the unique ability to foster cooperation and collaborations among scientists of very different disciplines, to which his numerous collaborative publications attest. At scientific meetings, in the work place, at his home, as well as at various social functions, Arnold's conviviality, energy, and pleasant personality invariably created an inspiring atmosphere for everyone to excel at whatever they do. He left an indelible mark on our science and on all who have had the good fortune to have met him.

Among his many hobbies and interests Arnold was an avid fisherman, bridge player, mushroom collector, and gourmand.

Arnold was predeceased by his wife, Hanni, of 49 years in 1999, and is survived by his son Mario, daughters Angela Kindhauser and Franca Alpin, five grandchildren, and two great-grandsons.

GABRIEL G. SAUCY
PERCY S. MANCHAND

ROBERT E. IRELAND
1929–2012

Robert Ellsworth Ireland, Emeritus Professor of Chemistry at the University of Virginia, Charlottesville, died on February 4th, 2012 in Sarasota, Florida. He is survived by his wife, Margaret, his brother Andrew and his sons Mark and Richard.

Bob Ireland was one of the legends of organic synthesis, and he will be remembered by his many contributions to synthetic methodology and natural product total synthesis, as well as by his dedicated training of several generations of influential industrial, government, and academic scientists.

Bob was born in Ohio in 1929 and almost became a professional baseball player. After receiving his A.B. degree from Amherst College in 1951, he moved to the University of Wisconsin-Madison and on to a stellar career in chemistry.

Under the direction of Professor William S. Johnson, he studied the formation of steroids and completed a total synthesis of estrone and 14-isoestrone. He graduated with a Ph.D. in 1954 and joined the research group of Professor William G. Young at the University of California, Los Angeles as an NSF postdoctoral fellow. During this time, he continued his research on steroids and investigated the rearrangement of allylic alcohols in the presence of thionyl chloride. The final term of his Midwest period began in 1956 when he accepted a faculty position at the University of Michigan.

In his nine years in Ann Arbor, Bob attracted his first cohort of outstanding coworkers and established the foundation of his unique academic career. A mix of synthetic methods, for example the use of the N-butylthiomethylene blocking group, and approaches to terpenoid natural products such as 9-isopimaradiene resulted from this period. These activities expanded and increased after the move of the Ireland group to the California Institute of Technology in Pasadena in 1965. In particular, the "Ireland-Claisen Rearrangement" was reported in seminal papers in 1972 and 1976. Furthermore, major natural product targets such as monensin, lasalocid, tirandamycic acid, chlorothricolide and aphidicolin were successfully tackled. Many of these syntheses showcased the tremendous utility and versatility of the Ireland-Claisen methodology and introduced new variants of the sigmatropic shift, such as the "Straight-Chain Claisen" and the "Black-Belt Claisen". He also introduced new methods for the preparation of pyranoid glucals.

At Caltech, Bob wrote the first textbook on synthetic strategy. *Organic Synthesis* was published by Prentice Hall in 1969 and included memorable quotes such as "Stereochemistry Raises its Ugly Head" or "Multistage Synthesis: Logistics and Stereochemistry Combine to Produce Nightmares". In this treatise, Bob introduced strategic thinking, not unlike a battlefield plan, by facing the complex target molecule head-on as well as circumventing its defenses.

In 1985, Bob surprised many when, in spite of his love of Orange County, Caltech, and the Athenaeum, he accepted the directorship of the Merrell-Dow Research Institute in Strasbourg, France. However, Bob had always been a Francophile and an adventurer, and he was never afraid to take risks. While he returned to the United States in 1986 as Chair and Commonwealth Professor of Chemistry at the University of Virginia in Charlottesville, his time in Strasbourg reinforced his dedication to chemistry and his appreciation of the many frontiers of science. In 1987, he was honored with the newly created Thomas Jefferson Chair in Chemistry at UVA. In addition to the syntheses of FK-506 and carba-monensin, several new processes emanated from Charlottesville, not the least of these including a highly cited and improved preparation of the Dess-Martin periodinane. Unfortunately, his health forced him into an early retirement in 1995, which nonetheless, he greatly enjoyed with his wife Margaret, his enthusiasm for Macintosh computers, and his continued vivid interest in new developments in synthesis.

Bob's work earned him many accolades, including the prestigious ACS Ernest Guenther Award in the Chemistry of Natural Products in 1977 and the ACS Award for Creative Work in Synthetic Organic

Chemistry in 1988. He chaired the ACS Division of Organic Chemistry in 1980 and was the Editor of Volume 54 of *Organic Syntheses* in 1974. He and his group contributed as submitters or checkers to 34 procedures published in *Organic Syntheses*.

Over the course of his illustrious academic research career at Michigan, Caltech, and Virginia, Bob mentored numerous graduate students and postdoctoral associates who have moved on to key positions in industry, government, and universities, and they continue to spread Bob's infectious love for synthesis across the globe. Bob was never afraid to speak his mind, and he enjoyed life inside and outside the laboratory. We will always remember him as a keen intellect, a towering educator and scientist, and a relentless, enthusiastic supporter of his students.

PETER WIPF

PREFACE

Like many readers and end-users of *Organic Syntheses*, my knowledge of the OS enterprise was more or less limited to following the procedures in the Volumes (Collected and Individual) when seeking reproducible and reliable reactions. Over the course of my undergraduate, graduate and academic career, I often turned to OS to find best practices concerning specific reactions and also to make much-needed intermediates that were subsequently used in synthetic sequences. In fact, OS even found its way into undergraduate advanced labs, because students could learn so much from the careful instructions that help to teach the next generation of researchers in organic chemistry how to properly do an organic reaction.

My other point of contact with OS was that we had submitted a procedure that was successfully checked and ultimately published several years ago. It was a proud moment to see one of our reactions appear and be available for others to use. Little did I know all that went into the process, behind the scenes. I only learned that when I became a member of the Board and started selecting, approving and checking procedures.

When I was called by Ed Grabowski, then at Merck, on behalf of the Board, to join the OS family, I looked over the list of those who had served as Members and was in awe of the many who had for 80 years (at the time) maintained and built on the spirit of the "brand" as the Gold Standard for reliable procedures. I subsequently learned of the incredible dedication of the members of the Board to identifying, checking and finally publishing procedures that would be so valuable to the community. My role changed from Foreign Editor, a five-year term, to Editor, an eight-year term, and the honor to be a volume editor. What you see before you is the combined work of many authors and checkers and it is my pleasure to share diverse transformations and products, as you turn the pages (literally or electronically). I have grouped the contributions into the areas that are covered in the Preface, though the list of procedures follows a chronological order. As was mentioned in a previous Preface, OS now publishes Discussion Addenda, wherein classic and important procedures get an update by the original submitters. In this way, readers can learn how the procedure has been used and if it has evolved in any way since the original date of publication.

Asymmetric synthesis and organofluorine chemistry combine in a procedure by Shaw, Kalow and Doyle entitled "Fluoride Ring Opening Kinetic

Resolution of Terminal Epoxides" on page 9. Introduction of the trifluoromethyl group is presented by Song, Yee and Senanayake in "Synthesis of Trifluoromethylketones from Carboxylic Acids: 4-(3,4-Dibromophenyl)-1,1,1-trifluoromethylpentan-2-one" on page 210. On page 374, a Discussion Addendum for introduction of the CF_3 group is presented by Sato, Tarui, Omote, Kumadaki and Ando.

Oxidation continues to hold a special place in OS and Ratnikov, Goldmann, McLaughlin and Doyle report on "Allylic Oxidation Catalyzed by Dirhodium (II) Tetrakis [epsilon-caprolactamate] of *tert*-Butyldimethylsilyl-protected *trans*-Dehydroandrosterone", page 19. Oxidation of benzylic alcohols is also presented by Uyanik and Ishihara in "2-Iodoxy-5-Methylbenzenesulfonic Acid-Catalyzed Selective Oxidation of 4-Bromobenzyl Alcohol to 4-Bromobenzaldehyde or 4-Bromobenzoic Acid with Oxone" on page 105. Oxidation using Oxone is also a key step in "Oxone Mediated Synthesis of Benimidazoles from 1,2-Phenylenediamines with Aldehydes: Preparation of 2-(4-Cyanophenyl)-1-[2-(3,4-Dimethoxyphenyl)-Ethyl]-1H Benzimidazole-5 Carboxylic Acid Ethyl Ester" as reported by by Gillard and Beulieu on page 131. The Discussion Addendum on "Oxidation of Nerol to Neral with Iodosobenzene and TEMPO" by Leonelli, Margarita and Piancatelli gives readers an update on this method of oxidation.

Similarly, reduction remains an important area for investigation and Willwacher, Ahmad and Movassaghi describe the "Preparation of n-Isopropylidene-N'-2-Nitrobenzenesulfonyl Hydrazine (IPNBSH) and its Use in Palladium Catalyzed Synthesis of Monoalkyl Diazenes Synthesis of 9-Allylanthracene" on page 230. Asymmetric hydroformylation is reported by Wong, Adint and Landis in "Synthesis of (2*R*)-3-[[1,1-Dimethylethyl)dimethylsilyl]oxy]-2-methylpropanal by Rhodium Catalyzed Asymmetric Hydroformylation" on page 243. Another asymmetric transformation, which produces enantiomerically pure protected amines, is reported by Thalen, Roesch and Bäckvall in the DKR "Synthesis of (R)-2-Methoxy-*N*-(1-Phenylethyl)Acetamide via Dynamic Kinetic Resolution" on page 255.

Heterocycles remain a key component of OS and Yang, Ulysse, McLaws, Keefe, Guzzo and Haney report on the "Preparation of Tetrahydroisoquinoline-3-ones via Cyclization of Phenyl Acetamindes using Eaton's Reagent," page 44.

C-H Activation is emerging as a new strategy in organic synthesis and a contribution by Martin and Flores-Gaspar on the "Synthesis of 8,8-Dipropylbicyclo [4.2.0] octa-1,3,5-trien-7-one via Pd-Catalzyed Intramolecular C-H Bond Acylation," on page 159.

Each volume of *Organic Syntheses* could not be completed without authors submitting procedures (voluntarily or when invited), without the Board members and the students and post-doctoral fellows who do the checking, and without the efforts of Rick Danheiser and Chuck Zercher, who as Editor-in-Chief and Associate Editor, keep the trains running on time, which is no small feat. I thank each and every one of you, as well as Carl Johnson and the other members of the Board of Directors, whose careful management has made Organic Synthesis, Inc. a model organization and one I feel privileged to have served for the past 8 years.

MARK LAUTENS
Toronto, Ontario

CONTENTS

$$Me_3Si\diagup\!\!\diagdown \xrightarrow[\substack{aq.\ MeOH \\ 50\ ^\circ C}]{\substack{NaHSO_3 \\ PhCO_3t\text{-}Bu}} Me_3Si\diagup\!\diagdown SO_3Na \xrightarrow[0\ ^\circ C\ to\ rt]{\substack{SOCl_2 \\ cat.\ DMF}} Me_3Si\diagup\!\diagdown SO_2Cl$$

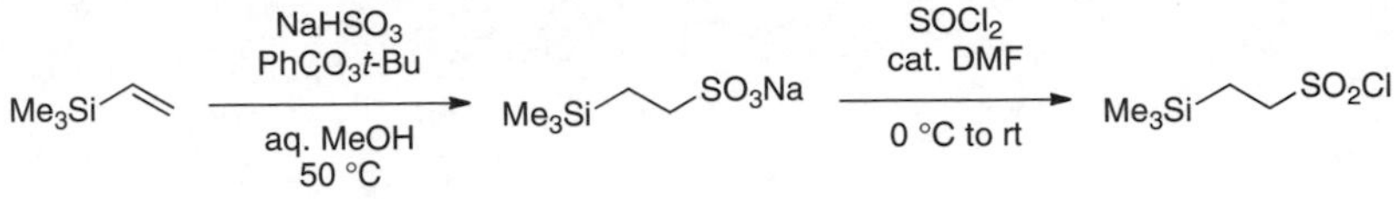

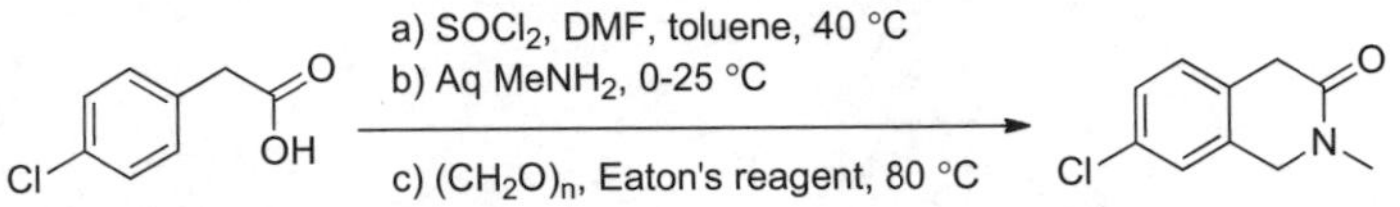

a) SOCl₂, DMF, toluene, 40 °C
b) Aq MeNH₂, 0-25 °C
c) (CH₂O)ₙ, Eaton's reagent, 80 °C

a) Li, THF, reflux
b) NBS, CCl₄, reflux
c) KOtBu, THF, 0 °C to rt

MeO₂C—⟨ ⟩—CN + (N-OH / Me, H) → RhCl(PPh₃)₃ (0.5 mol %), 110 °C, toluene, 4 h → MeO₂C—⟨ ⟩—C(=O)NH₂

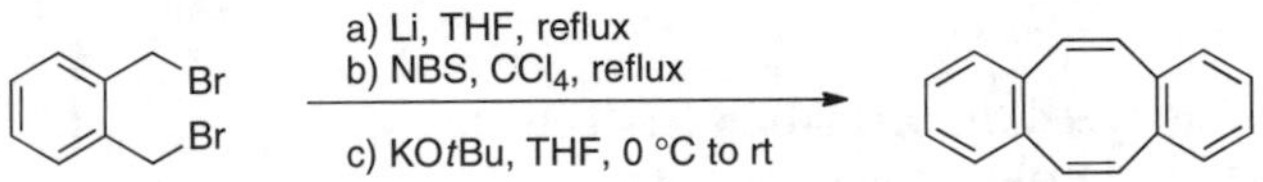

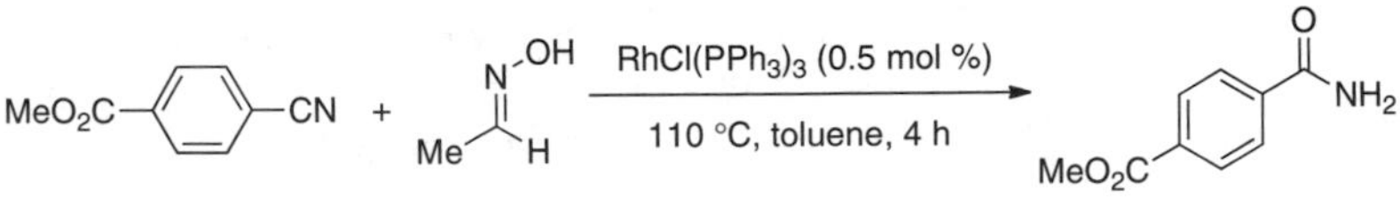

Preparation of Enantioenriched Homoallylic Primary Amines   **88**
José C. González-Gómez, Francisco Foubelo and
Miguel Yus*

Discussion Addendum for: **98**
(Phenyl)[2-(trimethylsilyl)phenyl]iodonium Triflate.
An Efficient and Mild Benzyne Precursor
Tsugio Kitamura*

Muhammet Uyanik and Kazuaki Ishihara*

Tiantong Lou, E-Ting Liao, Ashraf Wilsily, and Eric Fillion*

Teruyuki Hayashi*

Oxone®-Mediated Synthesis of Benzimidazoles from 1,2-Phenylenediamines and Aldehydes: Preparation of 2-(4-Cyano-Phenyl)-1-[2-(3,4-Dimethoxyphenyl)-Ethyl]-1H-Benzimidazole-5-Carboxylic Acid Ethyl Ester

James R. Gillard and Pierre L. Beaulieu*

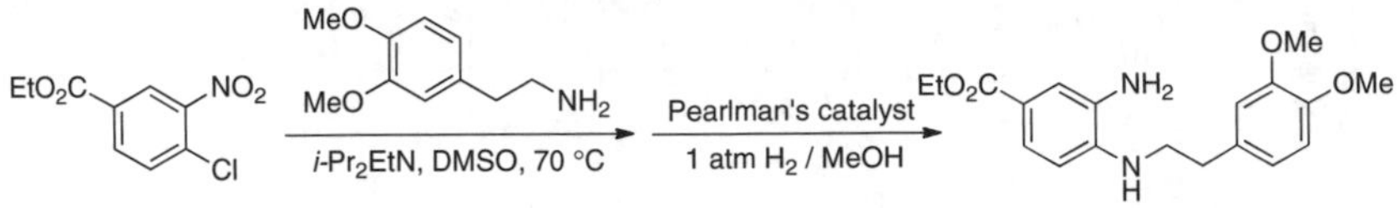

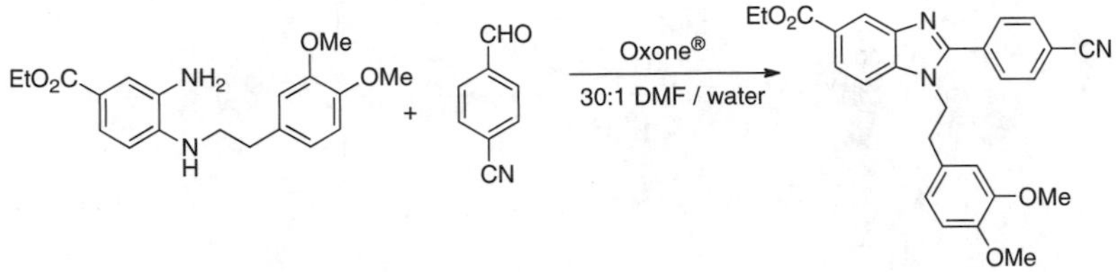

131

Discussion Addendum for:
Preparation of α- Acetoxy Ethers by the Reductive Acetylation of Esters: *endo*-1-Bornyloxyethyl Acetate

143

Nicholas Sizemore and Scott D. Rychnovsky*

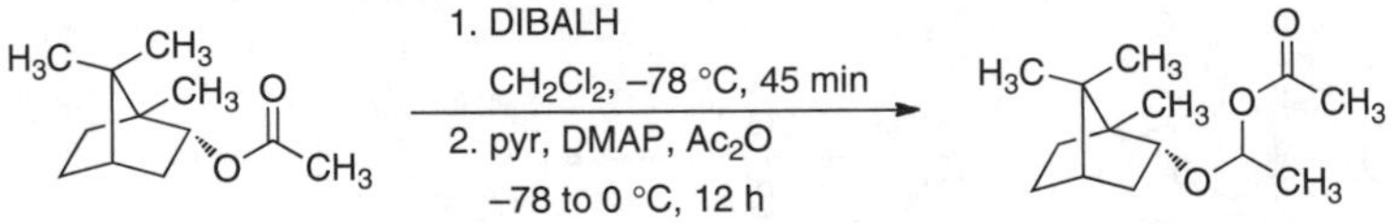

xl

Preparation of 1-Monoacylglycerols via the Suzuki-Miyaura   183
Reaction: 2,3-Dihydroxypropyl (*Z*)-tetradec-7-enoate

Dexi Yang, Valerie A. Cwynar, David J. Hart,* Jakkam Madanmohan,
Jean Lee, Joseph Lyons and Martin Caffrey

Discussion Addendum for: 202
Suzuki-Miyaura Cross-Coupling:Preparation of 2'-Vinylacetanilide

Donal F. O'Shea*

Synthesis of Trifluoromethyl Ketones from Carboxylic Acids: 210
4-(3,4-Dibromophenyl)- 1,1,1-trifluoro-4-methylpentan-2-one

Jonathan T. Reeves,* Zhulin Tan, Daniel R. Fandrick,
Jinhua J. Song, Nathan K. Yee and Chris H. Senanayake

Discussion Addendum for:
ortho-Formylations of Phenols; Preparation of 3-Bromosalicylaldehyde

Trond Vidar Hansen and Lars Skattebøl*

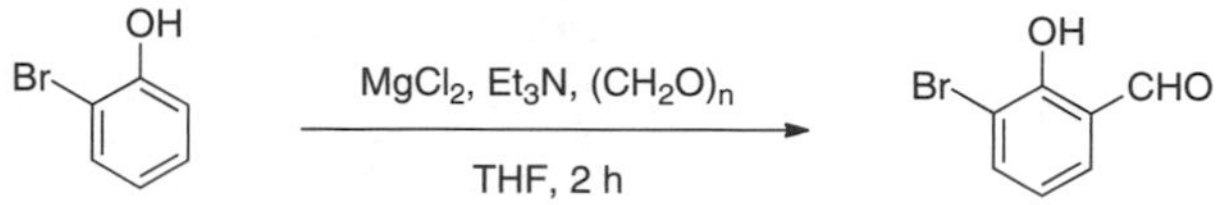

220

Br—(OH) phenol → [MgCl$_2$, Et$_3$N, (CH$_2$O)$_n$ / THF, 2 h] → Br—(OH)(CHO) phenol

Preparation of n-Isopropylidene-n'-2- Nitrobenzenesulfonyl Hydrazine (IPNBSH) and Its Use in Palladium–catalyzed Synthesis of Monoalkyl Diazenes Synthesis of 9- Allylanthracene

Jens Willwacher, Omar K. Ahmad, and Mohammad Movassaghi*

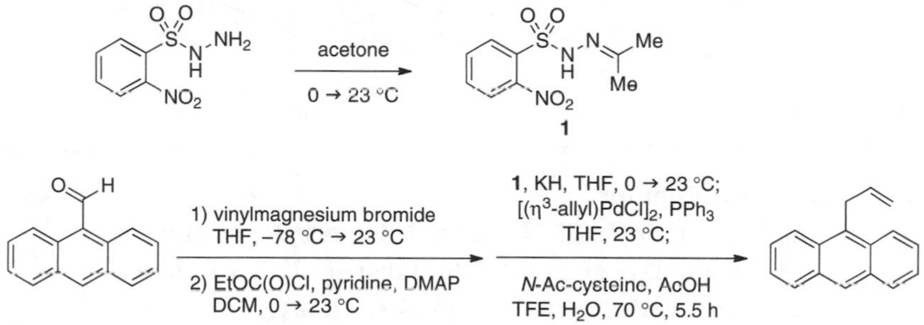

230

(benzenesulfonyl hydrazine with NO$_2$) → [acetone, 0 → 23 °C] → (hydrazone, Me, Me, NO$_2$) **1**

(anthracene-9-carbaldehyde) → 1) vinylmagnesium bromide, THF, −78 °C → 23 °C; 2) EtOC(O)Cl, pyridine, DMAP, DCM, 0 → 23 °C → **1**, KH, THF, 0 → 23 °C; [(η^3-allyl)PdCl]$_2$, PPh$_3$, THF, 23 °C; N-Ac-cysteine, AcOH, TFE, H$_2$O, 70 °C, 5.5 h → (9-allylanthracene)

Synthesis of (2R)-3-[[(1,1-Dimethylethyl) dimethylsilyl]oxy]-2-methylpropanal by Rhodium-Catalyzed Asymmetric Hydroformylation

Gene W. Wong, Tyler T. Adint, Clark R. Landis*

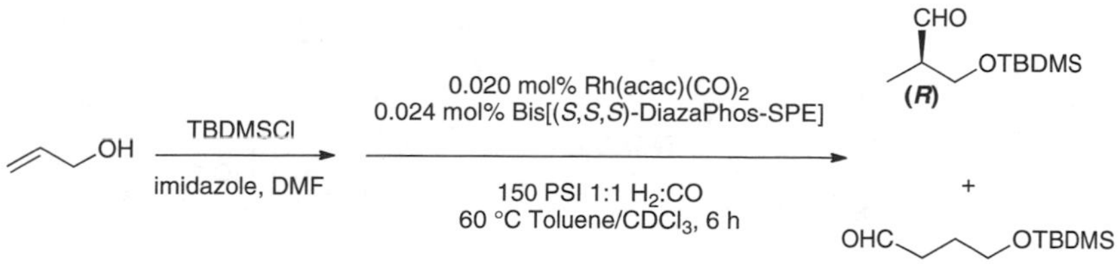

243

(allyl alcohol, OH) → TBDMSCl, imidazole, DMF → 0.020 mol% Rh(acac)(CO)$_2$, 0.024 mol% Bis[(S,S,S)-DiazaPhos-SPE], 150 PSI 1:1 H$_2$:CO, 60 °C Toluene/CDCl$_3$, 6 h → CHO, OTBDMS (R) + OHC, OTBDMS

Lisa Kanupp Thalén, Christine Rösch, and Jan-Erling Bäckvall*

Ar = *p*-MeO-C₆H₄

Discussion Addendum for: **267**

[3 + 4] Annulation Using a [β-(Trimethylsilyl) acryloyl]silane
and the Lithium Enolate of an α,β-Unsaturated Methyl Ketone:
(1*R*,6*S*,7*S*)-4-(*tert*-Butyldimethylsiloxy)-6-(trimethylsilyl)-
bicyclo [5.4.0]undec-4-en-2-one

Kei Takeda* and Michiko Sasaki

o-Aminobenzaldehyde, Redox-Neutral Aminal Formation and Synthesis of Deoxyvasicinone

Chen Zhang, Chandra Kanta De, and Daniel Seidel*

Multicomponent Synthesis of Tertiary Diarylmethylamines: 1-((4-Fluorophenyl) (4-methoxyphenyl)methyl)piperidine

Erwan Le Gall* and Thierry Martens

Synthesis of Quinolines by Electrophilic Cyclization of N-(2-Alkynyl)Anilines: 3-Iodo-4-Phenylquinoline

Yu Chen, Anton Dubrovskiy, and Richard C. Larock*

Discussion Addendum for:
Bicyclopropylidene
Armin de Meijere* and Sergei I. Kozhushkov

$\triangleright$—CO$_2$Me $\xrightarrow[\text{Et}_2\text{O, 20 °C}]{\text{EtMgBr, Ti}(i\text{-PrO})_4}$ $\triangleright\triangleleft$ OH $\xrightarrow[\text{CH}_2\text{Cl}_2, -15 \text{ to } 20 \text{ °C}]{\text{Ph}_3\text{P}\cdot\text{Br}_2, \text{Py, CH}_2\text{Cl}_2}$ $\triangleright\triangleleft$ Br

$\triangleright\triangleleft$ Br $\xrightarrow[\text{20 °C}]{t\text{-BuOK, DMSO}}$ $\triangleright\!=\!\triangleleft$

Discussion Addendum for: - 350
Synthesis of 1,2:4,5- Di-*o* isopropylidene–D-erythro-2,3-hexodiulo-2,6-pyranose. A Highly Enantioselective Ketone Catalyst for Epoxidation/Asymmetric Epoxidation of *trans*-β-Methylstyrene and 1-Phenylcyclohexene using a D-Fructose-derived Ketone: (*R,R*)-*trans*-β–Methylstyrene Oxide and (*R,R*)-1-Phenylcyclohexene Oxide
Thomas A. Ramirez, O. Andrea Wong, and Yian Shi*

[Reaction scheme: D-fructose derivative with OH groups, reacted with MeO—C(CH₃)₂—OMe, (CH₃)₂CO, HClO₄, then PCC, 3 Å MS, CH₂Cl₂ to give diisopropylidene ketone **1**]

Ph $\diagdown$ $\xrightarrow[\substack{\text{Oxone - KOH}\\\text{CH}_3\text{CN-DMM}}]{\mathbf{1}}$ Ph $\triangleleft$O (epoxide), 91-92% ee

[1-phenylcyclohexene] $\xrightarrow[\substack{\text{H}_2\text{O}_2\text{-K}_2\text{CO}_3\\\text{CH}_3\text{CN}}]{\mathbf{1}}$ [1-phenylcyclohexene oxide], 96-98% ee

Preparation of (R_a)-Methyl 3-(dimethyl(phenyl)silyl)   **420**
hexa-3,4-dienoate
Ryan A. Brawn and James S. Panek*

Discussion Addendum for: **432**
Boric Acid Catalyzed Amide Formation from
Carboxylic Acids and Amines: *n*-Benzyl-4-phenylbutyramide
Pingwah Tang*

Preparation of 3-Alkylated Oxindoles from *N*-Benzyl   **438**
Aniline via a Cu(II)-Mediated Anilide Cyclization Process
David S. Pugh and Richard J. K. Taylor*

Gold(I)-Catalyzed Decarboxylative Allylic Amination of Allylic *N*-Tosylcarbamates

Dong Xing and Dan Yang*

Preparation of *t*-Butyl-3-Bromo-5-Formylbenzoate Through Selective Metal-Halogen Exchange Reactions

Juan D. Arredondo, Hongmei Li and Jaume Balsells*

Reductive Radical Decarboxylation of Aliphatic Carboxylic Acids

Eun Jung Ko, Craig M. Williams, G. Paul Savage,* and John Tsanaktsidis

Efficient Synthesis of Oxime Using *O*-TBS-*N*-   **480**
Tosylhydroxylamine: Preparation of (2*Z*)-4-
(Benzyloxy)but-2-enal Oxime

Katsushi Kitahara, Tatsuya Toma, Jun Shimokawa and
Tohru Fukuyama*

Preparation of Cyclobutenone   **491**

Audrey G. Ross, Xiaohua Li, and Samuel J. Danishefsky*

Convenient Synthesis of α-Diazoacetates from α-Bromoacetates   **501**
and *N,N'*-Ditosylhydrazine: Preparation of Benzyl Diazoacetate

Eiji Ideue, Tatsuya Toma, Jun Shimokawa and Tohru Fukuyama*

Handling and Disposal of Hazardous Chemicals

The procedures in this volume of *Organic Syntheses* are intended for use only by persons with prior training in experimental organic chemistry. All hazardous materials should be handled using the standard procedures for work with chemicals described in references such as *Prudent Practices in the Laboratory* (The National Academies Press, Washington, D.C., 2011 www.nap.edu). All chemical waste should be disposed of in accordance with local regulations. For general guidelines for the management of chemical waste, see Chapter 8 of Prudent Practices and the discussion below.

The procedures described in this volume must be conducted at one's own risk. Organic Syntheses, Inc., its Editors, and its Board of Directors do not warrant or guarantee the safety of individuals using these procedures and hereby disclaim any liability for any injuries or damages claimed to have resulted from or related in any way to the procedures herein.

Handling Hazardous Chemicals

A Brief Introduction

An excellent general reference covering all aspects of work with hazardous chemicals is *Prudent Practices in the Laboratory* (The National Academies Press, Washington, D.C., 2011 www.nap.edu). Readers are strongly encouraged to refer to this volume for information on evaluating hazards, the storage of chemicals, and work with toxic, flammable, and highly reactive substances.

The following sections provide a brief introduction to some of the most important aspects of work with hazardous chemicals.

1

Several general principles apply to all work conducted in laboratories where hazardous substances are stored or are in use.

1. **Preparation**. A cardinal rule of laboratory research is that researchers must determine the potential hazards associated with an experiment before beginning it. *Before working with any chemical, researchers should determine what physical and health hazards are associated with the substance.* This determination may require consulting library references and material safety data sheets, and may involve discussions with supervisors and safety professionals.

2. **Minimize exposure to chemicals**. All skin contact with chemicals in the laboratory should be avoided. Laboratory hoods and other ventilation devices should be used to prevent exposure to airborne substances whenever possible (note that the use of hoods should be regarded as **required** for work with many hazardous substances). Eye protection and protective clothing (lab coat and appropriate gloves) should be worn at all times when working in laboratories where hazardous chemicals are stored or are in use.

3. **Do not underestimate risks.** Assume that any mixture of chemicals will be more toxic than its most toxic component. All new compounds and substances of unknown toxicity should be treated as toxic substances.

4. **Be prepared for accidents.** Before beginning an experiment, know what specific action you will take in the event of the accidental release of any hazardous substances involved. Know the location of all safety equipment including fire extinguishers, fire blankets, eye washes, safety showers, spill carts and spill control materials, be familiar with the location of the nearest fire alarm and telephone, and know what telephone numbers to call in the event of an emergency. Know the location of the circuit breakers for your laboratory. Researchers should avoid working alone in the laboratory. Work with certain hazardous

2

substances and carrying out certain hazardous experiments when alone in the lab may be strictly forbidden.

Chemical Hazards

Researchers should be familiar with the different classes of hazardous substances encountered in their laboratory.

a. Acute Effects. These effects occur soon after exposure. Acute effects include burn, inflammation, allergic responses, damage to the eyes, lungs, or nervous system (e.g., dizziness), and unconsciousness or death (as from overexposure to HCN). The effect and its cause are usually obvious and so are the methods to prevent it. They generally arise from inhalation or skin contact, and can be avoided if one follows prudent practices of carrying out experiments in an effective fume hood, wearing appropriate protective clothing, and avoiding all skin contact with hazardous chemicals by wearing gloves. Ingestion is a rare route of exposure in synthetic laboratories, and can be avoided if one follows the prudent practice of never eating or drinking in the laboratory, wearing gloves and a lab coat that are removed when leaving the lab, and always washing hands after leaving the lab.

b. Chronic Effects. These effects occur after a long period of exposure or after a long latency period and may show up in any of numerous organs. Carcinogenic substances are frequently encountered in research laboratories, and the carcinogenic effects of many substances are unknown. Reproductive and developmental toxins are other classes of chronic toxins that may be encountered in the laboratory.

Because all chemicals are toxic under some conditions, and relatively few have been thoroughly tested, it is prudent to minimize exposure to all chemicals. In practice this means avoiding all skin contact by wearing appropriate impermeable gloves and lab coats, and avoiding inhalation of toxic vapors, mists, and dusts by carrying out experiments in an effective fume hood. Eye protection should be worn at all times in laboratories where hazardous chemicals are stored or are in use. The type of eye protection that is required will vary with the nature of the potential hazards and can range

3

from impact resistant safety glasses with side shields, to goggles, to full-face shields.

Fire Hazards

Flammable substances are among the most common hazardous materials found in research laboratories. Flammable substances are materials that readily catch fire and burn in air. A flammable liquid does not itself burn; it is the vapors from the liquid that burn. The rate at which different liquids produce flammable vapors depends on their vapor pressure, which increases with temperature. The degree of fire hazard depends also on the ability to form combustible or explosive mixtures with air, the ease of ignition of these mixtures, and the relative densities of the liquid with respect to water and of the gas with respect to air.

The following basic precautions should be followed when working with flammable substances.

1. Flammable substances should be handled only in areas free of any ignition sources. Ignition sources include open flames, electrical equipment (especially motors), static electricity, and for some highly flammable materials (e.g. carbon disulfide), even hot surfaces.

2. Never heat a flammable substance with an open flame.

3. When transferring flammable liquids in metal equipment, static-generated sparks should be avoided by bonding and the use of ground straps.

4. Ventilation is one of the most effective ways to prevent the formation of flammable mixtures. A laboratory hood should be used whenever significant quantities of flammable substances are transferred from one container to another, allowed to stand or heated in open containers, or handled in any other way. Be sure that the hood is free of all ignition sources including, in particular, variacs.

4

5. Special precautions are required when working with pyrophoric substances, which should only be handled by experienced researchers trained in their safe use.

Explosion Hazards

Explosive substances are materials that decompose under conditions of mechanical shock, elevated temperature, or chemical action, with the release of large volumes of gases and heat. Special precautions are required for the safe use of explosive materials. For example, work with explosive materials will generally require the use of special protective apparel (face shields, gloves, lab coats) and protective devices such as explosion shields and barriers.

Organic peroxides are one of the most common classes of potentially explosive substances encountered in synthetic organic chemistry research laboratories. Researchers should be familiar with the methods for testing for the presence of peroxides and should be aware of the classes of compounds that are prone to form peroxides by autoxidation, including aldehydes, ethers, hydrocarbons with allylic, benzylic, or propargylic hydrogen atoms, and conjugated dienes, enynes, and diynes.

Disposal of Chemical Waste

A Brief Introduction

General guidelines for the management of chemical waste can be found in Chapter 8 of *Prudent Practices in the Laboratory* (The National Academies Press, Washington, D.C., 2011 www.nap.edu). However, researchers should be familiar with the regulations governing the storage and disposal of chemical waste in the local area and should carry out their research in accord with these laws and regulations.

Effluents from synthetic organic chemistry fall into the following categories:

1. **Gases**
    1a. Gaseous materials either used or generated in an organic reaction.
    1b. Solvent vapors generated in reactions swept with an inert gas and during solvent stripping operations.
    1c. Vapors formed from volatile reagents, intermediates, and products.

2. **Liquids**
    2a. Waste solvents and solvent solutions of organic solids (see item 3b below).
    2b. Aqueous layers from reaction work-up containing volatile organic solvents.
    2c. Aqueous waste containing non-volatile organic materials.
    2d. Aqueous waste containing inorganic materials.

3. **Solids**
    3a. Metal salts and other inorganic materials.
    3b. Organic residues (tars) and other unwanted organic materials.
    3c. Used silica gel, charcoal, filter aids, spent catalysts, and the like.

6

Effluent Gases

Some reactions generate gases and vapors from volatile compounds that as part of the experimental design can be trapped at the point of generation by "scrubbing" the effluent gas. The gas being swept from a reaction set-up is led through tubing to a large trap to prevent suck-back and into a sintered glass gas dispersion tube immersed in the scrubbing fluid. A bleach container can be conveniently used as a vessel for the scrubbing fluid. The nature of the effluent determines which of four common fluids should be used: dilute sulfuric acid, dilute alkali or sodium carbonate solution, laundry bleach when an oxidizing scrubber is needed, and sodium thiosulfate solution or diluted alkaline sodium borohydride when a reducing scrubber is needed. Ice should be added to the scrubbing fluid if an exotherm is anticipated.

Larger scale operations may require the use of a pH meter or starch/iodide test paper to ensure that the scrubbing capacity is not being exceeded.

At the end of the experiment, the solution is then disposed of as chemical waste according to the local regulations.

Fluoride Ring-Opening Kinetic Resolution of Terminal Epoxides: Preparation of (*S*)-2-Fluoro-1-Phenylethanol

Submitted by Travis W. Shaw, Julia A. Kalow, and Abigail G. Doyle.[1]
Checked by Florian Vogt and Brian M. Stoltz.

1. Procedure

Caution! This procedure should be carried out in an efficient fume hood. Appropriate precautions should be taken to avoid inhalation or direct contact with styrene oxide, which is a known carcinogen, and benzoyl fluoride, which is toxic and a potent lachrymator.

(S)-2-Fluoro-1-phenylethanol. A 500-mL, single-necked, round-bottomed flask equipped with a magnetic stir bar (24 x 8 mm), (Note 1) is charged with (*R,R*)-(−)-*N,N'*-bis(3,5-di-*tert*butylsalicylidene)-1,2-cyclohexanediaminocobalt(II) (1.51 g, 2.5 mmol, 0.025 equiv (Note 2). 1,5-Diazabicyclo[4.3.0]non-5-ene (247 μL, 2 mmol, 0.02 equiv(Note 3) and 1,1,1,3,3,3-hexafluoroisopropanol (8.66 mL, 82.5 mmol, 0.825 equiv) are added to the flask via syringe. The mixture is diluted with diethyl ether (250 mL (Note 4) to afford a dark brown solution and stirring is initiated at 500 rpm. Styrene oxide (11.4 mL, 100 mmol, 1.0 equiv (Note 5) and benzoyl fluoride (5.99 mL, 55 mmol, 0.55 equiv) are both added via syringe. The reaction is allowed to stir open to air for 15 min (Note 6) at room temperature (Note 7). The flask is then capped with a polyethylene cap and secured with Parafilm wrap (Note 8). The reaction is stirred at 500 rpm for 13–14 h at room temperature (Note 9), and then quenched by the addition of methanolic ammonia (7 N, 150 mL) via syringe (Note 10). The solution is

allowed to stir for 2–3 h or until the disappearance of 1,1,1,3,3,3-hexafluoropropan-2-yl benzoate, as monitored by TLC (Notes 11 and 12). The reaction is then transferred to a 1-L separatory funnel (Note 13) and washed with aqueous 1M NaOH (3 x 200 mL). The combined aqueous portions are then extracted with diethyl ether (3 x 100 mL). The combined organic portions are dried over 10 g of anhydrous Na_2SO_4, filtered, and concentrated by rotary evaporator (25 °C, 100 mmHg, then 60 mmHg) to afford a dark brown oil. The crude product is charged onto a column (75 mm x 300 mm) packed with 200 g SiO_2 (Note 14) as a slurry with hexanes. The loaded column is topped with a 1 cm layer of sand and 75 mL fractions are collected, eluting with 500 mL hexanes, 625 mL of 5% diethyl ether-hexanes, 625 mL of 10% diethyl ether-hexanes, 375 mL of 15% diethyl ether-hexanes, and then 1625 mL of 20% diethyl ether-hexanes. TLC is used to monitor the course of the column (Note 15). Fractions 39–56 are concentrated by rotary evaporation under vacuum (25 °C, 100 mmHg, then 60 mmHg) to provide (S)-2-fluoro-1-phenylethanol (5.8–5.9 g, 41–42% yield) as a brown oil (Note 16) in 98–98% enantiomeric excess (Note 17).

2. Notes

1. No special precautions were taken to exclude air or moisture from the reaction.

2. The (salen)Co(II) precatalyst was purchased from Strem Chemicals, Inc. and stored in a dessicator. 1,5-Diazabicyclo[4.3.0]non-5-ene (98%) (DBN), styrene oxide (97%) and methanolic ammonia (7 N) were purchased from Aldrich Chemical Company, Inc. Benzoyl fluoride (98%) was purchased from Alfa Aesar. 1,1,1,3,3,3-Hexafluoro-2-propanol (>99%) (HFIP) was purchased from TCI America. All chemicals, except for DBN (see Note 3) and styrene oxide (see Note 5), were used as received. All chemicals were stored at room temperature on the benchtop except for the methanolic ammonia, which was stored in a refrigerator at 0 °C. However, we have found that open bottles of benzoyl fluoride slowly hydrolyze under air, generating benzoic acid (7.4 wt% over 4 months), the presence of which leads to depressed rates in the fluorination reactions.

3. DBN was distilled under reduced pressure (12 mmHg, bp = 101-102 °C) from calcium hydride into a Schlenk flask containing 3Å molecular sieves and stored under a nitrogen atmosphere on the benchtop. Use of DBN that had been aged under air led to slower rates of reaction (see Note 9).

10

Molecular sieves 3Å 8-12 mesh were purchased from Acros Organics and were flame dried under reduced pressure (0.1 mmHg) and the flask was backfilled with argon (three cycles) prior to use.

4. Diethyl ether (99.9%) was obtained from Fisher Scientific and purified by a solvent delivery system, with passage through a packed dry neutral alumina column.[2]

5. Commercial styrene oxide is contaminated with approximately 1% carbonyl-containing impurities that lead to depressed rates in the kinetic resolutions. On a 0.3 mmol scale, after 5 h, the reaction reaches 50% conversion with pure styrene oxide and only 35% conversion using unpurified material. Therefore, styrene oxide is purified prior to use as follows: A 1-L, 1-necked, round-bottomed flask is equipped with a magnetic stir bar (50 x 8.5 mm) and charged with styrene oxide (16.0 mL, 140 mmol). HPLC-grade methanol (700 mL, 99.8%, EMD Chemicals Inc.) is added and the flask is held in an ice-water bath with stirring (600 rpm) for the remaining duration of the procedure. After the solution has cooled for 45 min, NaBH₄ (529.6 mg, 14 mmol, 98%, J.T. Baker Chemical Co.) is added in a single portion. The reaction is allowed to stir for 1 h or until disappearance of acetophenone by TLC (see Note 11). The R_f of the acetophenone is 0.48 (1:1 hexanes:diethyl ether) visualizing with a UV light source (254 nm). The methanol is then removed under reduced pressure on the rotary evaporator (25 °C, 100 mmHg) and the remaining material is purified by column chromatography using 130 g silica gel (see Note 14) on a 70 mm x 350 mm glass column, eluting with 9:1 hexanes:diethyl ether (250 mL) and then 4:1 hexanes:diethyl ether (750 mL). The eluent containing the product is concentrated under reduced pressure (25 °C, 100 mmHg) to yield pure styrene oxide (12.3 g, 102 mmol, 73%).

In addition to the purification described above, the checkers alternatively purified styrene oxide as follows: A 500-mL, 1-necked, round-bottomed flask is equipped with a magnetic stir bar (24 x 8 mm) and charged with styrene oxide (28.6 mL, 250 mmol, 1.0 equiv) and diethyl ether (250 mL) (Note 4). The flask is sealed with a rubber septum, connected to a nitrogen/vacuum manifold and brought to 0 °C in an ice-water bath with stirring (600 rpm) for the remaining duration of the procedure. After the solution has been cooled for 15 min, LiBH₄ (10 mL, 10 mmol, 0.025 equiv, 1 M in THF, Aldrich Chemical Company, Inc.) is added with a syringe over a period of 10 min. The water-ice bath is removed after the addition is completed and the reaction is allowed to stir for 1 h at ambient temperature.

Ethanol (1 mL) is added and the reaction mixture further allowed to stir until the evolution of hydrogen gas stops. Sodium sulfate (2 g) is added and the solids are then removed by filtration through a short column (2 cm) of Celite® (washed with 50 mL of diethyl ether) and the remaining solution is concentrated by rotary evaporation under reduced pressure (25 °C, 100 mmHg). The remaining oil is transferred into a 50-mL round-bottomed flask and short path distillation (a forerun of about 2 mL is collected) yields pure styrene oxide as a clear, colorless oil (22.7 g, 76%).

6. The color of the solution changes to a dark brown upon addition of the liquid reagents, indicative of aerobic oxidation of the Co(II) precatalyst (red) to the active Co(III) species. Rapid stirring of the reaction open to air facilitates this oxidation.

7. The ambient temperature was recorded as 25 °C by the submitters. The checkers recorded it as 23 °C.

8. A polyethylene cap is preferred to a rubber septum, as polyethylene does not absorb solvent.

9. Reactions with opened bottles of DBN that had been stored under air were observed to take up to 20 h to reach completion. However, the selectivity of the fluoride ring-opening reactions remained the same in all cases. Thus, to obtain optimal results, the course of the reaction should be monitored and the reaction quenched at the specified end-point: the reaction was found to be complete when the enantiomeric excess of the styrene oxide was approximately 90% by chiral GC analysis on an Agilent 6850 using a Supelco BetaDex 120 column (30.0 m x 0.25 mm), and a flame ionization detector. The assay conditions are 76 °C for 35 min, a gradient of 15 °C/min until 150 °C is reached, followed by a 2 min hold at 150 °C. The retention times for the two enantiomers of styrene oxide are t_R(major) = 28.3 min, t_R(minor) = 31.0 min). The submitters reported using an Agilent 7890 with a J&W Scientific Cyclodex–β column (30.0 m x 0.25 mm), and a flame ionization detector. The assay conditions then are 100 °C for 15 min, a gradient of 15 °C/min until 150 °C is reached, followed by a 4 min hold at 150 °C. The retention times for the two enantiomers of styrene oxide are t_R(major) = 13.4 min, t_R(minor) = 13.9 min. GC samples were obtained by passing a small aliquot (~0.1 mL) of the reaction through a Pasteur pipette containing a cotton plug and 350 mg silica gel and eluting with diethyl ether.

10. Ammonia converts the 1,1,1,3,3,3-hexafluoropropan–2–yl benzoate, derived from the solvolysis of benzoyl fluoride by HFIP, to benzamide, which may be removed by alkaline aqueous workup. Ammonia also

12

converts the (salen)Co catalyst to a species that does not elute on silica gel. This procedure simplifies purification of the product by column chromatography. The rate at which the methanolic ammonia is added does not affect yield or selectivity.

11. All TLC measurements were performed using plates pre-coated with silica Gel 60 F_{254} purchased from EMD Chemicals Inc.

12. 1,1,1,3,3,3-Hexafluoropropan–2–yl benzoate has an R_f of 0.76 (4:1 hexanes:diethyl ether) and can be visualized with a UV light source (254 nm)

13. The reaction flask is rinsed with three 20 mL portions of diethyl ether.

14. SiliCycle SiliaFlash F60 (40–53 μm, 60 Å) silica gel was used.

15. The R_f value of the title compound is 0.46 (1:1 hexanes:diethyl ether) and the starting material has an R_f of 0.55 (1:1 hexanes:diethyl ether). Both compounds can be visualized with a UV light source (254 nm).

16. The brown oil was found to be clean judged by ^{1}H NMR spectroscopy (purity >95%). A colorless oil is obtained by short path distillation of the title compound at 110 °C under 16 mmHg of pressure with 91.4% (5.3 g from 5.8 g) of the compound recovered. This material was submitted to elemental analysis (Note 17).

17. Enantiomeric excess was determined by chiral GC analysis on a Agilent 6850 using a Supelco Beta DEX^{TM} 120 column (30.0 m x 0.25 mm), and a flame ionization detector. The assay conditions are 102 °C for 40 min, a gradient of 80 °C/min until 180 °C is reached, followed by a 1-minute hold at 180 °C. The retention times for the two enantiomers of the title compound are t_R(major) = 32.4 min, t_R(minor) = 35.6 min. (S)-2-Fluoro-1-phenylethanol has the following spectroscopic data: ^{1}H NMR (500 MHz, CDCl$_3$) δ: 2.44 (br s, 1 H), 4.48 (dddd, J = 3.2, 5.8, 9.6, 48.5 Hz, 2 H), 5.03 (ddd, J = 3.2, 8.3, 13.9 Hz, 1 H), 7.32–7.41 (m, 5 H); ^{13}C NMR (125.7 MHz, CDCl$_3$) δ: 72.9 (d, J = 19.7 Hz), 87.1 (d, J = 174.4 Hz), 126.3 (s), 128.4 (s), 128.6 (s), 138.1 (d, J = 8.2 Hz); FTIR (neat, cm^{-1}) 3391 (br), 3064 (m), 3033 (m), 2979 (m), 2950 (m), 2891 (m), 1494 (s), 1455 (s), 1311 (m), 1199 (s), 1098 (s), 1072 (s), 1010 (s), 916 (m), 897 (s), 835 (m), 759 (s), 701 (s); $[α]^{20}_D$ = +55.6 (CHCl$_3$, c = 1.01); HRMS (m/z) (GC-EI$^+$) calc. for C$_8$H$_9$FO (M$^+$ 140.0637, found 140.0639; TLC (Hex/Et$_2$O = 1:1) Rf = 0.46; Anal calcd for C$_8$H$_9$FO: C 68.56, H 6.47, found C 68.60, H 6.49.

Safety and Waste Disposal Information

All hazardous materials should be handled and disposed of in accordance with "Prudent Practices in the Laboratory"; National Academies Press; Washington, DC, 2011.

3. Discussion

The selective incorporation of fluorine atoms into organic molecules profoundly affects their physical properties and is a common strategy for the optimization of pharmaceuticals, agrochemicals, materials, and catalysts. Accordingly, approximately 20–25% of drugs in the pharmaceutical pipeline contain at least one fluorine atom.[3] While catalytic asymmetric methods for C–F bond formation using electrophilic "F+" equivalents have been identified, complementary strategies using abundant, inexpensive nucleophilic fluoride sources are scarce.[4] This procedure describes an application of the first highly enantioselective catalytic method for nucleophilic fluorination to the kinetic resolution of racemic styrene oxide, providing (S)-2-fluoro-1-phenylethanol in >98% ee.[5] Previous methods to access such fluorohydrins from the enantiopure epoxide proceed with variable regioselectivity and require elevated temperatures, prolonged reaction times, and rigorous exclusion of water.[6]

The protocol is operationally simple, and may be performed in standard glassware without the exclusion of air or moisture. The chiral (salen)Co precatalyst, amine cocatalyst, and fluoride source are commercially available. Notably, unlike many metal fluorides and HF-containing reagents, benzoyl fluoride is a bench-stable, non-hydroscopic liquid that can be handled using standard techniques. The combination of benzoyl fluoride and 1,1,1,3,3,3-hexafluoroisopropanol (HFIP) serves as a latent source of HF, allowing mild conditions and efficient catalysis.

Under these conditions, aliphatic and aromatic epoxides undergo kinetic resolution by fluoride ring-opening exclusively at the terminal position (Table 1).[7] A silyl glycidyl ether is tolerated, highlighting the chemoselectivity of the nucleophilic fluorine species generated in the co-catalytic reaction (Table 1, entry 3). In the above procedure, purification of the styrene oxide prior to the kinetic resolution allows the catalyst loadings to be lowered to 2.5 and 2.0 mol%, and the stoichiometries of benzoyl

14

fluoride and HFIP to be reduced. The kinetic resolution offers complementary scope to well-developed strategies for the organocatalytic electrophilic fluorination of aldehydes, which can provide the alternative regioisomer in high enantiomeric excess ((R)-2-fluoro-2-phenyl-1-ethanol, 54% yield, 99% ee).[8]

Using the commercially available chiral isothiourea (–)-tetramisole in place of DBN, meso epoxides undergo efficient desymmetrization, yielding trans fluorohydrins in 55–88% yield and 58–95% ee (Table 2). Intriguingly, a significant mismatched effect on rate and enantioselectivity is observed when the opposite enantiomer of (salen)Co is used with (–)-tetramisole. Based on this and other preliminary mechanistic evidence, the reactivity and selectivity are ascribed to a cooperative effect between the chiral Lewis acid and amine co-catalysts.

Table 1. Kinetic resolution of terminal epoxides by fluoride opening.[7]

entry	R	Co (mol %)	yield (%)	ee (%)	k_{rel}
1	n-Bu	2	36	99	>300
2[a]	Ph	2.5	42	98.2	>200
3	TBSOCH$_2$	5	44	88	32

[a] This work.

Table 2. Desymmetrization of meso epoxides by fluoride opening.[7]

product	yield (%)	ee (%)	product	yield (%)	ee (%)
	77	85		75	90
	65	93		88	86
	82	90		84	80
	87	95		55	58

1. Department of Chemistry, Frick Laboratory, Princeton University, Princeton, NJ 08544; agdoyle@princeton.edu; Financial support was provided by Princeton University and the donors of the American Chemical Society Petroleum Research Fund.
2. Pangborn, A. B.; Giardello, M. A.; Grubbs, R. H.; Rosen, R. K.; Timmers, F. J. *Organometallics* **1996**, *15*, 1518–1520.
3. For recent reviews, see: (a) Hagmann, W. K. *J. Med. Chem.* **2008**, *51*, 4359–4369. (b) Müller, K., Faeh, C. and Diederich, F. *Science* **2007**, *317*, 1881– 1886. (c) Purser, S., Moore, P. R., Swallow, S. and Gouverneur, V. *Chem. Soc. Rev.* **2008**, *37*, 320–330.
4. For a recent review, see: Cahard, D.; Xu, X.; Couve-Bonnaire, S.; Pannecoucke, X. *Chem. Soc. Rev.* **2010**, *31*, 558–568.
5. Haufe and Bruns have effected the fluorination of styrene oxide in 43% yield and 74% ee using 50 mol% (salen)CrCl: Haufe, G. and Bruns, S. *Adv. Synth. Catal.* **2002**, *344*, 165–171.
6. Akiyama, Y.; Fukuhara, T.; Hara, S. *Synlett* **2003**, *10*, 1530–1532.

7. Kalow, J. A.; Doyle, A. G. *J. Am. Chem. Soc.* **2010**, *132*, 3268–3269.

8. Beeson, T. D.; MacMillan, D. W. C. *J. Am. Chem. Soc.* **2005**, *127*, 8826–8828.

Appendix
Chemical Abstract Nomenclature; (Registry Number)

(*R,R*)-(–)-*N,N'*-Bis(3,5-di-*tert*-butylsalicylidene)-1,2-cyclohexanediaminocobalt(II): Cobalt, [[2,2'-[(1*R*,2*R*)-1,2-cyclohexanediylbis[(nitrilo-κ*N*)methylidyne]]bis[4,6-bis(1,1 dimethylethyl)phenolato-κ*O*]](2-)]-, (SP-4-2)-; (176763-62-5)

Styrene oxide: 2-Phenyloxirane; (96-09-3)

1,5-Diazabicyclo[4.3.0]non-5-ene: Pyrrolo[1,2-a]pyrimidine, 2,3,4,6,7,8-hexahydro-; (3001-72-7)

Benzoyl fluoride: Benzoyl fluoride; (455-32-3)

1,1,1,3,3,3-Hexafluoro-2-propanol: 2-Propanol, 1,1,1,3,3,3-hexafluoro-; (920-66-1)

Sodium borohydride: Borate(1-), tetrahydro-, sodium (1:1); (16971-29-2)

Abigail Doyle received her A.B. and A.M. summa cum laude in Chemistry and Chemical Biology from Harvard University in 2002. From 2003-2008, she was an NDESG and NSF graduate fellow working with Professor Eric Jacobsen at Harvard University. Her graduate research included the development of transition metal-catalyzed enantioselective alkylation reactions of enolates with alkyl halides. Abby began as an Assistant Professor in the Department of Chemistry at Princeton University in July 2008. Her research group is working in the area of catalysis and asymmetric synthesis. One area of interest is the use of transition metal catalysts to prepare chiral building blocks that contain stereogenic C–F bonds using nucleophilic fluorine sources.

Travis Shaw was born in 1987 in Corpus Christi, TX. He graduated from The University of Texas at Austin in 2010 with a B.S. in biochemistry. While at UT Austin he worked on the total synthesis of a sesquiterpene-neolignan natural product under the advisory of Professor Dionicio Siegel. Travis also spent a summer performing synthetic studies in alkaloid total synthesis in the lab of Professor Karl Gademann at the École polytechnique fédérale de Lausanne in Lausanne, Switzerland. He joined the Doyle group in the summer of 2010 and his research is currently focused on the asymmetric hydrohalogenation of olefins.

Julia Kalow was born in Newton, MA (US). She graduated from Columbia University in 2008 (B.A., Chemistry and Creative Writing), where she did undergraduate research in the laboratory of Professor James Leighton. She is currently pursuing her Ph.D. at Princeton University under the supervision of Professor Abigail Doyle, where her research focuses on the development of asymmetric catalytic methods for nucleophilic fluorination. Julia is a National Science Foundation Predoctoral Fellow and Bristol-Myers Squibb Endowed Graduate Fellow in Synthetic Organic Chemistry.

Forian Vogt was born in Feuchtwangen (Germany) in 1981 and studied chemistry at the Technical University Munich (Germany). He obtained his M.Sc. degree under the supervision of Prof. Thorsten Bach (Munich) and Prof. Martin Banwell (Australian National University) in 2006. He then joined the lab of Prof. Thorsten Bach as a Ph.D. student in June 2006 focusing on the synthesis of hydrogen bond mediated ligands and catalysts and their application in stereoselective synthesis. After receiving his Ph.D. degree in March 2010, he moved to California for postdoctoral studies with Prof. Brian Stoltz where his research interests include the development of catalytic, asymmetric methods and their utility in natural product total syntheses

Allylic Oxidation Catalyzed by Dirhodium(II) Tetrakis[ε-caprolactamate] of *tert*-Butyldimethylsilyl-protected *trans*-Dehydroandrosterone

Submitted by Maxim O. Ratnikov,[1*] Petra L. Goldmann,[1] Emily C. McLaughlin,[2] and Michael P. Doyle.[1]

Checked by Kainan Zhang, Supriyo Majumder, and Kay Brummond.

1. Procedure

A. *Dirhodium(II) tetrakis[ε-caprolactamate] acetonitrile solvate,* $Rh_2(cap)_4(CH_3CN)_2$ (**1**). An oven-dried 250-mL, single-necked (24/40 joint) round-bottomed flask containing a Teflon-coated stir-bar is charged with dirhodium(II) tetraacetate (2.00 g, 4.5 mmol) (Note 1), ε-caprolactam (11.84 g, 104.6 mmol) (Note 2), and dry chlorobenzene (110 mL) (Note 3). The flask is fitted with a Soxhlet extraction apparatus (Note 4) containing a thimble with 30 g of an oven-dried Na_2CO_3 and sand mixture in a 3:1 ratio (Note 5). An additional 50 mL of chlorobenzene is added to partially fill the thimble-chamber of the Soxhlet apparatus.   The mixture is refluxed at 185 °C (external temperature) open to the atmosphere (Note 6), and after only 15 min the color of the solution turns from green to dark blue. After 10 h the vessel is removed from the oil bath, the Soxhlet extraction apparatus is removed, and the solvent is evaporated under reduced pressure using a rotary evaporator to yield a blue, glass-like solid residue. 2-Propanol (100 mL) (Note 7) is added to the same flask, and the solid residue is mixed vigorously with 2-propanol for 30 seconds using a metal spatula to dissolve chlorobenzene that is entrapped in the residue (Note 8). The 2-propanol is evaporated from the solid residue under reduced pressure to give a purple powder suitable for filtration (Note 9). Acetone (100 mL) is added to dissolve excess ε-caprolactam contained in the powder, and the solid is gravity-filtered through a Büchner funnel and then washed with an additional 50 mL of acetone (Note 10). The filtered purple solid is transferred to a pre-weighed, 250-mL round-bottomed flask and dried under high vacuum (0.1 mmHg) for 1 h to yield 3.52 g of a dull purple powder (Note 11). To render the powder catalytically active (Note 12) the solid is recrystallized from 66 mL of boiling acetonitrile/methanol (10:1) (Note 13) to give 1.65 g (50% yield) of $Rh_2(cap)_4(CH_3CN)_2$ (1) (Note 14) as shiny deep purple crystals.

B. *Androst-5-en-17-one, 3-[[(1,1-dimethylethyl)dimethylsilyl]oxy]-, (3β)- preparation* (3). A 250-mL, single-necked (24/40 joint) Erlenmeyer flask is dried at about 200 °C under vacuum (0.01 mmHg) for approximately 10 min. The flask is charged under the atmosphere of dry nitrogen with 2 (5.00 g, 17.3 mmol) (Note 15) and 1*H*-imidazole (1.77 g, 26.0 mmol) (Note 16), followed by *N,N*-dimethylformamide (25.0 mL) (Note 17). The resulting suspension is sonicated at 20 °C with gentle swirling until homogeneous (Note 18). *tert*-Butyldimethylsilyl chloride (3.39 g, 22.5 mmol) (Note 19) is added as a solid to the reaction mixture. The flask is equipped with a magnetic stir bar (Note 20). After about 3 min of stirring, a white precipitate begins to form. The reaction is stirred for 30 min, at which time water (100 mL) is added to ensure complete product precipitation. The white clay-like substance is scraped from the glass walls of the Erlenmeyer flask, filtered under vacuum, and then dried overnight under vacuum (0.01 mmHg) at 20 °C to yield 6.83 g (98%) of 3 as a white powder (Note 21).

C. *(3β)-3-[[(1,1-dimethylethyl)dimethylsilyl]-oxy]-Androst-5-ene-7,17-dione* (4). A 200-mL, Erlenmeyer flask (24/40 joint) containing a Teflon-coated magnetic stir bar is charged with 3 (8.04 g, 20.0 mmol), 1 (146 mg, 0.20 mmol, 1.00 mol %), and 1,2-dichloromethane (40 mL) (Note 22). The suspension is sonicated for 5 min at 20 °C with occasional gentle

20

swirling (Note 18). After addition of T-HYDRO (14.3 mL, 99.5 mmol, 5.0 equiv) in one portion (Note 23), the flask containing the reaction mixture is equipped with a septum containing an outlet needle (Note 24) and its contents are stirred at 40 °C (external temperature) (Note 25). The color of the solution turns from purple to dark red after about 2 min following addition of T-HYDRO. After 24 h the mixture is concentrated on a rotary evaporator (30 °C, 10 mmHg) (Note 26). The product is purified by column chromatography (Note 27). Fractions containing the product are combined, and the solvent is evaporated under reduced pressure on a rotary evaporator (30 °C, 30 mmHg). The remaining white solid is dissolved in diethyl ether (20 mL) at room temperature. The solution is heated to 35 °C and then cooled to room temperature whereupon a white powder crystallizes from the solution. Diethyl ether is removed using a pipette. The above washing procedure is repeated twice. The resulting white powder is collected and dried overnight under reduced pressure at room temperature (60 °C, 4 mmHg) to yield **4** (5.50 g, 66 %) (Notes 28 and 29).

2. Notes

1.   Dirhodium(II) tetraacetate was obtained from Pressure Chemicals Co. and used as received.

2.   ε-Caprolactam, 99%, was obtained from Alfa Aesar and used as received.

3.   Chlorobenzene, 99+%, was obtained from Acros and stored over activated 3Å molecular sieves prior to use.

4.   A Soxhlet apparatus (PYREX®, 31 mm internal diameter × 135 mm external length), which contained a cellulose single thickness extraction thimble obtained from Whatman® (33 mm internal diameter × 94 mm external length), was used.

5.   Sodium carbonate was used to trap acetic acid liberated in the ligand exchange reaction as sodium acetate/sodium bicarbonate. A piece of glass wool was placed at the top and the bottom of the carbonate-sand mixture to prevent its entry into the reaction flask.

6.   The temperature of the oil bath was maintained at 185–195 °C (external temperature). The Soxhlet extractor was covered with aluminum foil for insulation to provide a higher rate for solvent turnover. Solvent turnover in the Soxhlet extractor (about 60 mL each time) occurred every 30–40 minutes.

7.   Commercial ACS grade 2-propanol was used as received.

8.   Complex **1** has low solubility in 2-propanol.

9.   If the solid remains a slurry, two additional 2-propanol washes of 100 mL each are performed.

10.   Vacuum filtration is applied when washing with commercial ACS-grade acetone.

11.   The dull purple powder was identified as $Rh_2(cap)_6$ with two axially bound ε-caprolactam ligands:

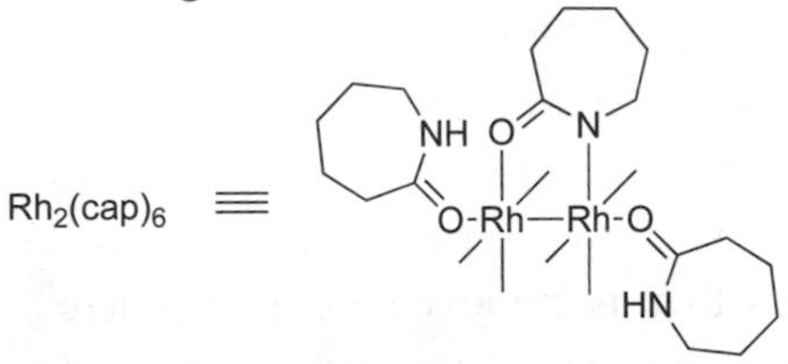

Calcd. for $C_{36}H_{62}N_6O_6Rh_2$: C, 49.09; H, 7.10; N, 9.54. Found: C, 49.55; H, 6.82; N, 9.29. The analysis was not confirmed by the checkers.

12.   The two axially-ligated ε-caprolactam molecules are tightly bound; they inhibit formation of the catalytically active dirhodium complex.

13.   Recrystallization procedure: In a 100-mL, single-necked (24/40 joint) round-bottomed flask, equipped with a refluxing condenser and a Teflon-coated stir-bar, crude $Rh_2(cap)_4$ (701.1 mg) is stirred in acetonitrile (40 mL) and methanol (4 mL). The flask is placed in an oil bath maintained at 100 °C (external temperature). When the mixture reaches boiling temperature, additional methanol is added dropwise via Pasteur pipette from the top of the refluxing condenser until a deep purple homogenous solution is obtained. The flask is removed from the oil bath, and the hot solution is vacuum filtered through a funnel containing a small cotton plug. (The submitters caution against the use of vacuum filtration. The recommend gravity filtration with minimal cooling until filtration is complete.) The solution is allowed to cool to room temperature and then placed into a freezer overnight. A shiny purple crystalline solid is isolated by removal of the supernatant liquid (using a pipette). The solid is washed with acetonitrile (3 x 5 mL) and dried under vacuum at 4 mmHg for 8 h.

14.   Data for $Rh_2(cap)_4(CH_3CN)_2$: ^{1}H NMR (400 MHz, $CDCl_3$) δ: 1.39–1.67 (m, 26 H), 2.07 (s, 11 H), 2.31–2.43 (m, 8 H), 3.34–3.23 (m, 8 H), The ^{1}H NMR spectral regions 1.39-1.67 and 2.07 do not integrate to the correct number of protons; however, the data, including relative integration values, are in agreement with those provided by the submitter; ^{13}C NMR (100 MHz, $CDCl_3$) δ: 2.3, 24.3, 29.0, 30.7, 36.8, 53.3, 115.9, 185.7; IR

22

(film) 3383, 2958, 2918, 2847, 2359, 2339, 1683, 1596, 1434, 1348 cm^{-1}; HRMS (ESI) Calcd. for $C_{24}H_{40}N_4O_4Rh_2$: 654.1160, found 654.1180 [M$^+$– 2(CH$_3$CN)]; Anal. Calcd. for $C_{28}H_{46}N_6O_4Rh_2$: C, 45.66; H, 6.30; N, 11.41; Found: C, 45.65; H, 6.44; N, 11.90. The checkers were unable to obtain satisfactory CHN analysis (checker's results: Anal. Calcd. for $C_{28}H_{46}N_6O_4Rh_2$: C, 45.66; H, 6.30; N, 11.41; Found: C, 42.31; H, 6.48; N, 8.23); however, the oxidation in Step C proceeded efficiently and without difficulty using the catalyst prepared by the checkers.

15. *trans*-Dehydroandrosterone (99%) was obtained from Sigma-Aldrich. and used as received.

16. 1*H*-Imidazole (ACS reagent, ≥99%) was obtained from Sigma-Aldrich and used as received.

17. *N,N*-Dimethylformamide (99.8 %, anhydrous) was obtained from Aldrich, stored under nitrogen, and used as received.

18. Sonication facilitates dissolution of starting materials. Steroids **2** and **3** form solid clumps upon addition that have a low dissolution rate when stirring with a magnetic stir bar.

19. *tert*-Butyldimethylsilyl chloride (reagent grade, 97 %) was obtained from Aldrich and used as received.

20. Use of a mechanical stirrer is necessitated by the formation of a large amount precipitate that significantly increases the viscosity of the reaction mixture.

21. The product has following properties. mp 148–149 °C, lit.[3] value 146–147 °C; $[\alpha]_D^{18}$ +3.4° (*c* 1.0, CHCl$_3$); lit.[3] $[\alpha]_D^{23}$ = +7.7° (*c* 1.0, CHCl$_3$); IR (CH$_2$Cl$_2$): 2949, 2925, 2888, 2855, 1744, 1372, 1258, 1094, 841 cm^{-1}; R_f = 0.58 (EtOAc/hexane 1:4); ^{1}H NMR (400 MHz, CDCl$_3$) δ: 0.05(s, 6 H), 0.88 (s, 9 H + 3 H), 0.95–1.09 (m, 5 H), 1.24–1.30 (m, 2 H), 1.41–1.74 (m, 7 H), 1.79–1.97 (m, 3 H), 2.03–2.27 (m, 4 H), 2.45 (dd, *J* = 19.2, 8.4 Hz, 1 H), 3.47 (tt, *J* = 10.8, 4.8 Hz, 1 H), 5.34 (d, *J* = 5.2 Hz, 1 H); ^{13}C NMR (100 MHz, CDCl$_3$) δ: –4.6, 13.5, 18.2, 19.4, 20.3, 21.8, 25.9, 30.8, 31.4, 31.5, 32.0, 35.8, 36.7, 37.3, 42.7, 47.5, 50.3, 51.7, 72.4, 120.4, 141.7, 221.0; HRMS (ESI): Calcd. for $C_{25}H_{42}O_2SiNa$ (M+Na): 425.2852. Found: 425.2846; Anal. Calcd. for $C_{25}H_{42}O_2Si$: C, 74.56; H, 10.51. Found: C, 74.78; H, 10.70.

22. 1,2-Dichloroethane (spectrophotometric grade, ≥99%) was obtained from Aldrich and used as received.

23. T-HYDRO, a 70% solution of *tert*-butyl hydroperoxide in water, was obtained from Aldrich and used as received. A smaller excess of T-HYDRO leads to an incomplete conversion.

24. The needle provides a gas outlet that prevents a pressure built-up in the reaction mixture. The septum limits solvent evaporation.

25. The reaction can be monitored by silica gel TLC (Whatman, 200 µm, with F-254) with EtOAc-hexane (1:4) as an eluent and visualized with $KMnO_4$. Compound **4** has an R_f of 0.49 in EtOAc-hexane (1:4).

26. If the residue is a syrup, the residue should be diluted with a dichloromethane prior to transfer onto the chromatography column.

27. The column (5 cm diameter × 20 cm height) of silica gel (95 g of silica gel; Sorbent Technologies, 40–63 µm) is packed as slurry in EtOAc-hexane (1:60) and a layer of sand (1 cm) is added on top. The concentrated reaction mixture is placed on the top of the column and is eluted with EtOAc-hexane mixture of the following ratios and volumes:

EtOAc-hexane Ratio	Volume, mL
1:60	160
1:30	40
1:2	400

The product is collected in fractions 15 to 28 in 13*100 Fisher test tubes. The presence of **4** can be monitored by TLC on silica and visualized with UV light (254 nm) or $KMnO_4$ (see note 25).

28. The product has following properties: mp 124–125 °C, lit.[4] value 127–129 °C; $[\alpha]_D^{17} = -53.5°$ (*c* 1.0, $CHCl_3$), lit.[4] $[\alpha]_D^{23} = -51.6°$ (*c* 1.86, $CHCl_3$); IR (solid): 2949, 2880, 2857, 1740, 1671, 1629, 1462, 1407, 1379, 1295, 1254, 1190, 1096, 1008, 957, 869, 837, 806, 777 cm^{-1}; $R_f = 0.49$ (EtOAc/hexane 1:4); ^1H NMR (400 MHz, $CDCl_3$) δ: 0.05 (s, 6 H), 0.88 (s, 13 H), 1.20 (m, 6 H), 1.48–1.87 (m, 9 H), 1.92 (dt, $J = 13.5, 3.4$ Hz, 1 H), 2.07–2.17 (m, 1 H), 2.35–2.50 (m, 4 H), 2.77–2.85 (m, 1 H) 3.60 (tt, $J = 10.4, 5.2$ Hz, 1 H), 5.71 (s, 1 H); ^{13}C NMR (75 MHz, $CDCl_3$) δ: –4.4, –4.3, 14.1, 17.8, 18.5, 20.9, 24.5, 26.1, 31.1, 32.0, 36.0, 36.7, 38.8, 43.0, 44.7, 46.1, 48.2, 50.5, 71.5, 126.0, 167.2, 201.5, 220.7; HRMS (ESI): Calcd. for (M+H): 417.2825. Found: 417.2798; Anal. Calcd. For $C_{25}H_{40}O_3Si$: C, 72.06; H, 9.68. Found: C, 71.83; H, 9.73.

29. To address the differences in specific rotation between that reported in footnote 28 and literature values, the *tert*-butyldimethylsilyl

24

group is removed with 1.0 M tetrabutylammonium fluoride solution in THF to prepare corresponding hydroxyl derivative and compare its specific rotation with that reported in the literature:

$[\alpha]_D^{21} = -82.4°$ (*c* 1.0, DCE), lit.[5] $[\alpha]_D^{23} = -82.8°$ (methanol).

Safety and Waste Disposal Information

All hazardous materials should be handled and disposed of in accordance with "Prudent Practices in the Laboratory"; National Academies Press; Washington, DC, 2011.

3. Discussion

The selective oxidation of hydrocarbons is a topic of increasing importance in organic chemistry and desirability for organic synthesis.[6-22] Allylic oxidation is particularly useful because it generates an α,β-unsaturated carbonyl moiety that is broadly functional in organic syntheses.[23-25] Steroids are prevalent substrates for oxyfunctionalization because the hydrocarbon steroidal core is easily accessible from natural materials, and selective functionalization can be easily achieved.

An oxidative system of great potential has been recently reported: rhodium(II) caprolactamate-catalyzed oxidations by *tert*-butyl hydroperoxide (TBHP). The detailed procedure for allylic oxidation and a synthesis of the catalyst are presented in this article. This methodology has proven to be superior in allylic oxidation[26-28] as well as in benzylic[29] and propargylic[30] oxidations, oxidative Mannich reactions of tertiary anilines[31] and oxidations of secondary amines to imines.[32] Some examples of allylic oxidation are presented in Table 1. These reactions give high product yields even when electron poor alkenes (entries 1-3, Table 1) are employed, and this methodology is particularly advantageous in oxidations of steroidal substrates (entries 4-5, Table 1). Although allylic oxidation of 1-substituted-cycloalkenes occurs exclusively at the 3-position in most cases,[24-26] an exception is that of limonene oxidation (Figure 1).[25] Although allylic oxidation of acyclic α,β-unsaturated carbonyl compounds and cyclic alkenes occur in good yield, allylic oxidations of simple acyclic alkenes have not been reported. The mechanism of the reaction is reported to involve formation of allylic radical intermediates that undergo trapping by molecular oxygen.[5]

Table 1. Examples of allylic oxidation with $Rh_2(cap)_4$ / T-HYDRO system.

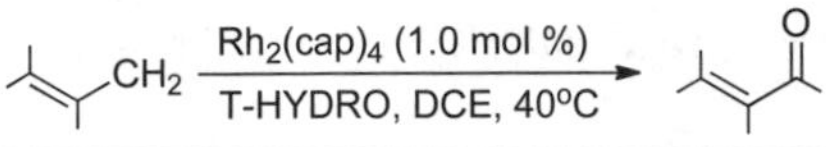

Entry	Substrate	Equiv of T-HYDRO	Time (h)	Product	Yield (%)
1	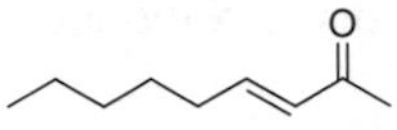	10	40	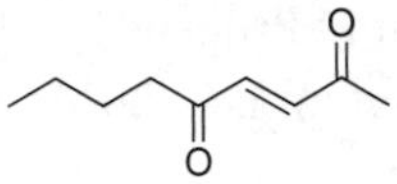	84[27]
2	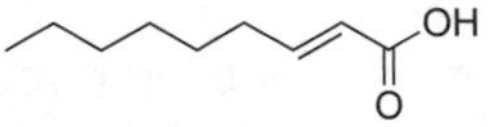	10	40	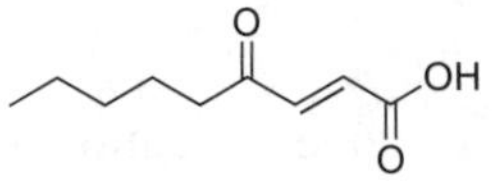	87[27]
3	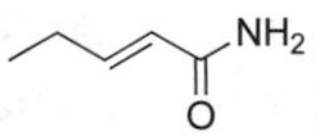	8	40	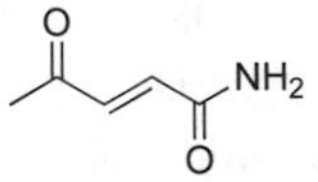	65[27]
4	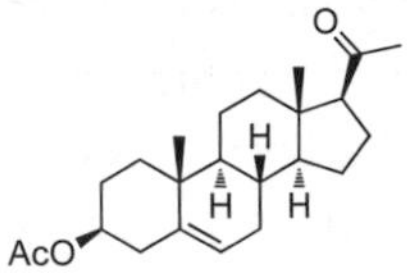	5	20	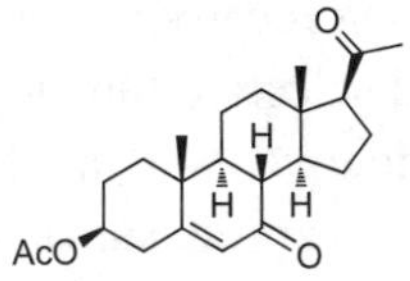	80[28]
5	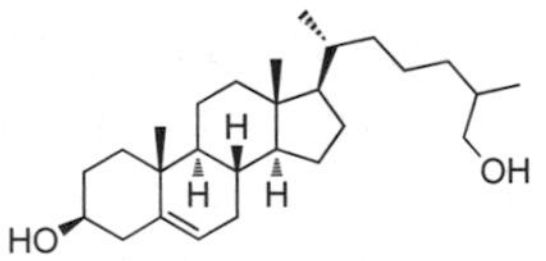	5	20	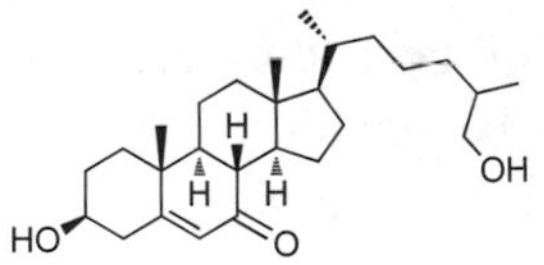	60[28]

Figure 1.[27] Limonene oxidation

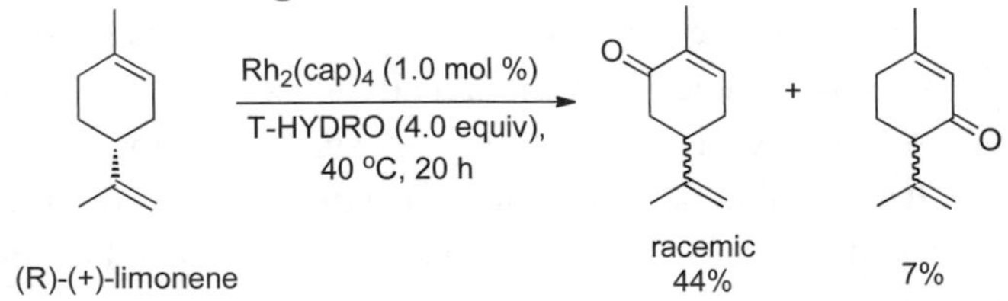

(R)-(+)-limonene   racemic 44%   7%

Org. Synth. **2012**, *89*, 19-33

Several methodologies have been developed to achieve efficient allylic oxidation as alternatives to stoichiometric oxidants selenium dioxide[33,34] and chromium(VI).[35-41] Selenium(IV) oxide has the opposite regioselectivity for carbonyl formation of chromium(VI) or of *tert*-butyl hydroperoxide oxidations, and both selenium dioxide and chromium(VI) have well-known disadvantages in their applications.[42] As an alternative, procedures have been introduced that employ the mild oxidant *tert*-butyl hydroperoxide (TBHP), catalyzed by CrO_3,[43] (n-Bu$_3$SnO)$_2$CrO$_2$,[44] Cr(CO)$_6$,[45,46] benzotriazole $\times$ CrO$_3$,[47] ZrOH$_2^+$ $\times$ Cr$_2$O$_7^{2-}$,[48] 2PyH$^+$ $\times$ Cr$_2$O$_7^{2-}$,[49,50] C$_5$H$_8$N$_2$H[CrO$_3$F],[36] RuCl$_3$,[51] Co(OAc)$_2$,[52-54] CuI,[55,56] BiCl$_3$,[57] Mn$_3$O(OAc)$_9$,[4] NaClO$_2$,[58] NaOCl,[59] and Pd(OH)$_2$.[58,59] Most often excess *tert*-butyl hydroperoxide is required because of its catalytic conversion to di-*tert*-butyl peroxide, water, molecular oxygen.[25] Typical conditions and reaction outcomes are presented for comparison in Table 2 for the oxidation of cholesteryl acetate by those reactants for which this process has been reported. Chromium reagents and chromium-based catalysts (entries 9-11, Table 2) have limited applicability especially on an industrial scale due to a well-established high toxicity and carcinogenicity of chromium(VI)[42] Reactions catalyzed by chlorine-containing salts (entries 1-2, Table 2) have limited attractiveness for large scale processes due to moderate yields obtained in these oxidation reactions. Some copper- and manganese-based systems (entries 3, 5, Table 2) require large loadings of heavy metals that complicate waste treatment procedures. Several oxidation systems are based on the use of a TBHP solution in decane as an oxidant (entries 3-4, 6-7, Table 2) that is significantly more expensive than the 70% aqueous solution of TBHP (T-HYDRO). Hydrogen peroxide has been reported to effect allylic oxidation promoted by an iron(III) complex (entries 12-13, Table 2), but yields are low and further refinement will be required in order for this oxidant to be effectively employed.[60,61] Use of over 10 equiv of *tert*-butyl hydroperoxide per mole of substrate (entries 4, 5, and 8, Table 2) is to be avoided since the increased explosive potential of hydroperoxides must be taken into consideration during scale-up. The rhodium-based process occurs in a high yield with low catalyst and peroxide loadings under mild conditions. The oxidant, T-HYDRO, is inexpensive, and the dirhodium-catalyzed oxidation process is relatively free of competing reactions that include both oxidation of alcohols and alcohol formation.

Table 2. Cholesteryl acetate oxidation

Entry	Catalyst/Additive (mol %)	Oxidant	Equivalents of Oxidant	Time (h)	Temp (°C)	Yield (%)
1	NaClO$_2$ (120%)	T-HYDRO	5	60	80	66[58]
2	NaOCl (200%)	T-HYDRO	6	10	2 - 5	68[59]
3	Mn$_3$O(OAc)$_3$ (10%)	TBHP in decane	5	48	20	85[4]
4	BiCl$_3$ (10%)	TBHP in decane	11	22	70	82[57]
5	CuI (68%) / n-Bu$_4$N$^+$Br$^-$ (12%)	T-HYDRO	30	24	40	76[55]
6	CuI (2.6%)	TBHP in decane	7	24	70	80[56]
7	Co(OAc)$_2$ immob. on silica (2.2%)	TBHP in decane	7	48	70	70[53]
8	RuCl$_3$ × H$_2$O (0.7%)	T-HYDRO	10	24	20	75[51]
9	Cr$_2$O$_7^{2-}$ (immob. on SiO$_2$ / ZrO$_2$)	T-HYDRO	2	n/a	20	48[48]
10	Cr$_2$O$_7^{2-}$ × 2PyH$^+$ (2.5%)	T-HYDRO	7	50	40	78[50]
11	Cr(CO)$_6$ (50%)	TBHP (90% in water)	3	15	80	80[45]
12	Fe(o-C$_5$H$_4$CO$_2$)$_3$ (50%)	H$_2$O$_2$ (30% aq.)	3	4.5	20	20[60]
13	Fe(PA)$_3$ (50%)	H$_2$O$_2$ (30% aq.)	5	3	20	34[61]
14	Rh$_2$(cap)$_4$ (1.0%)	T-HYDRO	5	20	40	80[28]

Org. Synth. **2012**, *89*, 19-33

1. University of Maryland, Department of Chemistry, College Park, MD 20742-4454. moratn@umd.edu, mdoyle3@umd.edu. Support for this report was generously provided by the National Science Foundation (CHE-0456911).-Hudson, NY 12504.

2. Bard College, Annandale-on-Hudson, NY 12504.

3. Hosoda, H.; Fukushima, D.K.; Fishman, J. *J. Org. Chem.* **1973**, *38*, 4209-4211.

4. Shing, T. K. M.; Yeung, Y.-Y.; Su, P. L. *Org. Lett.* **2006**, *8*, 3149-3151.

5. Dodson, R. M.; Nicholson, R. T.; Muir, R. D. *J. Am. Chem. Soc.* **1959**, *81*, 6295-6297.

6. Bulman-Page, P. C.; McCarthy, T. J. In *Comprehensive Organic Synthesis*; Trost, B. M., Ed.; Pergamon: Oxford, UK, 1991; Vol. 7.

7. Olah, G. A.; Molnar, A. *Hydrocarbon Chemistry, 2nd Edition*, 2003.

8. Meunier, B. *Chem. Rev.* **1992**, *92*, 1411-1456.

9. Reese, P. B. *Steroids* **2001**, *66*, 481-497.

10. Diaz-Requejo, M. M.; Perez, P. J. *Chem. Rev.* **2008**, *108*, 3379-3394.

11. Skouta, R.; Li, C.-J. *Tetrahedron* **2008**, *64*, 4917-4938.

12. Herrerias, C. I.; Yao, X.; Li, Z.; Li, C.-J. *Chem. Rev.* **2007**, *107*, 2546-2562.

13. Vedernikov, A. N. *Curr. Org. Chem.* **2007**, *11*, 1401-1416.

14. Dick, A. R.; Sanford, M. S. *Tetrahedron* **2006**, *62*, 2439-2463.

15. Conley, B. L.; Tenn, W. J.; Young, K. J. H.; Ganesh, S. K.; Meier, S. K.; Ziatdinov, V. R.; Mironov, O.; Oxgaard, J.; Gonzales, J.; Goddard, W. A.; Periana, R. A. *J. Mol. Catal. A: Chem.* **2006**, *251*, 8-23.

16. Punniyamurthy, T.; Velusamy, S.; Iqbal, J. *Chem. Rev.* **2005**, *105*, 2329-2363.

17. Shul'pin, G. B. In *Transition Metals for Organic Synthesis*; 2nd ed.; Beller, M., Bolm, C., Eds.; New York: Wiley-VHC, 2004; Vol. 2; pp. 215-241.

18. Stahl, S. S. *Angew. Chem., Int. Ed.* **2004**, *43*, 3400-3420.

19. Labinger, J. A. *J. Mol. Catal. A: Chem.* **2004**, *220*, 27-35.

20. Katsuki, T. In *Comprehensive Asymmetric Catalysis*; Jacobsen, E. N., Pfaltz, A., Yamamoto, H., Eds.; Springer: Berlin, 1999; Vol. 2; pp. 791-802.

21. Stahl, S.; Labinger, J. A.; Bercaw, J. E. *Angew. Chem., Int. Ed.* **1998**, *37*, 2181-2192.

22. Shilov, A. E.; Shul'pin, G. B. *Chem. Rev.* **1997**, *97*, 2879-2932.

23. Chen, M. S.; White, M. C. *J. Am. Chem. Soc.* **2004**, *126*, 1346-1347.

24. Delcamp, J. H.; White, M. C. *J. Am. Chem. Soc.* **2006**, *128*, 15076-15077.
25. Arends, I. W. C. E.; Kodama, T.; Sheldon, R. A. In *Ruthenium Catalysts and Fine Chemistry*; Bruneau, C., Dixneuf, P. H., Eds.; Springer-Verlag: Heidelberg, Germany, 2004; pp. 277-320.
26. Catino, A. J.; Forslund, R. E.; Doyle, M. P. *J. Am. Chem. Soc.* **2004**, *126*, 13622-13623.
27. McLaughlin, E. C.; Choi, H.; Wang, K.; Chiou, G.; Doyle, M. P. *J. Org. Chem.* **2009**, *74*, 730-738.
28. Choi, H.; Doyle, M. P. *Org. Lett.* **2007**, *9*, 5349-5352.
29. Catino, A. J.; Nichols, J. M.; Choi, H.; Gottipamula, S.; Doyle, M. P. *Org. Lett.* **2005**, *7*, 5167-5170.
30. McLaughlin, E. C.; Doyle, M. P. *J. Org. Chem.* **2008**, *73*, 4317-4319.
31. Catino, A. J.; Nichols, J. M.; Nettles, B. J.; Doyle, M. P. *J. Am. Chem. Soc.* **2006**, *128*, 5648-5649.
32. Choi, H.; Doyle, M. P. *Chem. Commun.* **2007**, 745-747.
33. Rabjohn, N. *Org. React.* **1976**, *24*, 261-415.
34. Crich, D.; Zou, Y. *Org. Lett.* **2004**, *6*, 775-777.
35. Hua, Z.; Carcache, D. A.; Tian, Y.; Li, Y.-M.; Danishefsky, S. J. *J. Org. Chem.* **2005**, *70*, 9849-9859.
36. Bora, U.; Chaudhuri, M. K.; Dey, D.; Kalita, D.; Kharmawphlang, W.; Mandal, G. C. *Tetrahedron* **2001**, *57*, 2445-2448.
37. Kasal, A. *Tetrahedron* **2000**, *56*, 3559-3565.
38. Li, D.; Spencer, T. A. *Steroids* **2000**, *65*, 529-535.
39. Pouzar, V.; Slavikova, T.; Cerny, I. *Steroids* **1998**, *63*, 454-458.
40. Parish, E. J.; Wei, T. Y. *Synth. Commun.* **1987**, *17*, 1227-1233.
41. Dauben, W. G.; Fullerton, D. S. *J. Org. Chem.* **1971**, *36*, 3277-3282.
42. Parish, E. J.; Kizito, S. A.; Qiu, Z. *Lipids* **2004**, *39*, 801-804.
43. Muzart, J. *Tetrahedron Lett.* **1987**, *28*, 4665-4668.
44. Muzart, J. *Synth. Commun.* **1989**, *19*, 2061-2067.
45. Pearson, A. J.; Chen, Y. S.; Han, G. R.; Hsu, S. Y.; Ray, T. *J. Chem. Soc., Perkin Trans.* **1985**, 267-273.
46. Pearson, A. J.; Chen, Y. S.; Hsu, S. Y.; Ray, T. *Tetrahedron Lett.* **1984**, *25*, 1235-1238.
47. Parish, E. J.; Honda, H.; Hileman, D. L. *Synth. Commun.* **1990**, *20*, 3359-3366.
48. Baptistella, L. H. B.; Sousa, I. M. O.; Gushikem, Y.; Aleixo, A. M. *Tetrahedron Lett.* **1999**, *40*, 2695-2698.

49. Chidambaram, N.; Chandrasekaran, S. *J. Org. Chem.* **1987**, *52*, 5048-5051.
50. Fousteris, M. A.; Koutsourea, A. I.; Nikolaropoulos, S. S.; Riahi, A.; Muzart, J. *J. Mol. Catal. A: Chem.* **2006**, *250*, 70-74.
51. Miller, R. A.; Li, W.; Humphrey, G. R. *Tetrahedron Lett.* **1996**, *37*, 3429-3432.
52. Jurado-Gonzalez, M.; Sullivan, A. C.; Wilson, J. R. H. *Tetrahedron Lett.* **2003**, *44*, 4283-4286.
53. Salvador, J. A. R.; Clark, J. H. *Chem. Commun.* **2001**, 33-34.
54. Salvador, J. A. R.; Clark, J. H. *Green Chem.* **2002**, *4*, 352-356.
55. Arsenou, E. S.; Koutsourea, A. I.; Fousteris, M. A.; Nikolaropoulos, S. S. *Steroids* **2003**, *68*, 407-414.
56. Salvador, J. A. R.; Sa e Melo, M. L.; Campos Neves, A. S. *Tetrahedron Lett.* **1997**, *38*, 119-122.
57. Salvador, J. A. R.; Silvestre, S. M. *Tetrahedron Lett.* **2005**, *46*, 2581-2584.
58. Silvestre, S. M.; Salvador, J. A. R. *Tetrahedron* **2007**, *63*, 2439-2445.
59. Marwah, P.; Marwah, A.; Lardy, H. A. *Green Chem.* **2004**, *6*, 570-577.
60. Kotani, E.; Takeya, T.; Egawa, H.; Tobinaga, S. *Chem. Pharm. Bull.* **1997**, *45*, 750-752.
61. Takeya, T.; Egawa, H.; Inoue, N.; Miyamoto, A.; Chuma, T.; Kotani, E. *Chem. Pharm. Bull.* **1999**, *47*, 64-70.

Appendix
Chemical Abstracts Nomenclature (Registry Number)

ε-Caprolactam: 2*H*-Azepin-2-one, hexahydro-; (105-60-2)
Dirhodium(II) tetrakis[ε-caprolactamate]: Rhodium, tetrakis[μ-(hexahydro-2H-azepin-2-onato-κN1:κO2)]di-, (Rh-Rh); (138984-26-6)
trans-Dehydroandrosterone: Androst-5-en-17-one, 3-hydroxy-, (3β)-; (53-43-0)
tert-Butyl hydroperoxide: Hydroperoxide, 1,1-dimethylethyl; (75-91-2)
1,2-Dichloroethane: Ethane, 1,2-dichloro-; (107-06-2)
Androst-5-ene-7,17-dione, 3-hydroxy-, (3β)-; (566-19-8)
TBDMSCl: Silane, chloro(1,1-dimethylethyl)dimethyl-; (18162-48-6)
1*H*-Imidazole; (288-32-4)

N,N-Dimethylformamide: Formamide, N,N-dimethyl-; (68-12-2)

Androst-5-en-17-one, 3-[[(1,1-dimethylethyl)dimethylsilyl]oxy]-, (3β)-;
  (42151-23-5)

Androst-5-ene-7,17-dione, 3-[[(1,1-dimethylethyl)dimethylsilyl]oxy]-, (3β)-;
  (157302-54-0)

Michael P. Doyle joined the faculty at Hope College in 1968 and moved to Trinity University in San Antonio, TX, in 1984. In 1997 he came to Tucson, AZ, as Vice President, then President, of Research Corporation and Professor of Chemistry at the University of Arizona. In 2003 he moved to the University of Maryland, College Park, as Professor and Chair of the Department of Chemistry and Biochemistry. His research interests are in catalysis, especially reactions catalyzed by paddlewheeled dirhodium compounds, and in the diverse chemistries that surround the dirhodium framework.

Maxim O. Ratnikov was born in Moscow, Russia, in 1984. In 2007 he received his M.S. degree from the Higher Chemical College in Moscow participating in a development of polyfluorinated alcohols as a new electrophilic media under Dr. William Smit's supervision. After receiving his degree, he moved to University of Kentucky where he worked on investigation of bio-inspired nanoparticles under Dr. Marc Knecht's supervision. In 2008, he joined Professor Michael Doyle in University of Maryland where he is currently working towards his doctoral degree. His research interests are in the development of novel oxidative methods and their applications.

Petra L. Goldmann was born in 1986 in Munich where she studied chemistry and biochemistry in Ludwig Maximilian University, Munich, Germany and received her B.Sc. degree in 2009. In spring 2009, she did her internship studies with Professor Michael Doyle in University of Maryland, College Park, USA in the development of allylic oxidation catalyzed by dirhodium (II) caprolactamate. Working towards her PhD studies, she is currently participating in elaboration of new organometallic-based synthetic methodology Professor Paul Knochel in Ludwig Maximilian University

Emily McLaughlin was born in Pennsylvania and received her B.S. degree from Ohio Northern University in 2001. She completed her Ph.D. studies in total synthesis under Professor Jeffrey Winkler at the University of Pennsylvania in 2006. Soon thereafter, she commenced her post-doctoral training at the University of Maryland with Professor Michael Doyle. In the Doyle lab, she worked on various oxidations catalyzed by dirhodium(II) caprolactamate. Emily is currently an Assistant Professor of Chemistry at Bard College, NY where she and her students are pursuing the design of an aziridination protocol to be applied toward the synthesis of amino acids and heterocyclic scaffolds.

Kainan Zhang received his Bachelor's degree in chemical engineering from Hebei University of Science and Technology in 2004. He then joined the graduate program at the Beijing Institute of Technology and obtained a master's degree where he is working on the synthesis of β-lactamase inhibitor Tazobactam. He then moved to the United States and obtained a Ph.D. in Chemistry. During his Ph.D. studies, he worked on transition metal-catalyzed cycloaddition reactions under the guidance of Dr. Janis Louie. He is now a postdoctoral associate and working under the direction of Professor Kay M. Brummond. His current research interests include Rh(I)-catalyzed cyclocarbonylation reactions of allene-ynes.

Supriyo Majumder was born in Kolkata, India in 1980. He received his B.Sc. in 2002 from University of Calcutta and M.Sc in chemistry in 2004 from IIT-Madras, India. In 2009 he earned his PhD in chemistry from Michigan State University under the supervision of Professor. Aaron L. Odom. Then he joined the research group of Professor. Kay M. Brummond at University of Pittsburgh and worked in the diversity oriented synthesis using an allene-containing Tryptophan scaffold. Currently, he is working as a research fellow with Professor. Edwin Vedejs at University of Michigan. His current research interests include ionic-hydrogenation with phosphine boranes and intramolecular hydroboration of olefins.

DISCUSSION ADDENDUM for:
2-Trimethylsilylethanesulfonyl Chloride (SES-Cl)

Prepared by Joshua R. Sacher and Steven M. Weinreb.*[1]
Original article: Weinreb, S. M.; Chase, C. E; Wipf, P.; Venkatraman, S.
Org. Synth. **1997**, *75*, 161.

Sulfonamides are among the most stable amine protecting groups that are able to tolerate a broad range of reaction conditions. This stability, however, can sometimes be problematic for protecting group removal, often requiring harsh conditions. The 2-(trimethylsilyl)ethanesulfonyl (SES) group provides a sulfonamide that combines the compatibility with many synthetic transformations with relatively benign removal conditions. The chemistry of reactions involving the SES group has previously been reviewed;[2] herein, we report some of the more recent examples of its use in methodology and total synthesis.

The SES group can be installed on a primary or secondary amine using 2-(trimethylsilyl)ethanesulfonyl chloride (SES-Cl) and a base, typically a tertiary amine or sodium hydride.[3] An alternate procedure using silver cyanide has been used to improve yields in difficult protections.[4] Generally, cesium fluoride in DMF or tetrabutylammonium fluoride (TBAF) in acetonitrile at elevated temperature is sufficient to remove the SES group,[3] producing TMS-F, ethylene, sulfur dioxide, and the free amine. Additionally, the SES group can be cleaved by HF,[5] tris(dimethylamino)sulfonium difluorotrimethylsilicate (TASF),[6] or refluxing 6 N HCl.[7] In certain special cases, judicious selection of the deprotection strategy can lead to different products, as demonstrated in the formation of either pyrrole **2** or pyrroline **3** from the corresponding SES-protected pyrroline **1** (Scheme 1).[8]

Org. Synth. **2012**, *89*, 34-43
Published on the Web 8/24/2011
© 2012 Organic Syntheses, Inc.

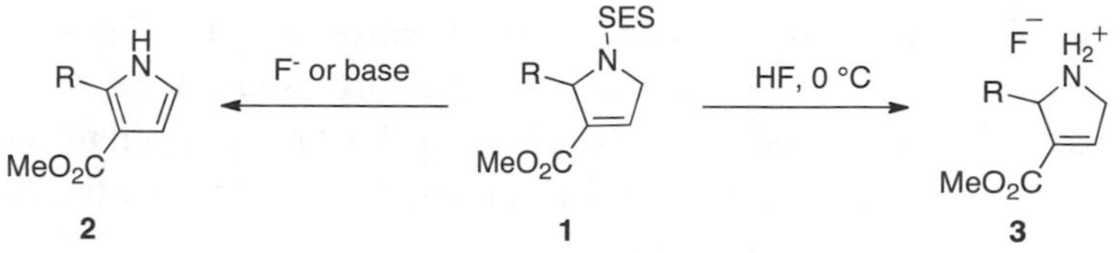

Scheme 1

In addition to its use in introducing a protecting group, SES-Cl can also effectively acidify an amine nitrogen and simultaneously provide a leaving group, as demonstrated in a synthesis of L-azetidine-2-carboxylic acid (**9**), a naturally occurring non-proteogenic amino acid.[9] *N*-Sulfonylation of amino alcohol **4** with SES-Cl, followed by treatment of the resulting bis-SES compound **5** with base, gave the orthogonally-protected cyclic amino acid **6** (Scheme 2). The SES and *t*-butyl ester protecting groups were removed selectively with TBAF or TFA, to yield amine **7** and acid **8**, respectively, while global deprotection to form amino acid **9** was accomplished with HF.

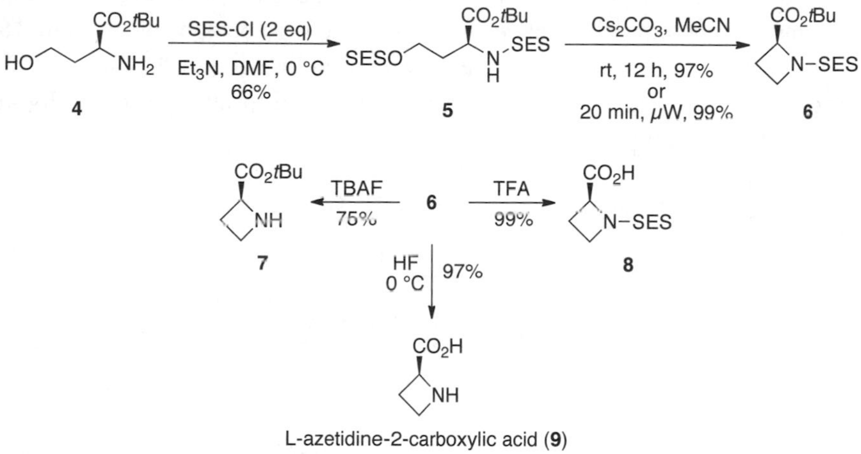

L-azetidine-2-carboxylic acid (**9**)

Scheme 2

Although the traditional manner of SES group installation involves protection of an existing amine, it has become increasingly common to use related derivatives to introduce a SES-protected nitrogen into a molecule. One of the most common SES derivatives, 2-(trimethylsilyl)ethanesulfonamide (SES-NH₂), is easily prepared from the chloride by reaction with anhydrous ammonia[10] or ammonium hydroxide.[11]

SES-NH$_2$ has been shown to be effective in palladium-catalyzed sulfonamidations involving electron deficient aryl and heteroaryl halides (Scheme 3).[12] Both aryl bromides **10** and chlorides **11** give good to excellent yields of the corresponding aryl sulfonamides **12** and **13**, and the reaction tolerates a wide range of functionality.

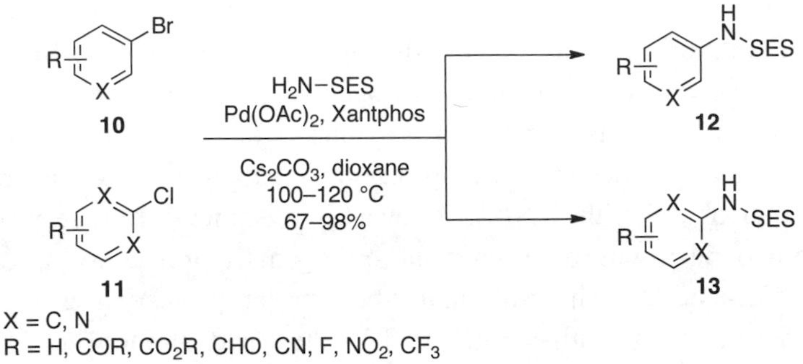

X = C, N
R = H, COR, CO$_2$R, CHO, CN, F, NO$_2$, CF$_3$

Scheme 3

SES-NH$_2$ has also been *N*-alkylated in high yield with benzyl alcohol (**14**) under ruthenium[13] and copper[14] catalysis to give benzyl sulfonamide **15** (Scheme 4). When benzylic acetate **16** was used in place of the alcohol, the sulfonamide alkylation to form **17** has been shown to take place rapidly at ambient temperature.[15]

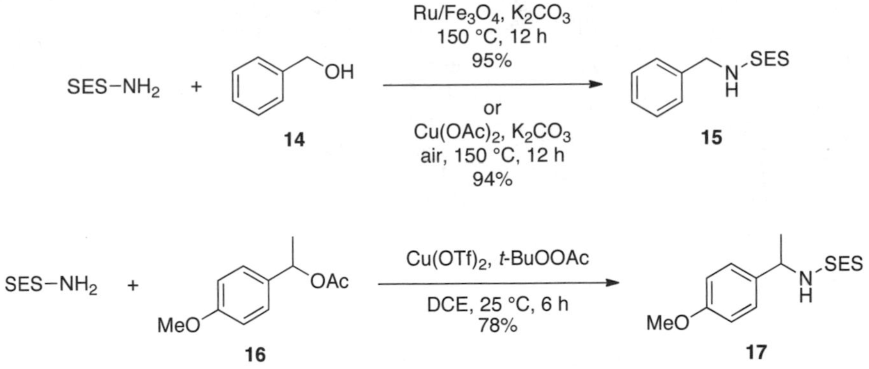

Scheme 4

SES-Sulfonamides bearing a terminal alkene moiety can undergo a cyclization in the presence of a metal catalyst or oxidant (Scheme 5). For example, utilizing a chiral copper catalyst, Chemler and coworkers

36

performed an enantioselective carboamination of **18**, which they proposed occurs via a single-electron mechanism involving radical **19** to form hexahydro-1*H*-benz[*f*]indole **20**.[16]

In contrast to the 5-*exo* cyclization observed with metal catalysts, Michael and coworkers demonstrated that oxidative cyclization of **21** with hypervalent iodine gave solely the 6-*endo* product **24**.[17] A plausible rationalization for this transformation involves opening of an oxidatively-formed aziridinium intermediate **22** with trifluoroacetate resulting in the observed *endo* selectivity. Hydrolysis of the trifluoroacetate **23** gave the SES-protected hydroxypiperidine **24**. In this system, the SES protecting group gave slightly better yields than the corresponding tosyl, or 2- and 4-nosylsulfonamides.

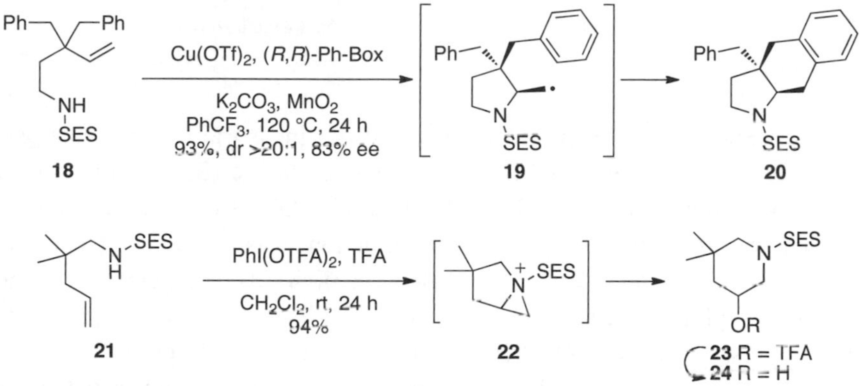

Scheme 5

Iodosulfonamidation methodology using SES-NH$_2$ and a cationic iodine complex developed by Danishefsky[18] was recently adopted by Jurczak and Chaladaj in a formal synthesis of galantinic acid (**25**) (Scheme 6).[19] Activation of the enol ether **26** with I(*sym*-coll)$_2$PF$_6$, followed by opening of the iodonium intermediate by SES-NH$_2$, gave the iodosulfonamide **27** with good diastereoselectivity. Under basic aqueous conditions, sulfonamidoalcohol **29** was produced by hydrolysis of the SES-aziridine intermediate **28**. Reductive opening of the pyran ring, protection of the resulting diol as the acetonide, and hydrogenolysis of the PMB group gave advanced intermediate **30** that could easily be converted to galantinic acid (**25**).

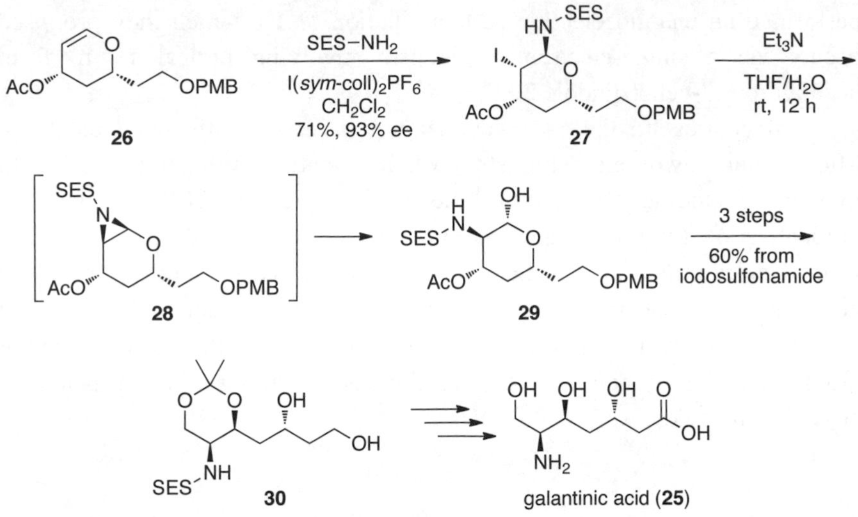

Scheme 6

SES-sulfonamide has also been used in aziridination reactions, typically as the derived ([N-SES]imino)phenyliodinane (SESN=IPh)[20] or in a one-pot procedure using SES-NH$_2$ and PhI=O.[21] Recently, a catalytic asymmetric modification of this aziridination was used by Trost and colleagues in a total synthesis of (–)-oseltamivir (**31**) (Scheme 7).[22] The choice of sulfonamide and catalyst in the aziridine formation was a key in this concise synthesis. Of the sulfonamides examined, only SES-NH$_2$, coupled with a bulky rhodium catalyst, were found to give satisfactory results in the conversion of diene **32** to aziridine **33**. Regioselective opening of **33** with 3-pentanol then gave the corresponding ether **34**, which was acetylated and deprotected to give the free base of (–)-oseltamivir (**31**).

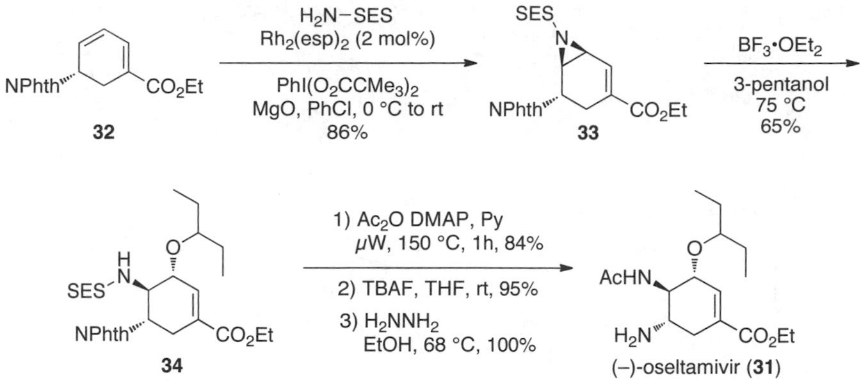

Scheme 7

Recently, Komatsu and coworkers developed a high yielding, metal-free sulfonamide aziridination method using SES-NH$_2$ and t-butyl hypoiodite to convert styrene (**35**) into SES-aziridine **36** (Scheme 8).[23] In a screening of sulfonamides, SES-NH$_2$ was found to be superior to 2-nitrophenyl-, n-butyl-, and p-toluenesulfonamides. The authors note that the SES group is particularly useful in aziridine chemistry because it can be removed without intervention of undesirable side reactions.

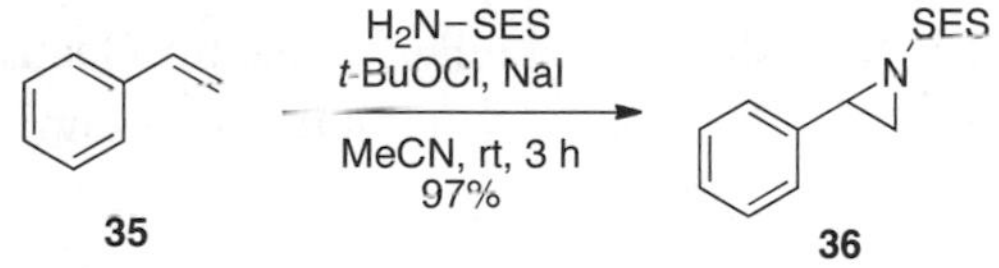

Scheme 8

Due to the electron-withdrawing nature of the sulfonamide moiety, SES aziridines are useful substrates for ring opening reactions. For example, in a synthesis of (+)-preussin (**37**), SES-aziridine **39** was opened by the lithium anion of allyl sulfone **38** to give SES-sulfonamide **40**.[24] Isomerization of the alkene to intermediate **41** by treatment with TBAF at elevated temperature promoted a stereoselective 5-*endo*-trig cyclization that took place with concomitant SES removal. The resulting pyrrolidine **42** was then converted to the natural product **37** in 3 steps.

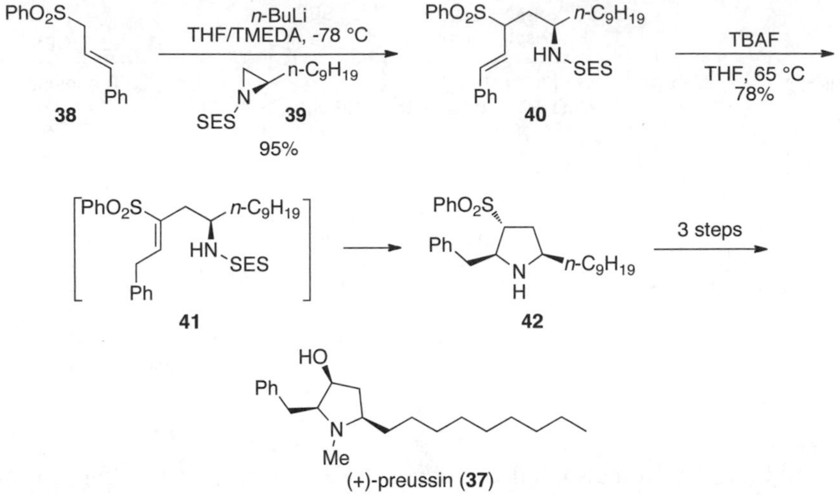

Scheme 9

2-(Trimethylsilyl)ethanesulfonyl azide (SES-N$_3$), which is easily prepared by reaction of SES-Cl with NaN$_3$ in acetone,[25a] has also been used in metal-catalyzed aziridination reactions. For instance, Katsuki and coworkers used SES-N$_3$ to effect an asymmetric transformation of alkenes **43** to aziridines **44** using a chiral ruthenium-salen catalyst (Scheme 10).[25] Interestingly, SES-N$_3$ showed higher enantioselectivity than the corresponding 2- or 4-nosylazides, while maintaining a similar yield.

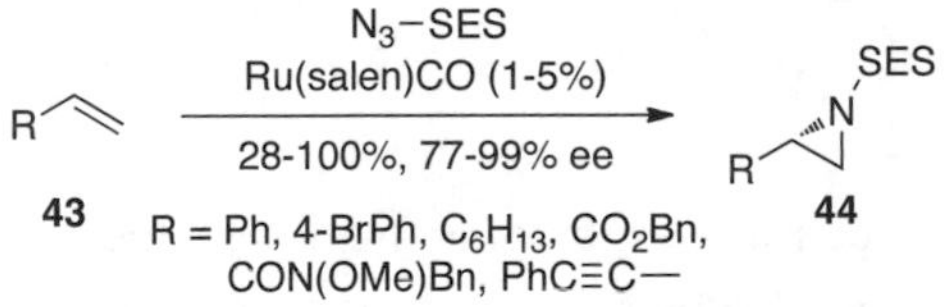

Scheme 10

SES-N$_3$ has also been shown to participate in alkyne-azide 1,3-dipolar cycloaddition reactions (Scheme 11). Thus, in the presence of a catalytic amount of CuI and 2,6-lutidine, the reaction of the azide with phenylacetylene (**45**) proceeds smoothly to give the 4-phenyl 1,2,3-triazole **46**.[26] By changing the base to triethylamine and adding an alcohol to the reaction mixture, the corresponding *N*-protected imidate **48** can be isolated.[27] If water is used, the acylated sulfonamide **49** is obtained in high yield.[28] The

40

reactions involving alcohol and water most likely proceed through addition to a transient ketenimine intermediate **47**.

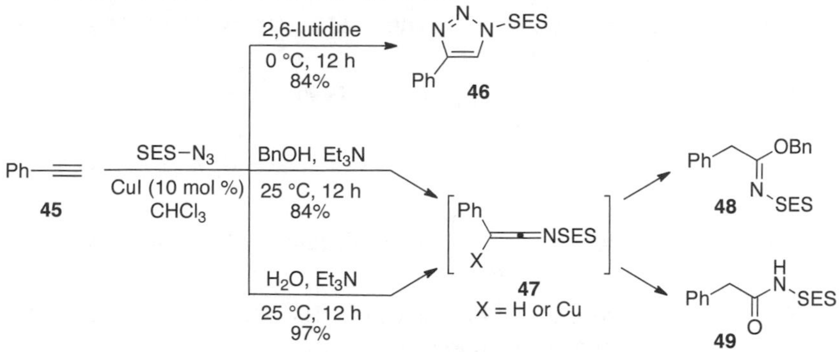

Scheme 11

N-Acyl-SES-sulfonimides can also be obtained from carboxylic acids and SES-N$_3$ via a one-pot, three-step sequence (Scheme 12).[29] For example, activation of acids **50** with *t*-butyl chloroformate, followed by treatment with lithio trimethylsilyl thiolate, and methanolysis of the resulting silylated compound gives the thioacid, which is immediately combined with SES-N$_3$ to give the N-acylated sulfonimides **51**. These sulfonimides can be N-alkylated and the SES group can be selectively removed under mild conditions.

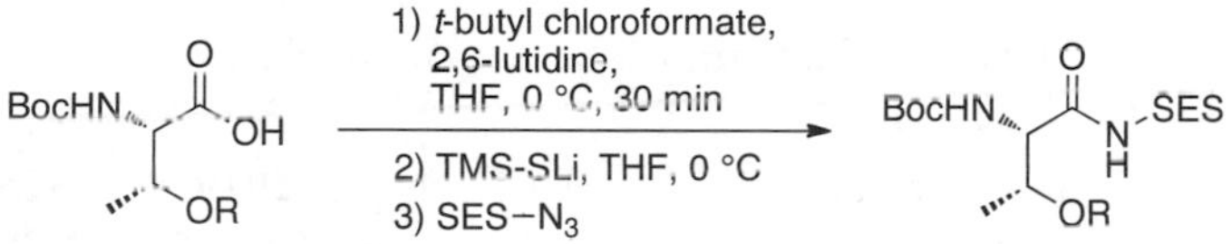

50a: R = H
50b: R = Bn

51a: R = H, 94%
51b: R = Bn, 96%

Scheme 12

1. Department of Chemistry, The Pennsylvania State University, University Park, PA, 16802. smw@chem.psu.edu
2. For recent reviews, see: a) Ribière, P.; Declerck, V.; Martinez, J.; Lamaty, F. *Chem. Rev.* **2006**, *106*, 2249; b) Chambert, S.; Désiré, J.; Décout, J.-L. *Synthesis* **2002**, 2319.
3. Weinreb, S. M.; Demko, D. M.; Lessen, T. A. *Tetrahedron Lett.* **1986**, *27*, 2099.

4. Hale, K. J.; Domostoj, M. M.; Tocher, D. A.; Irving, E.; Scheinmann, F. *Org. Lett.* **2003**, *5*, 2927.

5. Boger, D. L.; Ledeboer, M. W.; Kume, M.; Searcey, M.; Jin, Q. *J. Am. Chem. Soc.* **1999**, *121*, 11375.

6. Dauban, P.; Dodd, R. H. *J. Org. Chem.* **1999**, *64*, 5304.

7. Decicco, C. P.; Grover, P. *Synlett* **1997**, 529.

8. Declerck, V.; Allouchi, H.; Martinez, J.; Lamaty, F. *J. Org. Chem.* **2007**, *72*, 1518.

9. Bouazaoui, M.; Martinez, J.; Cavelier, F. *Eur. J. Org. Chem.* **2009**, 2729.

10. Stien, D.; Anderson, G. T.; Chase, C. E.; Koh, Y.-H., Weinreb, S. M. *J. Am. Chem. Soc.* **1999**, *121*, 9574.

11. Benohoud, M.; Leman, L.; Cardoso, S. H.; Retailleau, P.; Dauban, P.; Thierry, J.; Dodd, R. H. *J. Org. Chem.* **2009**, *74*, 5331.

12. Anjanappa, P.; Mullick, D.; Selvakumar, K.; Sivakumar, M. *Tetrahedron Lett.* **2008**, *49*, 4585.

13. Shi, F.; Tse, M. K.; Zhou, S.; Pohl, M.-M.; Radnik, J.; Hübner, S.; Jähnisch, K.; Brückner, A.; Beller, M. *J. Am. Chem. Soc.* **2009**, *131*, 1775.

14. Shi, F.; Tse, M. K.; Cui, X.; Gördes, D.; Michalik, D.; Thurow, K.; Deng, Y.; Beller, M. *Angew. Chem. Int. Ed.* **2009**, *48*, 5912.

15. Powell, D. A.; Pelletier, G. *Tetrahedron Lett.* **2008**, *49*, 2495.

16. Miao, L.; Haque, I.; Manzoni, M. R.; Tham, W. S.; Chemler, S. R. *Org. Lett.* **2010**, *12*, 4739.

17. Lovick, H. M.; Michael, F. E. *J. Am. Chem. Soc.* **2010**, *132*, 1294

18. Griffith, D. A.; Danishefsky, S. J. *J. Am. Chem. Soc.* **1990**, *112*, 5811

19. Chaladaj, W.; Jurczak, J. *Eur. J. Org. Chem.* **2011**, 1223.

20. Dauban, P.; Dodd, R. H. *J. Org. Chem.* **1999**, *64*, 5304.

21. Dauban, P.; Sanière, L.; Tarrade, A.; Dodd, R. H. *J. Am. Chem. Soc.* **2001**, *123*, 7707.

22. a) Trost, B. M.; Zhang, T. *Angew. Chem. Int. Ed.* **2008**, *47*, 3759; b) Trost, B. M.; Zhang, T. *Chem. Eur. J.* **2011**, *17*, 3630.

23. Minakata, S.; Morino, Y.; Oderaotoshi, Y.; Komatsu, M. *Chem. Commun.* **2006**, 3337.

24. Caldwell, J. J.; Craig, D.; East, S. P. *ARKIVOC* **2007**, *7*, 67.

25. a) Kawabata, H.; Omura, K.; Uchida, T.; Katsuki, T. *Chem. Asian. J.* **2007**, *2*, 248; b) Kawabata, H.; Omura, K.; Katsuki, T. *Tetrahedron Lett.* **2006**, *47*, 1571.

26. Yoo, E. J.; Ahlquist, M.; Kim, S. H.; Bae, I.; Fokin, V. V.; Sharpless, K. B.; Chang, S. *Angew. Chem. Int. Ed.* **2007**, *46*, 1730.
27. Yoo, E. J.; Bae, I.; Cho, S. H.; Han, H.; Chang, S. *Org. Lett.* **2006**, *8*, 1347.
28. Cho, S. H.; Yoo, E. J.; Bae, I.; Chang, S. *J. Am. Chem. Soc.* **2005**, *127*, 16046.
29. Barlett, K. N.; Kolakowski, R. V.; Katukojvala, S.; Williams, L. J. *Org. Lett.* **2006**, *8*, 823.

Steven M. Weinreb was born in Brooklyn, and grew up in Rochester, NY. He did his undergraduate work at Cornell University, and received a doctorate from the University of Rochester with Marshall Gates in 1967. He then held NIH Postdoctoral Fellowships at Columbia University during 1966-67 with Gilbert Stork and at MIT during 1967-70 with George Büchi. He began his independent scientific career at Fordham University in New York City as an Assistant Professor of Chemistry in 1970. In 1978, he joined the faculty at Penn State and was named Russell and Mildred Marker Professor of Natural Products Chemistry in 1987. Throughout his career he has been involved in research on the total synthesis of natural products, heterocyclic chemistry and the development of new synthetic methods

Joshua Sacher was born in Wilmington, Delaware in 1984. He obtained his B.S. degree in biochemistry from the University of Delaware in 2005. While at Delaware, he did undergraduate research with Douglass Taber and had an internship at Cephalon, Inc. He then joined the Weinreb group at Penn State University, where he is currently pursuing his Ph.D degree.

Preparation of Tetrahydroisoquinoline-3-ones Via Cyclization of Phenyl Acetamides Using Eaton's Reagent

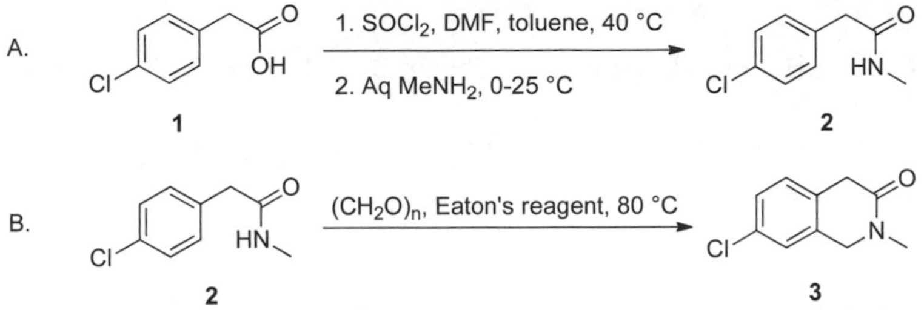

A.

1. SOCl$_2$, DMF, toluene, 40 °C

2. Aq MeNH$_2$, 0-25 °C

B.

(CH$_2$O)$_n$, Eaton's reagent, 80 °C

Submitted by Qiang Yang, Luckner G. Ulysse, Mark D. McLaws, Daniel K. Keefe, Peter R. Guzzo, and Brian P. Haney.[1]
Checked by Andrew S. Cosbie and Margaret M. Faul.

1. Procedure

A. *2-(4-Chlorophenyl)-N-methylacetamide (2)*. A three-necked, 500-mL round-bottomed flask is equipped with a 4-cm magnetic stir bar, a temperature probe, a 50-mL addition funnel, and a reflux condenser with a nitrogen gas inlet adaptor (Note 1). The apparatus is flushed with nitrogen for a minimum of 15 min and charged with 4-chlorophenyl acetic acid (50.0 g, 293 mmol) and 100 mL of anhydrous toluene (Note 2) under nitrogen. Anhydrous DMF (2.28 mL, 2.15 g, 29.3 mmol, 0.1 equiv) (Note 2) is added and the mixture is heated to 35 °C with a heating mantle. SOCl$_2$ (25.6 mL, 41.9 g, 352 mmol, 1.2 equiv) (Note 2) is added through the addition funnel over 30 min at such a rate that the internal temperature is maintained at <40 °C (Note 3). The resulting light brown solution is stirred at 40 °C and monitored by HPLC until the reaction is complete (Notes 4, 5, and 6).

A three-necked, 1-L round-bottomed flask is equipped with an 8-cm mechanical stirrer, a temperature probe, a 250-mL addition funnel, and a nitrogen gas inlet adaptor. The flask is charged with methylamine solution (40 wt% in water or 11.5M, 128 mL, 114 g, 1470 mmol, 5.0 equiv) (Note 7) and cooled to 5 °C with an ice bath.

The acid chloride solution is transferred from the 500-mL flask to the 250-mL addition funnel. The solution is added to the 1-L flask containing

44

Published on the Web 8/25/2011

the methylamine solution over 45 min while maintaining the internal temperature at <25 °C with an ice bath (Note 8). *Caution! The reaction is exothermic* (Note 9). The resulting white suspension is stirred at ambient temperature for a minimum of 30 min and filtered through a Büchner funnel. The filter cake is washed with water (2 × 100 mL) and dried under vacuum (30 mmHg) at 40 °C for a minimum of 24 h to afford 50.9–51.5 g (95–96%) of the desired product **2** as a white crystalline solid (Notes 4 and 10).

B. *7-Chloro-2-methyl-1,4-dihydro-2H-isoquinolin-3-one (3)*. A three-necked, 500-mL round-bottomed flask is equipped with a 3-cm magnetic stir bar, a temperature probe, a 50-mL addition funnel, and a reflux condenser with a nitrogen gas inlet adaptor (Note 1). The apparatus is flushed with nitrogen for a minimum of 15 min and charged with 50 mL of Eaton's reagent (Note 11). 2-(4-Chlorophenyl)-*N*-methylacetamide (10.0 g, 54.5 mmol) is charged in portions, resulting in an exotherm from 23 °C to 29 °C (Note 12). Paraformaldehyde (1.98 g, 65.3 mmol, 1.2 equiv) (Notes 13 and 14) is added and the reaction mixture is heated at 80 °C with a heating mantle for 2 h to afford a brown solution [*Caution! The reaction is exothermic* (Note 15)], at which point HPLC analysis indicates complete consumption of starting material **2** (Notes 4 and 16). The reaction mixture is cooled to 5 °C with an ice bath and water (50 mL) is added through the addition funnel over 30 min while maintaining the internal temperature at <25 °C. *Caution! The addition of water is exothermic.* Isopropyl acetate (IPAc) (50 mL) is added and the mixture is cooled to 5 °C with an ice bath. The pH of the mixture is adjusted to 8–8.5 using 19M NaOH solution (~48-49 mL) while maintaining the internal temperature at <25 °C with an ice bath. *Caution! The addition of NaOH is exothermic.* The solid precipitate is filtered off with a Büchner funnel. The filter cake is washed with IPAc (10 mL). The filtrate mixture is transferred to a 500-mL separatory funnel and the phases are separated. The aqueous phase is extracted with IPAc (50 mL) and the organic phases are combined. The combined organic phases are concentrated by rotary evaporation (35 °C, 30 mmHg) to afford a brown residue. The residue is dissolved in EtOAc (50 mL) and charged on a column (5 × 20 cm) of 200 g of silica gel (Note 17). The column is eluted with 1 L of EtOAc and fraction collection (50-mL fractions) is begun after 500 mL of solvent is eluted. Elution is continued with 1 L of 10% MeOH/EtOAc and the desired product is obtained in fractions 12-22 (Notes 18 and 19), which are concentrated by rotary evaporation (35 °C, 30 mmHg) to afford a yellow oil. The oil is dried under vacuum (30 mmHg) at ambient temperature (20–25 °C) for a minimum of 24 h to afford 9.00-9.07 g (85%) of a light yellow solid (Notes 4, 20, and 21).

2. Notes

1. The glassware was oven-dried at 80 °C for a minimum of 24 h.

2. 4-Chlorophenyl acetic acid (99%), anhydrous toluene, anhydrous DMF, and thionyl chloride (Reagent grade) were purchased from Aldrich and used as received.

3. Vigorous off-gassing was observed during and after $SOCl_2$ was added. Sufficient ventilation should be employed to avoid pressure accumulation during the reaction. On large scale (e.g. >300 g scale) a caustic scrubber was used to sequester acidic off-gas.

4. Reaction progress and product purity were evaluated by HPLC analysis using a Waters Sunfire C18 3.5 μm column (150 mm × 4.6 mm) with mobile phases A (water + 0.05% TFA) and B (acetonitrile + 0.05% TFA) and detection at 220 nm; flow: 1.0 mL/min; temp. 25 °C; and gradient: 0 min: A = 95%, B = 5%; 20 min: A = 5%, B = 95%; 20.1 min: A = 95%, B = 5%; and 22 min: A = 95%, B = 5%.

5. The reaction conversion samples were prepared as follows: A sample of reaction mixture (25 μL) was quenched into 50 μL of 40 wt% aqueous $MeNH_2$ and shaken for 5 min. It was diluted with 1:1 acetonitrile/water to 2 mL and 5 μL of the resulting solution was injected on HPLC. The retention time for starting material **1** was 12.3 min, and the retention time of product **2** was 10.5 min under the HPLC conditions in Note 4. Alternatively, the reaction can be monitored by TLC. Silica plates with glass backing were used. The plates were developed in 3:1 EtOAc/Hexanes and visualized with UV light. The R_F value for starting material **1** was 0.47. The R_F value for product **2** was 0.31.

6. Generally the reaction should be complete within 30 min.

7. Methylamine solution (40 wt% in water or 11.5M) was purchased from Sigma Aldrich and used as received. A total of five equiv of $MeNH_2$ were used to quench excess $SOCl_2$ and acids generated during the reaction.

8. The methylamine solution was diluted with an equal volume of water in cases when a thick suspension formed and became difficult to stir.

9. Significant exotherm was observed during the addition of acid chloride solution to methylamine solution. The exotherm was controlled by the addition rate of acid chloride. As confirmed by safety assessment

46

evaluated by RC1 calorimetry, the heat output of the amide formation stayed constant during the addition of the acyl chloride, characteristic of a fast reaction with no potential for latent reaction.[2]

10. Physical properties and spectral data for **2** are as follows: mp 106–107 °C; ^{1}H NMR (600 MHz, CDCl$_3$) δ: 2.77 (d, J = 4.9 Hz, 3 H), 3.53 (s, 2 H), 5.43 (bs, 1 H), 7.20 (d, J = 8.4 Hz, 2 H), 7.33 (d, J = 8.4 Hz, 2 H); ^{13}C NMR (150 MHz, CDCl$_3$) δ: 26.5, 42.9, 129.1, 130.8, 133.3, 133.4, 170.9; IR (film) [cm^{-1}]: 3281, 1646, 1557, 1492, 1408, 1260, 1085, 1017, 806, 734; HRMS (ES+) calculated for C$_9$H$_{11}$NOCl 184.05292, found 184.05252; HPLC >99% (t_R = 10.5 min).

11. Eaton's reagent was purchased from Aldrich (7.7 wt% of P$_2$O$_5$ in MsOH) and used as received. Alternatively, fresh Eaton's reagent was prepared in-house at a concentration of 7.5 wt% by adding P$_2$O$_5$ (12 g) to methanesulfonic acid (100 mL) at <25 °C. A slight exotherm occurred which was easily controlled by the rate of P$_2$O$_5$ addition. Once the addition was complete, the solution was stirred at ambient temperature for 18 h and then stored in airtight containers under nitrogen. Cleaner reaction profile was obtained when freshly prepared Eaton's reagent was used.

12. Safety assessment evaluated by RC1 calorimetry indicated that a mild thermal event took place upon addition of the phenyl acetamide (ΔH = –35 kJ/mol, ΔT_{ad} = 14 K) which rapidly subsided.[2]

13. Paraformaldehyde (95%) and sodium hydroxide solution (19M or 50 wt%) were purchased from Aldrich and used as received.

14. RC1 study revealed that addition of paraformaldehyde was marked by an initial endothermic event devoid of any significant thermal activity until the system was heated.[2]

15. The thermal profile observed during heating resulted in a net output of –120 kJ/mol (ΔT_{ad} = 43 K). Upon completion of the heat cycle, a decaying heat flow was observed which ended within an hour, indicating that no rapid temperature spike or uncontrollable heat accumulation occurred.[2]

16. The reaction progress was monitored using the HPLC method described in Note 4. The reaction conversion samples were prepared as follows: A sample of reaction mixture (25 µL) was quenched into 1 mL of water. It was diluted with acetonitrile to 2 mL and 5 µL of the resulting solution was injected on HPLC. The retention time for starting material **2** was 10.5 min, and the retention time of product **3** was 11.3 min under the HPLC conditions in Note 4.

17. Silica gel SiliaFlash® F60 (40–63 μm/230–400 mesh) was purchased from Silicycle.

18. The starting material **2** co-elutes with the product during column chromatography purification, thus it is necessary to ensure complete consumption of the starting material before work up.

19. The fractions were monitored by HPLC analysis for purity using the HPLC method described in Note 4. HPLC samples were prepared as follows: A sample of each fraction (100 μL) was evaporated under a flow of nitrogen to a residue and dissolved in 2 mL of 1:1 acetonitrile/water. The typical injection volume was 10 μL, but the volume was adjusted for some fractions based on the concentration of each fraction.

20. Physical properties and spectral data for **3** are as follows: mp 51–54 °C; ^{1}H NMR (600 MHz, CDCl$_3$) δ: 3.11 (s, 3 H), 3.58 (s, 2 H), 4.47 (s, 2 H), 7.09 (d, J = 8.2 Hz, 1 H), 7.17 (d, J = 1.9 Hz, 1 H), 7.23 (dd, J = 8.2 Hz, 1 H); ^{13}C NMR (150 MHz, CDCl$_3$) δ: 34.4, 36.3, 52.4, 125.2, 127.7, 128.7, 130.7, 132.3, 132.6, 168.2; IR (film) [cm^{-1}]: 1626, 1488, 1389, 1331, 1244, 1085, 902, 810; HRMS (ES+) calculated for C$_{10}$H$_{11}$NOCl 196.05292, found 196.05257; HPLC 97.6-98.3% (t_R = 11.3 min).

21. The product needs to be stored under nitrogen at –20 °C as slow decomposition was observed when it was stored at ambient temperature (20–25 °C).

Safety and Waste Disposal Information

All hazardous materials should be handled and disposed of in accordance with "Prudent Practices in the Laboratory"; National Academies Press; Washington, DC, 2011.

3. Discussion

Tetrahydroisoquinoline is a ubiquitous structural framework present in numerous pharmacologically relevant molecules.[3,4] Such structural framework was envisioned to be achieved via the Pictet-Spengler cyclization;[5] however, conditions commonly used for the preparation of tetrahydroisoquinoline-3-ones required the use of polyphosphoric acid (PPA).[6] The high temperature requirement of this reaction coupled with the high viscosity associated with the PPA at 160 °C rendered this condition unattractive on scale.

48

The present procedure is a simple, efficient, and economical method for the preparation of tetrahydroisoquinoline-3-ones that can be easily scaled up. The safety concerns are minimized with the use of Eaton's reagent since the resultant mixtures are more mobile solutions with temperature requirements much lower than that of PPA. Although the use of Eaton's reagent as a dehydrating agent in various reactions is well documented, Eaton's reagent-mediated cyclization of phenyl acetamide derivatives with formaldehyde was not available in the literature[7-15]. Cyclization of phenyl acetamides using Eaton's reagent was broadened to understand the scope and limitation of the process. Some representative examples of tetrahydroisoquinoline-3-ones prepared by this method are compiled in Table 1. For the phenyl acetamides with electron-withdrawing groups on 2- or 4- positions, the desired cyclized products were isolated in excellent yield by neutralization of the reaction to pH 8.0 – 8.5 with sodium hydroxide (NaOH), followed by extraction of the product with isopropyl acetate (IPAc) and isolation by concentration. A ~2:1 mixture favoring 6-chloro-2-methyl-1,4-dihydro-2*H*-isoquinolin-3-one was obtained when 2-(3-chlorophenyl)-*N*-methylacetamide was subjected to this condition, as the formation of 8-chloro-2-methyl-1,4-dihydro-2*H*-isoquinolin-3-one was more hindered by the chloro group. Complicated results were afforded when phenyl acetamides with electron-donating groups were treated with Eaton's reagent under the same conditions.

Table 1. Scope of cyclization using Eaton's reagent

Phenyl acetamide	Product	Isolated yield (%)
		95[a]
		80[a]
		89[a]
		96[a]
		98[a]
		77[a]
		79[a]
		71[a]
		41[b]
		18[b]

[a] Crude yield; [b] Purified yield.

1. Chemical Development Department, AMRI, 21 Corporate Circle, Albany, New York 12203, U.S.A. e-mail: qiang.yang@amriglobal.com.
2. Ulysse, L. G.; Yang, Q.; McLaws, M. D.; Keefe, D. K.; Guzzo, P. R.; Haney, B. P. *Org. Process Res. Dev.* **2010**, *14, 225–228.*
3. Wenner, W. *J. Med. Chem.* **1965**, *8*, 125–126.
4. Klutschko, S.; Blankley, C. J.; Fleming, R. W.; Hinkley, J. M.; Werner, A. E.; Nordin, I.; Holmes, A.; Hoefle, M. L.; Cohen, D. M.; Essenburg, A. D.; Kaplan, H. R. *J. Med. Chem.* **1986**, *29*, 1953–1961.
5. Pictet, A.; Spengler, T. *Ber. Dtsch Chem. Ges.* **1911**, *44*, 2030–2036.
6. Ivanov, I.; Venkov, A. *Heterocycles* **2001**, 55, 1569–1572.
7. Hansen, K. B.; Balsells, J.; Dreher, S.; Hsiao, Y.; Kubryk, M.; Palucki, M; Rivera, N.; Steinhuebel, D.; Armstrong, J. D.; Askin, D.; Grabowski, E. J. J. *Org. Process Res. Dev.* **2005**, *9*, 634–639.
8. Zewge, D.; Chen, C. Y.; Deer, C.; Dormer, P. G.; Hughes, D. L. *J. Org. Chem.* **2007**, *72*, 4276–4279.
9. Skalitzky, D. J.; Marakovits, J. T.; Maegley, K. A.; Ekker, A.; Yu, X. H.; Hostomsky, Z.; Webber, S. E.; Eastman, B. W.; Almassy, R.; Li, J. *J. Med. Chem.* **2003**, *46*, 210–213.
10. Dorow, R. L.; Herrinton, P. M.; Hohler, R. A.; Maloney, M. T.; Mauragis, M. A.; McGhee, W. E.; Moeslein, J. A.; Strohbach, J. W.; Veley, M. F. *Org. Process Res. Dev.* **2006**, *10*, 493–499.
11. Pandit, C. R.; Polniaszek, R. P.; Thottathil, J. K. *Synth. Commun.* **2002**, *32*, 2427–2432.
12. Zhao, D.; Hughes, D. L.; Bender, D. R.; DeMarco, A. M.; Reider, P. J. *J. Org. Chem.* **1991**, *56*, 3001–3006.
13. Kano, S.; Yokomatsu, T.; Yuasa, Y.; Shibuya, S. *Chem. Pharm. Bull.* **1985**, *33*, 340–346.
14. Gala, D.; Dahanukar, V. H.; Eckert, J. M.; Lucas, B. S.; Schumacher, D. P.; Zavialov, I. A. *Org. Process Res. Dev.* **2004**, *8*, 754–768.
15. Chae, J.; Buchwald, S. L. *J. Org. Chem.* **2004**, *69*, 3336–3339.

Appendix
Chemical Abstracts Nomenclature (Registry Number)

4-Chlorophenylacetic acid: *p*-Chlorophenylacetic acid, PCPA; (1878-66-6)
Thionyl chloride: sulfur chloride oxide, sulfurous oxychloride, sulfinyl chloride, sulfurous dichloride, thionyl dichloride; (7719-09-7)

Methylamine: monomethylamine, aminomethane, carbinamine, mercurialin; (74-89-5)

2-(4-Chlorophenyl)-N-methylacetamide; (60336-41-6)

Eaton's reagent: Phosphorus pentoxide, 7.7 wt% solution in methanesulfonic acid; (39394-84-8)

Paraformaldehyde: Polyformaldehyde; Polyoxymethylene; Formaldehyde polymer; Polyoxymethylene glycol; Trioxymethylene; (30525-89-4)

Qiang Yang received his B.S. in Chemistry from Beijing Normal University in 1996 and M.S. in Organic Chemistry from Institute of Chemistry, the Chinese Academy of Sciences in 1999. Qiang completed his Ph.D. study under the direction of Prof. David C. Baker at the University of Tennessee, Knoxville in 2002. His Ph.D. research focused on the total synthesis of C-/O- and S-/O- linked oligosaccharide mimetics of Hyaluronic Acid (HA). Qiang joined the Chemical Development group at AMRI in 2003 focusing on route scouting, process development, and technology transfer of processes for pilot manufacture of Active Pharmaceutical Ingredients intended for toxicological and clinical evaluations.

Luckner G. Ulysse Jr. (born in 1970) started his academic studies at the University of Louvain la Neuve, in Belgium transferring later to Walsh University in Canton, OH obtaining his BS degree Magna Cum Laude in 1991. As an NIH pre-doctoral fellow, he completed his Doctoral studies from Purdue University in Chemistry, graduating Cum Laude under the direction of professor Jean Chmielewski (1996). He joined Advanced ChemTech (1996), working on unnatural amino acids, and in 1999 started at AMRI as a Senior Scientist. After nearly 10 years at AMRI working on process development of various APIs, Luckner headed the global chemical development/GMP group as its vice president. In 2010, he joined Cherokee Pharmaceuticals, LLC as Director-Head of Technical Operations overseeing a portfolio of commercial and late stage products.

Mark McLaws (born in 1972) received his B.A. degree in chemistry from Southern Utah University in 1997. He continued his studies under the mentorship of Gary Keck at the University of Utah where he obtained a Ph.D. in 2003. His graduate research included work towards the total synthesis of the natural products Dolabelide B and Rhizoxin D. Upon completing his graduate work, Mark moved to Albany, NY and joined the Chemical Development at AMRI.

Daniel K. Keefe earned his B.S. in Chemistry from Rensselaer Polytechnic Institute and joined AMRI in 2001. He has worked in the areas of Chemical Development and Development Manufacturing, gaining experience in both non-cGMP and cGMP environments. Dan has successfully transferred numerous large-scale projects to the AMRI Rensselaer Pilot Plant and High Potency Suites.

Peter Guzzo obtained his Ph.D. in Organic Chemistry at the University of Notre Dame under Professor Marvin Miller and conducted post-doctoral studies with Professor Arthur Schultz at Rensselaer Polytechnic Institute. Currently, Pete is the Director of Discovery R&D, Chemistry at AMRI where he leads efforts on internal drug discovery programs with a particular focus on central nervous system and metabolic diseases.

Brian P. Haney was born in 1972 in Syracuse, NY. He received his B.S. in Chemistry from Allegheny College (PA.) in 1996 where he worked with Prof. Edward Walsh studying organosulfur compounds. He completed his Ph.D. at the University of Pittsburgh in 2000. His thesis work was performed under the direction of Prof. Dennis P. Curran and focused on tandem radical cyclizations and their application to the synthesis of Natural Products. He joined the Chemical Development group at AMRI in 2000, where he focused on route scouting, process development, technology transfer and Highly Potent Active Pharmaceutical Ingredient (HPAPI) development. Brian moved into AMRI's Rensselaer Large Scale Manufacturing team in 2009 and is currently the Section Head for Commercial Manufacturing.

Andrew Cosbie earned his B.S. in Chemical Engineering from the University of California, Santa Barbara in 2009. He joined Amgen later that year and has been working as a process engineer in the Chemical Process Research and Development group. Andrew has been gaining experience in process characterization, modeling of unit operations, and scale-up of chemical processes in a GMP setting.

Dibenzo[a,e]cyclooctene: Multi-gram Synthesis of a Bidentate Ligand

Submitted by Géraldine Franck, Marcel Brill and Günter Helmchen.[1]
Checked by Tohru Fukuyama and Takuya Nishimura.

Caution! Once α,α'-dibromo-o-xylene is dissolved in an organic solvent it becomes lachrymatory. The use of granular lithium on a larger scale requires special care due to the exothermic formation of lithium bromide.

1. Procedure

A. 5,6,11,12-Tetrahydrodibenzo[a,e]cyclooctene (**1**). A 1-L, three-necked round-bottomed flask, equipped with a 4-cm egg-shaped magnetic stirring bar, a reflux condenser, an internal thermometer and a 250-mL pressure-equalizing dropping funnel (capped with a rubber septum), is flame-dried and purged with argon. Degassed THF (200 mL) (Note 1) is added and subsequently granular lithium (6.63 g, 956 mmol, 2.5 equiv) (Note 2). The dropping funnel is charged with a solution of freshly sublimed α,α'-dibromo-*o*-xylene (101 g, 382 mmol) in degassed THF (150 mL)

(Note 3). This solution is then added dropwise to the stirred suspension of granular lithium over a period of 1.5 h, which begins to reflux after 20 min. During the period of addition the internal temperature of the mixture is kept between 65 °C and 70 °C (Note 4). After complete addition, the mixture is refluxed for 2 h (oil bath) until TLC-analysis indicates complete consumption of the substrate (Notes 5, 6 and 7).

The reaction mixture is allowed to cool to room temperature and is then, for removal of residual lithium, filtered through a fritted glass funnel (5.0 cm diameter, porosity 0) under an atmosphere of argon. The funnel is subsequently rinsed with THF (100 mL) (Note 1). The filtrate is concentrated by rotary evaporation at a bath temperature of 40 °C (255-225 mmHg) and then at 50 °C (7.5 mmHg) for 1 h in order to remove THF (Note 8). Dichloromethane (700 mL) is added to the residue. After stirring for 2 min, insoluble material (Note 9) is removed by filtration with a fritted glass funnel (8.0 cm diameter, porosity 4) filled with silica gel (100 g) (Note 10). The residue is washed with dichloromethane (700 mL) and the filtrate dried over Na_2SO_4 (50 g). The latter is removed by filtration and the filtrate concentrated at 40 °C by rotary evaporation (675-525 mmHg and then 15 mmHg) for 1 h. The crude product remains as a pale yellow semifluidic mass (40.5 g), which solidifies under vacuum overnight (25 °C, 0.3 mmHg). Finally, the product is purified by Kugelrohr distillation (air bath temperature 130–175 °C, 0.25 mmHg) (Note 11) to yield pure **1** (19.6 g, 49%) as colorless plates (Note 12).

B.   5,11-Dibromo-5,6,11,12-tetrahydrodibenzo[a,e]-cyclooctene (**2**). Under an atmosphere of argon in a flame-dried, 500-mL, three-necked round-bottomed flask, equipped with a reflux condensor and a 4-cm egg-shaped magnetic stirring bar, **1** (19.2 g, 92.3 mmol) is dissolved in carbon tetrachloride (200 mL) (Note 13). NBS (35.3 g, 198.4 mmol, 2.15 equiv) (Notes 14 and 15) is added, and the stirred mixture is then refluxed at 85 °C (oil bath temperature). After 2 h NMR analysis (Note 16) indicates complete consumption of the starting material **1** (Note 17). The hot suspension (65 °C) is filtered through a fritted glass funnel (8.0 cm diameter, porosity 3), and the remaining succinimide is washed with carbon tetrachloride (400 mL) (Note 18). The solvent of the filtrate is evaporated under gradually reduced pressure (40 °C, 225 to 18 mmHg). A pale yellow residue remains, which is scraped off the inner wall of the flask, poured into a fritted glass funnel (7.0 cm diameter, porosity 3), washed with distilled water (300 mL) and dried overnight under vacuum (25 °C, 0.3 mmHg) to give 35.4 g of crude **2** as a pale yellow solid (Note 19).

C. *Dibenzo[a,e]cyclooctene* (**3**). Under an atmosphere of argon in a flame-dried, 250-mL, single-necked round-bottomed flask equipped with a magnetic stirring bar, **2** (34.1 g crude product) is dissolved in THF (150 mL) (Note 1). Under an atmosphere of argon in a second flame-dried, 500-mL, single-necked round-bottomed flask equipped with a magnetic stirring bar, potassium *tert*-butoxide (41.6 g, 371 mmol, 4 equiv, assuming 100% purity of **2**) (Note 20) is dissolved in THF (100 mL) (Note 1). The resulting cloudy solution is then cooled with an ice bath. At 0 °C the solution of **2** is transferred through a teflon tube (1.5 mm internal diameter) over a period of 15 min into the 500-mL flask. During the period of addition the color changes from orange to brown and finally to black. Then the ice bath is removed, the mixture is allowed to warm to room temperature and is stirred for 2 h until TLC-analysis indicates complete consumption of the dibromide **2** (Notes 21 and 22). Distilled water (15 mL) is added and the black mixture is poured onto a pad of silica gel (diameter 10 cm x 4 cm height) that has already been wetted with diethyl ether. The pad is rinsed with 700 mL of diethyl ether, and the eluent is collected directly as a single fraction, dried over Na$_2$SO$_4$ (40 g), filtered and concentrated at 40 °C by rotary evaporation (675 to 75 mmHg) and then at 15 mmHg for 1 h. The crude product remains as a brown solid (18.5 g) (Note 23), which is scraped off the inner wall of the flask and dried overnight under vacuum (25 °C, 0.3 mmHg). Purification by Kugelrohr distillation (air bath temperature 130–145 °C, 0.2 mmHg) (Note 24, 25) affords 13.0 g (69%, two steps from **1**) of dibenzo[a,e]cyclooctene (**3**) as colorless plates (Note 26).

2. Notes

1. THF 99.9+% was purchased from Kanto Chemicals Co., Inc., and was purified using a Glass Contour Solvent System; The submitters purchased from Sigma-Aldrich, Inc. and processed through a Solvent Purification System (M. Braun SPS-800, filter material: MB-KOL-A, MB-KOL-M; catalyst: MB-KOL-C).

2. Granular lithium, 99.9+%, was purchased from Sigma-Aldrich, Inc. and was used as received.

3. α,α'-Dibromo-*o*-xylene, 96%, was purchased from Aldrich Chemical Co., Inc. and was sublimed in a Kugelrohr distillation apparatus, which was equipped according to Note 11 (air bath temperature 130–150 °C,

0.07 mmHg) yielding white crystals, which were rinsed out of the receiver flask with diethyl ether. The solvent was removed via rotary evaporation and the residual colorless crystals were dried under vacuo for 2 h (25 °C, 0.04 mmHg).

Upon use of α,α'-dibromo-o-xylene as purchased, i. e. without purification, the yield of **1** dropped to 44-46%, and the submitters have experienced problems with drying the crude product before purification by Kugelrohr distillation.

4.   The addition of α,α'-dibromo-o-xylene may cause a rapid rise of temperature. The addition may need to be temporarily suspended upon onset of the exothermic reaction.

5.   The TLC analysis was performed on Merck precoated analytical plates, 0.25 mm thick, silica gel 60 $F_{254}$and PE as eluent; detection: UV-light (254 nm) or aqueous $KMnO_4$ (1% $KMnO_4$ + 5% Na_2CO_3 in H_2O), R_f (α,α'-dibromo-o-xylene) = 0.21, R_f(**1**) = 0.18, R_f(**4**) = 0.12 (Note 7).

6.   The submitters also monitored the reaction by the GC/MS analysis with a Hewlett-Packard 5890 Series II gas chromatograph equipped with a 5972 series mass selective detector and a HP-1 column (25 m × 0.2 μm; 0.33 mm film thickness, carrier gas: helium). For reaction control the following parameters were used; temperature-program: 50 °C for 3 min; heating rate 20 °C/min; 250 °C for 10 min; injection-temperature 250 °C). GC/MS (EI): t_R(α,α'-dibromo-o-xylene) = 8.9 min, m/z = 264 [M^+]; t_R(**1**) = 10.9 min, m/z 208 [M^+]; t_R(**4**) = 11.5 min, m/z = 312 [M^+].

4

7.   The by-product 5,6,11,12,17,18-hexahydro-tribenzo[a,e,i]cyclodo-decene (**4**) was identified by the following data: mp = 184–185 °C (lit.,[2] 184.5 °C); ^1H NMR (CDCl$_3$, 300.13 MHz) δ: 3.04 (s, 12 H, CH$_2$), 7.20–7.36 (m, 12 H, Aryl); ^{13}C NMR (CDCl$_3$, 75.48 MHz) δ: 37.45 (t, CH$_2$), 126.87, 130.61 (2 d, C-Ar), 140.22 (s, $C_{quart.}$-Ar); IR (KBr): 3058, 3011, 2939, 2864, 1913, 1489, 1473, 1450, 1158, 1051, 1040 cm^{-1}; HRMS: Calcd. for $C_{24}H_{24}$:

312.1878. Found: 312.1888; MS (EI, 70 eV): m/z (%): 312 (40), 207 (100), 193 (30), 178 (30), 115 (40), 105 (35), 91 (25), 78 (25).

8.   At ca. 75 mmHg a white solid precipitates.

9.   The material is LiBr.

10. Silica gel (0.04-0.100 mm) was purchased from Kanto Chemical.

11. Purification by Kugelrohr distillation was carried out in an apparatus equipped with a straight oven tube (29 mm neck fit and 20 cm length) and a cooled 100-mL single bulb receiver flask. A high vacuum pump was used to reduce the pressure.

The submitters reported that fraction 1 (130–160 °C, 0.08–0.05 mmHg) yielded 17.3 g of **1**, fraction 2 (160–180 °C, 0.05 mmHg) yielded 4.2 g of **1** and the triple-coupled by-product **4** as a mixture and fraction 3 (185–250 °C, 0.06 mmHg) yielded 3.8 g of **4** as a pale brown solid. Fraction 2 was subjected to Kugelrohr distillation again to yield 3.3 g of pure **1**.

On a smaller scale run (6.6 g, 25 mmol, of α,α'-dibromo-o-xylene) **1** and **4** were isolated via chromatography on a column (2.5 x 50 cm) of 70 g silica gel with petroleum ether (35-60 °C) as eluent. The product **1** was obtained in fractions 5-15 (100-mL fractions) and **4** in fractions 20-33. After evaporation of the solvent by rotary evaporation (40 °C, 262 mmHg) **1** was obtained as colorless plate shaped crystals (1.09 g, 42%) and the by-product **4** as colorless needles (0.14 g, 6%).

12. Compound **1** has the following physicochemical properties: mp = 108–109 °C (lit.[10] mp 108.5–110 °C); ^{1}H NMR (CDCl$_3$, 400 MHz) δ: 3.06 (s, 8 H, CH$_2$), 6.99 (s, 8 H, Aryl-H); ^{13}C NMR (CDCl$_3$, 100 MHz) δ: 35.12 (t, CH$_2$), 126.07, 129.64 (d, C-Ar), 140.57 (s, C$_{quart.}$-Ar); IR (KBr): 3060, 3011, 2939, 2899, 2845, 2359, 1914, 1491, 1476, 1450, 1157, 1079, 1050.cm^{-1}; Anal. Calcd. for C$_{16}$H$_{16}$: C, 92.26; H, 7.74. Found: C, 92.18; H, 7.87. HRMS: Calcd. for C$_{16}$H$_{17}$: 209.1330. Found: 209.1321; MS (EI, 70 eV): m/z (%): 208 (54), 193 (100), 178 (40), 165 (11), 152 (5), 130 (7), 116 (24), 104 (29), 91 (13), 78 (27).

13. Carbon tetrachloride (99%, extra pure) was purchased from Kanto Chemical and was used as received.

14. NBS, 99%, purchased from Aldrich Chemical Co., Inc was used as received.

15. The use of 2.15 equiv of NBS was necessary for the complete consumption of the substrate. With only 2.0 equiv of NBS the product was obtained as a mixture of undesired monobrominated product and desired

product **2** in a ratio of 14/86, which was determined via its proton NMR spectrum.

16. Analysis via TLC monitoring was unsuccessful, because the R_f values of the dibrominated product **2** and the monobrominated intermediate are the same. Therefore, even if the spot of **1** has disappeared and only the spot of the dibrominated product **2** is visible, conversion might still be incomplete.

17. NMR analysis was conducted in the following manner; a portion of reaction mixture (approximately 10-15 μl) was removed by Pasteur pipette or capillary and dissolved in CDCl$_3$. Completion of the reaction was confirmed by disappearance of monobrominated intermediate peaks found in the ^{1}H NMR (400 MHz) spectrum at $\delta = 3.04$ (s), 3.06 (s), 3.48–3.76 (m), 3.96 (dd, $J = 11.0$, 14.2 Hz).

The submitters monitored the reaction by GC/MS, For reaction control the "default"-method was used (see Note 6). GC/MS (EI): t_R(succinimide) = 5.9 min, $m/z = 99$ [M$^+$]; t_R(**1**) = 10.9 min, m/z 208 [M$^+$]; t_R(monobrominated intermediate) = 12.5 min, $m/z = 286$ [M$^+$]; t_R(**2**) = 14.8 min, $m/z = 366$ [M$^+$].

18. The yield of crude product was significantly lower if the solution was filtered after allowing it to cool to room temperature.

19. Compound **2** was used in the next step without further purification. The compound can, however, be purified by recrystallization: The solid (3.0 g) is dissolved in 12 mL of hot carbon tetrachloride at 85 °C (oil bath temperature). The yellow solution is allowed to cool slowly to room temperature by turning off the heating of the oil bath. Then the flask is cooled to 6 °C (refrigerator) overnight, and the resulting crystals are collected by suction filtration through a small Büchner funnel (diameter 2 cm), washed with 4 mL of ice-cold dichloromethane, and then dried under vacuum (0.04 mmHg, 14 h) to provide 1.3 g of **2** as colorless rods with the following physicochemical properties: mp = 188–189 °C (lit.[10] mp 188–189 °C); ^{1}H NMR (CDCl$_3$, 400 MHz) δ: 3.65 (dd, $J = 8.2$, 14.2 Hz, 2 H), 4.29 (dd, $J = 11.0$, 14.2 Hz, 2 H), 5.33 (dd, $J = 11.0$, 8.2 Hz, 2 H), 6.95–7.11 (m, 8 H, Ar-H) ^{13}C NMR (CDCl$_3$, 100 MHz) δ: 43.65 (d, C-2), 52.98 (s, C-1), 127.88, 129.03, 130.78, 130.92 (4 d, C-Ar), 136.32, 138.33 (2 s, C$_{quart.}$); IR (KBr): 3062, 3018, 2892, 1961, 1600, 1579, 1490, 1449, 1312, 1110, 1084 cm^{-1}; Anal. Calcd. for C$_{16}$H$_{14}$Br$_2$: C, 52.49; H, 3.85; Br, 43.65. Found: C, 52.37; H, 4.03; Br, 43.72. HRMS: Calcd. for C$_{16}$H$_{14}$^{79}Br^{81}Br: 365.9442.

Found: 365.9457; MS (EI, 70 eV): 287 (22), 285 (23), 205 (100), 191 (20), 178 (15), 165 (10), 115 (11), 101 (16), 89 (23).

20. Potassium *tert*-butoxide (97%) was purchased from Fluka, stored in a Schlenk flask under argon and was used as received.

21. The TLC analysis was carried out according to Note 5, R_f (2) = 0.11, R_f (3) = 0.20.

22. The submitters also monitored reaction by GC/MS. For GC/MS-control the "default"-method was used (see Note 6). GC/MS (EI): t_R(3) = 10.9 min, m/z = 204 [M$^+$]; t_R(2) = 14.8 min, m/z = 366 [M$^+$].

23. If after rotary evaporation the crude product remains as a semifluid mass and cannot be scraped off the inner walls of the flask, it is dissolved in diethyl ether and the solvent is evaporated once more.

24. Kugelrohr distillation was carried out in an apparatus equipped according to Note 11. The first fraction, up to 85 °C, is *tert*-butanol.

25. From a smaller scale run (0.62 g, 2.98 mmol of 1) 3 was isolated via chromatography on a column (3.0 x 15 cm) of 25 g silica gel with petroleum ether (35-60 °C) as eluent. The desired product 3 was obtained in fractions 2-3 (50-mL fractions), which were concentrated by rotary evaporation (40 °C, 262 mmHg) to yield 3 as colorless plates (0.44 g, 72% over two steps).

26. Physicochemical properties of dibenzo[a,e]cyclooctene 3: mp 108–109 °C (lit.[10] 109.5–110 °C); ^{1}H NMR (CDCl$_3$, 300.13 MHz) δ: 6.72 (s, 4 H, CH), 7.02–7.11 (m, 8 H, Ar-H); ^{13}C NMR (CDCl$_3$, 100 MHz) δ: 126.76 (d, CH), 129.03, 133.18 (2 d, C-Ar), 137.00 (s, C$_{quart.}$-Ar); IR (KBr): 3054, 3010, 1922, 1815, 1650, 1490, 1432, 1400, 1153, 1088, 1039 cm^{-1}; Anal. Calcd. for C$_{16}$H$_{12}$: C, 94.08; H, 5.92. Found C, 94.14; H, 6.14. HRMS: Calcd. for C$_{16}$H$_{13}$: 205.1017. Found: 205.1017. MS (EI, 70 eV): 204 (79), 203 (100), 202 (63), 176 (5); 150 (4), 101 (30), 88 (10), 76 (7).

Safety and Waste Disposal Information

All hazardous materials should be handled and disposed of in accordance with "Prudent Practices in the Laboratory"; National Academies Press; Washington, DC, 2011.

3. Discussion

Dibenzo[a,e]cyclooctene (dbcot = **3**) is a strongly binding diene ligand for transition metal ions, and its transition metal complexes show interesting properties for the use in catalysis.[3, 4, 5, 6] However, applications are limited because of poor accessibility of **3**. Commercial sources until now only offer very small amounts of the compound (milligram scale).

The first synthesis of **3** was reported in 1946 by Fieser et al.,[7] who condensed o-phthalaldehyde and o-phenylenediacetonitrile in the presence of sodium ethoxide to obtain a dicyano derivative of **3**. Three more steps afforded the title compound in an overall yield of 6%. A Wittig approach described by Griffin et al. starting from o-phthalaldehyde led to an overall yield of 15%.[8] Other routes led either to low yields,[9] as well, and/or required inconvenient reaction conditions, such as the use of sodium powder,[10] Ni(CO)$_4$ and mercury amalgam.[11] Another approach is the photolytic-induced isomerization of dibenzobarrelene (9,10-dihydro-9,10-ethenoanthracene) to **3**.[12, 13] Dibenzobarrelene can be prepared by Diels-Alder cycloaddition of anthracene and trans-1,2-dichloroethylene,[13] dimethylacetylene carboxylate[14] or phenyl vinyl sulfoxide[15] as dienophile in the key step.

The procedure described herein in detail is a modification of a method published by Wudl et al.[16] Though in principal straightforward and elegant, this synthesis in our hands suffered from poor reproducibility of the first step due to the use of a large excess of hazardous material like powdered lithium sand and difficult purification of the hydrocarbon **1** by chromatography. Our aim was the development of a protocol that could be run reliably on a multigram scale. This aim was reached by lowering the amounts of reagents and solvents needed and the application of Kugelrohr distillation instead of column chromatography for purification.

Concerning the properties of **3** as a ligand for transition metal complexes,[3] preorganization of the double bonds leads to very stable transition metal complexes. From 1953 onwards, coordination compounds with **3** as ligand have been prepared (silver,[17] palladium,[11] molybdenum, chromium).[18] In 1983, Crabtree et al. demonstrated exceptional stability for several molybdenum, rhodium and iridium complexes.[19] Their work showed that **3** is an excellent ligand, which binds more strongly to these metal centers than cod and resists further transformations such as hydrogenation. Because of its tight binding **3** was employed as a catalyst poison in order to

62

distinguish between homogeneously and heterogeneously-catalyzed reactions.[20] More recent publications describe syntheses and organometallic coordination chemistry of a variety of Ir-,[21] Rh-,[21] Pd-,[22, 23] and Pt-[23] complexes of **3**.

1. Organisch-Chemisches Institut der Universität Heidelberg, Im Neuenheimer Feld 270, D-69120 Heidelberg, Germany. E-mail: g.helmchen@oci.uni-heidelberg.de. Support of this work by the Deutsche Forschungsgemeinschaft (SFB 623) is gratefully acknowledged.
2. Baker, W.; Banks, R.; Lyon, D. R.; Mann, F. G. *J. Chem. Soc.* **1945**, 27-30.
3. Defieber, C.; Grützmacher, H.; Carreira, E. M. *Angew. Chem. Int. Ed.* **2008**, *47*, 4482-4502.
4. Läng, F.; Breher, F.; Stein, D.; Grützmacher, H. *Organometallics* **2005**, *24*, 2997-3007.
5. Spiess, S.; Welter, C.; Franck, G.; Taquet, J.-P.; Helmchen, G. *Angew. Chem. Int. Ed.* **2008**, *47*, 7652-7655.
6. Franck, G.; Brödner, K.; Helmchen, G. *Org. Lett.* **2010**, *12*, 3886-3889.
7. Fieser, L. F.; Pechet, M. *J. Am. Chem. Soc.* **1946**, *68*, 2577-2580.
8. Griffin, C. E.; Peters, J. A. *J. Org. Chem.* **1963**, *28*, 1715-1716.
9. Kagan, J.; Chen, S.; Agdeppa, D.; Watson, W.; Zabel, V. *Tetrahedron Lett.* **1977**, *18*, 4469-4470.
10. Cope, A. C.; Fenton, S. W. *J. Am. Chem. Soc.* **1951**, *73*, 1668-1673.
11. Avram, M.; Dinu, D.; Mateescu, G.; Nenitzescu, C. D. *Chem. Ber.* **1960**, *93*, 1789-1794.
12. Rabideau, P. W.; Hamilton, J. B.; Friedman, L. *J. Am. Chem. Soc.* **1968**, *90*, 4465-4466.
13. Wang, X.; Hou, X.; Zhou, Z.; Mak, T.; Wong, H. *J. Org. Chem.* **1993**, *58*, 7498-7506.
14. Figeys, H. P.; Dralants, A. *Tetrahedron* **1972**, *28*, 3031-3036.
15. Doecke, C. W.; Klein, G.; Paquette, L. A. *J. Am. Chem. Soc.* **1978**, *100*, 1597-1599.
16. Chaffins, S.; Brettreich, M.; Wudl, F. *Synthesis* **2002**, 1191-1194.
17. Wittig, G.; Eggers, H.; Duffner, P. *Liebigs Ann. Chem.* **1958**, *619*, 10-27.

18. Müller, J.; Göser, P.; Elian, M. *Angew. Chem. Int. Ed. Eng.* **1969**, *8*, 374-375. Brown, M.; Zubkowski, J. D.; Valente, E. J.; Yang, G.; Henry, W. P. *J. Organomet. Chem.* **2000**, *613*, 111-118.
19. Anton, D. R.; Crabtree, R. H. *Organometallics* **1983**, *2*, 621-627.
20. Anton, D. R.; Crabtree, R. H. *Organometallics* **1983**, *2*, 855-859.
21. Singh, A.; Sharp, P. R. *Inorg. Chim. Acta* **2008**, *361*, 3159-3164.
22. DeGray, J. A.; Geiger, W. E.; Lane, G. A.; Rieger, P. H. *Inorg. Chem.* **1991**, *30*, 4100-4102.
23. Singh, A.; Sharp, P. R. *Organometallics* **2006**, *25*, 678-683.

Appendix
Chemical Abstracts Nomenclature; (Registry Number)

5,6,11,12-Tetrahydrodibenzo[a,e]cyclooctene; (1460-59-9)
α,α'-Dibromo-*o*-xylene, 1,2-Bis(bromomethyl)benzene; (91-13-4)
Tetrahydrofuran; (109-99-9)
Lithium; (7439-93-2)
5,6,11,12,17,18-Hexahydro-tribenzo[a,e,i]cyclododecene; (4730-57-8)
5,11-Dibromo-5,6,11,12-tetrahydrodibenzo[a,e]-cyclooctene; (94275-22-6)
N-Bromosuccinimide; (128-08-5)
Carbon tetrachloride; (56-23-5)
Dibenzo[a,e]cyclooctene; (262-89-5)
Potassium *tert*-butoxide; (865-47-4)

G. Helmchen (b. 1940) is a Professor Emeritus at the Institute of Organic Chemistry of the Ruprecht-Karls-Universität Heidelberg. He pursued undergraduate studies at the TH Hannover (Dipl.-Chem. 1965) and graduate work under the guidance of Prof. V. Prelog at the ETH Zürich (Dr. sc. techn. 1971). He then carried out a Habilitationsarbeit at the TU Stuttgart (1975-1980). In 1980 he was appointed Professor C3 at the Universität Würzburg. In 1985 he moved to Heidelberg His interest in catalysis dates back to ca. 1990.

Géraldine Franck, was born in Munich (Germany) in 1979. She studied chemistry in Heidelberg and Bristol and received her diploma degree from the Ruprecht-Karls-Universität Heidelberg in 2006. Under the guidance of Prof. G. Helmchen she completed her Ph.D. in 2010. Her research interests are the application of Ir-catalyzed allylic substitution reactions in syntheses of chiral building blocks and natural products.

Marcel Brill was born in 1986 in Göttingen, Germany. He started studying chemistry at the Ruprecht-Karls-Universität in Heidelberg in 2005. During a practical course of advanced organic chemistry he worked on the synthesis of dbcot in the group of Prof. G. Helmchen. In 2010 he received his diploma degree in chemistry under the supervision of Prof. P. Hofmann. The same year he started his Ph.D. studies where he is currently working on the development of chiral NHC ligands and their application in enantioselective catalysis.

Takuya Nishimura was born in 1983 in Kanagawa, Japan. He received his B.S. in 2007 from the University of Showa Pharmaceutical Science. He is pursuing his Ph.D. at the Graduate School of Pharmaceutical Sciences, The University of Tokyo, under the guidance of Professor Tohru Fukuyama. His research interests are in the area of natural product synthesis.

Anhydrous Hydration of Nitriles to Amides:
p-Carbomethoxybenzamide

Submitted by Dahye Kang, Jinwoo Lee and Hee-Yoon Lee.[1]
Checked by Hiroshi Kumazaki and Tohru Fukuyama.

1. Procedure

A.   p-Carbomethoxylbenzamide. A 100-mL single-necked, round-bottomed flask that is equipped with a magnetic stirring bar (round, 5 x 15 mm), a reflux condenser and Argon bubbler is charged with the following order of the reagents: methyl *p*-cyanobenzoate (5.0 g, 31 mmol) (Note 1), toluene (9.5 mL) (Note 2), Wilkinson's catalyst (145 mg, 0.16 mmol, 0.5 mol%) (Note 3), and acetaldoxime (9.2 g, 9.5 mL, 156 mmol, 5.0 equiv) (Notes 4 and 5). The reaction mixture is heated to reflux with an oil bath (bath temperature 130 °C) for 4 h and then is allowed to cool down to room temperature (Notes 6 and 7). The reaction mixture is concentrated under reduced pressure on rotary evaporator (20–30 mmHg, bath temperature 40–45 °C). Water (20 mL) is added to the concentrate and the mixture is heated with an oil bath (90-100 °C bath temperature) under air for 1 h and cooled down to room temperature (Note 8). The resulting mixture is filtered through a Büchner funnel. The solid product is washed with water (40 ml). The solid is transferred to a 50-mL round-bottomed flask and dried under vacuum (0.1 mmHg) to afford 5.1 g (92%) of *p*-carbomethoxylbenzamide (Notes 9 and 10) as a white solid.

2. Notes

1.   Methyl *p*-cyanobenzoate (99%) was purchased from Aldrich and used as received.
2.   The submitters purchased toluene (CHROMASOLV[B] for HPLC, >99.9%) from Aldrich and it was used as received. The checkers purchased toluene (>99.5%) from Kanto Chemical Company and used it as received.

66

3. Wilkinson's catalyst (RhCl(PPh$_3$)$_3$) was purchased from Aldrich and used as received.

4. Acetaldoxime (>99%) was purchased from TCI and used as received.

5. Since acetonitrile produced from the reaction also reacts with acetaldoxime to form acetamide, at least five equivalents of acetaldoxime are required to complete the reaction.

6. As the reaction proceeds, solid begins to appear in the flask.

7. The reaction is monitored by TLC analysis on Merck silica gel 60 F254 plates (hexane/ethyl acetate = 3:2); visualization was accomplished with 254 nm UV light and ethanolic solution of phosphomolybdic acid with heat: R$_f$ (starting nitrite) = 0.69; R$_f$ (amide product) = 0.06 .

8. The crude product is heated in water to dissolve all the acetamide.

9. *p*-Carbomethoxylbenzamide exhibits the following physical and spectroscopic properties: mp = 206 °C (lit.[8] 206 °C); IR (neat) cm^{-1}: 3404, 3197, 1724, 1655, 1281, 1116; ^{1}H NMR (400 MHz, CD$_3$OD) δ: 3.85 (s, 3 H), 7.96 (d, 2 H), 8.10 (d, 2 H); ^{13}C NMR (100 MHz, CD$_3$OD) δ: 53.0, 129.0, 130.7, 134.3, 139.4, 167.9, 171.4; Anal. Calcd for C$_9$H$_9$NO$_3$: C, 60.33; H, 5.06; N, 7.82. Found: C, 60.48; H, 5.08; N, 7.76.

10. Recrystallization of the product can be performed according to the following procedure: A 500-mL, single-necked, round-bottomed flask is charged with *p*-carbomethoxybenzamide (5.1 g, 28 mmol) and ethanol (150 mL). The suspension is heated to reflux with an oil bath (bath temperature 90 °C) until complete dissolution occurs Then the oil bath is removed, and the solution is allowed to cool to room temperature overnight. The crystallized product is collected with a Büchner funnel, washed with cooled ethanol (10 mL), and dried *in vacuo* to afford 4.4 g (86%) of pure amide.

Safety and Waste Disposal Information

All hazardous materials should be handled and disposed of in accordance with "Prudent Practices in the Laboratory"; National Academies Press; Washington, DC, 2011.

3. Discussion

The amide is one of the important functional groups in chemical industry, as well as the pharmaceutical industry,[2] and can be readily prepared from the corresponding nitriles, since they are *isohypsic*.[3] However, hydration of nitriles to amides requires strong acidic or basic conditions, and is accompanied often by the formation of the corresponding carboxylic acids through further hydrolysis.[4] To circumvent side reactions and harsh reaction conditions, transition metal-catalyzed hydrolysis reactions of nitriles or rearrangement reactions of oximes have been developed.[5] During the study of Rh-catalyzed rearrangement of oximes into amides[6] we found out that the transformation not only proceeded through the intermediacy of nitriles, but the reaction was also catalyzed by the same nitrile intermediates. A brief mechanistic study of this transformation strongly indicated that dehydration of oximes into nitriles or hydration of nitriles into amides did not produce or use water molecules. Instead, oxygen and hydrogen atoms of the oximes are delivered directly to the nitriles to convert the nitriles into amides. Thus, the concentration of nitriles stays constant. This nitrile-catalyzed transformation of oximes into amides prompted the development of a new method of hydrolyzing nitriles into amides without using or generating water molecules.[7] (Scheme 1) Acetaldoxime was selected as the reagent along with $RhCl(PPh_3)_3$ as the catalyst for nitrile hydrolysis and the reaction was optimized with five equivalent of acetaldoxime and 0.5 mol% of the catalyst in toluene.[8]

Scheme 1. Mechanistic proposal for the aldoxime promoted amide formation.

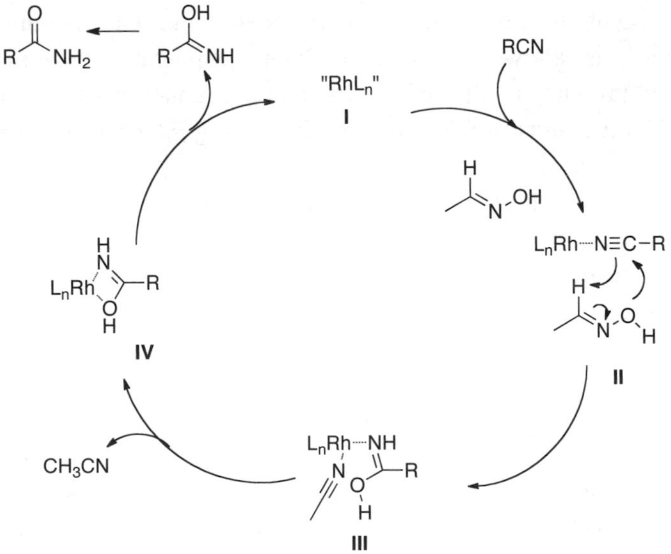

The procedure described here provides not only a mild and practical way of preparation of amides from nitriles, but also provides a method of hydration without utilizing water as the nucleophile during the reaction. The choice of acetaldoxime as the hydrating agent makes all the byproducts from this reagent to be either volatile or soluble in water, and thus simplifies the work-up process and eliminates the purification step of the amide products.

The scope of functional group tolerance is summarized in the Table 1. Aliphatic and aromatic nitriles, regardless of the substitution patterns, are hydrolyzed efficiently. Wide range of functional group tolerability indicates that the reaction is non-nucleophilic as well as neutral, since the hydrolysis of esters, the isomerization of olefins, or the deprotection of protecting group is not observed.

Table 1. Substrate scope of the hydration reaction[a]

Entry	Nitrile	Yield[b]	Entry	Nitrile	Yield[b]
1		95	7		81
2		99	8		82
3		79	9		92
4		96	10		81
5		90	11		76
6		87	12		75
			13		80

[a] 0.5 mmol scale. [b] Isolated yields.

In summary, water-free selective hydrolysis of nitriles into amides using acetaldoxime as the water surrogate with Wilkinson's catalyst is demonstrated as a practical and mild procedure compatible with various functional groups.

1. Department of Chemistry, Korea Advanced Institute of Science and Technology (KAIST), Daejeon, 305-701, Republic of Korea, E-mail: leehy@kaist.ac.kr.
2. a) *The Chemistry of Functional Groups, The Chemistry of Cyano Group* (Eds.: Rappoport, Z.; Patai, S.), Wiley, London, **1970**. b) Fatiadi, A. J. in *The Chemistry of Functional Groups, Supplement C, The Chemistry of Triple-Bonded Functional Groups*, Part 2 (Eds.: Rappoport, Z.; Patai, S.), Wiley, London, **1983**, pp. 1057–1303.
3. Burns, N. Z.; Baran, P. S.; Hoffmann, R. W. *Angew. Chem. Int. Ed.* **2009**, *48*, 2854–2867.
4. Kukushikin, V. Y.; Pombeiro, A. J. L. *Inorg. Chim. Acta* **2005**, *358*, 1–21 and references therein.
5. (a) Kim, H. S.; Kim, S. H.; Kim, J. N. *Tetrahedron Lett.* **2009**, *50*, 1717–1719. (b) Goto, A.; Endo, K.; Saito, S. *Angew. Chem. Int. Ed.* **2008**, *47*, 3607–3609. (c) Crestani, M. G.; Garcia, J. J.; *J. Mol. Cat. A: Chem.* **2009**, *299*, 26–36. (d) Solhy, A.; Smahi, A.; Badaoui, H. E.; Elaabar, B.; Amoukal, A.; Tikad, A.; Sebtia, S.; Macquarrie, D. J.; *Tetrahedron Lett.* **2003**, *44*, 4031–4033. (e) Thallaj, N. K.; Przybilla, J.; Welter, R.; Mandon, D. *J. Am. Chem. Soc.* **2008**, *130*, 2414–2415. (f) Ramon, R. S.; Marion, N.; Nolan, S. P. *Chem. Eur. J.* **2009**, *15*, 8695–8697.
6. Kim, M.; Lee, J.; Lee, H.-Y.; Chang, S. *Adv. Synth. Catal.* **2009**, *351*, 1807–1812.
7. Lee, J.; Kim, M.; Chang, S.; Lee, H.-Y. *Org. Lett.* **2009**, *11*, 5598–5601.
8. Renckhoff, G.; Huelsmann, H. L. US 3,252,977, May 24, 1966.

Appendix
Chemical Abstracts Nomenclature; (Registry Number)

p-Carbomethoxylbenzamide: Benzoic acid, 4-(aminocarbonyl)-, methyl ester; (6757-31-9)
Methyl *p*-cyanobenzoate: Benzoic acid, 4-cyano-, methyl ester; (1129-35-7)

Wilkinson's catalyst: Rhodium, chlorotris(triphenylphosphine)-; (14694-95-2)

Acetaldoxime: Acetaldehyde, oxime; (107-29-9)

Hee-Yoon Lee received his B.S. degree in Chemistry from Seoul National University in 1980, his M.S. 1982 with Eun Lee, and his Ph.D. 1988 in Chemistry from Stanford University with Paul A. Wender. After two years of postdoctoral work at the Department of Chemistry, Columbia University with Gilbert Stork, he worked at Merck Research Laboratories at West Point in 1990 as a senior research chemist. Then, he took up the professorship at KAIST in 1994, where he is focusing on the development of new synthetic strategies and application to total synthesis of natural products.

Dahye Kang was born in Incheon, Republic of Korea, in 1991. After graduating from Bupyong Girls High School in 2009 skipping her senior year, she entered KAIST in 2009. She is currently an undergraduate student in the Department of Chemistry at KAIST, studying organic synthesis in the laboratory of Professor Hee-Yoon Lee.

Jinwoo Lee was born in Daegu, Republic of Korea, in 1982. He received B.S. in 2003 from KAIST, Ph.D. in 2010 under the supervision of Hee-Yoon Lee. He is currently postdoctoral research fellow at Tuberculosis Research Section (National Institute of Allergy and Infectious Diseases) at the NIH, where he participates in the design, chemical synthesis and biological evaluation of novel compounds as anti-tubercular agents with Clifton E. Barry, III.

Hiroshi Kumazaki was born in Gifu, Japan in 1988. He received his B.S. in 2010 from the University of Tokyo. In the same year, he began his graduate studies at the Graduate School of Pharmaceutical Sciences, the University of Tokyo, under the direction of Professor Tohru Fukuyama. His current interest is total synthesis of natural products.

Discussion Addendum for:
dl-Selective Pinacol-Type Coupling Using Zinc, Chlorosilane, and Catalytic Amounts of Cp₂VCl₂; *dl*-1,2-Dicyclohexylethanediol (1,2-Ethanediol,1,2-dicyclohexyl-)

Prepared by Toru Amaya and Toshikazu Hirao*. [1]
Original article: Hirao, T.; Ogawa, A.; Asahara, M.; Muguruma, Y.; Sakurai, H. *Org. Synth.* **2005**, *81*, 26.

The pinacol coupling reaction achieves reductive carbon-carbon bond formation from carbonyl compounds by low-valent early transition metals to give 1,2-diols. A catalytic system for the pinacol coupling was developed for the first time in our group.[2] Catalytic, diastereoselective, and/or enantioselective methods have been reported recently.[3] Organic reaction in water or aqueous media is of interest in organic synthesis from the vantage points of its cost, safety, and environmental concern.[4] Therefore, the development of an efficient synthetic methodology to form a carbon-carbon bond in water or aqueous media appears to be of importance. Here, a vanadium-catalyzed pinacol coupling reaction in water is described.

Vanadium-Catalyzed Pinacol Coupling Reaction in Water

Metallic salts and metallic co-reductant were investigated for the pinacol coupling reaction in water (Scheme 1).[5] After optimization using benzaldehyde, 100 mol% of VCl₃ as a metallic salt was revealed to promote the reaction in the presence of 300 mol% of Zn as a co-reductant at room temperature to give 1,2-diphenylethane-1,2-diol in 86% yield with 64/36 ratio of the *dl* and *meso* isomers. VCl₃ is essential for this reaction because no reaction was observed in the absence of VCl₃. Three metallic Al, Mg, and Mn were employed in the presence of VCl₃ to show that metallic Al was more efficient as a co-reductant (92%, *dl/meso* : 65/35, Table 1, entry 1). Several co-solvents, such as H_2O-DMF, H_2O-THF, and H_2O-MeOH, can be used in these reaction conditions. As compared with VCl_3, VBr_3 gave a slightly better yield.

Reducing the amount of VCl_3 to 33 mol% to the substrate was not a problem (isolated yield 72%, NMR yield 92%, *dl/meso* : 56/44, Table 1, entry 2). It should be noted that the catalytic pinacol coupling reaction successfully proceeded in water even in the absence of a chlorosilane. It is in sharp contrast to the reaction in organic solvent, which requires a chlorosilane as an essential additive. These findings provide a synthetically versatile method. Various aromatic aldehydes (R = 4-MePh, 4-MeOPh, 2-ClPh, 3-ClPh, 4-ClPh, 2-BrPh, and 2-furyl) underwent the reductive coupling with the catalytic VCl_3/Al system in water to give the corresponding 1,2-diols in moderate to good yields. Acetophenone was not reduced under the similar conditions.

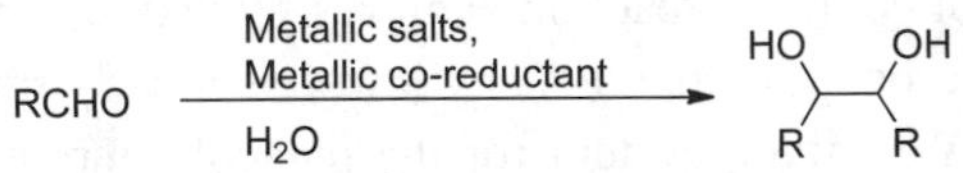

$$RCHO \xrightarrow[\text{H}_2\text{O}]{\substack{\text{Metallic salts,} \\ \text{Metallic co-reductant}}} \underset{R \quad R}{\overset{HO \quad OH}{\diagdown}}$$

Scheme 1. Pinacol coupling reaction in water.

Table 1. Vanadium-Promoted Pinacol Coupling Reaction of Benzaldehyde in Water

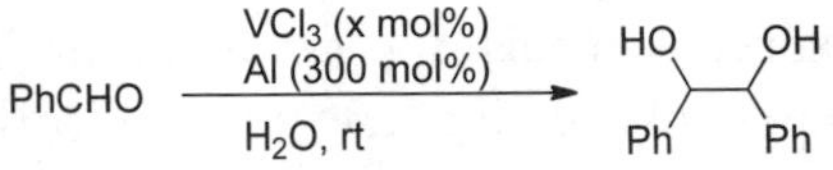

$$PhCHO \xrightarrow[\text{H}_2\text{O, rt}]{\substack{\text{VCl}_3 \ (x \ \text{mol\%}) \\ \text{Al} \ (300 \ \text{mol\%})}} \underset{Ph \quad Ph}{\overset{HO \quad OH}{\diagdown}}$$

Entry	VCl_3 (mol%)	Isolated yield (%)	Selectivity (*dl/meso*)
1	100	92	65/35
2	33	72 (92)[a]	56/44

[a] ^{1}H-NMR yield.

Although the detailed mechanism of the catalytic reaction requires more investigation, a low-valent vanadium generated by treatment with Al is likely to be involved in a catalytic cycle involving one-electron transfer from the catalyst to the carbonyl group. Another reaction path where vanadium hydroxide formed by *in situ* hydrolysis acting as a Lewis acid and Al metal serves as a reducing agent might be possible.

74

1. Department of Applied Chemistry, Graduate School of Engineering, Osaka University, Yamada-oka, Suita, Osaka 565-0871, Japan.
2. (a) Hirao, T.; Hasegawa, T.; Muguruma, Y.; Ikeda, I. *J. Org. Chem.* **1996**, *61*, 366-367. (b) For a review: Hirao, T. *Chem. Rev.* **1997**, *97*, 2707-2724. (c) For an account: Hirao, T. *Synlett* **1999**, 175-181.
3. For a review: Chatterjee, A.; Joshi, N. N. *Tetrahedron* **2006**, *62*, 12137-12158.
4. (a) *Organic Synthesis in Water*; Grieco, P. A., Ed.; Blackie Academic & Professional: Glasgow, 1998. (b) Li, C.-J.; Chan, T.-H. *Organic Reactions in Aqueous Media*; Wiley & Sons: New York, 1997.
5. Xu, X.; Hirao, T. *J. Org. Chem.* **2005**, *70*, 8594-8596.

Toru Amaya was born in Japan, in 1976. He received his B.S., M.S., and Ph.D. from Tokyo Institute of Technology. After completion of his Ph.D. in 2003 under the direction of Professor Takashi Takahashi, he studied as a postdoctoral fellow with Professor Julius Rebek, Jr. at Scripps Research Institute in USA until August of 2004. In September 2004, he joined Professor Toshikazu Hirao's group at Osaka University as an Assistant Professor, which is the current position. His research interests include functional organic chemistry and synthetic organic chemistry.

Toshikazu Hirao graduated from Kyoto University in 1973, where he obtained his doctorate in 1978. He became Assistant Professor at Osaka University. Dr. Hirao was promoted to Associate Professor in 1992 and Professor in 1994. His research interests lie in the development of new methods in organic synthesis, and redox-active systems consisting of transition metals and/or planar and nonplanar π-conjugated compounds. These research efforts are correlated to bioorganometallic conjugate systems. He is vice president of the Chemical Society of Japan (2010-2012). He received international awards, as exemplified by "Award for outstanding achievements in bioorganometallic chemistry" and "Vanadis award" in 2008.

Discussion Addendum for:
Synthesis of 4-, 5-, and 6-Methyl-2,2'-Bipyridine by a Negishi Cross-Coupling Strategy: 5-Methyl-2,2'-Bipyridine

Prepared by Tiandong Liu and Cassandra L. Fraser*.[1]
Original article: Smith, A. P.; Savage, S. A.; Love, J. C. and Fraser, C. L. *Org. Synth.* **2002**, *78*, 51.

Negishi coupling is a powerful tool in the preparation of bipyridines due to its high yield, mild conditions and relatively low cost.[2,3] It shows good tolerance to various substituents and the corresponding pyridyl zinc halide precursors can efficiently couple with many halogen-substituted heterocycles in addition to pyridine.[4] In recent years, some other palladium-catalyzed reactions, such as Stille and Suzuki coupling, have emerged to make pyridine derivatives.[5-7] Recent progress in palladium-catalyzed coupling reactions between pyridyl and other heterocycles will be discussed.

Advances in Bipyridine Synthesis via Negishi Coupling

Negishi reactions require aryl zinc halides and suitable coupling partners (Figure 1). The pyridyl zinc halides can be achieved by transmetallation with pyridyl lithium[2,3,8-10] or direct reaction between pyridyl halides and active zinc (Zn*).[11] For nearly quantitative yields, (trifluoromethane-sulfonyl)oxypyridine reagents are an excellent choice for coupling partners.[2,3] Many recent reports, however, employ pyridyl halides,

Org. Synth. **2012**, *89*, 76-81
Published on the Web 9/26/2011

which are more accessible. Chloro,[8,9,12] bromo[10,11,13] and iodo substituted compounds[11,14] are good coupling agents in Negishi[11] reactions; whereas, fluoro reagents are typically inert.

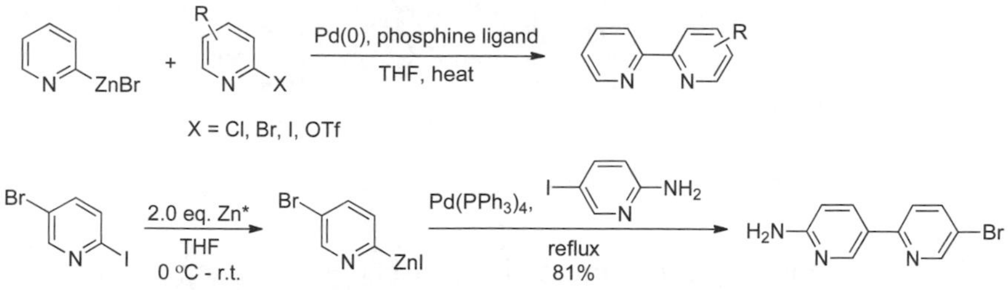

Figure 1. General example of preparation of bipyridines by Negishi coupling between pyridyl zinc reagents and pyridyl halides or triflates, and illustration of selectivity in zinc reagent preparation and halide coupling.

Generally, aromatic iodides are more reactive than bromides and chlorides, however bromides are most commonly employed.[15] Halides at the 3-position can also participate in the coupling reactions to provide 2,3'-bipyridines, but their reactivity is less than that of 2-halopyridines.[11] This allows for selective coupling at the 2-halo site in the presence of dihalo-substituted pyridines (Figure 2).

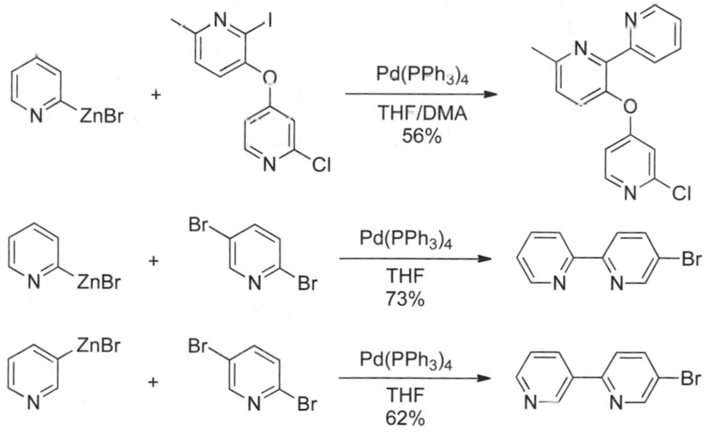

Figure 2. Chemoselectivity and regioselectivity in Negishi coupling in the preparation of bipyridine complexes.

Negishi coupling shows impressive tolerance of various functional groups, including alkyne, CN, COOR, NO₂, NR₂, OR, OH, and TMS.[8,9,11] This enables further functionalization of 2,2′-bipyridines (Figure 3).

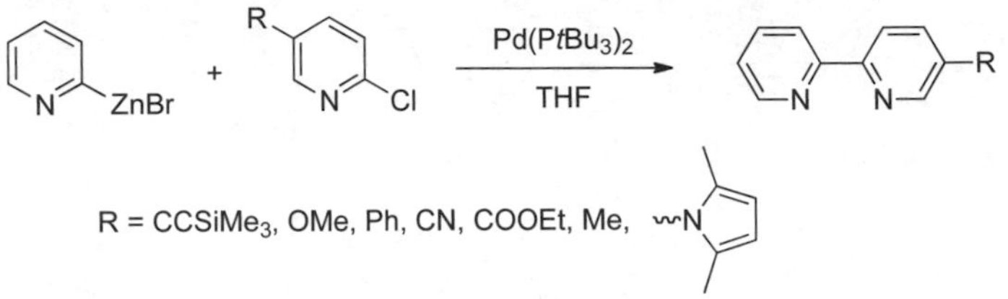

R = CCSiMe₃, OMe, Ph, CN, COOEt, Me, ⌇N⟨ ⟩

Figure 3. Functional group tolerance in Negishi cross coupling.

Bipyridines Prepared by Stille and Suzuki Coupling

Other palladium catalyzed coupling reactions can also be used to prepare 2,2′-bipyridine derivatives (Figure 4).[16] Stille coupling can provide various bipyridine compounds with moderate to good yields. The challenge with this method is that coupling reactions usually have to be carried out in toluene under refluxing conditions for a couple days. Heat sensitive compounds may not be tolerant of this method.[17-20] Additionally, toxicity is a concern for tin reagents. Due to the difficulty of obtaining stable 2-pyridylboron coupling precursors,[21] Suzuki coupling was not used in making 2,2′-bipyridine compounds until recent years. The relatively high catalyst loadings could also limit its applications in organic synthesis.[22]

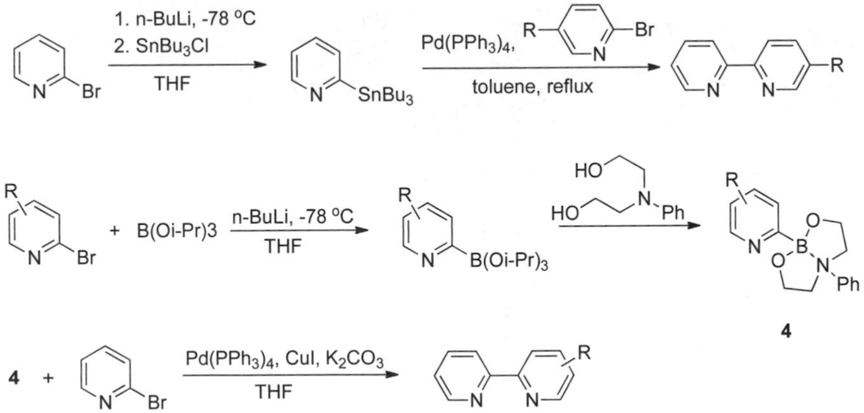

Figure 4. Preparation of 2,2′-bipyridine by Stille and Suzuki coupling.

Org. Synth. **2012**, *89*, 76-81

Negishi Coupling of Other Pyridyl Heterocyclic Compounds

Among heterocycles containing pyridine units, terpyridine complexes have received considerable attention due to their strong binding affinity with many metal ions.[23] Negishi coupling is a common method to prepare terpyridine derivatives.[24] Pyridyl zinc halide can react with a great many halide heterocycles to generate products with pyridine moieties.[11,14] Sometimes high regioselectivity was observed[25] (Figure 5).

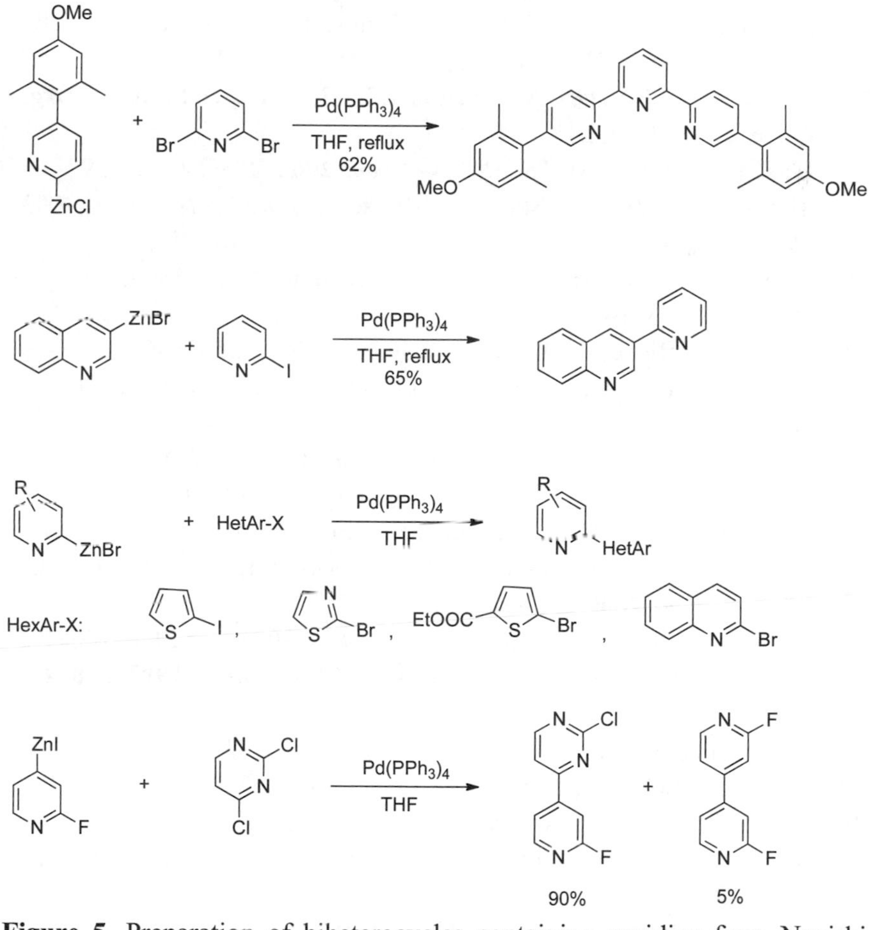

Figure 5. Preparation of biheterocycles containing pyridine from Negishi coupling.

1. Department of Chemistry, University of Virginia, Charlottesville, VA 22904, fraser@virginia.edu. We thank the National Science Foundation (CHE 0718879) for support for this work.
2. Savage, S. A.; Smith, A. P.; Fraser, C. L. *J. Org. Chem.* **1998**, *63*, 10048–10051.
3. Smith, A. P. S., S. A.; Love, J. C.; Fraser, C. L. *Org. Synth.* **2002**, *78*, 51–56.
4. Heller, M.; Schubert, U. S. *Eur. J. Org. Chem.* **2003**, *2003*, 947–961.
5. Collins, I. *J. Chem. Soc., Perkin Trans. 1* **2000**, 2845–2861.
6. Donnici, C. L.; Oliveira, I. M. F. d.; Temba, E. S. C.; Castro, M. C. R. d. *Quim. Nova.* **2002**, *25*, 668–675.
7. Newkome, G. R.; Patri, A. K.; Holder, E.; Schubert, U. S. *Eur. J. Org. Chem.* **2004**, *2004*, 235–254.
8. Lützen, A.; Hapke, M. *Eur. J. Org. Chem.* **2002**, *2002*, 2292–2297.
9. Lützen, A.; Hapke, M.; Staats, H.; Bunzen, J. *Eur. J. Org. Chem.* **2003**, *2003*, 3948–3957.
10. Trécourt, F.; Gervais, B.; Mallet, M.; Quéguiner, G. *J. Org. Chem.* **1996**, *61*, 1673–1676.
11. Kim, S.-H.; Rieke, R. D. *Tetrahedron* **2010**, *66*, 3135–3146.
12. Kiehne, U.; Bunzen, J.; Staats, H.; Lützen, A. *Synthesis* **2007**, 1061–1069.
13. Petzold, H.; Heider, S. *Eur. J. Inorg. Chem.* **2011**, *2011*, 1249–1254.
14. Rieke, R. D.; Kim, S.-H. *Tetrahedron Lett.* **2011**, *52*, 244–247.
15. Ward, R. A.; Powell, S. J.; Debreczeni, J. E.; Norman, R. A.; Roberts, N. J.; Dishington, A. P.; Gingell, H. J.; Wickson, K. F.; Roberts, A. L. *J. Med. Chem.* **2009**, *52*, 7901–7905.
16. Campeau, L.-C.; Fagnou, K. *Chem. Soc. Rev.* **2007**, *36*, 1058–1068.
17. Lehmann, U.; Henze, O.; Schlüter, A. D. *Chem. Eur. J.* **1999**, *5*, 854–859.
18. Eschbaumer, C.; Heller, M. *Org. Lett.* **2000**, *2*, 3373–3376.
19. Heller, M.; Schubert, U. S. *J. Org. Chem.* **2002**, *67*, 8269–8272.
20. Mathieu, J.; Marsura, A. *Synth. Commun.* **2003**, *33*, 409–414.
21. Kudo, N.; Perseghini, M.; Fu, G. C. *Angew. Chem. Int. Ed.* **2006**, *45*, 1282–1284.
22. Gütz, C.; Lützen, A. *Synthesis* **2010**, 85–90.
23. Sauvage, J. P.; Collin, J. P.; Chambron, J. C.; Guillerez, S.; Coudret, C.; Balzani, V.; Barigelletti, F.; De Cola, L.; Flamigni, L. *Chem. Rev.* **1994**, *94*, 993–1019.

24. Loren, J. C.; Siegel, J. S. *Angew. Chem. Int. Ed.* **2001**, *40*, 754–757.
25. Hattinger, G.; Schnérch, M.; Mihovilovic, M. D. *J. Org. Chem.* **2005**, *70*, 5215–5220.

Cassandra L. Fraser was born in Norfolk, Virginia in 1962 and grew up in Michigan. She obtained a B.A. degree from Kalamazoo College in 1984, an M.T.S. degree from Harvard Divinity School in 1988, and a Ph.D. from The University of Chicago in 1993 with Brice Bosnich. After postdoctoral research with Robert Grubbs at Caltech from 1993-5, she joined the faculty in the Department of Chemistry at the University of Virginia, where she was promoted to Associate Professor in 2001 and Professor in 2005, now with joint appointments in Biomedical Engineering and Architecture. Her research involves luminescent materials for imaging and sensing in biomedicine and sustainable design.

Tiandong Liu was born in Nanjing, China in 1979. He obtained a B.S. degree from Nanjing University in 2001. He began his graduate work at the University of Virginia with Lin Pu exploring enantioselective luminescent sensors for chiral carboxylic acids in 2006 and in 2010 joined the Fraser group to explore boron complexes with optical oxygen sensing and mechanochromic luminescence properties.

Discussion Addendum for:
Efficient Synthesis of Halomethyl-2,2′-Bipyridines: 4,4′-Bis(Chloromethyl)-2,2′-Bipyridine

Prepared by Tiandong Liu and Cassandra L. Fraser*.[1]
Original article: Smith, A. P.; Lamba, J. J. S.; Fraser, C. L. *Org. Synth.* **2002**, *78*, 82.

Halogenation of dimethyl-2, 2′-bipyridines via trimethylsilyl (TMS) intermediates and electrophilic halide reagents, CX_3CX_3 (X = Cl, Br or F), is a very useful method for bipyridine (bpy) derivation.[2-4] This approach benefits from nearly quantitative yields, few by-products, and low cost reagents. In the past decade, there have been only a few improvements reported for this reaction, but many bipyridine derivatives have been prepared using this method. Many bipyridine analogues have been accessed via TMS-bpy intermediates.

Monohalogenation of Dimethyl-Bipyridine

Other common methods for methylpyridine halogenation involve radical reactions or hydroxymethylbipyridine precursors. The radical method suffers from high reactivity and poor selectivity, often generating mixtures of halogenated species, whereas hydroxyl halogenations typically involve multi-step processes.

82

Org. Synth. **2012**, *89*, 82-87
Published on the Web 9/26/2011

Monohalogenation of dimethyl-bipyridine may be performed by the TMS procedure with a reduced amount of lithium diisopropylamide (LDA). The reported yields were nearly quantitative (~99%)[6] for 5-bromomethyl-5'-methyl-2,2'-bipyridine and over 60% for 5-chloromethyl-5'-methyl-2,2'-bipyridine[5] and. However, the same reaction with 4,4'-dimethyl-2,2'-bipyridine generated the monobromo product in only 35% yield.[7]

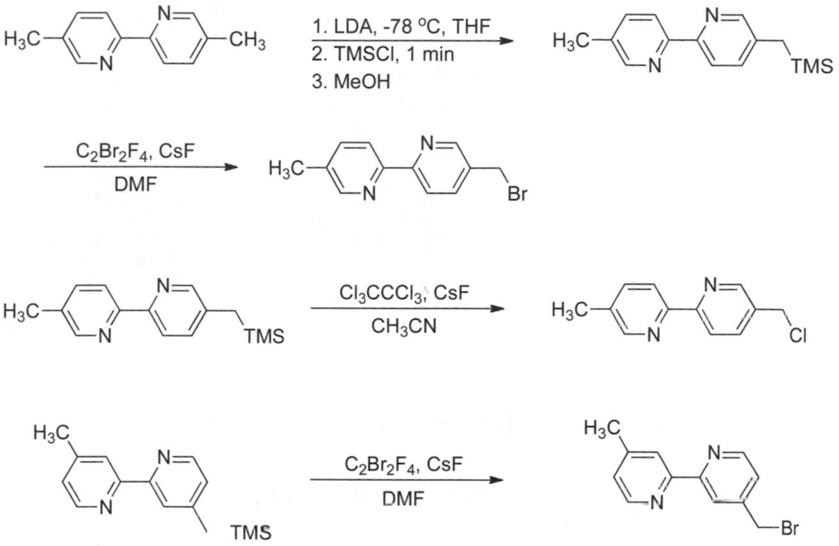

Figure 1. Monohalogenation of dimethyl-bipyridines.

Bipyridine Derivatization via Silane Reagents

Given that the trimethylsilyl-bpy reagents were good nucleophile precursors in the halogenation of dimethylbipyridine, their reactivity was extended to other electrophilic substrates, such as aldehydes[2] and alkylhalides.[8] Tetrachlorosilane can also be used to trap lithium intermediates and generate reactive trichlorosilane species, which have been reacted with alcohols to generate trialkoxysilylmethyl bipyridines or hydroxyl groups on zeolite surfaces for catalysis.[9] New derivatives are accessible via these approaches using various reagents and substrates.

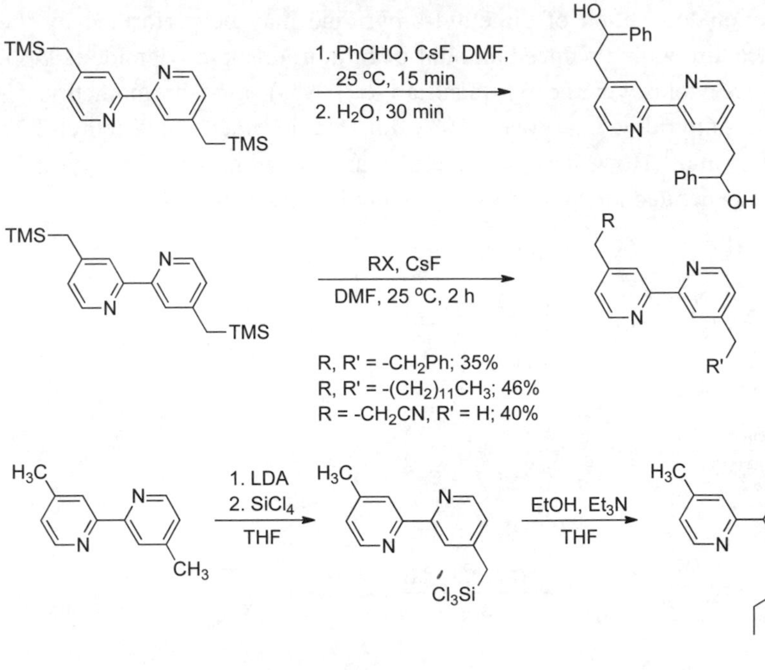

Figure 2. Compounds derived from trimethylsilylmethyl bipyridine.

Polymers Containing Bipyridines

Bipyridyl units appear frequently in functionalized polymeric materials as ligands for metal atoms or to modify electronic or optical properties.[10-14] The bis(halomethyl)-bipyridine complexes and their corresponding compounds with ruthenium or iron were utilized as initiators to generate metallopolymers, including block polymers.[15-21]

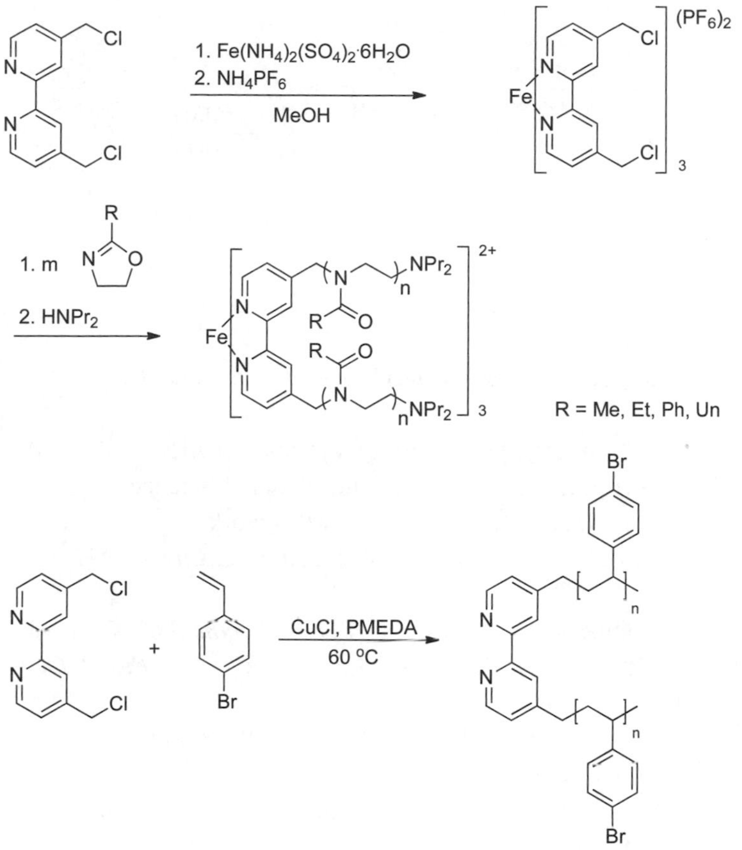

Figure 3. Bis(halomethyl) bipyridine in polymerization.

Bisphosphonate Bipyridine Complexes

Le Bozec, et al. used 4,4′-bis(bromomethyl)-2,2′-bipyridine to prepare 4,4′-bis(phosphonate)-2,2′-bipyridine, which can be used to produce many new derivatives via Wadsworth-Emmons reactions with aldehydes.[22,23] A similar approach was adopted by Grätzel and coworkers to generate bipyridines with extended π conjugation for solar cell applications.[24]

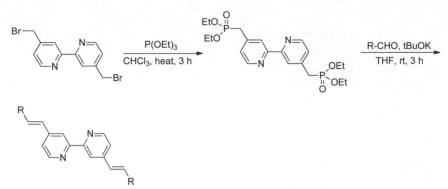

Figure 4. Preparation of bis(phosphonate) bipyridine compounds.

1. Department of Chemistry, University of Virginia, Charlottesville, VA 22904, fraser@virginia.edu. We thank the National Science Foundation (CHE 0718879) for support for this work.
2. Fraser, C. L.; Anastasi, N. R.; Lamba, J. J. S. *J. Org. Chem.* **1997**, *62*, 9314-9317.
3. Smith, A. P.; Lamba, J. J. S.; Fraser, C. L. *Org. Synth.* **2002**, *78*, 82-87.
4. Newkome, G. R.; Patri, A. K.; Holder, E.; Schubert, U. S. *Eur. J. Org. Chem.* **2004**, *2004*, 235-254.
5. Laïb, S.; Petit, M.; Bodio, E.; Fatimi, A.; Weiss, P.; Bujoli, B. *C. R. Chim.* **2008**, *11*, 641-649.
6. Schubert, U. S.; Eschbaumer, C.; Hochwimmer, G. *Tetrahedron Lett.* **1998**, *39*, 8643-8644.
7. Burton, J. W.; Anderson, E. A.; O'Sullivan, P. T.; Collins, I.; Davies, J. E.; Bond, A. D.; Feeder, N.; Holmes, A. B. *Org. Biomol. Chem.* **2008**, *6*, 693-702.
8. Smith, A. P.; Corbin, P. S.; Fraser, C. L. *Tetrahedron Lett.* **2000**, *41*, 2787-2789.
9. Dutta, P. K.; Vaidyalingam, A. S. *Microporous Mesoporous Mater.* **2003**, *62*, 107-120.
10. Whittell, G. R.; Hager, M. D.; Schubert, U. S.; Manners, I. *Nat. Mater.* **2011**, *10*, 176-188.
11. Lucht, B. L.; Tilley, T. D. *Chem. Commun.* **1998**, 1645-1645.
12. Chan, S. H.; Lam, L. S. M.; Tse, C. W.; Man, K. Y. K.; Wong, W. T.; Djurišić, A. B.; Chan, W. K. *Macromolecules* **2003**, *36*, 5482-5490.
13. Liu, Y.; Zhang, S.; Miao, Q.; Zheng, L.; Zong, L.; Cheng, Y. *Macromolecules* **2007**, *40*, 4839-4847.

14. Cheng, Y.; Zou, X.; Zhu, D.; Zhu, T.; Liu, Y.; Zhang, S.; Huang, H. *J. Polym. Sci., Part A: Polym. Chem.* **2007**, *45*, 650-660.
15. Peter, K.; Thelakkat, M. *Macromolecules* **2003**, *36*, 1779-1785.
16. Collins, J. E.; Fraser, C. L. *Macromolecules* **1998**, *31*, 6715-6717.
17. Collins, J. E.; Lamba, J. J. S.; Love, J. C.; McAlvin, J. E.; Ng, C.; Peters, B. P.; Wu, X.; Fraser, C. L. *Inorg. Chem.* **1999**, *38*, 2020-2024.
18. McAlvin, J. E.; Fraser, C. L. *Macromolecules* **1999**, *32*, 6925-6932.
19. McAlvin, J. E.; Scott, S. B.; Fraser, C. L. *Macromolecules* **2000**, *33*, 6953-6964.
20. Wu, X.; Fraser, C. L. *Macromolecules* **2000**, *33*, 7776-7785.
21. Fraser, C. L.; Smith, A. P.; Wu, X. *J. Am. Chem. Soc.* **2000**, *122*, 9026-9027.
22. Viau, L.; Maury, O.; Le Bozec, H. *Tetrahedron Lett.* **2004**, *45*, 125-128.
23. Lohio, O.; Viau, L.; Maury, O.; Le Bozec, H. *Tetrahedron Lett.* **2007**, *48*, 1229-1232.
24. Klein, C.; Baranoff, E.; Nazeeruddin, M. K.; Grätzel, M. *Tetrahedron Lett.* **2010**, *51*, 6161-6165.

Cassandra L. Fraser was born in Norfolk, Virginia in 1962 and grew up in Michigan. She obtained a B.A. degree from Kalamazoo College in 1984, an M.T.S. degree from Harvard Divinity School in 1988, and a Ph.D. from The University of Chicago in 1993 with Brice Bosnich. After postdoctoral research with Robert Grubbs at Caltech from 1993-5, she joined the faculty in the Department of Chemistry at the University of Virginia, where she was promoted to Associate Professor in 2001 and Professor in 2005, now with joint appointments in Biomedical Engineering and Architecture. Her research involves luminescent materials for imaging and sensing in biomedicine and sustainable design.

Tiandong Liu was born in Nanjing, China in 1979. He obtained a B.S. degree from Nanjing University in 2001. He began his graduate work at the University of Virginia with Lin Pu exploring enantioselective luminescent sensors for chiral carboxylic acids in 2006 and in 2010 joined the Fraser group to explore boron complexes with optical oxygen sensing and mechanochromic luminescence properties.

Preparation of Enantioenriched Homoallylic Primary Amines

A.

B.

Submitted by José C. González-Gómez, Francisco Foubelo and Miguel Yus.[1]

Checked by Rhia M. Martin and Jonathan A. Ellman.

1. Procedure

A. *(4S,S$_S$)-N-(tert-Butanesulfinyl)tridec-1-en-4-amine* (**4**). An oven-dried 100-mL three-necked round-bottomed flask is charged with (S$_S$)-N-*tert*-butanesulfinamide (**2**, 2.42 g, 20 mmol) and indium powder (2.90 g, 25 mmol) (Note 1). The flask is fitted with a condenser and equipped with a 0.8 × 2.5 cm magnetic stirring bar, a thermometer, a rubber septum, and a nitrogen inlet. After flushing with nitrogen, the flask is charged with THF (40 mL) by syringe, followed by decanal (**1**, 4.2 mL, 22 mmol) and Ti(OEt)$_4$ (9.0 mL, 43 mmol), which are also added by syringe (Note 2). The slurry is stirred at 23–25 °C for 1 h, after which time (Note 3) allyl bromide (**3**, 2.65 mL, 30 mmol) (Note 4) is added via syringe over 5 min at 23–25 °C (Note 5). The mixture is heated to 60 °C for 5 h (Notes 6 and 7). After cooling to room temperature, the reaction mixture is carefully added to a mixture of ethyl acetate:brine (200 mL:50 mL) with stirring. The resulting white suspension is filtered through a short plug (5 cm × 2 cm) of Celite, which washed with ethyl acetate (about 600 mL total) until no more product is eluted. The organic layer is concentrated under reduced pressure and the resulting white suspension is diluted with hexanes:ethyl acetate (1:1) (50 mL). The suspension is filtered through a short plug (5 cm × 2 cm) of Celite, which is then washed with hexanes:ethyl acetate (1:1) (3 × 50 mL). The

88

organic layer is concentrated under reduced pressure (40–80 mmHg) to afford crude protected amine **4** as a yellow oil (Note 8). The crude material is purified by flash column chromatography (33 × 5.5 cm column containing 320 g of silica gel) (Note 9). A mixture of hexanes:*tert*-butyl methyl ether (2:1) is used as the eluent, and fractions of 25 mL are collected using 30-mL vials (Notes 10 and 11). Fractions containing pure product are evaporated under reduced pressure to afford amine (4*S*,*S$_S$*)-**4** (3.21 g, 53%, >99:1 dr by ^{1}H-NMR and HPLC) as a colorless oil (Notes 12, 13 and 14). Mixed fractions are purified by flash column chromatography (25 × 4.5 cm column containing 160 g of silica gel). A mixture of hexanes:*tert*-butyl methyl ether (2:1) is used as the eluent, and fractions of 20 mL are collected using 30 mL vials. Fractions containing pure product are evaporated under reduced pressure to afford additional amine (4*S*,*S$_S$*)-**4** (935 mg, 15%, >99:1 dr by ^{1}H-NMR and HPLC) as a colorless oil to provide a 69% overall yield of diastereomerically pure product **4** (Note 15).

B.   (4S)-1-Tridecen-4-amine (5). In a 125-mL round-bottomed flask, equipped with a 0.8 x 2.5 cm magnetic stirring bar, is placed a solution of compound (4*S*,*S$_S$*)-**4** (3.20 g, 21.4 mmol) in dry MeOH (100 mL) (Note 16), and the solution is cooled to 0–5 °C using an ice-water bath. To this cooled solution is added a 4M solution of HCl in dioxane (25 mL). After 10 min, the cooling bath is removed, and the reaction mixture is stirred for 2 h at room temperature. The resulting mixture is carefully concentrated to dryness under vacuum. The residue is dissolved in ethyl acetate (50 mL), cooled to 0–5 °C using an ice-water bath and treated with aqueous 6M NaOH (50 mL). After stirring for 10 min (Note 17), the phases are separated, and the aqueous layer is extracted with ethyl acetate (2 × 50 mL). The combined organic layers are successively washed with 2M NaOH (1 × 5 mL), H$_2$O (1 × 5 mL) and brine (1 × 5 mL). The organic layer is dried over anhydrous sodium sulfate, filtered, and concentrated under reduced pressure (40–80 mBar). The residue is carefully dried under vacuum (ca 0.7 mmHg) until constant weight to afford pure amine **5** (1.94 g, 93%) as a yellow oil (Notes 18 and 19).

2. Notes

1.   The checkers received (*S$_S$*)-*N*-*tert*-butanesulfinamide (**2**) as a gift from AllyChem (>99% *ee* by chiral HPLC on a Chiracel AS column, 90:10

hexanes/i-PrOH, 1mL/min, λ = 222 nm). Indium powder (>99%) was purchased from Alfa Aesar. The submitters received (S_S)-N-tert-butanesulfinamide (2) as a gift from Medalchemy. (S_S)-N-tert-butanesulfinamide (2) is also available from many commercial suppliers.

2.   THF (HPLC grade, 99.9% purity) was purchased from Scharlau and used as received. A new bottle of decanal (1) (96 %, Alfa Aesar) was used without further purification. Titanium(IV) tetraethoxide (33–35% TiO_2) was purchased from Across Organics.

3.   At this point, a sample was removed and submitted to ^{1}H NMR analysis, confirming full conversion of 2 into the corresponding N-tert-butanesulfinyl imine 6 intermediate. ^{1}H NMR (300 MHz, $CDCl_3$) δ: 0.88 (t, J = 6.6 Hz, 3 H), 1.20 (s, 9 H), 1.23–1.39 (m, 12 H), 1.56–1.69 (m, 2 H), 2.52 (td, J = 7.4, 4.8 Hz, 2 H), 8.07 (t, J = 4.8 Hz, 1 H),

6

4.   Allyl bromide (3), 99% was purchased from Alfa Aesar and filtered through a short plug (0.7 × 2.5 cm) of alumina (activity 1).

5.   *Caution:* Approximately 5 min after the addition of allyl bromide (3), upon beginning to heat the reaction mixture, the reaction became very exothermic with the internal temperature rapidly rising to ~65 °C. It was therefore necessary to remove the heating bath for about 5 min.

6.   It is important to control the temperature in the reaction mixture. At 40 °C the reaction is significantly slower.

7.   The reaction progress can be monitored by TLC analysis on Merck silica gel 60 F_{254} plates and visualization with phosphomolybdic acid (5% in EtOH). Using hexanes:ethyl acetate (3:1) as eluent, compound 2 has an R_f = 0, N-tert-butanesulfinyl imine 6 has an R_f = 0.77, diastereomer ($4R,S_S$)-4 (minor) has an R_f = 0.32, diastereomer ($4R,S_S$)-4 (major) has an R_f = 0.26, and the homoallyl alcohol by-product 7 has an R_f = 0.85.

7

8.   The crude oil contained a 10:1 mixture of $(4R,S_S)$-**4** : $(4S,S_S)$-**4** with the diastereomeric purity determined by ^{1}H-NMR analysis based on the NH signals: major diastereomer 3.22 (d, J = 6.1 Hz); minor diastereomer 3.07 (d, J = 5.9 Hz) (see attached NMR).

9.   Silica gel 60, 230-400 mesh (Sorbent Technologies) was used.

10. The column fractions were checked by TLC analysis on Merck silica gel 60 F_{254} plates with hexanes:ethyl acetate (3:1) as eluent and visualization with phosphomolybdic acid. R_f values are given in Note 7.

11. The minor diastereomer $(4R,S_S)$-**4** elutes before the major one. Concentration of fractions 58-94 furnished 2.18 g (36%) of a 78:22 mixture of $(4R,S_S)$-**4**/$(4S,S_S)$-**4** (according to ^{1}H NMR). Concentration of fractions 95-168 furnished 3.21 g (53%) of $(4S,S_S)$-**4** (>99:1 according to ^{1}H NMR).

12. The physical properties of $(4S,S_S)$-**4** are as follows: ^{1}H NMR (400 MHz, CDCl$_3$) δ: 0.88 (t, J = 6.9 Hz, 3 H), 1.20 (s, 9 H), 1.22–1.39 (m, 14 H), 1.43–1.53 (m, 2 H), 2.30 (dt, J = 13.9, 6.9 Hz, 1 H), 2.41 (dt, J = 12.8, 6.4 Hz, 1 H), 3.19 (d, J = 6.0 Hz, 1 H), 3.25–3.36 (m, 1 H), 5.13 (d, J = 1.7 Hz, 1 H), 5.17 (d, J = 2.0 Hz, 1 H), 5.70–5.85 (m, 1 H); ^{13}C NMR (101 MHz, CDCl$_3$) δ: 14.2, 22.7, 25.5, 29.4, 29.56, 29.61, 29.63, 32.0, 35.0, 40.5, 54.9, 55.8, 118.9, 134.3; IR (neat, cm^{-1}) 3208 (br), 2923, 2854, 1640, 1457, 1389, 1363, 1176, 1051. Anal. calcd. for C$_{17}$H$_{35}$NOS: C, 67.72; H, 11.70; N, 4.65. Found: C, 67.39; H, 11.58; N, 4.69. The submitters report the specific rotation of a 97:3 mixture of $(4S,S_S)$-**4**/$(4R,S_S)$-**4** as $[\alpha]^{20}_D$ +45.8 (c 2.6, CH$_2$Cl$_2$).

13. The diastereomeric purity was determined by ^{1}H NMR analysis (see Note 8).

14. The submitters further confirmed the diastereomeric purity of compound **4** by converting it to the corresponding benzamide according to the following procedure. To a solution of **4** (0.1 mmol) in MeOH (1 mL) was added 4M of HCl in dioxane (0.3 mL), and the solution was stirred 1.5 h at room temperature. After cooling to 0 °C, 2M NaOH (5 mL) was added, followed by benzoyl chloride (32 µL, 0.25 mmol), and the resulting mixture was stirred over 2 h at room temperature (23 °C). After aqueous workup, a sample [GC-MS (EI): 301 (M$^+$, 1%), 260 ([M-C$_3$H$_5$]$^+$, 39%); 105 ([M-Bz]$^+$, 100%); 77 (18%)] was submitted to chiral HPLC analysis. The chromatography was performed on a JASCO 200-series apparatus equipped with a Chiralcel OD-H column 25 cm × 0.46 cm (isocratic elution with 10%

i-PrOH in *n*-hexane, 1.0 mL/min, UV detection at 254 nm); t_R = 6.3 min (*R*), t_R = 7.7 min (*S*).

15. Concentration of fractions 48-72 furnished 977 mg (16%) of a 66:34 mixture of (4*R*,S_S)-4/(4*S*,S_S)-4. Concentration of fractions 73-108 furnished 935 mg (15%) of (4*S*,S_S)-4 (>99:1 dr).

16. MeOH (99.8%) was purchased from J. T. Baker and distilled from magnesium.

17. At this point the pH of aqueous phase was >12. Importantly, at pH 8–9 the corresponding ammonium hydrochloride is isolated using the same procedure.

18. The submitters report that according to GC-MS (EI): 136 ([M]$^+$, 0.6%); 79 ([M-C$_4$H$_9$]$^+$, 5%); 57 ([C$_4$H$_9$]$^+$, 23%); the main impurity is likely *t*-BuSO$_2$Me (t_R = 4.9 min), which is volatile. After keeping under house vacuum (0.7–3.5 mmHg) until constant weight, only the expected compound **5** was observed by GC (t_R = 10.5 min). GC analyses were obtained on an Agilent 5973N equipped with an HP-5 column (30 m × 0.25 mm, ID × 0.25 μm) and an EI (70 EV) detector. The temperature program: hold at 60 °C for 3 min, ramp from 60 °C to 270 °C at 15 °C/min, hold at 270 °C for 10 min.

19. The physical properties of **5** are as follows: ^{1}H NMR (500 MHz, CDCl$_3$) δ: 0.86 (t, *J* = 6.7 Hz, 3 H), 1.20 (s, 2 H), 1.44–1.21 (m, 17 H), 1.88–2.01 (m, 1 H), 2.13–2.27 (m, 1 H), 2.75 (s, 1 H), 5.05 (s, 1 H), 5.08 (d, *J* = 8.3 Hz, 1 H), 5.69–5.83 (m, 1 H); ^{13}C NMR (126 MHz, CDCl$_3$) δ: 14.4, 23.0, 26.6, 29.6, 29.86, 29.92, 30.0, 32.2, 38.0, 42.9, 50.9, 77.1, 77.3, 77.6, 117.4, 136.3; IR (neat, cm^{-1}) 3354 (br), 3076, 2956, 2922, 2853, 1640, 1466, 1439, 1378, 1354, 1130. The submitters report the specific rotation as $[\alpha]^{20}_D$ –6.2 (c 1.1, CH$_2$Cl$_2$)

Safety and Waste Disposal Information

All hazardous materials should be handled and disposed of in accordance with "Prudent Practices in the Laboratory"; National Academy Press; Washington, DC, 1995.

3. Discussion

Addition of allylmetals to the C=N bond of enantiomerically pure *N*-*tert*-butanesulfinyl imines are among the most widely used approaches for the asymmetric synthesis of homoallylic amine derivatives.[2]

Allylmagnesium,[3] allylzinc,[4] and allylindium[5] species have been used; however, the synthesis and isolation of the corresponding *N*-sulfinyl imines has been required. In contrast, we describe in this procedure a method based on *in situ* formation of *N-tert*-butanesulfinyl imines and subsequent indium-mediated allylation.[6] Thus, enantioenriched homoallylamine derivatives are directly obtained from commercially available aldehydes, allyl bromides, and chiral *tert*-butanesulfinamide. It was found that two equivalents of Ti(OEt)$_4$ were required to promote the reaction efficiently,[7] and good results for a range of aldehydes were obtained when allyl bromide (**3**) was added after *in situ* pre-formation of the *N*-sulfinyl imine for 1 hour at room temperature. Notably, the reaction works better when aliphatic aldehydes are used, but aromatic and α,β-unsaturated aldehydes also afford the corresponding homoallylic amine derivative in good yields and stereoselectivities (Table 1).[5] It is worth noting that although

Table 1. Other homoallylic amine derivatives prepared using this protocol

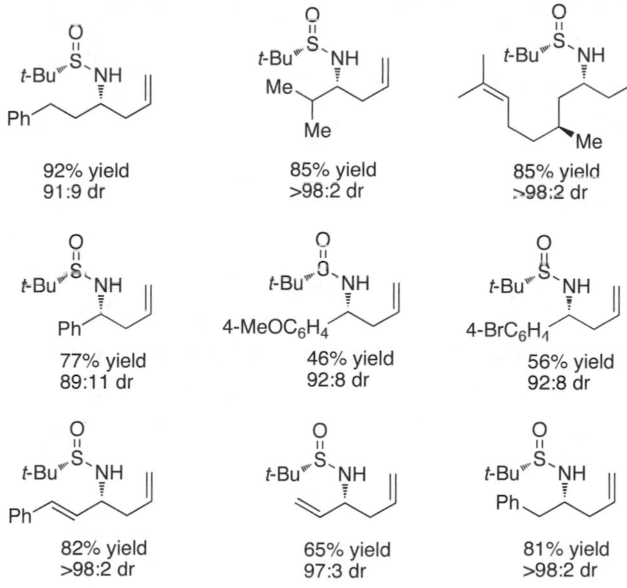

92% yield
91:9 dr

85% yield
>98:2 dr

85% yield
>98:2 dr

77% yield
89:11 dr

46% yield
92:8 dr

56% yield
92:8 dr

82% yield
>98:2 dr

65% yield
97:3 dr

81% yield
>98:2 dr

enantioenriched aromatic homoallylic amines currently can be prepared using a large number of protocols, methods to prepare the aliphatic derivatives with good selectivity are scarce. Since conversion of *tert*-butanesulfinamide to the *N*-sulfinyl imine with Ti(OEt)$_4$ is not complete within 1 hour for all of the aldehydes explored, it is possible that indium(III) salts formed *in situ* accelerate both the formation of the *N*-sulfinyl imine and its allylation.[8] Importantly, despite the presence of Lewis acid (indium and titanium salts) and Lewis base species (water/ethanol), we observed the same sense and degree of stereoselectivity[9] that we have previously observed for the indium-mediated allylation of pre-formed *N*-sulfinyl imines.[4b]

Additionally, the reaction was investigated using 3-phenylpropanal and other allyl bromides, such as methallyl bromide, prenyl bromide and racemic cyclohexenyl bromide, each of which proceeded with very good chemo- and stereoselectivities (Table 2).

Table 2. Reaction with other allylic bromides

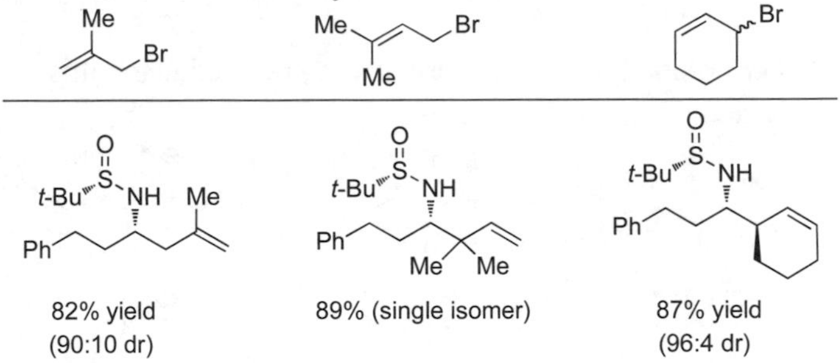

| 82% yield (90:10 dr) | 89% (single isomer) | 87% yield (96:4 dr) |

Moreover, it is worth mentioning that the synthetic utility of homoallylamine derivative (4*S*,*S*_S)-**4** has already been illustrated in a short synthesis of (+)-isosolenopsin,[10] while (4*R*,*R*_S)-**4** was used to prepare (+)-preussin and several analogs.[11] On the other hand, the *N*-Cbz protected amine **5** has been recently used in a four-step synthesis of the alkaloid (+)-241D (Figure 1).[12]

Figure 1.

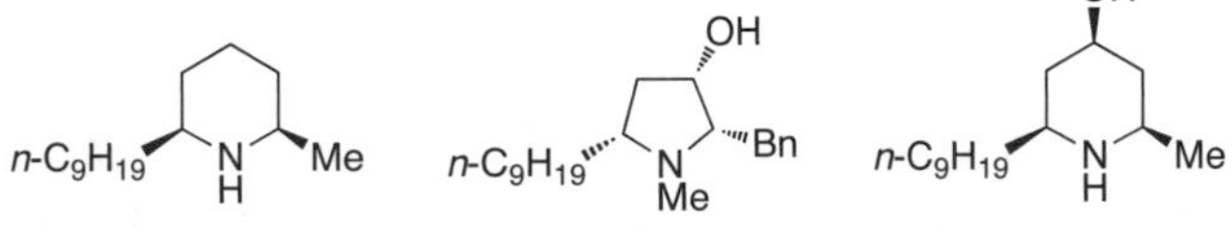

(+)-isosolenopsin   (+)-preussin   (+)-241D

1. Departamento de Química Orgánica, Facultad de Ciencias and Instituto de Síntesis Orgánica (ISO), Universidad de Alicante, Apdo. 99, 03080 Alicante, Spain. E-mail: josecarlos.gonzalez@ua.es; foubelo@ua.es; yus@ua.es. We thank the Spanish Ministerio de Ciencia e Innovación (Grant No. CTQ2007-65218 and Consolider Ingenio 2010-CSD-2007-00006), the Generalitat Valenciana (Grant No. PROMETEO/2009/039 and FEDER), and the University of Alicante for generous and continuous financial support, as well as MEDALCHEMY for a gift of chemicals.

2. For a recent review on the asymmetric synthesis of amines via *N-tert*-butanesulfinyl imines, see: Robak, M. T.; Herbage, M. A.; Ellman, J. A. *Chem. Rev.* **2010**, *110*, 3600.

3. (a) Cogan, D. A.; Liu, G.; Ellman, J. A. *Tetrahedron* **1999**, *55*, 8883. (b) Grainger, R. S.; Welsh, E. J. *Angew. Chem. Int. Ed.* **2007**, *46*, 5377.

4. (a) Sun, X.-W.; Xu, M.-H.; Lin, G.-Q. *Org. Lett.* **2006**, *8*, 4979. (b) Kolodney, G.; Sklute, G.; Perrone, S.; Knochel, P.; Marek, I. *Angew. Chem. Int. Ed.* **2007**, *46*, 9291.

5. (a) Cooper, I. R.; Grigg, R.; MacLachlan, W. S.; Thornton-Pett, M.; Sridharan, V. *Chem. Commun.* **2002**, 1372. (b) Foubelo, F.; Yus, M. *Tetrahedron: Asymmetry* **2004**, *15*, 3823. (c) Sun, X.-W.; Liu, M.; Xu, M.-H.; Lin, G.-Q. *Org. Lett.* **2008**, *10*, 1259.

6. González-Gómez, J. C.; Medjahdi, M.; Foubelo, F.; Yus, M. *J. Org. Chem.* **2010**, *75*, 6308.

7. (a) Ti(OEt)$_4$ it is an excellent water scavenger and catalyst for the formation of *N-tert*-butanesulfinyl imines, see: Liu, G.; Cogan, D. A.; Owens, T. D.; Tang, T. P.; Ellman, J. A. *J. Org. Chem.* **1999**, *64*, 1278. (b) Ti(OEt)$_4$ has also been successfully used in a one-pot protocol for the formation and reduction of *N-tert*-butanesulfinyl ketimines, see: Tanuwidjaja, J.; Peltier, H. M.; Ellman, J. A. *J. Org. Chem.* **2007**, *72*, 626.

8. It is known that In(OTf)$_3$ increases the rate of the indium-mediated allylation of chiral hydrazones, see: Cook, G. R.; Maity, B. C.; Kargbo, R. *Org. Lett.* **2004**, *6*, 1741.

9. It is reported that the stereochemical outcome of the Zn/In-mediated allylation of *N-tert*-butanesulfinyl imines can be highly dependent on the presence of Lewis acids or bases. See: Lin, G.-Q.; Xu, M.-H.; Zhong, Y. W.; Sun, X.-W. *Acc. Chem. Res.* **2008**, *41*, 831.

10. González-Gómez, J. C.; Foubelo, F.; Yus, M. *Synlett* **2008**, 2777.
11. Bertrand, M. B.; Wolfe, J. P. *Org. Lett.* **2006**, *8*, 2353.
12. Kumar-G, R. S. C.; Reddy, V.; Shankaraiah, G.; Babu, K. S.; Rao, J. M. *Tetrahedron Lett.* **2010**, *51*, 1114.

Appendix
Chemical Abstracts Nomenclature; (Registry Number)

(S_s)-*N*-*tert*-Butanesulfinamide: (S)-(–)-2-Methyl-2-propanesulfinamide; (343338-28-3)

Indium; (7440-74-6)

Decanal: Decyl aldehyde; (112-31-2)

Titanium(IV) ethoxide: Tetraethyl titanate; (3087-36-3)

Allyl bromide: 3-Bromo-1-propene; (106-95-6)

(*rac*)-1-Tridecen-4-amine; (893402-88-5)

Miguel Yus was born in Zaragoza (Spain) in 1947, and received his BSc (1969), MSc (1971) and PhD (1973) degrees from the University of Zaragoza. After spending two years as a postdoctoral fellow at the Max Planck Institut für Kohlenforschung in Mülheim a.d. Ruhr he returned to Spain to the University of Oviedo where he became associate professor in 1977, being promoted to full professor in 1987 at the same university. In 1988 he moved to a chair in Organic Chemistry at the University of Alicante where he is currently the head of the newly created Organic Synthesis Institute (ISO). His current research interest is focused on the preparation of very reactive functionalized organometallic compounds and their use in synthetic organic chemistry, arene-catalyzed activation of different metals, preparation of new metal-based catalysts, including metallic nanoparticles for homogeneous and heterogeneous selective reactions, and asymmetric catalysis.

José Carlos González Gómez was born in 1971 and grew up in Havana, Cuba. He obtained his B.S. (1994) and M.S (1998) in chemistry at Havana University. In 1999 he moved to the University of Santiago de Compostela (Spain), where he got his Ph.D degree (2003) working on the synthesis of bioactive psoralen derivatives. After a postdoctoral stay (2005-2007) at the ETH in Zurich, he started as "Juan de la Cierva" researcher at the University of Alicante, where he became Assistant Professor in 2008. Dr. González-Gómez has co-authored about 30 papers and his research interest is currently focused on the development and application of stereoselective reactions for the synthesis of bioactive compounds, mainly involving chiral sulfinimines.

Francisco Foubelo was born in 1961 and grew up in Eastern Asturias. He studied chemistry at the University of Oviedo from which he received B.S. (1984), M.S. (1986), and Ph.D. (1989) degrees. After a postdoctoral stay (1989–1991) as a Fulbright fellow at Princeton University, he moved to the University of Alicante where he became Associate Professor in 1995 and Full Professor in 2002. Dr. Foubelo has co-authored more than 100 papers, and his current research interests are focused on the development of new synthetic methodologies involving chiral sulfinimines and on metal-promoted functionalization of alkenes and alkynes.

Rhia M. Martin received her B.S. degree at Georgetown University in Washington, DC in 2007. She then joined Professor Jonathan A. Ellman's and Professor Robert G. Bergman's research groups at University of California, Berkeley, where she is pursuing her Ph. D. degree. Her graduate research focuses on the development of methodologies to access nitrogenous heterocycles through C-H functionalization.

Discussion Addendum for:
(Phenyl)[2-(trimethylsilyl)phenyl]iodonium Triflate. An Efficient and Mild Benzyne Precursor

Prepared by Tsugio Kitamura.*[1]
Original article: Kitamura, T.; Todaka, M.; Fujiwara, Y. *Org. Synth.* **2002**, *78*, 104.

Although many convenient and reliable precursors of benzyne have been prepared and utilized in organic synthesis,[2] a hypervalent iodine-containing molecule, (phenyl)[2-(trimethylsilyl)phenyl]iodonium triflate (**2**), has been developed for service as an efficient and mild benzyne precursor. The major developments over the past years have been the improvement of methods of synthesis of the hypervalent iodine benzyne precursor and the applications to related reactive intermediates. The summary will be reported in this Discussion Addendum.

Improvement of Hypervalent Iodine Precursor

2-(Trimethylsilyl)phenyliodonium triflate **2** generates benzyne more efficiently than other benzyne precursors such as benzenediazonium-2-carboxylate, *o*-dihalobenzenes (1,2-bromofluorobenzene and 1,2-dibromobenzene), and 2-(trimethylsilyl)phenyl triflate. The efficiency of

98

Org. Synth. **2012**, *89*, 98-104
Published on the Web 9/28/2011
© 2012 Organic Syntheses, Inc.

this benzyne precursor **2** is attributable to the high leaving group ability of hypervalent iodine groups, which has been discussed previously.[3] Hypervalent iodine benzyne precursor **2** was prepared in two steps from inexpensive and commercially available 1,2-dichlorobenzene. Initial preparation of 1,2-bis(trimethylsilyl)benzene (**1**) by reaction of 1,2-dichlorobenzene and Me$_3$SiCl in the presence of Mg and a catalytic amount of I$_2$ in HMPA is followed by transformation into **2** by the reaction with PhI(OAc)$_2$ activated with TfOH in CH$_2$Cl$_2$. However, several drawbacks remained: (1) use of toxic HMPA solvent and severe conditions for the preparation of **1**, (2) difficulty for introduction of functional groups, and (3) low solubility of the hypervalent iodine species in most organic solvents.

To improve the reaction procedure, an alternative benzyne precursor, 2-(hydroxydimethylsilyl)phenyliodonium triflate (**5**), has been prepared in three steps starting with 1,2-dibromobenzene.[4] This process avoids the use of carcinogenic HMPA and severe reaction conditions (high temperature and long reaction time). The reaction of **5** with Bu$_4$NF under mild conditions generates benzyne efficiently, which provides the benzyne adducts in high yields. Therefore, **5** is a convenient alternative benzyne precursor. Incorporation of synthetically important carbonyl groups into hypervalent iodine benzyne precursors is difficult, because carbonyl groups cannot tolerate Grignard conditions. Synthesis of benzyne precursors bearing the carbonyl groups has been accomplished using pyran-2-ones as starting materials. Reaction of 1,2-bis(trimethylsilyl)acetylene with methyl 2-pyrone-5-carboxylate or 5-acylpyran-2-ones at 200 °C in a sealed tube affords the corresponding methyl 3,4-bis(trimethylsilyl)benzoate[5] or 1-acyl-3,4-bis(trimethylsilyl)-benzenes[6] in good to high yields, which can be converted to 2-(trimethylsilyl)-phenyliodonium triflates **6**, which contain carbonyl groups, by treatment with PhI(OAc)$_2$ and TfOH. Treatment of **6** with Bu$_4$NF in CH$_2$Cl$_2$ generates benzynes that contain ketone functionality, and reaction with furan provides the benzyne adducts in high yields. In general, hypervalent iodine benzyne precursors are stable solids and easy to handle, but they do not dissolve in most organic solvents. To improve the solubility of the hypervalent iodine benzyne precursors, long-chained precursors (**7**) have been synthesized.[7] As the alkyl chain of benzyne precursors **7** is lengthened, the solubility in non-polar organic solvents and the yield of the benzyne adduct with furan gradually increases.

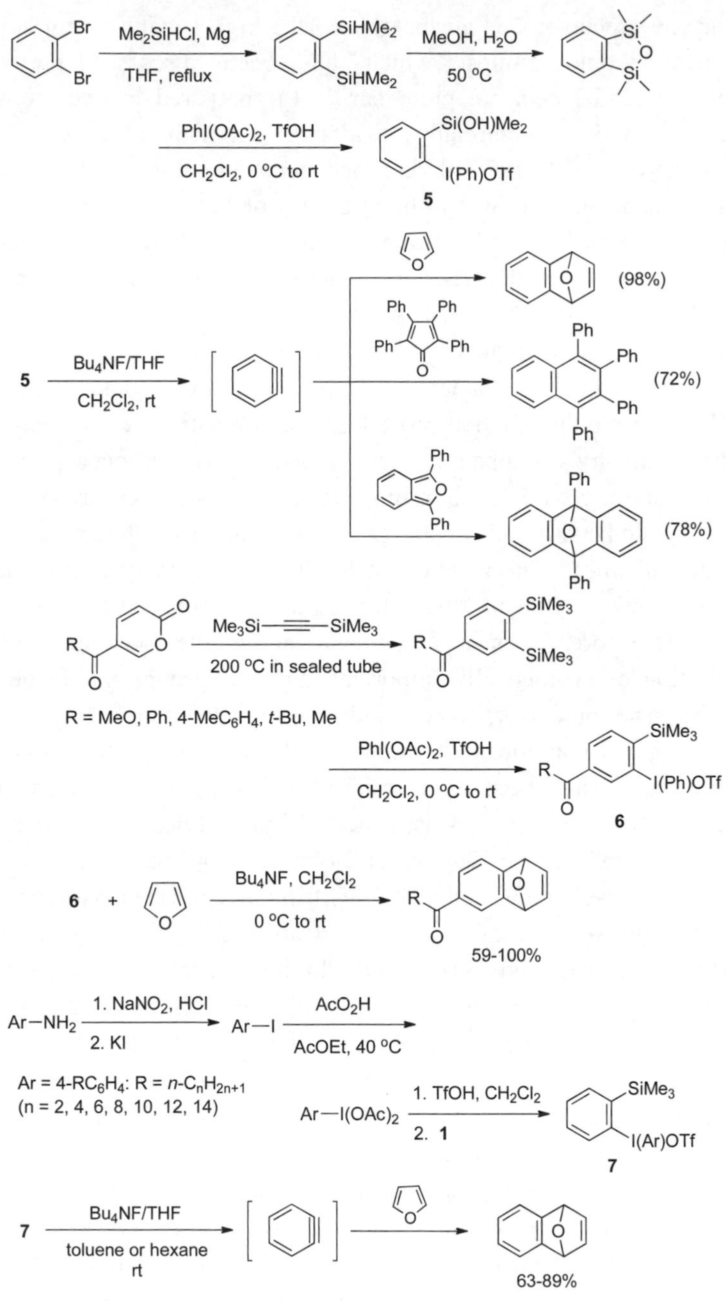

Figure 1. Improvements of synthetic procedures for hypervalent iodine precursors.

Org. Synth. **2012**, *89*, 98-104

1,2-Bis(trimethylsilyl)benzene (**1**) is the key starting material for the synthesis of 2-(trimethylsilyl)phenyliodonium triflate **2**. Recent improvement on the synthesis of **1** have been reported in which 1,2-dibromobenzene is used instead of 1,2-dichlorobenzene. The use of Rieke magnesium or Mg turnings/1,2-dibromoethane[8] and Fe-catalyzed DIBAL-H activated Grignard reaction have led to improved reactions.[9] Through these variations, severe reaction conditions and the use of toxic HMPA solvent can be avoided.

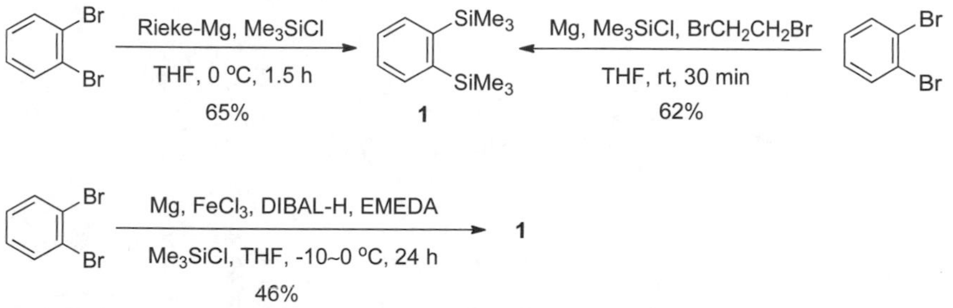

Figure 2. Improvements of synthesis of **1**.

Other Applications Using Hypervalent Iodine

When hypervalent iodine and silyl groups are contiguous, fluoride ion can trigger the concerted elimination reaction of these two groups. This method is applicable not only to the generation of benzyne, but also to the generation of reactive intermediates related to benzyne. Thus, 2,3-didehydronaphthalene (**8**),[10] bicycle[2.2.1]hept-2-en-5-yne (**9**),[11] dehydro-*o*-carborene (**10**),[12] 2,3-didehydrothiphene (**11**),[13] and 2,3-didehydropyrrole (**12**)[14] have been generated by this method. Benzobisoxadisiloles have been applied to synthesis of hypervalent iodine benzyne precursors, which can serve as the synthetic equivalents of 1,4-benzdiyne (**13**) and related species.[15]

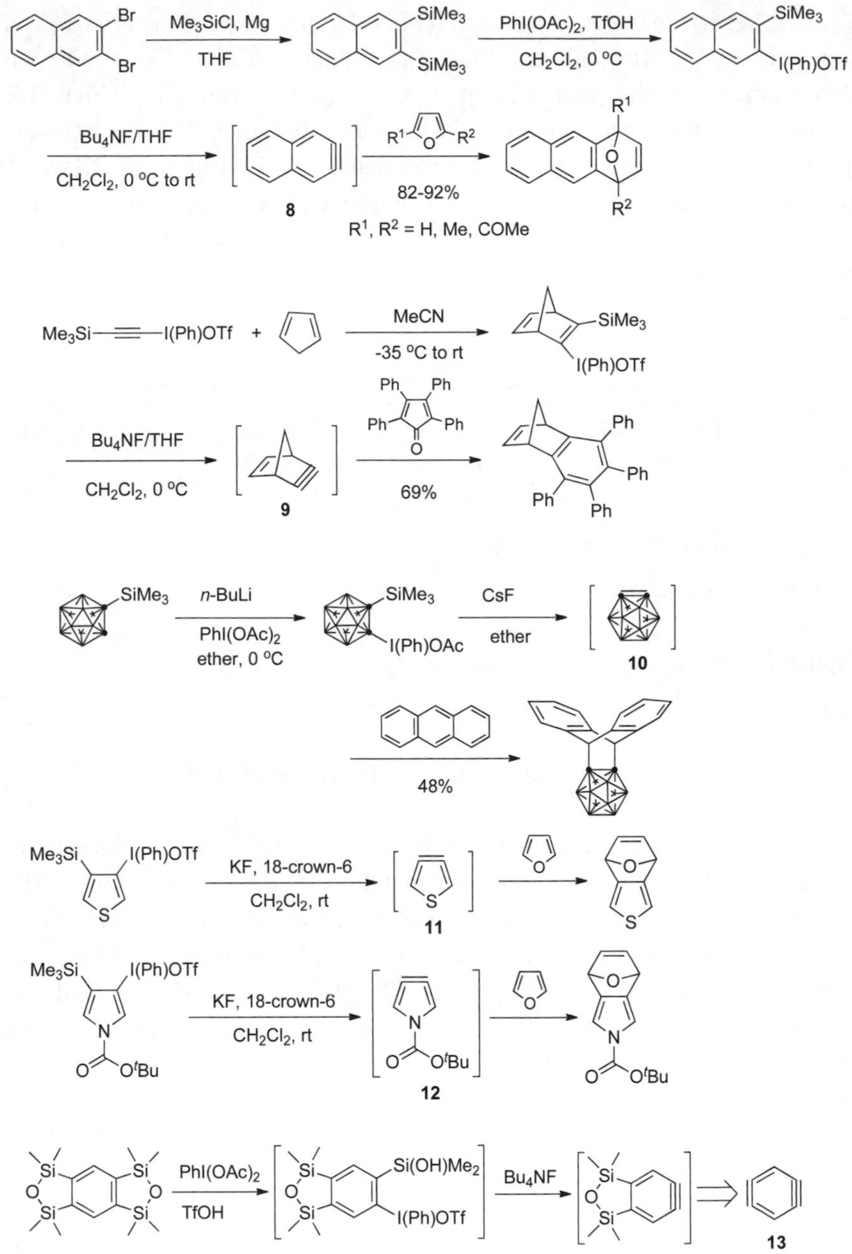

Figure 3. Generation of benzyne-related species using hypervalent iodine/silane methodology.

Org. Synth. **2012**, *89*, 98-104

1. Department of Chemistry and Applied Chemistry, Graduate School of Science and Engineering, Saga University, Honjo-machi, Saga 840-8502, Japan.

2. (a) Winkler, M.; Wenk, H. H.; Sander, W. In *Reactive Intermediate Chemistry*; Moss, R. A.; Platz, M. S.; Jones, M. J., Eds.; John Wiley & Sons: Hoboken, 2004, 741-794. (b) Wenk, H. H.; Winkler, M.; Sander, W. *Angew. Chem. Int. Ed.* **2003**, *42*, 502-527. (c) Pellissier, H.; Santelli, M. *Tetrahedron* **2003**, *59*, 701-830. (d) Hart, H. In *The Chemistry of Triple-bonded Functional Groups, Supplement C2*; Patai, S., Ed.; John Wiley and Sons: Chichester, 1994, Chapt. 18, 1017-1134. (e) Kessar, S. V. In *Comprehensive Organic Synthesis*; Trost, B. M.; Fleming, I., Eds.; Pergamon Press: New York, 1991, Vol. 4, 483-515. (f) Gilchrist, T. L. In *The Chemistry of Triple-bonded Functional Groups, Supplement C*; Patai, S; Rappoport, Z., Eds.; John Wiley and Sons: Chichester, 1983, Chapt. 11, 383-419. (g) Hoffmann, R. W. In *Chemistry of Acetylenes*; Viehe, H. G., Ed.; Dekker: New York, 1969, 1063-1148. (h) Hoffmann, R. W. *Dehydrobenzene and Cycloalkynes*; Academic Press: New York, 1967. (h) Kitamura, T. *Aus. J. Chem.* **2010**, *63*, 987-1101.

3. (a) Kitamura, T.; Todaka, M.; Fujiwara, Y. *Org. Synth.* **2002**, *78*, 104-112. (b) Kitamura, T.; Yamane, M.; Inoue, K.; Todaka, M.; Fukatsu, N.; Meng, Z.; Fujiwara, Y. *J. Am. Chem. Soc.* **1999**, *121*, 11674-11679. (c) Kitamura, T.; Yamane, M. *J. Chem. Soc. Chem. Commun.* **1995**, 983-984.

4. Kitamura, T.; Meng, Z.; Fujiwara, Y. *Tetrahedron Lett.* **2000**, *41*, 6611-6614.

5. Kitamura, T.; Wasai, K.; Todaka, M.; Fujiwara, Y. *Synlett* **1999**, 731-732.

6. Kitamura, T.; Aoki, Y.; Isshiki, S.; Wasai, K.; Fujiwara, Y. *Tetrahedron Lett.* **2006**, *47*, 1709-1712.

7. (a) Kitamura, T.; Abe, T.; Fujiwara, Y.; Yamaji, T. *Synthesis* **2003**, 213-126. (b) Abe, T.; Yamaji, T.; Kitamura, T. *Bull. Chem. Soc. Jpn.* **2003**, *76*, 2175-2178.

8. Lorbach, A.; Reus, C.; Bolte, M.; Lemer, H.-W.; Wagner, M. *Adv. Synth. Cat.* **2010**, *352*, 3443-3449.

9. Bader, S. L.; Kessler, S. N.; Wegner, H. A. *Synthesis* **2010**, 2759-2762.

10. Kitamura, T.; Fukatsu, N.; Fujiwara, Y. *J. Org. Chem.* **1998**, *63*, 8579-8581.

11. Kitamura, T.; Kotani, M.; Yokoyama, T.; Fujiwara, Y. *J. Org. Chem.* **1999**, *64*, 680-681.
12. Jeon, J.; Kitamura, T.; Yoo, B.-W.; Kang, S. O.; Ko, J. *Chem. Commun.* **2001**, 2110-2111.
13. (a) Ye. X.-S.; Li, W.-K.; Wong, H. N. C. *J. Am. Chem. Soc.* **1996**, *118*, 2511-2512. (b) Ye, X.-S.; Wong, H. N. C. *J. Org. Chem.* **1997**, *62*, 1940-1954.
14. Liu, J.-H.; Chan, H.-W.; Xue, F.; Wang, Q.-G.; Mak, T. C. W.; W. H. N. C. *J. Org. Chem.* **1999**, *64*, 1630-1634.
15. (a) Chen, Y.-L.; Zhang, H.-K.; Wong, W.-Y.; Lee, A. W. M. *Tetrahedron Lett.* **2002**, *43*, 2259-22262. (b) Chen, Y.-L.; Sun, J.-Q.; Wei, X.; Wong, W.-Y.; Lee, A. W. M. *J. Org. Chem.* **2004**, *69*, 7190-7197. (c) Chen, Y.-L.; Hau, C.-K.; Wang, H.; He, H.; Wong, M.-S.; Lee, A. W. M. *J. Org. Chem.* **2006**, *71*, 3512-3517. (d) Chen, Y.-L.; Wong, M.-S.; Wong, W.-Y.; Lee, A. W. M. *Tetrahedron Lett.* **2007**, *48*, 2421-2425.

Tsugio Kitamura was born in Nagasaki, Japan in 1954. He received his B. Eng. and Dr. Eng. degrees from Kyushu University in 1977 and 1982, respectively, under the direction of Prof. Hiroshi Taniguchi. He joined the faculty as assistant professor at Kyushu University in 1982 and was promoted to associate professor in 1993. In 1986-1988 he worked as a postdoctoral fellow with Professor Peter J. Stang at University of Utah, USA. In 2002 he moved to Saga University as full professor. His research interests are in the areas of synthetic organic chemistry including vinyl cations, hypervalent iodines, and transition-metal catalysts.

104

2-Iodoxy-5-Methylbenzenesulfonic Acid-Catalyzed Selective Oxidation of 4-Bromobenzyl Alcohol to 4-Bromobenzaldehyde or 4-Bromobenzoic Acid with Oxone

Submitted by Muhammet Uyanik and Kazuaki Ishihara.*[1]
Checked by Margaret Faul and Richard Crockett.

1. Procedure

Caution! Although no problems were encountered with the explosion of Oxone or other chemicals in the use of this oxidation procedure, prudence dictates that all operations should be conducted in a hood.

A. *4-Bromobenzaldehyde.* A 500-mL, three-necked, round-bottomed flask equipped with an overhead mechanical stirrer, a reflux condenser fitted with a nitrogen inlet adapter, and a glass stopper (Note 1) is charged with Oxone (12.8 g, 20.9 mmol, 0.65 equiv) (Notes 2 and 3) and acetonitrile (70 mL) (Note 4). The mixture is stirred vigorously (1000 rpm) (Notes 5 and 6) at room temperature for 2 h. To the resulting white suspension are added potassium 2-iodo-5-methylbenzenesulfonate (0.11 g, 0.32 mmol, 1 mol%) (Note 7) and 4-bromobenzyl alcohol (6.00 g, 32.1 mmol) (Note 8), and the wall of the flask is washed with acetonitrile (10 mL) (Note 9). The resulting mixture is heated to 70 °C in an oil bath and stirred vigorously (1000 rpm) (Note 5). After 5 h (Note 10), the resulting white suspension is cooled to room temperature and filtered through a plug of tightly packed Celite (Note 11) on a sintered glass funnel, which is successively washed with diethyl ether (100 mL). To the filtrate is added water (150 mL), and the aqueous layer is separated and extracted with EtOAc (2 x 100 mL). The combined organic layers are washed with saturated aqueous NaHCO$_3$ (100 mL), saturated aqueous Na$_2$SO$_3$ (100 mL), water (100 mL) and saturated brine (100 mL). The resulting organic layers are dried over anhydrous Na$_2$SO$_4$, filtered and concentrated by rotary evaporation (30 °C, 20 mmHg) to give a white solid (5.78–6.25 g) (Note 12). The solid is transferred to a 250-mL, round-bottomed flask. Methanol (40 mL) and deionized water (20 mL) are added, and the mixture is heated to 70 °C until all solids are completely dissolved (Note 13). To the resulting solution is added additional deionized water (30 mL) all at once. The resulting suspension is cooled to 25 °C over 1.5 h and allowed to stir at room temperature (> 3 h) to give a white precipitate. The solids are collected by suction filtration on a Büchner funnel. The reaction flask is rinsed with 50 mL water. This rinse is used to wash the wet cake and collected with the filtrate. The cake is allowed to dry on the funnel for 1 h under vacuum suction with a nitrogen bag fitted over the funnel. The cake is then transferred to a 150-mL, round-bottomed flask and further dried for 12 h (25 °C, 1 mmHg) to provide 4-bromobenzaldehyde (4.45–4.73 g, 24.0–25.6 mmol, 76–80% yield) as a white powder (Notes 14 and 15). Solids that formed in the first filtrate are collected by suction filtration on a Büchner funnel, washed with water (30 mL), and then transferred to a 150-mL, round-bottomed flask and dried for 12 h (25 °C, 1 mmHg) to provide further 4-bromobenzaldehyde (0.25–0.41 g, 0.97–1.35 mmol, 3–4% yield) as a white powder. The combined yield of 4-bromobenzaldehyde (4.63–4.98 g, 25.0–26.9 mmol) is 79–85% (Notes 14 and 15).

B. 4-Bromobenzoic acid. A 500-mL, three-necked, round-bottomed flask equipped with an overhead mechanical stirrer, a reflux condenser fitted with a nitrogen inlet adapter, and a glass stopper (Note 1) is charged with Oxone (25.7 g, 41.7 mmol, 1.3 equiv) (Notes 2 and 3) and acetonitrile (70 mL) (Note 4). The mixture is stirred vigorously (1000 rpm) (Notes 5 and 6) at room temperature for 2 h. To the resulting white suspension are added potassium 2-iodo-5-methylbenzenesulfonate (0.11 g, 0.32 mmol, 1 mol%) (Note 7) and 4-bromobenzyl alcohol (6.00 g, 32.1 mmol) (Note 8), and the wall of the flask is washed with acetonitrile (10 mL) (Note 9). The resulting mixture is heated to 70 °C in an oil bath and stirred vigorously (1000 rpm) (Notes 5). After 2.6 h (Note 16), the starting materials are consumed completely. To the mixture is added deionized water (100 mL), and the

106

resulting mixture is stirred vigorously (1000 rpm) at 70 °C. After 6.5 h, TLC showed no aldehyde and HPLC analysis showed < 5 area % of aldehyde. The resulting pale yellow mixture is cooled to room temperature. EtOAc (200 mL) and water (100 mL) are added, and the mixture is transferred to a 1-L separatory funnel. The aqueous layer is separated and extracted with EtOAc (3 x 100 mL). The combined organic layers are washed with saturated aqueous NaHSO₃ (100 mL), water (100 mL) and saturated brine (100 mL). The resulting organic layers are dried over anhydrous Na₂SO₄, filtered and concentrated by rotary evaporation (30 °C, 20 mmHg) to give a pale yellow solid (6.14–6.37 g). The solid is transferred to a 250-mL, round-bottomed flask. Methanol (120 mL) and deionized water (5 mL) are added, and the mixture is heated to reflux (66 °C) to afford a turbid solution. The solution is cooled to room temperature over 1.5 h and aged for 15 min. To the resulting suspension is added additional deionized water (95 mL) over 2 h and the mixture is stirred at room temperature overnight to give pale yellow crystals. The crystals are collected by suction filtration on a Büchner funnel. The reaction flask is rinsed with 50 mL water. This rinse used to wash the wet cake and collected with the filtrate. The cake is allowed to dry on the funnel for 1 h under vacuum suction with a nitrogen bag fitted over the funnel. The cake is then transferred to a 150-mL, round-bottomed flask and further dried for 12 h (25 °C, 1 mmHg) to provide 4-bromobenzoic acid (5.28–5.5 g, 26.3–27.6 mmol, 83–87% yield) as pale yellow powder (Notes 13 and 16). Solids that formed in the first filtrate are collected by suction filtration on a Büchner funnel, washed with water (50 mL), and then transferred to a 100-mL, round-bottomed flask and dried for 12 h (25 °C, 1 mmHg) to provide further 4-bromobenzoic acid (0.18–0.19 g, 0.89–0.94 mmol, 2–3% yield) as pale yellow powder. The combined yield of 4-bromobenzoic acid (5.47–5.73 g, 27.2–28.5 mmol) is 86–90% (Note 14 and 17).

2. Notes

1.   A glass stir bearing with a ground glass stir shaft (10 mm diameter) and stir lube was used. To the stir shaft was affixed a semi circular, Teflon stir paddle (19 x 60 mm).

2.   Oxone (2KHSO₅·KHSO₄·K₂SO₄) was purchased from Aldrich Chemical Company, Inc., and used as received.

3.   Although no problems were encountered with the explosion of Oxone or other chemicals in the use of this oxidation procedure, prudence

dictates that all operations should be conducted in a hood.

4. Analytical reagent-grade <u>acetonitrile</u> was purchased from the Aldrich Chemical Company and used as received.

5. Oxone was almost insoluble in acetonitrile. Although Oxone adhered to the wall of the flask during the reaction, vigorous stirring ($\geq$800 rpm) is preferred for efficient grinding of Oxone and efficient alcohol oxidation.

6. Due to very turbulent mechanical stirring and vibrations, and to prevent leakage of solvent and compounds, it is necessary prepare all flask joints as follows: apply grease and wrap in Teflon tape followed by parafilm. The stir bearing was secured to the flask with copper wire to prevent separation from flask.

7. Potassium 2-iodo-5-methylbenzenesulfonate (99.6%) was obtained from the Aldrich Chemical Company and used as received.

8. 4-Bromobenzyl alcohol (99%) was purchased from the Aldrich Chemical Company and used as received.

9. The starting material and *pre*-catalyst adhered to the wall of the flask. Thus, the wall was washed with additional acetonitrile.

10. The progress of the reaction is followed by TLC analysis on silica gel with 20% EtOAc–hexane as an eluent and visualization with a UV lamp at 254 nm and molybdatophosphoric acid. The alcohol starting material has $R_f = 0.18$ and the aldehyde product has $R_f = 0.54$.

11. Celite (~6 g) was used pre-wetted with acetonitrile.

12. Evaluation of aldehyde volatility: The aldehyde (961 mg) was placed under house vacuum (20~40 mmHg) for 3 days, after which time only 22 mg (2.3%) remained.

13. At this point a turbid solution formed.

14. Yields are adjusted for 99 wt% starting material and do not account for purity of products.

15. 4-Bromobenzaldehyde has the following properties: mp (DSC) 58–59 °C; ^{1}H NMR (400 MHz, DMSO-d_6) δ: 7.81-7.86 (m, 4 H), 9.99 (s, 1 H); ^{13}C NMR (100 MHz, DMSO-d_6) δ: 128.9, 131.4, 132.5, 135.3, 192.5; Anal. Calcd. for C_7H_5BrO: C, 45.44; H, 2.72. Found: C, 45.33; H, 2.82.

16. The progress of the reaction is followed by TLC analysis on silica gel with 50% EtOAc–hexane as an eluent and visualization with a UV lamp at 254 nm and molybdatophosphoric acid. The alcohol starting material has $R_f = 0.45$, the aldehyde intermediate has $R_f = 0.74$ and the acid product has $R_f = 0.12$. The reaction progress can also be monitored at $\lambda = 260$ nm using

an Agilent 1100 HPLC equipped with a diode array detector and an Agilent ZORBAX SB-Phenyl column (3.5 μm, 4.6 x 150 mm). Mobile phase: water / acetonitrile (0.1% TFA each). Linear Gradient: Aqueous phase from 30% to 95% over 15 min, then back to 30% over 0.5 min, and holding at 30% for an additional 4.5 min (total method time of 20 min isothermally at 30 °C). COOH λ_{max} = 245 nm, CHO λ_{max} = 260 nm. Elution times: ROH - 5.6 min , RCOOH – 6.4 min, RCHO – 8.2 min.

17. 4-Bromobenzoic acid has the following properties: mp 247–250 °C; ^{1}H NMR (400 MHz, DMSO-d_6) δ: 7.70 (d, J = 8.6 Hz, 2 H), 7.86 (d, J = 8.6 Hz, 2 H), 13.16 (s, 1 H); ^{13}C NMR (400 MHz, DMSO-d_6) δ: 127.1, 130.2, 131.5, 131.9, 166.8; Anal. Calcd. For $C_7H_5BrO_2$: C, 41.82; H, 2.51. Found: C, 41.83; H, 2.58.

Safety and Waste Disposal Information

All hazardous materials should be handled and disposed of in accordance with "Prudent Practice in the Laboratory"; National Academy Press; Washington, DC, 1995.

3. Discussion

Many excellent methods have been reported for the alcohol oxidation reaction, which is one of the simplest transformations in synthetic organic chemistry.[2] However, there is a strong need for more efficient, chemoselective and greener methods that do not require heavy metallic species for such transformations, particularly in the pharmaceutical industry.[3] Also, selective cascade oxidative transformations of alcohols to carbonyl compounds (i.e. primary alcohols → aldehydes → carboxylic acids, etc.) would be powerful tools, since the target molecule can be obtained directly in a one-pot sequence. The transition metal- or nitroxyl radical-catalyzed oxidation of alcohols to ketones or aldehydes has attracted great attention because aqueous H_2O_2 or gaseous O_2 can be used as a stoichiometric oxidant.[2] However, it is technically difficult to control the amount of gaseous O_2 added as an oxidant. Moreover, aqueous H_2O_2 and gaseous O_2 are often concentrated under evaporation and high pressure to increase their respective reactivities, but such treatments can be dangerous because of their explosiveness due to static electricity, heating, shock, etc. In contrast, Oxone ($2KHSO_5 \cdot KHSO_4 \cdot K_2SO_4$) offers several advantages with regard to safety, stability, ease of transport, simple handling, controllable addition, nontoxic nature, etc., although aqueous H_2O_2 and gaseous O_2 are more atom-economical than Oxone.

Recently, we reported a highly efficient and chemoselective oxidation of various alcohols to carbonyl compounds such as aldehydes, carboxylic acids, and ketones with powdered Oxone in the presence of catalytic amounts of 2-iodobenzenesulfonic acid or its sodium salt under nonaqueous conditions.[4] Cycloalkanones can be further oxidized to cycloalkenones and lactones by controlling the amount of Oxone used under the same conditions as above. The oxidation rate in IBS-catalyzed oxidation was further accelerated by the use of *powdered* Oxone due to its increased surface area. 2-Iodoxybenzenesulfonic acid (IBS)[5] as iodine(V), which is generated in situ from 2-iodobenzenesulfonic acid and Oxone, serves as the actual catalyst for the alcohol oxidation (Scheme 1).[4, 6]

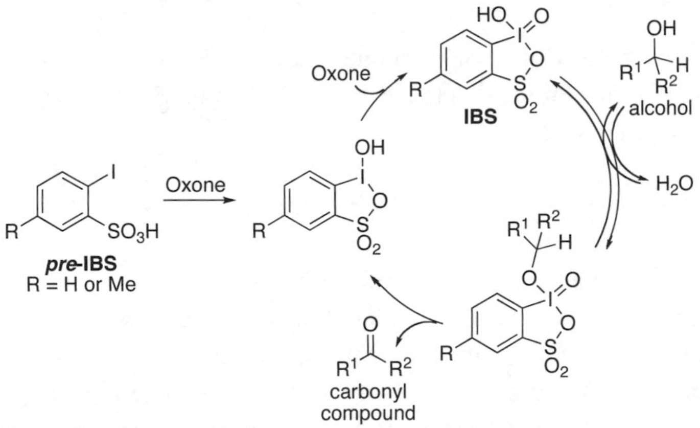

Scheme 1. Proposed Mechanism of *In Situ*-Generated IBS-Catalyzed Alcohol Oxidation

Various structurally diverse secondary and primary alcohols could be oxidized with IBS under optimized conditions (Table 1). Aldehydes readily react with water to give hydrates, which are oxidized to carboxylic acids in the presence of Oxone.[7] According to previously reported IBX-catalyzed alcohol oxidations,[8] it is difficult to selectively oxidize primary alcohols to the corresponding aldehydes or carboxylic acids. Fortunately, as shown in Table 1, not only primary α,β-unsaturated alcohols, such as propargylic and benzylic alcohols, but also aliphatic alcohols could be selectively oxidized to

110

Table 1. IBS-Catalyzed Oxidation of Primary and Secondary Alcohols[a]

Entry	Alcohol	Carbonyl Compound	pre-IBS / Oxone	Conditions	Yield (%)
1[b,c]	(menthol)	(menthone)	1 mol% / 0.6 equiv	CH$_3$CN, 70 °C, 7 h	88[b]
2	Et$_3$SiO—cyclohexyl—OH	Et$_3$SiO—cyclohexyl=O	2 mol% / 0.8 equiv	EtOAc, Na$_2$SO$_4$, 70 °C, 7 h	78
3	BnO—cyclohexyl—OH	BnO—cyclohexyl=O	2 mol% / 0.8 equiv	EtOAc, Na$_2$SO$_4$, 70 °C, 8 h	94
4[c,d]	(2-octanol)	(2-octanone)	2 mol% / 0.8 equiv	CH$_3$CN, 70 °C, 7 h	89[d]
5	Me—C$_6$H$_4$—CH$_2$OH	Me—C$_6$H$_4$—CHO	1 mol % / 0.6 equiv	CH$_3$CN, 70 °C, 3 h	95
6	Me—C$_6$H$_4$—CH$_2$OH	Me—C$_6$H$_4$—CO$_2$H	1 mol % / 1.2 equiv	CH$_3$CN, 70 °C, 4 h	94
7[e]	O$_2$N—C$_6$H$_4$—CH$_2$OH	O$_2$N—C$_6$H$_4$—CHO	1 mol% / 0.65 equiv	CH$_3$CN, 70 °C, 6 h	90
8[e]	O$_2$N—C$_6$H$_4$—CH$_2$OH	O$_2$N—C$_6$H$_4$—CO$_2$H	1 mol% / 1.3 equiv	CH$_3$CN, 70 °C, 9 h	91
9	C$_5$H$_{11}$—≡—CH$_2$OH	C$_5$H$_{11}$—≡—CHO	1 mol% / 0.6 equiv	CH$_3$CN, 70 °C, 3 h	84
10	C$_5$H$_{11}$—≡—CH$_2$OH	C$_5$H$_{11}$—≡—CO$_2$H	1 mol% / 1.2 equiv	CH$_3$CN, 70 °C, 4 h	93
11[f]	Ph—CH$_2$OH	Ph—CHO	5 mol % / 0.6 equiv	CH$_3$CN, 70 °C, 3 h	79[e]
12[f]	Ph—CH$_2$OH	Ph—CO$_2$H	1 mol % / 1.2 equiv	CH$_3$CN, 70 °C, 4 h	90[e]

[a] Unless otherwise noted, results in reference 4a are shown. After unsaturated primary alcohol was consumed, H$_2$O was added to accelerate the further oxidation of aldehyde to the carboxylic acid with Oxone (entries 6, 8 and 10). [b] The reaction was carried out using (−)-menthol (40.4 g, 256 mmol) in the presence of potassium 2-iodo-5-methylbenzenesulfonate. The yield was determined by GC analysis and included epimer (ca. 4%). [c] The experiments were performed at Nissan Chemical Industries, Ltd. [d] The reaction was carried out using 2-octanol (10.0 g, 76.8 mmol) in the presence of potassium 2-iodo-5-methylbenzenesulfonate. The yield was determined by GC analysis. [e] The reaction was carried out as described in experimental procedures above using 4-nitrobenzyl alcohol (6.13 g, 40.0 mmol) in the presence of potassium 2-iodo-5-methylbenzenesulfonate. [f] The reaction was carried out using 4-phenyl-1-butanol (6.01 g, 40.0 mmol) in the presence of potassium 2-iodo-5-methylbenzenesulfonate (5 mol%) and Na$_2$SO$_4$.[4a] Aliphatic primary alcohols were slowly added to the reaction mixture to prevent ester formation in the oxidation.

the corresponding aldehydes and carboxylic acids in excellent yield by controlling the amount of Oxone added in the presence of the precatalyst sodium 2-iodobenzenesulfonate (1–4 mol%): 0.6–0.8 equiv and 1.2 equiv of Oxone were used for selective oxidation to aldehydes and carboxylic acids, respectively (entries 5–12). The present protocol could be applied to the chemoselective oxidation of alcohols bearing several functional or protective groups such as silyloxy, benzyloxy, ketal, alkenyl, alkynyl, halo, pyridinyl, and thiophene groups.[4]

1. Graduate School of Engineering, Nagoya University, Furo-cho, Chikusa, Nagoya, 464-8603 Japan; TEL: +81-52-789-3331; FAX: +81-52-789-3222; E-mail: ishihara@cc.nagoya-u.ac.jp; We gratefully acknowledge support from Nissan Chemical Industries, Ltd.

2. (a) Tojo, G.; Fernandez, M. *Oxidation of Primary Alcohols to Carboxylic Acids*; Springer, Berlin, 2006. (b) Tojo, G.; Fernandez, M. *Oxidation of Alcohols to Aldehydes and Ketones*; Springer, Berlin, 2006. (c) Backwall, J. E.; *Modern Oxidation Methods*; Wiley-VCH, New York, 2004. (d) Ley, S. V. in *Comprehensive Organic Synthesis;* Pergamon, Oxford, ed. 3, 1999, vol. 7, chap. 2, pp. 251-327. (e) Marko, I. E.; Giles, P. R.; Tsukazaki, M.; Brown, S. M.; Urch, C. J. *Science* **1996**, *274* (5295), 2044-2046. (f) ten Brink, G.-J.; Arends, I. W. C. E.; Sheldon, R. A. *Science* **2000**, *287* (5458), 1636-1639. (g) Enache, D. I.; Edwards, J. K.; Landoin, P.; Solsona-Espriu, B.; Carley, A. F.; Herzing, A. A.; Watanabe, M.; Kiely, C. J.; Knight D. W.; Hutchings G. J. *Science* **2006**, *311* (5275), 362-365. (h) Liu, R.; Liang, X.; Dong, C.; Hu, X. *J. Am. Chem. Soc.* **2004**, *126* (13), 4112-4113.

3. Caron, S.; Dugger, R. W.; Ruggeri, S. G.; Ragan, J. A.; Ripin, D. H. B. *Chem. Rev.* **2006**, *106* (7), 2943-2989.

4. (a) Uyanik, M.; Akakura, M.; Ishihara, K. *J. Am. Chem. Soc.* **2009**, *131* (1), 251-262. (b) Uyanik, M.; Fukatsu, R.; Ishihara, K. *Org. Lett.* **2009**, *11* (15), 3470-3473. (c) Uyanik, M.; Ishihara, K. *Chem. Commun.* **2009**, (16), 2086-2099. (d) Uyanik, M.; Ishihara, K. *Aldrichim. Acta,* **2010**, *43*, 83-91.

5. For the synthesis of 2-iodoxybenzoic acid (IBX), an analogue of IBS with Oxone see: Frigerio, M.; Santagostino, M.; Sputore, S. J. *J. Org. Chem.* **1999**, *64 (12)*, 4537-4538.

6. Although IBS has been synthesized from 2-iodobenzenesulfonic acid

and Oxone in water by Zhdankin and co-workers, its oxidative ability has not been investigated due to its low stability: Koposov, A. Y.; Litvinov, D. N.; Zhdankin, V. V.; Ferguson, M. J.; McDonald, R.; Tykwinski, R. R. *Eur. J. Org. Chem.* **2006**, (21), 4791-4795.

7. Travis, B. R.; Sivakumar, M.; Hollist, G. O.; Borhan, B. *Org. Lett.* **2003**, *5* (7), 1031-1034.
8. (a) Thottumkara, A. P.; Bowsher, M. S.; Vinod, T. K. *Org. Lett.* **2005**, *7* (14), 2933-2936. (b) Schulze, A.; Giannis, A. *Synthesis* **2006**, (2), 257-260. (c) Page, P. C. B.; Appleby, L. F.; Buckley, B. R.; Allin, S. M.; McKenzie, M. J. *Synlett*, **2007**, (10), 1565-1568.

Appendix
Chemical Abstracts Nomenclature; (Registry Number)

Oxone: Potassium peroxymonosulfate sulfate; (37222-66-5)
Potassium 2-iodo-5-methylbenzenesulfonate; (1093215-92-9)
4-Bromobenzyl alcoholBenzenemethanol, 4-bromo-873-75-6
4-Bromobenzaldehyde; (1122-91-4)
4-Bromobenzoic acid; (586-76-5)

Kazuaki Ishihara was born in Aichi, Japan, in 1963, and received his PhD from Nagoya University in 1991 under the direction of Professor Hisashi Yamamoto. He had the opportunity to work under the direction of Professor Clayton H. Heathcock at the University of California, Berkeley, as a visiting graduate student for three months in 1988. He was a JSPS Fellow under the Japanese Junior Scientists Program from 1989 to 1991. After completing his postdoctoral studies with Professor E. J. Corey at Harvard University, he returned to Japan and joined Professor H. Yamamoto's group at Nagoya University as an assistant professor in 1992, and became Associate Professor in 1997. In 2002, he was appointed to his current position as a Full Professor at Nagoya University. His research interests include asymmetric catalysis, biomimetic catalysis induced by artificial enzymes, dehydrative condensation catalysis towards green and sustainable chemistry, and acid–base combination chemistry.

Muhammet Uyanik was born in Samsun, Turkey, in 1981 and received his PhD from Nagoya University in 2007 under the direction of Professor Kazuaki Ishihara. He was appointed as an Assistant Professor at Nagoya University in 2007. His research interests includes oxidation reactions and asymmetric catalysis.

Richard Crockett was born in North Carolina in 1969 and received his B.S. from University of California, Berkeley in 2002. In the winter of 2003 he began working at Amgen, Inc. as a Process Chemist in the Chemical Process Research and Development group. He is currently a Senior Research Associate for Amgen.

Catalytic Intramolecular Friedel-Crafts Reaction of Benzyl Meldrum's Acid Derivatives: Preparation of 5,6-Dimethoxy-2-Methyl-1-Indanone

Submitted by Tiantong Lou, E-Ting Liao, Ashraf Wilsily, and Eric Fillion.[1]
Checked by Maurizio Bernasconi and Andreas Pfaltz.

1. Procedure

A.   *5-(3,4-Dimethoxybenzyl)-2,2-dimethyl-1,3-dioxane-4,6-dione* (**1**). An oven-dried 1-L, two-necked, round-bottomed flask equipped with an oval (Note 1) magnetic stir bar, a rubber septum and a two-tap Schlenk adaptor connected to a bubbler and an nitrogen/vacuum manifold (Note 2) is charged with 3,4-dimethoxybenzaldehyde (20.0 g, 120 mmol, 1.0 equiv) (Note 3), 2,2-dimethyl-1,3-dioxane-4,6-dione (Meldrum's acid) (17.7 g, 123 mmol, 1.0 equiv) (Note 4) and ethanol (250 mL) (Notes 5 and 6) by temporary removal of the septum. After complete dissolution of the reactants, piperidine (1.2 mL, 12.3 mmol, 0.1 equiv) (Note 7) and acetic acid (0.7 mL, 12.3 mmol, 0.1 equiv) (Note 7) are added sequentially by syringe (Note 8) and the solution is stirred at room temperature for 30 min. Sodium triacetoxyborohydride (52.1 g, 246 mmol, 2.0 equiv) (Note 9) is then added in four equal portions, one after the other on a 30 min interval by temporarily removing the septum. The reaction mixture is stirred for another

60 min at room temperature (Note 10), after which it is cooled using an ice bath and quenched by the addition of saturated ammonium chloride solution (60 mL).

The reaction mixture is then transferred into a 1-L separatory funnel using dichloromethane (3 x 80 mL) and saturated ammonium chloride (3 x 80 mL). Deionized water (50 mL) is added to facilitate the separation of the two layers. The layers are separated and the organic phase is collected into a 2-L Erlenmeyer flask. The aqueous phase is extracted with dichloromethane (2 x 150 mL) and the combined organic layers are dried over $MgSO_4$ (20 g), filtered through a coarse glass frit (Note 11) into a 1-L round-bottomed flask, concentrated by rotary evaporation (35 °C, approximately 2 mmHg) and dried under reduced pressure (room temperature, 0.15 mmHg) over 30 min. The crude product is transferred into a 500-mL round-bottomed flask equipped with an oval (Note 1) stir bar, suspended in methanol (80 mL) (Notes 6 and 12) and warmed at 55 °C for 20 min (Note 13). After allowing the solution to cool to room temperature (~20 min), the reaction vessel is stored in a freezer at −25 °C for 17 h. The yellow solid is collected by suction filtration (room temperature, approximately 10 mmHg), through a coarse frit (Note 11), washed with methanol (10 mL, −25 °C), dried for 30 min under reduced pressure (0.15 mmHg, room temperature, 1 h) to yield 29.9 g (85 % yield) of product **1** as pale yellow crystals (Notes 14 and 15).

B.   *5-(3,4-Dimethoxybenzyl)-2,2,5-trimethyl-1,3-dioxane-4,6-dione (2)*. An oven-dried 500-mL, three-necked, round-bottomed flask equipped with an oval (Note 1) magnetic stir bar, a 50-mL addition funnel, a two-tap Schlenk adaptor connected to a bubbler and an nitrogen/vacuum manifold (Note 2) and a rubber septum is charged with **1** (21.9 g, 74.5 mmol, 1.0 equiv) and potassium carbonate (15.4 g, 112 mmol, 1.5 equiv) (Note 16) by temporary removal of the septum. Freshly distilled *N,N*-dimethylformamide (74 mL) (Note 17) is added into the reaction flask (Note 18). After complete dissolution of the reactants, iodomethane (9.3 mL, 149 mmol, 2.0 equiv) (Note 19) is added via syringe to the addition funnel, and added dropwise to the reaction mixture (Note 20).

After 3.5 h (Note 21), the reaction mixture is transferred with deionized water (2 x 30 mL) and dichloromethane (2 x 30 mL) into a 1-L Erlenmeyer flask equipped with an oval (Note 1) stir bar, filled with deionized water (100 mL) and cooled using an ice bath. After stirring for 5 min, the solution is transferred into a 1-L separatory funnel using deionized water (30 mL) and dichloromethane (30 mL). The layers are partitioned and the organic phase is collected into a 1-L Erlenmeyer flask. The aqueous phase is further extracted with 10% dichloromethane in hexanes (3 x 150 mL). The combined organic layers are washed with water (200 mL) and brine (200 mL), dried over $MgSO_4$ (20 g), filtered through a coarse glass frit (Note 11) into a 1-L round bottom flask, and concentrated by rotary evaporation (40 °C, approximately 10 mmHg). The crude product is dried under reduced pressure (room temperature, 0.2 mmHg) for 30 min prior being transferred to a 250-mL round-bottomed flask, dissolved in methanol (30 mL) (Notes 6 and 12) and warmed at 55 °C for 20 min (Note 13). After allowing the solution to cool to room temperature (~20 min), it is stored in the freezer at –25 °C for 17 h. The crystals are collected by suction filtration through a coarse frit (Note 11) under vacuum (room temperature, approximately 10 mmHg) and washed with methanol (10 mL, –25 °C). Product **2** is then collected into a 250-mL round-bottom flask and dried under reduced pressure (room temperature, 0.15 mmHg) for 30 min to give 20.21 g (88 % yield) of pale yellow crystals (Note 22).

C. *5,6-Dimethoxy-2-methyl-1-indanone (3).* An oven-dried 1-L, three-necked, round-bottomed flask equipped with an oval (Note 1) magnetic stir bar, a 500-mL addition funnel, a reflux condenser with a two-tap Schlenk adaptor connected to a bubbler and an nitrogen/vacuum manifold (Note 2) on the top, and a glass stopper is charged with **2** (12.0 g, 38.9 mmol, 1.0 equiv) by temporary removal of the glass stopper. Freshly distilled nitromethane (390 mL) (Notes 23 and 24) is added into the reaction flask (Note 25) via the addition funnel. When the reaction mixture begins to reflux at 100 °C (Note 13), TMSOTf (0.7 mL, 0.1 equiv) (Note 26) is added via the addition funnel, which is rinsed with nitromethane (approximately 5 mL). After 60 min (Note 27) the reaction mixture is allowed to cool for 15 min before it is placed in an ice bath. After stirring for 5 min, saturated ammonium chloride solution (10 mL) is added.

After 15 min of stirring, the reaction mixture is then transferred into a 1-L round-bottom flask with dichloromethane (100 mL) before it is concentrated by rotary evaporation (45 °C, approximately 4 mmHg). The resulting crude product is transferred into a 2-L separatory funnel using deionized water (2 x 50 mL) and ethyl acetate (3 x 100 mL). The layers are separated and the organic phase is collected into a 2-L Erlenmeyer flask. The aqueous phase is then extracted with ethyl acetate (2 x 200 mL) and the combined organic phases are washed with brine (200 mL), dried over $MgSO_4$ (20 g), filtered through a coarse glass frit (Note 11) into a 2-L round-bottomed flask and concentrated by rotary evaporation (40 °C, approximately 4 mmHg). The crude product is transferred using dichloromethane (2 x 50 mL) into a 250-mL round-bottomed flask equipped with an oval (Note 1) magnetic stir bar, then concentrated by rotary evaporation (40 °C, approximately 4 mmHg) and dried under reduced pressure (room temperature, 0.2 mmHg) for 30 min to yield a slightly purple powder. To the solid crude product is added ethanol (30 mL) (Notes 5 and 6) and the suspension is heated to 90 °C for 20 min (Note 13). The solution is allowed to cool to room temperature for 40 min. The precipitated colorless crystals are filtered off through a coarse frit (Note 11), washed with cold ethanol (5 °C, 10 mL) (Note 5 and 6) collected into a 250-mL round-bottomed flask and dried under reduced pressure overnight (room temperature, 0.08 mmHg), to give 5.36 g (67 % yield) of white crystals (Notes 28, 29 and 30).

2. Notes

1. Length: 4 cm, diameter: 2 cm
2. A picture of this adaptor can be found in: *Org. Synth.* **2008**, *85*, 64–71.
3. 3,4-Dimethoxybenzaldehyde (99%) was purchased from Aldrich Chemical Company, Inc. and was ground to a fine powder before use.
4. 2,2-Dimethyl-1,3-dioxane-4,6-dione (98%) was purchased from Aldrich Chemical Company and was used as received.
5. Ethanol (ACS reagent) was obtained from Fluka and was used as received.
6. The graduated cylinder was oven-dried.
7. Piperidine (99%, ReagentPlus) and acetic acid ($\geq$ 99.7%, ACS reagent) were obtained from Aldrich Chemical Company, Inc. and both reagents were used as received. The rate of addition of those compounds does not influence the outcome of the reaction.
8. During the condensation, the reaction changed from a clear yellow mixture to a cloudy yellow mixture.
9. Sodium triacetoxyborohydride (95%) was purchased from Aldrich Chemical Company, Inc. and was used as received. **Caution**: After the addition of each portion of the reducing agent (mainly after the first two portions) strong hydrogen gas evolution is observed and the internal temperature of the reaction increases (from 25 °C to 36 °C). The checkers

118

observed in some cases that upon the first addition of the reducing agent the reaction could not be stirred properly due to the increased viscosity of the reaction mixture. In this case an additional 80 mL of ethanol were added.

10. The reaction was monitored via TLC with 3:1 n-hexane/ethyl acetate as eluent (R_f value of 0.31). The color of the reaction mixture changed from cloudy yellow to cloudy white as sodium triacetoxyborohydride was added.

11. A 125-mL filter funnel with a coarse frit, porosity 3, was used.

12. Methanol (Exceeds ACS Specifications) was obtained from J.T. Baker and used as received.

13. A silicon oil bath was used.

14. The product displays the following physicochemical properties: pale yellow crystals; mp 136 °C. Lit. 142–144 °C; ^{1}H NMR (CDCl$_3$, 400 MHz) δ: 1.48 (s, 3 H), 1.72 (s, 3 H), 3.45 (d, J = 4.8 Hz, 2 H), 3.72 (t, J = 4.8 Hz, 1 H), 3.84 (s, 3 H), 3.86 (s, 3 H), 6.77 (m, 1 H), 6.85–6.87 (m, 2 H); ^{13}C NMR (CDCl$_3$, 400 MHz) δ: 27.6, 28.6, 32.1, 48.5, 56.0 (2 C), 105.4, 111.3, 113.3, 122.2, 129.6, 148.3, 149.0, 165.6 (2 C); IR (ATM): 2993, 2943, 2860, 1782, 1747, 1516, 1447, 1259, 1155, 1020, 862 cm^{-1}; Anal. Calcd. for C$_{15}$H$_{18}$O$_6$: C, 61.20; H, 6.16, found: C, 61.07; H, 6.39.

15. In some cases the product contained 5–8% of a side product. This side product was identified as the alkylidene Meldrum's acid (Knoevenagel condensation intermediate). This product has the following analytical data: ^{1}H NMR (CDCl$_3$, 400 MHz) δ: 1.79 (s, 6 H), 3.96 (s, 3 H), 3.99 (s, 3 H), 6.96 (d, J = 8.5 Hz, 1 H), 7.65 (dd, J = 8.5, 2.0 Hz, 1 H), 8.31 (d, J = 2.1 Hz, 1H), 8.37 (s, 1 H). The side product does not influence the outcome of the subsequent reactions, and it is removed in the recrystallization in Step B.

16. Potassium carbonate ($\geq$ 99.0%, ACS reagent) was purchased from Aldrich Chemical Company, Inc. and used as received.

17. N,N-Dimethylformamide (certified ACS) was purchased from Aldrich Chemical Company and was freshly distilled over calcium hydride.

18. The solvent was added using a syringe.

19. Iodomethane (99%, stabilized with copper, ReagentPlus) was purchased from Aldrich Chemical Company, Inc. and used as received.

20. Iodomethane was added over a period of 9 min (1 mL/minute). The internal temperature raised from 28 °C to 33 °C.

21. The reaction was monitored via TLC with 3:1 n-hexane/ethyl acetate as eluent (R_f value of 0.30). The reaction mixture was a cloudy white solution.

22. The product displays the following physicochemical properties: pale yellow crystals; mp 88–89 °C. ^{1}H NMR (CDCl$_3$, 400 MHz) δ: 0.99 (s, 3 H), 1.61 (s, 3 H), 1.73 (s, 3 H), 3.28 (s, 2 H), 3.83 (s, 3 H), 3.84 (s, 3 H), 6.69–6.77 (m, 3 H); ^{13}C NMR (CDCl$_3$, 400 MHz) δ: 26.1, 28.7, 29.5, 44.7, 52.6, 56.0 (2 C), 105.4, 111.3, 113.2, 122.5, 128.0, 148.7, 149.1, 170.2 (2 C); IR (ATM): 3006, 2937, 2840, 1774, 1738, 1512, 1379, 1261, 1238, 1201, 1145, 1022, 980, 825, 680 cm^{-1}. Anal. Calcd for C$_{16}$H$_{20}$O$_6$: C, 62.33; H, 6.54. Found: C, 62.33; H, 6.42.

23. Nitromethane (96%, Reagent Grade) was purchased from Aldrich Chemical Company, Inc. and was freshly distilled under nitrogen over calcium hydride behind a safety shield.

24. The use of nitromethane gave optimal selectivity, yielding product **3** over its regioisomer, 6,7-dimethoxy-2-methyl-1-indanone (**4**), in a >20:1 ratio. Since regioisomer **4** is an oil at room temperature, it was easily separated from product **3** through a simple recrystallization. The Friedel-Crafts acylation was also carried out at reflux in acetonitrile (**3:4** ratio of 9:1), toluene (**3:4** ratio of 8:1), and chlorobenzene (**3:4** ratio of 7:1).

25. The addition funnel was charged with the solvent using a cannula.

26. TMSOTf (99%) was purchased from Aldrich Chemical Company, Inc. and distilled under vacuum at approximately 10 mmHg and stored in a Schlenk tube protected from light. TMSOTf was added over a period of 2 min (0.35 mL/min).

27. The reaction was monitored via TLC with 2:1 *n*-hexane/ethyl acetate as eluent (R$_f$ value of 0.31). The reaction mixture changed from yellow to dark purple over the course of one hour.

28. The product displays the following physicochemical properties: white crystals; mp 133 °C. Lit. 132–133 °C; ^{1}H NMR (CDCl$_3$, 500 MHz) δ: 1.29 (d, J = 7.4 Hz, 3 H), 2.63 (dd, J = 16.8, 3.4 Hz, 1 H), 2.69 (ddq, J = 7.4, 7.4 , 3.6 Hz, 1 H), 3.30 (dd, J = 16.8, 7.4 Hz, 1 H), 3.90 (s, 3 H), 3.96 (s, 3 H), 6.86 (s, 1 H), 7.17 (s, 1 H); ^{13}C NMR (CDCl$_3$, 500 MHz) δ 16.8, 34.9, 42.3, 56.2, 56.3, 104.6, 107.5, 129.1, 148.9, 149.5, 155.6, 208.4; IR (ATM): 2963, 2925, 2838, 1679, 1588, 1496, 1461, 1427, 1364, 1318, 1239, 998, 834 cm^{-1}; Anal. Calcd. for C$_{12}$H$_{14}$O$_3$: C, 69.88; H, 6.84, found: C, 69.83; H, 6.89.

29. To isolate the regioisomer, the reaction was performed in toluene and regioisomer **4** was purified using flash silica gel chromatography eluting with 2:1 *n*-hexane/ethyl acetate.

30. The regioisomer, 6,7-dimethoxy-2-methyl-1-indanone, displays the following physicochemical properties: light yellow oil; ^{1}H NMR (CDCl$_3$, 300 MHz) δ: 1.27 (d, J = 7.2 Hz, 3 H), 2.57–2.72 (m, 2 H), 3.27 (dd, J = 16.3, 7.7 Hz, 1 H), 3.86 (s, 3 H), 3.99 (s, 3 H), 7.04 (d, J = 8.2 Hz, 1 H), 7.15 (d, J = 8.2 Hz, 1 H); ^{13}C NMR (CDCl$_3$, 75 MHz) δ: 16.1, 33.6, 43.0, 58.8, 61.5, 119.9, 120.8, 128.4, 145.9, 146.9, 150.9, 206.6; IR (CH$_2$Cl$_2$): 1708 cm^{-1}; HRMS m/z calcd for C$_{12}$H$_{14}$O$_3$ (M$^+$): 206.0943. Found: 206.0945.

Waste Disposal Information

All hazardous materials should be handled and disposed of in accordance with "Prudent Practices in the Laboratory"; National Academy Press; Washington, DC, 1995.

3. Discussion

1-Indanones have proven synthetic utility in numerous biologically active natural products and play a major role in medicinal chemistry and the development of pharmaceuticals.[2] Although the intramolecular Friedel-Crafts acylation is the most powerful means of preparing 1-indanones, it suffers from significant drawbacks.[3] Typical procedures rely on the reaction of carboxylic acids or acid chlorides with stoichiometric amounts of Brønsted or Lewis acids, which poses problems for cyclization precursor preparation, product isolation, and functional group compatibility. Our group has reported that the use of Meldrum's acid as acylating agent provides a solution that overcomes the problems associated with the carbocyclization of carboxylic acids or acid chlorides. Furthermore, the cyclization of Meldrum's acid derivatives provides an expedient and efficient entry into 2-substituted 1-indanones.[4] Enolizable benzyl Meldrum's acid derivatives are readily functionalized under mild reactions conditions. Such facile α-alkylations are significantly more difficult for carboxylic acids and acid chlorides, and therefore synthesis of 2-substituted 1-indanones is typically performed after Friedel-Crafts acylation, where the standard difficulties of mono-alkylating ketones present themselves. In addition, compared to traditional acylation procedures, the generation of volatile and inert side products, acetone and CO$_2$, simplified product isolation.[5,6,7]

Table 1. Preparation of 2-substituted 1-indanones

Entry	Quaternized Meldrum's acid	Sc(OTf)$_3$ (mol%)	Time (min)	Indanone	Yield (%)
1	R = Me	10	45	R = Me	77
2	R = Ph	10	60	R = Ph	67
3	R = Bn	10	45	R = Bn	80
4	R = CHCH=CH$_2$	10	45	R = CHCH=CH$_2$	76
5	R = CH$_2$CCH	10	45	R = CH$_2$CCH	80
6	R = CH$_2$C$_6$H$_4$ (4-CN)	10	320	R = CH$_2$C$_6$H$_4$ (4-CN)	78
7	R = CH$_2$C$_6$H$_4$ (4-NO$_2$)	10	250	R = CH$_2$C$_6$H$_4$ (4-NO$_2$)	81
8	R = CH$_2$C$_6$H$_5$	7	85	R = CH$_2$C$_6$H$_5$	80
9	R = OMe	10	45	R = OMe	80
10	R = Me	11	30	R = Me	87
11	R = Me	10	20	R = Me	69
12	R = TBDPS	12	20	R = TBDPS	94
13	R = Me	9	20	R = Me	75
14	R = TBDPS	9	30	R = TBDPS	86
15	R = TIPS	10	40	R = TIPS	77

Org. Synth. **2012**, *89*, 115-125

As shown in Table 1, reactions of quaternized benzyl Meldrum's acids gave 2-substituted 1-indanones in excellent yields. A variety of groups could be added to C(5) of the starting materials, leading to diverse 2-substituents in the products while demonstrating further functional group compatibility (alkenes, alkynes, nitro and cyano groups).

1. Department of Chemistry, University of Waterloo, Waterloo, Ontario N2L 3G1, CANADA, efillion@uwaterloo.ca. This work was supported by the Government of Ontario (Early Researcher Award to E. F.), the Natural Sciences and Engineering Research Council of Canada (NSERC), the Canadian Foundation for Innovation (CFI), the Ontario Innovation Trust (OIT), and the University of Waterloo. A.W. is indebted to NSERC for CGS-M and CGS-D scholarships. T. L. and E-T. L. thank NSERC for USRA scholarship.
2. Catoen-Chackal, S.; Facompré, M.; Houssin, R.; Pommery, N.; Goossens, J.-F.; Colson, P.; Bailly, C.; Hénichart, J.-P. *J. Med. Chem.* **2004**, *47*, 3665–3674.
3. Trost, B. M., Fleming, I., Eds.; Pergamon Press: Oxford, UK, **1991**; Vol. 2, pp 733–752.
4. Mane, R.; Krishna Rao, G. S. *Chem. Ind. (London)* **1976**, 786 – 787.
5. Larock, R. *Comprehensive Organic Transformations*, 2nd ed.; Wiley-VCH: New York, **1999**; pp 1422–1433.
6. Fillion, E.; Fishlock, D. *Org. Lett.* **2003**, *5*, 4653–4656.
7. Fillion, E.; Fishlock, D.; Wilsily, A.; Goll, J. *J. Org. Chem.* **2005**, *70*, 1316–1327.

Appendix
Chemical Abstracts Nomenclature; (Registry Number)

3,4-Dimethoxybenzaldehyde; (120-14-9)
2,2-Dimethyl-1,3-dioxane-4,6-dione; (2033-24-1)
Piperidine; (110-89-4)
Sodium Triacetoxyborohydride; (56553-60-7)
Iodomethane; (74-88-4)
Trimethylsilyl Trifluoromethanesulfonate; (27607-77-8)

Eric Fillion received his undergraduate degree in biochemistry at the Universite de Sherbrooke. After completing his M. Sc. in medicinal chemistry at the Université de Montreal with Professor Denis Gravel, he began his doctoral studies at the University of Toronto under the direction of Professor Mark Lautens. From 1998-2000, he was an NSERC post-doctoral fellow in the laboratories of Professor Larry E. Overman at the University of California, Irvine. In August 2000, he joined the Department of Chemistry at the University of Waterloo, where he is currently a Professor of Chemistry.

Tiantong (Tim) Lou was born in 1987 in Nanjing, China and moved with his family to Canada in 1997. Starting in the summer of 2008, he joined the Fillion group as an NSERC USRA and worked on the asymmetric methylation of alkylidene Meldrum's acids.

E-Ting Liao was born in 1985 in Taiwan. She graduated with a Bachelor's degree in Biomedical Sciences in 2008 from the University of Waterloo. She is now in the university's School of Pharmacy. As an NSERC USRA student, E-Ting has worked with Professor Eric Fillion's group for three summers, dealing with Meldrum's Acid derivatives. Her projects involved 1,6-conjugate addition of dialkylzinc reagents to 5-(3-aryl-2-propenylidene) Meldrum's Acids, as well as Friedel-Crafts acylation.

Ashraf Wilsily (born 1982) received a B.Sc. (Hons.) in Chemistry (2004) from the University of Waterloo. In 2005, he pursued a doctoral studies with Professor Eric Fillion in the area of asymmetric conjugate addition reactions and persistent intramolecular C–H•••X (X = O, S, Br, Cl, F) hydrogen bonds involving benzyl Meldrum acid derivatives. In January 2010, he joined the group of Professor Gregory C. Fu at MIT to pursue his postdoctoral studies.

Maurizio Bernasconi was born in 1985 in Mendrisio, Switzerland. He studied chemistry at the ETH Zürich where he received his M.Sc degree in 2009 after completing his Master thesis under the direction of F. D. Toste at the University of California, Berkeley. In 2009 he joined the lab of Andreas Pfaltz at the University of Basel, where he is currently carrying out his Ph.D. thesis working on Iridium catalyzed asymmetric hydrogenation.

Discussion Addendum for:
Au(I)-Catalyzed Hydration of Alkynes: 2,8-Nonanedione

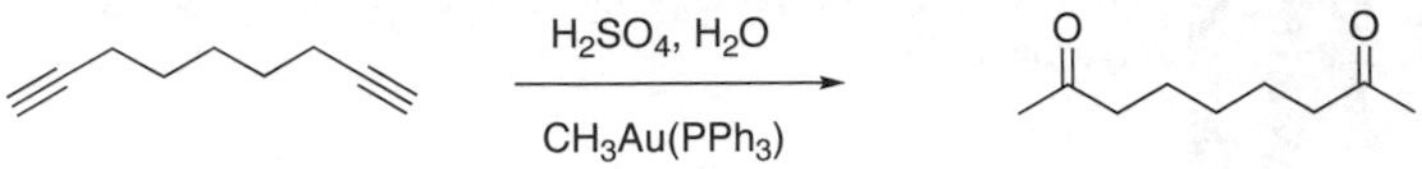

H_2SO_4, H_2O
$CH_3Au(PPh_3)$
O
O

Prepared by Teruyuki Hayashi.[1]
Original article: Eiichiro Mizushima, Dong-Mei Cui, Dilip Chandra Deb Nath, Teruyuki Hayashi,* and Masato Tanaka; *Org. Synth.* **2006**, *83*, 55.

The authors have applied hydration conditions for the reaction of 1,6-heptadiynes to form cyclohexenones.[2] Various substituents at the 4-position of 1,6-heptadiynes are tolerated to give 5-substituted 3-methylcyclohex-2-enones.

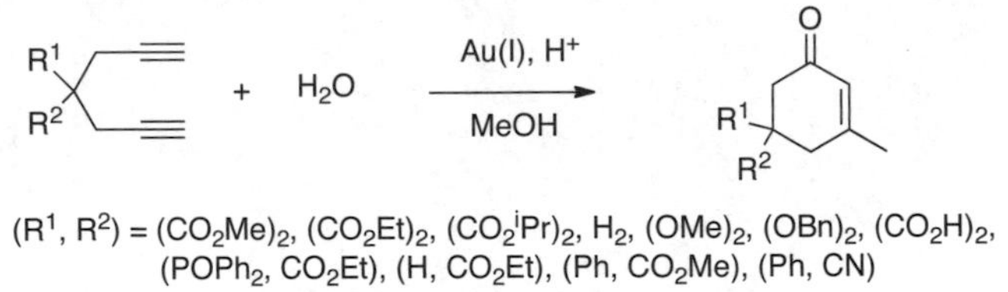

R1
R2
+ H_2O
Au(I), H^+
MeOH
O
R1
R2
(R^1, R^2) = $(CO_2Me)_2$, $(CO_2Et)_2$, $(CO_2{}^iPr)_2$, H_2, $(OMe)_2$, $(OBn)_2$, $(CO_2H)_2$,
($POPh_2$, CO_2Et), (H, CO_2Et), (Ph, CO_2Me), (Ph, CN)

Scheme 1 Hydrative Cyclization of 1,6-Diynes

The formation of the six-membered ring instead of the dihydration is attributed to the elimination of the ring from the monohydrated intermediate

Figure 1 Catalytic Cycle of Hydrative Cyclization

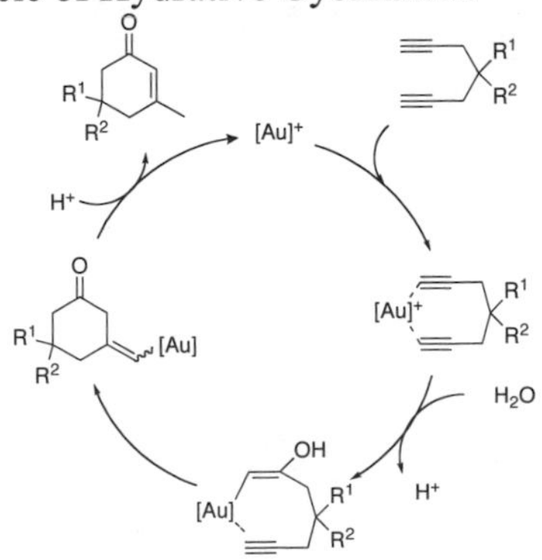

O
R1
R2
[Au]+
R1
R2
H^+
O
R1
R2
[Au]
[Au]+
R1
R2
H_2O
OH
[Au]
R1
R2
H^+

Org. Synth. **2012**, *89*, 126-130
Published on the Web 10/4/2011

complex. A similar reaction has been reported in supported mercury-catalyzed hydration of 1,6-heptadiyne.[3]

2,6-Heptanedione is prepared through reduction of 2,6-lutidine[4] or methylation of methylperoxycyclopentanone.[5]

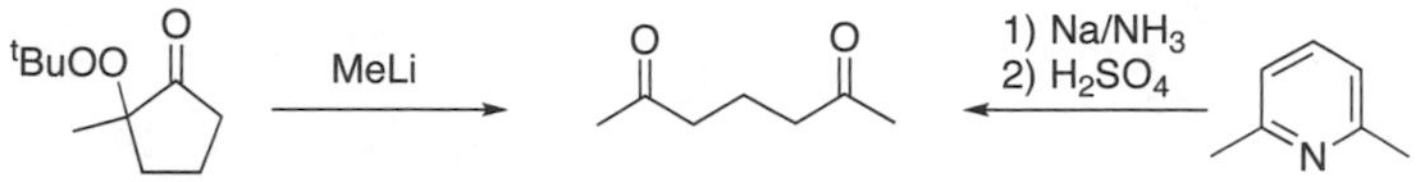

Scheme 2 Syntheses of 2,6-Heptanedione

α,ω-Diynes of C8 to C10 are dihydrated to give the corresponding diketones. Recently developed gold-carbene complexes,[6,7] as well as cationic iridium complexes with the aid of GdCl$_3$,[8] catalyze the reaction. Indium complexes also dihydrate α,ω-diynes of C9, C10, and C14 in the presence of diphenyldiselenide.[9]

Figure 2 New Dihydration Catalysts for α,ω-Diynes

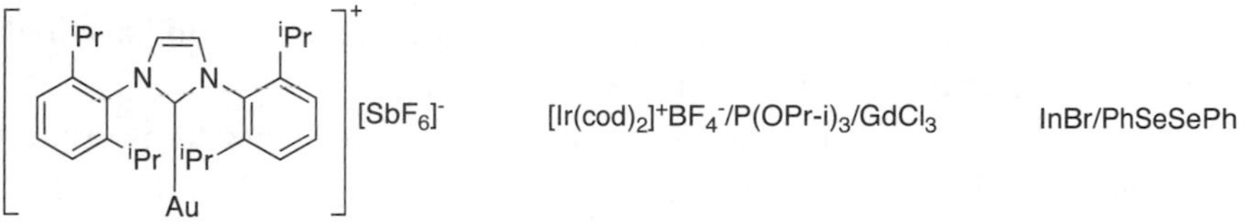

[SbF$_6$]$^-$ [Ir(cod)$_2$]$^+$BF$_4^-$/P(OPr-i)$_3$/GdCl$_3$ InBr/PhSeSePh

Ref. 6,7   **Ref. 8**   **Ref. 9**

Wacker-type oxidation of α,ω-dienes should be an industrially better process to form the corresponding diketones. However, the selectivity and reactivity are not high enough, with the diketone selectivities of Pd(OAc)$_2$/molybdovanadophosphate system[10] being 40-60%. PdCl$_2$/N,N-dimethylacetamide system[11] gives the mixture of diketone and monoketone.

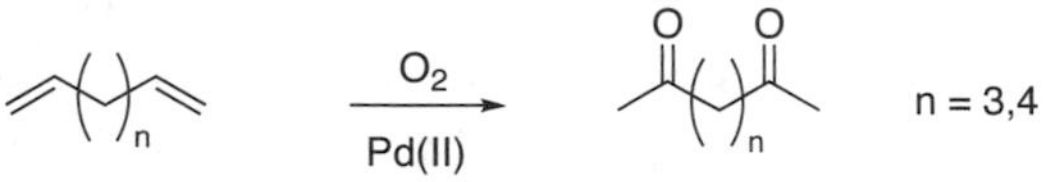

n = 3,4

Scheme 3 Wacker-Type Oxidation of α,ω-Dienes

Recent progress in the catalysis of alkyne hydration utilizes cationic gold complexes,[6,7,12,13] gold-carbene complexes,[6,14,15] and water-soluble gold complexes.[16] The application of these catalyst systems to dihydration of diynes gives the diketones.

Figure 3 New Hydration Catalysts for Alkynes

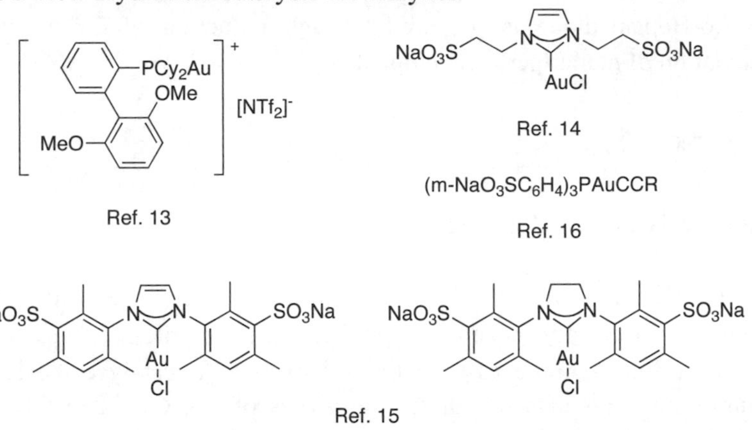
PCy2Au
OMe
MeO
[NTf2]-
Ref. 13
NaO3S
SO3Na
AuCl
Ref. 14
(m-NaO3SC6H4)3PAuCCR
Ref. 16
NaO3S
SO3Na
Au
Cl
NaO3S
SO3Na
Au
Cl
Ref. 15

For industrial application, heterogeneous catalysts are required. Those including $\{[NP(O_2C_{12}H_8)]_{0.85}[NP(OC_6H_4PPh_2)_2(AuPF_6)_{0.5}]_{0.15}\}$,[17] tin-tungsten mixed oxide,[18] and Hg(II)/silica[3] have been developed to have high activity and selectivity for the hydration of alkynes.

Neutral, metal-free hydration catalysis has been discovered by Abbott researchers.[19] They have reported that the hydration of alkynes proceeds with microwave irradiation in superheated water. On the other hand, graphene oxide has been reported to catalyze the reaction without metal components as an example of carbocatalysts.[20] These new systems have not been applied to the dihydration of diynes.

As for dihydration of α,ω-diynes, mononuclear gold complex systems with some Brønsted acids still seem the catalysts of choice at the present moment.

1. National Institute of Advanced Industrial Science and Technology (AIST). Present address is Department of Chemistry, the University of Tokyo, Hongo, Bunkyo-ku, Tokyo 113-0033, Japan. The author thanks AIST for financial support to this research.
2. Zhang, C.; Cui, D. -M.; Yao, L. -Y.; Wang, B. -S., Hu, Y. -Z.; Hayashi, T. *J. Org. Chem.* **2008**, *73*, 7811-7813.
3. Mello, R.; Alcalde-Aragonés, A.; González-Núñez, M. E. *Tetrahedron Lett.* **2010**, *51*, 4281-4283.
4. Zhou, J.; List, B. *J. Am. Chem. Soc.* **2007**, *129*, 7498-7499.

5. Nishinaga, A.; Rindo, K.; Matsuura, T. *Synthesis* **1986**, 1038-1041.
6. Marion, N.; Ramón, R. S.; Nolan, S. P. *J. Am. Chem. Soc.* **2009**, *131*, 448-449.
7. Nun, P.; Ramón, R. S.; Gaillard, S.; Nolan, S. P. *J. Organomet. Chem.* **2011**, *696*, 7-11.
8. Hirabayashi, T.; Okimoto, Y.; Saito, A.; Morita, M.; Sakaguchi, S.; Ishii, Y. *Tetrahedron* **2006**, *62*, 2231-2234.
9. Peppe, C.; Lang, E. S.; Ledesma, G. N.; de Castro, L. B.; Barros, O. S. D.; de Azevedo Mello, P. *Synlett* **2005**, 3091-3094.
10. Yokota, T.; Sakakura, A.; Tani, M.; Sakaguchi, S.; Ishii, Y. *Tetrahedron Lett.* **2002**, *43*, 8887-8891.
11. Mitsudome, T.; Umetani, T.; Nosaka, N.; Mori, K.; Mizugaki, T.; Ebitani, K.; Kaneda, K. *Angew. Chem. Int. Ed.* **2006**, *45*, 481-485.
12. Roembke, P.; Schmidbaur, H.; Cronje, S.; Raubenheimer, H. *J. Mol. Catal. A: Chem.* **2004**, *212*, 35-42.
13. Leyva, A.; Corma, A.; *J. Org. Chem.* **2009**, *74*, 2067-2074.
14. Almássy, A.; Nagy, C. E.; Bényei, A. C.; Joó, F. *Organometallics* **2010**, *29*, 2484-2490.
15. Czégéni, C. E.; Papp, G.; Kathó, Á.; Joó, F. *J. Mol. Catal. A: Chem.* **2011**, *340*, 1-8.
16. Sanz, S.; Jones, L. A.; Mohr, F.; Laguna, M. *Organometallics* **2007**, *26*, 952-957.
17. Carriedo, G. A.; López, S.; Suárez-Suález, S.; Presa-Soto, D.; Presa-Soto, A. *Eur. J. Inorg. Chem.* **2011**, 1442-1447.
18. Jin, X.; Oishi, T.; Yamaguchi, K.; Mizuno, N. *Chem. Eur. J.* **2011**, *17*, 1261-1267.
19. Vasudevan, A.; Verzal, M. K. *Synlett* **2004**, 631-634.
20. Dreyer, D. R.; Jia, H. -P.; Bielawski, C. W. *Angew. Chem. Int. Ed.* **2010**, *49*, 6813-6816.

Teruyuki Hayashi was born in Tokyo, Japan in 1949. He obtained B.Sc. at the University of Tokyo in 1973. He then joined National Institute of Advanced Industrial Science and Technology (AIST). He obtained his D.Sc. in 1982 from the University of Tokyo, was promoted to senior researcher in 1983, to group leader in 1994, and to deputy director in 2001. His research interests include invention of new catalytic methods for organic and/or material syntheses. Since 2008, he has been project professor at Department of Chemistry, the University of Tokyo.

Oxone®-Mediated Synthesis of Benzimidazoles from 1,2-Phenylenediamines and Aldehydes: Preparation of 2-(4-Cyano-Phenyl)-1-[2-(3,4-Dimethoxyphenyl)-Ethyl]-1H-Benzimidazole-5-Carboxylic Acid Ethyl Ester

Submitted by James R. Gillard and Pierre L. Beaulieu.[1]
Checked by York Schramm and Andreas Pfaltz.

1. Procedure

A. 4-[2-(3,4-Dimethoxyphenyl)-ethylamino]-3-nitrobenzoic acid ethyl ester (3). A 250-mL, single-necked, round-bottomed flask with a 24/40 ground-glass joint is equipped with a large egg-shaped magnetic stir bar (40 mm x 18 mm) (Note 1). The flask is charged with ethyl-4-chloro-3-nitrobenzoate **1** (5.00 g, 21.8 mmol, 1.00 equiv) (Note 2) and anhydrous DMSO (20 mL) (Note 3). To this homogeneous yellow mixture, 3,4-dimethoxyphenethylamine **2** (4.04 mL, 23.9 mmol, 1.10 equiv) (Note 4) is added dropwise over 1–2 min using a Pasteur pipette. As soon has the first drop is added, the reaction mixture quickly changes to a dark reddish-orange color, but the mixture remains homogeneous throughout the addition without detectable exotherm. *N,N*-Diisopropylethylamine (5.31 mL, 30.5 mmol,

Published on the Web 10/5/2011
© 2012 Organic Syntheses, Inc.

1.4 equiv) (Note 5) is added dropwise (1–2 min), which results in a slight cloudiness, but no exotherm is detected during or after addition. The flask is closed with a septum (Note 6) and the reaction mixture is stirred in an oil bath heated to 70 °C. During the warming process, the dark reddish-orange color lightens somewhat, but the mixture remains cloudy. The reaction is stirred for 18 h at 70 °C after which time the dark reddish-orange reaction mixture is homogeneous. The reaction is judged to be complete by the disappearance of the starting ester by TLC (Note 7). The mixture is cooled to room temperature (Note 8) and de-ionized water (2 mL) is added dropwise with a Pasteur pipette over a period of ca. 2–3 min with efficient stirring while a persistent cloudiness develops. Precipitation/crystallization of the product occurs rapidly after the addition, giving rise to a very thick dark-red suspension (Note 1). After formation of the suspension, water (150 mL) is added in 10 mL portions over a period of ca. 5 min. During the addition of water the mixture becomes slightly warm to the touch. The resulting dark-orange to dark-yellow suspension of product is stirred at room temperature for approximately 1 h. The solids are collected by suction-filtration on filter paper (Note 9) using a ceramic Büchner funnel and the solids are washed with four portions of water (20 - 30 mL each). The solids are allowed to dry for ca. 1–2 h under air suction on the filter funnel. The solids are then collected and placed under vacuum (2 mmHg) at ambient temperature until constant mass is achieved to provide 7.92–7.98 g (97–98% yield) of a red-orange to dark yellow solid (Note 10) which is suitable for use in the next step without further purification.

B. *Ethyl 3-amino-4-((3,4-dimethoxyphenethyl)amino)benzoate (4).* A 500-mL, single-necked, round-bottomed flask with a 24/40 ground-glass joint is equipped with a large egg-shaped magnetic stir bar (40 mm x 18 mm). The flask is charged with nitro compound **3** (7.00 g, 18.7 mmol, 1.00 equiv) and methanol (285 mL) (Note 12). The flask is then fitted with a 3-way gas inlet adapter with an argon-filled balloon attached to one of the arms. While stirring the resultant yellow suspension, the flask is evacuated and back-filled with argon gas three times (Note 13). While stirring under a blanket atmosphere of argon, the adapter on the flask is briefly removed and Pearlman's catalyst (500 mg, Note 14) is added as a suspension in methanol (10 mL + 5 mL rinse). The argon balloon is then replaced with another balloon filled with hydrogen gas and, while stirring the mixture, the flask is evacuated and back-filled with hydrogen gas three times (Note 13). The dark yellow-black suspension is allowed to stir under the hydrogen atmosphere

132

for a period of 24 h, after which TLC analysis indicate the disappearance of compound **3** (Note 15). The hydrogen balloon is once again replaced with an argon-filled balloon and the flask is evacuated and back-filled with argon gas three times (Note 13). The 3-way stopper is then removed, Celite 545® (10g) (Note 16) is added and the suspension stirred under the argon blanket for 10 min. The solid is removed by suction filtration over filter paper (Note 17) using a ceramic Büchner funnel and the solids are washed with two portions of methanol (20 - 30 mL each). The clear, dark brown filtrate and washings are collected and combined and the volatiles removed under reduced pressure (30 °C, 5 - 10 mmHg, then at room temperature, 0.1 mmHg) to give 6.0–6.3 g (93–98% yield) of compound **4** as a dark gray-black solid (Note 18) which is suitable for use in the next step without further purification.

C. *2-(4-Cyano-phenyl)-1-[2-(3,4- dimethoxyphenyl)- ethyl]-1 H-benzimidazole-5-carboxylic acid ethyl ester (6).* A 50-mL, 2-necked round-bottomed flask with 10/14 ground glass joints is equipped with an oval-shaped magnetic stir bar, a thermometer adapter and an alcohol thermometer. The flask is charged with diamine **4** (5.00 g, 14.5 mmol), 4-cyanobenzaldehyde **5** (2.10 g, 16.0 mmol, 1.10 equiv) (Note 19), DMF (30 mL) (Note 20) and de-ionized water (1 mL). The resulting dark brown homogeneous mixture is cooled to an internal temperature of 0–5 °C with the aid of an ice-water bath. Oxone® (5.80 g, 9.44 mmol, 0.65 equiv) (Note 21) is added in one portion and the resulting heterogeneous mixture is stirred while immersed in the ice-water bath. After addition of Oxone®, the following observations are made: the internal temperature of the mixture rises to ca. 6-8 °C and stays at this temperature for about 1 h; the granular Oxone® slowly loses its consistency and the mixture becomes a loose, easily stirred light brown slurry. After a period of 2 h, HPLC analysis indicates the disappearance of compound **4** (Note 11). The stirred mixture (Note 22) is allowed to warm to room temperature and is then transferred in small portions using a Pasteur pipette over ca. 15 min to a vigorously stirred solution of potassium carbonate (2.76 g, 20 mmol) (Note 23) in water (450 mL). During the addition, the desired benzimidazole product **6** precipitates as a light brown-beige solid (Note 25). The resulting slurry is stirred for an additional hour, then the solids are collected by suction filtration on a filter paper (Note 9) using a ceramic Büchner funnel. The damp filter cake is allowed to partially air-dry overnight under suction in the filter funnel and is then collected and placed under vacuum (0.03 mmHg)

(Note 26) at ambient temperature until constant mass is achieved to provide 5.88–6.04 g (89–91% yield) of a light tan solid (Notes 24 and 27). The material was further purified by dissolving the crude product in dichloromethane (250 mL), washing with diluted sodium hydroxide solution (1.0 M, 3 x 100 mL), then washing with water (100 mL) and brine (100 mL). The organic layer was dried with anhydrous $MgSO_4$, filtered over a filter paper (Note 28) and solvents removed on the rotary evaporator (Note 29). The residue was dried *in vacuo* (0.03 mmHg) for 48 h to provide 4.9 g (74% yield) of a beige solid (Note 30).

2. Notes

1.   It is important to use a large stir bar. In the subsequent precipitation step, an undersized stir bar may not adequately stir the resulting slurry, or may become jammed. On a larger scale, the authors would suggest the use of a stirring motor and paddle assembly.

2.   Ethyl-4-chloro-3-nitrobenzoate was purchased from Alfa Aesar (powder, 97%, cat. no. A19338) and was used as received.

3.   Anhydrous DMSO was purchased from Sigma-Aldrich Company Inc. and was used as received.

4.   3,4-Dimethoxyphenethylamine was purchased from Sigma-Aldrich Company Inc. (liquid, 97%, cat. no. D136204-100G) and was used as received.

5.   *N,N*-Diisopropylethylamine was purchased from Sigma-Aldrich Company Inc. (ReagentPlus®, 99%, cat. no. D125806) and was used as received.

6.   In this experiment, it is sufficient to loosely stopper the flask with a septum since all reagents and starting materials have boiling points above 70 °C. In cases where more volatile amines are used as substrate, the authors recommend the use of a reflux condenser and a larger excess of amine substrate (1.3–2 equiv) to ensure that step **A** proceeds to completion.

7.   The progress of the reaction was monitored by TLC analysis on glass plates precoated (250 µm thickness) with silica gel 60 F_{254} (purchased from EMD Chemicals Inc.). The eluent was 10% EtOAc in hexane and visualization performed with UV light (254 nm). The nitroarene starting material **1** has $R_f = 0.50$ while phenethylamine **2** and product **3** remain at baseline.

134

8.   After cooling to room temperature, the product may begin to crystallize to give a thick slurry or solid mass if allowed to stand. The product can be brought back into solution by reheating to 70 °C and then recooling to room temperature before proceeding as described.

9.   Whatman N° 1 filter paper (7 cm diameter) was used.

10.   Compound **3** has the following characteristics: R_f = 0.37 (30% EtOAc in hexane); HPLC t_R = 6.7 min, 98.5 to >99.5% homogeneity (Note 11); mp 113.6 – 115.9 °C; FTIR (thin film NaCl cm^{-1}): 3367, 2936, 1710, 1625, 1517, 1287, 1222, 1155, 1027; ^{1}H NMR (500 MHz, DMSO) δ: 1.31 (t, J = 7.1 Hz, 3 H), 2.89 (t, J = 7.0 Hz, 2 H), 3.63 (dd, J = 12.8, 6.5 Hz, 2 H), 3.73 (s, 3 H), 3.76 (s, 3 H), 4.28 (q, J = 7.0 Hz, 2 H), 6.82 (d, J = 8.1 Hz, 1 H), 6.89 (d, J = 8.1 Hz, 1 H), 6.93 (s, 1 H), 7.16 (d, J = 9.1 Hz, 1 H), 7.95 (d, J = 9.0 Hz, 1 H), 8.49 (t, J = 5.2 Hz, 1 H), 8.58 (s, 1 H); ^{13}C NMR (126 MHz, DMSO) δ: 14.1, 33.8, 44.1, 55.3, 55.3, 60.6, 111.9, 112.6, 114.8, 116.0, 120.6, 128.2, 130.2, 130.9, 135.6, 147.2, 147.4, 148.7, 164.3; MS (FAB, EI, 70 eV, 250°C) m/z (relative intensity): 374.2 (13.85%), 223.1 (30.54), 152.1 (30.15%), 151.1 (100%). HRMS (APPI) calcd for $C_{19}H_{22}N_2O_6$: 374.1478, found: 374.1485; Anal. calcd for $C_{19}H_{22}N_2O_6$: C, 60.95; H, 5.92; N, 7.48. Found: C, 60.80; H, 5.98; N, 7.52.

11.   HPLC chromatography was performed on a Waters 2695 Separation Module fitted with a Waters^{TM} 2487 Dual Wavelength detector monitoring at 220 and 254 nm. Separations were achieved using a Waters Sunfire^{TM} brand C18, 3.5μm, 4.6 mm x 30 mm column, employing a flow rate of 3mL/min. with a 5% to 100% water/acetonitrile gradient in 10 min with 0.10% trifluoroacetic acid added as modifier.

12.   Methanol was purchased from EMD (OmniSolv® HPLC grade) and used as received.

13.   CAUTION: There may be some initial frothing and/or bubbling during the evacuation.

14.   Pearlman's catalyst (20% Palladium hydroxide on carbon; Pd content 20% dry weight basis. Moisture content ~50%. cat. no. 330094-10G) was purchased from Sigma-Aldrich Company Inc. and used as received.

15.   The progress of the reaction was monitored by TLC analysis on silica gel (Note 7) with 30% EtOAc in hexane as eluent and visualization with UV light (254 nm). Nitroarene **3** has R_f = 0.37 while product **4** has R_f = 0.12.

16.   Celite 545® was purchased from EMD and used as received.

17.   Whatman N° 1 filter paper (5.5 cm diameter) was used.

18. Compound **4** has the following characteristics: R_f = 0.12 (30% EtOAc in hexane); HPLC t_R = 4.0 min (Note 11); mp 100.8 – 103.0 °C; FTIR (thin film NaCl cm^{-1}): 3397, 2935, 1694, 1600, 1516, 1294, 1143, 1027; ^{1}H NMR (500 MHz, DMSO) δ: 1.26 (t, J = 7.1 Hz, 3 H), 2.83 (t, J = 7.3 Hz, 2 H), 3.32 (dd, J = 13.1, 6.9 Hz, 2 H), 3.71 (s, 3 H), 3.75 (s, 3 H), 4.19 (q, J = 7.1 Hz, 2 H), 4.76 (s, 2 H), 5.27 (t, J = 5.2 Hz, 1 H), 6.50 (d, J = 8.4 Hz, 1 H), 6.79 (dd, J = 8.1, 1.5 Hz, 1 H), 6.86 (d, J = 8.2 Hz, 1 H), 6.89 (d, J = 1.5 Hz, 1 H), 7.19 (d, J = 1.9 Hz, 1 H), 7.24 (dd, J = 8.3, 1.7 Hz, 1 H); ^{13}C NMR (126 MHz, DMSO) δ: 14.4, 34.2, 44.9, 55.4, 55.5, 59.4, 108.0, 111.9, 112.7, 114.1, 117.0, 120.5, 120.7, 132.2, 134.2, 140.3, 147.3, 148.7, 166.4. MS (FAB, EI, 70 ev, 250°C) m/z (relative intensity): 345.2 (4.46%), 344.2 (20.57%), 194.1 (11.19%), 193.1 (100%), 165.1 (11.06%), 152.1 (11.24%), 120.1 (4.11%). Compound **4** can be partially purified in the following manner: a 50-mL Erlenmeyer flask equipped with a magnetic stir bar is charged with crude compound **4** (1.32 g) and MeOH (15 mL). The dark purple solution with a beige-colored suspension is stirred for 1 h at room temperature. The solid is collected by filtration and washed with MeOH (3 x 5 mL) at room temperature. The solid is partially dried under air suction and then dried to constant weight under vacuum (2 mmHg) to give 0.75 g of a light beige-colored solid which had comparable spectroscopic properties to the crude material (see above) in addition to: mp 101.6 – 102.8 °C; Anal. calcd for $C_{19}H_{24}N_2O_4$: C, 66.26; H, 7.02; N, 8.13. Found: C, 66.20; H, 7.19; N, 8.12.

19. 4-Cyanobenzaldehyde was purchased from Alfa Aesar (98% cat. no. A14914) and used as received.

20. DMF (ACS grade) was purchased from EMD and used as received.

21. Oxone® (potassium peroxymonosulfate; $2KHSO_5.KHSO_4.K_2SO_4$) was purchased from Sigma-Aldrich Company Inc. (cat. no. 228036-100G) and used as received.

22. The authors suggest stirring the reaction mixture continuously during the transfer to avoid the settling of particles.

23. Potassium carbonate (ACS grade) was purchased from EMD and used as received.

24. Compound **6** has the following characteristics: R_f = 0.32 (50% EtOAc in hexane); HPLC t_R = 4.9 min, >98% homogeneity (Note 11); mp 125 – 130 °C; FTIR (thin film NaCl cm^{-1}): 2937, 1710, 1611, 1516, 1299, 1223, 1027; ^{1}H NMR (500 MHz, DMSO) δ: 1.36 (t, J = 7.1 H, 3 H), 2.82 (s,

2 H), 3.45 (s, 3 H), 3.67 (s, 3 H), 4.35 (q, J = 7.1 Hz, 2 H), 4.63 (t, J = 6.3 Hz, 2 H), 6.08 (d, J = 8.1 Hz, 1 H), 6.21 (d, J = 1.2 Hz, 1 H), 6.58 (d, J = 8.1 Hz, 1 H), 7.63 (d, J = 8.2 Hz, 2 H), 7.89 (dd, J = 12.8, 8.4 Hz, 3 H), 7.96 (d, J = 8.5 Hz, 1 H), 8.29 (s, 1 H); ^{13}C NMR (126 MHz, DMSO) δ: 14.3, 34.4, 46.1, 54.9, 55.5, 60.6, 111.5, 111.8, 112.1, 118.4, 120.4, 121.1, 123.8, 124.2, 129.5, 129.8, 132.2, 134.1, 138.8, 142.2, 147.6, 148.4, 153.8, 166.2; MS (FAB, EI, 70 eV, 250°C) m/z (relative intensity): 456.1 (8.8%), 455.1 (28.91%), 276.0 (4.41%), 151.1 (100%). HRMS (APPI) calcd for $C_{27}H_{26}N_3O_4$: 455.1845, found: 455.1858; EA by submitters: Anal. calcd for $C_{27}H_{25}N_3O_4$: C, 71.19; H, 5.53; N, 9.23. Found: C, 70.88; H, 5.45; N, 9.19.

25. The submitters report observation of a light grey solid.

26. The submitters applied a vacuum of 2 mmHg.

27. The submitters reported that the material is analytically pure at this stage, as determined by CHN analysis. In order to obtain analytically pure product, the checkers, in consultation with the submitters, developed the purification protocol described above.

28. Whatman LS 14 1/2 filter paper (18.5 cm diameter) was used.

29. The water bath was warmed to 40 °C and a reduced pressure of 300 mmHg applied.

30. The material has the same characteristics as the crude product. mp = 132–134.5 °C. Anal. calcd for $C_{27}H_{25}N_3O_4$: C, 71.19; H, 5.53; N, 9.22. Found: C, 70.96; H, 5.48; N, 9.20.

Safety and Waste Disposal Information

All hazardous materials should be handled and disposed of in accordance with "Prudent Practices in Laboratory"; National Academy Press; Washington, DC, 1995.

3. Discussion

The benzimidazole scaffold constitutes a "privileged structure" in material sciences and for drug design. It is a central constituent of many biologically active substances from both natural and synthetic origin.[2] Several methods have been described in the literature for the synthesis of substituted benzimidazole derivatives.[3] They often involve condensation of 1,2-phenylenediamine derivatives with aldehydes or carboxylic acids, performed under a variety of reaction conditions that vary in scope, yield, harshness of reaction conditions and ease of isolation and purification of the final product. Recently, we described an Oxone®-mediated oxidative condensation of aldehydes and 1,2-phenylediamines that is suitable not only for the high-throughput synthesis of benzimidazole libraries, but can also be carried out on a large scale.[4] The method that is illustrated in this procedure is highly versatile, simple to perform, economical, metal-free and environmentally friendly. Furthermore, intermediates and the final product from this sequence are isolated in pure form and high yields by simple precipitation. While several methods for the synthesis of benzimidazoles have continued to appear in the literature since publication of the original report in 2003, attributes of this Oxone®-mediated synthesis of benzimidazoles still make the method described herein a procedure of choice for the preparation of such derivatives on small or large scale.

Table 1. Preparation of 1,2-phenylenediamine substrates[a]

entry	starting material	R^{2}NH$_2$	% yield	% yield
1	EtOOC — NO$_2$ / Cl	cyclohexyl–NH$_2$	98%	99%
2	NO$_2$ / F	MeO, MeO — CH$_2$CH$_2$NH$_2$	100%	98%
3	EtOOC — NO$_2$ / Cl	(CH$_3$)$_3$C–NH$_2$	99%	70%
4	F$_3$C — NO$_2$ / Cl	4-methylphenyl–NH$_2$	98%	95%

[a] See reference 4 for details

The required 1,2-phenylenediamine starting materials can be obtained from commercial sources (a search in the Symyx ACD® 2009.3 catalog

Org. Synth. **2012**, *89*, 131-142

reveals >6000 hits) or prepared through a two-step sequence which involves S_NAr displacement on a 2-chloro or 2-fluoronitroarene derivative followed by reduction of the nitro functionality to the aniline. A few examples of these transformations are provided in Table 1.[4]

Condensation of 1,2-phenylenediamines with aldehydes tolerates a variety of functional groups on either substrates. The reaction can be performed on electron rich or poor systems, aromatic, heterocyclic (e.g. quinolines, pyridines, thiophenes, furans) or aliphatic aldehydes, and tolerates unprotected functionalities such as phenol hydroxyl groups, free carboxylic acids, nitriles and amides. The reaction is usually very rapid and provides good yields of pure products (>90 % homogeneity) after simple precipitation from the reaction medium with aqueous base. Table 2 is illustrative of the broad applicability of this methodology.

In the original report, the authors described a limitation of the methodology when aldehydes are sensitive to Oxone® in the acidic reaction medium (e.g. pivaldehyde, some heterocyclic aldehydes and α,ß-unsaturated aldehydes that may be susceptible to Baeyer-Villiger oxidation).[4] The authors have discovered since that in some cases, if such sensitive aldehydes are allowed to pre-form the presumed aminal intermediate by stirring with the diamine for a few hours before addition of Oxone®, these substrates can be engaged in the reaction and provide the desired benzimidazole in improved yields.

Table 2. Scope of the oxone®-mediated preparation of benzimidazoles from 1,2-phenylenediamines and aldehydes[a]

Entry	Starting material	R³CHO	Reaction time (h)	% yield[b] (crude)	% homogeneity[c] (crude)
1	EtOOC— ... NH₂ / NH-cyclohexyl	2-pyridyl—CHO	1	85	>99
2	NH₂ / NH-CH₂CH₂-(3-MeO,4-OMe-phenyl)	HOOC— ... —CHO (meta)	0.5	66	98
3		HO—⟨ ⟩—CHO	1	84	>99
4		2-quinolinyl—CHO	1	73	>99
5		cyclohexyl—CHO	2	80	96
6		(CH₃)₂CH—CHO	0.5	83	>99
7	EtOOC— ... NH₂ / NH-t-Bu	O₂N—⟨ ⟩—CHO	2.5	79	99
8		NC—⟨ ⟩—CHO	18	89	97
9		MeO,MeO—⟨ ⟩—CHO	19	70	92
10		(methylenedioxyphenyl)—CHO	5	95	86
11		(N-methyl-3-chloroindol-2-yl)—CHO	22	59	97
12	F₃C— ... NH₂ / NH-(4-methylphenyl)	Br—⟨ ⟩—CHO	1.5	89	97
13		AcHN—⟨ ⟩—CHO	20	86	96
14		2-thienyl—CHO	1.5	88	96
15		(CH₃)₂CHCH₂—CHO	1.5	68	94

[a]See reference 4 for details. [b]Yield of crude material obtained after dilution of reaction mixture with 20 volumes of water and collection of the precipitate by filtration or extraction ith EtOAc. [c]Reversed-phase HPLC homogeneity of the crude products.

140

1. Boehringer Ingelheim (Canada) Ltd.; Research and Development; 2100 Cunard street, Laval (Québec) Canada, H7S 2G5; pierre.beaulieu@boehringer-ingelheim.com.
2. Boiani, M.; Gonzalez, M. *Mini-Reviews Med. Chem.* **2005**, *5*, 409-424.
3. Grimmet, M. R.; In *Comprehensive Heterocyclic Chemistry*; Katritzky, A. R., Rees, C. W., Eds.; Pergamon: Oxford, 1984; Vol. 5, pp 457-498.
4. Beaulieu, P. L.; Haché, B.; von Moos, E. *Synthesis* **2003**, 1683-1692.

Appendix
Chemical Abstracts Nomenclature; (Registry Number)

Ethyl-4-chloro-3-nitrobenzoate; (16588-16-2)
3,4-Dimethoxyphenethylamine: 2-(3,4-dimethoxyphenyl)ethylamine; (120-20-7)
4-Cyanobenzaldehyde; (105-07-7)
Oxone®: potassium peroxymonosulfate; (37222-66-5)

Pierre Beaulieu was born in 1957 in Canada and received his BSc and MSc degrees in chemistry from the University of Ottawa (Canada) in 1978 and 1980 respectively. He received his PhD degree in 1983 from the University of Alberta in Edmonton (Canada) working under the supervision of Professor Derrick Clive. Following Post-doctoral studies with Professor Stephen Hanessian at the University of Montréal (Canada), he was appointed Associate Director of the chemical biology and polypeptide unit at the Clinical Research Institute of Montréal from 1986 to 1988. He then joined the Research and Development division of Boehringer Ingelheim (Canada) Ltd. in 1988 as a medicinal chemist. He is currently Highly Distinguished Scientist working on the design and development of antiviral drugs for the treatment of hepatitis C virus infection.

James Gillard was born in 1965 in Nova Scotia, Canada. He received his BSc degree in chemistry from Saint Francis Xavier University, Antigonish (Canada) in 1987, and his MSc from Memorial University of Newfoundland, St. John's (Canada) in 1992, under the supervision of Professor D. Jean Burnell. He joined the Research and Development division of Boehringer Ingelheim (Canada) Ltd. in 1992 as a medicinal chemist. He is currently Research Scientist working on the development of antiviral drugs for the treatment of hepatitis C virus infection.

York Schramm was born in 1984 and received his MSc at the University of Basel (Switzerland) carrying out his master thesis at the University of Bristol (United Kingdom) under the supervision of Professor Guy Lloyd-Jones in 2009. He subsequently started his graduate studies in Basel under the supervision of Professor Andreas Pfaltz working on Mechanistic Investigations of Asymmetric Hydrogenation.

Discussion Addendum for:
Preparation of α-Acetoxy Ethers by the Reductive Acetylation of Esters: *endo*-1-Bornyloxyethyl Acetate

Prepared by Nicholas Sizemore and Scott D. Rychnovsky.*[1]
Original Article: Kopecky, D. J.; Rychnovsky, S. D. *Org. Synth.* **2003**, *80*, 177.

Reductive acetylation has proven to be a powerful methodology primarily because its products are excellent substrates for carbon-carbon bond forming reactions. Since 2003, the *Organic Syntheses* article and the primary references have been cited over 100 times. This discussion addendum aims to provide a brief overview of recent developments in this methodology. Improvements and variations of α-acetoxy ether formation, as well as the synthetic utility of α-acetoxy ethers will be discussed. Recent applications in the context of the total synthesis of complex natural products are also included.

Improvements and Variations of α-Acetoxy Ether Formation

The original reductive acetylation protocol involves the addition of diisobutylaluminum hydride (DIBALH) as a 1.0 M solution in hexanes to the ester at –78 °C. The resulting aluminum alkoxide is then treated with pyridine, 4-dimethylaminopyridine (DMAP), and acetic anhydride (Ac$_2$O) sequentially. The internal temperature during these additions should not exceed –72 °C in order to maximize the yield and prevent undesired side reactions.[2] While this method is general, recent developments include the use of asymmetric electrophiles in order to generate more complex acyl ethers and the extension of reductive acetylation to lactams and imides.

Shortly after the original article, Rovis and Zhang reported the use of acid fluorides as competent reagents for the trapping of such aluminum alkoxides, the results of which are summarized in Figure 1.[3] Aryl, vinyl, and alkyl acid fluorides are all effective trapping reagents (67-97% yield) and steric bulk is well tolerated (pivaloyl fluoride = 92% yield). This development allows access to more complex acyl acetal derivatives without sacrificing a full equivalent of the acid, which represents an improvement in atom economy over the use of symmetric anhydrides.

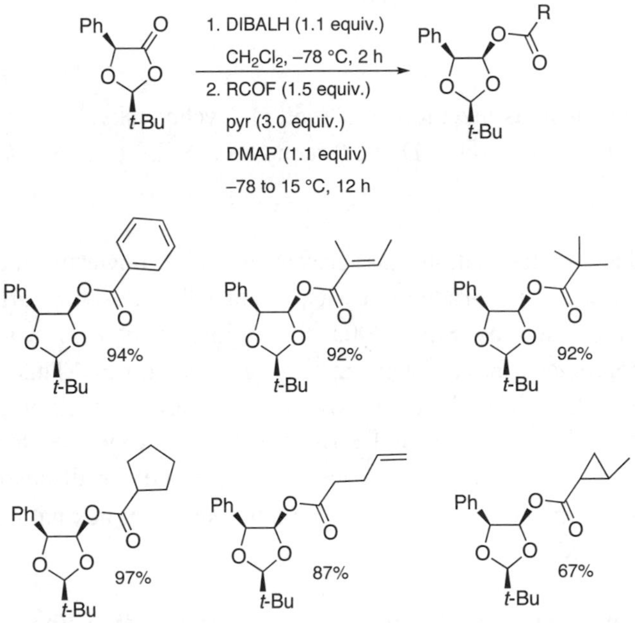

Figure 1. Acid Fluorides as Trapping Reagents.

More recently, Dalla and co-workers have extended reductive acetylation strategy to the formation of α-acetoxy pyrrolidines (Figure 2).[4] While DIBALH will reduce imides, the resultant aluminum alkoxide is not nucleophilic enough to react with acetic anhydride. Therefore, the reductant of choice for the preparation of α-acetoxy amines is lithium triethylborohydride (LiEt$_3$BH), whose lithium alkoxide is sufficiently nucleophilic for effective trapping with acetic anhydride.

144

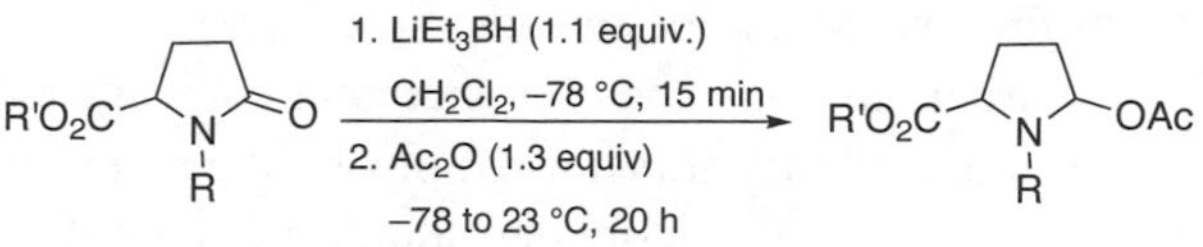
R'O2C
N
R
O
1. LiEt3BH (1.1 equiv.)
CH2Cl2, –78 °C, 15 min
2. Ac2O (1.3 equiv)
–78 to 23 °C, 20 h
R'O2C
N
R
OAc

R	R'	% yield
Boc	Me	100
Boc	*t*-Bu	96
CBz	*t*-Bu	77

Figure 2. Formation of α-Acetoxy Pyrrolidines.

A variety of *N*-alkylated imides, shown in Figure 3, afford α-acetoxy lactams in high yields (75-100%), with the notable exception of cases where there is α-branching with respect to the nitrogen substituent (e.g. the imide of α-methylbenzylamine, MeBn). Steric bulk on the imide also limits the conversion to the α-acetoxy lactams, as shown by the modest yield for the *bis*-OTBS imide. Despite these limitations, this reductive acetylation strategy offers rapid entry into *N*-acylimium ions for use in Mannich reactions.

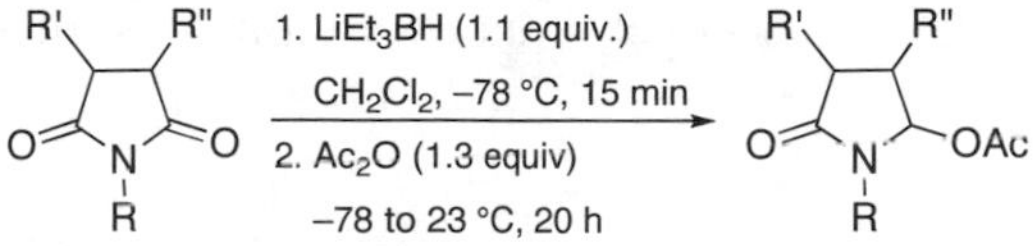
R'
R"
O
N
R
O
1. LiEt3BH (1.1 equiv.)
CH2Cl2, –78 °C, 15 min
2. Ac2O (1.3 equiv)
–78 to 23 °C, 20 h
R'
R"
O
N
R
OAc

R	R'	R"	% yield	cis:trans[a]
PMB	H	OAc	88	100.0
Allyl	H	OAc	79	100:0
Bn	H	OTBS	75	100:0
PMB	H	OTBS	75	8:92
Bn	OAc	OAc	95	0:100
Allyl	OAc	OAc	80	0:100
Bn	OTBS	OTBS	30	0:100
MeBn	H	H	0	

a: refers to stereochemistry between R" and the newly formed acetate.

Figure 3. α-Acetoxy Lactams from Imides.

Synthetic Utility of α-Acetoxy Ethers

Treatment of α-acetoxy ethers with a Lewis acid (commonly $BF_3{\cdot}OEt_2$, $MgBr_2{\cdot}OEt_2$, TMSOTf, $SnCl_4$ or $TiCl_4$) results in the formation of an oxocarbenium ion, which can be intercepted by a variety of nucleophiles to afford a new carbon-carbon or carbon-heteroatom bond adjacent to an ether linkage. The scope of such transformations has been investigated in the context of diastereoselective nucleophilic additions. α-Acetoxy ethers are also useful substrates for cascade sequences, such as oxonia Cope–Prins and Sakurai–Prins–Ritter cyclizations. The use of α-acetoxy ethers in enantioselective organocatalytic oxa-Pictet Spengler reactions have also been examined.

The use of α-(trimethylsilyl)benzyl alcohol as a chiral auxiliary has been investigated by Rychnovsky and Crossrow for diastereoselective additions to oxocarbenium ions generated from α-acetoxy ethers.[5] This chiral auxiliary was chosen due to the ease of preparation of both enantiomers and facile deprotection or conversion to the benzyl ether. Treatment of the α-acetoxy ether with a variety of nucleophiles, including allyl silanes, silyl enol ethers, and silyl ketene acetals, in the presence of TMSOTf proceed in high yields (87-98%) and high diastereoselectivities (20:1 to 80:1), as shown in Figure 4. Similarly, (E)-crotyltrimethylsilane affords the *syn* product in high yield and diastereoselectivity (82%, 27:2.4:1). The addition of cyanide and ethyl nucleophiles proceed in high yields (97% and 75% respectively), albeit with modest diastereoselectivity (5:1 and 4:1).

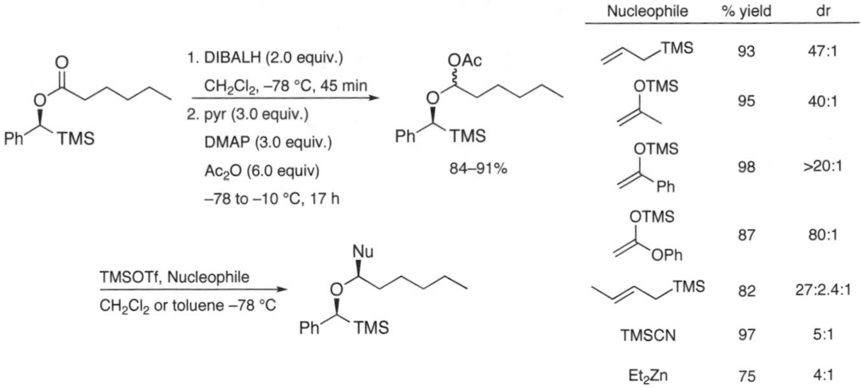

Nucleophile	% yield	dr
⌇TMS	93	47:1
OTMS	95	40:1
OTMS ⌇Ph	98	>20:1
OTMS ⌇OPh	87	80:1
⌇TMS	82	27:2.4:1
TMSCN	97	5:1
Et₂Zn	75	4:1

Figure 4. Diastereoselective Additions of Nucleophiles.

Rychnovsky and Dalgard have reported on the expedient synthesis of tetrahydropyranones through an oxonia Cope–Prins cyclization.[6] This strategy, outlined in Figure 5, involves a reductive acetylation of an ester **1**,

146

followed by treatment with TMSOTf to generate oxocarbenium ion **3**. The pendant allyl moiety participates in a 2-oxonia Cope rearrangement to generate oxocarbenium ion **4**, which is poised for a Prins cyclization (intramolecular Mukaiyama aldol reaction) with the silyl enol ether to afford tetrahydropyranone **5**. This transformation is effective for a variety of substrates (Figure 6) affording the tetrahydropyranone product in high yields (77-99%), including those bearing quaternary centers adjacent to the ketone.

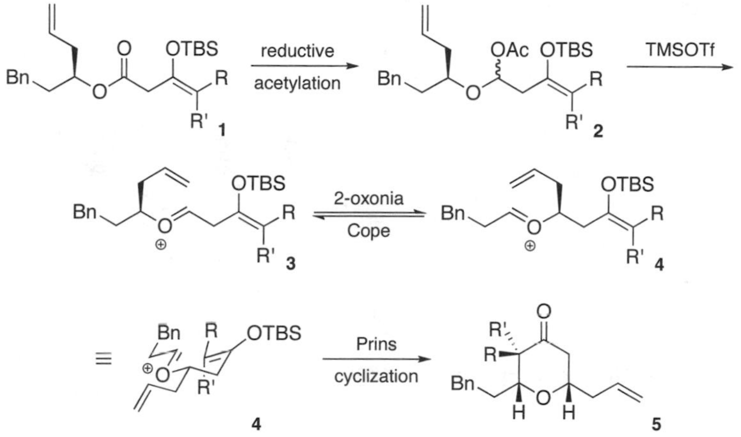

Figure 5. 2-Oxonia Cope–Prins Cyclization.

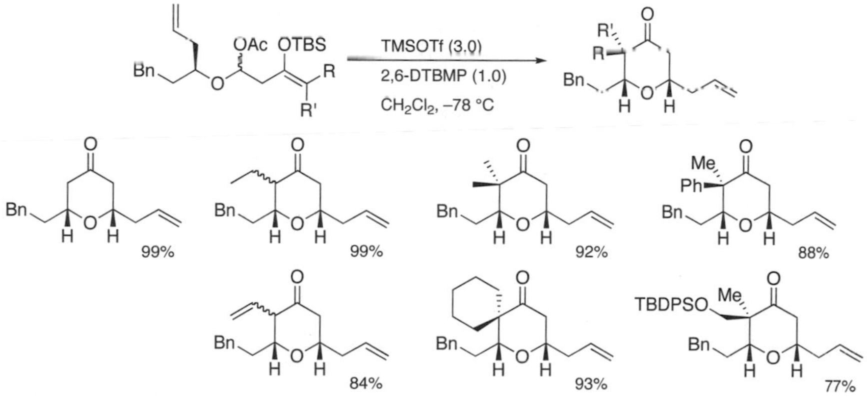

Figure 6. Synthesis of Tetrahydropyranones.

Rovis and Epstein have also utilized α-acetoxy ethers in a cascade sequence to provide 4-amino tetrahydropyrans through a Sakurai–Prins–Ritter cyclization.[7] As shown in **Figure 7**, treatment of 4-acetoxy-1,3-

dioxane **6** with TMSOTf generates oxocarbenium ion **7**, which upon treatment with allyl silane affords alkene **8**. In the presence of TfOH, acetal cleavage results in formation of oxocarbenium ion **9**, which undergoes a Prins cyclization to give carbocation **10**. Addition of a nitrile to the carbocation affords 4-amino tetrahydropyran **11** after hydrolysis.

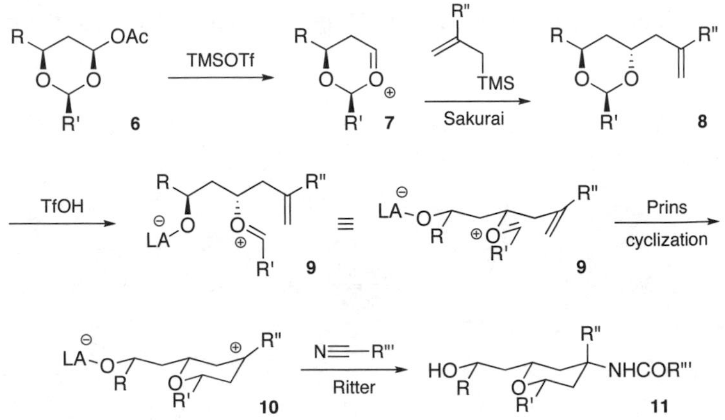

Figure 7. Sakurai–Ritter–Prins Cyclization.

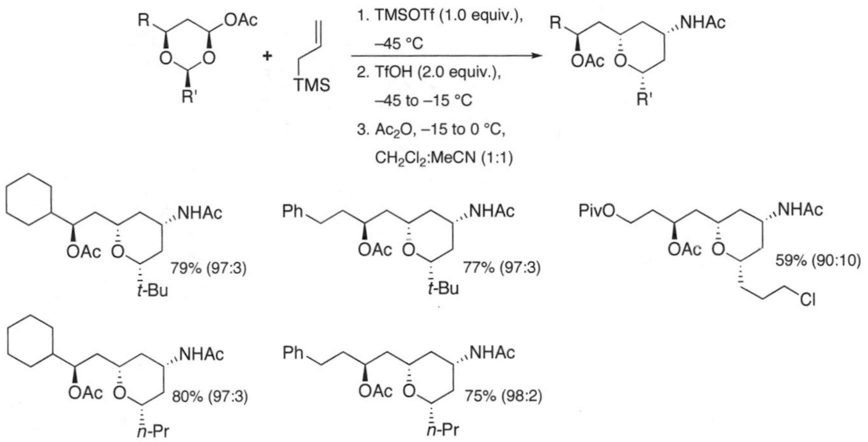

Figure 8. Synthesis of 4-Amino Tetrahydropyrans.

The Sakurai–Ritter–Prins reaction is effective for a variety of dioxane derivatives, as shown in Figure 8. This transformation affords the 4-amino tetrahydropyran products in good yields (59-88%) and high

148

diastereoselectivies (90:10 to 99:1). It should be noted that 2-substituted allyl silanes (Figure 9) and a variety of nitriles are effective partners in this transformation.[7]

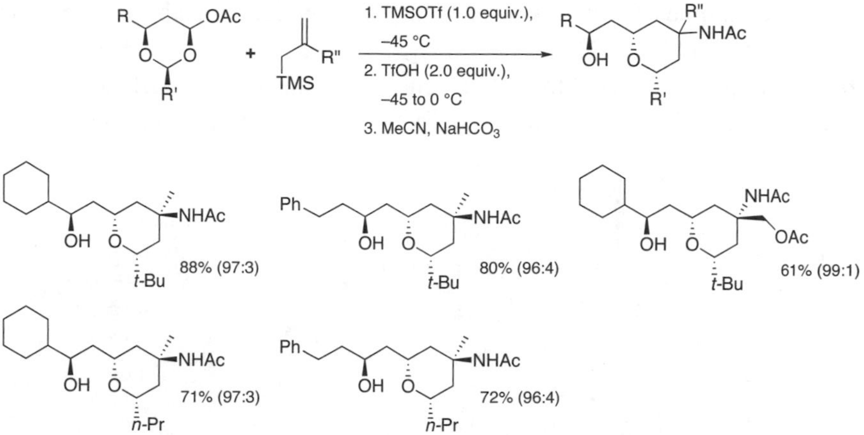

Figure 9. Synthesis of Quaternary 4-Amino Tetrahydropyrans.

Jacobsen and Doyle have utilized α-acetoxy ethers as substrates in the development of enantioselective organocatalytic oxa-Pictet–Spengler reactions (Figure 10).[8] An ester derivative of tryptophol was reduced and trapped to give α-acetoxy ether **12**. In the presence of a thiourea catalyst **13**, TMSCl, and BaO, oxocarbenium ion formation and subsequent ring closure afforded tetrahydropyrano-indole **14** in good yield (82%) and good enantiomeric excess (79%).

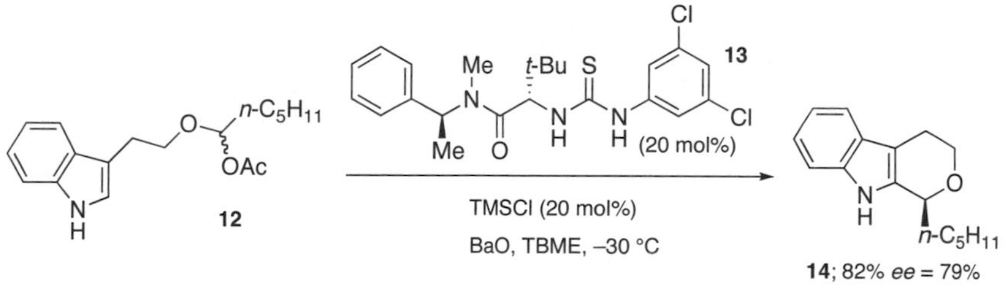

Figure 10. Organocatalytic Asymmetric Oxa-Pictet–Spengler Cyclization.

Highlighted Applications in the Total Synthesis of Complex Natural Products: Intramolecular Allylation of α-Acetoxy Ethers

Kadota, Yamamoto and co-workers utilized reductive acetylation to unite two complex fragments in their highly convergent synthesis of marine neurotoxin gambierol (**19**).[9] The strategy, shown in Scheme 1, begins with the coupling of ABC acid to FGH alcohol *via* Yamaguchi esterification. The resulting ester **15** is then subjected to reductive acetylation conditions. While

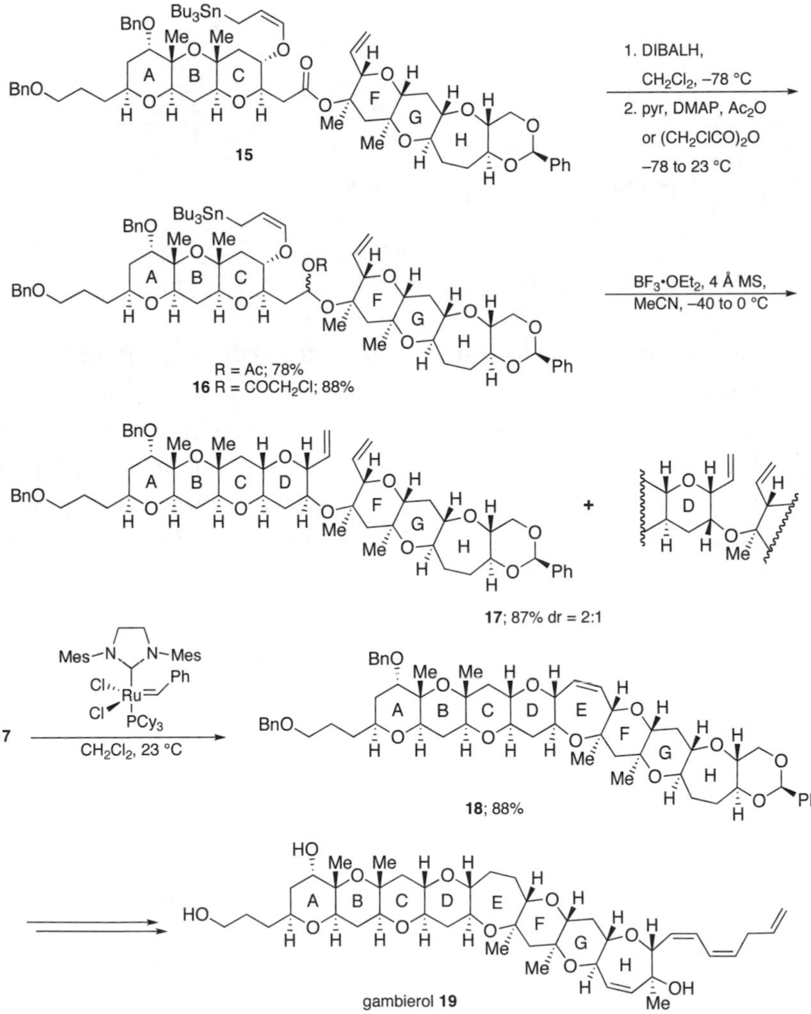

Scheme 1. Kadota and Yamamoto's Total Synthesis of Gambierol.

Org. Synth. **2012**, *89*, 143-158

Ac$_2$O is an effective trapping reagent, chloroacetic anhydride provided increased yield and proper diastereoselectivity for the subsequent ring closure. Treatment of the resulting α-acetoxy ether **16** with boron trifluoride etherate in the presence of 4 Å molecular sieves afforded effective closure of the 6-membered D ring in high yield (87%) and modest diastereoselectivity (2:1) with respect to the acyclic ether. The resulting diene **17** is primed for a ring closing metathesis to forge the 7-membered E ring. The complex polycyclic ether framework **18** was then further elaborated to gambierol (**19**).

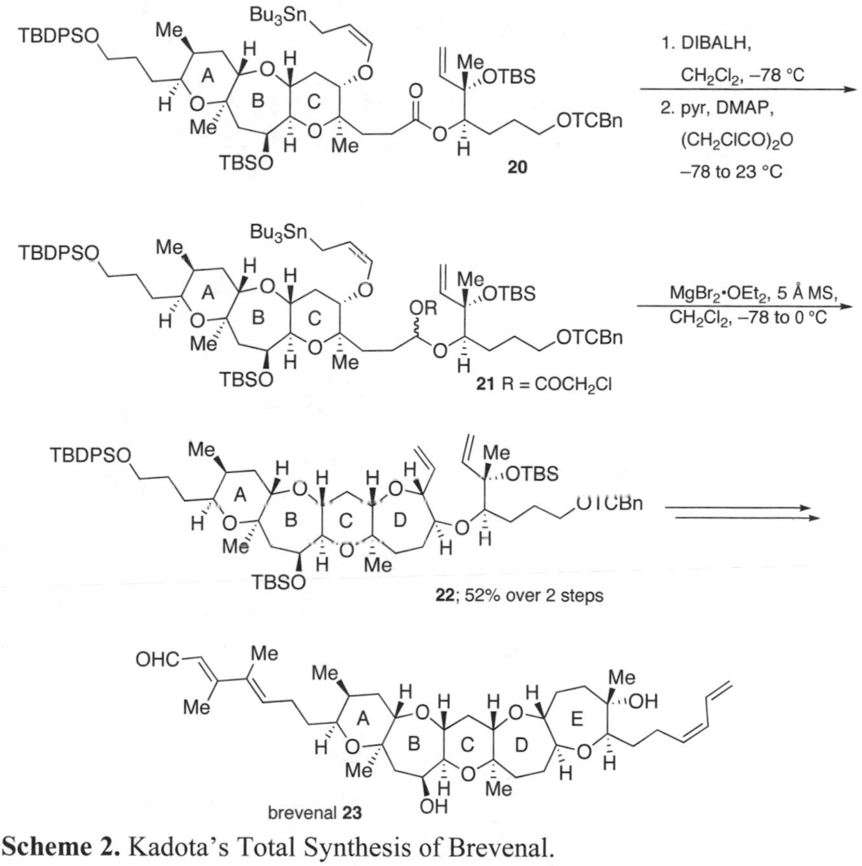

Scheme 2. Kadota's Total Synthesis of Brevenal.

A similar approach was employed in Kadota and co-worker's total synthesis of the nontoxic brevetoxin antagonist brevenal (**23**).[10] As shown in Scheme 2, the 7-membered D ring arose from the reductive chloroacetylation of ester **20**, followed by treatment of α-acetoxy ether **21** with magnesium bromide etherate and 5 Å molecular sieves. The resulting oxocarbenium ion was then trapped by the allyl stannane to afford the tetracycle **22** in good yield as a single diastereomer (52% from the ester **20**). This cyclization again sets the stage for a ring closing metathesis sequence to afford a pentacycle, which was elaborated to brevenal (**23**).

The authors also reported a second-generation strategy (Scheme 3) where acyclic acid **24** is coupled to ABC alcohol **25**.[10b] This route employs the same reduction/cyclization sequence to forge the E ring in good yield as the single desired diastereomer (57% from the ester **26**) and sets up the closure of the D ring in an analogous fashion.

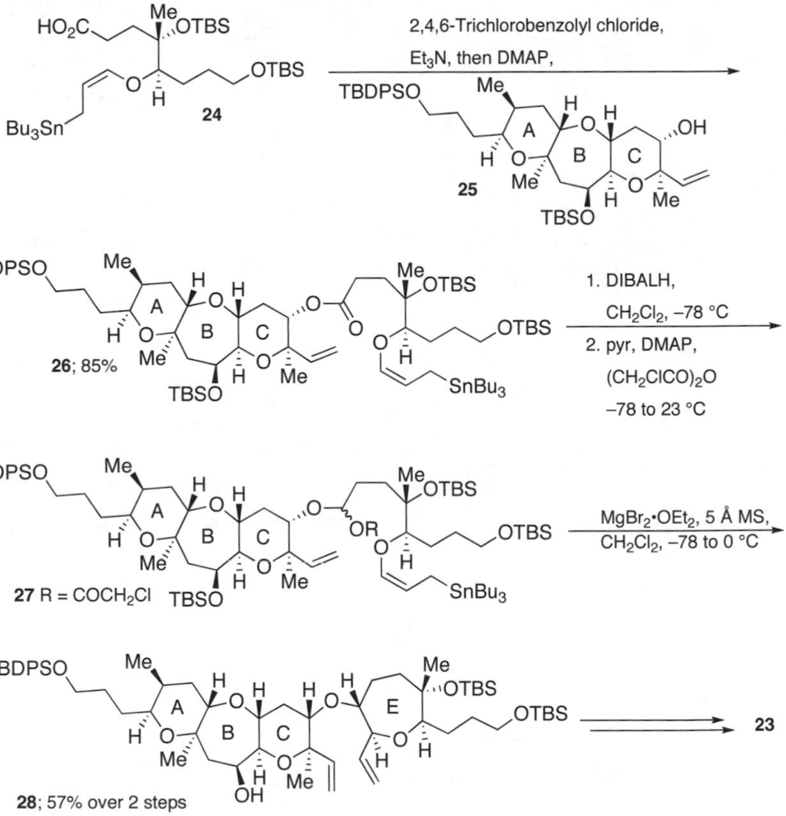

Scheme 3. Kadota's Second Generation Strategy to Brevenal.

Recently, the total synthesis of ciguatoxin CTX3C (**34**) by Yamashita and co-workers employed reductive acetylation to facilitate a radical

152

cyclization to give the C ring of tridecacyclic natural product (Scheme 4).[11] The AB acid and E alcohol were subjected to Yamaguchi esterification conditions to afford the ester **29**. Following reductive acetylation, α-acetoxy ether **30** underwent transacetylation in the presence of diisobutylaluminum phenylselenide to afford the α-seleno ether **31** in high yield (83% from the ester). The α-seleno ether **31** was carried through 3 steps before C-ring radical cyclization was initiated by triethylborane and oxygen in the presence of tributyltin hydride to provide sulfoxide **33** in good yield as a single diastereomer (86%). This approach offered increased stereoselectivity over a direct allylation strategy (as in gamberiol). Sulfoxide **33** was then carried forward to afford ciguatoxin CTX3C (**34**).

Reductive acetylation has proven to be a powerful method for the construction of complex cyclic ethers *via* intramolecular allylation. The strategy has played a key role in the preparation of brevetoxin-like ladder polyethers.[12] The method's utility is largely due to the ease with which substrates can be brought together by a Yamaguchi esterification. This protocol allows for a highly convergent and flexible approach to these marine natural products.

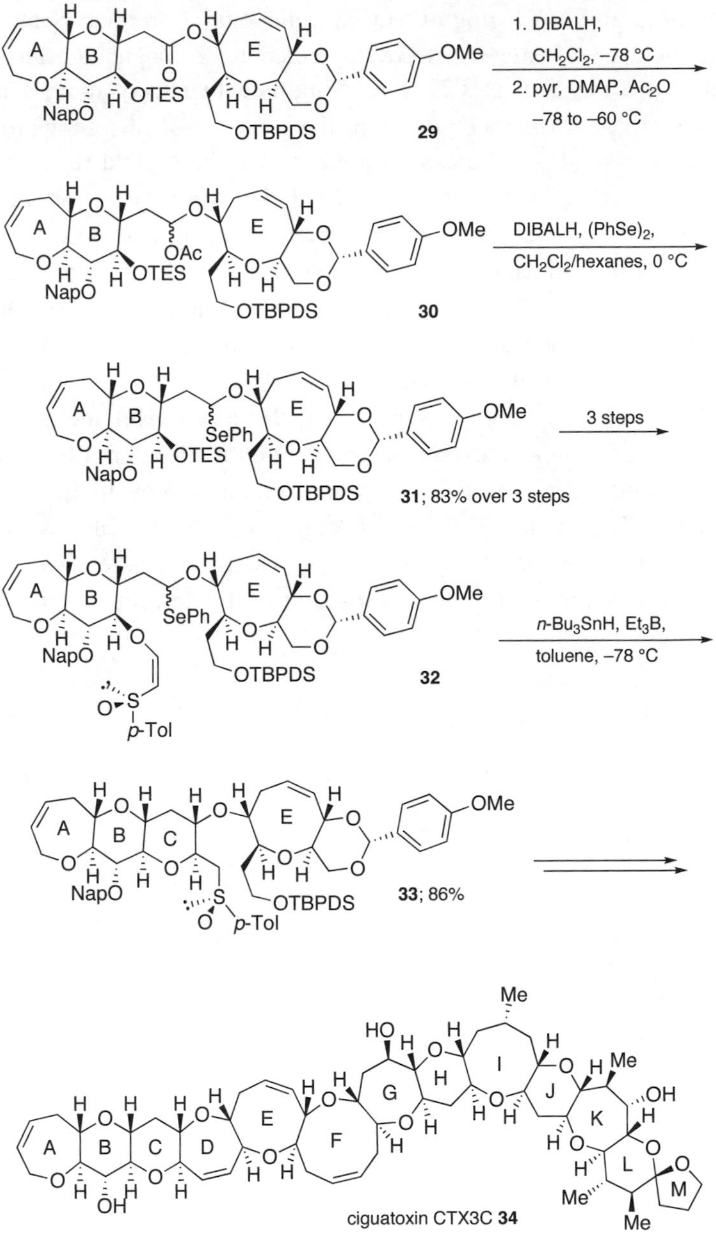

Scheme 4. Yamashita's Total Synthesis of Ciguatoxin CTX3C.

Intermolecular Couplings of α-Acetoxy Ethers and Silyl Enol Ethers

Crimmins and DeBaillie utilized reductive acetylation to install the α,β-unsaturated ketone moiety of the tetrahydropyran fragment of the marine cytotoxin bistramide A (**38**).[13] The synthetic strategy is outlined in Scheme 5. The lactone **35** was reduced with DIBALH followed by trapping with Ac$_2$O to give the α-acetoxy ether **36** in high yield (87%) and good diastereoselectivity (7:1). Treatment of α-acetoxy ether **36** with (*E*)-3-penten-2-one in the presence of excess TMSOTf led to formation of tetrahydropyran **37**, a key building block for the synthesis of bistramide A (**38**), in high yield (87%) and good diastereoselectivity (9:1).

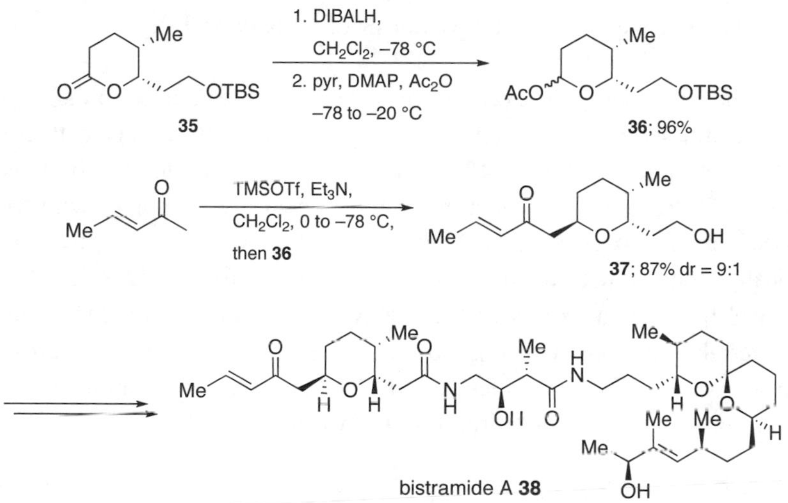

Scheme 5. Crimmins' Total Synthesis of Bistramide A.

Rychnovsky and co-workers used a related strategy to construct the western tetrahydropyran fragment of marine cytotoxin leucascandrolide A (**43**).[14] As shown in Scheme 6, the lactone **39** was reduced with DIBALH followed by trapping with Ac$_2$O to give the α-acetoxy ether **40** in high yield (91%) and high diastereoselectivity (15:1). Treatment of α-acetoxy ether **40** with silyl enol ether **41** in the presence of catalytic zinc bromide led to formation of tetrahydropyran **42** in high yield (92%) and good diastereoselectivity (20:1). Tetrahydropyran **42** was then further elaborated to leucascandrolide A (**43**).

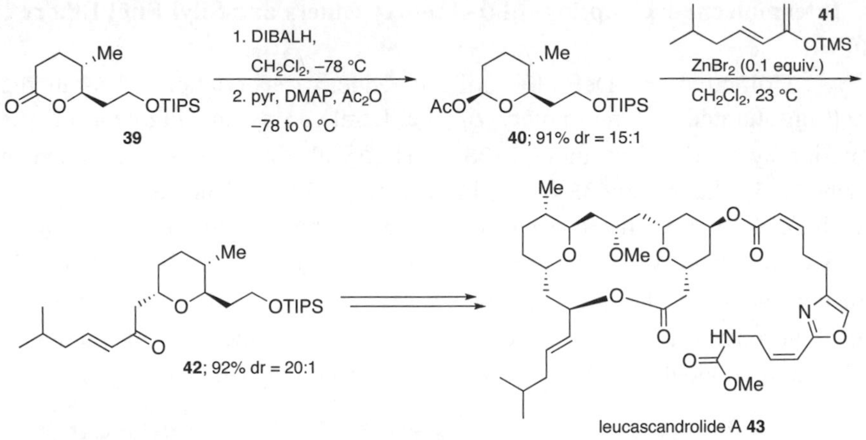

Scheme 6. Rychnovsky's Total Synthesis of Leuscandrolide A.

The most complex example of such intermolecular couplings comes from Crimmins and co-workers recent synthesis of the cancer cell growth inhibitor (+)-irciniastatin A (48).[15] This convergent strategy is outlined in Scheme 7. Reduction of lactone 44 followed by trapping with Ac₂O proceeds to give α-acetoxy ether 45 in quantitative yield. Treatment of the sterically hindered α-acetoxy ether 45 with boron trifluoride etherate followed by a slight excess of the highly functionalized silyl enol ether 46 gave tetrahydropyran 47 in good yield (59%; 85% based on recovered ketone) and high diastereoselectivity (>20:1). Tetrahydropyran 47 was then carried forward to afford (+)-irciniastatin A (48).

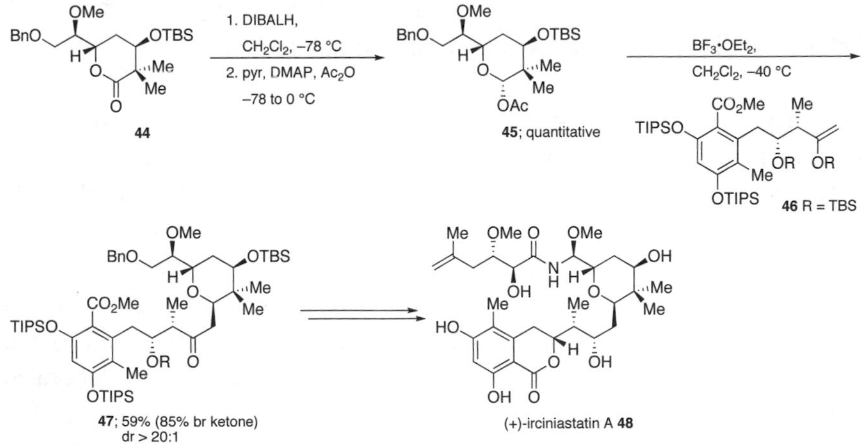

Scheme 7. Crimmins' Total Synthesis of Irciniastatin A.

Org. Synth. **2012**, *89*, 143-158

Reductive acetylation followed by intermolecular coupling has proven to be a useful strategy for coupling complex fragments in total synthesis.[16] The success of this methodology is largely due to the functional group compatibility of the reduction and mild conditions for oxocarbenium ion formation. Furthermore, the coupling itself often proceeds in high yield and high diastereoselectivity. These features validate this method as a highly convergent process in complex natural product synthesis.

1. Department of Chemistry, 1102 Natural Sciences II, University of California–Irvine, Irvine, CA 92697-2025. Support was provided by the National Institute of General Medicine (GM-43854) and the Vertex Scholar Fellowship (N.S.).
2. (a) Dahanukar, V. H.; Rychnovsky, S. D. *J. Org. Chem.* **1996**, *61*, 8317. (b) Kopecky, D. J.; Rychnovsky, S. D. *J. Org. Chem.* **2000**, *65*, 191. (c) Kopecky, D. J.; Rychnovsky, S. D. *Org. Synth.* **2003**, *80*, 177.
3. Zhang, Y.; Rovis, T. *Tetrahedron* **2003**, *59*, 8979.
4. Szemes, F., Jr.; Fousse, A.; Othman, R. B.; Bousquet, T.; Othman, M.; Dalla, V. *Synthesis* **2006**, 875.
5. Rychnovsky, S. D.; Crossrow, J. *Org. Lett.* **2003**, *5*, 2367.
6. Dalgard, J. E.; Rychnovsky, S. D. *J. Am. Chem. Soc*, **2004**, *126*, 15662.
7. Epstein, O. L.; Rovis, T. *J. Am. Chem. Soc*, **2006**, *128*, 16480.
8. (a) Doyle, A. G.; Jacobsen, E. N. *Abstract of Papers*, 232nd National Meeting of the American Chemical Society, San Francisco, CA, Sept 10-14, 2006; American Chemical Society: Washington, DC, 2006, ORGN-076. (b) Doyle, A. G. Ph.D. Thesis, Harvard University, Cambridge, MA, 2008. (c) Larghi, E. L.; Kaufman, T. S. *Eur. J. Org. Chem.* **2011**, 5195.
9. (a) Kadota, I.; Takamura, H.; Sato, K.; Ohno, A.; Matsuda, K.; Yamamoto, Y. *J. Am. Chem. Soc.* **2003**, *125*, 46. (b) Kadota, I.; Takamura, H.; Sato, K.; Ohno, A.; Matsuda, K.; Satake, M.; Yamamoto, Y. *J. Am. Chem. Soc.* **2003**, *125*, 11893.
10. (a) Takamura, H.; Kikuchi, S.; Nakamura, Y.; Yamagami, Y.; Kishi, T.; Kadota, I.; Yamamoto, Y. *Org. Lett.* **2009**, *11*, 2531. (b) Takamura, H.; Yamagami, Y.; Kishi, T.; Kikuchi, S.; Nakamura, Y.; Kadota, I.; Yamamoto, Y. *Tetrahedron* **2010**, *66*, 5329.
11. Yamashita, S.; Ishihara, Y.; Morita, H.; Uchiyama, J.; Takeuchi, K.; Inoue, M.; Hirama, M. *J. Nat. Prod.* **2011**, *74*, 357.

12. For reviews of convergent strategies in polycyclic ether synthesis see: (a) Inoue, M. *Org. Biomol. Chem.* **2004**, *2*, 1811. (b) Inoue, M. *Chem. Rev.* **2005**, *105*, 4379. (c) Sasaki, M.; Fuwa, H. *Nat. Prod. Rep.* **2008**, *25*, 401.

13. Crimmins, M. T.; DeBaillie, A. C. *J. Am. Chem. Soc.* **2006**, *128*, 4936.

14. Van Orden, L. J.; Patterson, B. D.; Rychnovsky, S. D. *J. Org. Chem.* **2007**, *72*, 5784.

15. Crimmins, M. T.; Stevens, J. M.; Schaaf, G. M. *Org. Lett.* **2009**, *11*, 3990.

16. Other examples include: Total synthesis of (+)-muconin (a) Yoshimitsu, T.; Makino, T.; Nagaoka, H. *J. Org. Chem.* **2004**, *69*, 1993. Total synthesis of (–)-panacene (b) Boukouvalas, J.; Pouliot, M.; Robichaud, J.; MacNeil, S.; Snieckus, V. *Org. Lett.* **2006**, *8*, 3597. Total synthesis of candicanoside A (c) Tang, P.; Yu, B. *Angew. Chem. Int. Ed.* **2007**, *46*, 2527. (d) Tang, P.; Yu, B. *Eur. J. Org. Chem.* **2009**, 259.

Professor Rychnovsky earned his B.S. from UC Berkeley in 1981 under the guidance of Paul Bartlett. He obtained his Ph.D. from Columbia University in 1986 with Gilbert Stork. NIH Postdoctoral Fellowships with David Evans at Harvard and Stuart Schreiber at Yale University preceded his academic career at the University of Minnesota in 1988. He was promoted to Associate Professor in 1994, and moved to UC Irvine as a full Professor in 1995. Professor Rychnovsky has received numerous honors including an Arthur C. Cope Scholar and was recently named an ACS Fellow. His research interests include natural product synthesis and the development of new synthetic methods.

Nicholas Sizemore was born in Akron, Ohio in 1981. He obtained a B.S. in Chemistry from Case Western Reserve University under the guidance of Prof. Philip Garner in 2005. He began his career as a medicinal chemist at Sanofi-Aventis in Exploratory and Internal Medicine working with Dr. James Pribish and Dr. Michael Fennie. In 2009, he joined the research group of Prof. Scott Rychnovsky at the University of California, Irvine. His research interests include total synthesis of complex natural products, computational chemistry and synthetic methodology.

Synthesis of 8,8-Dipropylbicyclo[4.2.0]octa-1,3,5-trien-7-one via Pd-catalyzed Intramolecular C-H Bond-Acylation

Submitted by Ruben Martin and Areli Flores-Gaspar.[1]
Checked by Hongqiang Liu and Mark Lautens.

1. Procedure

A. *2-(2-Bromophenyl)-2-propylpentanenitrile* (**1**). An oven-dried, 500-mL, round-bottomed flask containing 2-bromophenylacetonitrile (8.0 g, 40.8 mmol, 1.0 equiv) (Note 1) is equipped with an oval magnetic stirring bar (32 mm x 15 mm), argon inlet, and a rubber septum. An argon atmosphere is maintained throughout the reaction using an argon manifold system. The flask is charged through the septum via syringe with anhydrous THF (150 mL) (Note 2) and NaHMDS (122.4 mL, 122.4 mmol, 3.0 equiv) (Note 3) is added dropwise over a 4 min period, which results in the solution becoming brown. After stirring for 20 min, the flask is immersed in a room temperature water bath and 1-iodopropane (8.8 mL, 89.8 mmol, 2.2 equiv) (Note 4) is added dropwise over a 3 min period, resulting in a pale brown slurry. The reaction is followed by TLC analysis (Note 5). After stirring for 2.5 h, the septum is removed and saturated NH$_4$Cl solution (100 mL) is added (Note 6). The organic layer is separated using a 500-mL separatory funnel and the aqueous solution is extracted with diethyl ether (3 x 20 mL) (Note 7). The combined organic layers are dried over MgSO$_4$ (10 g) (Note 8), filtered and concentrated by rotary evaporation (from 760 mmHg to 26 mmHg, 40 °C). The residue is transferred to a 500-mL round-bottomed flask and dried for 4 h under vacuum (10 mmHg). The crude compound **1** thus obtained is used directly in the next step without further purification (Note 9).

2-(2-Bromophenyl)-2-propylpentanal (**2**). The previous 500-mL round-bottomed flask with the crude 2-(2-bromophenyl)-2-propylpentanenitrile (**1**) is equipped with an oval stirring bar (32 mm x 15 mm), argon inlet and a rubber septum. An argon atmosphere is maintained throughout the reaction using an argon manifold system. The flask is charged through the septum (*via* syringe) with anhydrous CH_2Cl_2 (150 mL) (Note 10) and immersed in a previously cooled dry ice/acetone bath at −78 °C (internal temperature) (Note 11). Then, DIBAL-H (45 mL, 45.0 mmol, 1.1 equiv) is added dropwise via syringe over a 15 min period (Note 12), resulting in a pale orange solution. The reaction progress is followed by TLC analysis (Note 13) or GC analysis (Note 14). After 2 h reaction time, additional DIBAL-H (20.4 mL, 20.4 mmol, 0.5 equiv) is added dropwise via syringe; after stirring for an additional 1 h, no more starting material is observed by TLC analysis (Note 15). The flask is removed from the cooled bath, and the reaction is then quenched by slow addition of ethyl acetate (150 mL) (Note 16) and 2M HCl (100 mL) (Note 17) over a 10 min period. The organic phase is separated using a 1-L separatory funnel and the aqueous solution is extracted with ethyl acetate (2 x 50 mL). The combined organic layers are dried over magnesium sulfate (8 g) and concentrated by rotary evaporation (from 760 mmHg to 26 mmHg, 40 °C). The residue is purified by column chromatography on silica gel (Note 18). The title compound is thus obtained as a yellow oil (8.00–8.09 g, 69–70% yield) (Note 19).

B. *8,8-Dipropylbicyclo[4.2.0]octa-1,3,5-trien-7-one* (**3**). An oven-dried 500-mL Schlenk flask equipped with a reflux condenser and a stirring bar (32 mm x 15 mm) is charged with $Pd(OAc)_2$ (134.7 mg, 2.0 mol%) (Note 20), *rac*-BINAP (560.4 mg, 3.0 mol%) (Note 21), Cs_2CO_3 (11.73 g, 36.01 mmol, 1.2 equiv) (Note 22). The Schlenk flask is evacuated and back-filled with argon (this sequence was repeated three times over a period of 3 min). Under an argon atmosphere, the 1,4-dioxane (150 mL) solution of *2-(2-bromophenyl)-2-propylpentanal* (**2**) (8.50 g, 30.01 mmol), (Note 23) is then added by syringe. The mixture is then placed in a pre-heated oil bath (Note 24) at 110 °C for 22 h under argon atmosphere, resulting in a black slurry. The mixture is then allowed to cool to room temperature, diluted with EtOAc (3 x 50 mL) and filtered through a Celite® plug (19.4 g, 50 mL) (Note 25) eluting with additional EtOAc (2 x 30 mL). The filtrate is

160

concentrated and purified by column chromatography on silica gel (Note 26), obtaining 4.24 g (20.96 mmol, 70% yield) of the title compound as a yellow oil (Notes 27 and 28).

2. Notes

1.   2-Bromophenylacetonitrile (97%) was purchased from Alfa Aesar and used as received.

2.   THF was distilled from Na/benzophenone ketyl. Submitters used THF anhydrous (content in H_2O <10 ppm) that was dried from an Instrument Solvent Purification System (MBraun-SPS).

3.   NaHMDS (1.0 M in THF) was purchased from Aldrich and used as received.

4.   1-Iodopropane (99%) was purchased from Aldrich and used as received.

5.   TLC analysis (performed using EMD TLC silica gel 60 F254 plates thin-layer chromatography) using hexanes:EtOAc (95:5) as the eluent; visualization with $KMnO_4$ stain; 2-bromophenylacetonitrile: R_f= 0.49, mono-alkylated product: R_f=0.79 and compound 1: R_f=0.89

6. NH_4Cl was purchased from ACP; the solution was prepared using 110 g of NH_4Cl and 100 mL of distilled water.

7.   Diethyl ether (stabilized with ~1 ppm of 2,6 di-*tert* butyl-*p*-cresol) was purchased from Caledon and used as received.

8.   Magnesium sulfate anhydrous was purchased from ACP and used as received.

9.   Crude compound 1 has the following properties: Brown oil; ^{1}H NMR (400 MHz, $CDCl_3$) δ: 0.92 (t, *J* = 7.2 Hz, 6 H), 1.06–1.19 (m, 2 H), 1.39–1.52 (m, 2 H), 1.97 (ddd, *J* = 14.0, 12.4, 4.4 Hz, 2 H), 2.61 (ddd, *J* = 14.0, 12.0, 4.4 Hz, 2 H), 7.16 (ddd, *J* = 8.0, 7.2, 1.2 Hz, 1 H), 7.32 (ddd, *J* = 8.0, 7.2, 1.2 Hz, 1 H), 7.61 (dd, *J* = 8.0, 1.2 Hz, 1 H), 7.72 (dd, *J* = 8.0, 1.6 Hz, 1 H). ^{13}C NMR (100 MHz, $CDCl_3$) δ: 13.9, 18.9, 39.3, 50.9, 120.3, 123.0, 127.6, 129.2, 131.5, 135.3, 135.9.

The pure compound 1 was prepared following the procedure *A* using 2-bromophenylacetonitrile (1.46 g, 7.44 mmol, 1.0 equiv), 20 mL of anhydrous THF, NaHMDS (22 mL, 22 mmol, 3.0 equiv), 1-iodopropane (1.60 mL, 16.4 mmol, 2.2 equiv). Column chromatography was performed on 75 mL of Silica gel 230-400 mesh SiliaFlash®P60, purchased from Silicycle. It was wet packed in a 3 cm diameter column using hexanes/ethyl

Acetate: 90/10 and the crude material was directly loaded to the column (the remaining residue was loaded in the minimal amount of hexanes/ethyl acetate 90/10). 10 mL fractions were collected at 0.15 mL/s rate, eluting with hexanes/ethyl Acetate: 90/10. All the fractions (9 to 18) containing the desired product were combined and concentrated by rotary evaporation (from 760 mmHg to 26 mmHg, 40 °C), and dried under high vacuum (10 mmHg), to yield 1.77 g (9.23 mmol, 85% yield) of the title compound as a brown oil. ^{1}H NMR (300 MHz, CDCl$_3$) δ: 0.92 (t, J = 7.2 Hz, 6 H), 1.04–1.21 (m, 2 H), 1.37–1.54 (m, 2 H), 1.97 (ddd, J = 16.8, 12.0, 4.5 Hz, 2 H), 2.61 (ddd, J = 16.8, 12.3, 4.5 Hz, 2 H), 7.16 (ddd, J = 8.7, 7.5, 1.5 Hz, 1 H), 7.32 (ddd, J = 8.7, 7.5, 1.2 Hz, 1 H), 7.60 (dd, J = 7.8, 1.2 Hz, 1 H), 7.71 (dd, J = 7.8, 1.5 Hz, 1 H). ^{13}C NMR (75 MHz, CDCl$_3$) δ: 13.9, 18.9, 39.2, 50.8, 120.2, 123.0, 127.5, 129.2, 131.5, 135.2, 135.9. IR (neat, cm^{-1}): 2961, 2932, 2874, 2359, 2233, 1629, 1470, 1391, 1021, 758. HRMS calcd. for (C$_{14}$H$_{18}$BrN+NH$_4$): 297.0966, Found 297.0953. Anal. calcd. for C$_{14}$H$_{18}$BrN: C, 60.01; H, 6.47; N, 5.00 Found: C, 59.80; H, 6.46; N, 4.89.

10. Dichloromethane anhydrous (content in H$_2$O <10 ppm) was dried from an Instrument Solvent Purification System (MBraun-SPS).

11. Submitter used an immersion cooler HAAKE EK90 with methanol bath for –78 °C.

12. Diisobutyl aluminiumhydride (DIBAL-H), 1M solution in hexane, Sureseal ^{TM} was purchased from Aldrich and used as received. Submitters used DIBAL-H, 1M solution in hexane, Acroseal^{TM} that was purchased from Acros Organics and titrated before use with the following procedure: Hoye, T. R.; Aspaas, A. W.; Eklov, B. M.; Ryba, T. D. *Org. Lett.* **2005**, *7*, 2205.

13. TLC is run twice using a mixture of Hexanes:EtOAc (40:1) as eluent (compound **1**: R_f= 0.50, compound **2** R_f=0.46), using a KMnO$_4$ stain.

14. Submitters also used GC to monitor the reaction progress. Compounds **1** and **2** are easily distinguished by GC: t_{r1}=5.04 min; t_{r2}=5.24 min.

GC-method (Agilent 19091J-413): Initial Temp: 70°C; Maximum Temp: 300°C; Initial Time: 1.0 min, Equilibration Time: 3.0 min; Ramp: Rate = 50.0 °/min; Final Temp = 250 °C, Final Hold Time= 1.50 min; Run Time: 6.10 min; Pressure: 10.10 psi; Split flow: 97.1 mL/min; Gas type: Helium; Capillary column: HP-5, 5% phenyl methyl siloxane

15. Addition of DIBAL-H at the start of experiment rather than semi-batch addition led to incomplete reduction in a 3 h period.

16. Ethyl acetate was purchased from Fisher Scientific and used as received.

17. HCl (37-38%) was purchased from Fisher Scientific; a 2M HCl solution is preparing by adding 16.7 mL of HCl (37-38%) to a 250-mL volumetric flask containing 83.3 mL of distilled water.

18. Column chromatography was performed on 260 mL of silica gel (230-400 mesh SiliaFlash®P60), purchased from Silicycle. It was wet packed in a 5-cm diameter column using hexanes/ethyl acetate (96/4) and the crude material was directly loaded to the column (The remaining residue was loaded in the minimal amount of hexanes/ethyl acetate (96/4)). Fractions of 30 mL were collected at 0.5 mL/s rate, eluting with hexanes/ethyl acetate (96/4). All fractions (10-18) containing the desired product were combined, concentrated by rotary evaporation (from 760 mmHg to 26 mmHg, 40 °C), and dried overnight at 10 mmHg. In order to avoid any decomposition, compound 2 was kept under argon atmosphere.

19. Compound 2 has the following physical properties: ^{1}H NMR (400 MHz, CDCl$_3$) δ: 0.88 (t, J = 7.2 Hz, 6 H), 0.97–1.11 (m, 2 H), 1.16–1.30 (m, 2 H), 2.00 (ddd, J = 26.0, 12.0, 4.8 Hz, 4 H), 7.13–7.19 (m, 1 H), 7.33–7.38 (m, 2 H), 7.60 (d, J = 8.0 Hz, 1 H), 9.86 (s, 1 H). ^{13}C NMR (100 MHz, CDCl$_3$) δ: 14.9, 17.0, 35.1, 58.6, 123.9, 127.5, 129.1, 130.4, 135.0, 140.2, 204.4. IR (neat, cm^{-1}): 2957, 2872, 1716, 1564, 1466, 1432, 1380, 1264, 1167, 1113, 1067, 1029, 971. HRMS calcd. for (C$_{14}$H$_{19}$BrO+NH$_4$): 300.0963, Found 300.0959. Anal. calcd. for C$_{14}$H$_{19}$BrO: C, 59.37; H, 6.76. Found: C, 59.42; H, 6.91. Submitters also determined the purity of 2 using GC analysis. The range of yield for different runs is from 63% to 70%.

20. Pd(OAc)$_2$ (min. 98%; 99.9% Pd) was purchased from Strem Chemicals and used as received. Submitters noted that Pd(OAc)$_2$ (99.98% (metal basis); Pd 47% min) purchased from Alfa-Aesar gave similar efficiency.

21. Racemic 2,2'-bis(diphenylphosphino)-1,1'-binaphthyl 98% was purchased from Strem Chemicals and used as received. Submitters noted that racemic 2,2'-bis(diphenylphosphino)-1,1'-binaphthyl 98% from Atomax chemicals gave similar efficiency.

22. Cs$_2$CO$_3$ 99.9% [metal basis] was purchased from Aldrich and was stored in the glove box. The exact amount of cesium carbonate was weighed out inside the glove box and then added to the reaction mixture under an

argon stream outside the glove box. Submitters used Cs_2CO_3 99% [metal basis] purchased from Alfa Aesar.

23. 1,4-Dioxane was distilled over sodium, and used directly without degassing. Submitters used dioxane anhydrous, 99.8% that was purchased from Sigma Aldrich. Instead of the addition of a dioxane solution of 2-(2-bromophenyl)-2-propylpentanal (2), submitters added 2 to the flask at the start of experiment followed by air exclusion and addition of dioxane.

24. Oil Bath: silicone oil δ=0.97, was purchased from Fisher Scientific and used as received (working temperature from –40 °C to +200 °C).

25. Celite® 545 coarse was purchased from Sigma-Aldrich and used as received.

26. Column chromatography was performed on 500 mL of silica gel 230-400 mesh SiliaFlash®P60, purchased from Silicycle. The column was wet packed in a 8-cm diameter column with hexanes and the crude material was directly loaded to the column (The remaining residue was loaded in the minimal amount of hexanes/ethyl acetate (10/1)). Fractions of 30 mL were collected at 0.8 mL/s rate eluting with the following gradient: 500 mL hexane, 850 mL hexanes/ethyl acetate: 30/1, 300 mL hexanes/ethyl acetate: 20:1, 400 mL hexanes/ethyl acetate: 10/1, 300 mL hexanes/ethyl acetate: 5/1. The fractions (10-17) containing compound 3 (R_f=0.65; hexanes:EtOAc (90:10)) are collected, combined and concentrated by rotary evaporation (from 760 mmHg to 26 mmHg, 40 °C). In order to avoid any decomposition, compound 3 was kept under argon atmosphere.

27. Compound 3 has the following physical properties: ^{1}H NMR (400 MHz, $CDCl_3$) δ: 0.83 (t, J = 7.2 Hz, 6 H), 1.13–1.30 (m, 4 H), 1.74 (ddd, J = 8.8, 6.4, 2.4 Hz, 4 H), 7.32 (dt, J = 6.8, 0.8 Hz, 1 H), 7.37 (td, J = 6.8, 0.8 Hz, 1 H), 7.42 (dt, J = 6.8, 0.8 Hz, 1 H), 7.48 (td, J = 6.8, 0.8 Hz, 1 H). ^{13}C-NMR ($CDCl_3$, 100 MHz) δ: 14.7, 19.1, 37.4, 74.2, 120.9, 123.2, 129.2, 135.1, 146.1, 160.6, 197.1. IR (neat, cm^{-1}): 3064, 2958, 2873, 2845, 1754, 1582, 1461, 1441, 1379, 1274, 1142, 1092, 926. HRMS calcd. for ($C_{14}H_{19}O+NH_4$): 203.1436, Found 203.1433. Anal. calcd. for $C_{14}H_{18}O$: C, 83.12; H, 8.97. Found: C, 82.97; H, 8.79. The range of yields for different runs is from 67% to 71%.

28. The only by-product generated in the reaction is [(1E)-1-propylbut-1-enyl]benzene (4) as a mixture of diastereoisomers (16:1, favoring E isomer, as judged by NOESY) in 9.0% yield. This by-product elutes prior to the main fraction in the column chromatography and is readily removed (R_f=0.90; hexanes:EtOAc (90:10)). Compound 4 has the following physical

properties: Colorless oil; ^{1}H-NMR (400 MHz, CDCl$_3$) δ: 1.06 (t, J = 7.6 Hz, 3 H), 1.23 (t, J = 7.6 Hz, 3 H), 1.51–1.60 (m, 2 H), 2.38 (q, J = 7.2 Hz, 2 H), 2.65 (t, J = 7.2 Hz, 2 H), 5.83 (t, J = 7.2 Hz, 1 H), 7.33–7.37 (m, 1 H), 7.42–7.46 (m, 2 H), 7.49–7.52 (m, 2 H); ^{13}C NMR (100 MHz, CDCl$_3$) δ: 13.9, 14.4, 21.8, 21.9, 31.6, 126.32, 126.34, 128.1, 130.9, 139.4, 143.4. IR (neat, cm^{-1}): 3058, 2930, 2871, 1599, 1491, 1457, 1443, 1377, 1074, 1030, 754, 697. HRMS Calcd for (C$_{13}$H$_{19}$): 175.14868, Found 175.14834. Anal. Calcd for C$_{13}$H$_{18}$: C, 89.59; H, 10.41. Found: C, 89.39; H, 10.33.

Waste Disposal Information

All hazardous materials should be handled and disposed of in accordance with "Prudent Practices in the Laboratory"; National Academy Press; Washington, DC, 1995

3. Discussion

Benzocyclobutenones are an intriguing class of four-membered ring ketones that have been used extensively as powerful synthetic intermediates in organic synthesis[2]. The reactivity of benzocyclobutenones is primarily associated to their unique high electrophilicity of the carbonyl unit, allowing

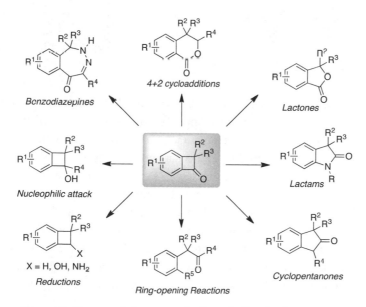

Figure 1. Benzocyclobutenones in Organic Synthesis

a myriad of different transformations, ranging from classical 1,2-additions, ring-expansion, ring-opening reactions, cycloadditions, heterocycle synthesis or preparation of complex benzocyclobutanes via reduction of the carbonyl backbone, among many others (Figure 1).

Despite their potential as synthetic intermediates, benzocyclobutenones are elusive compounds to prepare in a straightforward and general fashion. To the best of our knowledge, the synthesis of benzocyclobutenones is usually accomplished via two different routes: (1) intramolecular addition of organolithium or Grignard reagents to Weinreb amides (route 1, Figure 2)[3] or (2) [2+2]-cycloaddition of silyl enol ethers and benzyne (route 2, Figure 2),[4] as elegantly described by Suzuki and coworkers. The application profile of these methods, unfortunately, is quite limited, as only a limited set of substitution patterns can be accessed; additionally, these procedures do not tolerate the presence of functional groups, as stochiometric amounts of organolithium derivatives are required.

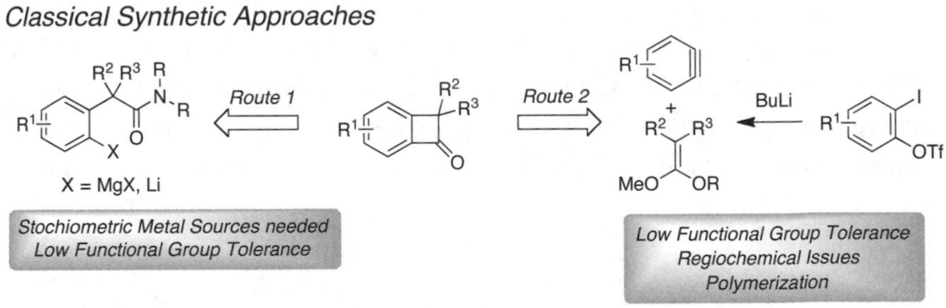

Classical Synthetic Approaches

Figure 2. Alternative Pathways to Benzocyclobutenones

In recent years, metal-catalyzed C-H bond-functionalization strategies have become one of the most popular areas of research in organic (organometallic) chemistry.[5] The attractiveness of such methodologies is based on the ability to build up molecular complexity from rather inert and abundant C-H bonds, thus allowing unconventional and elegant bond disconnection strategies for assembling valuable organic structures. Our group has recently reported that benzocyclobutenones can be prepared by intramolecular Pd-catalyzed acylation via C-H bond-functionalization.[6] Such an approach has the advantage of using readily available precursors and controlling the substitution pattern over the aryl backbone; additionally, the method tolerates a wide range of functional groups, allowing for the first

166

time, the preparation of highly complex benzocyclobutenones in a straightforward manner. The scope of this procedure is illustrated in Table 1.

Table 1. Pd-catalyzed Synthesis of Benzocyclobutenones via C-H Bond-Functionalization

Herein, we describe the preparation of the model compound [8,8-dipropylbicyclo[4.2.0]octa-1,3,5-trien-7-one], thus illustrating the simplicity of our new protocol for the direct conversion of commonly employed and readily available α-aryl aldehydes into benzocyclobutenones. This user-friendly methodology nicely complements the existing routes to benzocyclobutenones in the literature.[3,4] Given the practicality and flexibility, it is expected that this method will find immediate application in advanced organic synthesis.

1.   Institute of Chemical Research of Catalonia (ICIQ), AV. Països Catalans 16, 43007 Tarragona, Spain, E-mail: rmartinromo@iciq.es; We thank ICIQ Foundation, Consolider Ingenio 2010 (CSD2006-0003) and MICINN (CTQ2009-13840) for financial support. Johnson Matthey, Umicore, and Nippon Chemical Industrial are acknowledged for gifts of metal and ligand sources. R. M and A. F-G thank MICINN for RyC and FPU fellowships.
2.   (a) Sadana, A. K.; Saini, R. K.; Billups, W. E. *Chem. Rev.* **2003**, *103*, 1539. (b) Bellus, D.; Ernst, B. *Angew. Chem., Int. Ed.* **1988**, *27*, 797.
3.   Aidhen, I. S.; Ahuja, J. R. *Tetrahedron Lett.* **1992**, *33*, 5431.

4. (a) Hosoya, T.; Hasegawa, T.; Kuriyama, Y.; Matsumoto, T.; Suzuki, K. *Synlett* **1995**, 177. (b) Stevens, R. V.; Bisacchi, G. S. *J. Org. Chem.* **1982**, *47*, 2393.

5. For reviews, see: (a) Daugulis, O.; Do, H. –Q.; Shabashov, D. *Acc. Chem. Res.* **2009**, *42*, 1074. (b) Chen, X.; Engle, K. M.; Wang, D. –H.; Yu, J. –Q. *Angew. Chem., Int. Ed.* **2009**, *48*, 5094. (c) Ackermann, L.; Vicente, R.; Kapdi, A. R. *Angew. Chem. Int. Ed.* **2009**, *48*, 9792. (d) Colby, D. A.; Bergman, R. G.; Ellman, J. A. *Chem. Rev.* **2010**, *110*, 624.

6. Álvarez-Bercedo, P.; Flores-Gaspar, A.; Correa, A.; Martin, R. *J. Am. Chem. Soc.* **2009**, *132*, 466.

Appendix
Chemical Abstracts Nomenclature; (Registry Number)

2-Bromophenylacetonitrile: Benzeneacetonitrile, 2-bromo-; (19472-74-3)

NaHMDS: Silanamine, 1,1,1-trimethyl-N-(trimethylsilyl)-, sodium salt (1:1); (1070-89-9)

1-Iodopropane: Propane, 1-iodo-; (107-08-4)

DIBAL-H: Aluminum, hydrobis(2-methylpropyl)-; (1191-15-7)

Pd(OAc)$_2$: Acetic acid, palladium (2+) salt (2:1); (3375-31-3)

Rac-BINAP: Phosphine, 1, 1'-[1,1'binaphthalene]-2,2'-diylbis[1,1-diphenyl- ; (98327-87-8)

Cs$_2$CO$_3$: Carbonic acid, cesium salt (1:2); (534-17-8)

2-(2-Brommophenyl)-2-propylpentanal: Benzeneacetaldehyde, 2-bromo-α,α-dipropyl-; (1206450-98-7)

[8,8-Dipropylbicyclo[4.2.0]octa-1,3,5-trien-7-one]: Bicyclo[4.2.0]octa-1,3,5-trien-7-one, 8,8-dipropyl-; (1206451-50-4)

[(1*E*)-1-Propylbutyl-1-enyl]benzene: Benzene, [(1*E*)-1-propyl-1-buten-1-yl]-; (1151654-13-5)

Ruben Martin was born in 1976. He received his Ph.D in 2003 at the Universitat de Barcelona with Prof. Antoni Riera. After two postdoctoral stages at the Max-Planck-Institut für Kohlenforschung with Prof. Alois Fürstner and at the Massachusetts Institute of Technology with Prof. Stephen L. Buchwald, he initiated his independent career in 2008 at the Institute of Chemical Research of Catalonia (ICIQ). His interests are primarily focused on the metal-catalyzed activation of inert bonds.

Areli Flores-Gaspar was born in Mexico, City. She did her Bachelor studies and a M.Sc. at the Universidad Nacional Automoma de México. She is currently a Ph.D. student in Dr. Ruben Martin's group at Institut Català d'Investigació Química in Tarragona, Spain. Her work is focused on the development of novel synthetic transformations based upon C-H bond-activation protocols.

Hongqiang Liu was born in China. He did his B. Sc. and M. Sc, at Peking University and Peking Union Medical College respectively. He finished his Ph.D. degree in 2010 under supervision of Dr. John Vederas at the University of Alberta. He is currently a postdoctoral fellow at Dr. Mark Lautens group at the University of Toronto. His work is focused on molecular motor synthesis using palladium-catalyzed and norbornene-mediated domino reactions.

Discussion Addendum for:
Ring-closing Metathesis Synthesis of
N-Boc-3-pyrroline

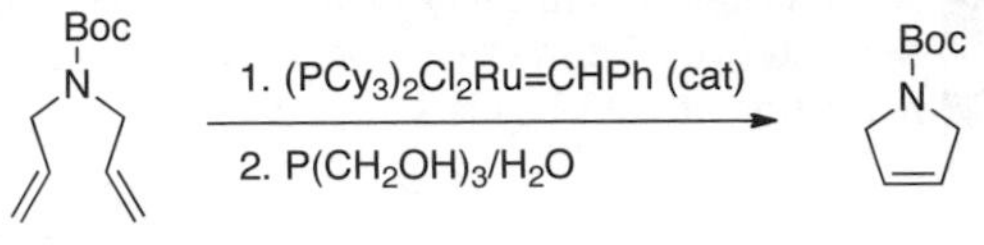

Prepared by Daniel J. O'Leary,[1] Richard Pederson,[2] and Robert H. Grubbs.*[3]
Original article: O'Leary, D. J.; Ferguson, M. L.; Grubbs, R. H. *Org. Synth.* **2003**, *80*, 85.

The olefin metathesis reaction has emerged as a widely used transformation in organic chemistry and materials science.[4] In keeping with the theme of our original article, this discussion addendum provides an overview of current large-scale ring-closing metathesis (RCM) applications of the Ru-based family of metathesis catalysts, the structures of which are shown in Figure 1.[5]

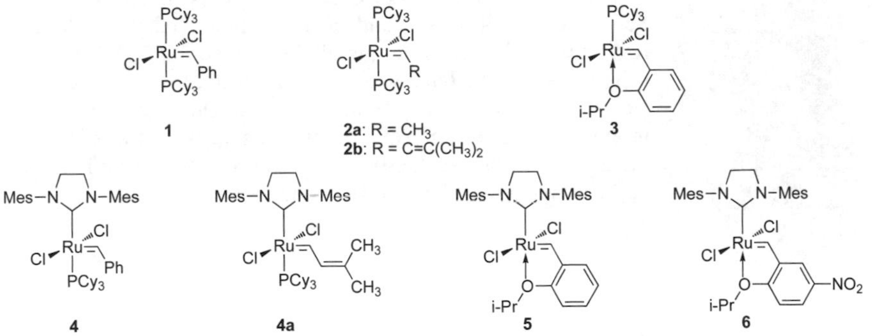

Figure 1. Commonly Used Olefin Metathesis Catalysts.

In our 2003 report for the preparation of *N*-Boc-3-pyrroline, we utilized 0.5 mol% of catalyst **1** in a refluxing (2.5 h) 0.4 M solution of *N*-Boc-diallylamine in CH_2Cl_2. The report also described an extractive method for removing Ru-derived catalytic species/impurities using the water-soluble

Org. Synth. **2012**, *89*, 170-182
Published on the Web 10/18/2011

P(CH$_2$OH)$_3$, readily prepared from a commercially available fire-retardant aqueous formulation of P(CH$_2$OH)$_4$Cl. Following Kugelrohr distillation, this method reliably provided ca. 30 g of crystalline N-Boc-3-pyrroline in 90-94% isolated yield. Concurrent with the development of our procedure, Helmchen reported[6] a 100 g preparation of Boc-3-pyrroline using 0.1 mol% of ethylidene **2a** in a room-temperature (15 h) 0.57 M CH$_2$Cl$_2$ solution of N-Boc-diallylamine. Catalyst deactivation or removal procedures were not taken and the product was isolated in 98% yield after vacuum distillation. This transformation has also been reported to proceed in 87% yield when run neat and with catalyst **5** loading as low as 500 ppm.[7]

Since the time of our report, however, RCM has rapidly matured and has been used to prepare kilogram quantities of structurally complex macrocyclic synthetic intermediates within the pharmaceutical industry. Optimizing a large-scale RCM process requires that several conditions be studied, namely, substrate type and functional group compatibility, catalyst selection, and solvent/temperature/time requirements. In many applications, removal of residual ruthenium is also a concern. In this Discussion Addendum, large-scale industrial RCM processes will be used to frame the relevant issues. These examples include Boehringer-Ingelheim's Hepatitis C protease inhibitor BILN-2061, GlaxoSmithKline's cathepsin K inhibitor SB-462795, and RCM-tethered macrocyclic peptides. The discussion will conclude with an overview of developments in the area of ruthenium removal.

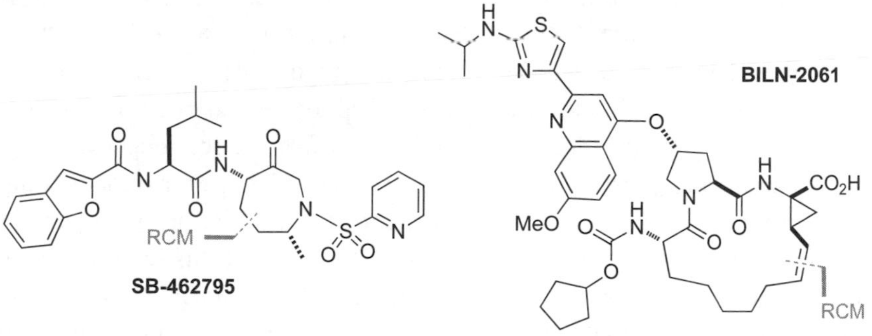

Figure 2. SB-462795 and BILN-2061 and their RCM disconnections.

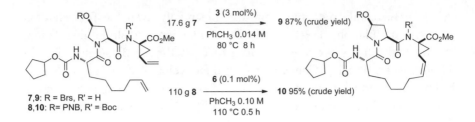

Figure 3. Comparison of the 2006 and 2009 RCM processes for construction of the BILN-2061 peptide macrocycle.

Boehringer-Ingelheim's 2006 disclosure[8-13] of a large-scale RCM approach to the hepatitis C viral protease inhibitor BILN-2061 provides an interesting case study for those contemplating a large-scale metathesis process. The BILN-2061 metathesis substrate is an unnatural tripeptide-derived diene, which closes to a (Z)-olefin-containing 15-membered ring. Two of the conformation-restraining amide bonds are endocyclic, and two additional macrocyclic torsional degrees of freedom are fixed by the presence of the five-membered proline ring and an endocyclic *trans*-cyclopropane. The optimized 2006 RCM process (Figure 3) was scaled up to produce >400 kg of cyclized product.[10] Three years later, the BI group reported a higher-yielding and order-of-magnitude greener procedure, one which dramatically reduced the amount of time, solvent, and catalyst loading.[13]

The first efforts to construct the BILN-2061 macrocycle used catalyst **1**.[9] These trials were met with problematic epimerization at the β-carbon of the vinylcyclopropyl amino acid residue (Figure 4). When a mono-phosphine catalyst such as **3** was used, the epimerization reaction was not observed. Analysis of model systems revealed that the rearrangement was facile when the bis-phosphine catalyst **1** was used. The rearrangement could also be promoted by adding one equivalent of phosphine to a mixture of catalyst **3** and the model system (Figure 4). The origin of this rearrangement was suggested to occur via the intermediacy of bis-phosphino or amino phosphino ruthenacyclopentene intermediates.[9,11] Further studies revealed that the site of catalyst initiation in the macrocycle precursor depended upon the substitution of the aminocylopropylcarboxylic acid nitrogen atom: acyl-type substituents such as *N*-Boc shifted the initiation to the rearrangement-stable nonenoic residue (Figure 4).[12] Before this detail was known, the reasonably efficient 2006 process using diene **7** and catalyst **3** was plagued

172

by problematic batch inconsistencies. Traces of morpholine (<20 ppm) in the toluene solvent were shown to be responsible for substrate isomerization and catalyst inhibition, with the former arising from bis-ligated Ru species of the type discussed earlier. These problems were alleviated by acid-washing the solvent prior to the metathesis reaction.

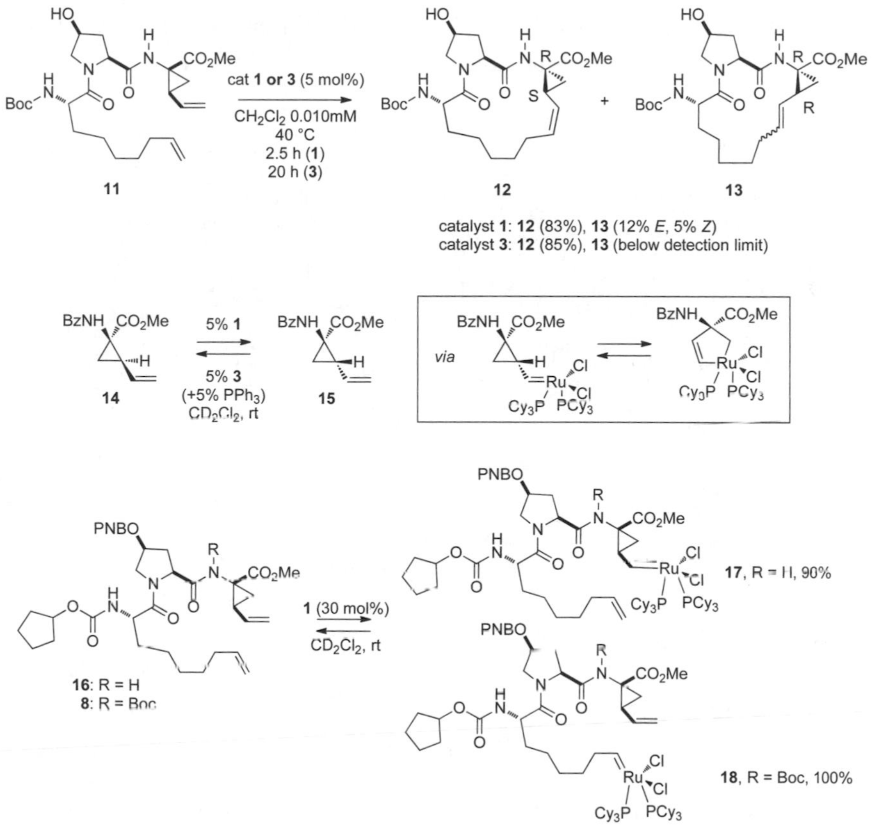

Figure 4. Catalyst-specific epimerization observed in RCM reactions of BILN-2061 substrates, suggested ruthenacyclopentene intermediates, and the effect of N-Boc substitution on site of catalyst initiation.

The aminocyclopropyl metathesis rearrangement encountered by the BI group was unexpected and highly substrate-specific. The most common side reactions in metathesis processes are alkene isomerization reactions.[14] Because of the reversible nature of olefin metathesis, isomerization can alter unreacted starting material or products. It can be problematic in reactions where the catalyst is stressed by temperature or time or in some product purifications involving direct distillation from Ru-containing pot residues.[15] For example, ring closure of diallyl ether **19** to **20** with catalyst **4** (Figure 5) is a rapid reaction, but significant product isomerization to vinyl ether **21** can occur at extended reaction times. Rearrangement is thought to arise from Ru-H species; it can be suppressed by solvent selection[16,17] or by additives such as tricyclohexylphosphine oxide,[18] acetic acid or 1,4-benzoquinones,[19] phenylphosphoric acid,[20] chlorocatechol borane,[21] or Cy_2BCl.[22]

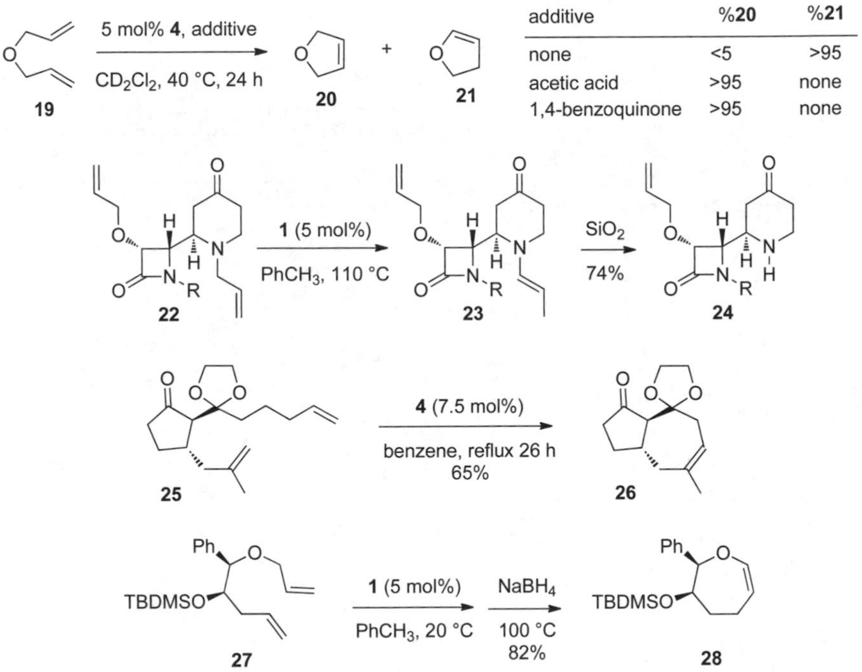

additive	%20	%21
none	<5	>95
acetic acid	>95	none
1,4-benzoquinone	>95	none

Figure 5. Select rearrangements observed in Ru-catalyzed metathesis reactions.

Rearrangement chemistry is the desired outcome in some applications. For example, Ru-catalyzed rearrangements are used for selective *O*-allyl[23] and *N*-allyl[24] deprotections. The latter, an allylamine to enamine transformation, was observed in lactam-piperidone **22** when treated with catalyst **1** in refluxing toluene (Figure 5). Olefin migration can also occur in hydrocarbon chains. For example, attempts to ring-close cyclopentanone-derived diene **25** with catalyst **4** did not return the [9.5.0] ring system.[25] Instead, the monosubstituted double bond was found to migrate at a rate faster than 9- or 8-membered ring closure, leading to exclusive generation of the [7.5.0] product **26**. Another strategy purposefully adds a hydride source to promote rearrangement after ring closure. For example, enol ether **28** is formed in 82% yield by first performing an RCM reaction on diene **27** to form the seven-membered ring product, which is then isomerized in a second step involving addition of $NaBH_4$ to the Ru-containing reaction mixture.[26]

Non-metathesis side reactions have also been reported in the context of Ru-promoted metathesis reactions. Hoye and Zhao have reported fragmentation reactions in RCM substrates containing secondary allylic alcohols; in sluggish metathesis reactions these systems can fragment to the methyl ketone.[27] Kharasch additions of $CHCl_3$ to olefins, as catalyzed by **1** at 65 °C, are also known.[28]

Although the problem of isomerization was largely solved and the RCM reaction was utilized at kg scale, the 2006 BILN-2061 process suffered from several other drawbacks. The first of these was the dilute substrate concentration (0.014 M), which required excessive solvent consumption. A study of the macrocyclization behavior of different acyclic dienes and dimeric intermediates revealed a 27-fold improvement in the effective molarity (EM) of cyclization, which translated to a 0.2 M RCM process.[12] A significant shortening of the reaction time was also realized. As mentioned earlier, an *N*-Boc group at the vinylcyclopropane amide residue was found to productively shift the site of catalyst initiation and led to a more facile and robust ring-closing reaction. While adding protection-deprotection steps, the improved 2009 process could therefore be run at higher concentrations and temperatures and with lower catalyst loadings (0.1 mol% **6** vs. 3 mol% **3**). Each of these factors contributed in a positive manner to the EM of cyclization. Although the details will not be considered here, one of the principle steps in optimizing the BILN-2061 macrocyclization was defining conditions which (i) minimized formation of dimeric compounds arising from acyclic diene metathesis (ADMET) processes,[29,30] and (ii) one in which these oligomers could be equilibrated to product.

The BILN-2061 RCM reaction optimization is best viewed in terms of green chemistry considerations. The 2009 process[13] represented a significant savings in terms of time, energy, and material. The RCM step was reduced to minutes instead of hours. Scrupulous degassing and solvent purification was necessary for the 2006 process, whereas the new process only required a brief degassing boil-out. The 2006 RCM process required as much as 150,000 L solvent per metric ton of diene, whereas the 2009 process reduced this amount to 7,500 L/MT. Because catalyst loading was greatly reduced in the 2009 process, Ru removal was likewise less resource-intensive. The 2006 process used excess 2-mercaptonicotinic acid (2 kg/1 kg diene) and aqueous bicarbonate to remove a portion of the soluble ruthenium, with subsequent silica gel and charcoal filtrations necessary to reduce Ru levels to 10 ppm in the active pharmaceutical ingredient (API). In contrast, the 2009 process used 50-fold less 2-mercaptonicotinic acid and did not utilize silica or charcoal filtrations. The Sheldon E-factor,[31] which is the ratio of waste mass to product mass, improved to 52 (computed for the 2009 Boc protection/RCM/Boc deprotection process) from 370 (computed for the 2006 RCM process).

Large-scale RCM was also used in GlaxoSmithKline's 2009 synthesis of the seven-membered azapane ring in cathepsin K inhibitor SB-462795 (Figure 6).[32] The metathesis step in this synthesis also required multiple optimization attempts. Initial efforts using diene **29** and catalyst **5** were stymied by the need for high catalyst loadings (5-10 mol%). Substrate-catalyst coordination was suspected to be the source of the problem, but reactions using Ti(O-iPr)$_4$ as a chelation inhibitor[33,34] were not successful due to substrate/product decomposition. Additionally, large residual Ru levels were deemed a safety hazard with downstream steps involving H$_2$O$_2$ (for chiral auxiliary removal) and an acyl azide (for nitrogen insertion via Curtius rearrangement). To complicate matters, the β-hydroxy ketone RCM product **30** decomposed via a retro-aldol reaction when the basic P(CH$_2$OH)$_3$/NaOH ruthenium removal procedure was employed. These setbacks led to the design of N/O-protected RCM precursor **31**. This substrate closed with lower catalyst loadings (1-2 mol%) but gave capricious levels of olefin migration when attempted with crude product coming out of a Curtius rearrangement employing hydrazine and t-butyl nitrite/HCl (Figure 6). Solvent swaps or isomerization inhibitors (acetic acid, styrene, or Cy$_3$PO) did not suppress the isomerization. The RCM process was made tractable by purifying precursor diene **31** as the HCl salt. When converted to the free base, it reliably provided **32** as a crystalline product in 90% yield using 1-2 mol% of catalyst **5**. A basic cysteine wash was used to remove Ru; this thiol-based approach was utilized after it was discovered that (i) formaldehyde, generated during *in situ* generation of P(CH$_2$OH)$_3$, added to

176

the carbamate nitrogen of product **32** to form hydroxymethylated **35**, and (ii) residual P reagents proved problematic for subsequent olefin hydrogenation.

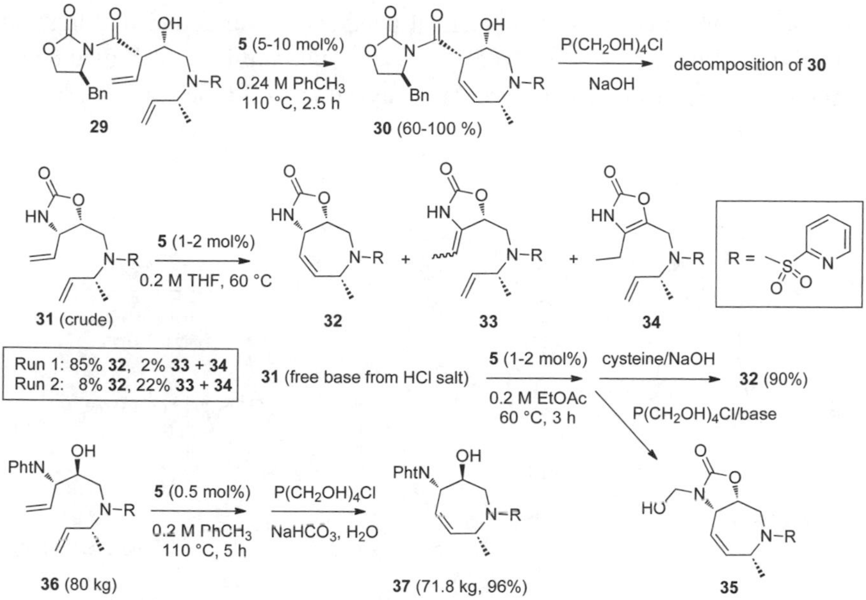

Figure 6. Selected RCM reactions used in the synthesis of SB-462795.

An alternative route to SB-462795 employed RCM precursor **36** containing a phthalimide-protected *trans*-1,2-aminoalcohol moiety. In comparison with diene **31**, this substrate offered a significant improvement in the RCM process, although here again significant optimization efforts were directed at identifying and minimizing trace impurity catalyst inhibitors.[35] Highly purified material could be closed in quantitative yields with as little as 0.25-0.5 mol% catalyst **5**. Olefin migration was not observed, even in reactions using crude material. When conducted in toluene, the product readily crystallized and provided material with low residual Ru content after the slurry was washed with $P(CH_2OH)_3/NaHCO_3$.[36] According to the published account, both approaches were amenable to large-scale manufacturing and the *trans*-1,2-aminoalcohol route was used to provide over 200 kg of SB-462795.

Other publicly disclosed pharmaceutical applications of RCM involve its use to conformationally constrain[37,38,39] and alter the metabolic profile of helical peptides by forming peptide macrocycles.[40] Scale-up aspects of this chemistry remain proprietary, but much of the exploratory work has been on solid support,[41] with microwave-assisted RCM macrocyclizations.[42,43] Many examples undergo efficient RCM reactions, and in a solution-phase study of minimal RCM constraints in 3_{10}-helical peptides, it was found that high *E*-selectivity could be realized in an unoptimized 18-membered macrocyclization (**38→39**, Figure 7) catalyzed by 5-7 mol% **4** in refluxing dichloromethane.[44]

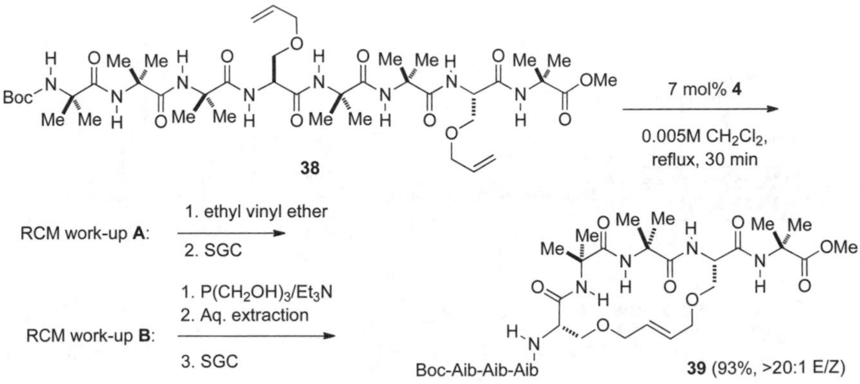

Figure 7. Representative helical peptide macrocyclization and work-up procedures.

The reactions shown in Figure 7 highlight two additional methods for post-RCM reaction processing: (i) the use of ethyl vinyl ether as a Ru catalyst poison (by formation of the stable Fischer carbene) prior to evaporation and silica gel chromatography (SGC), and (ii) a Ru removal procedure utilizing the commercially available, albeit expensive, tris(hydroxymethyl)-phosphine.[15,45]

The BI and GSK investigations show that large-scale RCM reactions can require significant optimization. It is expected that these paths will shorten as more applications are reported. These two examples reveal that at-scale process applications can utilize as little as 0.1-0.5 mol % Ru catalyst, which minimizes the waste stream and is in accord with sustainable green chemistry practice. Sustainable concepts in olefin metathesis have been reviewed elsewhere.[46] Pharmaceutical applications require that the API contain <10 ppm metal, and the two industrial examples discussed here utilized water-solubilizing procedures using 2-mercaptonicotinic acid,[13] cysteine,[32] or tris(hydroxy-methyl)phosphine.[36] Downstream processes such

178

as hydrogenation over charcoal-based catalysts can also be used to reduce Ru levels to <1 ppm.[35] Other approaches to remove Ru include amine-functionalized mesoporous silicates,[47] polar isocyanides,[48] lead tetraacetate,[49] activated carbon/silica gel,[50] and DMSO or triphenylphosphine oxide treatment of crude reaction mixtures.[51] DMSO/DMF mixtures have been used to remove colored Ru impurities in solid-phase RCM applications.[52] A recent report describes the use of 15% aqueous hydrogen peroxide as an efficient Ru removal agent; this process also oxidizes residual phosphine and carbene ligands to more polar entities, thus easing purification in some cases.[53]

1. Department of Chemistry, Pomona College, Claremont, CA 91711. D.J.O. thanks Pomona College and the National Science Foundation for their financial support.
2. Materia, Inc., 60 N. San Gabriel Blvd, Pasadena, CA 91107.
3. Division of Chemistry and Chemical Engineering, California Institute of Technology, Pasadena, CA 91125
4. Grubbs, R.H., Ed. *Handbook of Metathesis*; Wiley-VCH: Weinheim, 2003.
5. For a discussion of the properties and specificities of ruthenium-based metathesis catalysts, see: Schrodi, Y.; Pederson, R. L. *Aldrichimica Acta* **2007**, *40*, 45.
6. Sturmer, R., Schafer, B.; Wolfart, V.; Stahr, H.; Kazmaier, U.; Helmchen, G. *Synthesis* **2001**, *46*.
7. Kuhn, K. M.; Champagne, T. M.; Hong, S. H.; Wei, W.-H.; Nickel, A.; Lee, C. W.; Virgil, S. C.; Grubbs, R. H.; Pederson, R. L. *Org. Lett.* **2010**, *12*, 984.
8. Nicola, T.; Brenner, M.; Donsbach, K.; Kreye, P. *Org. Process Res. Dev.* **2005**, *9*, 513.
9. Poirer, M; Aubry, N.; Boucher, C.; Ferland, J-M; LaPlante, S.; Tsantrizos, Y.S. *J. Org. Chem.* **2005**, *70*, 10765.
10. Yee, N. K; Farina, V.; Houpis, I. N.; Haddad, N.; Frutos, R.P.; Gallou, F.; Wang, X-J.; Wei, X.; Simpson, R. D.; Feng, X.; Fuchs, V.; Xu, Y.; Tan, J.; Zhang, L.; Xu, J.; Smith-Keenan, L. L.; Vitous, J.; Ridges, M. D.; Spinelli, E. M.; Johnson, M. *J. Org. Chem.* **2006**, *71*, 7133.

11. Zeng, X.; Wei, X.; Farina, V.; Napolitano, E.; Xu, Y.; Zhang, L.; Haddad, N.; Yee, N. K.; Grinberg, N.; Shen, S.; Senanayake, C. H. *J. Org. Chem.* **2006**, *71*, 8864.

12. Shu, C.; Zeng, X.; Hao, M-H.; Wei, X.; Yee, N. K.; Busacca, C. A.; Han, Z.; Farina, V.; Senanayake, C. H. *Org. Lett.* **2008**, *10*, 1303.

13. Farina, V.; Shu, C.; Zeng, X.; Wei, X.; Han, Z.; Yee, N. K.; Senanayake, C. H. *Org. Process Res. Dev.* **2009**, *13*, 250.

14. Schmidt, B. *Eur. J. Org. Chem.* **2004**, 1865.

15. Maynard, H. D.; Grubbs, R. H. *Tetrahedron Lett.* **1999**, *40*, 4137.

16. Furstner, A.; Thiel, O. R.; Ackermann, L.; Schanze, H.-J.; Nolan, S. P. *J. Org. Chem.* **2000**, *65*, 2204.

17. Campbell, M.J.; Johnson, J. S. *J. Am. Chem. Soc.* **2009**, *131*, 10370.

18. Bourgeois, D.; Pancrazi, A; Nolan, S.P.; Prunet, J. *J. Organomet. Chem.* **2002**, *643-644,* 247.

19. Hong, S. H.; Sanders, D. P.; Lee, C. W.; Grubbs, R. H. *J. Am. Chem. Soc.* **2005**, *127*, 17160.

20. Gimeno, N.; Formentin, P.; Steinke, J.H.G.; Vilar, R. *Eur. J. Org. Chem.* **2007**, 918.

21. Moise, J.; Arseniyadis, S.; Cossy, J. *Org. Lett.* **2007**, *9*, 1695.

22. Vedrenne, E.; Dupont, H.; Qualef, S.; Elkaim, L.; Grimaud, L. *Synlett* **2005**, *4*, 670.

23. Cadot, C.; Dalko, P.I.; Cossy, J. *Tetrahedron Lett.* **2002**, *43*, 1839.

24. Alcaide, B.; Almendros, P.; Alonso, J. M. *Chem. Eur. J.* **2003**, *9*, 5793.

25. De Bo, G.; Marko, I. E. *Eur. J. Org. Chem.* **2011**, 1859.

26. Schmidt, B. *Eur. J. Org. Chem.* **2003**, 816.

27. Hoye, T. R.; Zhao, H. *Org. Lett.* **1999**, *1*, 1123.

28. Tallarico, J. A.; Malnick, L. M.; Snapper, M. L. *J. Org. Chem.* **1999**, *64*, 344.

29. Miller, S. J.; Kim, S.-H.; Chen, Z.-R.; Grubbs, R. H. *J. Am. Chem. Soc.* **1995**, *117*, 2108.

30. Conrad, J. C.; Eelman, M. D.; Duarte Silva, J. A.; Monfette, S.; Parnas, H. H.; Snelgrove, J. L.; Fogg, D. E. *J. Am. Chem. Soc.* **2007**, *129*, 1024.

31. Sheldon, R. *Green Chem.* **2007**, *9*, 1273.

32. Wang, H.; Matsuhashi, H.; Doan, B. D.; Goodman, S. N.; Ouyang, X.; Clark, Jr., W. M. *Tetrahedron* **2009**, *65*, 6291.

33. Furstner, A.; Langemann, K. *J. Am. Chem. Soc.* **1997**, *119*, 9130.

34. For an application in the synthesis of 3-pyrroline systems, see: Yang, Q.; Xiao, W.-J.; Yu, Z. *Org. Lett.* **2005**, *7*, 871.

35. Wang, H.; Goodman, S. N.; Dai, Q.; Stockdale, G. W.; Clark, Jr. *Org. Process Res. Dev.* **2008**, *12*, 226.

36. Pederson, R. L.; Fellows, I. M.; Ung, T. A.; Ishihara, H.; Hajela, S. P. *Adv. Synth. Catal.* **2002**, *344*, 728.

37. Miller, S. J.; Blackwell, H. E.; Grubbs, R. H. *J. Am. Chem. Soc.* **1996**, *118*, 9606.

38. Blackwell, H. E.; Grubbs, R. H. *Angew. Chem. Int. Ed.* **1998**, *37*, 3281.

39. Blackwell, H. E.; Sadowsky, J. D.; Howard, R. J.; Sampson, J. N.; Chao, J. A.; Steinmetz, W. E.; O'Leary, D. J.; Grubbs, R. H. *J. Org. Chem.* **2001**, *66*, 5291.

40. Van Arnum, P. *Pharm. Tech.* **2011**, *35*, 56.

41. Kim, Y.-W.; Grossmann, T. N.; Verdine, G. L. *Nature Protocols* **2011**, *6*, 761.

42. Patgiri, A.; Menzenski, M. Z.; Mahon, A. B.; Arora, P. S. *Nature Protocols* **2010**, *5*, 1857.

43. van Lierop, B. J.; Bornschein, C.; Jackson, W. R.; Robinson, A. J. *Aust. J. Chem.* **2011**, *64*, 806.

44. Boal, A. K.; Guryanov, I.; Moretto, A.; Crisma, M.; Lanni, E. L.; Toniolo, C.; Grubbs, R. H.; O'Leary, D. J. *J. Am. Chem. Soc.* **2007**, *129*, 6986.

45. Winkler, J. D.; Asselin, S. M.; Shepard, S.; Yuan, J. *Org. Lett.* **2004**, *6*, 3821.

46. Clavier, H.; Grela, K.; Kirschning, A.; Mauduit, M.; Nolan, S. P. *Angew. Chem. Int. Ed.* **2007**, *46*, 6786

47. McEleney, K; Allen, D. P.; Holliday, A. F.; Crudden, C. M. *Org. Lett.* **2006**, *8*, 2663.

48. Galan, B. R.; Kalbarczyk, K. P.; Szczepankiewicz, S.; Keister, J. B.; Diver, S. T. *Org. Lett.* **2007**, *9*, 1203.

49. Paquette, L. A.; Schloss, J. D.; Efremov, I.; Fabris, F.; Gallou, F.; Mendez-Andino, J.; Yang, J. *Org. Lett.* **2000**, *2*, 1259.

50. Cho, J. H.; Kim, B. M. *Org. Lett.* **2003**, *5*, 531.

51. Ahn, Y. M.; Yang, K.; Georg, G. I. *Org. Lett.* **2001**, *3*, 1411.

52. Hossain, M. A.; Guilhaudis, L.; Sonnevende, A.; Attoub, S.; van Lierop, B. J.; Robinson, A. J.; Wade, J. D.; Conlon, J. M. *Eur. Biophys. J.* **2011**, *40*, 555.

53. Knight, D. W.; Morgan, I. R.; Proctor, A. J. *Tetrahedron Lett.* **2010**, *51*, 638.

Daniel J. O'Leary was born in Seattle, Washington on 11 April 1964. He obtained B.A. degrees in chemistry and biology from Linfield College in 1986 and conducted undergraduate research at IBM Instruments (with Chuck Wade). He earned his Ph.D. (with Frank Anet) from UCLA in 1991. After an NSF post-doctoral fellowship (with Yoshito Kishi) at Harvard University, he joined the Pomona College faculty in 1994 and is presently Carnegie Professor of Chemistry. His research interests include conformational analysis, isotope effects, and applications of the olefin metathesis reaction to biologically relevant systems.

Richard L. Pederson was born in Albert Lea, Minnesota in November 1962. In 1990 he earned his Ph.D (with Chi-Huey Wong) from Texas A&M University and later that year joined Bend Research Inc, Bend OR, where in 1997 he and Professor Grubbs patented the production of insect pheromones using ruthenium metathesis. In January 2000, he joined Materia Inc., Pasadena CA, to start up the Fine Chemicals group and currently is the VP of R&D. His research interests include using metathesis to develop new products from renewable seed oils and the synthesis of insect pheromones. He is an inventor on 34 patents and patent applications.

Robert H. Grubbs was born in rural Kentucky on 27 February 1942. He earned his B.S. and M.S. degrees in chemistry in 1963 and 1965 from the University of Florida (working with Merle Battiste) and his Ph.D. (with Ron Breslow) from Columbia University in 1968. Following an NIH post-doctoral fellowship (with Jim Collman) at Stanford University, he joined the faculty at Michigan State University in 1969 and moved to the California Institute of Technology in 1978, where he is presently Victor and Elizabeth Atkins Professor of Chemistry. His research interests include mechanisms of metal-catalyzed reactions, polymer synthesis, and catalysis in organic synthesis.

Preparation of 1-Monoacylglycerols via the Suzuki-Miyaura Reaction: 2,3-Dihydroxypropyl (Z)-tetradec-7-enoate

Submitted by Dexi Yang,[1] Valerie A. Cwynar,[1] David J. Hart,[1] Jakkam Madanmohan,[1] Jean Lee,[2] Joseph Lyons[2] and Martin Caffrey.[2]
Checked by Zachary Gates and Viresh Rawal.

1. Procedure

A. *(2,2-Dimethyl-1,3-dioxolan-4-yl)methyl 5-Hexenoate* (**2**). A flame-dried, 250-mL, three-necked round-bottomed flask equipped with a oval-shaped magnetic stir bar (4 cm x 1.5 cm) and three rubber septa is charged via plastic syringe with 5-hexenoic acid (5.30 mL, 5.09 g, 44.6 mmol), dichloromethane (70 mL), and DL-1,2-isopropylideneglycerol (5.60 mL, 5.95 g, 45.0 mmol, 1.01 equiv) (Note 1). A septum is removed, solid 4-

(dimethylamino)pyridine (55 mg, 0.45 mmol, 0.01 equiv) (Note 1) is added through the open neck, the septum is replaced, and the flask is evacuated (<1 mmHg) and refilled with nitrogen before a septum is quickly replaced with a thermometer inlet adapter containing a low temperature thermometer. The flask is submersed in an ice/water bath, and the reaction mixture is allowed to stir gently until its temperature reaches 3–4 °C. Concurrently, a flame-dried, 100-mL recovery flask is charged with *N,N'*-dicyclohexylcarbodiimide (9.28 g, 45.0 mmol, 1.01 equiv) (Note 1). The flask is sealed with a rubber septum, evacuated (<1 mmHg) and refilled with nitrogen, then charged with dichloromethane (15 mL) using a 20-mL plastic syringe, and swirled gently to dissolve the solid. The resulting solution is added dropwise over 15 min to the stirred reaction mixture using a 20-mL plastic syringe (Note 2). A white precipitate appears about 5 min into the addition. At no point does the internal temperature rise above 6 °C. To achieve quantitative transfer, the recovery flask is rinsed with approximately 5 mL dichloromethane, and the resulting solution is added to the reaction mixture in one portion via a 20-mL plastic syringe. Upon completion of the addition, the reaction mixture is allowed to stir for 5 min before the cold bath is removed. After 3.5 h, the precipitate is removed via vacuum filtration (150 mmHg) through a Büchner funnel. The reaction flask is rinsed with an additional 50 mL of dichloromethane, which is poured into the Büchner funnel. The filtrate is concentrated by rotary evaporation (<30 °C, 20 mmHg) to give a clear, colorless oil that contains a small amount of white solid. The crude product is purified by flash chromatography on silica gel (Note 3) to give 9.25 g (91%) of ester **2** as a clear, colorless oil (Note 4).

B. *1-Iodo-1-octyne* (**4**). A flame-dried, 500-mL, three-necked round-bottomed flask under an atmosphere of nitrogen and equipped with a oval-shaped magnetic stir bar (4 cm x 1.5 cm), a 250-mL pressure-equalizing addition funnel, a low temperature thermometer, and a rubber septum is charged via 20-mL plastic syringes with 1-octyne (**3**) (14.75 mL, 11.02 g, 100.0 mmol) and hexanes (125 mL) (Notes 5 and 6). The resulting opaque solution is cooled with gentle stirring in an acetone/dry ice bath to an internal temperature of –52 °C (Note 7). The stirring rate is increased, and a solution of 2.5 M *n*-butyllithium in hexanes (40.0 mL, 100 mmol, 1.0 equiv) (Note 5) is added dropwise over 20 min via a 20-mL plastic syringe. A white precipitate begins to form less than 1 min into the addition. Upon completion of the addition, the flask is fully submersed in the dry ice bath to bring the internal temperature to –70 °C. The viscous reaction mixture is

184

stirred vigorously at this temperature for 1.5 h. On occasion, the stir bar stops stirring in the thick slurry and is dislodged using the thermometer. During this time, a flame-dried, 500-mL pear-shaped flask is charged with iodine (25.4 g, 100 mmol, 1.0 equiv) (Note 5). The flask is sealed with a rubber septum, evacuated (< 1 mmHg) and refilled with nitrogen, and diethyl ether (100 mL) (Note 5) is added via a 20-mL plastic syringe. The resulting suspension is swirled vigorously by hand for approximately 5 min before the deep red supernatant is transferred to the pressure-equalizing addition funnel via cannula. The residual solid iodine is taken up with vigorous stirring in 30 mL of diethyl ether, which is again transferred to the addition funnel via cannula. Finally, the sides of the addition funnel are rinsed with approximately 10 mL of ether using a 20-mL plastic syringe, and the ethereal solution of iodine is added dropwise over 25 min (Note 8) to the stirring slurry of acetylide. Upon completion of the addition, a substantial mass of iodine is scraped from the tip of the addition funnel into the reaction vessel using a spatula. The cold bath is removed, and the solution is allowed to warm to 20 °C over 2.5 h. A septum is removed, and distilled water (100 mL) is added cautiously through the open neck. The resulting biphasic mixture is transferred to a 1-L separatory funnel, and the organic phase is isolated. The aqueous phase is extracted with hexanes (100 mL), and the combined organic phases are washed with two 200-mL portions of water, followed by 50 mL of saturated aqueous sodium thiosulfate. The organic phase is dried over anhydrous magnesium sulfate (20 g), filtered through a Büchner funnel (150 mmHg), and concentrated by rotary evaporation (< 30 °C, 20 mmHg) to give a clear, light yellow oil that is transferred to a 50-mL round-bottom flask and distilled under vacuum using a short-path still (Note 9) to give 19.9 g (84%) of iodoalkyne 4 as a clear, colorless liquid (Note 10).

C. *(Z)-1-Iodo-1-octene* (**5**). A flame-dried, 250-mL, three-necked round-bottomed flask under an atmosphere of nitrogen, containing an oval-shaped magnetic stir bar (4 cm x 1.5 cm), and fitted with three rubber septa is charged via plastic syringes with borane dimethyl sulfide complex (7.51 mL, 6.02 g, 79.2 mmol, 1.06 equiv) and diethyl ether (75 mL) (Note 11). The resulting stirred solution is submersed in an ice-bath for 10 min before cyclohexene (16.0 mL, 13.0 g, 158 mmol, 2.11 equiv) (Note 11) is added dropwise over 10 min via a 20-mL plastic syringe. A white precipitate appears within the final minute of the addition. After 15 min, the ice bath is removed and stirring is continued for 1 h. The reaction flask is again placed in an ice bath and, after 5 min, treated with iodoalkyne 4 (75 mmol, 17.7 g), added dropwise over 10 min via a 10-mL plastic syringe. The solution is stirred for 30 min, the ice bath is removed, and stirring is continued for 1.25 h. The flask is once more immersed in the ice bath, and the mixture is allowed to stir for 15 min before a septum is removed and glacial acetic acid (30 mL) (Note 11) is added slowly via a 100 mL graduated cylinder. After replacing the septum, the ice bath is removed and the mixture is stirred for 2.25 h. Finally, the mixture is diluted with diethyl ether (30 mL) (Note 11) and deionized water (50 mL). The resulting biphasic mixture is transferred to a 500-mL separatory funnel, shaken, and the aqueous phase is discarded. The organic phase is washed with water (3 x 50 mL), and dried over magnesium sulfate (5 g). The drying agent is removed by vacuum filtration (150 mmHg), and the filtrate is concentrated by rotary evaporation (< 30 °C, 20 mmHg) to give a clear oil, containing small amounts of a white precipitate. This suspension is applied to a silica plug and eluted with hexanes (Note 12). Fractions containing the product are concentrated by rotary evaporation (< 30 °C, 20 mmHg) and the resulting yellow oil is distilled under vacuum using a short-path still to give 14.28 g (80%) of iodoalkene **5** as a clear, pale-yellow liquid (Note 13).

D. *(2,2-Dimethyl-1,3-dioxolan-4-yl)methyl (Z)-tetradec-7-enoate* (**6**). A flame-dried, 500-mL, three-necked round-bottomed flask equipped with a cylindrical-shaped magnetic stirbar (4 cm x 1 cm) and 2 rubber septa is charged through the open neck with ester **2** (5.0 g, 21.9 mmol). A third septum is added, the flask is evacuated (< 1 mmHg) and refilled with nitrogen by piercing a septum with a needle connected to a Schlenck manifold. Tetrahydrofuran (70 mL) (Note 14) is added via a 20-mL plastic syringe. The flask is submersed in an ice bath, and the solution is stirred for 5 min before it is treated with a 0.5 M solution of 9-borabicyclo[3.3.1]nonane [9-BBN] in THF (46 mL, 23.0 mmol, 1.05 equiv) (Note 14), added dropwise over 12 min via a 20-mL plastic syringe. Upon completion of the addition, the cold bath is removed and the reaction mixture is stirred for 2.5 h, then quenched with deionized water (70 mL), added slowly through an open neck via a 100 mL graduated cylinder. The septum is replaced, and the flask is evacuated (< 1 mmHg) and refilled with nitrogen. The resulting suspension is stirred for 2.5 h, after which time it is sparged with argon for 15 min (Note 15).

A separate flame-dried, 1-L, 3-necked round-bottomed flask equipped with a oval-shaped magnetic stir bar (4 cm x 1.5 cm) and two rubber septa is charged through the open neck with vinyl iodide **5** (6.25 g, 26.3 mmol, 1.2

186

equiv), cesium carbonate (7.14 g, 21.9 mmol, 1.0 equiv) and triphenylarsine (0.67 g, 2.2 mmol, 0.10 equiv) (Note 14). A third septum is added, the flask is evacuated (> 1 mmHg) and refilled with nitrogen, and DMF (130 mL) (Note 14) is added via a 20-mL plastic syringe. The resulting suspension is sparged with argon for 15 min (Note 15). Dichloro[1,1'-bis(diphenylphosphino)ferrocene]palladium(II)•dichloromethane adduct (Pd(dppf)Cl$_2$•CH$_2$Cl$_2$) (0.80 g, 1.1 mmol, 0.05 equiv) (Note 14) is added quickly to the flask through an open neck, the septum is quickly replaced, and the resulting red suspension is sparged with argon for 15 min (during this time, the color of the solution changes from red-orange to brown). The resulting degassed suspension is then treated with the aforementioned hydroboration mixture via cannula over 10 min (Note 16). Upon completion of the cannula transfer, the reaction vessel is evacuated (< 1 mmHg) and refilled with nitrogen, and the dark brown reaction mixture is stirred for 5 h before it is poured onto 500 mL saturated NaCl in a 2 L separatory funnel. The organic phase is isolated, and the aqueous phase is extracted with ether (3 x 300 mL). The combined organic phases are dried over magnesium sulfate (40 g), and the drying agent is removed via vacuum filtration through a Büchner funnel (150 mmHg). The filtrate is concentrated by rotary evaporation (< 30 °C, 20 mmHg) to give the product as a dark oil containing some residual DMF. Purification by flash chromatography on silica gel (Note 17) gives 7.12 g (88%) (Note 18) of acetonide 6 as a brown oil containing 7% (w/w) cyclooctanone (Note 19).

E. *2,3-Dihydroxypropyl (Z)-tetradec-7-enoate* (7). A 250-mL round-bottomed flask equipped with a oval-shaped magnetic stir bar (4 cm x 1.5 cm) is charged with acetonide 6 (4.80 g, 14.1 mmol), methanol (60 mL), and 2.0 M hydrochloric acid (4.0 mL). (Note 20) The flask is sealed with a rubber septum and the reaction mixture is allowed to stir for 6 h. After this time, the reaction mixture is poured into a 1-L separatory funnel containing a mixture of ether (350 mL) and saturated sodium bicarbonate (350 mL). The funnel is swirled gently to release carbon dioxide. The organic phase is separated, and the aqueous phase is extracted with ether (3 x 150 mL). The combined organic phases are dried over magnesium sulfate (20 g), which is removed by vacuum filtration through a Büchner funnel (150 mmHg). The filtrate is concentrated by rotary evaporation at (< 30 °C, 20 mmHg) to give an orange oil that is purified by flash chromatography on silica gel (Note 21) to give 3.02 g (71%) of monoacylglycerol 7 as an orange oil. This sample is a 20:1 mixture of 1-monoacylglycerol (1-MAG) 7 and the corresponding 2-

MAG, respectively (Notes 22, 23, and 24). If desired, the sample can be enriched further in the desired 1-MAG by recrystallization (Note 25).

2. Notes

1. 5-Hexenoic acid (98%) was purchased from TCI Americas and used as received. DL-1,2-Isopropylideneglycerol ("solketal," 98%), 4-dimethylaminopyridine (99%), and dicyclohexylcarbodiimide (99%) were purchased from Aldrich and used as received. Dichloromethane was purchased from Fisher (Certified ACS, stabilized), and dried by passage over activated alumina (Innovative Technology Inc. Pure Solv™ solvent purification system). The submitters purchased dichloromethane from Mallinckrodt (ACS reagent grade) and used as received. Submitters used glass syringes in experiments where checkers used plastic syringes.

2. The submitters used a 50-mL addition funnel equipped with a Newman stopcock.

3. Silica gel was purchased from Zeochem (ZEOprep 60, 40-63 micron particle size). A flash column is prepared by wet-packing a 7.5 cm diameter column with a slurry of 200 g silica in 15% ethyl acetate/hexanes [ethyl acetate and hexanes (both Certified ACS) were purchased from Fisher or Mallinckrodt and used as received]. The crude compound is loaded neat, and eluted with 15% ethyl acetate in hexanes. Fractions (50 mL) are collected and analyzed by TLC (Whatman 0.25 mm 60 Å silica gel plates; 4:1 hexanes:ethyl acetate eluent; potassium permanganate stain). The desired ester (**2**) begins to elute after ~180 mL of eluent is collected and continues to elute over ~1 L.

4. Ester **2** exhibits the following spectral properties: IR (neat) cm^{-1}: 1740; ^{1}H NMR (500 MHz, C$_6$D$_6$) δ: 1.24 (s, 3 H), 1.36 (s, 3 H), 1.57 (quintet, J = 8 Hz, 2 H), 1.86 (quartet, J = 7 Hz, 2 H), 2.07 (t, J = 7 Hz, 2 H), 3.45 (dd, J = 8 Hz, 6 Hz, 1 H), 3.66 (dd, J = 8 Hz, 6 Hz, 1 H), 3.95–4.05 (m, 3 H), 4.91–4.96 (m, 2 H), 5.56–5.64 (m, 1 H); ^{13}C NMR (125 MHz, C$_6$D$_6$) δ: 24.6, 25.9, 27.3, 33.6, 33.7, 65.0, 66.7, 74.3, 110.0, 115.8, 138.3, 173.0; Exact mass calcd for (C$_{12}$H$_{20}$O$_4$)$_2$Na$^+$: 479.2615; found (electrospray) m/z: 479.2615; Anal calcd for C$_{12}$H$_{20}$O$_4$: C, 63.14; H, 8.83; found: C, 62.94; H, 8.70.

5. 1-Octyne (97%) and iodine crystals (ACS reagent grade, ≥99.8%) were purchased from Aldrich and used as received. The n-butyllithium solution was purchased from Aldrich, transferred to a Schlenck flask upon

receipt, and titrated immediately prior to use by the method of Kofron (Kofron, W.G. and Baclawski, L.M. *J. Org. Chem.* **1976**. *41* (10), 1879). Anhydrous diethyl ether (Certified ACS, BHT stabilized) was purchased from Fisher and purified using the solvent purification system described in Note 1. Hexanes were purchased from Fisher (Certified ACS) and dried over microwave-activated 4 Å beaded molecular sieves (5 g/100 mL) for at least 12 h. The submitters purchased anhydrous diethyl ether (Certified ACS) and hexanes (ACS reagent grade) from Fisher and distilled them from sodium.

6. The submitters used 70 mL hexanes.

7. The checkers achieved the desired internal temperature by adjusting the depth of the flask in the dry ice/acetone bath. The submitters used a chloroform/dichloromethane-dry ice bath.

8. The submitters reported an addition time of 45 min.

9. The distillation must be carried out below 100 °C. On one occasion when the bath temperature exceeded 150 °C, the contents of the flask changed from light yellow to dark red in color, bumped vigorously, and poured over into the receiving flask. Submitters performed the distillation at 0.85 mmHg and recorded a bp of 63-65 °C.

10. The submitters reported a yield of 77%, and that the product turns pink on standing at room temperature due to residual iodine (color can be removed by an additional aqueous sodium thiosulfate wash). The checkers found the product to remain colorless over a period of several months when stored in a refrigerator under a nitrogen atmosphere. Iodoalkyne **4** exhibits the following properties: bp: 70–72 °C (2 mmHg); IR (neat) cm^{-1}: 2187; ^{1}H NMR (500 MHz, CDCl$_3$) δ: 0.89 (t, J = 7 Hz, 3 H), 1.30–1.40 (m, 6 H), 1.50 (quintet, J = 7 Hz, 2 H), 2.35 (t, J = 7 Hz, 2 H); ^{13}C NMR (125 MHz, CDCl$_3$) δ: -7.4, 14.3, 21.0, 22.7, 28.7, 31.5, 95.0; Anal calcd for C$_8$H$_{13}$I: C, 40.70; H, 5.55; found: C, 40.54; H, 5.44.

11. BH$_3$•Me$_2$S complex (purity not indicated) was purchased from Aldrich and used as received. Cyclohexene (inhibitor-free, 99%) was purchased from Aldrich and distilled prior to use. Acetic acid (ACS reagent grade) was purchased from Fisher and used as received. Anhydrous diethyl ether (Certified ACS, BHT stabilized) was purchased from Fisher and purified as described in Note 1. The submitters purchased anhydrous diethyl ether (Certified ACS, BHT stabilized) from Fisher and distilled it from sodium prior to use.

12. The checkers prepared the silica plug by wet-packing a 7.5-cm diameter column with a slurry of 100 g silica gel (purchased from Zeochem, ZEOprep 60, 40-63 micron particle size) in hexanes. The crude product is loaded neat, and eluted with hexanes. Fractions (50 mL) are collected and analyzed by TLC (Whatman 0.25 mm 60 Å silica gel plates; hexanes eluent; potassium permanganate stain). This procedure removed most, but not all, of the dicyclohexylborinic acid. The submitters used a 2-cm diameter column containing 30 g of silica gel, and eluted with pentanes. For an alternate method of removing the dicyclohexylborinic acid via precipitation, see Brown, H.C. *et al. J. Org. Chem.* **1989**. *54*, 6064-6067.

13. The submitters obtained a yield of 86%. Iodoalkene **5** exhibits the following properties: bp: 77–78 °C (3 mmHg); IR (neat) cm^{-1}: 1609; ^{1}H NMR (500 MHz, C$_6$D$_6$) δ: 0.86 (t, J = 7 Hz, 3 H), 1.14–1.23 (m, 8 H), 2.02 (quartet, J = 7 Hz, 2 H), 5.75 (quartet, J = 7 Hz, 1 H), 5.90 (dt, J = 7 Hz, 1 Hz, 1 H); ^{13}C NMR (125 MHz, C$_6$D$_6$) δ: 14.6, 23.3, 28.6, 29.4, 32.3, 35.3, 82.7, 141.8; Anal calcd for C$_8$H$_{15}$I: C, 40.35; H, 6.35; found: C, 40.47, 6.38.

14. The 9-borabicyclo[3.3.1]nonane [9-BBN] solution, cesium carbonate (99%) and triphenylarsine (97%) were purchased from Aldrich and used as received. Dichloro[1,1'-bis(diphenylphosphino)ferrocene]-palladium(II)-dichloromethane adduct (Pd(dppf)Cl$_2$•CH$_2$Cl$_2$) was purchased from Strem and used as received. THF (Optima, Inhibitor-free) was purchased from Fisher and purified using the solvent purification system described in Note 1. DMF (99.8%, extra dry over molecular sieves) was purchased from Acros Organics and purified using the solvent purification system described in Note 1.

15. Sparging is accomplished as follows: one of the septa is pierced with a 1.5 inch, 16.5 gauge needle, and a second with a 8 inch, 20 gauge needle connected to an argon cylinder. Gentle stirring is maintained as argon is bubbled through the suspension.

16. The transfer is accomplished by partially evacuating the receiving flask, with the hydroboration flask under a slightly positive pressure of nitrogen.

17. A flash column is prepared by wet-packing a 7.5-cm diameter column with a slurry of 400 g of silica gel (purchased from Zeochem, ZEOprep 60, 40-63 micron particle size) in 5% ethyl acetate/hexanes. The column is eluted with 5% ethyl acetate in hexanes; 50 mL fractions are collected and analyzed by TLC (Whatman 0.25 mm 60 Å silica gel plates; 19:1 hexanes:ethyl acetate eluent; potassium permanganate stain). The

190

desired compound begins to elute after ~900 mL, and elutes over 2.5 L. The submitters performed a second flash purification in order to obtain the product as pale yellow oil. The checkers found that a second purification yielded the product as a clear, light orange oil, in 81% yield.

18. The submitters reported a yield of 64% of **6** (twice chromatographed). The checkers found that when new bottles of Pd(dppf)Cl$_2$•CH$_2$Cl$_2$, Cs$_2$CO$_3$, and AsPh$_3$ were used, yields were reproducibly ~90%. If these reagents were used after being opened and stored for several weeks, however, yields dropped to as low as 50%. The use of fresh 9-BBN is also required. Ester **2** was recovered as the major side product in these low-yielding cases.

19. The presence of cyclooctanone[4] is inferred from ^{1}H NMR multiplets at 1.55, 1.87, and 2.40 ppm, and ^{13}C NMR signals at 24.9, 25.8, 27.4, and 42.1 ppm (the expected carbonyl resonance was not observed). The ratio of ester **6** to cyclooctanone is determined by integration of the resonances at 1.63 ppm (ester **6**, 2H) and 1.55 ppm (cyclooctanone, 4H). The percent yield of **6** takes into account the presence of cyclooctanone. An analytically pure sample of **6** was prepared by chemical reduction as follows: a 25-mL round-bottomed flask is charged with 500 mg of the ester/cyclooctanone mixture and 20 mg of sodium borohydride. Ethanol (10 mL) (Fisher, denatured) is added, and the resulting suspension is allowed to stir for 3 h. Saturated aqueous NaHCO$_3$ (10 mL) is added, the aqueous phase is extracted three times with ether, and the combined organic layers are dried over MgSO$_4$ and concentrated TLC analysis of the oil [silica gel, ethyl acetate-hexanes, 1:3 by volume] indicated the presence of three products with R$_f$ values of 0.06, 0.40 and 0.60. The lowest R$_f$ spot had the same TLC mobility as an authentic sample of cyclooctanol. This material was purified by chromatography over 30 g of silica gel, eluted with ethyl acetate-hexanes (1:15 by volume) to give (Z)-ethyl 7-tetradecenoate (R$_f$ = 0.60) and pure ester **6**, which exhibits the following spectral properties: IR (neat) cm^{-1}: 1742; ^{1}H NMR (500 MHz, CDCl$_3$) δ: 0.87 (t, J = 7 Hz, 3 H), 1.2–1.5 (m, 18 H), 1.63 (quintet, J = 7 Hz, 2 H), 2.01 (quintet, J = 7 Hz, 4 H), 2.34 (t, J = 7 Hz, 2 H), 3.73 (dd, J = 8 Hz, 6Hz, 1 H), 4.05–4.10 (m, 3 H), 4.31 (quintet, J = 6 Hz, 1 H), 5.29–5.38 (m, 2 H); ^{13}C NMR (125 MHz, CDCl$_3$) δ: 14.3, 22.9, 25.0, 25.6, 26.9, 27.2, 27.5, 29.0, 29.2, 29.6, 29.9, 32.0, 34.3, 64.8, 66.6, 74.9, 110.1, 129.6, 130.5, 173.8; Exact mass calcd for C$_{20}$H$_{36}$O$_4$Na$^+$: 363.2506; found (electrospray) m/z: 363.2503; Anal calcd for C$_{20}$H$_{36}$O$_4$: C, 70.55; H, 10.66; found: C, 70.21; H, 10.51.

20. Methanol (HPLC grade) and hydrochloric acid (Certified ACS) were purchased from Fisher and used as received.

21. A flash column is prepared by wet-packing a 7.5-cm diameter column with a slurry of 400 g of silica gel (purchased from Zeochem, ZEOprep 60, 40-63 micron particle size) in 37% ethyl acetate/hexanes. The column is eluted with 1 L of 37% ethyl acetate in hexanes, followed by 50% ethyl acetate in hexanes. Fractions (50 mL) are collected and analyzed by TLC (Whatman 0.25 mm 60 Å silica gel plates; 1:1 hexanes:ethyl acetate eluent; potassium permanganate stain). The desired compound begins to elute after ~1.95 L, and elutes over 1.4 L.

22. The submitters reported a yield of 78% using **6** that had been purified twice by column chromatography (see Note 17).

23. Purified **7** (containing 5% of the corresponding 2-MAG) exhibited the following spectral properties: IR (neat) cm^{-1}: 1740; ^{1}H NMR (500 MHz, CDCl$_3$) δ: 0.88 (t, J = 7 Hz, 3 H), 1.26–1.35 (m, 12 H), 1.64 (quintet, J = 8 Hz, 2 H), 2.02 (quintet, J = 7 Hz, 4 H), 2.35 (t, J = 8 Hz, 2 H), 3.59 (dd, J = 12 Hz, 6 Hz, 1 H), 3.69 (dd, J = 12 Hz, 4 Hz, 1 H), 3.93 (quintet, J = 6 Hz, 1 H), 4.11–4.20 (m, 2 H), 5.35 (m, 2 H); ^{13}C NMR (125 MHz, CDCl$_3$) δ: 14.3, 22.8, 25.0, 27.1, 27.4, 28.9, 29.1, 29.5, 29.9, 31.9, 34.3, 63.5, 65.3, 70.4. Signals at 3.82 (d, J = 5 Hz, 4 H) and 4.92 (quintet, J = 5 Hz, 1 H) in the ^{1}H spectrum and at 62.3, 75.1 in the ^{13}C spectrum are attributed to the 2-MAG; Exact mass calcd for C$_{17}$H$_{32}$O$_4$Na$^+$: 323.2193; found (electrospray) m/z: 323.2189; Anal calcd for C$_{17}$H$_{32}$O$_4$: C, 67.96; H, 10.74; found: C, 67.97; H, 10.59.

24. The submitters utilized DMSO-d$_6$ for the acquisition of NMR spectra. ^{1}H NMR (500 MHz, DMSO-d$_6$) δ: 0.86 (t, J = 7, 3 H), 1.26–1.30 (m, 12 H), 1.53 (quintet, J = 7, 2 H), 1.98–2.01 (m, 4 H), 2.38 (t, J = 7, 2 H), 3.31–3.38 (m, 2 H), 3.63 (m, 1 H), 3.91 (dd, J = 11.0, 6.5, 1 H), 4.04 (dd, J = 11.0, 4.5, 1 H), 4.59 (t, J = 5.5, 1 H, CH$_2$O$\underline{H}$), 4.82 (d, J = 5.5, 1H, CHO$\underline{H}$), 5.33, (m, 2 H); ^{13}C NMR (125 MHz, DMSO-d$_6$) δ: 14.4, 22.2, 24.8, 27.0, 27.1, 28.6, 28.8, 29.3, 29.6, 31.6, 33.9, 63.1, 66.0, 69.8, 129.9, 130.1, 173.3.

25. A sample of a mixture of 1-monoacylglycerol and 2-monoacylglycerol (1-MAG and 2-MAG) (3.3 g, 11.0 mmol) is added to an oven-dried 50 mL round-bottomed flask equipped with a sidearm, magnetic stir bar, and two rubber septa. The flask is evacuated and refilled with nitrogen, and a 1:9 solution of diethyl ether:petroleum ether (27 mL) is added via a 10-mL plastic syringe. The flask is submersed in a dry ice/acetonitrile bath, and the solution is allowed to stir for 10 min, during

which time a white solid forms. The solid is allowed to settle for 5 min before the flask is tilted to one side (approximately 30°), and as much supernatant as possible is removed via syringe. The residual white suspension is washed 3 times with a 9:1 solution of ether:petroleum ether pre-cooled in an ice/water bath (8 mL) (each time, the suspension is stirred for ~1 min, and allowed to settle for 5 min before withdrawing as much supernatant as possible through the side arm via syringe). Finally, the cold bath is removed (the precipitate dissolves quickly upon removal of the bath), and the resulting solution is concentrated to give 1.55 g (47%) of **7** as a clear, light yellow oil.

Safety and Waste Disposal Information

All hazardous materials should be handled and disposed of in accordance with "Prudent Practices in the Laboratory"; National Academies Press; Washington, DC, 2011.

3. Discussion

The use of unsaturated 1-monoacylglycerols (1-MAGs) of type **8** (Figure 1) for crystallizing membrane proteins by the *in meso* method has steadily increased.[5-7] Whereas several 1-MAGs of type **8** are commercially available, many 1-MAGs of interest for protein crystallization studies cannot be purchased.[8] Thus, there is a need for reliable methods for the preparation of 1-MAGs of type **8**.[9]

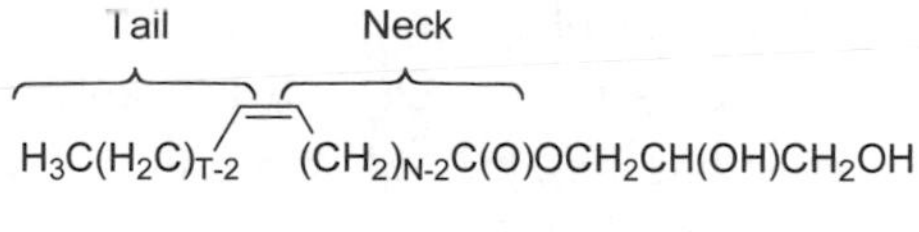

Tail   Neck

$$H_3C(H_2C)_{T-2} \quad (CH_2)_{N-2}C(O)OCH_2CH(OH)CH_2OH$$

N.T-MAG (**8**)

Figure 1. Neck (N) and Tail (T) represent the number of carbon atoms on either side of the double bond in the acyl chain. The neck and tail of 7.7-MAG (**7**) each have seven carbons.

Several procedures for preparing fatty acids that appear in **8** have been described. For example, hydrogenation of alkynoic acids,[10] or Wittig reactions between aldehydes and phosphoranes,[11] have been used to prepare fatty acids that might be useful for the preparation of commercially unavailable 1-MAGs.

An issue in any general synthetic approach to 1-MAGs of type **8** is preparation of the Z-olefin with good control of stereochemistry. The Suzuki-Miyaura reaction is a reliable method for the preparation of either E- or Z-olefins.[12] This reaction involves hydroboration of an alkene followed by Pd-mediated coupling of the intermediate borane with a 1-halogenated alkene with retention of haloalkene stereochemistry. This procedure describes use of the Suzuki-Miyaura reaction in a modular approach to 1-MAGs of type **8** (Figure 2) within the context of a synthesis of 7.7-MAG (**7**).

Scheme 1. Preparation of N.T-MAGs using Suzuki-Miyaura Coupling

$$
\begin{array}{c}
\text{1. 9-BBN} \\
\text{2. Pd(dppf)Cl}_2 \text{ (cat)} \\
\text{Ph}_3\text{As, Cs(CO}_3\text{)}_2 \\
\text{THF-H}_2\text{O}
\end{array}
$$

$H_2C=CH(CH_2)_{N-4}C(O)OCH_2CHCH_2$ **(9)** $\xrightarrow{\ \ \ \ \ H_3C(H_2C)_{T-2}\ \ \ I\ \ (10)\ \ \ }$ $H_3C(H_2C)_{T-2}$ $(CH_2)_{N-2}C(O)OCH_2CHCH_2$ **(11)**

$\xrightarrow{\text{MeOH-H}_2\text{O, HCl}}$ **8**

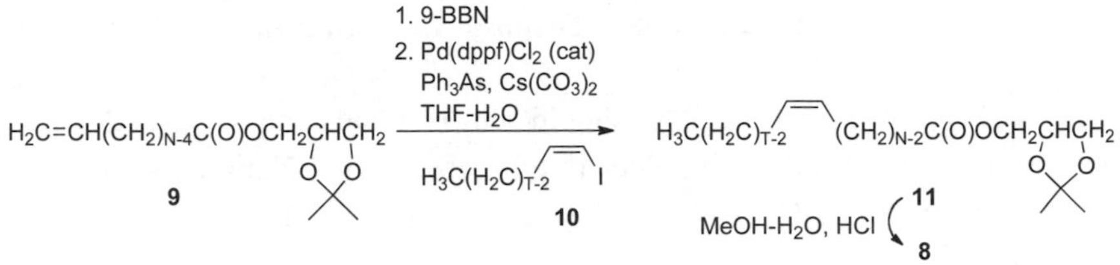

The coupling partners used in this synthesis of 1-MAGs are 1-iodoalkenes (**10**) and esters (**9**) derived from terminal alkenoic acids and solketal. Iodoalkenes can be prepared by a variety of methods.[13] This procedure uses the method of Brown involving conversion of a terminal alkyne to the corresponding 1-iodoalkyne, followed by a hydroboration-protonolysis to afford the terminal Z-1-iodoalkene.[14] We note that the preparation of Z-1-iodooctene presented here is modeled directly after the Denmark synthesis of Z-1-iodoheptene that appears in Volume 81 of *Organic Syntheses*.[15] The esters (**9**) can be prepared by DCC-mediated coupling of the corresponding unsaturated acid with solketal.[3] The Suzuki-Miyaura coupling is accomplished using Pd(dppf)Cl$_2$•CH$_2$Cl$_2$ as the catalyst with triphenylarsine as an additive (the "Johnson-Braun" modification).[17] This affords the 1-MAG acetonide (**11**), which is then hydrolyzed to the corresponding 1-MAG. Whereas the 1-MAG acetonides can be obtained in pure form from this procedure, the hydrolysis results in some isomerization of the 1-MAG to the corresponding 2-MAG (5-10%) in which the acyl group resides at the 2-position of glycerol.[16] The final 1-MAG, however, can usually be recrystallized to provide material contaminated with no more than 5% of the 2-MAG. Table 1 shows a variety of 1-MAG acetonides (**11**) and 1-MAGs (**8**) that have been prepared using this procedure.[18]

194

Table 1. Preparation of 1-Monoacylglycerols **8** via Scheme 1

1-MAG [N, T]	Yield 1-MAG Acetonide (**11**)	Yield 1-MAG[a] (**8**)
[5.7]	45%	**8a** 70%[b]
[5.9]	40%	**8b** 56%[b]
[5.13]	78%	**8c** 45%[b]
[6.7]	63%	**8d** 85%[b]
[6.12]	34%	**8e** 58%
[7.7]	64%	**8f** 78%[b]
[7.9]	77%	**8g** 45%
[7.10]	70%	**8h** 58%
[7.11]	76%	**8i** 85%
[8.7]	68%	**8j** 78%[b]
[8.8]	68%	**8k** 46%
[8.9]	64%	**8l** 63%[b]
[8.10]	53%	**8m** 53%[c]
[9.10]	68%	**8o** 50%[c]
[10.8]	61%	**8p** 55%
[11.9]	47%	**8q** 76%
[12.6]	53%	**8r** 64%
[13.5]	52%	**8s** 48%

[a]Yield of colorless recrystallized material unless noted otherwise. [b]Yield of 1-MAG from chromatography. The amount of 2-MAG in the 1-MAG follows: **8a** (6%), **8b** (5%), **8c** (0%), **8d** (trace), **8f** (7-9%), **8j** (0%), **8l** (0%). [c]Recrystallized 1-MAG contaminated with 2-MAG: **8m**, (3%), **8o** (5%).

1. The Ohio State University, Department of Chemistry, 100 W. 18th Avenue, Columbus, OH 43210, USA. E-mail: hart@chemistry.ohiostate.edu. The authors acknowledge the National Institutes of Health and Science Foundation Ireland for financial support.
2. Department of Chemical and Environmental Sciences, University of Limerick, Limerick, and Schools of Medicine and Biochemistry and Immunology, Trinity College, Dublin, Ireland.
3. Coleman, B. E.; Cwynar, V.; Hart, D. J.; Havas, F.; Madanmohan, J.; Patterson, S.; Ridenour, S.; Schmidt, M.; Smith, E.; Wells, A. J. *Synlett* **2004**, 1339–1342.
4. Fürstner, A.; Seidel, G. *Synlett* **1998**, 161–162. This paper describes cyclooctanone as a minor product (7%) in Suzuki reactions of the title reagent.
5. (a) Landau, E. M.; Rosenbusch, J. P. *Proc. Natl. Acad. Sci. USA* **1996**, *93*, 14532–14535. (b) Caffrey, M. *Curr. Opin. Stru. Biol.* **2000**, *10*, 486–497. (c) Caffrey, M. *J. Struct Biol.* **2003**, *142*, 108–132. (d) Caffrey, M. *Ann. Rev. Biophys.* **2009**, *38*, 29–51. (e) Cherezov, V.; Clogston, J.; Papiz, M. Z.; Caffrey, M. *J. Mol. Biol.* **2006**, *357*, 1605–1618. (f) Caffrey, M.; Lyons, J.; Smyth, T.; Hart, D. J. in *Current Topics in Membranes*, DeLucas, L., Ed.; Burlington: Academic Press, **2009**, *63*, 83-108.
6. Cherezov, V.; Caffrey, M. *J. Appl. Cryst.* **2005**, *38*, 398–400.
7. (a) Misquitta, Y.; Cherezov, V.; Havas, F.; Patterson, S.; Madanmohan, J.; Wells, A. J.; Hart, D. J.; Caffrey, M. *J. Struct. Biol.* **2004**, *148*, 169–175. (b) Misquitta, L. V.; Misquitta, Y.; Cherezov, V.; Slattery, O.; Madanmohan, J.; Hart, D.; Zhalnina, M.; Cramer, W. A.; Caffrey, M. *Structure* **2004**, *12*, 2113–2124. (c) Rosenbaum, D. M.; Zhang, C.; Lyons, J. A.; Holl, R.; Aragao, D.; Arlow, D. H.; Rasmussen, S. G. F.; Choi, H-J.; DeVree, B. T.; Sunahara, R. K.; Chae, P. S.; Gellman, S. H.; Dror, R. O.; Shaw, D. E.; Weis, W. I.; Caffrey, M.; Gmeiner, P.; Kobilka, B. K. *Nature* **2011**, *469*, 236-240. (d) Li, D.; Lee, J.; Caffrey, M.; *Cryst. Growth Des.* **2011**, *11*, 530-537. (e) Hofer, N.; Aragao, D.; Lyons, J. A.; Caffrey, M. *Cryst. Growth Des.* **2011**, *11*, 1182-1192. (f) Rasmussen, S. G. F.; DeVree, B. T.; Zou, Y.; Kruse, A. C.; Chung, K. Y.; Kobilka, T. S.; Thian, F. S.; Chae, P. S.; Pardon, E.; Calinski, D.; Mathiesen, J. M.; Shah, S. T. A.; Lyons, J. A.; Caffrey, M.; Gellman, S.

Org. Synth. **2012**, *89*, 183-201

H.; Steyaert, J.; Skiniotis, G.; Weis, W. I.; Sunahara, R. K.; Kobilka, B. K. *Nature* **2011**, *in press*.

8. Monoacylglycerols such as monoolein [9.9], monopalmitolein [9.7] and monovaccenin [11.7] are commercially available. **Note added in proof:** Several additional 1-MAGs, including [7.7], became commercially available from Avanti Polar Lipids, Inc. late in the checking of this procedure.

9. Jensen, R. G.; Pitas, R. E. in *Advances in Lipid Research* **1977**, *14*, 213–247.

10. Valicenti, A. J.; Pusch, F. M.; Holman, R. T. *Lipids* **1985**, *20*, 234–242.

11. (a) Gehring, L.; Haase, D.; Habben, K.; Kerkhoff, C.; Meyer, H. H.; Kaever, V. *J. Lipid Res.* **1998**, *39*, 1118–1126. (b) Zhang, X.; Schlosser, M. *Tetrahedron Lett.* **1993**, *34*, 1925–1928.

12. (a) Miyaura, N.; Ishiyama, T.; Ishikawa, M.; Suzuki, A. *Tetrahedron Lett.* **1986**, *27*, 6369–6372. (b) Miyaura, N.; Ishiyama, T.; Sasaki, H.; Ishikawa, M.; Satoh, M.; Suzuki, A. *J. Am. Chem. Soc.* **1989**, *111*, 314–321. (c) Trauner, D.; Chemler, S. R.; Danishefsky, S. J. *Angew. Chem. Int. Ed.* **2001**, *40*, 4544–4568.

13. (a) Alexakis, A.; Cahiez, G.; Normant, J. F. *Organomet. Chem.* **1979**, *177*, 293–298. (b) Miller, R. B.; McGarvey, G. *Syn. Comm.* **1978**, *8*, 291–299. (c) Stewart, S. K.; Whiting, A. *Tetrahedron Lett.* **1995**, *36*, 3932–3932. (d) Stork, G.; Zhao, K. *Tetrahedron Lett.* **1989**, *30*, 2173–2174. (e) Bestmann, H. J.; Rippel, H. C.; Dostalek, R. *Tetrahedron Lett.* **1989**, *30*, 5261–5262.

14. (a) Brown, H. C.; Blue, C. D.; Nelson, D. J.; Bhat, N. G. *J. Org. Chem.* **1989**, *54*, 6064–6067. (b) Svatos, A.; Saman, D. *Collect. Czech. Chem. Comm.* **1997**, *62*, 1457–1467.

15. Denmark, S. E.; Wang, Z. *Org. Syn.* **2004**, *81*, 42–53.

16. (a) Murgia, S.; Caboi, F.; Monduzzi, M.; Ljusberg-Wahren, H.; Nylander, T.; *Progr. Colloid Polym. Sci.* **2002**, *20*, 41–46. (b) Lyubachevskaya, G.; Boyle-Roden, E. *Lipids* **2000**, *35*, 1353–1358.

17. Johnson, C.R. and Braun, M.P. *J. Am. Chem. Soc.* **1993**, *115*, 11014–11015.

18. An alternative approach to the synthesis of 1-monoacylglycerols was reported while this work was in the checking process. See Fu, Y.; Weng, Y.; Hong W-X.; Zhang, Q. *Synlett* **2011**, 809–812.

Appendix
Chemical Abstracts Nomenclature; (Registry Number)

(2,2-Dimethyl-1,3-dioxolan-4-yl)methyl 5-hexenoate: 5-Hexenoic acid, (2,2-dimethyl-1,3-dioxolan-4-yl)methyl ester; (749266-07-7)

5-Hexenoic acid; (1577-22-6)

4-(Dimethylamino)pyridine: 4-Pyridinamine, *N,N*-dimethyl-; (1122-58-3)

N,N'-Dicyclohexylcarbodiimide; (538-75-0)

DL-1,2-Isopropylideneglycerol: 1,3-Dioxolane-4-methanol, 2,2-dimethyl-; (100-79-8)

1-Iodo-1-octyne; (81438-46-2)

1-Octyne; (629-05-0)

Iodine; (7553-56-2)

n-Butyllithium; (106-72-8)

(Z)-1-Iodo-1-octene; (52356-93-1)

Borane dimethyl sulfide complex: Boron, trihydro[thiobis[methane]]-, (T-4)-; (13292-78-0)

Cyclohexene; (110-83-8)

(2,2-Dimethyl-1,3-dioxolan-4-yl)methyl (*Z*)-tetradec-7-enoate: 7-Tetradecenoic acid, (2,2-dimethyl-1,3-dioxolan-4-yl)methyl ester, (7*Z*)-; (749266-24-8)

9-Borabicyclo[3.3.1]nonane: 9-BBN; (280-64-8)

Cesium carbonate; (534-17-8)

Triphenylarsine; (603-32-7)

Dichloro[1,1'-bis(diphenylphosphino)ferrocene]palladium(II)•dichloromethane; (95464-05-4)

2,3-Dihydroxypropyl (*Z*)-tetradec-7-enoate: 7-Tetradecenoic acid, 2,3-dihydroxypropyl ester, (7*Z*)-; (749266-35-1)

Org. Synth. **2012**, *89*, 183-201

David J. Hart was born in Lansing, Michigan, USA in 1948. He received his BS degree from the University of Michigan in 1972 after serving as a conscientious objector for two years (1967-1969) during the Vietnam War. He received his PhD in 1976 with William G. Dauben, and pursued postdoctoral studies at Caltech with David A. Evans. He moved to The Ohio State University Department of Chemistry in 1978 where he remained on the faculty for 31 years. He is now Professor Emeritus at OSU and is continuing research at a leisurely pace in the field of organic synthesis.

Dexi Yang was born and raised in China. He received his BS from Nanjing University in 2000 where he continued studies in bioinorganic chemistry (MS in 2003 with Wenxia Tang). He pursued graduate studies at The Ohio State University (PhD in 2008 with David J. Hart). His PhD studies focused on application of a radical cyclization strategy toward total synthesis of polyandranes and chaparrinone. He continued postdoctoral studies at Scripps Florida with Glenn C. Micalizio. Currently, he is working on application of titanium coupling reaction to stereoselective synthesis of highly substituted piperidines, indolizidines, quinolizidines and alkaloids.

Valerie Cwynar grew up in New Castle, PA. After receiving her B.S. in 2001 from the University of Pittsburgh, she continued her studies at The Ohio State University under the direction of Professor David J. Hart. Initially, her graduate work focused on the preparation of synthetic lipids, and transitioned to efforts toward the synthesis of (5*R*)- and (5*S*)-polyandrane. After receiving her Ph.D. in 2007, she moved to the University of Michigan to begin her postdoctoral studies under the direction of Professor Edwin Vedejs focusing on development of synthetic methodology inspired by (*R*)-telomestatin. Currently, she is a Scientist I at Vertex Pharmaceuticals Inc.

Dr. Madanmohan Jakkam was born in Andhrapradesh, India. He received his MS degree from the University of Pune, India in 1990, and his PhD at the National Chemical Laboratory in Pune. His doctoral studies were directed toward the synthesis of taxanoids and use of heterogeneous catalysts for organic transformations. He performed postdoctoral studies at National Taiwan University (synthesis of calexarines), at New Mexico State University (synthesis of perfluoro(bis)phosphonic acids), and at The Ohio State University (synthesis of monoacylglycerols). He then worked on polyamines at Medigen Biosciences in Madison, Wisconsin, and is currently a Research Scientist at GL Synthesis in Worcester, MA.

Jean P. Lee was born in Newport, Co. Tipperary, Ireland in 1978. She received her B.Sc. degree in Applied Chemistry from Kingston University, Surrey in 2000. She pursued graduate studies at the Royal College of Surgeons, Dublin (Ph.D. in 2005 with Marc Devocelle and John G. Kelly) and postdoctoral studies at the University of Limerick with Martin Caffrey from 2006-2008. She then moved from academia to industry in 2008 to a Process Specialist position for Abbott Ireland Diagnostic Division, Sligo where she is as a member of the Technical Operations Team working on process improvement projects.

Joseph A. Lyons was born in Ennis, Co. Clare, Ireland in 1982. He received his B. Sc. degree in Industrial Chemistry from the University of Limerick in 2004. He began his graduate studies towards a PhD between The Ohio State University and University of Limerick with Prof. Martin Caffrey. He is currently completing his graduate thesis on the use of synthetic monoacylglycerols as hosting matrices for the crystallization of integral membrane proteins.

Martin Caffrey grew up in Dublin, Ireland and was awarded a first-class honours degree in Agricultural Science at University College Dublin in 1972. With an MS in Food Science and a PhD in Biochemistry from Cornell University, Ithaca, NY, he embarked on a professorial career in the Chemistry Department at the Ohio State University, Columbus, Ohio. In 2003 he returned to Ireland to establish a multi-disciplinary program in Membrane Structural and Functional Biology at the University of Limerick with funding from Science Foundation Ireland and the USA National Institutes of Health. Its mission is to establish the molecular bases for biomembrane assembly and stability and to understand how membranes transform and transmit in health and disease. In 2009, his research group moved to Dublin when Prof. Caffrey received a Personal Chair at Trinity College Dublin with joint appointments in the School of Medicine and the School of Biochemistry and Immunology.

Zachary Gates received his B.S. degree in chemistry from The University of Chicago (2007), where he conducted undergraduate research in the laboratory of Stephen Kent. During this time he worked closely with graduate students Duhee Bang and Brad Pentelute (now Assistant professors at Yonsei University and Massachusetts Institute of Technology, respectively). Currently, Zak is a NSF predoctoral fellow in the laboratory of Viresh Rawal. Zak is broadly interested in the chemical sciences, chemical education, and the intersection of science, technology, and society. Hobbies include music, martial arts, small business, and community outreach.

Discussion Addendum for:
Suzuki-Miyaura Cross-Coupling: Preparation of 2'-Vinylacetanilide

Prepared by Donal F. O'Shea.*[1]
Original article: Cottineau, B.; Kessler, A.; O'Shea, D. F. *Org. Synth.* **2006**, *83*, 45.

Functionalized styrenes are of considerable importance as versatile synthetic intermediates and for the generation of new polymeric materials. The diverse spectrum of transformations achievable from the two carbon unit of the vinyl functional group has given rise to the development of numerous approaches for their synthesis.[2] Some of the more salient features of any reagent are bench-stability, a general set of reaction conditions in which it can be employed and substrate functional group tolerance. At the outset we felt an attractive approach to achieving vinylations would be to utilize a Suzuki-Miyaura cross coupling protocol which would necessitate a vinyl boronic acid reagent. But an immediate stumbling block to this approach was the instability of vinyl boronic acid which undergoes polymerization upon attempted isolation.[3a,c] Consequently to address this issue we turned to a pyridine promoted cyclotrimerization as an atom economical approach to stabilizing vinyl boronic acid as its trivinylboroxane-pyridine complex (Figure 1).[3a,b] While the trivinylboroxane is a bench stable solid, under the aqueous basic conditions of a Suzuki-Miyaura cross coupling it can hydrolyze to the vinyl boronic acid *in situ*.[3a] Since our initial report on use of this reagent it has become commercially available from several vendors and has been successfully exploited for C-C, C-O and C-N bond forming transformations.

Org. Synth. **2012**, *89*, 202-209
Published on the Web 10/25/2011

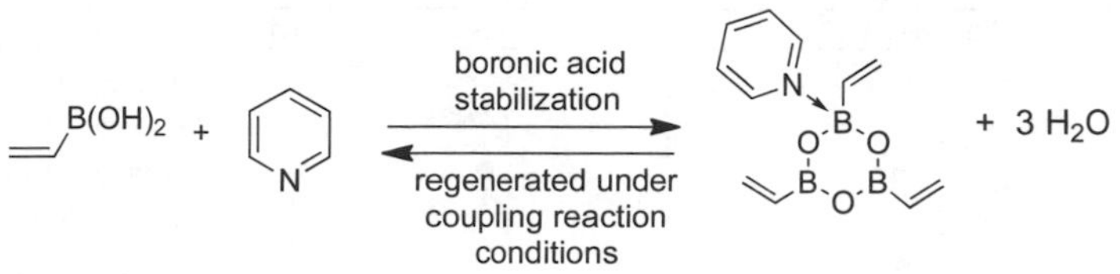

Figure 1. Stabilization of vinyl boronic acid and *in situ* reaction regeneration.

Vinylation of arenes with trivinylboroxane has been reported for a diverse range of *ortho- meta-* and *para-* substituted aryl rings (Table 1). The coupling reaction has shown an impressive range of functional group tolerance with good yields for many electronically and sterically challenging substrates. In the majority of examples the arylbromide substrate was utilized in conjunction with our general set of reaction conditions of Pd(PPh₃)₄ as catalyst and a carbonate base with dimethoxyethane / water as solvent (Table 1, entries 1-11). In several cases the trivinylboroxane was used in a 0.5 equiv. ratio to the coupling substrate illustrating that the vinyl-trimer can hydrolyze *in situ* to generate three equiv. of vinyl boronic acid.[5,18] It was shown that a recoverable polymer bound palladacycle could be used for vinylation reactions of aryl bromides (Table 1, entry 12) in addition to coupling of cinnamyl- and benzyl-chloride to provide 1-phenylpenta-1,4-diene and allylbenzene respectively (Table 1, entry 13). Multiple vinylations of substrates have also been accomplished with this reagent. For example, the di-vinylation of two different aryl rings of a HIV-1 protease inhibitor substrate (Table 1, entry 14) and of 1,4-dibromo-2-fluorobenzene were successfully accomplished (Table 1, entry 15). Remarkably, a tetra-vinylation provided 1,2,4,5-tetravinylbenzene in almost quantitative yields from the corresponding tetrabromobenzene (Table 1, entry 16). The palladium-catalyzed carbonylative cross-coupling of iodobenzene with 0.34 equiv. of trivinylboroxane gave phenylpropenone illustrating that each vinyl moiety of the reagent is available for reaction in this transformation (Table 1, entry 17).

Table 1. Representative examples of aryl, allyl and carbonylative vinylations.

Entry	Substrate	Product	Yield	Conditions
1[3a]			93%	Pd(PPh$_3$)$_4$, K$_2$CO$_3$, DME / H$_2$O, reflux, 15h
2[3a]			91%	Pd(PPh$_3$)$_4$, K$_2$CO$_3$, DME / H$_2$O, reflux, 15h
3[4]			90%	Pd(PPh$_3$)$_4$, K$_2$CO$_3$, DME / H$_2$O, reflux, 20h
4[5]			78%	Pd(PPh$_3$)$_4$, K$_2$CO$_3$, DME / H$_2$O, reflux, 20h
5[6]			77%	Pd(PPh$_3$)$_4$, K$_2$CO$_3$, DME / H$_2$O, reflux, 3h
6[7]			74%	Pd(PPh$_3$)$_4$, K$_2$CO$_3$, DME / H$_2$O, reflux, 20h
7[8]			95%	Pd(PPh$_3$)$_4$, K$_2$CO$_3$, DME / H$_2$O, reflux, 14h
8[9]			70%	Pd(PPh$_3$)$_4$, K$_2$CO$_3$, DME / H$_2$O, microwave, 130 °C, 1h
9[10]			78%	Pd(PPh$_3$)$_4$, K$_2$CO$_3$, DME / H$_2$O, reflux, 20h
10[11]			86%	Pd(PPh$_3$)$_4$, K$_2$CO$_3$, DME / H$_2$O, reflux, 20h
11[12]			70%	Pd(PPh$_3$)$_4$, K$_2$CO$_3$, DME / H$_2$O, reflux, 24h
12[13]			88%	PS-Pd, TBAB, K$_2$CO$_3$, H$_2$O, 100 °C, 2.5h
13[13]			80%	PS-Pd, TBAB, KOH, H$_2$O-acetone, 50 °C, 2.5h
14[14]			57%	Pd(OAc)$_2$, [(tBu$_3$)PH]BF$_4$, K$_2$CO$_3$, DME / H$_2$O, microwave, 100 °C, 15 min
15[15]			56%	Pd(PPh$_3$)$_4$, K$_2$CO$_3$, DME / H$_2$O, 85 °C, 24 h
16[16]			99%	Pd(OAc)$_2$, PPh$_3$ K$_2$CO$_3$, DME / H$_2$O, 100 °C, autoclave, 24h
17[17]	+ CO		67%	P$_{CO}$ 5 bar, Pd(OAc)$_2$, Ph$_3$P, THF, 80 °C, 20 h

204

Cross-coupling of heterocycles has also been extensively employed with the vinylboroxane reagent. Again, a uniformity of reaction conditions has been used across a diverse range of substrates often providing the products in excellent yields (Table 2). In addition to bromo-substituted heterocycles (entries 1-7), chloro (entry 8), triflate (entries 9-11) and tosylate (entry 12) derivatives have been successfully utilized as coupling partners.

Table 2. Representative examples of heterocycle vinylations.

Entry	Substrate	Product	Yield	Conditions
1[18]	NHCOtBu	NHCOtBu	85%	Pd(PPh$_3$)$_4$, K$_2$CO$_3$, DME / H$_2$O, reflux, 20 h
2[18]	NH	NH	86%	Pd(PPh$_3$)$_4$, K$_2$CO$_3$, DME / H$_2$O, reflux, 20 h
3[19]			93%	Pd(PPh$_3$)$_4$, K$_2$CO$_3$, DME / H$_2$O, reflux, 24 h
4[20]			90%	Pd(PPh$_3$)$_4$, K$_2$CO$_3$, DME / H$_2$O, 85 °C, 10 h
5[21]			88%	Pd(PPh$_3$)$_4$, Na$_2$CO$_3$, THF / H$_2$O, 70 °C, 12 h
6[22]	t-BuO$_2$C	t-BuO$_2$C	76%	Pd(PPh$_3$)$_4$, K$_2$CO$_3$, DME / H$_2$O, 85 °C, 2 h
7[23]			95%	Pd(PPh$_3$)$_4$, K$_2$CO$_3$, dioxane / H$_2$O, 75 °C
8[24]	OTHP	OTHP	89%	Pd(PPh$_3$)$_4$, K$_2$CO$_3$, DME / H$_2$O, reflux, 20 h
9[25]	OTf		76%	Pd(PPh$_3$)$_4$, K$_2$CO$_3$, DME / H$_2$O, reflux, 17 h
10[26]	TfO		71%	Pd(PPh$_3$)$_4$, K$_2$CO$_3$, DME / H$_2$O, microwave 150 °C, 30 min
11[27]	Bn, OTf, Ph	Bn, Ph	78%	Pd(PPh$_3$)$_4$, Na$_2$CO$_3$, dioxane / H$_2$O, microwave 90 °C, 5 min
12[28]	OTs, TBSO, TBSO OCH$_3$, NH$_2$	TBSO, TBSO OCH$_3$, NH$_2$	not reported	Pd(PPh$_3$)$_4$, LiBr, dioxane / H$_2$O

The scope of transformations utilizing this reagent has been broadened with both *O*-vinylation and *N*-vinylation methods reported (Table 3). Coupling conditions were mediated by copper acetate for both *O* and *N* heteroatoms at room temperature using CH_2Cl_2/pyridine as solvent system.[29] These mild conditions were tolerant of halo (entries 1-3), formyl (entry 4), phosphinate (entry 6) and aziridine (entries 9, 10) functional groups.

Table 3. Representative examples of *O* and *N* vinylations.

Entry	Substrate	Product	Yield	Conditions
1[29]			95%	Cu(OAc)$_2$, pyridine, CH$_2$Cl$_2$, rt, 24 h
2[29]			76%	Cu(OAc)$_2$, pyridine, CH$_2$Cl$_2$, rt, 24 h
3[29]			76%	Cu(OAc)$_2$, pyridine, CH$_2$Cl$_2$, rt, 24 h
4[29]			84%	Cu(OAc)$_2$, pyridine, CH$_2$Cl$_2$, rt, 24 h
5[30]			52%	Cu(OAc)$_2$, pyridine, CH$_2$Cl$_2$, rt, 24 h
6[31]			71%	Cu(OAc)$_2$, pyridine, CH$_2$Cl$_2$, O$_2$, rt, 40 h
7[32]			77%	Cu(OAc)$_2$, pyridine, CH$_2$Cl$_2$, O$_2$, rt, 40 h
8[32]			71%	Cu(OAc)$_2$, pyridine, CH$_2$Cl$_2$, O$_2$, rt, 40 h
9[33]			30%	Cu(OAc)$_2$, myristic acid, 2,6-lutidine, toluene, rt, 24 h
10[34]			27%	Cu(OAc)$_2$, pyridine, O$_2$, rt, 10 h

206

In summary, 2,4,6-trivinylcyclotriboroxane-pyridine complex continues to serve as a versatile and reliable vinyl source for a range of important coupling reactions.

1. School of Chemistry and Chemical Biology, University College Dublin, Belfield, Dublin 4, Ireland. Financial support from Science Foundation Ireland is gratefully acknowledged.
2. For a recent review see; Denmark, S. E.; Butler, C. R. *Chem. Commun.* **2009**, 20.
3. (a) Kerins F.; O'Shea D. F. *J. Org. Chem.* **2002**, *67*, 4968. (b) Matteson, D. S. *J. Org. Chem.* **1962**, *27*, 3712. (c) Matteson, D. S. *J. Am Chem. Soc.* **1960**, *82*, 4228.
4. Coleman, C. M.; O'Shea, D. F. *J. Am. Chem. Soc.* **2003**, *125*, 4054.
5. Kessler, A.; Coleman, C. M.; Charoenying, P.; O'Shea, D. F. *J. Org. Chem.* **2004**, *69*, 7836.
6. Charrier, N.; Demont, E.; Dunsdon, R.; Maile, G.; Naylor, A.; O'Brien, A.; Redshaw, S.; Theobald, P.; Vesey, D.; Walter, D. *Synthesis* **2006**, 3467.
7. Spraul, B. K.; Suresh, S.; Jin, J.; Smith Jr, D. W. *J. Am. Chem. Soc.* **2006**, *128*, 7055.
8. Murakami, M.; Kadowaki, S.; Fujimoto, A.; Ishibashi, M.; Matsuda, T. *Org. Lett.* **2005**, *7*, 2059.
9. Collier, S. J.; Wu, X.; Poh, Z.; Rajkumar, G. A.; Yet, L. *Synth. Comm.* **2011**, *41*, 1394.
10. Mennen, S. M.; Miller, S. J. *J. Org. Chem.* **2007**, *72*, 5260.
11. Bennasar, M. L.; Roca, T.; Monerris, M.; Garcia-Diaz, D. *J. Org. Chem.* **2006**, *71*, 7028.
12. Sengupta, S.; Mukhopadhyay, R.; Achari, B.; Banerjee, A. K. *Tetrahedron Lett.* **2005**, *46*, 1515.
13. Alacid, E.; Nájera, C. *J. Organomet. Chem.* **2009**, *694*, 1658.
14. Wannberg, J.; Sabnis, Y. A.; Vrang, L.; Samuelsson, B.; Karlén, A.; Hallberg, A.; Larhed, M. *Bioorg. Med. Chem.* **2006**, *14*, 5303.
15. Mantel, M. L. H.; Søbjerg, L. S.; Huynh, T. H. V.; Ebran, J.-P.; Lindhardt, A. T.; Nielsen, N. C.; Skrydstrup, T. *J. Org. Chem.* **2008**, *73*, 3570.
16. Su, K.-J.; Mieusset, J.-L.; Arion, V. B.; Knoll, W.; Brecker, L.; Brinker, U. H. *J. Org. Chem.* **2010**, *75*, 7494.
17. Pirez, C.; Dheur, J.; Sauthier, M.; Castanet, Y.; Mortreux, A. *Synlett*

2009, *11*, 1745.

18. (a) Cottineau, B.; O'Shea, D. F *Tetrahedron* **2007**, *63*, 10354. (b) Cottineau, B.; O'Shea, D. F. *Tetrahedron Lett.* **2005**, *46*, 1935.
19. Brown, P.; Dabbs, S. WO **2008**/128953 A1.
20. Miller, W. H.; Rouse, M. B.; Seefeld, M. A. US **2007**/0270417 A1.
21. Fackler, P.; Berthold, C.; Voss, F.; Bach, T. *J. Am. Chem. Soc.* **2010**, *132*, 15911.
22. Chao, R.; Zhang, W. WO **2010**/039719 A1.
23. Rouse, M. B.; Seefeld, M. A.; Leber, J. D.; McNulty, K. C.; Sun, L.; Miller, W. H.; Zhang, S.; Minthorn, E. A.; Concha, N. O.; Choudhry, A. E.; Schaber, M. D.; Heerding, D. A. *Bioorg. Med. Chem. Lett.* **2009**, *19*, 1508.
24. Gibson, C.; Schnatbaum, K.; Pfeifer, J. R.; Locardi, E.; Paschke, M.; Reimer, U.; Richter, U.; Scharn, D.; Faussner, A.; Tradler, T. *J. Med. Chem.* **2009**, *52*, 4370.
25. (a) Szadkowska, A.; Gstrein, X.; Burtscher, D.; Jarzembska, K.; Wozniak, K.; Slugovc, C.; Grela, K. *Organometallics* **2010**, *29*, 117. (b) Slugovc C.; Perner, B.; Stelzer, F.; Mereiter, K. *Organometallics* **2004**, *23*, 3622.
26. Harbottle, G. W.; Feeder, N.; Gibson, K. R.; Glossop, M.; Maw, G. N.; Million, W. A.; Morel, F. F.; Osborne, S.; Poinsard, C. *Tetrahedron Lett.* **2007**, *48*, 4293.
27. Bower, J. F.; Williams, A. J.; Woodward, H. L.; Szeto, P.; Lawrence, R. M.; Gallagher, T. *Org. Biomol. Chem.* **2007**, *5*, 2636.
28. Imoto, S.; Hori, T.; Hagihara, S.; Taniguchi, Y.; Sasaki, S.; Nagatsugi, F. *Bioorg. Med. Chem. Lett.* **2010**, *20*, 6121.
29. McKinley, N. F.; O'Shea, D. F. *J. Org. Chem.* **2004**, *69*, 5087.
30. Ward, A. F.; Wolfe, J. P. *Org. Lett.* **2011**, *13*, 4728.
31. Itoh, H.; Yamamoto, E.; Masaoka, S.; Sakai, K.; Tokunaga, M. *Adv. Synth. Catal.* **2009**, *351*, 1796.
32. Sakuma, T.; Yamamoto, E.; Aoyama, H.; Obora, Y.; Tsuji, Y.; Tokunaga, M. *Tetrahedron: Asymmetry*, **2008**, *19*, 1593.
33. Dalili, S.; Yudin, A. K. *Org. Lett.* **2005**, *7*, 1161.
34. Tsang, D. S.; Yang, S.; Alphonse, F.-A.; Yudin, A. K. *Chem. Eur. J.* **2008**, *14*, 886.

Donal O'Shea received his Ph.D. degree in Chemistry from University College Galway (Ireland) in 1994. He held post-doctoral positions in the University of Edinburgh (Scotland) and Carnegie Mellon University, Pittsburgh (USA) following which he was a research scientist at Eastman Kodak Company in Rochester, New York. In 1999 he returned to academia to a position in University College Dublin and was promoted to Associate Professor of Chemistry in 2007. His current research interests include development of new synthetic strategies utilizing mixed Li/K metal amides, automated flow micro-reactors, the development of light activated therapeutics and near-infrared fluorochromes.

Synthesis of Trifluoromethyl Ketones from Carboxylic Acids:
4-(3,4-Dibromophenyl)-1,1,1-trifluoro-4-methylpentan-2-one

Submitted by Jonathan T. Reeves[1], Zhulin Tan, Daniel R. Fandrick, Jinhua J. Song, Nathan K. Yee and Chris H. Senanayake.
Checked by David Hughes.

1. Procedure

A. 3-(3,4-Dibromophenyl)-3-methylbutanoic acid (**3**). A 250-mL, oven-dried three-necked round-bottomed flask, equipped with a 3-cm oval PTFE-coated magnetic stirring bar, is charged with aluminum chloride (13.4 g, 100 mmol, 2.0 equiv) (Note 1). The flask is fitted with a gas inlet adapter connected to a nitrogen line and a gas bubbler. The other two necks are capped with rubber septa; a thermocouple probe is inserted through one of the septa (Note 2). Dichloromethane (8 mL) and 1,2-dibromobenzene (8.0 mL, 15.7 g, 67 mmol, 1.3 equiv) are added via pipette and stirring is initiated. To the resultant slurry is charged a solution of 3,3-dimethylacrylic acid (5.00 g, 50 mmol, 1.00 equiv) in dichloromethane (10 mL) via syringe over 10 min, maintaining the internal temperature ≤ 32 °C. The reaction mixture is stirred at room temperature for 2.5 h (Note 3). A 1-L Erlenmeyer flask equipped with a 5 cm flat stir bar and a thermocouple probe (Note 2) is charged with water (150 mL) followed by concentrated hydrochloric acid (150 mL) and cooled to 2 °C in an ice-water bath. The Friedel-Crafts reaction mixture is transferred portion-wise via a 9-inch disposable glass pipette into the stirred, cold aqueous HCl solution over 5 min (Note 4). Ethyl

210

acetate (200 mL) is used to rinse residues from the reaction flask into the aqueous HCl. The biphasic mixture is then transferred to a 1-L separatory funnel and the lower aqueous phase is discarded. The organic phase is washed with water (100 mL), then brine (50 mL). The organic phase is filtered through a bed of sodium sulfate (30 g) in a medium porosity sintered glass funnel into a 500-mL round-bottomed flask, using ethyl acetate (2 x 50 mL) to rinse the filter cake. The solution is concentrated by rotary evaporation (50 °C, 20 mmHg) to a yellow oil (21 g, Note 5). The flask is charged with heptane (50 mL) and warmed to 60 °C using a water bath to complete dissolution (Note 6). The solution is allowed to cool to room temperature over about 2 h, during which time crystallization occurs. The slurry is further cooled to 0–5 °C by placement in a refrigerator overnight. The slurry is filtered through a medium porosity sintered glass funnel, washed with heptane (2 x 20 mL) pre-cooled to 0–5 °C, and then air-dried to constant weight to provide 3-(3,4-dibromophenyl)-3-methylbutanoic acid **3** as an off-white solid (12.1–12.8 g, 72–76% yield) (Note 7).

B. *4-(3,4-Dibromophenyl)-1,1,1-trifluoro-4-methylpentan-2-one* (**4**). A 250-mL three-necked, round-bottomed flask, equipped with a 3-cm oval PTFE-coated magnetic stirring bar and a reflux condenser fitted at the top with a gas inlet adapter connected to a nitrogen line and a gas bubbler, is charged with 3-(3,4-dibromophenyl)-3-methylbutanoic acid (**3**) (10.0 g, 29.7 mmol, 1.00 equiv) and toluene (70 mL) (Note 8). The other two necks are capped with rubber septa; a thermocouple probe is inserted through one of the septa (Note 2). To the resulting stirred slurry is charged trifluoroacetic anhydride (18.5 mL, 4.50 equiv) (Note 9). Then pyridine (14.5 mL, 6.0 equiv) is added dropwise (exothermic) via syringe over 5 min keeping the internal temperature of the reaction mixture ≤ 40 °C. The flask is heated using a heating mantle for 4 h, maintaining the internal temperature at 60–65 °C (Note 10). The reaction mixture is cooled to 2 °C in an ice-water bath. The reflux condenser is replaced with a 125-mL addition funnel through which is charged water (50 mL) to the reaction mixture over 5 min, maintaining the internal temperature of the reaction mixture < 40 °C (Note 11). The ice-water bath is removed and replaced with a heating mantle and the mixture is heated for 1 h, maintaining an internal temperature of 57–60 °C (Note 12). The reaction mixture is cooled to room temperature and transferred to a 500-mL separatory funnel using hexanes (150 mL) and water (100 mL). The layers are separated and the aq. layer is back extracted with 100 mL of 2:1 hexanes:toluene. The combined organic layers are washed with water (100 mL), then filtered through a bed of sodium sulfate (30 g) in a medium porosity sintered glass funnel into a 500-mL round-bottomed flask, using hexanes (2 x 50 mL) to rinse the filter cake. The filtrate is concentrated by rotary evaporation (50 °C, 20 mmHg) to an orange oil (13 g) (Note 13). The oil is purified by flash column chromatography (Note 14) to provide 4-(3,4-dibromophenyl)-1,1,1-trifluoro-4-methylpentan-2-one (**4**) as a light yellow oil (9.04–9.68 g, 78–84% yield) (Note 15).

2. Notes

1.   The following reagents and solvents in step A were used as received: aluminum chloride (Acros, 99%, powder), 1,2-dibromobenzene (98%, Sigma-Aldrich), 3,3-dimethylacrylic acid (Sigma-Aldrich, 97%), dichloromethane (Fisher, ACS reagent, 99.5%), ethyl acetate (Fisher, ACS reagent, 99%), 37% HCl (Fisher), and heptane (Sigma Aldrich, Chromasolv, 99%).   Deionized water was used throughout.

2.   The internal temperature was monitored using a J-Kem Gemini digital thermometer with a Teflon-coated T-Type thermocouple probe (12-inch length, 1/8 inch outer diameter, temperature range –200 to +250 °C).

3.   The checker monitored the reaction by ^{1}H NMR as follows. A 0.1 mL aliquot of the reaction mixture was quenched into 1 mL CDCl$_3$ and 1 mL 6N HCl.   The CDCl$_3$ layer was analyzed by NMR.   Diagnostic resonances were δ 5.7 (1 H), 2.2 (3 H), and 2.0 (3 H) for **2** and 2.63 (2 H) and 1.43 (6 H) for **3**. At the 90 min time point, approx 4% unreacted **2** remained. The submitters monitored both reactions using reverse phase HPLC employing the following conditions: column: TSK-gel SuperODS column (ID 4.6 mm, length 5.0 cm, particle size 2 μm); detection at 220 nm; mobile phase of 10% ramping to 90% MeCN in water (with 0.05 volume% TFA in both phases) over 3.5 min, then hold at 90% MeCN in water for 1.5 min; flow rate: 2.0 mL/min.   The starting materials, products and solvents had the following retention times: **1** (3.1 min); **2** (1.2 min); **3** (2.9 min); **4** (3.7 min); toluene (2.7 min). TLC may also be used.   The starting materials and products have the following R$_f$ values in 60:40 hexanes/EtOAc: **1** (0.9); **2** (0.7); **3** (0.35); **4** (0.9). In 95:5 hexanes/MTBE, **4** has an R$_f$ value of 0.5.

4.   The addition of the Friedel-Crafts reaction mixture into the aqueous HCl is exothermic. The temperature at the end of transfer was 20 °C.

212

5.   The composition of the crude material by ^{1}H NMR (by wt%) was approx. 83% product **3**, 11% 1,2-dibromobenzene, 5 % EtOAc, and 1% unreacted **2** (3 mol %).

6.   In most experiments the yellow oil crystallized prior to addition of heptane. Heating the heptane mixture to 60 °C effects dissolution of the solid, therefore a true recrystallization is achieved.

7.   3-(3,4-Dibromophenyl)-3-methylbutanoic acid **3** has the following physical and spectroscopic properties: off-white solid; mp 89–91 °C; IR (neat) 2966, 1710, 1648, 1473, 1377, 1250, 1166, 1011 cm^{-1}; MS m/z (relative intensity), 320.9 (21), 318.8 (42), 316.8 (21) [M-OH]; 278.9 (49), 276.8 (100), 274.8 (50) [M-CH$_2$CO$_2$H]; 215.0 (43), 141.0 (64), 127.0 (59), 118.0 (54), 102.0 (63); ^{1}H NMR (CDCl$_3$, 400 MHz) δ: 1.43 (s, 6 H), 2.63 (s, 2 H), 7.16 (dd, J = 8.5, 2.3 Hz, 1 H), 7.54 (d, J = 8.4 Hz, 1 H), 7.60 (d, J = 2.3 Hz, 1 H), 11.1 (broad s, 1 H); ^{13}C NMR (CDCl$_3$, 100 MHz) δ: 29.0, 37.0, 47.7, 122.4, 124.9, 126.7, 131.3, 133.5, 149.3, 177.4; Anal. Calcd. For C$_{11}$H$_{12}$Br$_2$O$_2$: C, 39.32; H, 3.60. Found: C, 39.49; H, 3.29; HPLC area % purity: 95 % (HPLC method: column, fused-core C-18, 4.6x100 mm, 2.7 μm particle size; gradient elution using a mobile phase of 90% aq. 0.1% H$_3$PO$_4$/10% MeCN to 10% aq. 0.1% H$_3$PO$_4$/90% MeCN over 6 min, holding the latter for 2 min; flow rate 1.8 mL per min; temperature 40 °C, detection at 210 nm; t$_R$ for **3** was 4.8 min.)

8.   The following reagents and solvents in step B were used as received: toluene (Sigma-Aldrich, Chromasolv, 99.9%), trifluoroacetic anhydride (Sigma-Aldrich, Reagent Plus, ≥ 99%), pyridine (Sigma-Aldrich, 99%), hexanes (Fisher, ACS reagent, >98.5%), methyl t-butyl ether (Sigma-Aldrich), and silica gel (Fisher, 230-400 mesh, 60 Å).

9.   The addition of TFAA is slightly endothermic. The temperature of the reaction mixture generally decreases by 2–4 °C after the TFAA is added.

10.   The submitters monitored the reaction for consumption of starting acid **3** by HPLC and TLC (Note 3). The checker monitored the reaction by ^{1}H NMR by quenching a 0.1 mL aliquot of the reaction mixture in 1 mL CDCl$_3$ and 1 mL water, concentrating the CDCl$_3$ to remove residual toluene, then redissolving in CDCl$_3$. The reaction samples at the 4 h time point had no detectable starting material (monitoring resonances at δ 2.63 and 1.43). Product **4** was the major species present.

11.   The addition of water is highly exothermic, particularly for the first few mL. The addition is also accompanied by the evolution of CO$_2$ gas.

12. Heating the reaction mixture at 60 °C for 1 hour drives the decarboxylation to completion.

13. The crude material by ^{1}H NMR was approx. 87 wt % product, 13 wt % toluene.

14. The checker used flash chromatography to purify product as follows: 150 g silica gel with an eluent of 300 mL 2% MTBE/hexanes, 300 mL 3% MTBE/hexanes, 600 mL MTBE/hexanes; 50 mL fractions. Product eluting in fractions 8-21 were combined and concentrated by rotary evaporation (40 °C, 20 mmHg), then held under vacuum (20 mmHg) for 14 h at room temperature to afford product as a yellow oil. Submitters used a Combi-Flash apparatus using a 150 g silica gel column using hexanes/methyl t-butyl ether (100:0 to 95:5 v/v) as eluent, collecting 25 mL fractions at a flow rate of 85 mL/min, and the UV detector was set at 220 nm. Residual toluene eluted first (2–4 min) with 100% hexanes. The eluent was then ramped over 13 min from 0 to 5% MTBE in hexanes. The product began eluting at 10 min and elution was complete after 26 min (50 fractions of 25 mL were required to collect the product.

15. 4-(3,4-Dibromophenyl)-1,1,1-trifluoro-4-methylpentan-2-one **4** has the following physical and spectroscopic properties: light yellow oil; IR (neat) 2971, 1765, 1469, 1380, 1285, 1200, 1149, 1025 cm^{-1}; GC-MS m/z (relative intensity), 390 (31), 388 (62), 386 (31), [M$^+$], 279 (84), 277 (100), 275 (86), [M – CH$_2$COCF$_3$], 251 (20), 249 (38), 247 (20), 115 (45); ^{1}H NMR (CDCl$_3$, 400 MHz) δ: 1.46 (s, 6 H), 3.05 (s, 2 H), 7.13 (dd, J = 8.5, 2.4 Hz, 1 H), 7.55 (d, J = 8.5 Hz, 1 H), 7.59 (d, J = 2.4 Hz, 1 H); ^{13}C NMR (CDCl$_3$, 100 MHz) δ: 28.8, 36.8, 48.5, 115.3 (q, J = 293 Hz, 1 C), 122.7, 125.2, 125.9, 131.1, 133.7, 148.5, 189.1 (q, J = 35 Hz, 1 C); Purity by GC: 97% (t_R = 11.7 min; conditions: Agilent DB35MS column; 30 m x 0.25 mm; initial temp 60 °C, ramp at 20 °C/min to 280 °C, hold 15 min); two peaks were present by reverse phase HPLC (conditions outlined in Note 7): t_R 5.3 min (30%, ketone hydrate of **4**) and 6.2 min (70%, ketone product **4**).

Safety and Waste Disposal Information

All hazardous materials should be handled and disposed of in accordance with "Prudent Practices in the Laboratory"; National Academies Press; Washington, DC, 2011.

3. Discussion

Trifluoromethyl ketones are useful intermediates for the introduction of the trifluoromethyl group into organic molecules.[2] While several methods for the synthesis of trifluoromethyl ketones have been reported, the procedure reported by Zard and co-workers is attractive in terms of simplicity.[3] This pioneering work involves treating primary (α-CH$_2$) acid chlorides at room temperature with trifluoroacetic anhydride and pyridine in CH$_2$Cl$_2$ for 0.5 – 8 hours, followed by hydrolysis/decarboxylation on addition of water. The application of the Zard procedure to more hindered primary acid chlorides, however, gives the trifluoromethyl ketone products in low conversions and yields, even after extended reaction times. It was found that by changing the solvent from CH$_2$Cl$_2$ to toluene and increasing the reaction temperature to 60 °C, more hindered substrates underwent full conversion and provided products in good yields. Furthermore, this process proved to be effective for the direct conversion of carboxylic acids to trifluoromethyl ketones (Table 1).[4] This avoids the need for conversion of the carboxylic acid to the corresponding acid chloride prior to trifluoromethyl ketone formation. Secondary (α-branched) carboxylic acids may also be converted to trifluoromethyl ketones if the reaction temperature is further increased to 100 °C. We have found this procedure is particularly well suited to 3,3-dimethyl-3-arylpropionic acid substrates, such as acid **3**, (entry 1) which contain aryl bromide functionality that is not stable to alternative methods employing strong bases. The starting acid **3** is conveniently prepared by a Friedel-Crafts reaction of 3,3-dimethylacrylic acid **2** with 1,2-dibromobenzene **1**.

Table 1. Examples of Direct Conversion of Carboxylic Acids to Trifluoromethyl Ketones

Entry	Acid	CF₃ Ketone	Yield[a]
1			80%
2			82%
3			55%
4			78%
5			63%
6			58%
7			52%
8			41%
9			66%

[a] Isolated Yield

Org. Synth. **2012**, *89*, 210-219

The proposed mechanism of the reaction involves α-trifluoroacetylation of a pyridinium enolate.[3] The pyridinium enolate may be derived from an acid chloride precursor in the case of Zard's reaction conditions[3] or a mixed anhydride of the starting carboxylic acid and trifluoroacetic acid in the present case (Scheme 1).[4]

Scheme 1. Proposed Mechanism for Acid to Trifluoromethyl Ketone Conversion

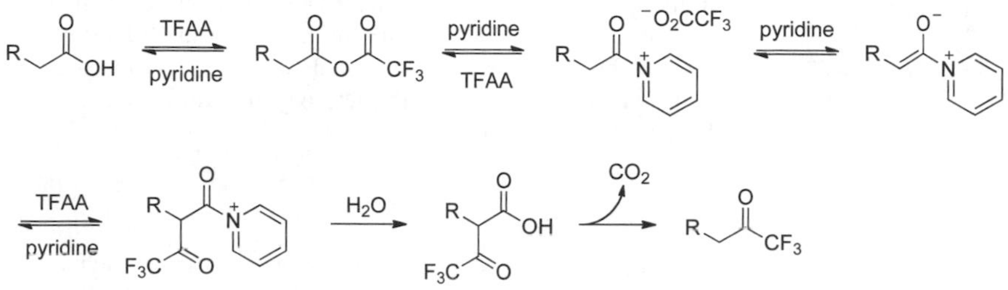

1. Department of Chemical Development, Boehringer Ingelheim Pharmaceuticals, Inc., 900 Ridgebury Road/P.O. Box 368, Ridgefield, Connecticut 06877-0368, USA, E-mail: jonathan.reeves@boehringer-ingelheim.com.
2. Reviews: (a) Bégué, J. P.; Bonnet-Delpon, D. *Tetrahedron* **1991**, *47*, 3207–3258; (b) Nenaidenko, V. G.; Sanin, A. V.; Balenkova, E. S. *Russian Chem. Rev.* **1999**, *68*, 437–458.
3. (a) Boivin, J.; El Kaim, L.; Zard, S. Z. *Tetrahedron Lett.* **1992**, *33*, 1285–1288; (b) Boivin, J.; El Kaim, L.; Zard, S. Z. *Tetrahedron* **1995**, *51*, 2573–2584; (c) Boivin, J.; El Kaim, L.; Zard, S. Z. *Tetrahedron* **1995**, *51*, 2585–2592.
4. Reeves, J. T.; Gallou, F.; Song, J. J.; Tan, Z.; Lee, H.; Yee, N. K.; Senanayake, C. H. *Tetrahedron Lett.* **2007**, *48*, 189–192.

Appendix
Chemical Abstracts Nomenclature; (Registry Number)

Aluminum Chloride; (7446-70-0)
1,2-Dibromobenzene; (583-53-9)

3,3-Dimethylacrylic acid: 2-Butenoic acid, 3-methyl- ; (541-47-9)
Trifluoroacetic anhydride; (407-25-0)

Jonathan T. Reeves was born in 1975 in Michigan, USA. He received a B.S. degree in chemistry from Hope College, Holland, MI in 1997. He then received a Ph.D. in organic chemistry under the guidance of Professor Peter Wipf at the University of Pittsburgh in 2002. After two years as an NIH postdoctoral fellow at Indiana University with Professor David R. Williams, he joined the Chemical Development department at Boehringer Ingelheim Pharmaceuticals, Inc., in Ridgefield, CT, where he is currently a Principal Scientist. His research interests include organometallic, heterocyclic and asymmetric methodology development.

Zhulin Tan received a B.S. degree in chemistry from Sichuan University in Chengdu, China (1988). He then joined Sichuan Chemical Engineering Institute as a research associate in Chengdu, China. After a brief stay at the University of Delaware, he moved to Worcester Polytechnic Institute in 1996 and obtained his M.S. in organic chemistry in 1999 under the guidance of Professor Stephen Weininger. He then joined the Department of Chemical Development at Boehringer Ingelheim Pharmaceuticals in Ridgefield, CT, where he is currently a Senior Research Associate. His research interests include heterocyle synthesis and organocatalysis using N-heterocyclic carbenes.

Daniel R. Fandrick received his B.S. degree in chemistry from the University of California, San Diego under the guidance of Professor Joseph M. O'Conner. In 2006, Daniel received his PhD degree in organic chemistry at Stanford University under the mentorship of Professor Barry M. Trost. His graduate studies focused on the development of the dynamic kinetic asymmetric transformations of vinyl aziridines and allenes with applications to total synthesis. After graduation, he joined the chemical development group at Boehringer-Ingelheim. His main interests are in the development of transition metal catalyzed and sustainable methodologies to provide efficient access to pharmaceutically useful scaffolds.

Dr. Jinhua Jeff Song received his undergraduate education at Nankai University in Tianjin, China (1992). After a brief stay at Rice University in Houston, TX, he moved to the Massachusetts Institute of Technology in 1993 and obtained his Ph.D. degree in 1998 under the supervision of Prof. Satoru Masamune. Subsequently, he joined the Department of Chemical Development at Boehringer Ingelheim Pharmaceuticals, where he is currently Senior Associate Director. His research areas encompass natural product synthesis, asymmetric synthesis of chiral biologically active compounds, efficient methodologies for heterocycle synthesis, and novel N-heterocyclic carbene-catalyzed reactions. Some of his work also received media attention and has been highlighted in the *Chemical and Engineering News*. Additionally, Dr. Song holds >15 patents on efficient synthesis of pharmaceutical agents.

Dr. Nathan Yee received his B.S. degree from the University of New York at Stony Brook in 1985, and then moved to the University of Illinois at Urbana-Champaign and obtained his Ph.D. degree in 1991 under the supervision of Prof. Robert Coates. He was a postdoctoral fellow at Yale University with Prof. Frederick Ziegler in 1991-1992. In 1992 he joined Boehringer Ingelheim Pharmaceuticals, Inc. as a Senior Scientist in Chemical Development, and he is currently Director of Chemical Development. He has been leading multi-disciplined teams engaged in Process Research, Analytical Research, and Solid State Physicochemical Characterization. His research interests include the development of new synthetic methodologies and their applications to chemical processes for drug substance manufacturing.

Dr. Chris H. Senanayake was born in Sri Lanka and received a BS degree (First Class) in Sri Lanka. He completed his MS at Bowling Green State University and his Ph.D. under the guidance of Professor James H. Rigby at Wayne State University in 1987. After postdoctoral studies with Professor Carl R. Johnson he joined the Department of Process Development at Dow Chemical Company in 1989. Currently, he is the Vice President of Chemical Development for Boehringer Ingelheim Pharmaceuticals and leading a group of highly talented scientists, engineers, and administrative staff located in Ridgefield, CT. Dr. Senanayake's research interests focus on the development of new asymmetric methods for the synthesis of bioactive molecules and heterocycles and on catalytic, enzymatic, and mechanistic studies.

Discussion Addendum for:
ortho-Formylations of Phenols;
Preparation of 3-Bromosalicylaldehyde

Trond Vidar Hansen[1] and Lars Skattebøl.[2*]
Original article: Hansen, T. V.; Skattebøl, L. *Org. Synth.* **2005**, *82*, 64.

Several methods are available for the direct formylation of phenols, but most of them suffer from lack of regioselectivity. Moreover, several of the methods involve the use of noxious reagents and harsh reaction conditions resulting in moderate to good yields of aldehydes.[3] An exception is the formylation of phenols using the acceptable reagents $MgCl_2$, Et_3N and paraformaldehyde giving high yields of salicylaldehydes from reactions exclusively at the *ortho*-position.[4] Additional examples and applications of this method since its appearance in 1999 are presented here. We have emphasized the scope of the reaction and its participation in one-pot preparations.

Scope of the *ortho*-Formylation Reaction

Numerous applications of the formylation reaction have been reported in the past twelve years. A number of substituted phenols have been successfully formylated (Figure 1). Alkyl and alkoxy substituents promote the formylation, with high to excellent yields of salicylaldehydes being obtained (Figure 1).[4,5] Highly substituted phenols undergo the reaction; for example, reaction of the phenol **1** furnished the salicylaldehyde **2** in 62% yield.[6] Surprisingly, however, α,ω–*bis*(*p*-hydroxyphenyl)alkanes were unreactive towards the method.[7]

Org. Synth. **2012**, *89*, 220-229
Published on the Web 11/29/2011
© 2012 Organic Syntheses, Inc.

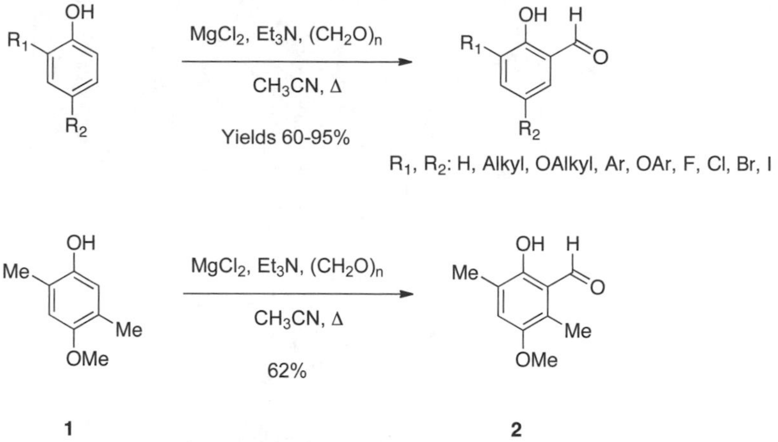

Figure 1. *Ortho*-formylation of phenols using $MgCl_2$, Et_3N and para-formaldehyde.

The method has been used for the formylation of estrogens.[8,9] The reactions occurred to provide excellent yields of regioisomeric aldehydes, with high preference for the 2-isomer (Table 1, Figure 2).

Table 1. Results from *ortho*-formylation of estrogens.

R_1	R_2	R_3	Regioisomeric ratio of 2- and 4-isomers	Yield %
OH	H	H	13:1	92
OH	CCH	H	6:1	86
OH	CH_2CH_3	H	12:1	90
H	OH	H	13:1	85
OAc	H	H	8:1	90
OH	H	OH	not determined	15
=O		H	9:1	90
OCH_2CH_2O		H	12:1	74

The practical use of this method was demonstrated with the synthesis of the anti-cancer agent 2-methoxyestradiol (**3**) from estradiol in 36% overall yield (Figure 2).[9]

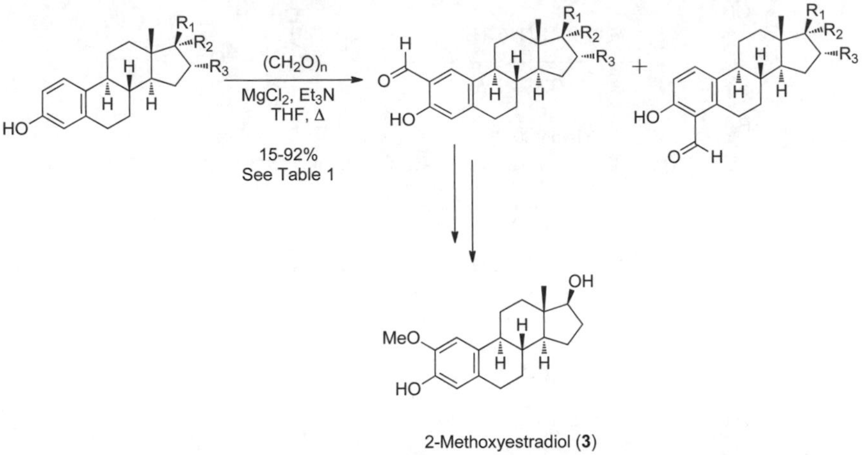
(CH2O)n
MgCl2, Et3N
THF, Δ
15-92%
See Table 1
2-Methoxyestradiol (3)

Figure 2. *ortho*-Formylation of estrogens.

Aromatic substituents have no detrimental effect on the reaction. Phenyl-,[10] perfluorophenyl-[11] and 2-nitrophenyl-substituted[11] phenols were formylated in good yields, and the same was the case with mono-protected 2,2´-bisphenol.[12] Naphthol derivatives behave similarly giving *ortho*-formylation products,[13,14] and of special merit is the preparation of 3-formyloctahydro-1,1´-binaphthol.[14b] 5-Methoxy-2-naphthol was formylated in 75% yield as part of the total synthesis of racemic juglomycin A.[15]

Oxygen functions are well tolerated; several mono protected resorcinols and methylenedioxy-substituted phenols have been formylated in excellent yields and high regioselectivity.[16,17] Moreover, several methylthio- and phenylthiophenol derivatives have been formylated in high yields.[18] Halogen-substituted phenols are good substrates in the reaction,[4] and this has been thoroughly substantiated in the last decade.[19]

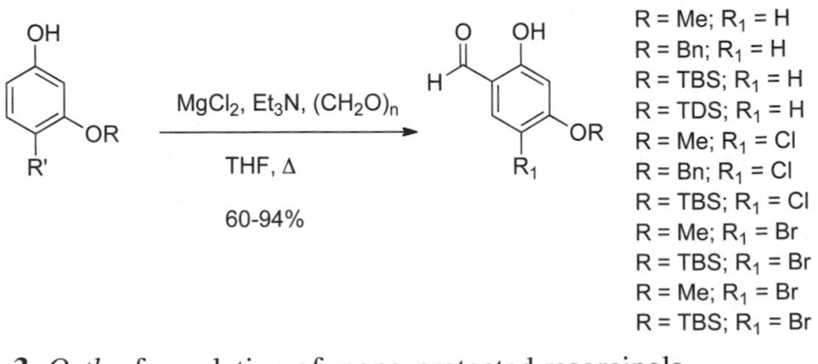
MgCl2, Et3N, (CH2O)n
THF, Δ
60-94%
R = Me; R1 = H
R = Bn; R1 = H
R = TBS; R1 = H
R = TDS; R1 = H
R = Me; R1 = Cl
R = Bn; R1 = Cl
R = TBS; R1 = Cl
R = Me; R1 = Br
R = TBS; R1 = Br
R = Me; R1 = Br
R = TBS; R1 = Br

Figure 3. *Ortho*-formylation of mono-protected resorcinols.

The method also tolerates electron attracting groups like cyano and ester, but the reaction proceeds at a slower rate when these groups are attached to the benzene ring, and more byproducts may appear as well.[4] As anticipated, faster reactions were encountered when the ester group was further removed from the benzene ring. Formylation of **4** gave **5** in 94% yield[20] and reaction of the phenol **6** furnished aldehyde **7** in high yield (Figure 4).[21] The presence of two ester groups was accepted; formylation of dimethyl 2-(4-hydroxyphenyl)succinate proceeded with excellent yield.[22] Moreover, protected tyrosine, *viz.* methyl *N*-Cbz-tyrosinate was successfully formylated in 43% yield.[23] Since the formylation of 4-cyanophenol was successful,[4,24] it is interesting to note that (4-hydroxyphenyl)acetonitrile could not be formylated by the different methods tried.[25]

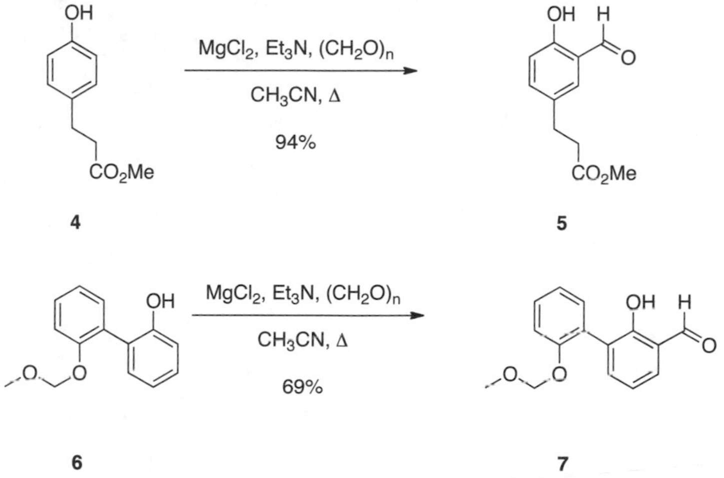

Figure 4. Examples of *ortho*-formylations.

The Formylation Method as Part of One-pot Reactions

Much can be gained with respect to efficiency and yields by using multi-component one-pot reactions, since several reactions can be carried out successively without isolation of intermediates. The reagents of this method appeared compatible for such a set-up, and several applications have been reported.

The catechol structural entity is present in natural products, many of biological interest. A one-pot preparation of catechols from the corresponding phenols has been reported[26] involving the formylation reaction followed by Dakin or Baeyer-Villliger oxidations. The yields are very good (Figure 5). The method has been used in the synthesis of combretastatins A-1 and B-1.[27]

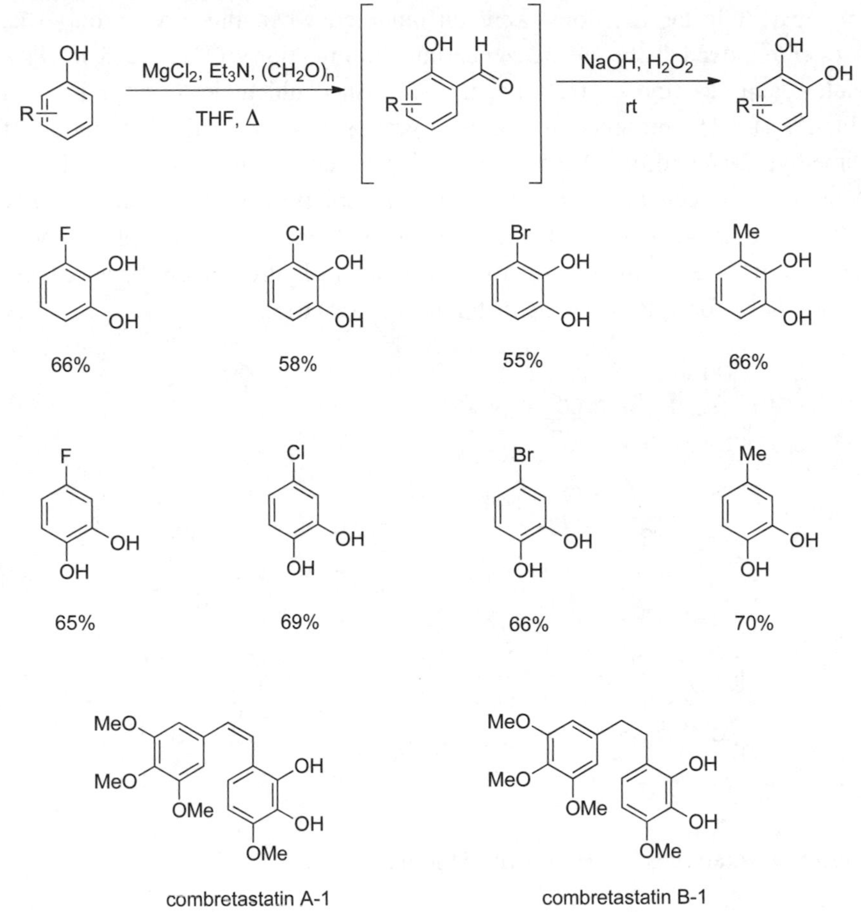

Figure 5. One-pot synthesis of catechols.

Salen ligands, that form part of the Jacobsen catalyst, were prepared in excellent yields by a one–pot reaction starting from the appropriate phenol (Figure 6).[28] The yields are considerably better than those previously obtained by a two-step protocol.

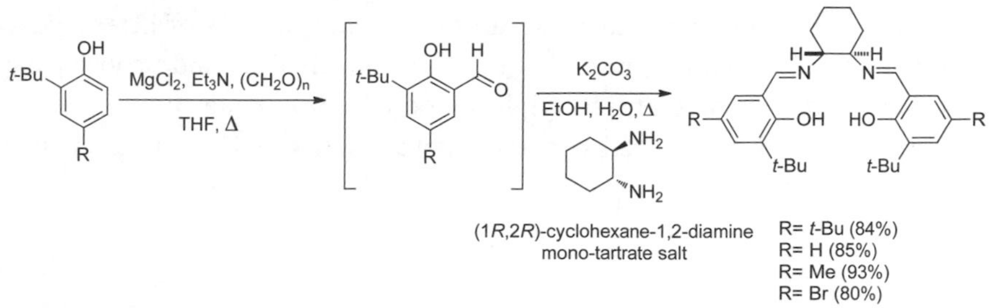

Figure 6. One-pot synthesis of salens.

By another one-pot reaction, salicylamines were prepared in good yields by reductive amination of the corresponding salicylaldehydes as outlined in Figure 7. The amines were further manipulated without isolation of intermediates to dihydro-2*H*-1,3-benzoxazines in good overall yields, considering the four reactions involved.[29] In a related one-pot procedure salicylnitriles were prepared.[30] In this case the imines formed from the respective salicylaldehyde and ammonia were oxidized with *o*-iodoxybenzoic acid to give good yields of the corresponding nitriles.

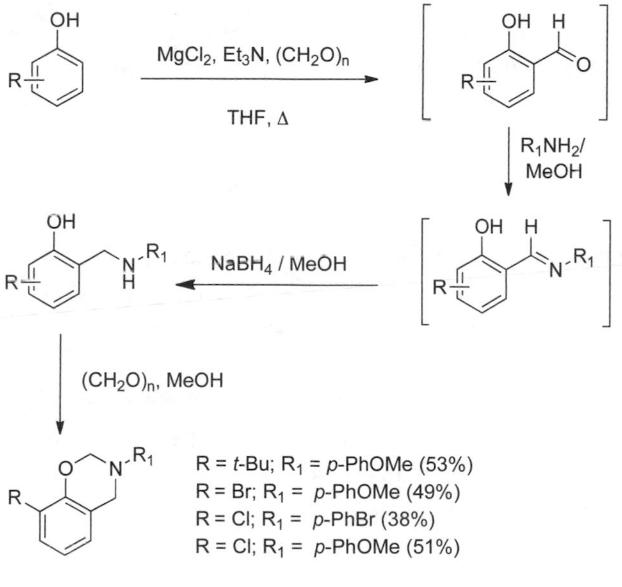

Figure 7. One-pot synthesis of salicylamines and dihydro-2*H*-1,3-benzoxazines.

Cinnamic acid derivatives are useful for the synthesis of heterocyclic compounds. Converting phenols to salicylaldehydes and subsequent treatment with methyl (triphenylphosporanylidene)acetate in a one-pot reaction afforded methyl 2-hydroxycinnamate derivatives in acceptable yields (Figure 8).[31]

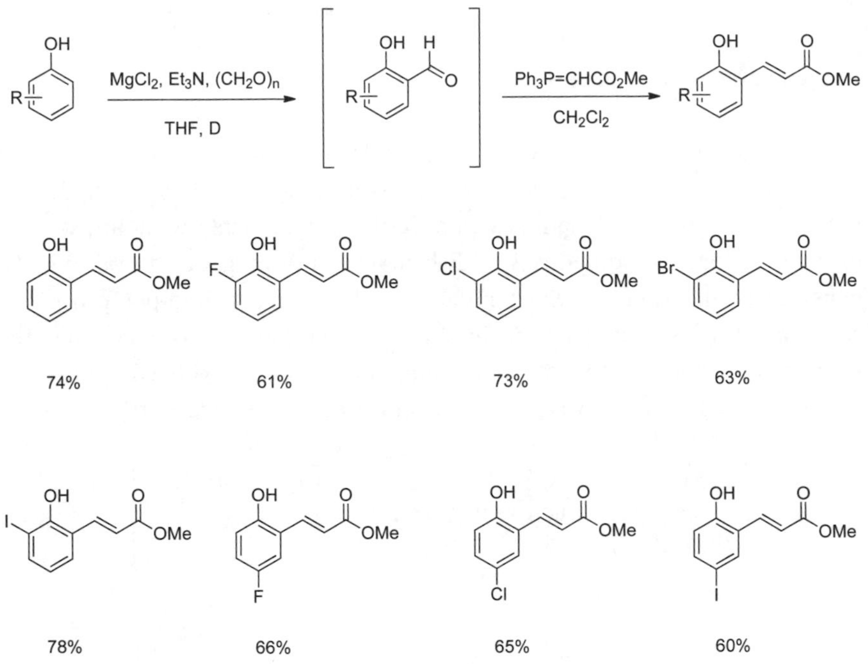

| 74% | 61% | 73% | 63% |

| 78% | 66% | 65% | 60% |

Figure 8. One-pot synthesis of methyl cinnamates.

Conclusion

The results have demonstrated that the method described above fits well into the assembly of phenol formylation reactions. It may prove to be the method of choice in many cases because of the ease of execution, regioselectivity, and good yields.

1. Department of Medicinal Chemistry, School of Pharmacy, University of Oslo, PO BOX 1068, 0316 Oslo, Norway.
2. Department of Chemistry, University of Oslo, PO BOX 1033, 0315 Oslo, Norway. E-mail: lars.skattebol@kjemi.uio.no.
3. (a) Olah, G. A.; Ohannasian, L.; Arvanaghi, M. *Chem. Rev.* **1987**, *87*, 671–686. (b) Laird, T. In *Comprehensive Organic Chemistry;* Stoddart, J. F., Ed., Pergamon, Oxford 1979, Vol. 1; p 1105–1160. (c) Kantlehner, W. *Eur. J. Org. Chem.* **2003**, 2530–2546.
4. (a) Hofsløkken, N. U.; Skattebøl, L. *Acta Chem. Scand.* **1999**, *53*, 258–262; (b) Hansen, T. V.; Skattebøl, L. *Org. Synth.* **2005**, *82*, 64–68.
5. Knight, P. D.; Clarkson, G.; Hammond, M. L.; Kimberley, B. S.; Scott, P. *J. Organomet. Chem.* **2005**, *690*, 5124–5144.
6. (a) Chen, Y.; Steinmetz, M. G. *Org. Lett.* **2005**, *7*, 3729–3732. (b) Chen, Y.; Steinmetz, M. G. *J. Org. Chem.* **2006**, *71*, 6053–6060.
7. Cort, A. D.; Mandolini, L.; Pasquini, C.; Schiaffino, L. *J. Org. Chem.* **2005**, 9814–9821.
8. (a) Liu, Y.; Lien, I.-F.; Ruttgaizer, S.; Dove, P.; Taylor, S. D. *Org. Lett.* **2004**, *6*, 209–212. (b) Liu, Y.; Kim, B.; Taylor, S. D. *J. Org. Chem.* **2007**, *72*, 8824–8830.
9. Akselsen, Ø. W.; Hansen, T. V. *Tetrahedron* **2011**, *67*, 7738–7742.
10. Doshi, J. M.; Tian, D.; Xing, C. *J. Med. Chem.* **2006**, *49*, 7731–7739.
11. Mu, H-L; Ye, W-P; Song, D-P; Li, Y-S. *Organometallics* **2010**, *29*, 6282–6290.
12. Riggs, J. R.; Kolesnikov, A.; Hendrix, J.; Young, W. B.; Shrader, W.D.; Vijaykumar, D.; Stephens, R.; Liu, L.; Pan, L.; Mordenti, J.; Green, M. J.; Sukbuntherng, J. *Bioorg. Med. Chem. Lett.* **2006**, *16*, 2224–2228.
13. Wright, B. J. D.; Hartung, J.; Peng, F.; Van de Water, R.; Liu, H.; Tan, Q-H.; Chou, T-C.; Danishefsky, S. J. *J. Am. Chem. Soc.* **2008**, *130*, 16786–16790.
14. (a) Zaichenko, N. L.; Kol´tsova, L. S.; Shcherbakova, I. M.; Os´kina, O. Yu.; Voznesenskii, V. N.; Mardaleishvili, I. R.; Shienok, A. I. *Russ. Chem. Bull. Int. Ed.* **2011**, *60*, 533–535. (b) Jung, H.; Nandhakumar, R.; Yoon, H.-J.; Lee, S.; Kim, K. M. *Bull. Korean Chem. Soc.* **2010**, *31*, 1289–1294.
15. Min, J.-P.; Kim, J-C.; Park, O-S. *Synth. Commun.* **2004**, 383–390.
16. Akselsen, Ø. W.; Skattebøl, L.; Hansen, T. V. *Tetrahedron Lett.* **2009**, *50*, 6339–6341.

17. Juhász, L.; Docsa, T.; Brunyászki, A.; Gergely, P.; Antus, S. *Bioorg. Med. Chem. Lett.* **2007**, *15*, 4048–4056.

18. (a) Sylvestre, I.; Wolowska, J.; Kilner, C. A.; McInnes, E. J. L.; Halcrow, M. A. *J. Chem. Soc. Dalton Trans.* **2005**, 3241–3249. (b) Hirano, K.; Biju, A. T.; Piel, I.; Glorius, F. *J. Am. Chem. Soc.* **2009**, *131*, 14190–14191.

19. (a) Feldman, K. S.; Eastman, K. J.; Laéssene, G. *Org. Lett.* **2002**, *4*, 3525–3528. (b) Marsh, G.; Stenutz, R.; Bergman, Å. *Eur. J. Org. Chem.* **2003**, 2566–2576. (c) Clever, G.; Polborn, K.; Curell, T. *Angew. Chem. Int. Ed.* **2005**, *44*, 7204–7208. (d) Charrier, C.; Bertrand, P. *J. Chem. Sci.* **2011**, *123*, 459–466.

20. Sun Z-N.; Wang, H-L.; Liu, F-Q.; Tam, P. K. H.; Yang, D. *Org. Lett.* **2009**, *11*, 1887–1890.

21. Hu, H.; Kolesnikov, A.; Riggs, J. R.; Wesson, K. E.; Stephens, R.; Leahy, E. M.; Shrader, W. D.; Sprengeler, P. A.; Green, M. J.; Sanford, E.; Nguyen, M.; Gjerstad, E.; Cabuslay, R.; Young, W. B. *Bioorg. Med. Chem. Lett.* **2006**, *16*, 2224–2228.

22. Shrader, W. D.; Kolesnikov, A.; Burgess-Henry J.; Rai, R.; Hendrix, J.; Hu, H.; Torkelson, S.; Ton, T.; Young, W. B.; Katz, B. A.; Yu, C.; Tang, J.; Cabuslay, R.; Sanford, E.; Janc, J. W; Sprengeler, P. A. *Bioorg. Med. Chem. Lett.* **2006**, *16*, 1596–1600.

23. Shen, K.: Qi, L.; Ravula, M. *Synthesis* **2009**, 3765–3768.

24. Becker, J. M.; Barker, J.; Clarkson, G. J.; van Gorkum, R.; Johal, G. K.; Walton, R. J.; Scott, P. *Dalton Trans.* **2010**, *39*, 2309–2326.

25. Shetty, R.; Moffet, K. K. *Tetrahedron Lett.* **2006**, *47*, 8021–8024.

26. Hansen T. V.; Skattebøl, L. *Tetrahedron Lett.* **2005**, *46*, 3357–3358.

27. Odlo, K.; Klaveness, J.; Rongved, P.; Hansen, T. V. *Tetrahedron Lett.* **2006**, *47*, 1101–1103.

28. Hansen, T. V.; Skattebøl, L. *Tetrahedron Lett.* **2005**, *46*, 3829–3830.

29. Anwar, H. F.; Skattebøl, L.; Hansen, T. V. *Tetrahedron* **2007**, *63*, 9997–10002.

30. Anwar, H. F.; Hansen, T. V. *Tetrahedron Lett.* **2008**, *49*, 4443–4445.

31. Anwar, H. F.; Skattebøl, L.; Skramstad, J.; Hansen, T. V. *Tetrahedron Lett.* **2005**, *46*, 5285–5287.

Trond Vidar Hansen obtained his MSc degree from the University of Trondheim in 1995 and his Ph.D. degree from the Agricultural University of Norway in 2001 (Professors Yngve Stenstrøm and Lars Skattebøl as supervisors). He was a postdoctoral fellow (2001-2003) at the University of Oslo with Professor Skattebøl and then at The Scripps Research Institute, La Jolla, with Professor K. Barry Sharpless (2004). In the same year he became Associate Professor at the University of Oslo, School of Pharmacy. His research interests are synthetic methodology, medicinal chemistry, lipid chemistry and synthesis of natural products.

Lars Skattebøl obtained a BSc from the Technical University of Norway in 1951 and a PhD in 1956 from the University of Manchester (Professor E. R. H. Jones as supervisor). After a postdoctoral year with Professor J. D. Roberts at California Institute of Technology, he spent eleven years with Union Carbide, Corp. at their corporate research institutes in Brussels and Tarrytown, N.Y. In 1970 he was appointed Professor at the University of Oslo, Department of Chemistry. His research interests are carbenes, lipids, synthetic methodology and natural product synthesis.

Preparation of *n*-Isopropylidene-*n'*-2-Nitrobenzenesulfonyl Hydrazine (IPNBSH) and Its Use in Palladium–catalyzed Synthesis of Monoalkyl Diazenes.
Synthesis of 9-Allylanthracene

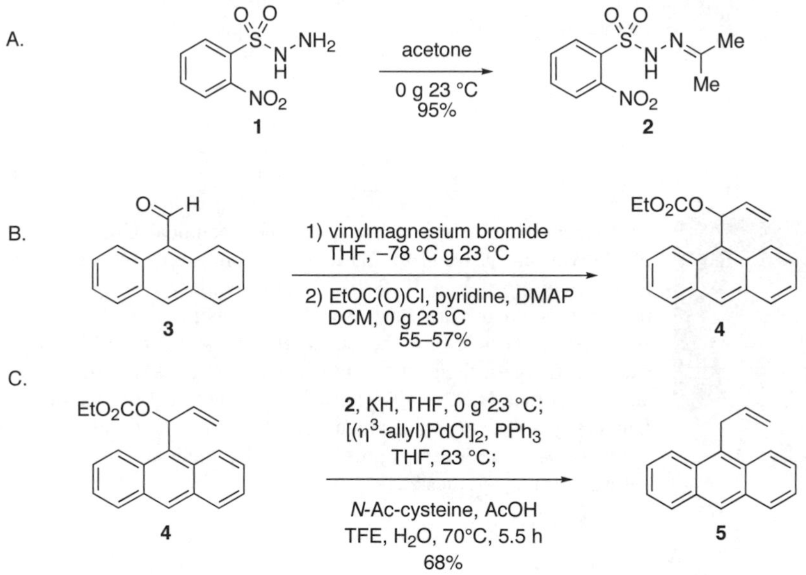

A.

B.

C.

Submitted by Jens Willwacher, Omar K. Ahmad, and Mohammad Movassaghi.[1]
Checked by Juliane Keilitz and Mark Lautens.

Potassium hydride is a pyrophoric solid and must not be allowed to come into contact with the atmosphere. This reagent should only be handled by individuals trained in its proper and safe use.

---

1. Procedure

A. N-Isopropylidene-N'-2-nitrobenzenesulfonyl hydrazine (2). An oven-dried 1-L, two-necked, round-bottomed flask, equipped with a 3 cm football-shaped stir bar, a rubber septum and inert gas inlet connected to a manifold, is charged with *o*-nitrobenzenesulfonyl hydrazide (**1**) (Note 1) (20.0 g,

Org. Synth. **2012**, *89*, 230-242
Published on the Web 12/2/2011
© 2012 Organic Syntheses, Inc.

92.1 mmol, 1 equiv) under an argon atmosphere. Acetone (Note 2) (300 mL) is added via cannula and the resulting solution stirred at room temperature (24 °C). After 10 min, TLC (Note 3) analysis of the reaction mixture indicated full conversion of the starting material (hexanes:ethyl acetate = 1:2; Rf_{SM} = 0.2, $Rf_{product}$ = 0.5, visualized with ceric ammonium molybdate). Aliquots of the solution are transferred to a 500-mL one-necked flask and the solvent is removed with a rotary evaporator (200 mmHg, 40 °C) to afford a yellow powder. Hexanes (100 mL), acetone (20 mL), and a 3-cm football-shaped stir bar are added and the suspension stirred for 10 min at room temperature. The white solid is collected by vacuum filtration (40 mm, Büchner funnel with fritted disc, medium porosity) and washed with hexanes (2 × 60 mL, 24 °C). The white solid is then transferred to a 250-mL flask and dried in vacuo (7 mmHg, 24 °C) for 24 h to afford *N*-isopropylidene-*N*'-2-nitrobenzenesulfonyl hydrazine (**2**) (Note 4) (22.5 g, 87.4 mmol, 95%) as an off-white solid.

B. 1-(Anthracen-9-yl)allyl ethyl carbonate (4). A flame-dried 1-L, three-necked round-bottomed flask, equipped with a thermometer, a rubber septum, a 3-cm football-shaped stir bar, and an inert gas inlet connected to a manifold, is charged with anthracene-9-carboxaldehyde (Note 5) (18.0 g, 87.3 mmol, 1 equiv). Anhydrous THF (Note 6) (360 mL) is added via cannula. The resulting dark brown solution is cooled to –78 °C (dry-ice–acetone bath, internal temperature) and stirred for 15 min at –78 °C. Vinyl magnesium bromide (Note 5) (1.0 M in THF, 96.0 mmol, 96.0 mL, 1.10 equiv) is added to the reaction mixture slowly via cannula over 14 min during which the internal temperature increased to –65 °C. The reaction mixture is stirred for one hour in the dry-ice–acetone bath and another hour after removal of the cooling bath (internal temperature: 15 °C). TLC analysis (hexanes:ethyl acetate = 17:3; Rf_{SM} = 0.3, $Rf_{product}$ = 0.2, visualized with ceric ammonium molybdate) indicated complete conversion of the starting material. The reaction mixture is cooled to 0 °C (ice–water bath, internal temperature), excess Grignard reagent is quenched with saturated aqueous ammonium chloride solution (500 mL, 23 °C) and the reaction mixture is diluted with CH_2Cl_2 (600 mL). The aqueous and organic layers are separated in a 3-L separatory funnel and the aqueous layer is further extracted with CH_2Cl_2 (2 × 350 mL). The combined organic layers are washed with water (700 mL) and brine (700 mL), dried over anhydrous sodium sulfate (110 g), and concentrated with a rotary evaporator (200 mmHg, 40 °C) in a 1-L round-bottomed flask. The flask is equipped with a magnetic stir bar for continuous agitation, and the residue is further dried in vacuo (6.5 mmHg, 23 °C) for 10 h. The crude alcohol is dissolved in anhydrous CH_2Cl_2 (Note 5) (360 mL) and the flask is fitted with a rubber septum and connected to a manifold via a needle. After addition of pyridine (Note 5) (10.6 mL, 131 mmol, 1.50 equiv) and N,N-dimethylpyridine-4-amine (Note 7) (1.1 g, 8.7 mmol, 0.10 equiv), the solution is cooled to 0 °C (ice–water bath, internal temperature). After 30 min, ethyl chloroformate (Note 7) (10.4 mL, 109 mmol, 1.25 equiv) is added dropwise via a syringe over 17 min, resulting in a dark red reaction mixture and a temperature increase to 6 °C. After 30 min, the ice–water bath is removed and the reaction mixture is allowed to warm to 23 °C. After 90 min, TLC analysis (hexanes:ethyl acetate = 17:3, Rf_{SM} = 0.2, $Rf_{product}$ = 0.35 (Note 8), visualized with ceric ammonium molybdate) indicated complete consumption of the starting material. The reaction mixture is diluted with a mixture of water and saturated aqueous ammonium chloride solution (1:1, 400 mL) and the aqueous and organic layers are separated in a 2-L separatory funnel. The aqueous layer is further extracted with CH_2Cl_2 (2 × 300 mL). The combined organic layers are washed with water (550 mL) and brine (550 mL), dried over anhydrous sodium sulfate (110 g), filtered and concentrated with a rotary evaporator (200 mmHg, 40 °C). The brown residue is then dissolved in a mixture of hexanes–ethyl acetate (6:1, 100 mL) and CH_2Cl_2 (40 mL), loaded on a short pad of silica gel (Note 9) (70 g, 9 × 5 cm) and flushed with hexanes–ethyl acetate (6:1, 400 mL). The resulting yellow solution is concentrated on a rotary evaporator (200 mmHg, 40 °C), and transferred to a 250-mL flask, and equipped with a magnetic stir bar. The residue is dried in vacuo (6.5 mmHg, 23 °C, 30 min) and the flask is fitted with a reflux condenser. The orange solid is dissolved in a mixture of hexanes and ethyl acetate (1:1, 42 mL, 70 °C, oil bath, external temperature), the stir bar is removed and the sealed flask is allowed to cool to –20 °C in a freezer for 16 h to allow crystallization of the desired product. The solvent is decanted and the yellow crystals are washed with hexanes–ethyl acetate (3 × 15 mL, 0 °C) and dried in vacuo (6.5 mmHg, 23 °C, 24 h) to afford carbonate **4** (Note 10) (14.7–15.2 g, 48.0–49.7 mmol, 55–57% yield) (Note 11).

C. 9-Allylanthracene (5). A flame-dried 500-mL, two-necked round-bottomed flask, equipped with an inert gas inlet connected to a manifold, a rubber septum and a 3-cm football-shaped stir bar, is charged with IPNBSH (**2**) (11.8 g, 45.8 mmol, 1.10 equiv). Anhydrous THF (Note 6) (160 mL) is added via cannula and the resulting solution is cooled to 0 °C (ice–water

232

bath, external temperature). A flame-dried 1-L, three-necked round-bottomed flask, equipped with a thermometer, an inert gas inlet, a rubber septum and magnetic stir bar is charged with potassium hydride (Note 12) (5.83 g of the suspension in oil, 1.75 g, 43.7 mmol, 1.05 equiv) under an argon atmosphere. Potassium hydride is washed with anhydrous hexane (3 x 20 mL) and dried in vacuo (6.5 mmHg, 23 °C, 30 min). Anhydrous THF (Note 6) (50 mL) is added via cannula and the resulting suspension is cooled to 0 °C (ice–water bath, internal temperature). The cold solution of IPNBSH is transferred to the suspension of potassium hydride via cannula over 15 min at 0 °C under an argon atmosphere, resulting in formation of a beige precipitate. The IPNBSH source flask is rinsed with anhydrous THF (Note 6) (25 mL, 0 °C) to complete the transfer to the reaction mixture. After 15 min, the ice–water bath is removed, and the reaction mixture is allowed to warm to 16 °C over 20 min. A flame-dried 250-mL, two-necked flask, equipped with an inert gas inlet, rubber septum and magnetic stir bar, is charged with carbonate **4** (12.7 g, 41.6 mmol, 1 equiv), allylpalladiumchloride dimer (Note 13) (228 mg, 0.624 mmol, 0.015 equiv) and triphenylphosphine (Note 13) (655 mg, 2.50 mmol, 0.06 equiv). Anhydrous THF (Note 6) (100 mL) is added via cannula and the resulting dark red solution is transferred slowly via cannula over 12 min to the solution of the potassium sulfonamide prepared above. The flask is rinsed with anhydrous THF (Note 6) (25 mL) to complete the transfer. After 30 min, TLC analysis indicated the complete consumption of the carbonate and formation of the intermediate arenesulfonyl hydrazone (hexanes:ethyl acetate = 17:3, Rf_{SM} = 0.35, $Rf_{hydrazone}$ = 0.05, visualized with ceric ammonium molybdate). *N*-Acetyl cysteine (Note 13) (543 mg, 3.33 mmol, 0.08 equiv) is added as a solid and glacial acetic acid (Note 14) (5.2 mL, 91.5 mmol, 2.2 equiv) is added via syringe. After 10 min, a mixture of 2,2,2-trifluoroethanol (Note 15) and deionized water (1:1, 180 mL, deoxygenated by purging with argon for 10 min) is added via cannula. The flask is fitted with a reflux condenser and heated to 70 °C (oil bath, external temperature). After 5.5 h, TLC analysis indicated consumption of the hydrazone–adduct (hexanes:ethyl acetate = 17:3, $Rf_{hydrazone}$ = 0.05, $Rf_{product}$ = 0.6, visualized with ceric ammonium molybdate). The reaction mixture is cooled to 23 °C, diluted with hexanes (250 mL) and excess acid is quenched with a mixture of saturated aqueous bicarbonate solution and water (1:1, 500 mL). The aqueous and organic layers are separated in a 3-L separatory funnel and the aqueous layer is further extracted with hexanes (2 × 350 mL). The combined organic layers are washed with water (450 mL) and brine (450 mL), dried over anhydrous sodium sulfate (110 g) and filtered. Volatiles are removed with a rotary evaporator (200 mmHg, 40 °C) and the residue dried in vacuo (6.5 mmHg, 10 h). The remaining orange residue is purified by flash column chromatography (Note 16). All product fractions are combined, concentrated with a rotary evaporator (200 mmHg, 40 °C) and dried in vacuo (6.5 mmHg, 23 °C, 5 h) to afford the reduction product **5** as a pale green solid (Note 17) (6.21 g, 28.4 mmol, 68% yield).

2. Notes

1. For the preparation of *o*-nitrobenzenesulfonyl hydrazide, see: *Organic Syntheses*, **1999**, *76*, 178; *Coll. Vol. 10*, **2004**, 165.

2. Acetone (Chromasolv for HPLC, ≥ 99%) and hexanes (mixture of isomers, Chromasolv for HPLC, ≥98.5%) were purchased from Aldrich and used as received.

3. TLC was performed on EMD Chemicals Inc. Silica Gel 60 F_{254} plates.

4. In different runs of this step, the yield varied from 95 to 96%. Analytical data for IPNBSH (**2**) are as follows: ^{1}H NMR (400 MHz, CDCl$_3$) δ: 1.90 (s, 3 H), 1.93 (s, 3 H), 7.73–7.79 (m, 2 H), 7.81–7.87 (m, 2 H), 8.22–8.28 (m, 1 H); ^{13}C NMR (100 MHz, CDCl$_3$) δ: 16.9, 25.2, 125.1, 131.7, 132.7, 133.1, 134.1, 148.2, 158.7; IR (neat): 3265, 1553, 1374, 1347, 1177 cm^{-1}; HRMS calc'd for C$_9$H$_{12}$N$_3$O$_4$S [M+H]: 258.05485, found: 258.05382; mp 133–134 °C (with decomposition); Anal. calcd. for C$_9$H$_{11}$N$_3$O$_4$S: C, 42.02; H, 4.31; N, 16.33, found: C, 42.06; H, 4.56; N, 16.32. IPNBSH (**2**) can be stored under an argon atmosphere at 23 °C for several months.

5. Anthracene-9-carbaldehyde (97%) and vinyl magnesium bromide in THF (1.0 M) were purchased from Aldrich Chemical Company, Inc. and used as received.

6. Anhydrous tetrahydrofuran was distilled from Na/benzophenone, anhydrous dichloromethane was obtained from a Solvent Purification System Mbraun MB-SPS, and anhydrous pyridine was purchased from Aldrich Chemical Company, Inc. and used as received. The submitters purchased tetrahydrofuran, dichloromethane, and pyridine from J.T. Baker (Cycletainer™) and dried the solvents by the method reported by Grubbs[2] et al. under positive argon pressure.

7. *N*,*N*-Dimethylpyridine-4-amine (99 %) and ethyl chloroformate (97%) were purchased from Aldrich Chemical Company, Inc. and used as received.

8. It is recommended to dilute the sample for TLC analysis due to decomposition of the carbonate on silica gel at high concentrations; one major decomposition product can be seen with an $Rf = 0.2$.

9. SiliCycle Silia*Flash*® P60 silica gel, 60-63 μm, 60 Å; the submitters used Zeochem® Silicagel Zeoprep 60HYD, 40–63 μm was used for flash column chromatography.

10. Analytical data for carbonate **4** are as follows: ^{1}H NMR (400 MHz, CDCl$_3$) δ: 1.26 (t, $J = 7.0$ Hz, 3 H), 4.10 (dq, $J = 10.5$, 7.0 Hz, 1 H), 4.21 (dq, $J = 10.5$, 7.0 Hz, 1 H), 5.26–5.34 (m, 2 H), 6.51 (ddd, $J = 17.2$, 10.6, 4.3 Hz, 1 H), 7.46–7.52 (m, 2 H), 7.57 (ddd, $J = 9.0$, 6.7, 1.6 Hz, 2 H), 7.74 (dt, $J = 4.7$, 2.0 Hz, 1 H), 8.01–8.05 (m, 2 H), 8.49 (s, 1 H), 8.57 (d, $J = 9.0$ Hz, 2 H); ^{13}C NMR (100 MHz, CDCl$_3$) δ: 14.2, 64.2, 75.6, 117.6, 124.9, 124.9, 126.2, 128.8, 129.2, 129.3, 129.9, 131.6, 136.1, 154.9; IR (neat): 3053, 2985, 1742, 1448, 1369, 1256, 1007, 877, 789, 731 cm^{-1}; HRMS calcd. for C$_{20}$H$_{18}$O$_3$ [M]$^+$: 306.1256, found 306.1259; mp 92–93 °C; Anal. calcd. for C$_{20}$H$_{18}$O$_3$: C, 78.41; H, 5.92, found: C, 78.03; H, 5.99.

11. The submitters reported a higher yield (17.9 g, 58.4 mmol, 67%), but the checkers were not able to reproduce this result. However, the checkers noted that after decantation of the solvent and washing of the crystals, the residual solution contained a significant amount of carbonate **4** (determined by ^{1}H NMR) which could not be crystallized, but accounts for the missing product.

12. Potassium hydride (30% by weight in mineral oil) was purchased from Aldrich Chemical Company, Inc and was used as received. In order to use the correct amount of KH for the reaction, the flask was weighed before addition of the KH suspension in oil and after the washing procedure and drying in high vacuum. If necessary, the amount of the starting material and reagents was adjusted to the amount of KH used. The submitters report that KH was washed with anhydrous hexanes prior to use and stored in a glovebox. *Note that potassium hydride is a pyrophoric solid and must be handled with care.*

13. Allylpalladiumchloride dimer (98%), triphenylphosphine (99%, Reagent Plus) and *N*-acetyl-L-cysteine (Sigma grade) were purchased from Aldrich Chemical Company, Inc. and used as received.

14. Glacial acetic acid (99.7 %) was purchased from Mallinckrodt Chemicals and used as received.

15. 2,2,2-Trifluoroethanol was purchased from Aldrich Chemical Company, Inc. and used as received.

16. Flash column chromatography (7 cm diameter, 20 cm height) was performed using silica gel (350 g). The residue was dissolved in DCM (100 mL) and silica gel (10 g) was added. The solvent was removed with a rotary evaporator (200 mmHg, 40 °C) and the silica gel with the product was loaded on the column. It was first eluted with hexanes (3 L), continued with hexanes–acetone (99:1, 2 L; then 98:2, 1 L). The first 1 L was discarded, collection of 50-mL fractions was begun. Fractions #39–100 contained the desired product. Mixed fractions #24–38 were collected separately, concentrated, and purified again by flash column chromatography (200 g SiO$_2$, 5 × 18 cm); loaded with hexanes (5 mL). While eluting with hexanes (1500 mL), collection of 20-mL fractions was begun, fractions #61-95 contained the desired product. TLC analysis during flash column chromatography was performed using 100% hexanes as the eluent ($Rf_{byproduct}$ = 0.4 (only visible under UV light), $Rf_{product}$ = 0.3, visualized with ceric ammonium molybdate).

17. In different runs of this step, the yield varied from 63 to 68%. Analytical data for 9-allylanthracene (5) are as follows: ^{1}H NMR (400 MHz, CDCl$_3$) δ: 4.41 (dt, J = 5.5, 2.0 Hz, 2 H), 5.03 (dq, J = 17.2, 2.0 Hz, 1 H), 5.12 (dq, J = 10.2, 1.6 Hz, 1 H), 6.25 (ddt, J = 17.2, 10.2, 5.5 Hz, 1 H), 7.47–7.58 (m, 4 H), 8.02–8.07 (m, 2 H), 8.26–8.31 (m, 2 H), 8.41 (s, 1 H); ^{13}C NMR (100 MHz, CDCl$_3$) δ: 31.9, 115.9, 124.5, 124.8, 125.6, 126.2, 129.1, 130.0, 131.5, 131.6, 136.4; IR (neat): 3081, 3052, 3004, 1638, 1623, 1446, 1157, 912, 883, 789, 730 cm^{-1}; HRMS calcd for C$_{17}$H$_{15}$ [M+H]: 219.11738, found: 219.11749; mp 43–44 °C; Anal. calcd for C$_{17}$H$_{14}$: C, 93.54; H, 6.46, found: C, 93.29; H 6.75.

Safety and Waste Disposal Information

All hazardous materials should be handled and disposed of in accordance with "Prudent Practices in the Laboratory"; National Academies Press; Washington, DC, 2011.

3. Discussion

Monoalkyl diazenes have been used as key intermediates in a wide range of organic transformations.[3] Many of these methods involve condensation of carbonyl compounds with arenesulfonyl hydrazides as a prelude to monoalkyl diazene formation. Myers discovered an efficient, mild, and stereospecific method for the reduction of allylic, benzylic and saturated alcohols using the reagent 2-nitrobenzenesulfonyl hydrazide (NBSH).[4] Inspired by these reports we developed the complementary reagent N-isopropylidene-N'-2-nitrobenzene-sulfonyl hydrazine (IPNBSH)[5] that offers thermal stability both in the solid state and in solution. We initially used IPNBSH in a key reductive transposition for the total synthesis of (–)-acylfulvene and (–)-irofulvene (Scheme 1).[6]

Scheme 1

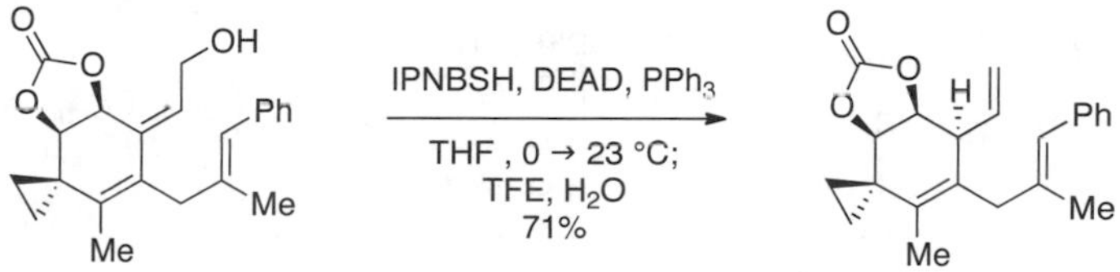

IPNBSH, DEAD, PPh₃

THF , 0 → 23 °C;
TFE, H₂O
71%

We recognized the utility of IPNBSH for generation of monoalkyl diazenes under basic reaction conditions.[5a] These observations prompted our development of IPNBSH as a diimide surrogate in palladium catalyzed allylic substitution reaction.[5b] Scheme 2 illustrates the proposed mechanism for the reduction of allylic carbonates as described in this procedure. The formation of a π-allyl complex[7] is followed by reaction with the potassium salt of IPNBSH to give the corresponding hydrazone-adduct. In situ hydrolysis and elimination of the arenesulfinic acid leads to the regioselective formation of the transient allylic monoalkyl diazene, that upon sigmatropic loss of dinitrogen affords the desired reduction product.

The reduction of allylic carbonates via the corresponding allylic monoalkyl diazene as described here is applicable to a broad range of substrates (Table 1). Both primary and secondary allylic carbonates undergo the reductive transformation to afford the desired products in good yields and high selectivity (E/Z, >96:4).[5]

Scheme 2

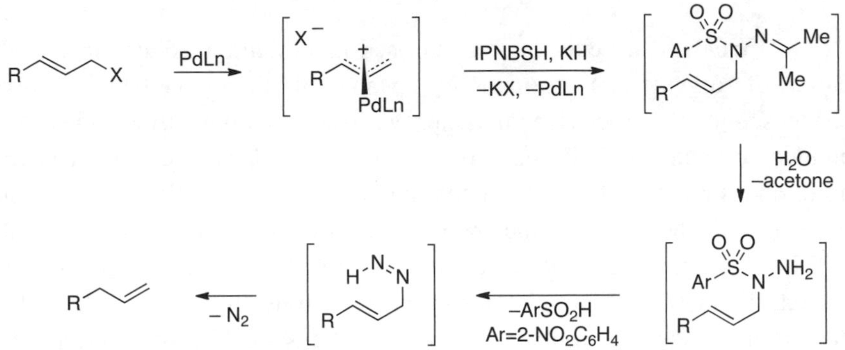

Table 1 Selected examples of IPNBSH-mediated reduction of allylic carbonates

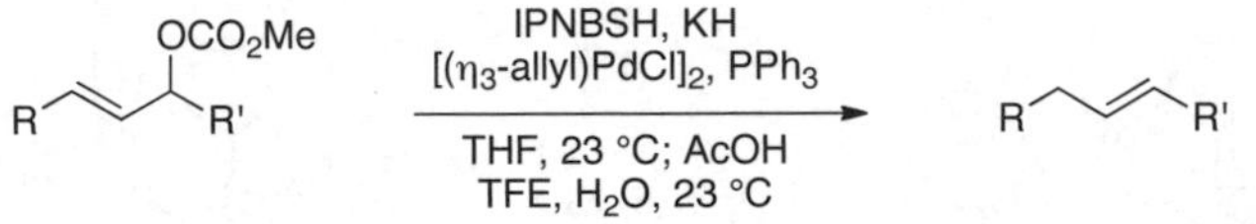

Entry	Substrate	Product	Yield
1	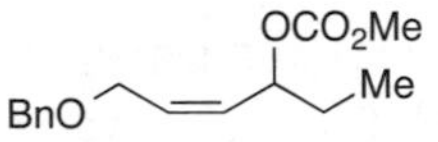	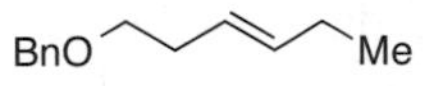	76
2	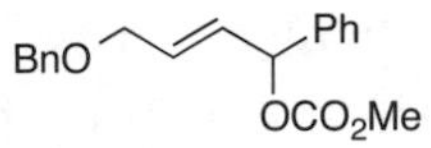	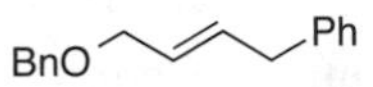	82
3	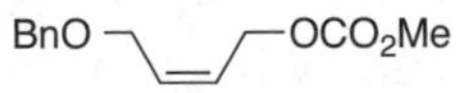	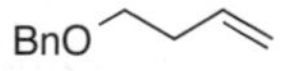	91
4	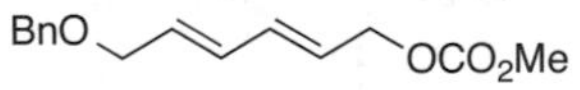	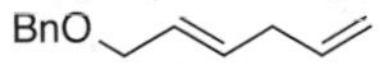	68
5	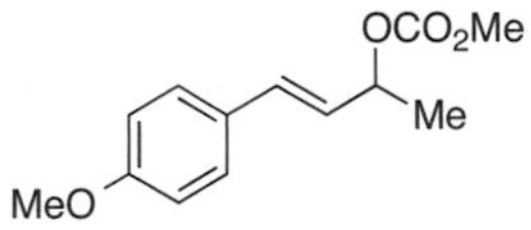	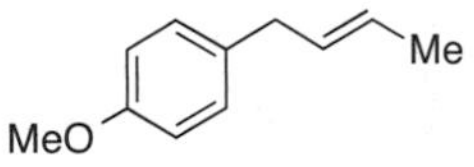	59

IPNBSH is further used for the reduction of vinyl epoxides to afford the corresponding homoallylic alcohols.[5b] Use of enantioenriched substrates did not lead to loss of enantiomeric excess during the reduction (Scheme 3).

238

Scheme 3

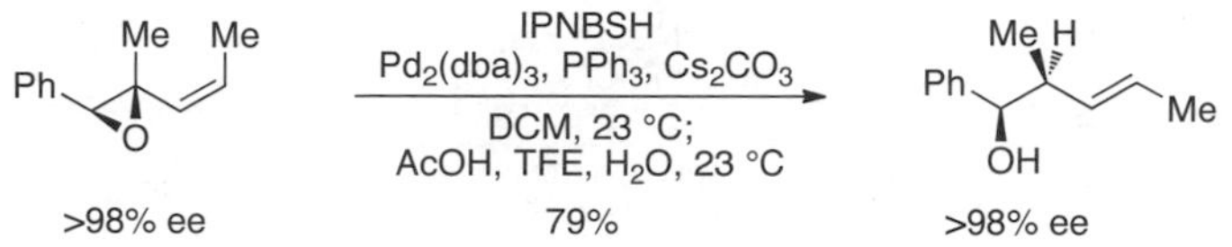

Me Me Ph⌒⟨O⟩⌒Me >98% ee	IPNBSH Pd₂(dba)₃, PPh₃, Cs₂CO₃ ————————————→ DCM, 23 °C; AcOH, TFE, H₂O, 23 °C 79%	Me H Ph⌒⟨OH⟩⌒Me >98% ee

Modification of this palladium-catalyzed diazene synthesis led to the development of an asymmetric version by using a chiral catalyst system. Racemic carbonates were converted into enantiomerically enriched sulfonyl hydrazones using Trost's[8] ligand (*1S,2S*)-(–)-1,2-diaminocyclohexane-*N,N'*-bis(2'-diphenylphospinobenzoyl) ((–)-6) and [(η³-allyl)PdCl]₂. Mild hydrolysis of these hydrazones provided enantiomerically enriched reductively transposed products (Scheme 4).

Scheme 4

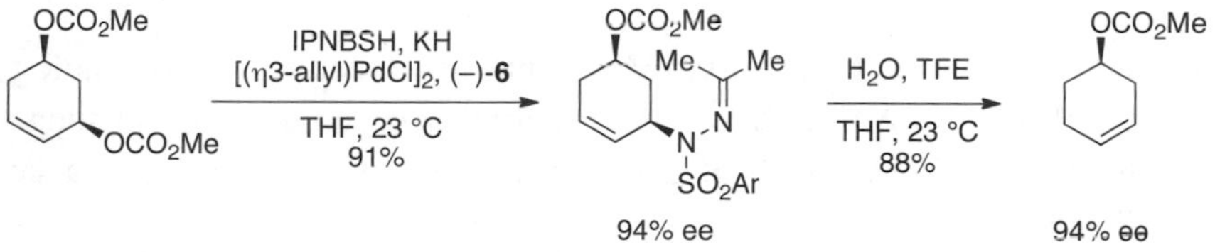

During the development of the procedure described here, we observed lower yields for the one-pot hydrazone formation and hydrolysis using 1-(anthracen-9-yl)allyl ethyl carbonate (**4**) as a substrate on large scale. After much experimentation, we recognized that this is due to interference of an active palladium species with the hydrazone adduct, resulting in formation of undesired by-products during the one-pot hydrolysis. We therefore developed a modification to the procedure to deactivate the palladium species after the formation of the hydrazone adduct. The best result was obtained using *N*-acetyl-L-cysteine as the palladium scavenger as described in the procedure above (Table 2). This additive prevents further activity of the palladium catalyst during the hydrazone hydrolysis step.

Table 2 Optimization studies for the synthesis of 9-allylanthracene

Entry	Additive	Yield
1	none	31%
2	Me-N-CH₂CH₂-NH-Me (diamine)	65%
3	Ph₂P-CH₂CH₂-PPh₂	68%
4	AcHN-CH(CH₂SH)-CO₂H	77%

1. Department of Chemistry, Massachusetts Institute of Technology, Cambridge, MA 02139. We are grateful for generous financial support by NIH-NIGMS (GM074825). M.M. is a Camille Dreyfus Teacher-Scholar.

2. Pangborn, A.B.; Giardello, M.A.; Grubbs, R.H.; Rosen, R. K.; Timmers, F. J. *Organometallics,* **1996**, *15*, 1518.

3. For representative examples, see: (a) Szmant, H. H.; Harnsberger, H. F.; Butler, T. J.; Barie, W. P. *J. Am. Chem. Soc.* **1952**, *74*, 2724. (b) Nickon, A.; Hill, A. S. *J. Am. Chem. Soc.* **1964**, *86*, 1152. (c) Corey, E. J.; Cane, D. E.; Libit, L. *J. Am. Chem. Soc.* **1971**, *93*, 7016. (d) Hutchins, R. O.; Kacher, M.; Rua, L. *J. Org. Chem.* **1975**, *40*, 923. (e) Kabalka, G. W.; Chandler, J. H. *Synth. Commun.* **1979**, *9*, 275. (f) Corey, E. J.; Wess, G.; Xiang, Y. B.; Singh, A. K. *J. Am. Chem. Soc.* **1987**, *109*, 4717. (g) Myers, A. G.; Kukkola, P. J. *J. Am. Chem. Soc.* **1990**, *112*, 8208. (h) Myers, A. G.; Finney, N. S. *J. Am. Chem. Soc.* **1990**, *112*, 9641. (i) Guziec, F. S., Jr.; Wei, D. *J. Org. Chem.* **1992**, *57*, 3772. (j) Wood, J. L.; Porco, J. A., Jr.; Taunton, J.; Lee, A. Y.; Clardy, J.; Schreiber, S. L. *J. Am. Chem. Soc.* **1992**, *114*, 5898. (k) Bregant, T. M.; Groppe, J.; Little, R. D. *J. Am. Chem. Soc.* **1994**, *116*, 3635. (l) Ott, G. R.; Heathcock, C. H. *Org. Lett.* **1999**, *1*, 1475. (m) Chai, Y.; Vicic, D. A.; McIntosh, M. C. *Org. Lett.* **2003**, *5*, 1039. (n) Sammis, G. M.;

Flamme, E. M.; Xie, H.; Ho, D. M.; Sorensen, E. J. *J. Am. Chem. Soc.* **2005**, *127*, 8612.

4. (a) Myers, A. G.; Zheng, B. *J. Am. Chem. Soc.* **1996**, *118*, 4492. (b) Myers, A. G.; Zheng, B. T*etrahedron Lett.* **1996**, *37*, 4841. (c) Myers, A. G.; Movassaghi, M.; Zheng, B. *J. Am. Chem. Soc.* **1997**, *119*, 8572.

5. (a) Movassaghi, M.; Ahmad, O. K. *J. Org. Chem.* **2007**, *72*, 1838. (b) Movassaghi, M.; Ahmad, O. K. *Angew. Chem., Int. Ed.* **2008**, *47*, 8909.

6. Movassaghi, M.; Piizzi, G.; Siegel, D. S.; Piersanti, G. *Angew. Chem., Int. Ed.* **2006**, *45*, 5859.

7. Trost, B. M.; Crawley, M. L. *Chem. Rev.* **2003**, *103*, 2921.

8. Trost, B. M.; Bunt, R. C. *J. Am. Chem. Soc.* **1994**, *116*, 4089.

Appendix
Chemical Abstracts Nomenclature; (Registry Number)

o-Nitrobenzenesulfonyl hydrazide; (5906-99-0)
N-Isopropylidene-*N'*-2-nitrobenzenesulfonyl hydrazine; (6655-27-2)
Anthracene-9-carboxaldehyde; (642-31-9)
Vinylmagnesium bromide; (1826-67-1)
N,N-Dimethylpyridine-4-amine; (1122-58-3)
Ethyl chloroformate; (541-41-3)
9-Allylanthracene: 9-(2-propeny-1-yl)anthracene; (23707-65-5)
Potassium hydride; (7693-26-7)
Allylpalladiumchloride dimer; (12012-95-2)
Triphenylphosphine; (603-35-0)
N-Acetyl cystcine; (616-91-1)

Mohammad Movassaghi carried out his undergraduate research with Professor Paul A. Bartlett at UC Berkeley, where he received his BS degree in chemistry in 1995. Mo then joined Professor Andrew G. Myers' group for his graduate studies and was a Roche Predoctoral Fellow at Harvard University. In 2001, Mo joined Professor Eric N. Jacobsen's group at Harvard University as a Damon Runyon–Walter Winchell Cancer Research Foundation postdoctoral fellow. In 2003, he joined the chemistry faculty at MIT where his research program has focused on the total synthesis of alkaloids in concert with the discovery and development of new reactions for organic synthesis.

Jens Willwacher was born in 1986 and grew up in Leverkusen, Germany. After receiving his high-school degree in 2005 and completing military service in 2006, he started studying chemistry at the Westfaelische Wilhems-Universitaet Muenster, Germany. During his undergraduate program, he joined the laboratories of Professor Mohammad Movassaghi at Massachusetts Institute of Technology during the summer 2010 as a visiting student. He conducted his diploma thesis with Professor Frank Glorius with the focus on rhodium-catalyzed C-H activation olefinations and recently joined Professor Alois Fürstner's group for his Ph.D. program.

Omar K. Ahmad pursued his undergraduate studies at Brown University, where he received his Sc.B. degree in chemistry in 2005. As an undergraduate he worked in the laboratories of Professor Dwight Sweigart and Professor Amit Basu. In 2005, Omar joined Professor Mohammad Movassaghi's research group at MIT for his graduate studies where his research has focused on the development of new methodologies for organic synthesis.

Juliane Keilitz was born in 1981 in Berlin, Germany. She studied chemistry at the Freie Universität Berlin, where she received her Ph.D. in 2010 under the supervision of Prof. Rainer Haag. After working on the development and application of polymer-supported catalysts, she is now pursuing post-doctoral research in the group of Prof. Mark Lautens at the University of Toronto. Her research focuses on the desymmetrization of cyclohexadienones.

Synthesis of (2*R*)-3-[[(1,1-Dimethylethyl)dimethylsilyl]oxy]-2-methylpropanal by Rhodium-Catalyzed Asymmetric Hydroformylation

A.

B.

Bis[(*S,S,S*)-DiazaPhos-SPE]

Submitted by Gene W. Wong, Tyler T. Adint, Clark R. Landis.[1]
Checked by Neil Strotman, James Cuff and David Hughes.

1. Procedure

> *Caution! Carbon monoxide is a highly toxic gas and manipulations should be conducted in a well-ventilated fume hood in the vicinity of a carbon monoxide detector. Hydrogen gas is highly flammable and explosive gas. Precautions should be taken when using synthesis gas (H_2/CO mixtures).*

A. Allyl (t-butyldimethyl)silyl ether 1. An oven-dried three-necked, 500-mL round-bottomed flask equipped with a 3-cm PTFE-coated oval stir bar is charged with allyl alcohol (8.6 g, 0.15 mol, 1.0 equiv), imidazole (21.1 g, 0.31 mol, 2.1 equiv), and dimethylformamide (100 mL). (Note 1) The flask is fitted with a gas inlet adapter connected to a nitrogen line and a gas bubbler. The other two necks are capped with rubber septa; a thermocouple probe is inserted through one of the septa. (Note 2) The stirred solution is cooled to 4 °C with an ice-water bath, then *t*-butyldimethylsilyl chloride

(25.5 g, 0.17 mol, 1.1 equiv) is added in portions over 2 min. (Note 3) The water bath is removed and the hazy solution is stirred 15 h at 22–23 °C. (Note 4) The reaction mixture is transferred to a 500-mL separatory funnel along with hexanes (200 mL) and water (60 mL). After mixing and settling, the aq. layer is removed and the organic layer washed with brine (75 mL). The hazy organic layer is filtered through a bed of sodium sulfate (50 g) in a medium porosity sintered glass funnel into a 500-mL round-bottomed flask, using hexanes (2 x 50 mL) to rinse the filter cake. The filtrate is concentrated by rotary evaporation (40 °C, 20 mmHg) to an oil (22 g). Since a substantial portion of the product co-distills during concentration, the distillate from the receiver flask is re-concentrated, providing additional product (2.8 g). The combined crude product is purified by column chromatography (Note 5) to afford allyl (*t*-butyldimethyl)silyl ether **1** (18.1–18.5 g, 71–73% yield) as a colorless oil. (Note 6)

   B.   *(2R)-3-[[(1,1-dimethylethyl)dimethylsilyl]oxy]-2-methylpropanal* **2**. In a glovebox, bis[(*S*,*S*,*S*)-DiazaPhos-SPE] (9.2 mg, 0.0070 mmol, 0.024 mol%) is added to a 20 mL vial followed by 0.22 mL CDCl₃. To a separate 4 mL vial is added Rh(acac)(CO)₂ (10 mg) followed by 1.94 mL toluene to prepare a 20 mM stock solution; 0.29 mL (1.5 mg, 0.0058 mmol, 0.020 mol%) of this stock solution is transferred to the 20 mL vial containing the ligand solution (Notes 7, 8). Allyl (*t*-butyldimethyl)silyl ether **1** (5.00 g, 6.17 mL, 29 mmol, 1.0 equiv) is added to the vial. The solution is divided into 4 x 4 mL vials, each containing a 5-mm glass bead, and loaded into a Symyx Heated Orbital Shaker System (HOSS) contained within the glove box. (Note 9) The system is taken through 3 cycles of pressurization (150 psig of 1:1 H_2:CO)/depressurization (0 psig) to replace the nitrogen atmosphere with synthesis gas. The system is pressurized with 150 psig 1:1 H_2:CO and heated to 60 °C for 16 h. (Note 10) The mixture is cooled and the system depressurized, then the vials are removed from the glove box. (Note 15) The hydroformylation reaction mixture is purified immediately (Note 16) by flash column chromatography (Note 17) to afford colorless oils of branched isomer (*R*)-**2** (3.24–3.43 g, 55–58% yield, Notes 18-20) and linear isomer **3** (1.62–1.70 g, 27–29% yield, Note 21).

2. Notes

1. The following reagents and solvents in step A were used as received: allyl alcohol (Sigma-Aldrich), imidazole (Acros, 99%), *t*-butyldimethylsilyl chloride (Acros, 98%), anhydrous DMF (Sigma-Aldrich, 99.8%), hexanes (Fisher, ACS reagent, >98.5%), ethyl acetate (Fisher, ACS reagent, >99.5%) and silica gel (Fisher, 230-400 mesh, 60 Å).

2. The internal temperature is monitored using a J-Kem Gemini digital thermometer with a Teflon-coated T-Type thermocouple probe (12-inch length, 1/8 inch outer diameter, temperature range –200 to +250 °C).

3. The reaction warmed to 25 °C over 5 min after addition of TBS-Cl.

4. The reaction was monitored by TLC, 5% EtOAc/hexanes, R_f 0.6, KMnO$_4$ stain.

5. Silica gel (250 g) was slurry-packed in a 5-cm diameter column using 2.5% EtOAc/hexanes. The product was eluted with 2.5% EtOAc/hexanes, collecting 100 mL fractions. Fractions 5-13 were combined and concentrated by rotary evaporation (40 °C, 20 mmHg) to afford **1** (14.5–15.4 g) as a colorless oil. The distillate from concentration was re-concentrated to provide an additional 2.7–4.0 g (combined yield, 18.1–18.5 g, 71–73%). The distillate from the final product concentration was assayed by ^{1}H NMR using toluene as an internal standard, indicating 2.4 g (10% yield) **1** was present in the distillate.

6. *Allyl (t-butyldimethyl)silyl ether 1* has the following physical and spectroscopic data: ^{1}H NMR (400 MHz, CDCl$_3$) δ: 0.09 (s, 6 H), 0.93 (s, 9 H), 4.18–4.20 (m, 2 H), 5.07–5.11 (m, 1 H), 5.25–5.30 (m, 1 H), 5.89–5.97 (m, 1 H); ^{13}C NMR (100 MHz, CDCl$_3$) δ: –5.0, 18.6, 26.2, 64.3, 114.1, 137.8; GC-MS (EI) *m/z*: 172 (6 %) [M$^+$], 157 (6%), [M - CH$_3$], 116 (31%), 115 (100%) [M - *t*-Bu], 99 (21%), 85 (69%) [M – Me$_2$,*t*-Bu], 75 (28%), 59 (48%); GC purity: 98% (t_R = 4.4 min; conditions: Agilent DB35MS column; 30 m x 0.25 mm; initial temp 60 °C, ramp at 20 °C/min to 280 °C, hold 15 min).

7. The following reagents and solvents in step B were used as received by the checkers: dicarbonylacetylacetonato rhodium(I) (Strem), toluene (Sigma Aldrich, anhydrous, >99.9%), CDCl$_3$ (Sigma-Aldrich, 99.8% atom % D), SynGas (49% carbon monoxide/51% hydrogen, Airgas) and bis[(*S*,*S*,*S*)-DiazaPhos-SPE] ligand: 2,2',2'',2'''-(1,2-phenylenebis[(1*S*,3*S*)-tetrahydro-5,8-dioxo-1*H*-[1,2,4]diazaphospholo[1,2-a]pyridazine-2,1,3(3*H*)-triyl])tetrakis(*N*-[(1*S*)-1-phenylethyl])benzamide (Sigma-Aldrich). The

ligand was prepared by the submitters according to their published procedure.[2] The submitters recrystallized dicarbonylacetylacetonato rhodium(I) from toluene and hexanes as fine green crystals. $CHCl_3$ may be used instead of $CDCl_3$.

8. Accurate volumes were measured and transferred using an Eppendoft® pipette.

9. Symyx HOSS equipment is described on the Symyx website. http://symyx.desantisbreindel.com/page.php?id=71

10. The submitter's equipment and experimental protocol are outlined in this Note and in Notes 11-14. A heavy wall reaction tube (Ace Glass #15 Ace-Tread®, 30 cm length x 38.1 mm O.D., 185 mL capacity) and a 0.5 x 0.125 inch magnetic stir bar are dried in a 125 °C oven overnight. In a glove box, the reaction tube is charged with stock solutions of $Rh(acac)(CO)_2$ and Bis[(S,S,S)-DiazaPhos-SPE] using a 1000 µL Eppendoft® pipette followed by 5 grams of substrate. The reaction tube is attached to the reactor head (Note 11). Notably, the addition of the alkene to the catalyst solution resulted in a yellow-white suspension due to partial precipitation of ligand and/or catalyst-ligand complex.

A blast shield must be used whenever the reactor is pressurized and safety procedures for using pressure tubes described in the Ace-Glass® catalog should be reviewed and followed.

The assembled reactor is removed from the glove box, placed in a fume hood, connected to the synthesis gas source and taken through 5 cycles of pressurization (150 psig of 1:1 H_2:CO)/depressurization (0 psig) to replace the nitrogen atmosphere with synthesis gas (Notes 12, 13, Figure 2). The reactor is then submerged in a heated silicon oil bath at the desired temperature. As synthesis gas is consumed, the reactor is repressurized to 150 psi to maintain approximately constant pressure (Note 14). After 2–3 hours (~30–40 psi of synthesis gas consumed) the suspension transforms to a homogeneous yellow solution. In six hours, ~90 psi of synthesis gas is consumed. At the end of the reaction time, the reactor is depressurized.

11. A custom-made reactor head used for hydroformylations is shown in Figure 1. The following parts were used to assemble the reactor head: **a**, Alltech® septum (High-temp, 3/8 in., AT79231) for aliquot-abstractions using a gas-tight syringe, **b**, Swagelok® Brass 1-Piece 40 Series Ball Valve (1.6 Cv, 1/4 in. MNPT x 1/4 in. Swagelok Tube Fitting; product #: B-43M4-

247S4), c, Swagelok® Brass Pipe Fitting, Cross (1/4 in. Female NPT; product #:B-4-CS), d, Brass Pipe Fitting, Street Tee, (1/4 in. Female NPT x 1/4 in. Male NPT x 1/4 in. Female NPT; product #:B-4-ST), e, Brass pressure release valve, 150 psi-350 psi (1/4 in. Female NPT; product #:B-4CPA-VI-150), f, Ashcroft® 0–160 psig pressure gauge (1/4 in. NPT, 3.5 in. Dial; McMaster-Carr 3846K311 0-160 psig range), g, Brass Pipe Fitting, Close Nipple (1/4 in. Male NPT), h, #15 Ace-Thred® (15 mm thread, 1/4 in. NPT PTFE Swagelok adapter; Prod. #: 5844-74), i, Kalrez® 6375 O-ring (9.30mm x 2.40 mm Part #: K31016K6375), j, #15 Ace Glass® pressure tube (30.5 cm L, 38.1 mm OD, Prod. #: 8648-33), k, Swagelok® Brass 1-Piece 40 Series 3-Way Ball Valve (0.75 Cv, 1/4 in. FNPT; product #: B-43XF4), l, Brass Pipe Fitting (1/4 in. male NPT to 1/4 in. male Swagelok Tube Fitting), m, SS tubing (1/4 in OD, 2 1/2 in. length), and n, Swagelok® SS Instrumentation Quick-Connect Stem w/ Valve, (0.2 Cv, 1/4 in. Swagelok Tube Fitting, Part #: SS-QC4-D-400). Threads b, d, f, g, and l were wrapped with PTFE tape prior to assembly. A thorough pressure check of reactor should be taken before conducting an experiment. The most common source of a leak is between the brass pipe fitting g and the plastic #15 Ace-Thread adapter h. Pressure release valve e is calibrated by pressurizing to 150 psig and loosening check valve until triggering, followed by tightening a quarter turn. Once assembled with the 185-mL pressure tube, the reactor is rather cumbersome to transport—the use of an 11.5" (W) x 13.5" (L) x 5.25"(D) Rubbermaid® dishpan with a 3"(D) x 1" (W) rectangle cut in the tub on the width side was used to hold the reactor.

12. A reverse threaded regulator is connected to a synthesis gas cylinder and Swagelok® Quick-Connects are used to attach to the reactor manifold. The synthesis gas cylinder was obtained from AirGas Inc. as a custom mixture (48.3±2% carbon monoxide balanced with hydrogen gas).

13. The reactor has two possible points of entry: Swagelok® Ball valve b fitted with a GC septum, for gas-tight syringe aliquots, and the Swagelok® 3-way Ball Valve k, for pressurizing and depressurizing the reactor. In Figure 2, k is opened carefully to the synthesis gas cylinder, charging the apparatus to 150 psig (it is advisable to set the regulator on the cylinder to ca. 150 psig and to have a safety shield in place). The valve on k is then opened to vent, releasing synthesis gas from the apparatus. After the pressure is reduced to <40 psi, the valve is turned back to the original closed position constituting one cycle. This procedure is repeated for five cycles and the reactor pressure is set at 150 psi. The glass tube of the reactor is

lowered into the oil bath for hydroformylation as seen in the far-right picture.

14. Synthesis gas is added manually to maintain at least 100 psig reactor pressure. It is not advisable to maintain reactor pressure by keeping the reactor open to the regulator on the synthesis gas cylinder because, in the event of a leak on the reactor or supply lines, large amounts of H_2 and CO could be released. A carbon monoxide detector is installed near the gas cylinder. Commonly, the synthesis gas line is detached from the reactor at the Swagelok® Quick-Connect during reaction and reconnected when adding more gas. However, if the synthesis gas line is not needed for other reactions, the Swagelok® Quick-Connect system can remain assembled throughout the reaction.

15. ^{1}H NMR of the crude product mixture indicated >99% conversion of alkene and a branched: linear (**2:3**) ratio of 2:1.

16. Aldehyde **2** is air-sensitive and flash chromatography should be performed immediately after depressurizing the reactor and the purified product stored in a freezer.

17. Silica gel (250 g) was slurry-packed in a 5-cm diameter column using 5% EtOAc/hexanes. The product was eluted with 5% EtOAc/hexanes, collecting 50 mL fractions, monitored by TLC. (10% EtOAc/hexanes, R_f **1** = 0.7, (R)-**2** = 0.38, **3** = 0.31, visualized with potassium permanganate stain, prepared as follows: 3 g $KMnO_4$, 20 g potassium carbonate, 5 mL of a 5% (w/w) solution of aqueous sodium hydroxide, and 300 mL of deionized water.) Fractions 9-26 were combined and concentrated by rotary evaporation (40 °C, 20 mmHg) to afford **2** (3.24–3.43 g) as a colorless oil. Fractions 29-38 were combined and concentrated by rotary evaporation (40 °C, 20 mmHg) to afford **3** (1.62–1.70 g) as a colorless oil.

18. *(2R)-3-[[(1,1-Dimethylethyl)dimethylsilyl]oxy]-2-methylpropanal* **2** has the following physical and spectroscopic data: $[\alpha]_D^{25}$ –34 (c 1.0, CH_2Cl_2); ^{1}H NMR (500 MHz, $CDCl_3$) δ: 0.06 (s, -Si(CH_3)$_2$C(CH$_3$)$_3$, 6 H), 0.89 (s, -Si(CH$_3$)$_2$C(CH_3)$_3$, 9 H), 1.10 (d, J = 6.9 Hz, -CHCH_3, 3 H), 2.52–2.56 (m, -CHCH$_3$, 1 H), 3.82 (dd, J = 6.4, 10.2 Hz, -CH_2OSi, 1 H), 3.86 (dd, J = 5.2 Hz, 10.2, -CH_2OSi, 1 H), 9.74 (d, J = 1.6 Hz, CHO-CH, 1 H); ^{13}C NMR (125 MHz, $CDCl_3$) δ: –5.33, –5.31, 10.5, 18.4, 26.0, 49.0, 63.7, 204.9; IR (neat): 2957, 2931, 2859, 1736 (C=O), 1473, 1258, 1101, 1033, 838, 778 cm^{-1}; GC-MS m/z (relative intensity): 145 (100) [M – t-Bu], 115 (95) [SiMe$_2$t-Bu], 101 (31), 85 (25) [Sit-Bu], 75 (54), 59 (25); GC purity: 98% (t$_R$ = 7.4 min, same conditions as in note 6); ee 94–96% determined by SFC

248

analysis of benzylamine reductive amination derivative as described in Note 20. The aldehyde oxidizes at a rate of about 1% per week when stored in a -20 °C freezer.

19. The (S)-enantiomer of **2** was prepared by the same procedure using Bis[(R,R,S)-DiazaPhos-SPE] as ligand; $[\alpha]_D^{25}$ +33 (c 1.0, CH_2Cl_2); ee 88%.

20. The submitters determined chiral purity by gas chromatographic analysis on a Varian Chrompack system using a β-DEX 225 capillary column from Supelco, 30 m x 0.25 mm ID x 0.25 um film thickness. The analytical method used to resolve the enantiomers as follow: 65 °C hold for 70 min, t_R(R)-**2**: 60.8 min, t_R(S)-**2**: 62.4 min. The checkers determined chiral purity by formation of the reductive amination product with benzylamine and analysis by supercritical fluid chromatography (SFC): tandem columns: 25 cm OZ : 25cm OZ, isocratic 8% 25mM i-butylamine in 2-propanol, 100 bar, 2.0 mL/min for 18 min. t_R (R)-**4** = 11.5 min, t_R (S)-**4** = 14 min.

Procedure for the preparation of the reductive amination product with benzylamine follows.

To a 20-mL vial equipped a 0.7 cm stir bar is added sequentially sodium triacetoxyborohydride (235 mg, 1.1 mmol), chloroform (2 mL), aldehyde **2** (73 mg, 0.36 mmol), and benzylamine (37 mg, 0.34 mmol). The heterogeneous mixture is stirred 16 h at ambient temperature, then quenched with 5 mL sat. $NaHCO_3$, stirring the biphasic mixture for 5 min. Dichloromethane (10 mL) is added, the layers are separated, and the organic layer dried by filtering through 2 g of sodium sulfate. The filtrate is concentrated by rotary evaporation (40 °C, 20 mmHg) to afford crude **4** (115 mg). The product is purified by silica gel chromatography using 10 g silica with an eluent of 97:2:0.5 CH_2Cl_2:MeOH:Et$_3$N, collecting 10 mL fractions. Fractions 7-10 were combined and concentrated by rotary evaporation to provide product **4** (68 mg, 67% yield) as a colorless oil having the following physical and spectroscopic data: TLC: R_f = 0.1 (97:2:0.5 CH_2Cl_2:MeOH:Et$_3$N); ^{1}H NMR (500 MHz, CDCl$_3$) δ: 0.04 (s, 6 H), 0.89 (s, 9 H), 0.91 (d, J = 6.7 Hz, 3 H), 1.68 (br s, 1 H), 1.85–1.91 (m, 1 H), 2.51 (dd, J = 6.0, 11.6 Hz, 1 H), 2.68 (dd, J = 6.8, 11.6 Hz, 1 H), 3.50–3.57 (m, 2 H), 3.77–3.82 (m, 2 H), 7.25–7.26 (m, 1 H), 7.31–7.33 (m, 4 H);

^{13}C NMR (125 MHz, CDCl$_3$) δ: –5.24, –5.21, 15.6, 18.5, 26.1, 36.1, 53.6, 54.5, 127.0, 128.3, 128.5, 140.9.

21. Linear product **3** has the following spectroscopic data: ^{1}H NMR (400 MHz, CDCl$_3$) δ: 0.04 (s, -Si(C*H$_3$*)$_2$C(CH$_3$)$_3$, 6 H), 0.88 (s, -Si(CH$_3$)$_2$C(C*H$_3$*)$_3$, 9 H), 1.86 (tt, *J* = 6.0, 7.1 Hz -CH$_2$C*H$_2$*CH$_2$-, 2 H), 2.50 (dt, *J* = 7.1, 1.8 Hz, C*H$_2$*CHO, 2 H), 3.65 (t, *J* = 6.0 Hz, -C*H$_2$*OSi, 3 H), 9.79 (t, *J* = 1.7 Hz, C*H*O, 1 H); ^{13}C NMR (100 MHz, CDCl$_3$) δ: –5.2, 18.5, 25.7, 26.1, 41.0, 62.3, 202.9; GC purity: 96% (t$_R$ = 7.9 min, same conditions as in Note 6)

Safety and Waste Disposal Information

All hazardous materials should be handled and disposed of in accordance with "Prudent Practices in the Laboratory"; National Academies Press; Washington, DC, 2011.

3. Discussion

Protected "Roche Aldehydes" (e.g., **2**, (2R)-3-[[(1,1-dimethylethyl)dimethylsilyl]oxy]-2-methylpropanal**)** are common starting materials for the synthesis of polyketides and related molecules.[3,4] Compared with the common reduction-to-alcohol-followed-by-selective-oxidation-to-aldehyde route to **2** from "Roche Ester",[4] hydroformylation of the protected commodity monomer, allyl alcohol, provides Roche Aldehyde derivatives rapidly, at low cost, and in an easily scalable process. For comparison purposes we have collected the following approximate costs of substrates, normalized to 25 g units, from a common supplier: Roche ester ($350/25g), allyl alcohol ($1.00/25g). The only byproducts of the enantioselective hydroformylation of **1** is the corresponding linear aldehyde; although achiral the linear aldehyde is isolated cleanly and constitutes a useful synthetic material also. On larger scales, it should be possible to separate the linear and branched aldehydes by careful vacuum distillation; we have not yet optimized the distillation conditions. An advantage of hydroformylation routes to chiral aldehydes is the absence of acids or bases in the reaction solution that catalyze racemization and condensation reactions. We note that although the Roche *Ester* has been synthesized by asymmetric hydrogenation of the methyl 2-(hydroxymethyl)-prop-2-enoate,[5] there is no report of a catalytic hydrogenation route to enantiopure Roche *Aldehyde*.

Org. Synth. **2012**, *89*, 243-254

Figure 1. The submitters assembled reactor with parts indicated

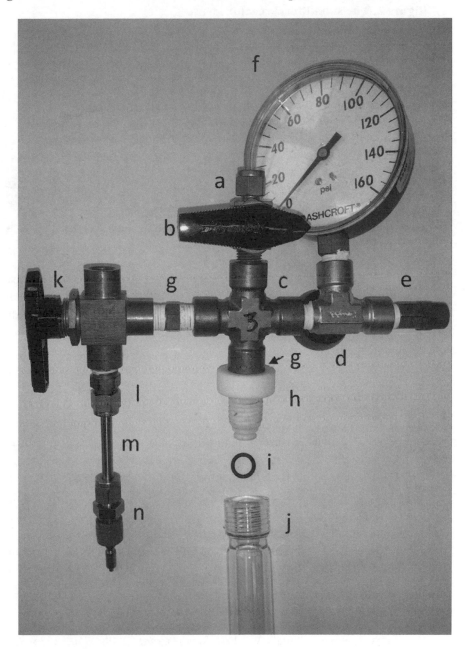

Figure 2. The submitters' reactor in-use

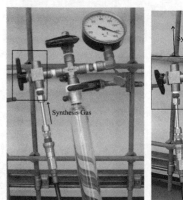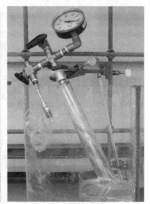

1. Department of Chemistry, University of Wisconsin, Madison, Wisconsin, 53706-1322. Email: Landis@chem.wisc.edu. We thank The Dow Chemical Company for their donation of Rh(acac)(CO)$_2$. Funding was provided by the National Science Foundation (CHE-0715491 and a graduate fellowship for G.W.W.).

2. Clark, T. P.; Landis, C. R.; Freed, S. L.; Klosin, K.; Abboud, K. A. *J. Am. Chem. Soc.* **2005**, *127*, 5040–5042.

3. For a general review on polyketides stereotetrads, see: Koskinen, A. M. P.; Karisalmi, K. *Chem. Soc. Rev.* **2005**, *34*, 677–690.

4. Roche Aldehyde for use in natural product syntheses see: (a) Früstner, A.; Nevado, C.; Waser, M.; Tremblay, M.; Chevrier, C.; Teplý, F.; Aïssa, C.; Moulin, E.; Müller, O. *J. Am. Chem. Soc.* **2007**, *129*, 9150–9161; (b) Canova, S.; Bellosta, V.; Bigot, A.; Mailliet, P.; Mignani, S.; Cossy, J. *Org. Lett.* **2007**, *9*, 145–148; (c) Ferrié, L.; Reymond, S.; Capdevielle, P.; Cossy, J. *Org. Lett.* **2006**, *8*, 3441–3443; (d) Ehrlich, G.; Kalesse, M. *Synlett* **2005**, 4, 655–657; (e) Smith, A. B., III, Brandt, B. M. *Org. Lett.* **2001**, *3*, 1685–1688.

5. (a) Qiu, M.; Wang, D.-Y.; Hu, X.-P.; Huang, J.-D.; Yu, S.-B.; Deng, J.; Duan, Z.-C.; Zheng, Z. *Tetrahedron: Asymmetry* **2009**, *20*, 210–213. (b) Pautigny, C.; Jeulin, S.; Ayad, T.; Zhang, Z.; Genêt, J.-P.; Ratovelomanana-Vidal, V. *Adv. Synth. Catal.* **2008**, *350*, 2525–2532.

252

(c) Holz, J.; Schäffner, B.; Zayas, O.; Spannenberg, A.; Börner A. *Adv. Synth. Catal.* **2008**, *350*, 2533–2545. (d) Wassenaar, J.; Kuil, M.; Reek J. N. H. *Adv. Synth. Catal.* **2008**, *350*, 1610–1614. (e) Jeulin, S.; Ayad, T.; Ratovelomanana-Vidal, V.; Genêt, J.-P. *Adv. Synth. Catal.* **2007**, *349*, 1592–1596. (f) Shimizu, H.; Saito, T.; Kumobayashi, H. *Adv. Synth. Catal.* **2003**, *345*, 185–189.

Appendix
Chemical Abstracts Nomenclature; (Registry Number)

Silane, (1,1-dimethylethyl)dimethyl(2-propen-1-yloxy)-; (85807-85-8)
Silane, chloro(1,1-dimethylethyl)dimethyl-; (18162-48-6)
Prop-2-en-1-ol; (107-18-6)
Rhodium, dicarbonyl(2,4-pentanedionato-κ-O2,κ-O4)-, (SP-4-2)-; (14874-82-9)
Benzamide, 2,2',2'',2'''-[1,2-phenylenebis[(1*S*,3*S*)-tetrahydro-5,8-dioxo-1H-[1,2,4]diazaphospholo[1,2-a]pyridazine-2,1,3(3H)-triyl]]tetrakis[*N*-[(1*S*)-1-phenylethyl]-; (851770-14-4)
Propanal, 3-[[(1,1-dimcthylethyl)dimethylsilyl]oxy]-2-methyl-, (2*R*)-; (97826-89-6),
Butanal, 4-[[(1,1-dimethylethyl)dimethylsilyl]oxy]-; (87184-81-4)

 Clark R. Landis was born in Aurora, IL in 1956. After completing his Ph.D. at the University of Chicago in 1983 under the direction of Jack Halpern, Professor Landis held professional positions at the Monsanto Company Corporate Research Lab, the University of Colorado-Boulder, and, since 1990, the University of Wisconsin-Madison. His research interests include bonding theory, computational methods, instrumentation development, chiral ligand synthesis, enantioselective catalysis, catalytic alkene polymerization, and the mechanisms of catalytic reactions.

Gene W. Wong was born in Reno, Nevada in 1985. He received his undergraduate chemistry degree from University of Nevada-Reno, where he conducted research with Prof. Brian J. Frost. He then moved to University of Wisconsin-Madison, where he is currently pursuing a Ph.D. in the research group of Prof. Clark R. Landis as a NSF Predoctoral Fellow. His research focuses on the development of bisdiazaphospholane libraries and rhodium catalyzed hydroformylation.

Tyler T. Adint was born in Fairbanks, Alaska in 1984. He received his undergraduate degree in 2003 from Lewis & Clark College, where he conducted research with Prof. Louis Y. Kuo. He is now pursuing his Ph.D. at the University of Wisconsin-Madison in the research group of Prof. Clark R. Landis. His current research interests concern the synthesis of bisdiazaphospholane ligands that utilize secondary interactions to control selectivity in rhodium catalyzed hydroformylations.

Synthesis of (*R*)-2-Methoxy-*N*-(1-Phenylethyl)Acetamide via Dynamic Kinetic Resolution

Submitted by Lisa Kanupp Thalén, Christine Rösch, and
Jan-Erling Bäckvall.[1]
Checked by Marc-André Müller and Andreas Pfaltz.

1. Procedure

A. (R)-2-Methoxy-N-(1-phenylethyl)acetamide (**2**). A flame-dried, 1-L two-necked round-bottomed flask is equipped with a two-tap Schlenk adapter connected to a bubbler and an argon/vacuum manifold (Note 1) in the middle neck. The flask is equipped with a stir bar (4.0 cm) and is charged with Novozym 435 (340 mg) (Note 2), Na$_2$CO$_3$ (900 mg, 8.49 mmol, 0.19 equiv) (Note 3), and Ru-complex **7** (745 mg, 0.56 mmol, 1.25 mol%) under a constant flow of argon. The other neck is equipped with a rubber septum and the vessel is evacuated and refilled with argon three times. Under an argon atmosphere, anhydrous toluene (225 mL) (Note 4) is added via syringe to afford a brown solution. (±)-1-Phenylethylamine (**1**, 5.80 mL, 5.45 g, 45.0 mmol, 1.00 equiv) (Note 5), 2,4-dimethyl-3-pentanol (8.00 mL, 6.63 g, 57.1 mmol, 1.27 equiv) (Note 6) and methyl methoxyacetate (3.40 mL, 3.57 g, 33.8 mmol, 0.75 equiv) (Note 7) are subsequently added via syringe. The septum is then changed to a glass stopper equipped with a teflon ring (Note 8) under constant argon flow. The glass stopper is secured with a metal clamp, the reaction mixture is stirred vigorously and the reaction mixture is heated to 100 °C. After the reaction mixture reaches 100 °C the two taps are closed (Notes 9 and 10). After 24 h the tap connected to the bubble counter is opened to release pressure. The second tap connected to the argon inlet is subsequently opened and the reaction vessel is allowed to cool for 10 min at rt. The adapter is then opened under argon flow, the glass stopper is exchanged for a rubber septa, and methyl methoxyacetate (1.60 mL, 1.68 g, 15.8 mmol, 0.35 equiv) (Note 7) is added via syringe. The septum is removed under an argon flow and Novozym 435 (110 mg) (Note 1) is quickly added. The flask is subsequently closed with a glass stopper secured with a metal clamp. The reaction mixture is heated up to 100 °C and the two taps are closed after reaching 100 °C. After a reaction time of 72 h, the reaction is cooled to rt, the solids are removed by filtration through a sintered glass funnel (poor size 3) (Note 11) and are subsequently washed with dichloromethane (4 × 10 mL) (Note 12). Half of the combined filtrate is transferred to a 250-mL round-bottomed flask and concentrated by rotary evaporation (40 °C, 20 mmHg). The remaining filtrate is also transferred to the same 250-mL round-bottomed flask and concentrated by rotary evaporation (40 °C, 20 mmHg). The crude material is further vacuum dried at rt (10 mmHg) until a solid is obtained (24 h). Purification by Kugelrohr distillation (Note 13) yields (*R*)-2-methoxy-N-(1-phenylethyl)acetamide (**2**) as a colorless solid (6.72–6.89 g, 34.8–35.7 mmol, 77–79%, 97–98% *ee*) (Note 14). Recrystallization from pentane yields (*R*)-2-methoxy-*N*-(1-phenylethyl)acetamide (**2**) as colorless crystals (6.03–6.18 g, 31.2–32.0 mmol, 69–71%, >99% *ee*) (Note 15).

B. *1,3-Bis(4-methoxyphenyl)propan-2-one* (**4**).[2] A 250-mL one-necked, round-bottomed flask equipped with a cross-shaped magnetic stir bar (2.5 cm) is charged with 2-(4-methoxyphenyl)acetic acid (8.30 g, 50.0 mmol, 1.00 equiv) (Note 16). The flask is not equipped with a stopper.

 Org. Synth. **2012**, *89*, 255-266

Upon addition of water (50 mL) a colorless suspension is obtained. Na_2CO_3 (5.30 g, 50.0 mmol, 1.00 equiv) (Note 17) is added in portions over a period of 5 min while the mixture is stirred vigorously to obtain a colorless solution. After the evolution of gas ceases , a solution of $MnCl_2$ (10.0 g, 50.0 mmol, 1.00 equiv) (Note 18) in water (15 mL) is slowly added over a period of 20 min. A white foam forms during the addition. Subsequently water (50 mL) is added resulting in the formation of a homogenous, white suspension. The reaction mixture is stirred vigorously for 3 h and subsequently filtered through a Büchner funnel. The resulting colorless solid is divided into three portions of an approximately equal amount and each portion is dried at 160 °C under reduced pressure (0.3 mmHg) for 1 h. The brown solid is finely ground using a mortar and pestle. Kugelrohr distillation (225 °C, 0.06 mmHg) affords 1,3-bis(4-methoxyphenyl)propan-2-one as light yellow solid (2.97–3.17 g, 11.0–11.7 mmol, 44–47%) (Note 19).

C. *2,3,4,5-Tetrakis(4-methoxyphenyl)cyclopenta-2,4-dienone* (**6**). A 250-mL, one-necked, round-bottomed flask equipped with an oval shaped magnetic stir bar (2.5 cm) is charged with 1,3-bis(4-methoxyphenyl)propan-2-one (2.00 g, 7.40 mmol, 1.00 equiv), 4,4-dimethoxybenzil (2.00 g, 7.40 mmol, 1.00 equiv) (Note 20) and tetramethylammonium hydroxide pentahydrate (4.00 g, 22.1 mmol, 1.00 equiv) (Note 21) is added. Upon addition of ethanol (100 mL) a yellow suspension is obtained. Subsequently, the flask is equipped with a reflux condenser and the mixture is refluxed for 3 h, during which time the suspension's color gradually changes from yellow to orange to black. The black mixture is allowed to cool to rt and is filtered through a sintered glass funnel (pore size 3) The resulting black powder is washed with ethanol until the washing solvent remains colorless (Note 22). Subsequently the solid is washed with pentane (3 × 10 mL). The black solid is dried under reduced pressure (10 mmHg) at rt to afford 2,3,4,5-tetrakis(4-methoxyphenyl)cyclopenta-2,4-dienone as black powder (2.85–3.06 g, 5.62–6.06 mmol, 76–82%) (Note 23).

D. *Ruthenium complex {{[2,3,4,5(4-OMe-$C_6H_4)_4$](η ^{5}-C_4CO)}$_2$H}-Ru$_2$(CO)$_4$(μ-H)* (**7**). A 250-mL oven-dried, two-necked, round-bottomed flask is equipped with a magnetic stir bar (2.5 cm), reflux condenser, a two-tap Schlenk adapter connected to a bubbler and an argon/vacuum manifold (Note 1) in the middle neck, and a rubber septum in the other neck. The apparatus is evacuated and refilled with argon three times. The flask is charged with 2,3,4,5-tetrakis(4-methoxyphenyl)cyclopenta-2,4-dienone (**6**, 1.00 g, 1.98 mmol, 1.00 equiv) and $Ru_3(CO)_{12}$ (428 mg, 0.67 mmol,

0.34 equiv) (Note 24) under a constant argon flow. Methanol (100 mL) (Note 25) is added. Argon is bubbled through the mixture for 5 min via a needle inserted through the septum. The septum is replaced with a glass stopper under a constant flow of argon. The mixture is heated to reflux (Note 26) for 24 h. During this time the color changes from black to mustard yellow. The mixture is cooled to rt and the precipitate is removed by filtration through a sintered glass funnel (pore size 3) with a filter paper. The collected solid is washed with pentane (3 × 10 mL) and subsequently dried under reduced pressure (10 mmHg) at rt to provide ruthenium complex $\{\{[2,3,4,5(4\text{-OMe-}C_6H_4)_4](\eta^5\text{-}C_4CO)\}_2H\}Ru_2(CO)_4(\mu\text{-H})$ (7) as a yellow powder (0.77–0.80 g, 581–603 μmol, 59–61%) (Note 27 and 28).

2. Notes

1. A two-tap Schlenk adapter connected to a bubbler and an argon/vacuum manifold is illustrated in Yu, J.; Truc, V.; Riebel, P.; Hierl, E.; Mudryk, B. *Org. Synth.* **2008**, *85*, 64–71.

2. *Candida antarctica* lipase B (CALB) was purchased from Sigma-Aldrich Chemicals Co., Inc. and was used as immobilized and thermostable Novozym 435.

3. Na_2CO_3 (anhydrous, extra pure) was purchased from Acros Organics and dried under reduced pressure (0.06 mmHg) at 120 °C for 1 h.

4. Toluene absolute, over molecular sieves ($H_2O \leq 0.005\%$), $\geq 99.7\%$ (GC) was obtained from Sigma-Aldrich Chemicals Co., Inc. and used as obtained.

5. (±)-1-Phenylethylamine (99%) was obtained from Sigma-Aldrich Chemicals Co., Inc. and distilled before use.

6. 2,4-Dimethyl-3-pentanol (99%) was obtained from Sigma-Aldrich Chemicals Co., Inc. and was stored over 3 Å molecular sieves. The alcohol is added as a hydrogen donor in the racemization process.[4]

7. Methyl methoxyacetate (99%) was obtained from Sigma-Aldrich Chemicals Co., Inc. and was stored over 3 Å molecular sieves.

8. The submitters used a greased glass stopper.

9. Upon heating the reaction mixture slowly changed from light brown to light yellow.

10. The reaction was stirred with a Heidolph MR 3001 K stirrer at 800 rpm.

11. The reaction vessel is washed with toluene (3 × 20 mL) to remove all of the residual solids.

12. In the submitter's protocol, prior to filtration, the grease is removed from the ground glass joint by wiping the joint with a tissue dampened with pentane.

13. The brown crude residue is then purified by Kugelrohr distillation (0.05 mmHg). Two bulbs are used for collection: one in the oven and one cooled with dry ice. The oven is heated to 40 °C and after 0.5 h, the bulb cooled with dry ice is exchanged for a clean bulb. The oven is then heated to 135 °C for 1 h. A white solid is collected in the dry ice cooled bulb. The solid is dissolved with DCM (10 mL) and concentrated under reduced pressure (40 °C, 530 mmHg) to afford (R)-2-methoxy-N-(1-phenylethyl)acetamide (2).

14. ^{1}H NMR (400 MHz, CDCl$_3$) δ: 1.54 (d, J = 6.9 Hz, 3 H), 3.42 (s, 3 H), 3.89 (d, J = 15.0 Hz, 1 H), 3.95 (d, J = 15.0 Hz, 1 H), 5.17–5.24 (m, 1 H), 6.77 (br s, 1 H), 7.26–7.31 (m, 1 H), 7.34–7.39 (m, 4 H); ^{13}C NMR (101 MHz, CDCl$_3$) δ: 22.1, 48.2, 59.3, 72.2, 126.3, 127.6, 143.2, 168.7; IR (ATR, cm^{-1}) 3324, 2975, 1643, 1521, 1448, 1197, 1110, 689; HRMS (ESI-MS) calc. (m/z) for C$_{11}$H$_{16}$NO$_2$ [M+H$^+$] 194.1176 found 194.1174; Chiral GC-analysis (CP-Chirasil-DEX CB column (25 m Ø x 0.25 mm)): Injector 250 °C; 60 kPa H$_2$, Program: 100 °C/ 5 min/ 3 °C × min^{-1}/ 155 °C/ 5 min/ 180 °C/ 10 min. t$_S$ = 25.5 min, t$_R$ = 26.3 min.

15. (R)-2-Methoxy-N-(1-phenylethyl)acetamide (2) was added to a round bottomed flask equipped with a stirring bar and pentane (75 mL/ 1.00 g compound) was added. A reflux condenser was connected and the suspension was stirred under reflux until a solution was obtained. Stirring was stopped and the oil bath was removed. The solution was allowed to cool to room temperature and allowed to stand for 14 h. Recrystallization is completed by storage in the freezer (- 25 °C) for 6 h. The crystals were collected by filtration and washed with cold pentane (3 × 10 mL) to afford (R)-2-methoxy-N-(1-phenylethyl)acetamide (2) as colorless crystals (90% recovery, >99% ee). mp 57–58 °C; [α]$^{23}_D$ = +102.4 (c = 1.00, CHCl$_3$).

16. 2-(4-Methoxyphenyl)acetic acid (99%) was obtained from Sigma-Aldrich Chemicals Co., Inc. and used as received.

17. Na$_2$CO$_3$ (anhydrous, extra pure) was obtained from Acros Organics and used as received.

18. MnCl$_2$ (tetrahydrate, extra pure) was obtained from Sigma-Aldrich Chemicals Co., Inc. and used as received.

19. mp 83–85 °C; ^{1}H NMR (400 MHz, CDCl$_3$) δ: 3.66 (s, 4 H), 3.80 (s, 6 H), 6.87 (d, J = 8.5 Hz, 4 H), 7.08 (d, J = 8.4 Hz, 4 H); ^{13}C NMR (101 MHz, CDCl$_3$) $δ$: 48.2, 114.4, 126.3, 130.6, 158.8, 206.6; IR (ART): ṽ/cm^{-1} = 3013, 2974, 2934, 2901, 2841, 2043, 1904, 1699, 1663, 1607, 1580, 1508, 1470, 1456, 1445, 1433, 1319, 1305, 1279, 1242, 1180, 1153, 1107, 1063, 1024, 966, 945, 932, 910, 878,789, 735, 716, 698, 669, 635; HRMS (ESI-MS) calc. (m/z) for C$_{17}$H$_{19}$O$_3$ [M+H$^+$] 271.1329; found 271.1324.

20. 4,4-Dimethoxybenzil (99%) was obtained from Acros Organics and used as received.

21. Tetramethylammonium hydroxide pentahydrate (97%) was obtained from Sigma-Aldrich Chemicals Co., Inc. and used as received.

22. The solid was rinsed in 10 mL aliquots of EtOH until the wash was colorless.

23. mp 257–260 °C; ^{1}H NMR (400 MHz, CDCl$_3$) δ: 3.79 (s, 12 H), 6.72 (d, J = 8.8 Hz, 4 H), 6.79 (d, J = 8.9 Hz, 4 H), 6.86 (d, J = 8.8 Hz, 4 H), 7.19 (d, J = 8.9 Hz, 4 H); ^{13}C NMR (101 MHz, CDCl$_3$) δ: 55.4, 113.6, 113.8, 123.9, 124.1, 125.9, 131.3, 131.6, 153.0, 159.0, 159.8, 201.5; IR (ART): ṽ/cm^{-1} = 3005, 2991, 2961, 2840, 1705, 1599, 1570, 1564, 1502, 1466, 1456, 1443, 1418, 1319, 1285, 1240, 1175, 1153, 1109, 1024, 1009, 960, 839, 810, 773, 694, 636; HRMS (ESI-MS) calc. (m/z) for C$_{33}$H$_{28}$O$_5$ [M$^+$] 504.1931; found 504.1917.

24. Ru$_3$(CO)$_{12}$ (99%) was obtained from Strem Chemicals Inc. and used as received.

25. Methanol (99.8%) was obtained from Sigma-Aldrich Chemicals Co., Inc.and used as received.

26. The oil bath is heated to 90 °C.

27. ^{1}H NMR (400 MHz, CDCl$_3$) δ: ⁻18.5 (s, 1 H), 3.69 (s, 12 H), 3.78 (s, 12 H), 6.54 (d, J = 8.8 Hz, 8 H), 6.57 (d, J = 8.8 Hz, 8 H), 6.93 (d, J = 8.8 Hz, 8 H), 7.05 (d, J = 8.8 Hz, 8 H); ^{13}C NMR (101 MHz, CDCl$_3$) δ: 55.2, 55.3, 87.7, 102.3, 113.2, 113.2, 122.8, 123.1, 132.5, 133.5, 153.9, 158.4, 201.5, ; IR (ATR, cm^{-1}) = 2958, 2933, 2904, 2832, 2028, 2004,1985, 1961, 1607, 1514, 1291, 1244, 1174, 1028, 824, 808; HRMS (ESI-MS) calc. (m/z) for C$_{70}$H$_{59}$O$_{14}$Ru$_2$ [M+H$^+$] 1327.1986; found 1327.1999.

28. The solid was stored in a desiccator under air.

Safety and Waste Disposal Information

All hazardous materials should be handled and disposed of in accordance with "Prudent Practices in the Laboratory"; National Academies Press; Washington, DC, 2011.

3. Discussion

Dynamic kinetic resolution (DKR) of primary amines has sparked a growing amount of interest over the past decade. DKR is an attractive method for the synthesis of chiral amines and combines the well-established and commonly used method of kinetic resolution (KR) with *in situ* racemization of the non-converted substrate. Thus the main drawback of KR (a theoretical maximum of 50% yield) is eliminated.

The first example of DKR of benzylic amines using a transition metal racemization catalyst and a lipase was demonstrated by the Reetz group in 1996.[3] The drawback of this method was the long reaction time of 8 days and the moderate yield.

In 2002, our group reported $\{\{[2,3,4,5(C_6H_4)_4](\eta^5-C_4CO)\}_2H\}$-$Ru_2(CO)_4(\mu\text{-H})$ (8) as a highly efficient amine racemization catalyst.[4] This catalyst was reported by Shvo and co-workers[5,6] in 1985 and has since been used in the hydrogenation of unsaturated substrates,[5] transfer hydrogenation of ketones and imines,[7] Oppenauer-type oxidations of alcohols[8,9] and amines,[10-12] and the racemization of alcohols[13,14] and amines.[4] A modified procedure for the synthesis of a tolyl-substituted analogue of 8 ($\{\{[2,5\text{-}(C_6H_4)_2\text{-}3,4\text{-}(4\text{-}CH_3\text{-}C_6H_4)_2](\eta^5-C_4CO)\}_2H\}Ru_2(CO)_4(\mu\text{-H})$) was reported by Casey et. al.[15] In 2005, our group reported 7 (anisyl-substituted analogue of 8) as a highly selective amine racemization catalyst.[16] Catalyst 7 was combined with *Candida antarctica* lipase B (CALB) and applied in the DKR of benzylic and aliphatic amines to obtain the corresponding amides in high yield and excellent ee.[16-18]

Independent work by the Jacobs-De Vos group identified Pd on alkaline earth supports as a viable racemization catalyst for use in chemoenzymatic DKR.[19,20] They observed that byproduct formation could be minimized by using Pd on alkaline supports instead of Pd/C. The group of Kim and Park later reported the DKR of primary amines combining a Pd nanocatalyst and CALB.[21] However, in both cases the substrate scope is limited to amines that are not susceptible to reduction under hydrogenation conditions.

Additional methods that combine racemization with enzymatic KR in the DKR of amines have been developed.[22-26]

The DKR protocol originally reported by our group utilizes 4 mol% **7**, 40 mg CALB per mmol amine substrate, and isopropyl acetate as the acyl donor.[16] A large excess of isopropyl acetate was used and upon concentrating the reaction mixture during the development of a large scale procedure, it was found that isopropyl acetate was not a suitable acyl donor under the desired conditions.[27] As isopropyl acetate became the primary solvent, a drop in enantiomeric excess was observed. Upon reducing the excess of isopropyl acetate to 1–3 equiv the reaction became sluggish. Therefore, an alternative acyl donor was desirable. Methyl methoxyacteate was chosen as the acyl donor and was added portion wise to avoid uncatalyzed chemical acylation of the substrate that occurs at elevated temperatures when using acyl donors of this type in the DKR of primary amines.[26,27]

For application on gram scale, the catalyst loadings were reduced to 1.25 mol% **7** and 10 mg CALB per mmol amine substrate. As seen in Table 1, a two-portion addition of CALB concurrent with a two-portion addition of methyl methoxyacteate provided amide **2** in the highest yield (83%, Table 1, entry 3).

Table 1. Dynamic kinetic resolution.

Entry	**1** (mmol)	Acyl donor (equiv)			Toluene (mL)	ee (%)	Yield (%)[b]
		0 h	24 h	48 h			
1	10	0.75	0.35	-	50	98	74
2[c]	45	0.4	0.35	0.35	225	98	68
3[d]	45	0.75	0.35	-	225	98	83

[a] Conditions: **1**, 1.25 mol% of **4**, 100 mg of CALB, 200 mg of Na$_2$CO$_3$, acyl donor, 1.25 equiv 2,4-dimethyl-3-pentanol, 100 °C, 72 h; [b] isolated yield; [c] 225 mg of CALB, addition of 113 mg of CALB after 24 and 48 h; [d] 340 mg of CALB, addition of 110 mg of CALB after 24 h.

262

1. Stockholm University, Department of Organic Chemistry, Arrhenius Laboratory, SE-106 91, Stockholm, Sweden, jeb@organ.su.se. This work was funded by the European Union (EU-project "INTENANT, Integrated Synthesis and Purification of Enantiomers"; NMP2-SL-2008-214129).

2. Diederich, F.; Carcanague, D. R. *Helv. Chim. Acta* **1994**, *77*, 800–818.

3. Reetz, M. T.; Schimmossek, K. *Chimia* **1996**, *50*, 668–669.

4. Pàmies, O.; Éll, A. H.; Samec, J. S. M.; Hermanns, N.; Bäckvall, J.-E. *Tetrahedron Lett.* **2002**, *43*, 4699–4702.

5. Blum, Y.; Czarkie, D.; Rahamim, Y.; Shvo, Y. *Organometallics* **1985**, *4*, 1459–1461.

6. Shvo, Y.; Czarkie, D.; Rahamim, Y.; Chodosh, D. F. *J. Am. Chem. Soc.* **1986**, *108*, 7400–7402.

7. Samec, J. S. M.; Bäckvall, J.-E. *Chem. Eur. J.* **2002**, *8*, 2955–2961.

8. Almeida, M. L. S.; Beller, M.; Wang, G.-Z.; Bäckvall, J.-E. *Chem. Eur. J.* **1996**, *2*, 1533–1536.

9. Csjernyik, G.; Éll, A. H.; Fadini, L.; Pugin, B.; Bäckvall, J.-E. *J. Org. Chem.* **2002**, *67*, 1657–1662.

10. Éll, A. H.; Samec, J. S. M.; Brasse, C.; Bäckvall, J.-E. *Chem. Commun.* **2002**, 1144–1145.

11. Éll, A. H.; Johnson, J. B.; Bäckvall, J.-E. *Chem. Commun.* **2003**, 1652–1653.

12. Samec J. S. M.; Éll A. H.; Bäckvall, J.-E. *Chem. Eur. J.* **2005**, *11*, 2327–2334.

13. Larsson, A. L. E.; Persson, B. A.; Bäckvall, J.-E. *Angew. Chem. Int. Ed. Engl.* **1997**, *36*, 1211–1212.

14. Persson, B. A.; Larsson, A. L. E.; Ray, M. L.; Bäckvall, J.-E. *J. Am. Chem. Soc.* **1999**, *121*, 1645–1650.

15. Casey, C. P.; Singer, S. W.; Powell, D. R.; Hayashi, R. K.; Kavana, M. *J. Am. Chem. Soc.* **2001**, *123*, 1090–1100.

16. Paetzold, J.; Bäckvall, J.-E. *J. Am. Chem. Soc.* **2005**, *127*, 17620–17621.

17. Hoben, C. E.; Kanupp, L.; Bäckvall, J.-E. *Tetrahedron Lett.* **2008**, *49*, 977–979.

18. Thalén, L. K.; Zhao, D.; Sortais, J.-B.; Paetzold, J.; Hoben, C.; Bäckvall, J.-E. *Chem. Eur. J.* **2009**, *15*, 3403–3410.

19. Parvulescu, A.; De Vos, D.; Jacobs, P. *Chem. Commun.* **2005**, 5307–5309.

20. Parvulescu, A. N.; Jacobs, P. A.; De Vos, D. E. *Chem. Eur. J.* **2007**, *13*, 2034–2043.

21. Kim, M. J.; Kim, W. H.; Han, K.; Choi, Y. K.; Park, J. *Org. Lett.* **2007**, *9*, 1157–1159.

22. Blacker, A. J.; Stirling, M. J.; Page, M. I. *Org. Process Res. Dev.* **2007**, *11*, 642–648.

23. Dunsmore, C. J.; Carr, R.; Fleming, T.; Turner, N. J. *J. Am. Chem. Soc.* **2006**, *128*, 2224–2225.

24. Gastaldi, S.; Escoubet, S.; Vanthuyne, N.; Gil, G.; Bertrand, M. P. *Org. Lett.* **2007**, *9*, 837–839.

25. Stirling, M.; Blacker, J.; Page, M. I. *Tetrahedron Lett.* **2007**, *48*, 1247–1250.

26. Veld, M. A. J.; Hult, K.; Palmans, A. R. A.; Meijer, E. W. *Eur. J. Org. Chem.* **2007**, *2007*, 5416–5421.

27. Thalén, L. K.; Bäckvall, J.-E. *Beilstein J. Org. Chem.* **2010**, *6*, 823–829.

Appendix
Chemical Abstracts Nomenclature; (Registry Number)

Novozym 435, Lipase acrylic resin from *Candida antárctica*; (9001-62-1)

Sodium carbonate; (497-19-8)

(±)-α-Methylbenzylamine; (618-36-0)

2,4-Dimethyl-3-pentanol; (600-36-2)

Methyl methoxyacetate; (6290-49-9)

4-Methoxyphenylacetic acid; (104-01-8)

Manganese(II) chloride tetrahydrate; (13446-34-9)

4,4'-Dimethoxybenzil; (1226-42-2)

Tetramethylammonium hydroxide pentahydrate; (10424-65-4)

(*R*)-2-Methoxy-*N*-(1-phenylethyl)acetamide; (162929-44-4)

Ruthenium carbonyl; (15243-33-1)

1,3-Bis(4-methoxyphenyl)propan-2-one; (29903-09-1)

2,3,4,5-Tetrakis(4-methoxyphenyl)cyclopenta-2,4-dienone; (49764-93-4)

Ruthenium complex {{[2,3,4,5(4-OMe-C_6H_4)$_4$](η^5-C_4CO)}$_2$H}-
$Ru_2(CO)_4(\mu$-H) ; (873815-22-6)

Jan-Erling Bäckvall was born in Malung, Sweden, in 1947. He received his Ph.D. from the Royal Institute of Technology, Stockholm, in 1975 under the guidance of Prof. B. Åkermark. After postdoctoral work (1975-76) with Prof. K. B. Sharpless at Massachusetts Institute of Technology he joined the faculty at the Royal Institute of Technology. He was appointed Professor of Organic Chemistry at Uppsala University in 1986. In 1997 he moved to Stockholm University where he is currently Professor of Organic Chemistry. He is a member of the Royal Swedish Academy of Sciences and the Finnish Academy of Science and Letters. He is also a member of the Nobel Committee for Chemistry. His current research interests include transition metal-catalyzed organic transformations, biomimetic oxidations, and enzyme chemistry.

Lisa Kanupp Thalén was born in Asheville, NC, in 1980. She received a B.Sc. from East Carolina University in 2002 and completed a research project in the group of Assoc. Prof. William E. Allen. She then received a M.Sc. from Uppsala University in 2005 while working in the group of Prof. Per I. Arvidsson. After completing her master's degree studies, she joined the group of Prof. Jan-Erling Bäckvall at Stockholm University and completed her doctoral studies in 2010. Her research focused on dynamic kinetic resolution and its application in the synthesis of biologically active small molecules.

Christine Rösch (née Höben) was born in 1978 in Mainz, Germany. She joined the group of Prof. H. Kunz at Johannes Gutenberg-Universität Mainz in 2002 and obtained her Ph.D. in 2006 while working on carbohydrate-based catalysts for the asymmetric Strecker-reaction. Afterwards she was a postdoctoral scholar in the group of Prof. J.-E. Bäckvall at the University of Stockholm, Sweden and worked on the dynamic kinetic resolution of amines. Since 2007 she has worked for BASF SE in Ludwigshafen, Germany.

Marc-André Müller was born in Bad Säckingen (Germany) in 1985 and did his chemistry studies at the University of Basel where he obtained his M. Sc. degree in 2010 under the supervision of Prof. Andreas Pfaltz. He began his Ph.D. work in May 2010 in the same group, where he is currently working in the field of Ir-catalyzed enantioselective hydrogenation of unfunctionalized olefins.

266

Discussion Addendum for:

[3 + 4] Annulation Using a [β-(Trimethylsilyl) acryloyl]silane and the Lithium Enolate of an α,β-Unsaturated Methyl Ketone: (1R,6S,7S)-4-(tert-Butyldimethylsiloxy)-6-(trimethylsilyl)bicyclo [5.4.0]undec-4-en-2-one

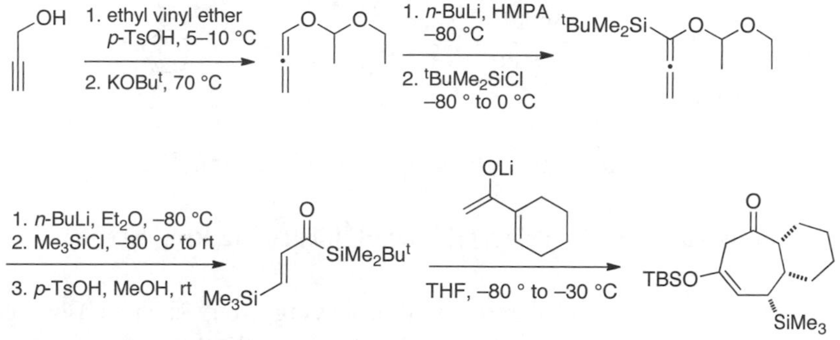

Prepared by Kei Takeda* and Michiko Sasaki.[1]

Original article: Takeda, K.; Nakajima, A.; Takeda, M.; Yoshii, E.; *Org. Synth.* **1999**, *76*, 199–213.

Brook rearrangement-mediated [3 + 4] annulation has evolved as a unique methodology for the construction of not only seven-membered carbocycles but also eight-membered carbo- and oxygen-heterocycles and has been applied to the synthesis of natural products after clarification[2] of the precise reaction mechanism accounting for the stereospecificity.

Synthesis of the Tricyclic Skeleton of Allocyathin B$_2$

The synthetic utility of annulation was first demonstrated by the synthesis of the unusual 5-6-7 tricyclic ring skeleton of allocyathin B$_2$, a compound that has been shown to have potent nerve growth factor synthesis-stimulating activity and to be a κ opioid receptor agonist (Scheme 1). The key [3 + 4] annulation proceeded smoothly even with a relatively complex four-carbon unit **1** to afford **2** as a single diastereomer in 50% yield, which was transformed to **3**.[3]

Published on the Web 1/6/2012
© Organic Syntheses, Inc.

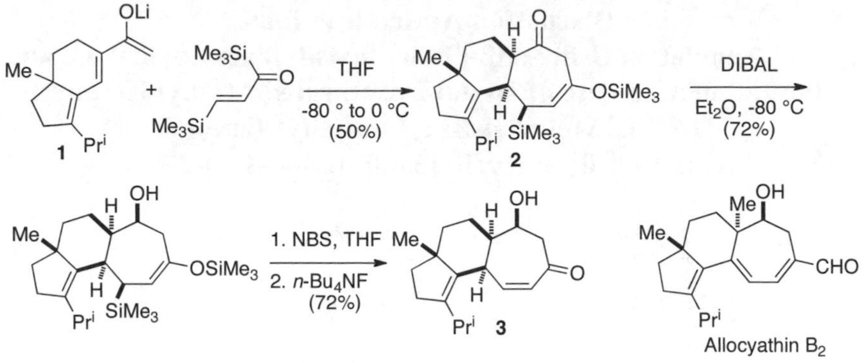

Scheme 1 Synthesis of the tricyclic skeleton of allocyathin B_2.

Construction of a Tricyclo[5.3.0.0^{1,4}]decenone Ring System

The use of acryloylsilanes **4** with a leaving group such as a halogen atom at the β-position as a three-carbon unit in the [3 + 4] annulation afforded tricyclic ketone derivatives **7a,b** in yields dependent upon the β-substituent of **4**, in addition to the [3 + 4] annulation–debromosilylation products **9a,b** (Scheme 2).[4] Small structural changes in the four-carbon unit significantly affect the product distribution. Thus, whereas cyclopentyl methyl ketone enolate gave **7a** in almost all cases, **9b** was formed as a byproduct in the case of the corresponding cyclohexyl derivative. Mechanistic studies including low–temperature quenching experiments suggested that **7a,b** can be formed via an S_N'-like intramolecular attack of the enolate at the C–4 position in the intermediate **6a,b**, and **9a,b** can be formed via tricyclic intermediate **8a,b**.

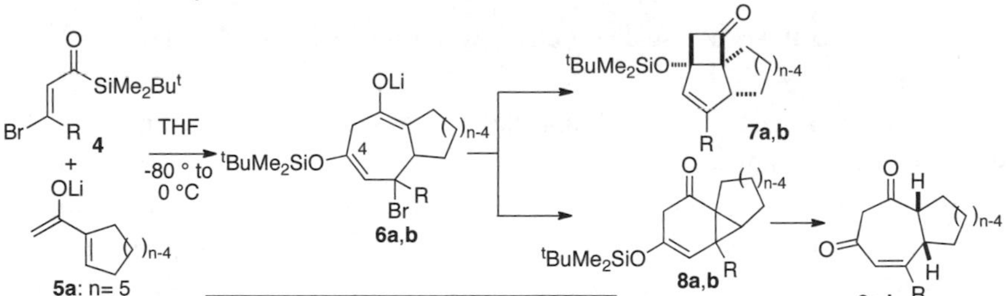

R	yield (%), n = 5		yield (%), n = 6	
	7a	**9a**	**7b**	**9b**
CH$_3$	68	0	27	45
n-Bu	49	0	27	14
n-hexyl	41	0	35	17
t-Bu	51	7	58	14
c-C$_3$H$_5$	56	0	39	32

Scheme 2 Construction of a tricyclo[5.3.0.0^{1,4}]decenone ring system.

268

BF₃·Et₂O-Mediated Intramolecular Allylstannane-Ketone Cyclizations

[3 + 4] annulation using a combination of (Z)-(β-(tributylstannnyl)acryloyl)silanes **10** and alkenyl methyl ketone enolate **11** proceeded in the same manner to give cycloheptenone derivative **12**, which upon treatment with BF₃·Et₂O, afforded bicyclo[4.1.0]heptenols **13**, an intramolecular addition product of the allylstannane system to the carbonyl group (Scheme 3).[5]

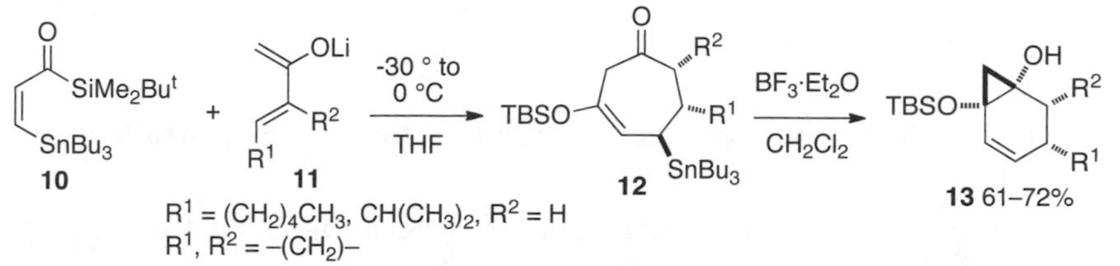

$R^1 = (CH_2)_4CH_3, CH(CH_3)_2, R^2 = H$
$R^1, R^2 = -(CH_2)-$

Scheme 3 BF₃·Et₂O-Mediated intramolecular allylstannane-ketone cyclizations.

Stereoselective Construction of Eight-Membered Carbocycles and Oxygen-Heterocycles

The use of the enolate **15** (X = CH₂) derived from 2-cycloheptenone as the four-carbon unit in [3 + 4] annulation instead of the enolates of alkenyl methyl ketones produced bicyclo[3.3.2]decenone derivatives **16**. The two-atom internal tether in these products could be oxidatively cleaved after conversion to α-hydroxy ketone **17** to give the cis-3,4,8-trisubstituted cyclooctenone enol silyl ethers **18** stereoselectively (Scheme 4).[6] This methodology has also been successfully applied to the construction of oxygen eight-membered heterocycles using enolates of 6-oxacyclohept-2-en-1-one **15** (X = O), affording eight-membered oxygen heterocycles **18** (X = O) possessing functionality that can easily be manipulated to generate other functionalized eight-membered ring products.[7]

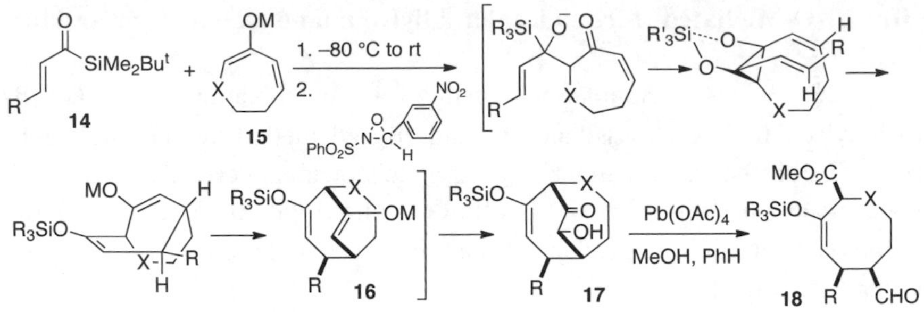

Scheme 4 Stereoselective construction of eight-membered carbocycles and oxygen-heterocycles.

Formal Total Syntheses of (+)-Prelaureatin and (+)-Laurallene

The versatility of the annulation has been highlighted through the formal total synthesis of (+)-prelaureatin, a biogenetic precursor of several members of the laurenan structural subclass (Scheme 5).[8] The annulation of **19** and sodium enolate **20** proceeded in a highly diastereoselective manner to afford exclusively **21** in 80% yield. The observed excellent selectivity could be explained in terms of the approach of the acryloylsilane from the same side as the C-7 substituent in **20** that is sterically less hindered because of pseudo equatorial disposition of the substituent on the seven-membered ring. The bicyclic derivative **21** was transformed into Crimmins' intermediate **22**[9] after oxidative cleavage of the two-carbon tether.

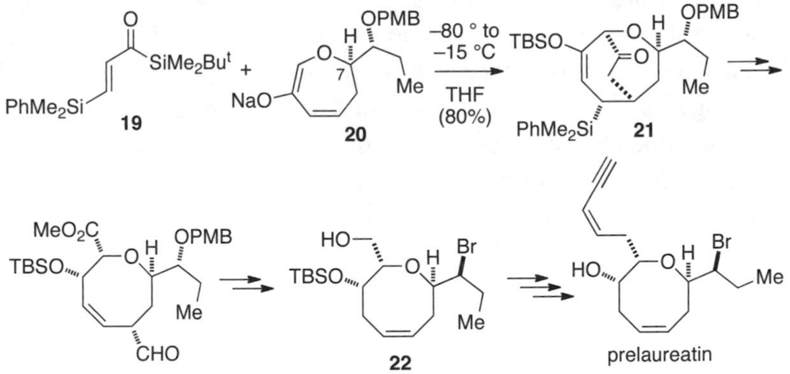

Scheme 5 Formal total syntheses of (+)-prelaureatin.

Stereocontrolled Construction of Seven- and Eight-Membered Carbocycles Using a Combination of Brook Rearrangement-Mediated [3 + 4] Annulation and Epoxysilane Rearrangement

The [3 + 4] annulation has also been expanded to include the construction of densely functionalized seven- and eight-membered carbocycles by combining it with an epoxysilane rearrangement,[10] which features a further extension of a stereocontrolled anion relay.[11] Reactions of δ-silyl-γ,δ-epoxy-α,β-unsaturated acylsilane **23** with alkenyl methyl ketone enolate **24** afforded highly functionalized cycloheptenone derivative **28** via a tandem process that involves Brook rearrangement followed by the resulting carbanion-induced ring-opening of the epoxide (**25** → **26**), a second Brook rearrangement, the formation of divinylcyclopropanediolate derivative **27** via internal carbonyl attack by the resulting carbanion, and an anionic oxy-Cope rearrangement (Scheme 6). The reactions using an alternative combination of three and four carbon units (**29** + **30**), in which an epoxysilane moiety was incorporated in the four-carbon unit, also give satisfactory results, via 1,4-*O*-to-*O* silyl migration (**31** → **32**). Use of enantioenriched acylsilane **33** and 2-cycloheptenone enolate **34** gave a moderate level (62% ee) of asymmetric induction in the bicyclic ketone **35**.

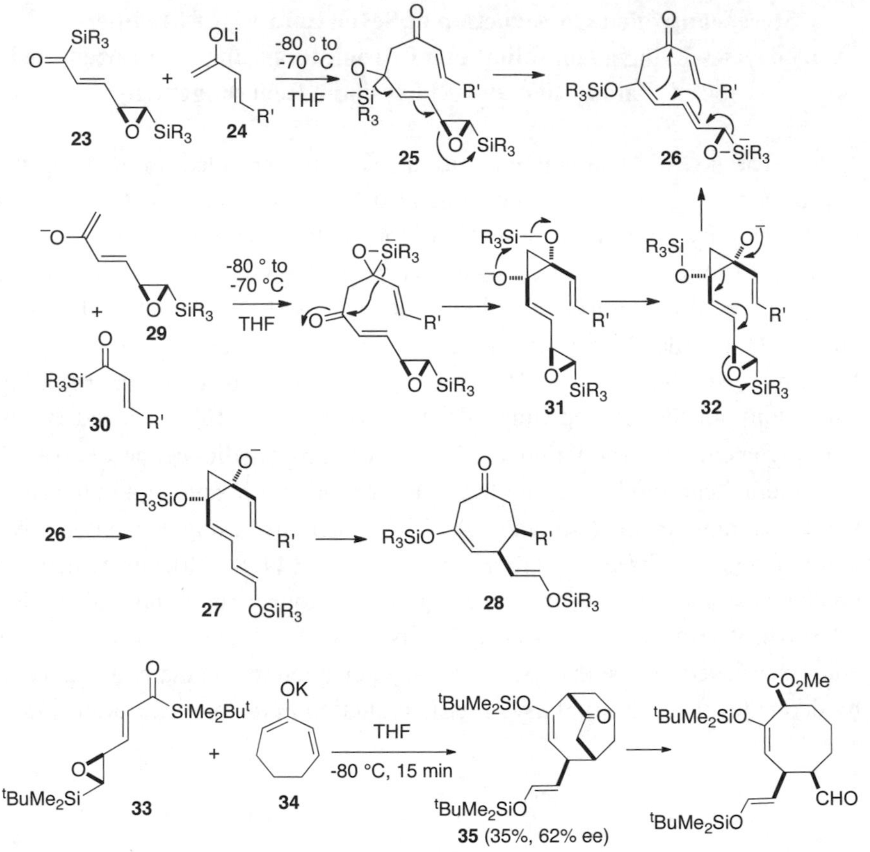

Scheme 6 Stereocontrolled construction of seven- and eight-membered carbocycles using a combination of Brook rearrangement-mediated [3 + 4] annulation and epoxysilane rearrangement.

1. Department of Synthetic Organic Chemistry, Graduate School of Medical Sciences, Hiroshima University, Hiroshima 734–8553, Japan.

2. Takeda, K.; Nakajima, A.; Takeda, M.; Okamoto, Y.; Sato, T.; Yoshii, E.; Koizumi, T.; Shiro, M. *J. Am. Chem. Soc.* **1998**, *120*, 4947–4959.

3. Takeda, K.; Nakane, D.; Takeda, M. *Org. Lett.* **2000**, *2*, 1903–1905.

4. Takeda, K.; Ohtani, Y. *Org. Lett.* **1999**, *1*, 677–679.

5. Takeda, K.; Nakajima, A.; Yoshii, E. *Synlett* **1996**, 753–754.

6. Takeda, K.; Sawada, Y.; Sumi, K. *Org. Lett.* **2002**, *4*, 1031–1033.

7. Sawada, Y.; Sasaki, S.; Takeda, K. *Org. Lett.* **2004**, *6*, 2277–2279.

8. (a) Sasaki, M.; Hashimoto, A.; Tanaka, K.; Kawahata, M.; Yamaguchi, K.; Takeda, K. *Org. Lett.* **2008**, *10*, 1803–1806. (b) Sasaki, M.; Oyamada, K.; Takeda, K. *J. Org. Chem.* **2010**, 75, 3941–3943.
9. Crimmins, M. T.; Elie, A. T. *J. Am. Chem. Soc.* **2000**, *122*, 5473–5476.
10. (a) Takeda, K.; Kawanishi, E.; Sasaki, M.; Takahashi, Y.; Yamaguchi, K. *Org. Lett.* **2002**, *4*, 1511–1514. (b) Sasaki, M.; Kawanishi, E.; Nakai, Y.; Matsumoto, T.; Yamaguchi, K.; Takeda, K. *J. Org. Chem.* **2003**, *68*, 9330–9339. (c) Sasaki, M.; Ikemoto, H.; Takeda, K. *Heterocycles* **2009**, *78*, 2919–2941.
11. Nakai, Y.; Kawahata, M.; Yamaguchi, K.; Takeda, K. *J. Org. Chem.* **2007**, *72*, 1379–1387.

Kei Takeda is professor of organic chemistry at Hiroshima University. He was born in 1952 and received both his B.S. degree (1975) and M.S. degree (1977) from Toyama University (with Eiichi Yoshii) and his Ph.D. degree (1980) from the University of Tokyo (with Toshihiko Okamoto). In 1980, he joined the faculty at Toyama Medical and Pharmaceutical University (Prof. Yoshii's group). After working at MIT with Professor Rick L. Danheiser (1988-1989), he was promoted to Lecturer (1989) and then to Associate Professor (1996). He became a professor of Hiroshima University in 2000. His research interests are in the invention of new synthetic reactions and chiral carbanion chemistry. He was the recipient of the Sato Memorial Award (1998) and the 41st Senji Miyata Foundation Award.

Michiko Sasaki is an assistant professor of organic chemistry at Hiroshima University. She obtained her B.S. degree (2002), M.S. degree (2003), and Ph.D. degree (2006) from Hiroshima University under the direction of Professor Kei Takeda and then remained in Takeda's group as an assistant professor. She spent one year (2010-2011) at MIT as a JSPS fellow (Excellent Young Researcher Overseas Visit Program) with Professor Rick L. Danheiser. Her current research interests lie in the area of development of new synthetic reactions. She was the recipient of the Chugoku-Shikoku Branch of Pharmaceutical Society of Japan Award for Young Scientists (2007).

o-Aminobenzaldehyde, Redox-Neutral Aminal Formation and Synthesis of Deoxyvasicinone

Submitted by Chen Zhang, Chandra Kanta De, and Daniel Seidel.[1]
Checked by Liang Huang, Brenda Burke, and Margaret Faul.

1. Procedure

A. *o-Aminobenzaldehyde* (**2**).   A 1-L, one-necked, round-bottomed flask, containing an egg-shaped magnetic stir bar (Note 1), is charged with o-nitrobenzaldehyde (**1**) (9.07 g, 60 mmol) (Note 2) and 170 mL of absolute ethanol (Note 3).   The resulting mixture is stirred for one min, at which point a yellow solution is formed.   Iron powder (10.05 g, 180 mmol, 3.0 equiv) (Note 4) is added to the stirring solution, followed by addition of diluted HCl (60 mL) (Note 5).   The flask is then equipped with a reflux condenser containing a nitrogen inlet, and heated under reflux for 60 min. Under continued stirring, the reaction mixture is allowed to cool to room temperature over a period of 35 min.   The reaction mixture is subsequently diluted with EtOAc (500 mL), followed by addition of anhydrous MgSO$_4$ (70 g) (Note 6).   The mixture is allowed to stir at room temperature for 5 min and then filtered through a Büchner funnel packed with celite (Note 7). The filter cake is washed with EtOAc (4 x 100 mL) and the combined

274

filtrates are concentrated to a volume of 15 mL (Notes 8 and 9). The resulting clear yellow solution is passed through a silica gel plug as quickly as possible, eluting with 20% EtOAc in hexanes (800 mL) under reduced pressure (Notes 10 and 11). The solvent is removed in vacuo (Note 8) and the residue dried under high vacuum (< 0.1 mmHg) for 2.5 h to obtain 6.95 g (74 wt% quantitative ^{1}H NMR, 57 mmol, 70%) of o-aminobenzaldehyde (**2**), which is used directly in the next step (Notes 12 and 13).

B. *1,2,3,3a,4,9-Hexahydropyrrolo[2,1-b]quinazoline* (**4**). An oven-dried, 500-mL one-necked, round-bottomed flask, containing a magnetic stir bar (Note 1), is charged with n-butanol (200 mL) (Note 14) and pyrrolidine (8.81 g, 10.17 mL, 123.8 mmol, 3.0 equiv) (Note 15). The flask is then equipped with a reflux condenser capped with a septum and a nitrogen inlet, and the reaction mixture is heated under reflux. Using a cannula that is inserted through the septum located on top of the reflux condenser, a solution of o-aminobenzaldehyde (**2**) (5.0 g, 41.3 mmol, 1.0 equiv) in n-butanol (24 mL) is added slowly via syringe pump over 24 h. Subsequent to the addition, the reaction mixture is heated under reflux for an additional 14 h. The reaction mixture is then allowed to cool to room temperature and is concentrated under reduced pressure (Note 16). The crude product is purified by flash column chromatography, eluting with 500 mL of 50% hexanes in EtOAc, followed by 4 L of 5% MeOH in EtOAc (Note 17). The solvent is removed in vacuo (Note 8) and the residue dried under high vacuum (< 0.1 mmHg) for 24 h to obtain 5.20–5.66 g (100 wt% by quantitative ^{1}H NMR, 32.1 mmol, 72–77%) of **4** as a light yellow microcrystalline solid (Notes 18 and 19).

C. *Deoxyvasicinone* (**5**). A 1-L one-necked, round-bottomed flask, containing a magnetic stir bar (Note 1), is charged with **4** (5.0 g, 28.7 mmol), and acetone (285 mL) (Note 20). The resulting mixture is stirred for one min at which point a light yellow solution is formed. KMnO$_4$ (22.67 g, 143 mmol) (Note 21) is added into the stirring solution. The flask is then equipped with a reflux condenser containing a nitrogen inlet, and the reaction mixture is heated under reflux for 3 h (Note 22). Under continued stirring, the reaction mixture is allowed to cool to room temperature over a period of 30 min and then filtered through a Büchner funnel packed with celite (Note 7). The filter cake is washed with acetone (2 x 50 mL) and the combined filtrates are concentrated under reduced pressure (Note 23). The combined filtrate is concentrated in vacuo (Note 8) and the residue dried under high vaccum (<0.1 mmHg) for 24 h to obtain 4.9–5.0 g (96 wt%, quantitative ^{1}H NMR, 87–89%) of (5) as a light yellow solid (Notes 24 and 25).

2. Notes

1.   The egg-shaped magnetic stir bar was of 3.8 cm length.   A stirring speed of 1000 rpm was maintained throughout the reaction.

2.   o-Nitrobenzaldehyde (98+%, Cat. No. 552-89-6) was obtained from Alfa Aesar and used as received.

3.   Ethanol (200 Proof - Absolute, Anhydrous ACS/USP grade, Cat No. 111ACS200) was obtained from Decon Laboratories, Inc and used as received.

4.   Baker analyzed reagent grade iron powder (99.3%, Cat. No. 7439-89-6) was obtained from Mallinckrodt Baker and used as received.

5.   Diluted HCl solution was prepared by dissolving 1 mL of concentrated HCl in 66 mL of distilled water.

6.   Magnesium sulfate certified anhydrous (Cat. No. 7487-88-9) was obtained from Fisher Chemicals and used as received.

7.   Celite (50 g) was used in a Büchner funnel with a 16.5 cm diameter.

8.   A standard rotary evaporator was used (25 °C bath temperature, 50 mmHg).

9.   Allowing the solution to go to dryness or remain on the rotary evaporator for prolonged periods of time will result in a significantly lower yield.

10.   Silica gel (standard grade, 230 x 400 mesh) was obtained from Sorbent Technologies and used as received.

11.   A coarse fritted glass funnel (diameter: 7 cm, height of silica gel: 4 cm) with a wet-packed silica gel plug was used in combination with a vacuum filter flask.

12.   The initially formed yellow oil turns into a yellow solid if kept in a –20 °C freezer.   However, 2 should not be stored for prolonged periods of time as it degrades gradually even at this temperature.   The reaction was also performed with 36.3 g of starting material, which resulted in a yield of 23.5 g (68%) of product with similar purity.

13.   Analytical data of 2: R_f = 0.48 in 20% EtOAc in hexanes; mp = 32–34 °C; ^{1}H NMR (500 MHz, CDCl$_3$) δ: 6.11 (br s, 2 H), 6.64–6.66

276

(m, 1 H), 6.76–6.77 (ddd, 1 H), 7.29–7.31 (ddd, 1 H), 7.47–7.49 (app dd, 1 H), 9.87 (s, 1 H); ^{13}C NMR (125 MHz, CDCl$_3$) δ: 116.0, 116.4, 118.9, 135.2, 135.7, 149.9, 194.0; IR (ATR) 3463, 3329, 3051, 2839, 2759, 1666, 1612, 1582, 1550, 1480, 1396, 1348, 1322, 1288, 1191, 1145, 1041, 881, 750 cm^{-1}; HRMS m/z calcd for C$_7$H$_8$N0 [M+H]$^+$: 122.05276, found 122.05957.

14. n-Butanol (99.9%, Cat. No. 71-36-3) was obtained from Sigma-Aldrich and used as received.

15. Pyrrolidine (99%, Cat. No. 123-75-1) was obtained from Alfa Aesar and was freshly distilled before use.

16. A standard rotary evaporator was used (60 °C bath temperature, 33 mmHg).

17. A wet-packed silica gel column was used (diameter: 8.0 cm, Height: 17 cm). The crude material was loaded as oil on top of the column.

18. The use of a syringe pump is not essential but allows for reproducibly higher yields. For instance, the submitters report that direct mixing of o-aminobenzaldehyde (6.06 g, 55 mmol) and three equivalents of pyrrolidine in n-butanol (220 mL) for 21 h under reflux gives rise to aminal 4 in 69% isolated yield.

19. Analytical data of 4: R$_f$ = 0.39 in 10% MeOH in EtOAc; mp = 63–64 °C; ^{1}H NMR (500 MHz, CDCl$_3$) δ: 1.61–1.68 (tdd, 1 H), 1.87–1.96 (m, 1 II), 1.97–2.07 (m, 1 H), 2.09–2.18 (m, 1 H), 2.66–2.70 (dt, 1H), 3.01–3.05 (dt, 1 H,), 3.67 (br s, 1 H), 3.90 (d, 1 H,), 4.04 (d, 1H), 4.13–4.17 (t, 1 H), 6.54 (d, 1 H), 6.70 (t, 1 H), 6.95 (d, 1 H), 7.01–7.02 (t, 1 H); ^{13}C NMR (125 MHz, CDCl$_3$) δ: 21.1, 31.8, 50.3, 50.5, 71.2, 114.9, 118.1, 119.4, 127.0, 127.2, 142.9; IR (ATR) 3040, 2967, 2826, 1608, 1585, 1476, 1380, 1252, 744 cm^{-1}; HRMS m/z calcd for C$_{11}$H$_{14}$N$_2$ [M+H]$^+$: 175.11570, found 175.12248.

20. Acetone CHROMASOLV for HPLC (> 99.9%, Cat. No. 67-64-1) was obtained from Sigma-Aldrich and used as received.

21. Potassium Permanganate, Baker Analyzed A.C.S. Reagent, J.T. Baker (99.0% min, Cat. No. 7722-64-7) was obtained from Mallinckrodt Baker and used as received.

22. The reaction is analyzed for complete consumption of starting material by ^{1}H NMR. The reaction will cleanly convert the starting material to product although the time required for the reaction may vary depending on the quality of KMnO$_4$. It is recommended that the reaction is driven to completion or final purification of the product is extremely difficult.

23. A standard rotary evaporator was used (25 °C bath temperature, 100 mmHg).

24. While it is recommended that efforts be made to ensure this reaction goes to completion, separation of the starting material and purification the product, although difficult, can be performed by flash column chromatography using a wet-packed silica gel column (diameter: 5.5 cm, height: 15 cm). The crude material was loaded as an oil on top of the column and the product eluted with 4 L of 3% MeOH in EtOAc.

25. Analytical data of **5**: R_f = 0.45 in 10% MeOH in EtOAc; mp = 99–102 °C; ^{1}H NMR (500 MHz, CDCl$_3$) δ: 2.22–2.34 (m, 2 H), 3.16 (t, 2 H), 4.18 (t, 2 H), 7.42 (t, 1 H), 7.62 (d, 1 H), 7.70 (t, 1 H), 8.25 (d, 1 H); ^{13}C NMR (125 MHz, CDCl$_3$) δ: 19.7, 32.7, 46.6, 120.6, 126.4, 126.5, 126.9, 134.3, 149.3, 159.6, 161.1; IR (ATR) 1669, 1609, 1558, 1464, 1424, 1383, 1334, 1322, 770, 693 cm^{-1}; HRMS m/z calcd for $C_{11}H_{10}N_2$ [M+H]$^+$: 187.07931, found 187.08591.

Safety and Waste Disposal Information

All hazardous materials should be handled and disposed of in accordance with "Prudent Practices in the Laboratory"; National Academies Press; Washington, DC, 2011.

3. Discussion

o-Aminobenzaldehydes are important building blocks that are widely used in the synthesis of quinolines via the classic Friedländer condensation.[2] The present procedure describes a convenient and improved approach for the preparation of o-aminobenzaldehyde from o-nitrobenzaldehyde.[3] In our evaluation of published methods for the reduction of o-nitrobenzaldehyde,[4] we found that the use of iron dust in combination with hydrochloric acid provides the best yields of o-aminobenzaldehyde.

As part of a program aimed at developing redox neutral functionalizations of amines,[5] we have recently reported novel reactions of o-aminobenzaldehydes and secondary amines that give rise to ring-fused aminals.[5a] In the course of our attempts to adapt this procedure to a larger scale production of these versatile building blocks, we found that reactions conducted in n-butanol (as opposed to the originally used ethanol) provide faster reaction rates and higher yields on larger scale. In addition, slow

278

addition of *o*-aminobenzaldehyde via syringe pump further improved the yield of this process. A low concentration of *o*-aminobenzaldehyde minimizes the well-known propensity of this substrate to undergo self-condensation.[6]

The aminal structural motif[7] is found in a number of natural products.[8] In addition, aminals such as **4** represent reduced versions of quinazolinone alkaloids, compounds that have attracted significant attention due to their diverse array of biological activities.[9] The selective oxidation of ring fused aminal **4**, reported herein, provides rapid access to deoxyvasicinone (**5**).[10,11] The latter has been reported to possess antimicrobial, anti-inflammatory, and antidepressant activities.[10]

1. Department of Chemistry and Chemical Biology, Rutgers, The State University of New Jersey, Piscataway, NJ 08854, USA, E-mail: seidel@rutchem.rutgers.edu. We gratefully acknowledge the National Science Foundation for support of this research (Grant CHE-0911192).

2. (a) Friedländer, P. *Ber.* **1882**, *15*, 2572. (b) Friedländer, P.; Gohring, C. F. *Ber.* **1883**, *16*, 1833. For a recent review on the Friedländer reaction, see: (c) Marco-Contelles, J.; Pe´rez-Mayoral, E.; Samadi, A.; Carrciras, M. C.; Soriano, E. *Chem. Rev.* **2009**, *109, 2652–2671.

3. Smith, L. I.; Opie, J. W. *Org. Synth.* **1955**, *Coll. Vol. 3*, 56–58; *Org. Synth.* **1948**, *28*, 11–13.

4. For example, see: (a) Hu, Y.-Z.; Zhang, G.; Thummel, R. P. *Org. Lett.* **2003**, *5*, 2251–2253. (b) Raitio, K. H.; Savinainen, J. R.; Vepsäläinen, J.; Laitinen, J. T.; Poso, A.; Järvinen, T.; Nevalainen, T. *J. Med. Chem.* **2006**, *49*, 2022–2027. (c) Diedrich, C. L.; Haase, D.; Saak, W.; Christoffers, J. *Eur. J. Org. Chem.* **2008**, 1811–1816 and references cited therein.

5. (a) Zhang, C.; De, C. K.; Mal, R.; Seidel, D. *J. Am. Chem. Soc.* **2008**, *130*, 416–417. (b) Murarka, S.; Zhang, C.; Konieczynska, M. D.; Seidel, D. *Org. Lett.* **2009**, *11*, 129–132. (c) Zhang, C.; Murarka, S.; Seidel, D. *J. Org. Chem.* **2009**, *74*, 419–422. (d) Murarka, S.; Deb, I.; Zhang, C.; Seidel, D. *J. Am. Chem. Soc.* **2009**, *131*, 13226–13227. (e) Zhang, C.; Seidel, D. *J. Am. Chem. Soc.* **2010**, *132*, 1798–1799. (f) Deb, I.; Seidel, D. *Tetrahedron Lett.* **2010**, *51*, 2945–2947.

6. For examples, see: (a) Jircitano, A. J.; Sommerer, S. O.; Shelley, J. J.;

Westcott, B. L., Jr. *Acta Crystallogr., Sect. C: Cryst. Struct. Commun.* **1994**, *C50*, 445–447. (b) Kolchinski, A. G. *Coord. Chem. Rev.* **1998**, *174*, 207–239 and references cited therein.

7. Hiersemann, M., "Functions bearing two nitrogens." In *Comprehensive Organic Functional Group Transformations II*, Katritzky, A. R. T., Richard J. K., Eds. Elsevier Ltd.: Oxford, UK, 2005; Vol. 4, pp 411–441.

8. For selected examples of aminal containing natural products, see: (a) Braekman, J. C.; Daloze, D.; Pasteels, J. M.; Van Hecke, P.; Declercq, J. P.; Sinnwell, V.; Francke, W. *Z. Naturforsch., C: Biosci.* **1987**, *42*, 627–630. (b) Hino, T.; Nakagawa, M. "Chemistry and reactions of cyclic tautomers of tryptamines and tryptophans." In *Alkaloids*, Academic Press: New York, 1988; Vol. 34, pp 1–75. (c) Hajicek, J.; Taimr, J.; Budesinsky, M. *Tetrahedron Lett.* **1998**, *39*, 505–508. (d) Crich, D.; Banerjee, A. *Acc. Chem. Res.* **2007**, *40*, 151–161.

9. For a review on quinazolinone alkaloids, see: Mhaske, S. B.; Argade, N. P. *Tetrahedron* **2006**, *62*, 9787–9826.

10. (a) Amin, A. H.; Mehta, D. R. *Nature* **1959**, *184*, 1317. (b) Jain, M. P.; Koul, S. K.; Dhar, K. L.; Atal, C. K. *Phytochemistry* **1980**, *19*, 1880–1882. (c) Al-Shamma, A.; Drake, S.; Flynn, D. L.; Mitscher, L. A.; Park, Y. H.; Rao, G. S. R.; Simpson, A.; Swayze, J. K.; Veysoglu, T.; Wu, S. T. S. *J. Nat. Prod.* **1981**, *44*, 745–747.

11. For selected examples of recent syntheses of deoxyvasicinone, see: (a) Mhaske, S. B.; Argade, N. P. *J. Org. Chem.* **2001**, *66*, 9038–9040. (b) Lee, E. S.; Park, J.-G.; Jahng, Y. *Tetrahedron Lett.* **2003**, *44*, 1883–1886. (c) Liu, J.-F.; Ye, P.; Sprague, K.; Sargent, K.; Yohannes, D.; Baldino, C. M.; Wilson, C. J.; Ng, S.-C. *Org. Lett.* **2005**, *7*, 3363–3366. (d) Hamid, A.; Elomri, A.; Daich, A. *Tetrahedron Lett.* **2006**, *47*, 1777–1781. (e) Bowman, W. R.; Elsegood, M. R. J.; Stein, T.; Weaver, G. W. *Org. Biomol. Chem.* **2007**, *5*, 103–113.

Appendix
Chemical Abstracts Nomenclature; (Registry Number)

o-Aminobenzaldehyde: Benzaldehyde, 2-amino-; (529-23-7)
o-Nitrobenzaldehyde: Benzaldehyde, 2-nitro-; (552-89-6)
Iron powder; (7439-896)

Pyrrolidine; (123-75-1)
1,2,3,3a,4,9-Hexahydropyrrolo[2,1-b]quinazoline: (495-58-9)
Deoxyvasicinone: Pyrrolo[2,1-b]quinazolin-9(1*H*)-one, 2,3-dihydro-;
   (530-53-0)
Potassium permanganate; (7722-64-7)

Daniel Seidel was born in Mühlhausen, Thüringen (Germany) in 1972. He obtained his Diplom at the Friedrich-Schiller-Universität Jena in 1998. He then joined the group of Prof. Jonathan L. Sessler at the University of Texas at Austin, where he received his Ph.D. in 2002. From 2002–2005, Daniel was an Ernst Schering Postdoctoral Fellow in the group of Prof. David A. Evans at Harvard University. He started his independent career at Rutgers University in August of 2005. Research in his group is focused on new concepts for asymmetric catalysis and the development of synthetic methods.

Chen Zhang was born in Fuzhou (P. R. of China) in 1982. He received his B.Sc. degree from the University of Science and Technology of China in 2005. In September of 2005, he joined the Seidel group as a Ph.D. student. His research is focused on the development of redox neutral reaction cascades.

Chandra Kanta De was born in West Bengal (India) in 1984. He obtained his B.Sc. (Honors in Chemistry) degree from RKMV Belur Math College (University of Calcutta, India) in 2004 and his M.Sc. degree in Chemistry from Indian Institute of Technology Bombay (India) in 2006. In September of 2006, he joined the Seidel group as a Ph.D student. His research is focused on asymmetric catalysis and the development of redox neutral reaction cascades.

Liang Huang obtained his M.S. degree in Chemistry from Murray State University in 1997. He then began working at AlliedSignal as a synthetic organic chemist. In 1998 he moved to CB Research and Development, Inc. taking a position as Scientist. He has been working in Amgen's Chemical Process Research and Development group as a Process Chemist since 1999.

Brenda J. Burke was born in Naples, NY in 1977. She graduated from the University of Rochester in 1999 *cum laude* with a B.S. degree in Chemistry. She completed her doctoral studies at the University of California, Irvine in 2004 under the direction of Professor Larry Overman after which time she moved to the University of Cambridge to begin her post-doctoral studies in the labs of Professor Steven Ley. In 2006, Brenda became a member of Amgen's Chemical Process Research and Development group, and in 2011 she started a position in the Process Research group Gilead Sciences.

Multicomponent Synthesis of
Tertiary Diarylmethylamines:
1-((4-Fluorophenyl)(4-methoxyphenyl)methyl)piperidine

Submitted by Erwan Le Gall and Thierry Martens.[1]
Checked by David Hughes.

1. Procedure

A. *1-((4 Fluorophenyl)(4-methoxyphenyl)methyl)piperidine* **1**. To an oven-dried 500-mL, 3-necked, round-bottomed flask equipped with a 4-cm oval PTFE-coated magnetic stir bar is added zinc dust (35 g, 0.54 mol, 9 equiv) and anhydrous acetonitrile (200 mL) (Note 1). The flask is fitted with a gas inlet adapter connected to a nitrogen line and a gas bubbler. The other two necks are capped with rubber septa; a thermocouple probe is inserted through one of the septa (Note 2). To the stirred slurry is added trifluoroacetic acid (1 mL, 13 mmol, 0.2 equiv) and 1,2-dibromoethane (1 mL, 12 mmol, 0.2 equiv). The stirred mixture (Note 3) is heated with a heating mantle to 50–55°C, held at this temperature for 5 min (Note 4), and then cooled to 25–30°C using a 20 °C water bath. Maintaining water bath cooling, 4-bromoanisole (33 g, 0.18 mol, 3 equiv) (Note 5) and cobalt bromide (4.0 g, 18 mmol, 0.3 equiv) (Note 6) are added sequentially. The resulting mixture is stirred for 30 min (Notes 7 and 8). Maintaining water bath cooling, piperidine (5.36 g, 63 mmol, 1.05 equiv) is added via syringe in one portion followed 2 min later with 4-fluorobenzaldehyde (7.45 g, 60 mmol, 1.0 equiv (Note 9), also added via syringe in one portion. The resulting mixture is stirred for 2 h (Note 10). The mixture is vacuum-filtered through a 110 mL disposable polyethylene fritted funnel (Note 11) into a 500-mL round-bottomed flask, using *t*-butyl methyl ether (3 x 50 mL) to rinse the filter cake. The filtrate is transferred to a 1-L separatory funnel and

Published on the Web 1/20/2012

washed with a saturated ammonium chloride solution (200 mL). The aqueous layer is extracted with *t*-butyl methyl ether (200 mL). The organic extracts are combined and washed with a saturated ammonium chloride solution (200 mL), then vacuum-filtered through a bed of sodium sulfate (50 g) in a 110 mL disposable polyethylene fritted funnel into a 1-L round-bottomed flask, using *t*-butyl methyl ether (2 x 75 mL) to rinse the filter cake. The filtrate is concentrated by rotary evaporation (50 °C, 20 mmHg), then held under vacuum (20 mmHg) for 24 h at 60 °C to afford a yellow thick oil containing solids (20–22 g) (Note 12). This material is purified by alumina chromatography, affording product **1** (16–17 g) that contains 5–7 wt % anisole and 10–12 wt % 4,4'-dimethoxybiphenyl (Notes 13-15). The highly crystalline biphenyl impurity is removed by crystallization as follows. To the product in a 500-mL round-bottomed flask is added a 3-cm PTFE coated oval stir bar and hexanes (50 mL). The slurry is stirred at ambient temperature for 15 min and then filtered through a 40-mL polyethylene fritted funnel into a 250-mL flask, rinsing the filter cake with hexanes (2 x 7 mL). The filtrate is concentrated by rotary evaporation (50 °C, 20 mmHg), then held under vacuum (20 mmHg) for 24–48 h at 60 °C (to remove anisole), affording 1-((4-fluorophenyl)(4-methoxyphenyl)methyl)piperidine **1** as a pale yellow oil (13.5–14.1 g, 75–78 % yield) (Notes 16 and 17).

2. Notes

1. The following reagents and solvents were used as received: zinc dust (Sigma-Aldrich, < 10 μm, ≥ 98%), acetonitrile (anhydrous, Fisher Optima), 1,2-dibromoethane (Sigma-Aldrich, 99%), trifluoroacetic acid (Fisher certified, 97%), cobalt bromide (Sigma-Aldrich, 99%), 4-fluorobenzaldehyde (Sigma-Aldrich, 99%), piperidine (Sigma-Aldrich, 99.5%), 4-bromoanisole (Acros, 98%, and Sigma-Aldrich, 99%), hexanes (Fisher, ACS reagent, 98.5%), *t*-butyl methyl ether (Sigma-Aldrich, >98%), triethylamine (Sigma-Aldrich, 99.5%).

2. The internal temperature was monitored using a J-Kem Gemini digital thermometer with a Teflon-coated T-Type thermocouple probe (12-inch length, 1/8 inch outer diameter, temperature range –200 to +250 °C).

3. The magnetic stir bar rotation speed was 500 rpm. Vigorous stirring is required to suspend Zn dust and the agglomerated zinc suspension after activation.

4. Gentle evolution of ethylene, resulting from the metallation-elimination of $BrCH_2CH_2Br$, generally starts between 45 °C and 50 °C,

284

along with the formation of ZnBr$_2$ which is highly favorable to the forthcoming generation of the organozinc reagent. Flocculation of the zinc occurs during the activation, constituting a visual confirmation the activation process has initiated.

5.   Three equiv of zinc vs 4-bromoanisole is required to obtain short reaction times. Lower amounts of zinc leads to incomplete conversion, resulting in contamination of the product with 4-bromoanisole, which is not readily removed by silica or alumina chromatography.

6.   Cobalt bromide is very hygroscopic, so contact with air must be minimized to avoid hydration (the hydration can be observed by the progressive modification of the powder color from intense green to mauve).

7.   Three equiv of 4-bromoanisole vs 4-fluorobenzaldehyde are required to counterbalance the formation of biaryl under these conditions and the loss of 1 equiv of the *in-situ* formed organozinc compound upon hydration. (Further detail on the mechanistic proposal is provided in the Discussion section). The yield for the formation of the arylzinc species was estimated to be 75–80 % based on quenching the reaction mixture with iodine to form 4-iodoanisole, as follows. A 0.1 mL aliquot of the reaction mixture was quenched into a solution of iodine (100 mg) in 2 mL THF. After standing for 2 min, hexanes (3 mL) and sat. sodium metabisulfite (3 mL) were added and the vial was shaken to mix the layers. The organic layer was separated, filtered through a plug of sodium sulfate and then concentrated by rotary evaporation (40 °C bath, 20 mmHg). The resulting solution was analyzed by ^{1}H NMR, providing the following molar ratio of products. 4-iodoanisole (75–80%), anisole (6–7%, equal to the trifluoroacetic acid charge), 4,4'-dimethoxybiphenyl (7–9 %, equivalent to 14–18 % of the 4-bromoanisole charge). Diagnostic resonances (CDCl$_3$, 400 MHz): 4-iodoanisole, δ 3.80; 4-bromoanisole, 3.81, anisole, 3.84; 4,4-dimethoxybiphenyl, δ 3.86. High levels of residual THF (δ 3.74–3.79) from incomplete evaporation of solvent may interfere with the integration of 4-iodoanisole.

8.   A 20 °C water bath was used to control the exothermic formation of the arylzinc species, maintaining the temperature <40 °C, as profiled in the graph below. This limited temperature rise is required for complete arylzinc formation.

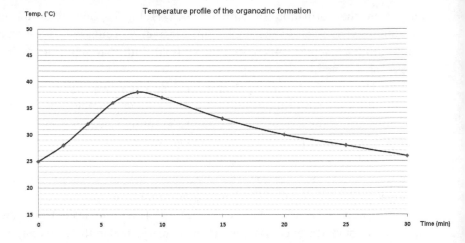

Temp. (°C)   Temperature profile of the organozinc formation

9. The reaction temperature increased from 26 to 29 °C on addition of piperidine, then to 36 °C on addition of the aldehyde. With water bath cooling, the reaction then slowly cools to 23–25 °C over the course of 20 min.

10. The checker monitored the reaction by ^{1}H NMR by quenching an aliquot of the reaction mixture in 1 mL CDCl$_3$/1 mL sat NH$_4$Cl. The organic layer was filtered through a plug of Na$_2$SO$_4$ and then concentrated to dryness. The residue was dissolved in CDCl$_3$ for NMR analysis. Diagnostic peaks: product **1**, 4.19 (s, 1H) and 3.77 (s, 3H), 4-bromoanisole (s, 3.80), anisole (s, 3.84), 4,4'-dimethoxybiphenyl (s, 3.87), and 4-fluorobenzaldehyde (s, 9.99). (The resonances for the methoxy group in these quenching experiments varied slightly from those reported in Note 7 due to differences in sample composition. Spiking experiments should be carried out for confirmation of individual components). The submitters monitored the reaction by gas chromatography using a SGE BP1 column, 5 m x 0.25 mm, initial temp 70 °C, ramp at 10 °C/min to 270 °C, hold 15 min.

11. The polyethylene fritted funnels were obtained from OptiChem.

12. The oil after rotary evaporation contains about 35 % anisole by weight, which can reduced to <10% prior to chromatography by applying high vacuum (3 mmHg) for 4 h at 60 °C or house vacuum (20 mmHg) at 60 °C for 24 h. Final weight of the crude material was 20–22 g.

13. Material was purified by either silica gel or alumina chromatography to remove polar impurities. Neither method was fully effective at separating the major by-products, anisole and 4,4'-

286

dimethoxybiphenyl. The best separation of 4,4'-dimethoxybiphenyl was obtained with alumina from E. Merck (sourced via Acros); however, E. Merck no longer supplies chromatography-grade alumina. Chromatography using alumina from either Aldrich (50–150 μm) or Selecto Scientific (N, Super 1, 63–150 μm) is described herein. A 6-cm glass column was wet-packed with alumina (450 g) using a 95:3.5:1.5 mixture of hexanes:MTBE:Et$_3$N. Sand (1 cm) was added to the top of the column. To the crude resinous product **1** was added 50 mL dichloromethane and 70 g alumina. The mixture was swirled for 2 min to adsorb thick resinous material on the walls of the flask onto the alumina. The slurry was concentrated to a powder by rotary evaporation (40 °C, 20 mmHg). This material was loaded onto the column using dichloromethane (50 mL). The column was eluted with a 95:3.5:1.5 hexanes:MTBE:Et$_3$N mixture (2 L), collecting 50-mL fractions. Fractions 7-32 were combined and concentrated in potions in a 500-mL round bottomed flask to afford 16–17 g of product **1** containing 5–7% anisole and 10–12 % 4,4'-dimethoxybiphenyl.

14. The product could also be chromatographed using silica gel (Fisher, 40–63 um, 60 Å) (300 g), wet-packed using a 95:3.5:1.5 mixture of hexanes:MTBE:Et$_3$N. The crude product was adsorbed onto silica (50 g) in dichloromethane (50 mL) and loaded on the column. Elution was carried out with a 95:3.5:1.5 mixture of hexanes:MTBE:Et$_3$N (1.5 L) then 92:6:2 hexanes:MTBE:Et$_3$N (1.5 L), collecting 50 mL fractions. Fractions 7-51 were combined and concentrated to afford 16–17 g of product **1** containing 5–7% anisole and 10–12 % 4,4'-dimethoxybiphenyl.

15. The checker monitored the chromatography using silica TLC, 90:8:2 mixture of hexanes:MTBE:Et$_3$N, monitored by UV, R$_f$ anisole = 0.5; product **1** = 0.3; 4,4-dimethoxybiphenyl = 0.3. Submitters carried out TLC on alumina using a 95.5:3:1.5 mixture of pentane:Et$_2$O:NEt$_3$ as eluent, R$_f$ product **1** = 0.65; anisole = 0.90.

16. Crystallization of **1** was carried out as follows. Diarylmethylamine **1** (5.0 g) was dissolved in 95:5 EtOH:water (40 mL) at ambient temperature in a 250-mL round bottom flask, then loosely stoppered and placed in a –20 °C freezer for 18 h. The resulting while solids were filtered through a 60-mL sintered glass fritted funnel and washed with 95:5 EtOH:water (10 mL, –20 °C), then air-dried at room temperature to afford 3.6 g (72 % recovery) of **1** as a white crystalline solid.

17. 1-((4-Fluorophenyl)(4-methoxyphenyl)methyl)piperidine has the following physical and spectroscopic data: white crystalline solid, mp 38–40 °C; IR, υ (cm^{-1}) : 2931, 2850, 2789, 2750, 1605, 1504, 1441, 1240, 1208,

1032, 820; ^{1}H NMR (CDCl$_3$, 400 MHz) δ: 1.43–1.46 (m, 2 H), 1.53–1.59 (m, 4 H), 2.29 (br s, 4 H), 3.77 (s, 3 H), 4.19 (s, 1 H), 6.80–6.84 (m, 2 H), 6.94–6.98 (m, 2 H), 7.25–7.29 (m, 2 H), 7.33–7.37 (m, 2 H); ^{13}C NMR (CDCl$_3$, 100 MHz) δ: 24.9, 26.5, 53.2, 55.4, 75.3, 113.9, 115.3 (d, J = 21.1 Hz), 129.1, 129.4 (d, J = 7.9 Hz), 135.2, 139.6 (d, J = 3.1 Hz), 158.6, 161.8 (d, J = 244.7 Hz); ^{19}F NMR (CDCl$_3$, 376 MHz) δ: –116.35; GC-MS, m/z (relative intensity): 299 (38) [M$^+$], 215 (100) [M - C$_5$H$_{10}$N], 204 (19) [M - C$_6$H$_4$F], 192 (13) [M – C$_6$H$_4$OMe], 171 (35); LC-MS, m/z (relative intensity): 300.1 (100) [MH$^+$], 215 (95) [M - C$_5$H$_{10}$N]; Anal. calcd. for C$_{19}$H$_{22}$FNO: C 76.22, H 7.41, N 4.68; found: C 76.16, H 7.80, N 4.68. GC area % of material prior to crystallization: 97% (t$_R$ = 14.7 min; conditions: Agilent DB35MS column; 30 m x 0.25 mm; initial temp 60 °C, ramp at 20 °C/min to 280 °C, hold 15 min; major impurities 0.5–1.0 % anisole, 0.5–0.6 % 4,4'-dimethoxybiphenyl, 0.1–0.2 % 4-bromoanisole, 0.1–0.2 % 4-fluorobenzaldehyde).

Safety and Waste Disposal Information

All hazardous materials should be handled and disposed of in accordance with "Prudent Practices in the Laboratory"; National Academies Press; Washington, DC, 2011.

3. Discussion

Diarylmethylamines are compounds of high interest, particularly as active ingredients of several pharmaceutical drugs such as Zyrtec® (cetirizine), Sibelium® (flunarizine), and Iperten® (manidipine) (Figure 1).

While these interesting properties have made them attractive synthetic targets,[2] the synthesis of tertiary diarylmethylamines by one-pot multicomponent procedures[3-4] are rare. Current methods for the synthesis of these compounds include the Petasis reaction[5] between arylboronic acids, aldehydes, and amines and the aromatic Mannich reaction[6] involving arenes, aldehydes, and amines. However, despite their important synthetic interest, these reactions present some limitations. The Petasis reaction requires the use of ortho-activated aldehydes such as salicylaldehydes[5a] and the aromatic Mannich reaction is efficient only with electron-rich aromatic compounds. In addition, in the latter case, control of the carbon that is involved in the C-C bond formation cannot be achieved.

288

Figure 1. Pharmaceutical drugs containing a diarylmethylamine scaffold

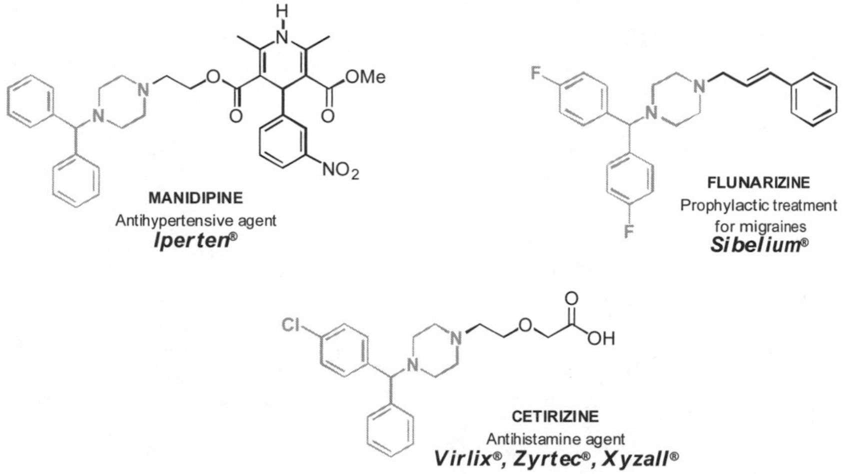

MANIDIPINE
Antihypertensive agent
Iperten®

FLUNARIZINE
Prophylactic treatment
for migraines
Sibelium®

CETIRIZINE
Antihistamine agent
Virlix®, Zyrtec®, Xyzall®

The procedure reported herein describes the synthesis of diarylmethylamines by a one-pot, zinc-mediated, cobalt-catalyzed three-component reaction between aryl bromides, secondary amines and aromatic or heteroaromatic aldehydes. This method provides convenient access to a range of tertiary diarylmethylamines from common and inexpensive commercial reagents and avoids the limitations of the above-mentioned reactions. The organozinc reagent, which is the nucleophile involved in this Mannich-like procedure, is pre-formed under conditions similar to those initially described by Gosmini and coworkers.[7] The final compound is obtained through the trapping of an iminium cation (or related species), formed *in situ* through the addition of the amine to the aldehyde, by the organozinc species. A possible reaction mechanism is depicted in Scheme 1.

Scheme 1. Possible reaction mechanism

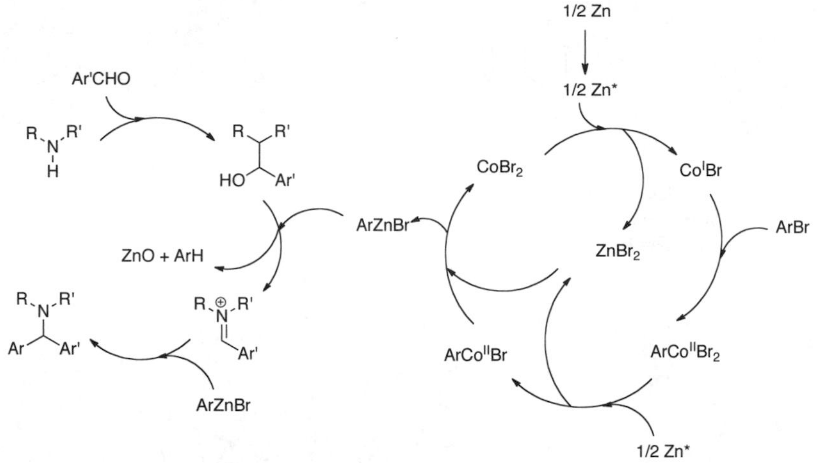

For efficient preparations of diarylmethylamines by this method, some experimental precautions should be taken: the reaction must be conducted with at least 3 equiv of the organozinc compound to both counterbalance the formation of biaryl, which systematically occurs under these conditions, and furnish an additional equivalent of organozinc compound to efficiently initiate the formation of an iminium equivalent by acid-base reaction. Zinc dust must also be used in large excess and a limited rise of the medium's temperature must be allowed to ensure a successful formation of the arylzinc species. The solvent and the catalyst must be anhydrous, otherwise the reaction will require an increased time for completion or not initiate.

This multicomponent procedure is illustrated by the description of a reaction providing 1-((4-fluorophenyl)(4-methoxyphenyl)methyl)piperidine. However, the scope includes other aldehydes, amines and halides (compounds **2-16**),[8] as depicted in Figure 2.

Org. Synth. **2012**, *89*, 283-293

Figure 2. Some structures accessible by the 3-component coupling procedure

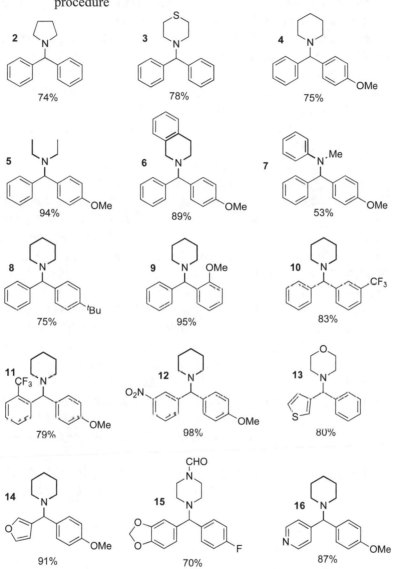

1. Électrochimie et Synthèse Organique, Institut de Chimie et des Matériaux Paris-Est, UMR 7182 CNRS - Université Paris-Est Créteil, 2 rue Henri Dunant, 94320 Thiais, France. E-mail: legall@glvt-cnrs.fr.

2. (a) Corey, E. J.; Helal, C. J. *Tetrahedron Lett.* **1996**, *37*, 4837-4840; (b) Opalka, C. J.; D'Ambra, T. E.; Faccone, J. J.; Bodson, S.; Cossement, E. *Synthesis* **1995**, 766-768; (c) Pflum, D. A.; Wilkinson, H. S.; Tanoury, G. J.; Kessler, D. W.; Kraus, H. B.; Senanayake, C. H.; Wald, S. A. *Org. Proc. Res. Dev.* **2001**, *5*, 110-115.

3. Zhu, J.; Bienaymé, H. in *Multicomponent Reactions*, Wiley–VCH, Weinheim, **2005**.

4. For recent reviews on multicomponent reactions (MCRs) and asymmetric MCRs (AMCRs), see: (a) Orru, R. V. A.; De Greef, M. *Synthesis* **2003**, 1471–1499; (b) Zhu, J. *Eur. J. Org. Chem.* **2003**, 1133–1144; (c) Hulme, C.; Gore, V. *Curr. Med. Chem.* **2003**, *10*, 51–80; (d) Balme, G.; Bossharth, E.; Monteiro, N. *Eur. J. Org. Chem.* **2003**, 4101–4111; (e) Simon, C.; Constantieux, T.; Rodriguez, J. *Eur. J. Org. Chem.* **2004**, 4957–4980; (f) Dömling, A. *Chem. Rev.* **2006**, *106*, 17–89; (g) D'Souza, D. M.; Müller, T. J. J. *Chem. Soc. Rev.* **2007**, *36*, 1095–1108; (h) Isambert, N.; Lavilla, R. *Chem. Eur. J.* **2008**, *14*, 8444–8454; (i) Arndtsen, B. A.; *Chem. Eur. J.* **2009**, *15*, 302–312; (j) Guillena, G.; Ramón, D. J.; Yus, M. *Tetrahedron: Asymmetry* **2007**, *18*, 693–700; (k) Ramón, D. J.; Yus, M. *Angew. Chem., Int. Ed.* **2005**, *44*, 1602–1634; (l) Dömling, A.; Ugi, I. *Angew. Chem., Int. Ed.* **2000**, *39*, 3168–3210.

5. (a) Petasis, N. A.; Boral, S. *Tetrahedron Lett.* **2001**, *42*, 539–542; (b) Petasis, N. A.; Zavialov, I. A. *J. Am. Chem. Soc.* **1998**, *120*, 11798–11799; (c) Petasis, N. A.; Akritopoulou, I. *Tetrahedron Lett.* **1993**, *34*, 583–586.

6. (a) Lubben, M.; Feringa, B. L. *J. Org. Chem.* **1994**, *59*, 2227–2233; (b) Cimarelli, C.; Mazzanti, A.; Palmieri, G.; Volpini, E. *J. Org. Chem.* **2001**, *66*, 4759–4765.

7. (a) Fillon, H.; Gosmini, C.; Périchon, J. *J. Am. Chem. Soc.* **2003**, *125*, 3867–3870; (b) Kazmierski, I.; Gosmini, C.; Paris, J. –M.; Périchon, J. *Tetrahedron Lett.* **2003**, *44*, 6417–6420. (c) Gosmini, C.; Amatore, M.; Claudel, S.; Périchon, J. *Synlett* **2005**, 2171–2174.

8. (a) Le Gall, E.; Haurena, C.; Sengmany, S.; Martens, T.; Troupel, M. *J. Org. Chem.* **2009**, *74*, 7970–7973; (b) Le Gall, E.; Troupel, M.; Nédélec, J.-Y. *Tetrahedron* **2006**, *62*, 9953–9965.

Erwan Le Gall received his PhD from the University of Rennes (France) in 1998. The work was focused on α-aminonitriles synthesis and their applications to the preparation of heterocyclic compounds. In 1999, he conducted his post-doctoral research at the LECSO group (Paris, France) under the guidance of Dr. Corinne Gosmini. He became Assistant Professor (2000) and Associate Professor (2009) at the Université Paris-Est Créteil. His current research focuses on the development of multicomponent reactions involving preformed or in-situ-generated organometallic reagents.

Thierry Martens received his PhD from the University Paris 6 (France) in 1988, under the direction of Pr. Maurice B. Fleury and Dr. Martine Largeron. In 1989, he conducted a post-doctoral work under the guidance of Dr. Nicole Moreau at the CERCOA CNRS, Thiais. He joined the Faculty of Pharmaceutical Sciences of the University Paris 5 in 1992 as an Assistant Professor. In 1994, he joined the group of Pr. Henri Philippe Husson and of Dr. Jacques Royer. His research topic was mainly focused on the development of electrochemical methods for the anodic oxidation of amines. In 2006, he became full Professor at the Université Paris-Est Créteil. His current research interests include synthetic electrochemistry and pharmaceutical chemistry.

Synthesis of Quinolines by Electrophilic Cyclization of *N*-(2-Alkynyl)Anilines: 3-Iodo-4-Phenylquinoline

Submitted by Yu Chen, Anton Dubrovskiy, and Richard C. Larock.[1,2]
Checked by Nicolas Armanino and Erick M. Carreira.

1. Procedure

A. *N-(2-Propynyl)aniline* (**3**). An oven-dried 500-mL, three-necked, round-bottomed flask equipped with a 25-mm egg-shaped magnetic stirring bar, a thermometer, a pressure-equalizing dropping funnel, and a rubber septum is charged with aniline (**1**, 22.35 g, 240.0 mmol, 4.0 equiv) (Notes 1 and 2), potassium carbonate (16.58 g, 120.0 mmol, 2.0 equiv) (Note 3), and *N,N*-dimethylformamide (DMF, 300 mL) (Note 4). The mixture is stirred for 5 min at room temperature. A solution of propargyl bromide (**2**, 7.14 g, 80% solution in toluene, 6.68 mL, 60.0 mmol, 1.0 equiv) (Note 5) in *N,N*-dimethylformamide (DMF, 25 mL) is added to the flask dropwise (Note 6). The reaction mixture is stirred at room temperature for 6 h (Note 7). The reaction mixture is filtered under reduced pressure (Note 8). The original three-necked round-bottomed flask and the Büchner funnel are rinsed with 150 mL of diethyl ether. The combined filtrate is transferred to a 1-L separatory funnel and washed with brine (300 mL). The aqueous phase is

Org. Synth. **2012**, *89*, 294-306
Published on the Web 2/1/2012
© 2012 Organic Syntheses, Inc.

extracted twice with diethyl ether (2 × 150 mL). The combined organic phases are washed with water (100 mL) and dried over anhydrous magnesium sulfate (MgSO₄) (Note 9), filtered through a fritted glass funnel, rinsed with diethyl ether (50 mL), and concentrated by rotary evaporation (25 °C, 40 mmHg) to give a dark brown oil. The residue is purified by flash column chromatography on silica gel (Note 10) to afford 6.52–6.85 g (83–87%) of *N*-(2-propynyl)aniline (**3**) as a light yellow oil (Note 11).

B. *N-(3-Phenyl-2-propynyl)aniline* (**5**). An oven-dried 500-mL, three-necked round-bottomed flask equipped with a 25-mm egg-shaped magnetic stirring bar, a thermometer, and two rubber septa (Note 12) is flushed with argon (Note 13) and charged with *N*-(2-propynyl)aniline (**3**, 4.76 g, 36.28 mmol, 1.0 equiv), triethylamine (225 mL) (Note 14), iodobenzene (**4**, 4.47 mL, 8.15 g, 40.0 mmol, 1.1 equiv) (Note 15) and bis(triphenylphosphine)palladium dichloride (510 mg, 0.73 mmol, 0.02 equiv) (Note 16). Copper iodide (69 mg, 0.36 mmol, 0.01 equiv) (Note 17) is then added in a single portion to the flask. The reaction mixture is stirred at room temperature for 6 h (Notes 18 and, 19). The reaction mixture is filtered under reduced pressure (Note 20). The original three-necked, round-bottomed flask and the Büchner funnel are rinsed with 150 mL of diethyl ether. The combined organic phases are transferred to a 1-L separatory funnel and washed with brine (300 mL). The aqueous phase is extracted twice with diethyl ether (2 × 100 mL). The combined organic phases are dried over anhydrous magnesium sulfate (MgSO₄) (Note 21), filtered through a fritted glass funnel, rinsed with diethyl ether (50 mL), and concentrated by rotary evaporation (25 °C, 40 mmHg) to give a dark brown oil. The residue is purified by flash column chromatography on silica gel (Note 22) to afford 6.48–6.61 g (86–88%) of *N*-(3-phenyl-2-propynyl)aniline (**5**) as a light yellow oil (Note 23).

C. *3-Iodo-4-phenylquinoline* (**6**). An oven-dried, 500-mL, three-necked, round-bottomed flask equipped with a 25-mm egg-shaped magnetic stirring bar, a thermometer, and two rubber septa is charged with *N*-(3-phenyl-2-propynyl)aniline (**5**, 5.23 g, 25.23 mmol, 1.0 equiv), sodium bicarbonate (2.24 g, 26.67 mmol, 1.06 equiv) (Note 24) and acetonitrile (126 mL) (Note 25). The reaction mixture is stirred at room temperature for 20 min. One rubber septum is replaced with a pressure-equalizing dropping funnel. A warm solution of iodine (19.2 g, 75.65 mmol, 3.0 equiv) (Note 26) in acetonitrile (126 mL) is added to the reaction mixture through the pressure-equalizing dropping funnel over 4 h with vigorous stirring (Notes

27, 28 and 29). An additional 35 mL of acetonitrile is added at the end of the addition to rinse the crystallized iodine from the pressure-equalizing dropping funnel into the reaction flask. The reaction mixture is stirred at room temperature for 18 h. The pressure-equalizing dropping funnel is replaced with a reflux condenser. The reaction flask is immersed in a preheated oil bath (at 70 °C) and the reaction mixture is stirred at 60 °C (temperature inside the reaction flask) for an additional 26 h (Note 30). The reaction mixture is cooled down to room temperature. The reflux condenser is replaced with a pressure-equalizing dropping funnel. An aqueous solution of $Na_2S_2O_3 \cdot 5H_2O$ (37.55 g, 151.29 mmol, 2 equiv with respect to iodine, in 150 mL water) is added to the reaction mixture through the pressure-equalizing dropping funnel over 6 min with stirring (Notes 31 and 32). After the addition, the reaction mixture is stirred at room temperature for 20 min. The resulting mixture is transferred into a 1-L separatory funnel and extracted with 100 mL of diethyl ether. The aqueous phase is further extracted three times with diethyl ether (100 mL, 50 mL, and 25 mL). The combined organic phases are dried over anhydrous magnesium sulfate ($MgSO_4$) (Note 33), filtered through a fritted glass funnel, rinsed with diethyl ether (4 × 50 mL), and concentrated by rotary evaporation (25 °C, 40 mm Hg) to give a dark orange powder. The residue is purified by recrystallization from hot 2-propanol (Notes 34, 35 and 36) or by flash column chromatography on silica gel (Note 37) to afford 5.46 g (65%) of 3-iodo-4-phenylquinoline (**6**) as a light orange powder (Notes 38 and 39).

2. Notes

1.   Aniline (**1**, 99.9%) was purchased from Fisher Scientific and used as received. The checkers used aniline (reagent plus 99%) purchased from Aldrich Chemical Co., Inc.

2.   A variety of ratios of aniline/propargyl bromide have been investigated. An excess of aniline (4 equiv to propargyl bromide) was found the most beneficial for the monoalkylation of aniline.

3.   Potassium carbonate (K_2CO_3, 99.5%, anhydrous) was purchased from Fisher Scientific and used as received. The checkers used potassium carbonate (ACS) from Merck.

4.   *N,N*-Dimethylformamide (DMF, 99.8%) was purchased from Fisher Scientific and used as received. The checkers used DMF (ACS reagent) from Aldrich Chemical Co., Inc.

Org. Synth. **2012**, *89*, 294-306

5. A solution of propargyl bromide (80% in toluene) was purchased from Aldrich Chemical Co., Inc. and used as received.

6. During the addition of propargyl bromide (**2**), the internal temperature of the reaction mixture increased from 25 to 29 °C and then gradually decreased to 25 °C. The addition took 10 min.

7. The authors recommend that the reaction be stopped after 6 hours in order to obtain a high yield of the desired monoalkylated aniline. Longer reaction times cause the generation of undesired dialkylated aniline.

8. The reaction mixture was filtered through the sintered glass Buchner funnel under 32 mmHg pressure.

9. Anhydrous magnesium sulfate was purchased from Fisher Scientific and used as received. To ensure proper dryness, 25 g of $MgSO_4$ were added to the organic phase and the resulting mixture was kept at room temperature for 20 min with occasional stirring.

10. Column chromatography was performed on a 5-cm diameter column, wet-packed with 300 g of silica gel (230-400 mesh) in hexanes. The length of silica gel was 35 cm. Dichloromethane (5 mL) was used to ensure the complete transfer of the crude compound onto the column. The following solvent systems were used: hexane (300 mL), hexane/ethyl acetate (30/1, 1000 mL), hexane/ethyl acetate (25/1, 1000 mL), hexane/ethyl acetate (20/1, 1500 mL). After elution of 1.2 L of solvent, 90 fractions of 25 mL were collected. Fractions 26 through 90 contained the desired product and were concentrated by rotary evaporation (40 °C bath, 20 mmHg), and dried under a vacuum (0.6 mmHg) at 25 °C with stirring for 20 h until a constant weight (6.52–6.85 g) was obtained. Fractions 20-25 contained a mixture of the desired product and dialkylated aniline, and fractions after 90 contained a mixture of the desired product and the unreacted aniline. These fractions were not collected.

11. The physical properties of *N*-(2-propynyl)aniline (**3**) follow: $R_f =$ 0.41 (TLC analysis performed on glass-backed silica gel TLC plates with a UV254 indicator, obtained from Sorbent Technologies; 10:1 hexane/ethyl acetate is used as the eluent; the product is visualized with a 254 nm UV lamp); IR (thin film): 3401, 3287, 1602, 1504, 1315, 1259, 751, 693 cm^{-1}; ^{1}H NMR (400 MHz, CDCl$_3$) δ: 2.25 (t, J = 2.4 Hz, 1 H), 3.92 (s, br, 1 H), 3.97 (d, J = 2.4 Hz, 2 H), 6.70 – 6.76 (m, 2 H), 6.82 (tt, J = 7.3, 1.1 Hz, 1 H),7.21 – 7.31 (m, 2 H); ^{13}C NMR (101 MHz, CDCl$_3$) δ: 33.7, 71.3, 81.0, 113.6, 118.7, 129.3, 146.8; HRMS *m/z* calcd. for C$_9$H$_{10}$N [M+H]$^+$, 132.0808,

found 132.0806; Anal. Calcd. for C$_9$H$_9$N: C, 82.41; H, 6.92; N, 10.68; found: C, 82.33; H, 6.97; N, 10.62.

12. One rubber septum is fitted with an argon inlet.

13. Dry argon (99.996%) was flushed through the flask for 2 min.

14. Triethylamine (Et$_3$N, 99.5%) was purchased from Aldrich Chemical Co., Inc. and used as received.

15. Iodobenzene (PhI, 98%) was purchased from Aldrich Chemical Co., Inc. and used as received.

16. Bis(triphenylphosphine)palladium dichloride [PdCl$_2$(PPh$_3$)$_2$] was donated by Kawaken Fine Chemicals Co., Ltd., and Johnson Matthey, Inc. and used as received. The checkers used bis(triphenylphosphine)palladium dichloride [PdCl$_2$(PPh$_3$)$_2$] (99%) from Aldrich Chemical Co., Inc..

17. Copper iodide (CuI, 99.5%) was purchased from Aldrich Chemical Co., Inc. and used as received.

18. One hour after the reaction had started, the internal temperature of the reaction mixture had increased from 25 to 30 °C with a suspension of yellow flakes present in the solution. After an additional hour, the internal temperature decreased to 25 °C.

19. Completion of the reaction was judged by the disappearance of the starting material (3) on TLC (SiO$_2$, eluent: 10:1 hexane/ethyl acetate, R$_f$ = 0.41).

20. The reaction mixture was filtered through the sintered glass Büchner funnel under 32 mmHg pressure.

21. Anhydrous magnesium sulfate was purchased from Fisher Scientific and used as received. To ensure proper dryness, 20 g of MgSO$_4$ were added to the organic phase and the resulting mixture was kept at room temperature for 20 min with occasional stirring.

22. Column chromatography was performed on a 5 cm diameter column, wet-packed with 385 g of silica gel (230-400 mesh) in hexanes. The length of silica gel was 41 cm. Dichloromethane (10 mL) was used to ensure the complete transfer of the material onto the column. The following solvent systems were used: hexane (125 mL), hexane/ethyl acetate (80/1, 400 mL), hexane/ethyl acetate (60/1, 300 mL), hexane/ethyl acetate (50/1, 300 mL), hexane/ethyl acetate (40/1, 300 mL), hexane/ethyl acetate (30/1, 600 mL), hexane/ethyl acetate (25/1, 500 mL), hexane/ethyl acetate (20/1, 700 mL). Among the 25 mL fractions collected, fractions 84 through 129 contained the desired product and were concentrated by rotary evaporation (40 °C bath, 20 mmHg), and then dried under vacuum (0.6 mmHg) at

25 °C with stirring for 20 h until a constant weight (6.48–6.61 g) was obtained.

23. The physical properties of *N*-(3-phenyl-2-propynyl)aniline (**5**) follow: R_f = 0.40 (TLC analysis performed on glass-backed silica gel TLC plates with a UV254 indicator, obtained from Sorbent Technologies; 10:1 hexane/ethyl acetate is used as the eluent; the product is visualized with a 254 nm UV lamp); IR (thin film): 3388, 3051, 2924, 1601, 1499, 1442, 1252, 752, 690 cm^{-1}; ^{1}H NMR (400 MHz, CDCl$_3$) δ: 4.01 (s, br, 1 H), 4.19 (s, 2 H), 6.75 – 6.80 (m, 2 H), 6.82 (tt, *J* = 7.3, 1.0 Hz, 1 H), 7.23 – 7.34 (m, 5 H), 7.39 – 7.47 (m, 2 H); ^{13}C NMR (101 MHz, CDCl$_3$) δ: 34.6, 83.3, 86.3, 113.6, 118.5, 122.9, 128.2, 128.3, 129.2, 131.7, 147.1; HRMS *m/z* calcd for C$_{15}$H$_{14}$N [M+H]$^+$, 208.1121, found 208.1125; Anal. calcd. for C$_{15}$H$_{13}$N: C, 86.92; H, 6.32; N, 6.76; found: C, 86.82; H, 6.37; N, 6.78.

24. Sodium bicarbonate (NaHCO$_3$, certified A.C.S.) was purchased from Fisher Scientific and used as received. The checkers used sodium hydrogen carbonate (ACS) from Merck.

25. Acetonitrile (CH$_3$CN, 99.8%, anhydrous) was purchased from Aldrich Chemical Co., Inc. and used as received.

26. Iodine (I$_2$, 99.9%, certified A.C.S.) was purchased from Fisher Scientific (submitters) or Aldrich Chemical Co., Inc. (checkers) and was purified by sublimation at 1.5 mmHg, 90 °C. The resublimed iodine was ground into a powder before use.

27. The iodine was dissolved in acetonitrile by heating with a heat gun. The latter was used to keep the iodine solution in the dropping funnel warm, thus preventing crystallization of the iodine and the funnel from plugging up. *Caution! Heat guns contain an electrically-heated filament that pose an ignition and spark hazard. Heat guns should only be used in a fume hood and care must be taken to ensure that no flammable vapors are present when heat guns are in use.*

28. During the addition of iodine to the reaction mixture, the internal temperature of the reaction mixture increased from 25 to 28 °C.

29. Slow addition of the solution of iodine in acetonitrile to the reaction mixture prevents the formation of undesired side products, which cause a significant problem in purification of the desired product.

30. Reaction has been monitored by executing ^{1}H NMR spectrum of ~0.10 mL of the reaction mixture extracted with diethyl ether from a saturated aqueous Na$_2$S$_2$O$_3$ solution. The conversion did not increase after

reaching a maximum of 86% as based on the integrations of the multiplet at 6.90 – 6.70 ppm (3H of **5**) and the doublet at 8.15 ppm (1H of **6**).

31. Sodium thiosulfate ($Na_2S_2O_3 \cdot 5H_2O$, certified A.C.S.) was purchased from Fisher Scientific and used as received.

32. During the addition of the sodium thiosulfate solution to the reaction mixture the internal temperature of the reaction mixture decreased from 25 to 19 °C. The solution turned from purple to slightly pink after the addition.

33. Anhydrous magnesium sulfate was purchased from Fisher Scientific and used as received. To ensure proper dryness, 20 g of $MgSO_4$ were added to the organic phase and the resulting mixture was kept at room temperature for 20 min with occasional stirring.

34. 2-Propanol (ACS) was purchased from Scharlab S.L. and used as received.

35. The crude material was charged in a 100-mL pear-shaped flask equipped with a 25-mm magnetic stirring bar and a reflux condenser and suspended in 2-propanol (65 mL). The mixture was then heated on an oil bath (80 °C) for 10 min until all the solids dissolved. The oil bath was removed. The mixture was allowed to slowly cool to room temperature with stirring and then placed in a freezer (–20 °C) for 12 h to complete crystallization. The solids were then filtered using a sintered glass Büchner funnel and washed with ice-cold 2-propanol (10 mL). The collected solids were dried under vacuum (0.6 mmHg) at 25 °C with stirring for 20 h until a constant weight (5.46 g) was obtained.

36. Recrystallization may also be performed from an ethanol (170 mL):water (120 mL) mixture with similar results.

37. The submitters provide conditions for purifying the crude compound by chromatography: Column chromatography was performed on a 7 cm diameter column, wet-packed with 367 g of silica gel (230-400 mesh) in hexanes. The length of silica gel was 24 cm. Dichloromethane (25 mL) was used to ensure the complete transfer of the crude material onto the column. The following solvent systems were used: chloroform/hexane (1/2, 3.5 L), chloroform/hexane (1/1, 9 L). Among the 50 mL fractions collected, fractions 111 through 250 contained the desired product and were concentrated by rotary evaporation (40 °C bath, 20 mmHg), and then dried under vacuum (0.6 mmHg) at 25 °C for 20 h until a constant weight (5.43 g) was obtained. Fractions 80 through 110 contained the mixture of

300

the desired product and unreacted *N*-(3-phenyl-2-propynyl)aniline (yields by crude ^{1}H NMR are 13 and 20%, respectively).

38. On smaller scale (13.0 mmol) the checkers obtained a higher yield (71 %) after chromatographic purification.

39. The physical properties of 3-iodo-4-phenylquinoline (6) follow: R_f = 0.34 (TLC analysis performed on glass-backed silica gel TLC plates with a UV254 indicator, obtained from Sorbent Technologies; 10:1 hexane/ethyl acetate is used as the eluent; the product is visualized with a 254 nm UV lamp); mp 130–132 °C; IR (thin film) cm^{-1}: 3063, 1679, 1566, 1483, 1437, 1375, 1230, 1095, 845, 758, 698, 668; ^{1}H NMR (400 MHz, CDCl$_3$) δ: 7.27 – 7.32 (m, 2 H), 7.41 – 7.51 (m, 2 H), 7.52 – 7.61 (m, 3 H), 7.74 (ddd, J = 8.4, 6.6, 1.7 Hz, 1 H), 8.15 (d, J = 8.4 Hz, 1 H), 9.27 (s, 1 H); ^{13}C NMR (101 MHz, CDCl$_3$) δ: 96.30, 126.71, 127.38, 128.64, 128.66, 128.95, 129.05, 129.47, 129.70, 140.33, 147.16, 152.35, 156.59; HRMS *m/z* calcd for C$_{15}$H$_{10}$IN [M]$^+$, 330.9858, found 330.9854; Anal. calcd for C$_{15}$H$_{10}$IN: C, 54.41; H, 3.04; N, 4.23; found: C, 54.13; H, 3.14; N, 4.13.

Safety and Waste Disposal Information

All hazardous materials should be handled and disposed of in accordance with "Prudent Practices in the Laboratory"; National Academies Press; Washington, DC, 2011.

3. Discussion

The electrophile-promoted intramolecular cyclization of alkynes containing proximate nucleophilic functional groups is an efficient synthetic method for the construction of a wide variety of carbocycles and heterocycles.[3] This synthetic method provides a facile protocol for the synthesis of functionalized cyclic compounds by regioselective addition of the nucleophile and the electrophile across the alkyne carbon-carbon triple bond. The preparation of 3-iodo-4-phenylquinoline described here illustrates a general protocol for the iodine promoted *6-endo-dig* electrophilic cyclization of *N*-(2-alkynyl)anilines.[4] The present protocol for the synthesis of substituted quinolines includes three steps, (1) monoalkylation of the aniline amine functional group, (2) Sonogashira coupling of the *N*-propargylaniline with an aryl iodide and (3) iodine-promoted intramolecular cyclization. The monoalkylation of aniline has employed a literature procedure with a slight modification.[5] A traditional Sonogashira coupling protocol has been used for the second step.[6] The third step is based on the protocol developed by the Larock group with a slight modification of the reaction conditions.[4]

The present procedure provides a convenient approach for the synthesis of 3-iodoquinolines under mild reaction conditions. It's noteworthy that the iodo functional group offers considerable potential for further structure elaboration, using well-known, palladium-catalyzed coupling protocols, such as Suzuki-Miyaura coupling[7] and Buchwald-Hartwig amination.[8] Besides iodine, other electrophiles have also been successfully employed in the present cyclization protocol, including iodine monochloride (Table 1, entries 2 and 4) and phenylselenyl bromide (Table 1, entry 5). A variety of functional groups are readily accommodated under the present conditions, including ester, nitro, and amino groups. (Table 1, entries 10, 11 and 13).

302

Table 1. Synthesis of quinolines by electrophilic cyclization of N-(2-alkynyl)anilines.

entry	propargylic aniline	electrophile	product(s)	% yield
1	R = Ph	I_2[a]	R = Ph	76
2		ICl[b]		83
3	R = p-MeOC$_6$H$_4$	I_2[a]	R = p-MeOC$_6$H$_4$	71
4		ICl[b]		73
5	R = p-MeOC$_6$H$_4$	PhSeBr[c]	R = p-MeOC$_6$H$_4$	74
6	R = p-FC$_6$H$_4$	I_2[a]	R = p-FC$_6$H$_4$	78
7	R = p-MeCOC$_6$H$_4$	I_2[a]	R = p-MeCOC$_6$H$_4$	57
8	R = n-Bu	I_2[a]	R = n-Bu	43
9		I_2[a]		80
10		I_2[a]		88
11		I_2[a]		8+71
12		I_2[a]		75
13		I_2[a]		55+0

[a] Reactions were run under the following conditions: 0.3 mmol of the propargylic aniline, 3 equiv of I_2, and 2 equiv of NaHCO$_3$ in 3 mL of CH$_3$CN were stirred at room temperature. [b] To 0.3 mmol of propargylic aniline and 2 equiv of NaHCO$_3$ in 2 mL of CH$_3$CN was added 2 equiv of ICl in 1 mL of CH$_3$CN dropwise at room temperature. [c] To 0.3 mmol of propargylic aniline and 2 equiv of NaHCO$_3$ in 2 mL of CH$_3$CN was added 2 equiv of PhSeBr in 1 mL of CH$_3$CN dropwise at room temperature.

This quinoline synthesis presumably proceeds by the mechanism illustrated in Scheme 1: (1) the carbon-carbon triple bond of the propargylic aniline coordinates to an iodine cation generating an iodonium intermediate **A**, (2) intramolecular electrophilic aromatic substitution by the activated alkyne at the *ortho* position of the aniline forms dihydroquinoline **B**, and (3) in the presence of I_2 or air, the dihydroquinoline **B** is oxidized to the corresponding quinoline.

Scheme 1. Proposed mechanism for the iodine-promoted, electrophilic cyclization of *N*-(2-alkynyl)anilines.

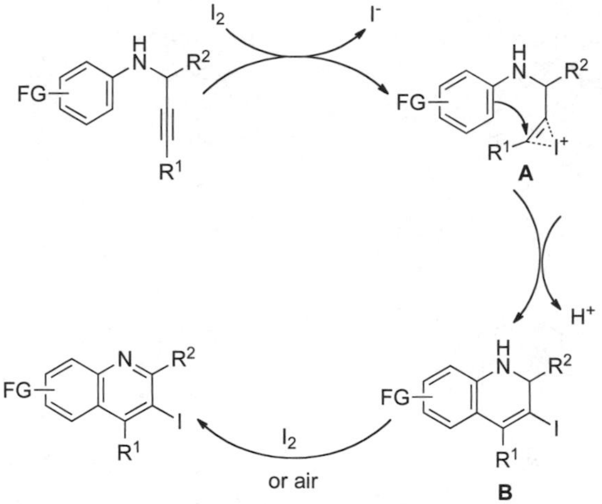

1. Department of Chemistry, Iowa State University, Ames, IA 50011; larock@iastate.edu. We gratefully acknowledge the National Institute of General Medical Sciences (GM 070620) for support of this research.
2. The non-corresponding authors' names are in alphabetical order.
3. For reviews, see: (a) Larock, R. C. In *Acetylene Chemistry. Chemistry, Biology, and Material Science;* Diederich, F.; Stang, P. J.; Tykwinski, R. R. Eds.; Wiley-VCH: New York, NY, 2005; Vol. 2, pp 51–99. (b) Godoi, B.; Schumacher, R. F.; Zeni, G. *Chem. Rev.* **2011**, *111*, 2937–2980.
4. (a) Zhang, X.; Campo, M. A.; Yao, T.; Larock, R. C. *Org. Lett.* **2005**, *7*, 763–766. (b) Zhang, X.; Yao, T.; Campo, M. A.; Larock, R. C. *Tetrahedron* **2010**, *66*, 1177–1187.
5. Marshall, J. A.; Wolf, M. A. *J. Org. Chem.* **1996**, *61*, 3238–3239.

6. (a) Sonogashira, K. In *Metal-Catalyzed Cross-Coupling Reactions;* Diederich, F.; Stang, P. J., Eds.; Wiley-VCH: Weinheim, Germany, 1998; Chapter 5, pp 203–229. (b) Sonogashira, K.; Tohda, Y.; Hagihara, N. *Tetrahedron Lett.* **1975**, *16*, 4467–4470.
7. (a) Miyaura, N.; Yamada, K.; Suzuki, A. *Tetrahedron Lett.* **1979**, *20*, 3437–3440. (b) For a review, see: Miyaura, N.; Suzuki, A. *Chem. Rev.* **1995**, *95*, 2457–2483.
8. (a) Guram, A. S.; Buchwald, S. L. *J. Am. Chem. Soc.* **1994**, *116*, 7901– 7902. (b) Paul, F.; Patt, J.; Hartwig, J. F. *J. Am. Chem. Soc.* **1994**, *116*, 5969–5970. For reviews, see: (c) Hartwig, J. F. *Acc. Chem. Res.* **1998**, *31*, 852–860. (d) Wolfe, J. P.; Wagaw, S.; Marcoux, J. F.; Buchwald, S. L. *Acc. Chem. Res.* **1998**, *31*, 805–818.

Appendix
Chemical Abstracts Nomenclature; (Registry Number)

Aniline; (62-53-3)
Propargyl bromide; (106-96-7)
Triethylamine; (121-44-8)
Iodobenzene; (591-50-4)
Bis(triphenylphosphine)palladium dichloride; (13965-03-2)
Copper iodide; (1335-23-5)
Iodine; (7553-56-2)
Sodium thiosulfate pentahydrate; (10102-17-7)

Richard C. Larock received his B.S. at the University of California, Davis in 1967. He then joined the group of Prof. Herbert C. Brown at Purdue University, where he received his Ph.D. in 1972. He worked as an NSF Postdoctoral Fellow at Harvard University in Prof. E. J. Corey's group and joined the Iowa State University faculty in 1972. His current research interests include aryne chemistry, electrophilic cyclization, palladium catalysis, and polymer chemistry based on biorenewable resources.

Yu Chen received his B.S. and M.S. degrees at Nankai University in China. He then joined Professor Andrei Yudin's research group at the University of Toronto working in the field of asymmetric catalysis, where he obtained his Ph.D. degree in 2005. After a two-year industrial appointment, he joined Professor Richard Larock's research group at Iowa State University as a postdoctoral fellow in 2007, working on an NIH-funded pilot-scale heterocyclic and carbocyclic library synthesis project. In 2009, he joined Queens College at City University of New York as an assistant professor. His research interests include late transition metal catalysis and asymmetric synthesis.

Anton V. Dubrovskiy received his Specialist (B.S./M.S.) degree at the Higher Chemical College of the Russian Academy of Sciences in Moscow, Russia in 2007. He then joined Professor Richard Larock's research group at Iowa State University, where he is currently pursuing his Ph.D. degree. His current research interest is in the development of new synthetic organic methodologies utilizing aryne intermediates.

Nicolas Armanino received his B.S. and M.S. degrees at the ETH Zürich, Switzerland in 2009. He then joined Professor Erick M. Carreira's research group at the same institution to pursue his Ph.D. degree. His current research is focused on the development of novel metal-catalyzed reaction cascades for the construction of complex organic molecules.

306

Discussion Addendum for:
Iridium-catalyzed Synthesis of Vinyl Ethers from Alcohols and Vinyl Acetate

Prepared by Yasushi Obora* and Yasutaka Ishii.*[1]
Original article: Hirabayashi, T.; Sakaguchi, S.; Ishii, Y. *Org. Synth.* **2005**, *82*, 55.

The synthesis of vinyl ethers by vinyl transfer from vinyl esters to alcohols constitutes the basis for much useful methodology. The originally reported iridium-catalyzed process provides a versatile and practical route to access vinyl alcohols.[2] This methodology was successfully expanded to facilitate allyl transfer from allyl acetates to alkyl alcohols, providing allyl ethers as products.[3] In addition, as a successful application of this protocol, the one-pot synthesis of γ,δ-unsaturated carbonyl compounds from allyl alcohols and vinyl acetates was achieved through in situ iridium-catalyzed formation of allylic vinyl ethers followed by Claisen rearrangement.[4] The above-mentioned advances will be summarized here.

Scope of Allylation of Alcohols from Allyl Acetates

Vinyl ethers and allyl ethers are important classes of compounds that have been used in polymer synthesis and in the pharmaceutical chemistry,[5] as well as starting materials for Claisen rearrangements,[6] cycloadditions,[7] hydroformylations,[8] and Mizoroki-Heck reactions,[9] etc.

We found that the iridium cationic complex $[Ir(cod)_2]^+BF_4^-$ catalyzed the allylation of alcohols with allyl acetate to afford allyl ethers as products.[3] For instance, the reaction of allyl acetate with *n*-octyl alcohol in the presence of a catalytic amount of $[Ir(cod)_2]^+BF_4^-$ complex afforded allyl octyl ether in quantitative yield (Figure 1).

Org. Synth. **2012**, *89*, 307-310
Published on the Web 2/6/2012

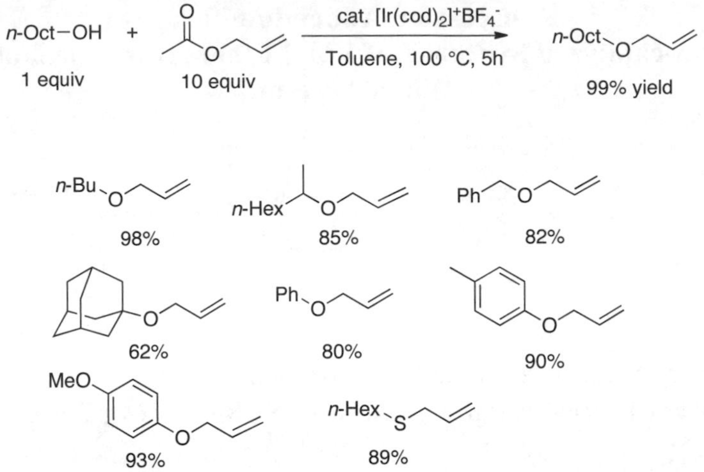

Figure 1. Ir-catalyzed reaction of *n*-octyl alcohol with allyl acetate to allyl octyl ether

Application of the Vinyl Transfer Methodology in Organic Synthesis

We previously reported the rearrangement of allyl homoallyl ethers to γ,δ-unsaturated aldehydes induced by the [IrCl(cod)]$_2$ complex.[10] Therefore, this protocol would serve to a useful route for the formation of γ,δ-unsaturated carbonyl compounds from allyl alcohols with vinyl or isopropenyl acetate through the formation of vinyl ethers as the key intermediate.[4] For example, the reaction of *trans*-2-methyl-3-phenyl-2-propen-1-ol with isopropenyl acetate in the presence of [IrCl(cod)]$_2$ catalyst combined with Cs$_2$CO$_3$ at 100 °C for 3 h followed by heating at 140 °C for 15 h afforded 5-methyl-4-phenyl-5-hexen-2-one in 83% yield (Figure 5). The reaction is thought to proceed through the Claisen rearrangement of the in situ generated allylic vinyl ether.

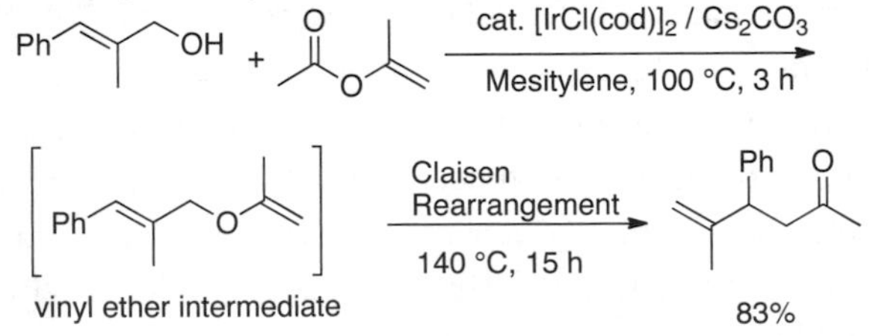

Figure 2. Ir-catalyzed one-pot synthesis of γ,δ-unsaturated carbonyl compounds

308

1. Department of Chemistry and Materials Engineering, Faculty of Chemistry, Materials and Bioengineering and ORDIST, Kansai University, Suita, Osaka 564-8680, Japan. E-mail: obora@kansai-u.ac.jp (YO), r091001@kansai-u.ac.jp (YI). This work was supported by a Grant-in-Aid for Scientific Research from the Ministry of Education, Culture, Sports, Science and Technology, Japan, the High-Tech Research Center Project for Private Universities, and the Strategic Project to Support the Formation of Research Bases at Private Universities, with a matching fund subsidy from the Ministry of Education, Culture, Sports, Science and Technology.

2. (a) Okimoto, Y.; Sakaguchi, S.; Ishii, Y. *J. Am. Chem. Soc.* **2002**, *124*, 1590–1591. (b) Ishii, Y.; Obora, Y.; Sakaguchi, S. *In Iridium Complex in Organic Synthesis*; Oro, L. A., Claver, C., Eds.; Wiley, New York, **2009**, 251–275. (c) Obora, Y.; Ishii, Y. *Synlett* **2011**, 30–51.

3. Nakagawa, H.; Hirabayashi, T.; Sakaguchi, S.; Ishii, Y. *J. Org. Chem.* **2004**, *69*, 3474–3477.

4. Morita, M.; Sakaguchi, S.; Ishii, Y. *J. Org. Chem.* **2006**, *71*, 6285–6286.

5. Yamaoka, T.; Watanabe, H. *J. Photopolym. Sci. Technol.* **2004**, *17*, 341–360.

6. Wei, X.; Lorenz, J. C.; Kapadia, S.; Saha, A.; Haddad, N.; Busacca, C. A.; Senanayake, C. H. *J. Org. Chem.* **2007**, *72*, 4250–4253 and references therein.

7. Esquivias, J.; Arrayas, R. G.; Carretero, J. C. *J. Am. Chem. Soc.* **2007**, *129*, 1480–1481 and references therein.

8. Kim, J. J.; Alper, H. *Chem. Commun.* **2005**, 3059–3061.

9. (a) *The Mizoroki Heck Reaction*,; Oestreich, M. Ed.; Wiley, Chichester, **2009**; (b) Obora, Y.; Ishii, Y. *Molecules* **2010**, *15*, 1487–1500.

10. Higashino, T.; Sakaguchi, S.; Ishii, Y. *Org. Lett.* **2000**, *2*, 4193–4195.

Yasutaka Ishii was born in Osaka, Japan, in 1941. He received his B.A. (1964) and M.S. (1967). In 1967, he was appointed Assistant Professor at Kansai University. In 1971, he received his Ph.D. degree under the supervision of Prof. Masaya Ogawa. He was a postdoctoral fellow with Professor Louis S. Hegedus at Colorado State University (1980-1981). In 1990 he was appointed as Professor at Kansai University. Since 2009, he has been an Emeritus Professor at Kansai University. His research interests include the development of practical oxidation reactions using molecular oxygen and hydrogen peroxide, homogeneous catalysis directed towards organic synthesis.

Yasushi Obora was born in 1969 in Shizuoka and received his B.Sc. (1991) and Ph.D. degree (1995) from Gifu University. After working as a postdoctoral fellow (1995-1997) at Northwestern University with Professor T. J. Marks, he moved to the National Institute of Materials and Chemical Research, AIST (1997-1999). In 1999, he moved to Catalysis Research Center at Hokkaido University as Research Associate. In 2006, he was appointed as an Associate Professor at Kansai University. His research interests include the development of new homogeneous catalysis and organometallic chemistry.

310

Discussion Addendum for:
Oxidation of Nerol to Neral with Iodosobenzene and TEMPO

Prepared by Francesca Leonelli,[1a] Roberto Margarita,[1b] and Giovanni Piancatelli.*[1c]
Original article: Piancatelli, G.; Leonelli, F. *Org. Synth.* **2006**, *83*, 18–23.

The oxidation of alcohols with IBD (iodosobenzene diacetate) with catalytic amounts of TEMPO (2,2,6,6-tetramethylpiperidin-1-oxyl), developed by Piancatelli, Margarita, and co-workers,[2] has become an established and effective protocol that has been continuously and successfully utilized, as reported in numerous literature reviews.[3] The methodology is characterized by a significantly high level of selectivity for the oxidation of primary alcohols to aldehydes[4] without over-oxidation to carboxylic acids, and a high chemoselectivity in the presence of secondary alcohols or other oxidizable functional groups.[5] The protocol has the additional benefits of being operationally simple and avoiding the use of heavy metals, odoriferous reagents, or expensive reagents.[3,5] Its robust nature is also described in a standardized protocol for the oxidation of nerol to neral in *Organic Syntheses*.[6]

Application in Total Synthesis

Beside the wide range usage of the Piancatelli/Margarita oxidation for classic reactions, it has become a standard method for efficient and clean conversion of alcohols to the aldehydes during the total synthesis of complex molecules. Some applications in this field will be highlighted.

A novel and chemoselective δ-lactonization was described in the stereocontrolled total synthesis of leucascandrolide A, a cytotoxic 18-membered macrolide (Scheme 1). This was conveniently achieved by selective oxidation of the primary alcohol to the lactol and further oxidation to generate the δ-lactone in 92% yield.[7]

Danishefsky completed the total synthesis of guanacastepene A by applying the highly selective properties of the IBD/TEMPO in the final step. The primary allyl alcohol was selectively oxidized in preference to the secondary one (Scheme 2).[8]

Scheme 1. Selective oxidation and δ-lactonization

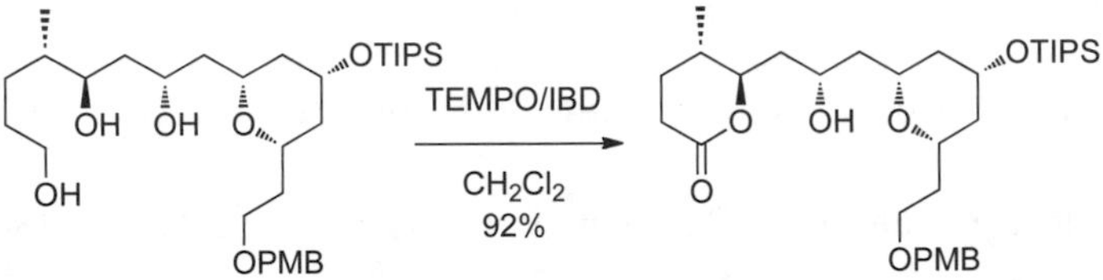

Scheme 2. Total Synthesis of guanacastepene A

guanacastepene A

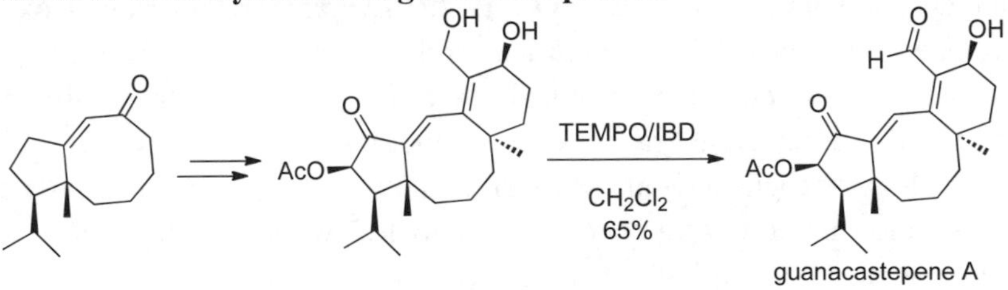

The TEMPO/IBD reagent combination allowed the oxidative lactonization of 1,6- and 1,7-diols, which formed seven- and eight-membered lactones, respectively, in good yields. These reactions have been performed in non-extreme-dilution conditions and on a large scale (up to 15 g) with high yields and without the formation of dimers or oligomers. In addition, the TEMPO/IBD-oxidative lactonization strategy avoided the use of protective groups, as well as separate oxidation steps. The TEMPO/IBD-mediated oxidative lactonization strategy was applied efficiently to the synthesis of isolaurepan (Scheme 3).[9]

Scheme 3. Total synthesis of isolaurepan.

isolaurepan

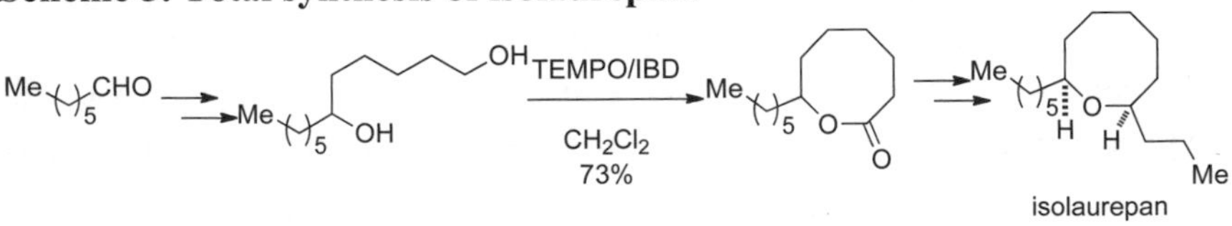

After testing several oxidation agents, the TEMPO/IBD protocol was selected for its efficiency and cleanliness in the preparation of (*R*)-3,4-dihydro-2*H*-pyran-2-carbaldehyde, a compound that is useful for the synthesis of potent adenosine agonist (Scheme 4).[10]

The selective oxidation of phenyl-substituted propargylic alcohols was attempted under many reaction conditions before the TEMPO/IBD methodology was identified as providing the best results in terms of stability, availability, reaction time and reaction yield. The aldehydes resulting from this reaction were used as building blocks in the preparation of porphyrins (Scheme 5).[11]

Scheme 4. Synthesis of (*R*)-3,4-dihydro-2H-pyran-2-carboxaldehyde.

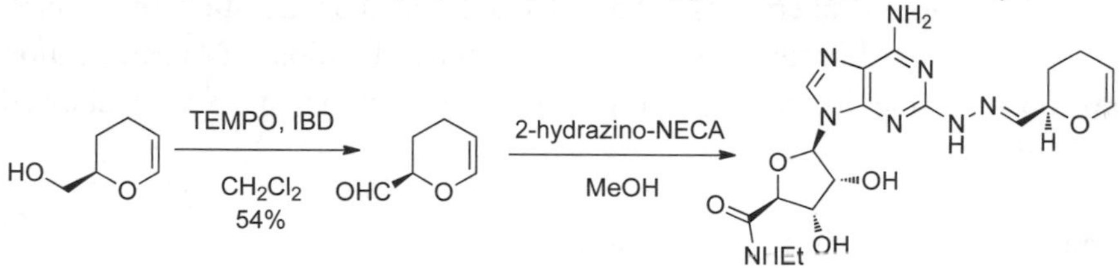

Scheme 5. Synthesis of *trans*-A₂B₂-porphyrins.

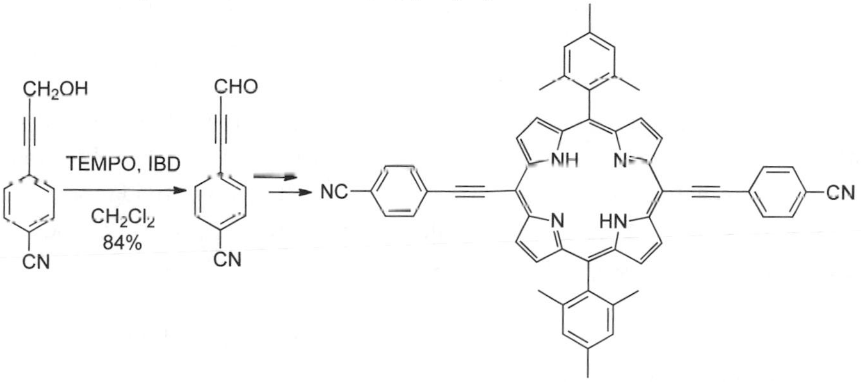

Capitalizing on the mild reaction conditions, the TEMPO/IBD procedure was successfully applied to the oxidation of a chiral primary alcohol. Other oxidants caused extensive and highly problematic epimerization of the aldehydes α-stereocenter. The resulting chiral aldehyde was a key intermediate in the synthesis of (+)-pumiliotoxin B (Scheme 6).[12]

Scheme 6. Total synthesis of (+)-pumiliotoxin B.

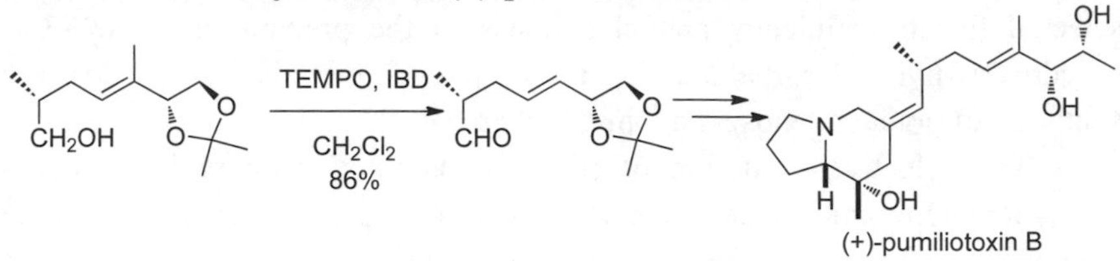

(+)-pumiliotoxin B

Industrial Applications

Paterson, together with Novartis Pharma, performed the oxidation of a polyol with TEMPO/IBD system to form an aldehyde in the course of a large-scale synthesis of the anti-cancer marine natural product (+)-discodermolide, (Scheme 7). The authors found that addition of a small amount of water resulted in the dramatic acceleration of the oxidation reaction and the formation of the product in high yields, which allowed scale-up of the reaction.[13]

Scheme 7. Synthesis of (+)-discodermolide.

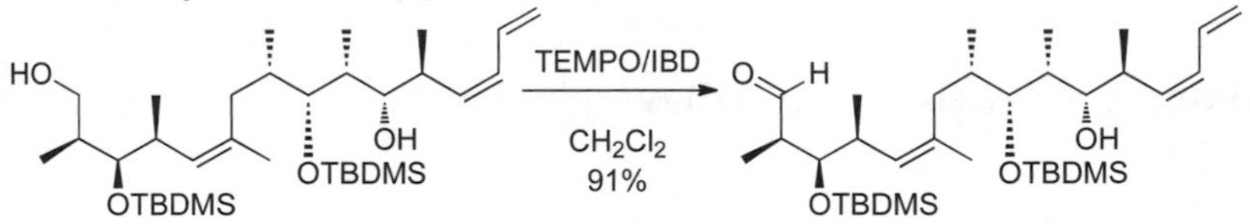

A group at Pfizer demonstrated that the oxidation process could be performed at a kg scale during their synthesis of the hepatitis C polymerase inhibitor (Scheme 8a).[14a] Similarly the IBD/TEMPO oxidation was employed industrially with a volatile aldehyde without workup and by telescoping to the next step (Scheme 8b).[14b]

Scheme 8. Multi-kilogram-scale synthetic application of the IBD/TEMPO oxidation methodology.

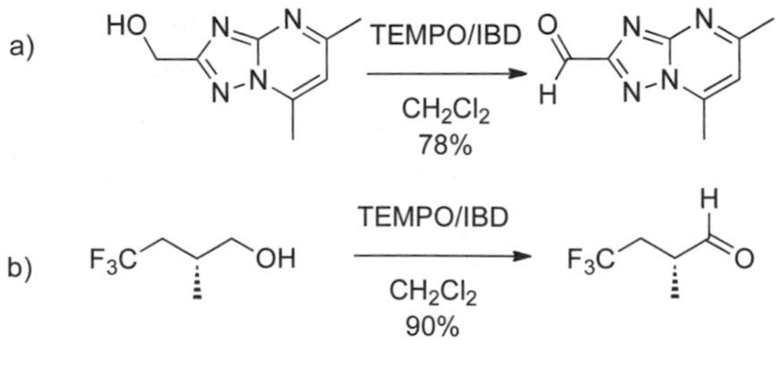

314

A Johnson and Johnson group, searching for a more scalable procedure for oxidation of a secondary alcohol, selected the TEMPO/IBD procedure (Scheme 9). For the product purification a simple switch of the solvent from dichloromethane to heptane, followed by cooling, produced a crystalline keto-derivative, an intermediate in the synthesis of (2'*R*)-2'-deoxy-2'-C-methyluridine, in 85-88.5% yield on scales that ranged from 100 g up to 10 kg.[15]

Scheme 9. Synthesis of (2'*R*)-2'-deoxy-2'-C-methyluridine.

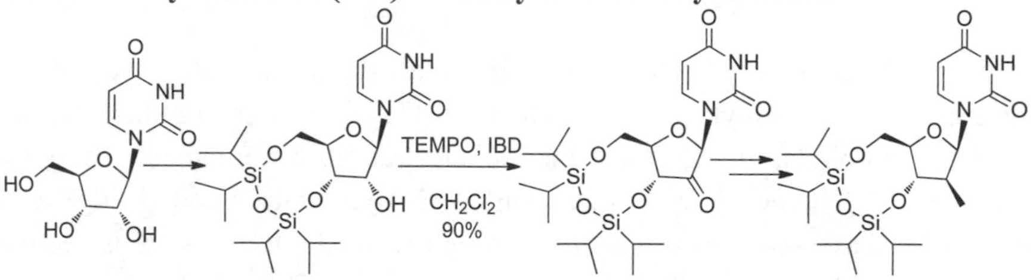

Development reactions

Interestingly, the mild reaction conditions and the tolerability of complex functional moieties have opened the route to combination reactions following the oxidation of alcohols. For example, Vatele has shown that the IBD/TEMPO system, in combination with stabilized phosphoranes, constitutes a mild and stereoselective one-pot method for the conversion of a variety of primary alcohols into their corresponding α,β-unsaturated esters (Scheme 10). The procedure was described by the author as having significant advantages over the existing ones for its chemoselectivity (oxidation of primary alcohols over secondary ones), general applicability to a broad range of complex molecules, and the employment of commercially available and safe reagents.[16]

Scheme 10. One-pot selective oxidation/olefination of primary alcohols.

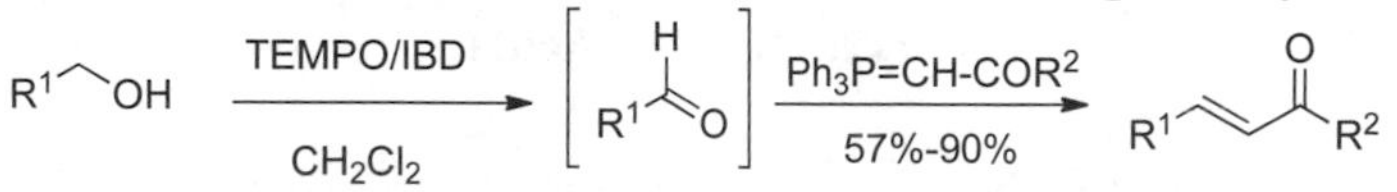

An additional example that demonstrates multiple one-pot reactions is the oxidation, under classical IBD/TEMPO conditions, of a primary alcohol in presence of an ynamide; the formation of the aldehyde triggers the ring-closing yne-carbonyl metathesis as a cascade reaction (Scheme 11).[17]

Scheme 11. One-pot oxidation/ring closing metathesis.

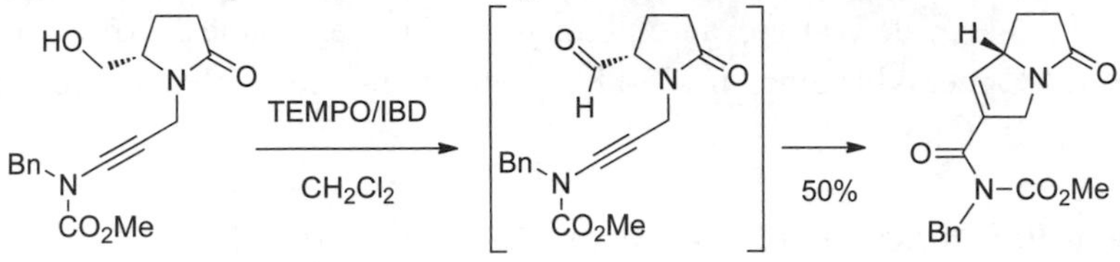

Modifications of the nitroxide catalyst, as in the use of 2-azaadamantane *N*-oxyl, have resulted in the ability to oxidize highly hindered alcohols, as well as the demonstration of enhanced catalytic efficiency (Scheme 12);[18a] meanwhile, C2-symmetric and bridgehead nitroxides have demonstrated that, together with IBD, stereoselective oxidations of secondary alcohols are at least conceptual feasibility.[18b]

Scheme 12. Oxidation with structurally less hindered class of nitroxyl radicals.

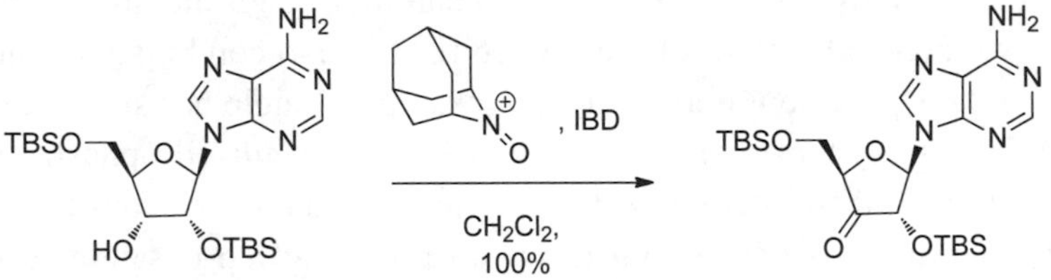

The simplicity of the reaction conditions have also triggered development of supported reagents (silica or polymer-based) or fluorous-tagged reagents, both for TEMPO and phenyliodoso acetate derivatives. The modifications improve the greenness of the oxidation process with more easily recoverable variations.[19]

Oxidation to Acids

Depending on the reaction conditions, primary alcohols can be oxidized to the corresponding carboxylic acids, usually in high yields.[3c] Observed initially by Epp and Widlansky in 1999,[20] this method has found its way into several natural product syntheses. For example, Sefton and coworkers reported that the oxidation of the primary alcohol with TEMPO

316

and IBD in aqueous acetonitrile led directly to the acid in very high yields (Scheme 13).[21]

Scheme 13. Oxidation of primary alcohol to carboxylic acid.

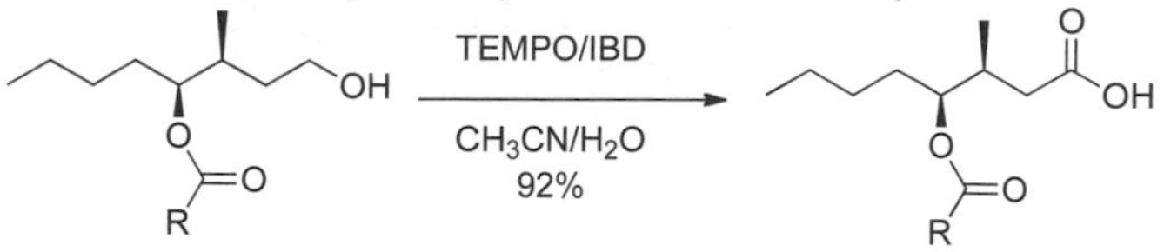

Van der Marel and his group disclosed TEMPO/IBD-mediated chemo- and regioselective oxidation of partially protected thioglycosides as a powerful means to obtain the corresponding thioglycuronic acids. The reaction was carried out in a mixture of dichloromethane and water (2:1) and proceeded in excellent yields. Interestingly, this oxidation protocol can be applied to alcohols containing thio and selenoethers (Scheme 14).[22]

Scheme 14. Oxidation of thioglycosides.

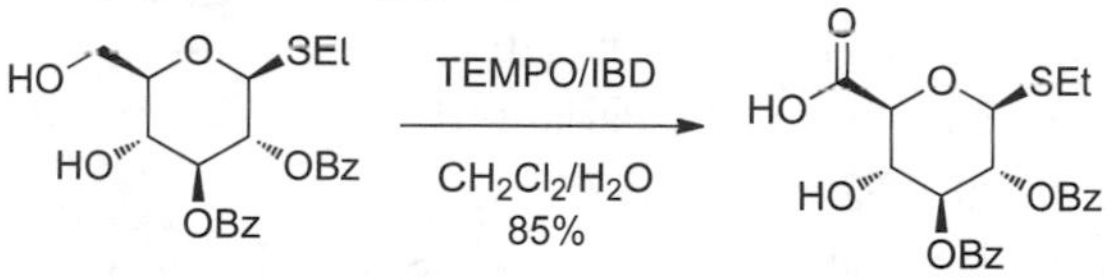

Optically active alcohols were oxidized to the corresponding acid with IBD in the presence of a catalytic amount of TEMPO in quantitative yields and without loss of enantiomeric purity (Scheme 15). These acids are key intermediates in the synthesis of several drugs, such as Tesaglitazar and Navaglitazar.[23]

After many attempts with other well-known oxidation procedures, the problematic oxidation of a primary alcohol to a carboxylic acid was achieved cleanly and in a good yield only when applying the TEMPO/IBD protocol. A series of 4,5-disubstituted γ-lactones, including whisky and cognac lactones (Scheme 16), were produced.[24] Further applications in this field are described in the reference 25.

Scheme 15. Stereospecific synthesis of Tesaglitazar and Navaglitazar precursors.

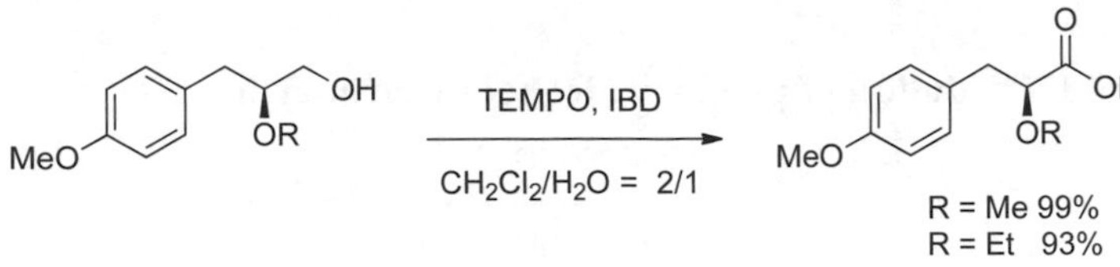

R = Me 99%
R = Et 93%

Scheme 16. Oxidation of primary alcohol to carboxylic acid.

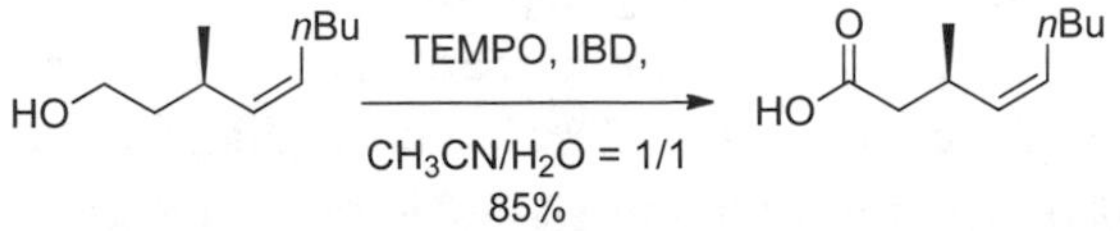

1. (a) Department of Chemistry, University "La Sapienza", P.le Aldo Moro, 5, 00185 Roma, Italy. (b) Corden Pharma, Via del Murillo Km 2.800 04013 Sermoneta-Latina Italy. (c) Retired, Private address: Via B.B. Spagnoli 57, 00144 Roma, Italy.

2. De Mico, A.; Margarita, R.; Parlanti, L.; Vescovi, A.; Piancatelli, G. *J. Org. Chem.* **1997**, *62*, 6974–6977.

3. For recent reviews see: (a) Adam, W; Saha-Moller, C. R.; Ganeshpure, P. A. *Chem. Rev.* **2001**, *101*, 3499–3548. (b) Zhdankin, V. V.; Stang, P. J. *Chem. Rev.* **2002**, *102*, 2523–2584. (c) Wirth, T. *Angew. Chem. Int. Ed.* **2005**, *44*, 3656–3665. (d) De Souza, M. V. N. *Mini-Rev. Org. Chem.* **2006**, *3*, 155–165. (e) Richardson, R. D.; Wirth, T. *Angew. Chem. Int. Ed.* **2006**, *45*, 4402–4404. (f) Zhdankin, V. V.; Stang, P. J. *Chem. Rev.* **2008**, *108*, 5299–5358. (g) Uyanik, M.; Ishihara, K. *Chem. Commun.* **2009**, 2086–2099. (h) Tebben, L.; Studer, A. *Angew. Chem. Int. Ed.* **2011**, *50*, 5034–5068.

4. *Inter alia*, see the references: (a) Nicolaou, K. C.; Barans, P. S.; Zhong, Y. -L.; Choi, H. -S.; Yoon, W. H.; He, Y. *Angew. Chem. Int. Ed.* **1999**, *38*, 1669–1675. (b) Jauch, J. *Angew. Chem. Int. Ed.* **2000**, *39*, 2764–2765. (c) Paterson, I.; Florence, G. J.; Gerlach, K.; Scott, J. P.; Sereinig, N. *J. Am. Chem. Soc.* **2001**, *123*, 9535–9544. (d) Nicolaou, K. C.; Baran, P. S. *Angew. Chem. Int. Ed.* **2002**, *41*, 2678–2720. (e) Paterson, I.; Delgado, O.; Florence, G. J.; Lyothier, I.; Scott, J. P.; Sereinig, N. *Org. Lett.* **2003**, *5*, 35–38. (f) Vong, B, G.; Kim, S. H.; Abraham, S.; Theodorakis, E. A. *Angew. Chem. Int. Ed.* **2004**, *43*, 3947–3951. (g)

318

Williams, D. R.; Kammler, D. C.; Donnell, A. F.; Goundry, W. R. F. *Angew. Chem. Int. Ed.* **2005**, *44*, 6715–6718. (h) Nicolaou, K. C.; Koftis, T. V.; Vyskocil, S.; Petrovic, G.; Tang, W.; Frederick, M. O.; Chen, D. Y. -K.; Li, Y.; Ling, T.; Yamada, Y. M. A. *J. Am. Chem. Soc.* **2006**, *128*, 2859–2872. (i) Paterson, I.; Ashton, K.; Britton, R.; Cecere, G.; Chouraqui, G.; Florence, G. J.; Stafford, J. *Angew. Chem. Int. Ed.* **2007**, *46*, 6167–6171. (l) Sarabia, F.; Martin-Galvez, F.; Garcia-Castro, M.; Chammaa, S.; Sanchez-Ruiz, A.; Tejon-Blanco, J. F. *J. Org. Chem.* **2008**, *73*, 8979–8986; (m) Muri, D.; Carreira, E. M. *J. Org. Chem.* **2009**, *74*, 8695–8712. (n) Fuwa, H.; Yamaguchi, H.; Sasaki, M. *Org. Lett.* **2010**, *12*, 1848–1851. (o) Butler, J. R.; Wang, C.; Bian, J.; Ready, J. M. *J. Am. Chem. Soc.* **2011**, *133*, 9956–9959.

5. For references, see for instance: (a) Hanessian S.; Claridge, S.; Johnstone, S. *J. Org. Chem.* **2002**, *67*, 4261–4274. (b) Paterson, I.; Tudge, M. *Angew. Chem. Int. Ed.* **2003**, *42*, 343–347. (c) Paterson, I.; Britton, R.; Delgado, O.; Meyer, A.; Poullennec, K. G. *Angew. Chem. Int. Ed.* **2004**, *43*, 4629–4633. (d) Muratake, H.; Natsume, M.; Nakai, H. *Tetrahedron* **2006**, *62*, 7093–7112. (e) Donohoe, T. J.; Chiu, J. Y. K.; Thpmas, R. F. *Org. Lett.* **2007**, *9*, 421–424. (f) Custar, D. W.; Zabawa, T. P.; Scheidt, K. A. *J. Am. Chem. Soc.* **2008**, *130*, 804–805. (g) Kleinbeck, F.; Carreira, E. M. *Angew. Chem. Int. Ed.* **2009**, *48*, 578–581. (h) Michalak, K.; Michalak, M.; Wicha, J. *J. Org. Chem.* **2010**, *75*, 8337–8350. (i) Yadav, J. S.; Pattanayak, M. R.; Das, P. P.; Mohapatra, D. K. *Org. Lett.* **2011**, *13*, 1710–1713.

6. Piancatelli G.; Leonelli F. *Org. Synth.* **2006**, *83*, 18–23. For reviews see: 3(f); Zhdankin, V. V., *Arkivoc* **2009**, 1–62.

7. Paterson, I.; Tudge, M. *Tetrahedron* **2003**, *59*, 6833–6849.

8. Siu, T.; Cox, C. D.; Danishefsky, S. J. *Angew. Chem. Int. Ed.* **2003**, *42*, 5629–5634.

9. Ebine, M.; Suga, Y.; Fuwa, H.; Sasaki, M. *Org. Biomol. Chem.* **2010**, *8*, 39–42.

10. Jagtap, P. G.; Chen, Z.; Koppetsch, K.; Piro, E.; Fronce, P.; Southan, G. J.; Klotz, K.-N. *Tetrahedron Lett.* **2009**, *50*, 2693–2696.

11. Nowak-Król, A.; Koszarna, B.; Yoo, S. Y.; Chromiński, J.; Węcławski, M. K.; Lee, C.-H.; Gryko, D. T. *J. Org. Chem.* **2011**, *76*, 2627–2634.

12. Manaviazar, S.; Hale, K. J.; LeFranc, A. *Tetrahedron Lett.* **2011**, *52*, 2080–2084.

13. Mickel, S. J.; Sedelmeier, G. H.; Niederer, D.; Schuerch, F.; Seger, M.; Schreiner, K.; Daeffler, R.; Osmani, A.; Bixel, D.; Loiseleur, O.; Cercus, J.; Stettler, H.; Schaer, K.; Gamboni, R.; Florence, G.; Paterson, I. *Org. Process Res. Dev.* **2004**, *8*, 113–121.

14. (a) Camp, D.; Matthews, C. F.; Neville, S. T.; Rouns, M.; Scott, R. W.; Truong, Y. *Org. Process Res. Dev.* **2006**, *10*, 814–821; (b) Wang, Z.; Resnick, L. *Tetrahedron* **2008**, *64*, 6440–6443.

15. Herrerias, C. I.; Zhang, T. Y.; Li, C.-J. *Tetrahedron Lett.* **2006**, *47*, 13–17.

16. Vatele, J.-M. *Tetrahedron Lett.* **2006**, *47*, 715–718.

17. Kurtz, K. C. M.; Hsung, R. P.; Zhang, Y. *Org. Lett.* **2006**, *8*, 231-234.

18. (a) Shibuya, M.; Tomizawa, M.; Suzuki, I.; Iwabuchi, Y. *J. Am. Chem. Soc.* **2006**, *128*, 8412–8413; (b) Graetz, B.; Rychnovsky, S.; Leu, W.-H.; Farmer, P.; Lin, R. *Tetrahedron: Asymmetry.* **2005**, *16*, 3584–3598.

19. (a) Moroda, A.; Togo, H. *Tetrahedron* **2006**, *62*, 12408–12414; (b) Holczknecht, O.; Cavazzini, M.; Quici, S.; Shepperson, I.; Pozzi, G. *Adv. Synth. Catal.* **2005**, *347*, 677–688; (c) Yamamoto, Y.; Kawano, Y.; Toy, P.H.; Togo, H. *Tetrahedron* **2007**, *63*, 4680–4687; (d) Subhani, M.A.; Beigi, M.; Eilbracht, P. *Adv. Synth. Catal.* **2008**, *350*, 2903–2909.

20. Epp, J. B.; Widlanski, T. S. *J. Org. Chem.* **1999**, *64*, 293–295.

21. Raunkjær, M.; Pedersen, D. S.; Elsey, G. M.; Mark A. Sefton, M. A.; Skouroumounis, G. K. *Tetrahedron Lett.* **2001**, *42*, 8717–8719.

22. (a) van den Bos, L. J.; Codée, J. D. C.; van der Toorn, J. C.; Boltje, T. J.; van Boom, J. H.; Overkleeft, H. S.; van der Marel, G. A. *Org. Lett.,* **2004**, *6*, 2165–2168. (b) Raunkjær, M.; El Oualid, F.; Gijs A. van der Marel, G. A.; Overkleeft, H. S.; Overhand, M. *Org. Lett.* **2004**, *6*, 3167–3170.

23. Brenna, E.; Fuganti, C.; Gatti, F. G.; Parmeggiani, F. *Tetrahedron: Asymmetry* **2009**, *20*, 2694-2698.

24. Adrio, L. A.; Hii, K. K. *Eur. J. Org. Chem.* **2011**, 1852–1857.

25. (a) Raunkjær, M.; El Oualid, F.; Gijs A. van der Marel, G. A.; Overkleeft, H. S.; Overhand, M. *Org. Lett.,* **2004**, *6*, 3167–3170; (b) Trost, B. M.; Dong, G. *Org. Lett.* **2007**, *9*, 2357–2359; (c) Rommel, M.; Enrst, A.; Koert, U. *E. J. Org. Chem.* **2007**, 4408–4430; (d) Vesely, J.; Rydner, L.; Oscarson, S. *Carb. Res.* **2008**, *343*, 2200–2208; (e) Chen, J.; Zhou, Y.; Chen, C.; Xu, W.; Yu, B. *Carbohydr. Res.* **2008**, *343*, 2853–2862; (f) Arungundram, S.; Al-Mafraji, K.; Asong, J.; Leach III,

320

F. E.; Amster, I. J.; Venot, A.; Turnbull, J. E.; Boons, G.-J. *J. Am. Chem. Soc.* **2009**, *131*, 17394–17405; (g) Yadav, J. S.; Bezawada, P.; Chenna, V. *Tetrahedron Lett.* **2009**, *50*, 3772–3775; (h) Reddy, G. V.; Kumar, R. S. C.; Sreedhar, E.; Babu, K. S.; Madhusudana Rao, J. *Tetrahedron Lett.* **2010**, *51*, 1723–1726; (i) Yadav, J. S.; Ramesh, K.; Subba Reddy, U. V.; Subba Reddy, B. V.; Ghamdi, A. A. K. A. *Tetrahedron Lett.* **2011**, *52*, 2943–2945. (j) Walvoort, M. T. C.; Moggré, G.-J.; Lodder, G.; Overkleeft, H. S.; Codée, J. D. C.; van der Marel, G. A. *J. Org. Chem.* **2011**, *76*, 7301–7315.

Giovanni Piancatelli was born in Roma and graduated in chemistry at the University "*La Sapienza*" of Roma in 1963. After postdoctoral activity, he joined the Research Center for the Chemistry of Natural Organic Compounds (1964, CNR), where he was Director from 1984 to 1993. He was appointed Associate Professor of Organic Chemistry at the Department of Chemistry of the University "*La Sapienza*" of Roma (1983) and then Full Professor of Organic Chemistry (1985). He retired on October 2009. His research interests include the synthesis of natural products, and the invention and development of new reagents and reactions in organic synthesis.

Roberto Margarita is the Business Development Director for Corden Pharma managing marketing, sales and strategic business planning. He previously held a similar position at Bristol Myers Squibb. He also developed a background in manufacturing as API Business Unit Director and in pharmaceutical development in the CMC area, including preparation of NDAs. His educational background includes a Ph.D. in Chemical Sciences from the University of Rome "la Sapienza", a Post-Doctoral experience at the Université de Montreal, Montreal, Quebec, Canada and a Diploma in General Management from the Henley Management College in Henley–UK. He is author of more than 20 publications.

Francesca Leonelli was born in Rome in 1974. She graduated in Chemistry at the University of Rome "*La Sapienza*" in 1999 and she received her Ph.D. in Chemistry in 2003 at the same university. Since 2008 she has been a researcher in the Chemistry Department in the same University. Francesca Leonelli's research is principally focused on the natural compound synthesis and on the preparation of molecules of biological interest.

Discussion Addendum for:
Diastereoselective Homologation of D-(*R*)-Glyceraldehyde Acetonide Using 2-(Trimethylsilyl)thiazole: 2-*O*-Benzyl-3,4-isopropylidene-D-erythrose

Prepared by Alessandro Dondoni*[1] and Alberto Marra.
Original article: Dondoni, A.; Merino, P. *Org. Synth.* **1995**, *72*, 21–31.

Introduction

Over the last several decades, numerous heterocycles have emerged as useful tools (precursors, reagents, chiral auxiliaries) in new methodologies for organic synthesis.[2] In that context, the five-membered heterocycle 1,3-thiazole attained considerable popularity by virtue of its stability under a wide range of reaction conditions, as well as its facile conversion into the formyl group by a very efficient three-step one-pot procedure.[3] The widespread use of the thiazole ring as masked formyl group led to the formulation of a general synthetic strategy that we have referred to as the "Thiazole-Aldehyde Synthesis".[4] Thus, the stereoselective synthesis of a chiral α-hydroxy aldehyde (*aldehydo*-D-erythrose) by one-carbon chain homologation of D-glyceraldehyde using 2-trimethylsilylthiazole (2-TST or TMST) as a formyl anion equivalent (see scheme above) represented the first example of diastereoselective thiazole-based synthesis reported from our laboratory in 1986. Notably, this reaction sequence was repeated over several consecutive cycles, with the result being that various chiral polyhydroxylated aldehydes (*aldehydo* sugars) with up to eight carbon atoms were prepared.[5] Moreover, this strategy was applied also to dialdoses to give, via elongation of the side chain, homologues with up to nine carbon atoms.[6] Thus, as 2-TST can be easily prepared by an efficient procedure (see the original paper in this journal), or can be purchased at reasonable price from a number of suppliers, its use in the Thiazole-Aldehyde Synthesis over the past 25 years became routine. Most notable is the fact that 2-TST reacts readily with aldehydes and other *C*-electrophiles (acyl chlorides, pyridinium chlorides, ketenes),[7] as well as sulfenyl halides,[8] without the need of any catalytic fluoride ion. These fluoride-free *C*- and *S*-desilylation processes had no precedents in silicon chemistry as they proceed by a special mechanism involving a thiazolium ylide intermediate.[9]

Scope of the 2-TST-Based Thiazole-Aldehyde Synthesis

The key role of 2-TST (often referred to as the *Dondoni reagent*) serving as a masked formyl anion equivalent has been reviewed in a comprehensive article[3] covering the period ranging from its first synthesis[10] in 1981 up to 2003. Nevertheless, some examples highlighting the role of 2-TST in synthetic routes leading to biologically active compounds are reported again below in Schemes 1-10 of this discussion addendum.

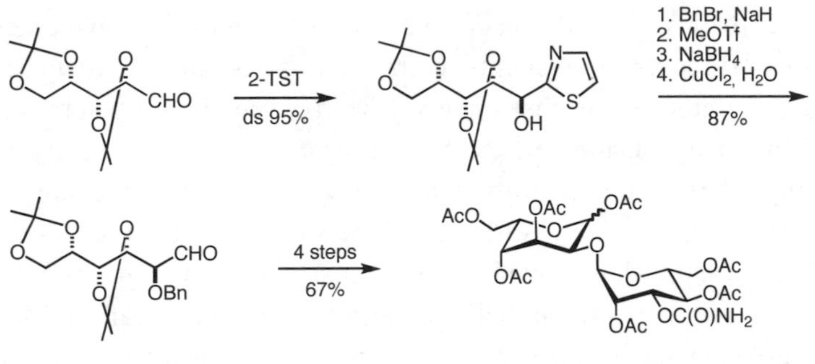

Scheme 1. Synthesis of the disaccharide subunit of bleomycin A_2 from *aldehydo*-L-xylose.[11]

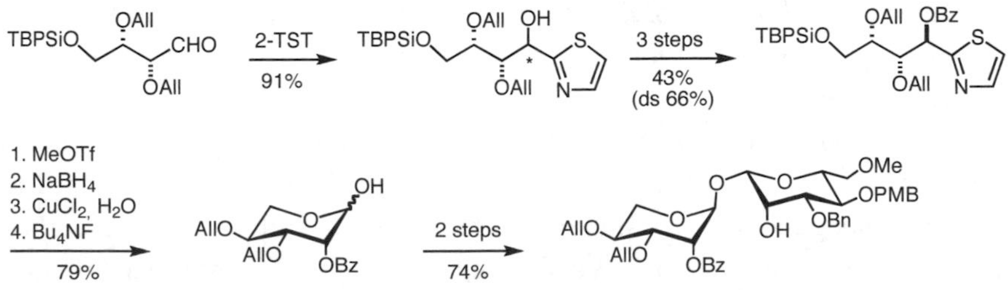

Scheme 2. Synthesis of the oligosaccharide portion of everninomicin 13,384-1 from *aldehydo*-L-threose.[12]

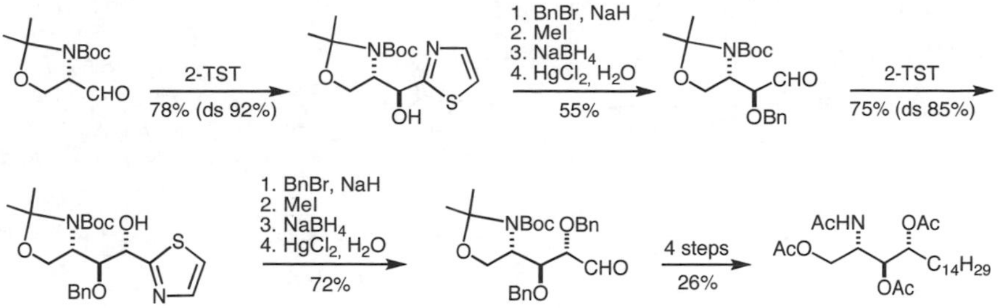

Scheme 3. Synthesis of an acetylated phytosphingosine from *N*-Boc serinal acetonide.[13]

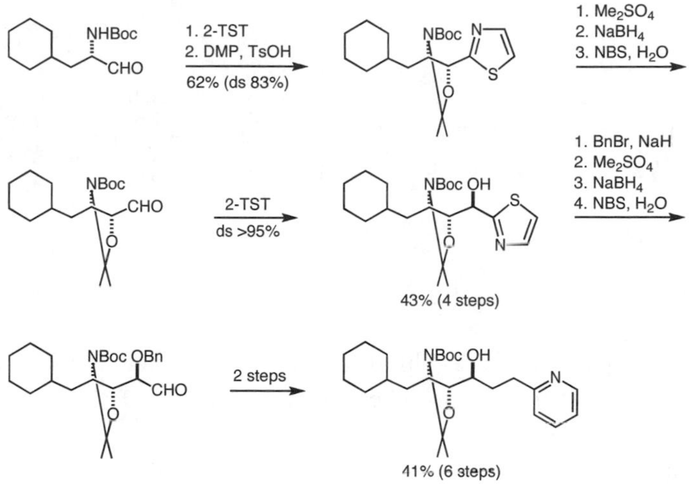

Scheme 4. Synthesis of a protected amino diol, a building block of renin inhibitors, from *N*-Boc cyclohexylalaninal.[14]

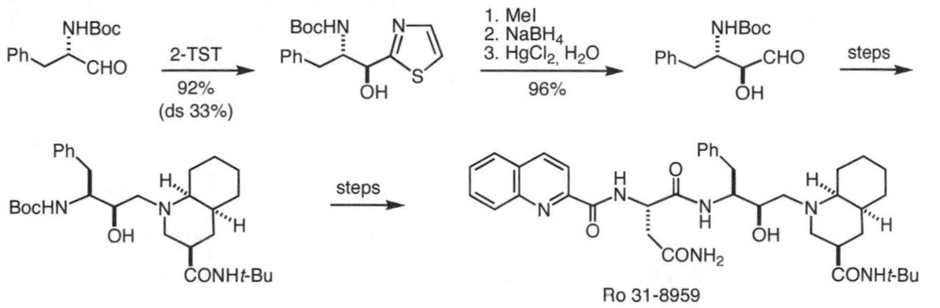

Scheme 5. Synthesis of a hydroxyethylamine isosteric dipeptide precursor of Saquinavir (Ro 31-8959) from *N*-Boc phenylalaninal (Hoffman-La Roche method).[15]

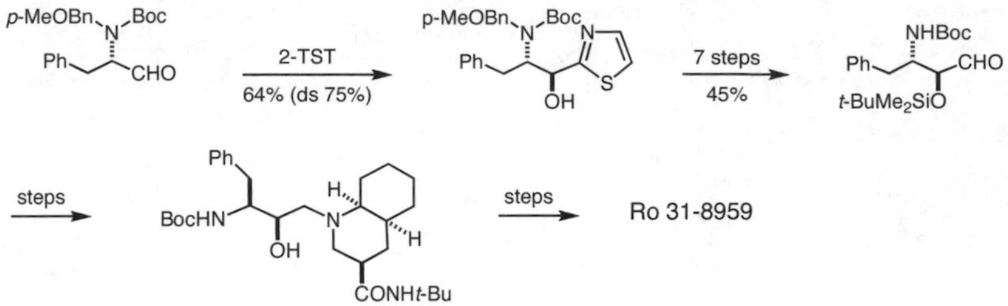

Scheme 6. Synthesis of a hydroxyethylamine isosteric dipeptide precursor of Saquinavir (Ro 31-8959) from N-Boc,N-PMB phenylalaninal (Dondoni method).[16]

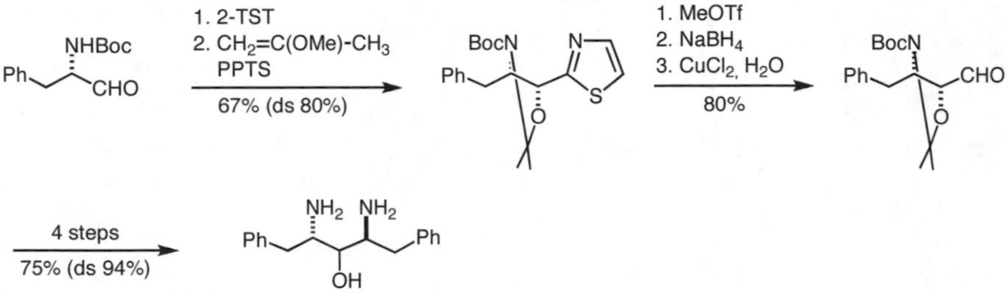

Scheme 7. Synthesis of a chiral pseudo C_2-symmetric 1,3-diamino-2-propanol derivative from N-Boc phenylalaninal.[17]

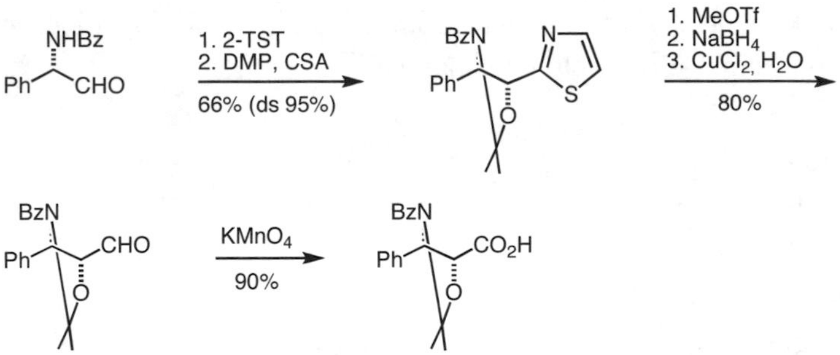

Scheme 8. Synthesis of N-benzoyl 3-phenylisoserine, viz the side chain of Taxol, from N-benzoyl phenylglycinal.[18]

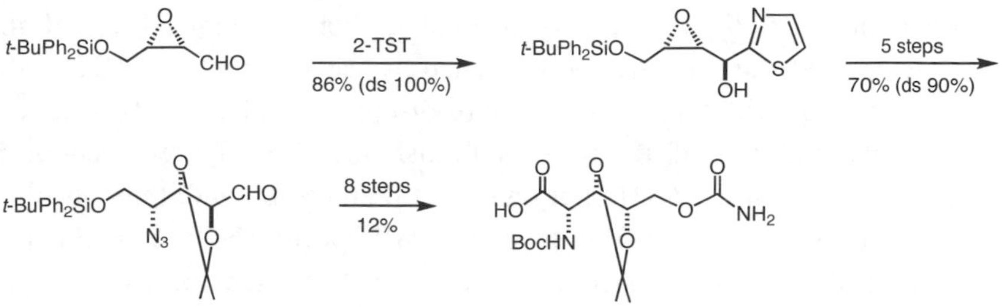

Scheme 9. Synthesis of *N*-Boc and acetonide protected 5-*O*-carbamoyl polyoxamic acid from chiral epoxy butanal.[19]

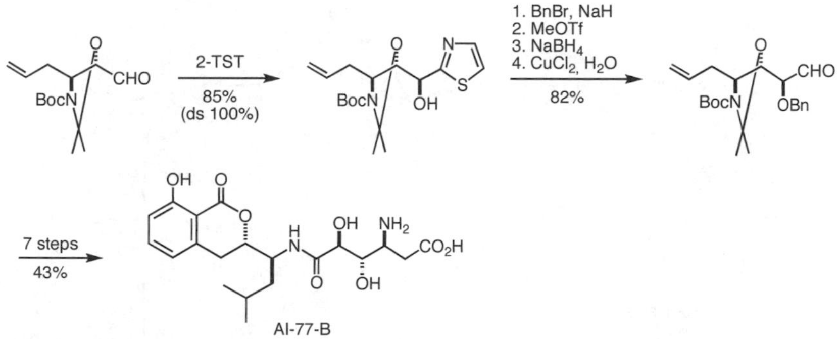

AI-77-B

Scheme 10. Synthesis of the amino acid side chain of the pseudopeptide microbial agent AI-77-B from a formyl oxazolidine.[20]

Even with the development of new research methods, the Thiazole-Aldehyde Synthesis based on the use of 2-TST continued to be an attractive synthetic tool. Thus, recent applications of this chemistry that have been carried out in our own and other laboratories are reported below.

In the context of a study on the synthesis of *C*-fucosides of biological relevance, we set out to develop the syntheses of *R*- and *S*-epimeric *C*-fucosyl hydroxyphenyl acetates[21] (Scheme 11). These glycoconjugates were expected to constitute the glycosidic part of new antibiotics similar to existing natural products.[22] The β-L-*C*-fucosyl aldehyde **1**, which was prepared in our laboratory, was utilized as the starting material for the synthesis of both epimers. The chiral alcohols **4** and **5** with *R* and *S* configuration at quaternary carbon atoms, respectively, were produced by using essentially the same reagents, but employing them in an inverse order. For example, aldehyde **1** was first reacted with phenylmagnesium bromide (PhMgBr) and the resulting mixture of diastereomeric alcohols was oxidized to give the ketone **2**. Addition of 2-lithiothiazole (2-LTT) to this ketone afforded the *R*-configured tertiary alcohol **4**. For the preparation of the alternate epimer, the sequence was initiated by the addition of 2-TST to the aldehyde **1** and oxidation of the mixture of epimeric alcohols to ketone **3** followed by addition of PhMgBr to furnish the *S*-configured alcohol **5**. Having both stereoisomeric tertiary alcohols **4** and **5** in hand, their absolute configuration was assigned from their ^{1}H NMR spectra (NOE experiments). Finally, both thiazolyl alcohols were readily transformed in the corresponding aldehydes **6** and **7** by the standard thiazole-to-formyl unmasking protocol and these aldehydes were oxidized to the representative esters **8** and **9**.

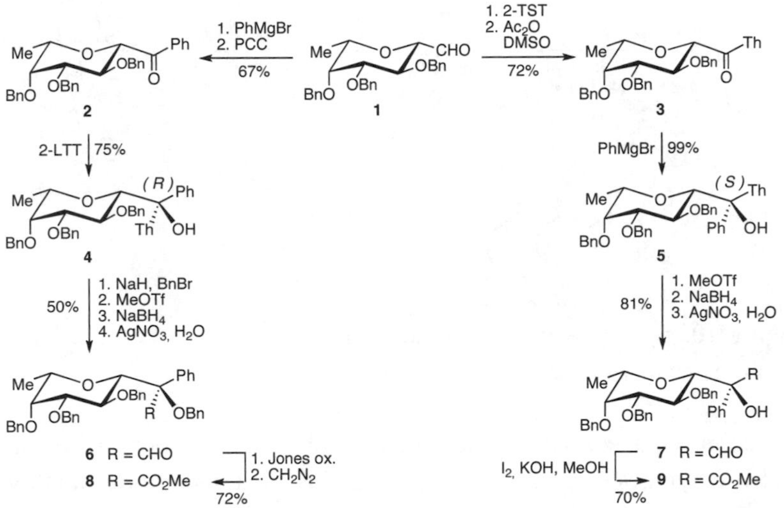

Scheme 11. Syntheses of *R*- and *S*-epimer *C*-fucosyl hydroxyphenyl acetaldehydes and acetates.

Another paper on the use of 2-TST in the Thiazole-Aldehyde Synthesis that appeared in mid-2005 was reported by Alcaide, Almendros, and their co-workers.[23] These authors developed a synthetic route leading to azabicyclo[4.3.0]nonane (indolidizinone) amino esters from β-lactams (Scheme 12) by using α-alkoxy β-lactam acetaldehydes as key intermediates. Interest in the synthesis of indolizidinone amino acids stemmed from their behavior as conformationally restricted dipeptide mimetics. Thus, the readily available enantiopure 4-oxoazetidine-2-carbaldehydes **10** and **11** were transformed into the one-carbon higher homologues via stereoselective addition of 2-TST to give the thiazole

Org. Synth. **2012**, *89*, 323-333

derivatives **12-13** and unmasking of the formyl group from the thiazole ring (Scheme 12). In all cases, well established conditions[4] for 2-TST addition to aldehydes, as well as for thiazole cleavage, were employed resulting in the isolation of the α-alkoxy β-lactam acetaldehydes **14** and **15** in good yields. These aldehydes were transformed into the target indolizidinone amino esters **16** and **17** via standard heterocyclic chemistry. A similar 2-TST-based strategy was followed by the same research group for the conversion of enantiopure *cis*-4-formyl-1-(3-methyl-2-butenyl)-3-methoxy-β-lactam into the corresponding higher homologue, an α-silyloxylated aldehyde that served as key intermediate for the synthesis of a mixture of diastereomeric dihydroxycarbacephams via an intramolecular carbonyl-ene reaction.[24] It was emphasized in both instances that the addition of 2-TST to 4-oxoazetidine 2-carbaldehydes to give the α-hydroxyalkylthiazoles was not a trivial process as the same authors had previously observed the reaction of 2-TST with *N*-aryl-4-formyl-β-lactams to give enantiopure α-alkoxy-γ-keto acid derivatives via N1-C4 bond cleavage of the β-lactam ring.[25]

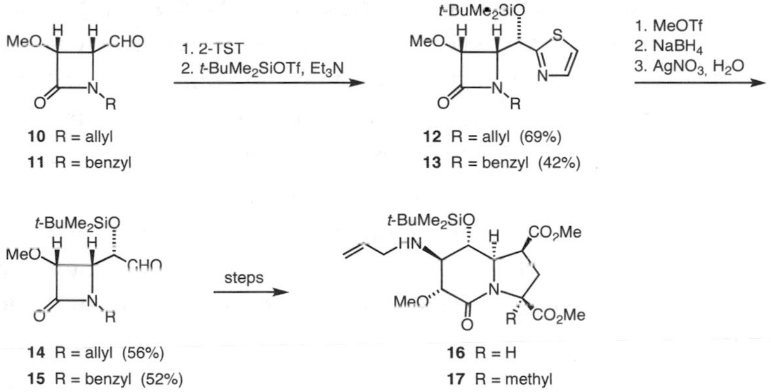

Scheme 12. Synthesis of α-alkoxy β-lactam acetaldehydes from 4-oxoazetidine-2-carbaldehydes en route to azabicyclo[4.3.0]nonane (indolidizinone) amino esters.

A fourth paper on the use of 2-TST was reported in 2007 by Cateni and co-workers in the effort to prepare a natural cerebroside from Euphorbiaceae.[26] These authors took advantage of the availability of the *O*- and *N*-protected amino sugar **18** (L-*ribo* configuration) that was previously prepared in our laboratory by 2-TST-based homologation of the Garner aldehyde.[13] Thus, the *aldehydo* sugar **18** was transformed into the one-carbon higher homologue **19** by the standard 2-TST-based methodology

(Scheme 13). Wittig reaction of this *aldehydo* sugar (L-*allo* configuration) with a long chain alkyl phosphorane afforded under suitable conditions the Z-olefin **20** displaying a chiral polar head and a lipophilic tail. Compound **20** was, in fact, the required C_{18}-sphingosine, which after glycosylation and acylation at the polar head, led to the targeted natural cerebroside.

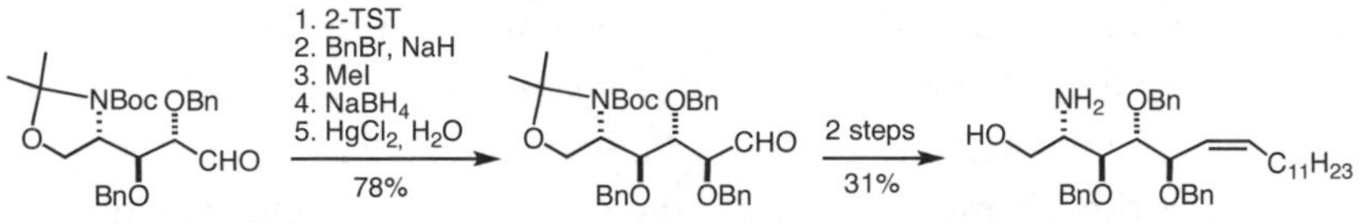

Scheme 13. Synthesis of a C_{18}-sphingosine intermediate in the synthesis of a natural cerebroside.

Conclusion

From the selected examples reported in the above schemes, it appears that since its discovery in 1981, 2-TST has found use as a formyl anion equivalent in various synthetic routes leading to compounds of biological relevance. The fidelity of this reagent to react *spontaneously* with aldehydes bearing a wide variety of substituents has been amply demonstrated. It has to be mentioned, however, that the synthetic utility of 2-TST extends beyond its use in the Thiazole-Aldehyde Synthesis, as in recent years numerous applications of this reagent have appeared in thiazole chemistry[27] and in synthetic routes leading to thiazole-containing compounds of biological relevance.[28]

1. Dipartimento di Chimica, Laboratorio di Chimica Organica, Università di Ferrara, Via L. Borsari 46, 44100 Ferrara, Italy. adn@unife.it
2. Meyers, A. I. In *Heterocyles in Organic Synthesis*; Taylor, E. C., Weissberger, A., Eds; Wiley: New York, 1974.
3. Dondoni, A. *Org. Biomol. Chem.* **2010**, *8*, 3366–3385.
4. Dondoni, A.; Marra, A. *Chem. Rev.* **2004**, *104*, 2557–2599.
5. Dondoni, A.; Fantin, G.; Fogagnolo, M.; Medici, A. *Angew. Chem. Int. Engl.* **1986**, *25*, 835–837

6. (a) Dondoni, A.; Fantin, G.; Fogagnolo, M.; Medici A.; Pedrini, P. *J. Org. Chem.* **1989**, *54*, 693–702. (b) Dondoni, A.; Fantin, G.; Fogagnolo, M.; Medici, A. *Tetrahedron* **1987**, *43*, 3533–3539.

7. (a) Dondoni, A.; Fantin, G.; Fogagnolo; Medici, A.; Pedrini, P. *J. Org. Chem.* **1988**, *53*, 1748–1761. (b) Dondoni, A. *Pure Appl. Chem.* **1990**, *62*, 643–652.

8. Munavalli, S.; Rohrbaugh, D. K.; Rossman, D. I.; Durst, H. D.; Dondoni, A. *Phosphorus, Sulfur, Silicon* **2002**, *177*, 2465–2470.

9. Dondoni, A.; Douglas, A. W.; Shinkai, I. *J. Org. Chem.* **1993**, *58*, 3196–3200.

10. Medici, A.; Pedrini, P.; Dondoni, A. *J. Chem. Soc., Chem Commun.* **1981**, 655–656.

11. Dondoni, A.; Marra, A.; Massi, A. *J. Org. Chem.* **1997**, *62*, 6261–6267.

12. (a) Nicolaou, K. C.; Rodriguez, R. M.; Fylaktakidou, K. C.; Suzuki, H.; Mitchell, H. J. *Angew. Chem., Int. Ed.* **1999**, *38*, 3340–3345. (b) Nicolaou, K. C.; Mitchell, H. J.; Fylaktakidou, K. C.; Rodriguez, R. M.; Suzuki, H. *Chem. Eur. J.* **2000**, *6*, 3116–3148.

13. (a) Dondoni, A.; Fantin, C.; Fogagnolo, M.; Pedrini, P. *J. Org. Chem.* **1990**, *55*, 1439-1446. (b) Dondoni, A.; Perrone, D. *Org. Synth.* **1999**, *77*, 64–77.

14. Wagner, A.; Mollath, M. *Tetrahedron Lett.* **1993**, *34*, 619–622.

15. Parkes, K. E. B.; Bushnell, D. J.; Crackett, P. H.; Dunsdon, S. J.; Freeman, A. C.; Gunn, M. P.; Hopkins, R. A.; Lambert, R. W.; Martin, J. A.; Merrett, J. H.; Redshaw, S.; Spurden, W. C.; Thomas, G. J. *J. Org. Chem.* **1994**, *59*, 3656–3664.

16. Dondoni, A.; Perrone, D.; Merino, P. *J. Org. Chem.* **1995**, *60*, 8074–8080.

17. Dondoni, A.; Perrone, D.; Rinaldi, M. *J. Org. Chem.* **1998**, *63*, 9252–9264.

18. Dondoni, A.; Perrone, D.; Semola, M. T. *Synthesis* **1995**, 181–186.

19. Dehoux, C.; Monthieu, C.; Baltas, M.; Gorrichon, L. *Synthesis* **2000**, 1409–1414.

20. (a) Ghosh, A. K.; Bischoff, A.; Capiello, J. *Org. Lett.* **2001**, *3*, 2677–2680. (b) Ghosh, A. K.; Bischoff, A.; Capiello, J. *Eur. J. Org. Chem.* **2003**, 821–832.

21. Dondoni, A.; Catozzi, N.; Marra, A. *J. Org. Chem.* **2004**, *69*, 5023–5036.

22. Pasetto, P.; Frank, R. W. *J. Org. Chem.* **2003**, *68*, 8042–8060.

23. Alcaide, B.; Almendros, P.; Redondo, M. C.; Ruiz, M. P. *J. Org. Chem.* **2005**, *70*, 8890–8894.

24. Alcaide, B.; Almendros, P.; Pardo, C.; Rodríguez-Ranera, C.; Rodríguez-Vicente, A. *Chem. Asian, J.* **2009**, *4*, 1604–1611.

25. Alcaide, B.; Almendros, P.; Redondo, M. C. *Org. Lett.* **2004**, *6*, 1765–1767.

26. Cateni, F.; Zacchigna, M.; Zilic, J.; Di Luca, G. *Helv. Chim. Acta* **2007**, *90*, 282–290.

27. (a) Zanirato, P.; Cerini, S. *Org. Biomol. Chem.* **2005**, *3*, 1508–1513. (b) Mamada, M.; Nishida, J.; Kumaki, D.; Tokito, S.; Yamashita, Y. *Chem. Mater.* **2007**, *19*, 5404–5409. (c) Zambon, A.; Borsato, G.; Brussolo, S.; Frascella, P.; Lucchini, V. *Tetrahedron Lett.* **2008**, *49*, 66–69. (d) Primas, N.; Bouillon, A.; Lancelot, J.-C.; El-Kashef, H.; Rault, S. *Tetrahedron* **2009**, *65*, 5739–5746. (e) Foerster, T.; Wann, D. A.; Robertson, H. E.; Rankin, D. W. H. *Dalton Trans.* **2009**, 3026–3033. (f) Daniels, M. H.; Hubbs, J. *Tetrahedron Lett.* **2011**, *52*, 3543–3546. (g) Amati, M.; Belviso, S.; D'Auria, M.; Lelj, F.; Racioppi, R.; Viggiani, L. *Eur. J. Org. Chem.* **2010**, 3416–3427.

28. (a) Costanzo, M. J.; Almond, H. R., Jr.; Hecker, L. R.; Schott, M. R.; Yabut, S. C.; Zhang, H.-C.; Andrade-Gordon, P.; Corcoran, T. W.; Giardino, E. C.; Kauffman, J. A.; Lewis, J. M.; de Garavilla, L.; Haertlein, B. J.; Maryanoff, B. E. *J. Med. Chem.* **2005**, *48*, 1984–2008. (b) Margiotta, N.; Ostuni, R.; Ranaldo, R.; Denora, N.; Laquintana, V.; Trapani, G.; Liso, G.; Natile, G. *J. Med. Chem.* **2007**, *50*, 1019–1027. (c) Spadafora, M.; Mehiri, M.; Burger, A.; Benhida, R. *Tetrahedron Lett.* **2008**, *49*, 3967–3971. (d) Faragalla, J.; Heaton, A.; Griffith, R.; Bremner, J. B. *Synlett* **2009**, 306–309. (e) Pedras, M. S. C.; Minic, Z.; Sarma-Mamillapalle, V. K. *J. Agric. Food Chem.* **2009**, *57*, 2429–2435.

Alessandro Dondoni was born and grew up in Ostiglia, Italy. He received a degree (laurea) in industrial chemistry from the University of Bologna in 1960. After postdoctoral studies at the IIT in Chicago with Sidney I. Miller, he returned to the University of Bologna in 1963 where he began his independent scientific career. In 1973 he joined the University of Ferrara and was named Professor of organic chemistry in 1975. He retired in 2008 and at present (2012) is Senior Research Scientist in the same University. At the beginning of his career he was involved in studies of the kinetics and mechanism of 1,3-dipolar cycloaddition reactions and physicochemical properties of sulfur compounds, while more recently he has focused on the development of new synthetic methods for heterocyclic and carbohydrate chemistry.

Alberto Marra graduated in Pharmaceutical Sciences from the University of Pisa (Italy) and obtained his Ph.D. degree in Organic Chemistry from the University Paris VI (France) under the supervision of Prof. P. Sinaÿ. After a postdoctoral fellowship at the University of Zurich (Switzerland) with Prof. A. Vasella, he was appointed to a lectureship in Organic Chemistry at the University of Ferrara, where he joined the group of Prof. A. Dondoni. In 1998 he was promoted to the position of Associate Professor at the same university. He was Visiting Professor at the *Université de Picardie* (2008) and the *Université de Cergy-Pontoise* (2010).

Discussion Addendum for:
Bicyclopropylidene

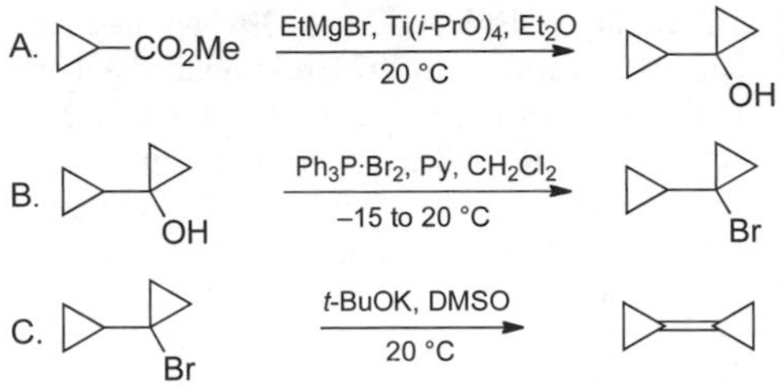

Prepared by Armin de Meijere* and Sergei I. Kozhushkov.[1]
Original article: de Meijere, A.; Kozhushkov, S. I.; Späth, T. *Org. Synth.*
2002, *78*, 142–151.

This facile and easily scalable synthesis[2] has made bicyclopropylidene
(BCP), a uniquely strained, tetrasubstituted alkene, accessible in bulk
quantities. Thus, various applications of BCP as a multifunctional C_6
building block, and in mechanistic terms, for the understanding of certain
reaction principles,[3] have been developed over the past decade.

Preparation and Functionalization

The method shown above has been extended to the preparation of
ring-annelated BCPs.[4] The best approach to functionally monosubstituted
BCPs is by deprotonation with *n*-butyllithium in THF at 0 °C followed by
electrophilic substitution of the lithiobicyclopropylidene with appropriate
reagents.[5] Some recent examples[6–8] are shown in Figure 1, while the
previously published ones are summarized in reviews.[3b,c]

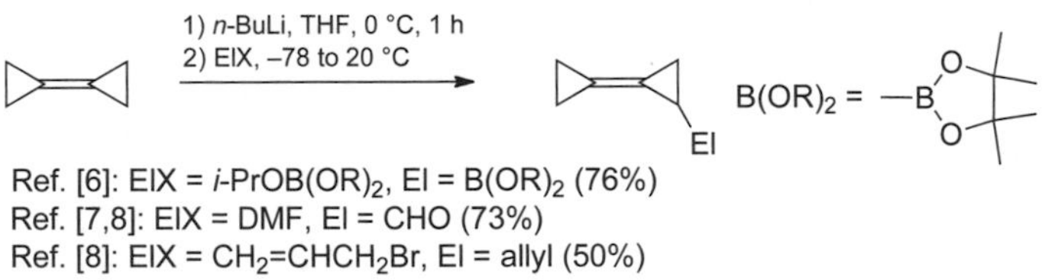

Ref. [6]: EIX = *i*-PrOB(OR)₂, EI = B(OR)₂ (76%)
Ref. [7,8]: EIX = DMF, EI = CHO (73%)
Ref. [8]: EIX = CH₂=CHCH₂Br, EI = allyl (50%)

Figure 1. Functionalization of BCP.[6–8]

On the other hand, oligosubstituted, as well as oligospirocyclo-
propanated BCPs, can better be prepared according to Neuenschwander et

334

Published on the Web 2/10/2012

al.[9] by dehalogenative coupling of 1-halo-1-lithiocyclopropanes generated from 1,1-dihalocyclopropanes upon treatment with alkyllithium reagents in the presence of copper(II) salts (see below).

Chemical Transformations

I. Isomerizations

BCP undergoes a clean rearrangement to methylenespiropentane when passed through a hot tube at 350 °C.[3a,c,e] In view of the ready availability of BCP, this rearrangement constitutes the most convenient preparative approach to methylenespiropentane.[8] When heated in a closed vessel as a pure compound or in solution (50% in toluene), a substantial fraction of BCP dimerizes to yield [4]rotane.[10] Theoretical aspects of the thermal,[3e,8,11] as well as the photochemical,[12] isomerizations of BCP have also been studied in detail.

II. Additions onto the Double Bond

While bromine addition to BCP and spirocyclopropanated BCPs is mainly of theoretical importance,[13] hydrogenation of substituted BCPs has been used several times to prepare specific compounds. Thus, a series of ω-trans-(bicyclopropyl)-substituted fatty acids for biological studies has been synthesized from BCP.[14] The key step in these preparations was a

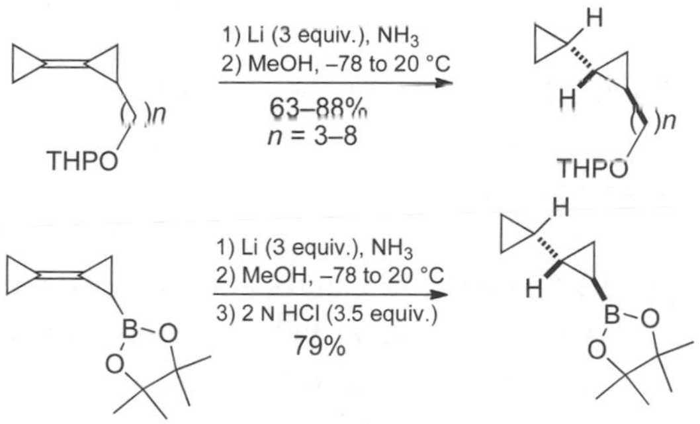

Figure 2. Reduction of the double bond in BCPs to yield bicyclopropyls.[14]

reasonably diastereoselective Birch reduction of the bicyclopropylidene moiety with lithium in liquid ammonia to afford THP-protected trans-(bicyclopropyl)alkanol derivatives in 63–88% isolated yields (Figure 2). 2-(trans-2'-Cyclopropylcyclopropyl)-4,4,5,5-tetramethyl-1,3-dioxa-2-borolane – an important intermediate for palladium-catalyzed cross-coupling reactions

– was also diastereoselectively prepared in 79% yield applying such a Birch reduction (Figure 2).[6]

On the other hand, hydrogenation of bicyclopropylidenes with diimide, in situ generated from 2-nitrobenzenesulfonyl hydrazide, furnished predominantly *cis*-bicyclopropyl derivatives. On the basis of these two reduction methods, five of the six possible diastereomeric [1,1';2',1";2",1"']quatercyclopropanes were obtained from diastereomeric *meso*- and *d,l*-bis(bicyclopropylidenyl).[15]

1-Cyclopropylcyclopropaneboronate – a starting material for the synthesis of derivatives of the skeletally interesting [1,1';1',1";1",1"']quater- and [1,1';1',1";1",1"';1"',1""] quinquecyclopropan-1-ol by a Matteson-type homologation[16] on dibromocyclopropane – could be obtained from the hydroboration product of BCP, yet, in only 27% yield (Figure 3).[17]

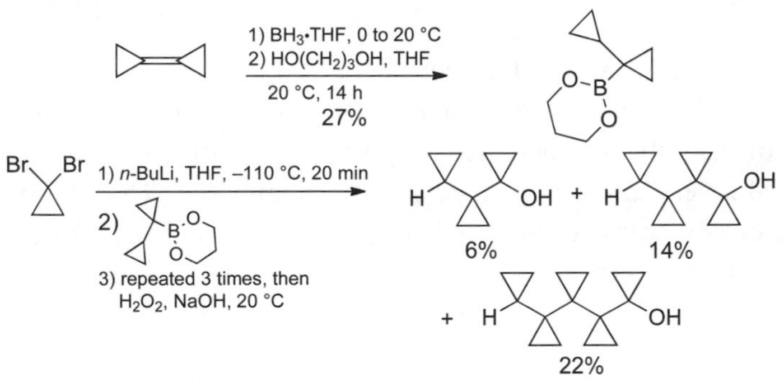

Figure 3. Hydroboration of the double bond in BCP and application of the product in the preparation of [1,1';1',1";1",1"']quater- and [1,1';1',1";1",-1"';1"',1""]quinquecyclopropanol.[17a]

Interestingly, the 1,1-linked cyclopropane moieties in dinitrobenzoates of these alcohols in the crystals adopt helical *all-gauche* conformations.[17a] Such helical overall structures have also been proved for higher 1,1-linked oligocyclopropyl hydrocarbons, which have consequently been termed [*n*]ivyanes[17b,c] by Sherburn et al.[17b]

III. Cycloadditions. Preparation of Triangulanes and Heterotriangulanes

Among other purposes, BCP serves as the best starting material for various linear and branched [*n*]triangulanes – hydrocarbons consisting exclusively of spiroannelated cyclopropane units.[3f,18] Thus, along with the

336

well-known 7,7-dibromo[3]triangulane,[3a–c] a new large family of 7-substituted [3]triangulanes (dispiro[2.0.2.1]heptanes) has been prepared during the last decade by halocarbenoid transfer to BCP under different conditions (Figure 4).[19,20] Among the chemical transformations of such [3]triangulanes, relative stabilities of spirocyclopropanated cyclopropyl cation intermediates in solvolytic processes[20] and unusual rearrangements of dibromotriangulanes in their reactions with alkyllithiums[19,21.22] have been studied in especially great detail.

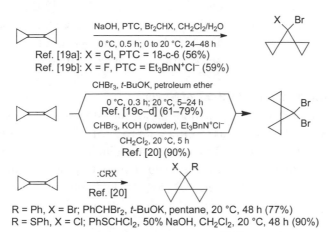

R = Ph, X = Br; PhCHBr₂, t-BuOK, pentane, 20 °C, 48 h (77%)
R = SPh, X = Cl; PhSCHCl₂, 50% NaOH, CH₂Cl₂, 20 °C, 48 h (90%)

Figure 4. Addition of dihalocarbenes onto BCP.[19–21]

On the other hand, 7,7-dibromo[3]triangulane can be subjected to dehalogenative coupling under the conditions of Neuenschwander et al.[9] in a repetitive fashion to afford perspirocyclopropanated bicyclopropylidenes. Along this route, the record-setting C_{2v}-[15]triangulane with 15 spirofused cyclopropane rings was built up starting from BCP.[23] An analogous key step – the dehalogenative coupling of a dibromocyclopropane to a bicyclopropylidene – was employed in the syntheses of enantiomerically pure linear [7]-, [9]- and [15]triangulanes,[24] while enantiomerically pure linear [4]- and [5]triangulanes were prepared from BCP itself.[25]

Epoxidations,[26] episulfidation,[27] reaction with a stable dialkylsilylene,[28] and CuCl-catalyzed phosphinidene addition[29] to BCP and spirocyclopropanated BCPs as well as their complexation[30] yielded 7-oxa-, 7-thia-, 7-sila-, and 7-phospha-, and 7-metalla[3]triangulane, respectively, as well as their spirocyclopropanated analogues (Figure 5).

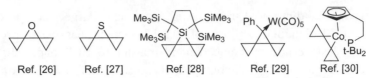

| Ref. [26] | Ref. [27] | Ref. [28] | Ref. [29] | Ref. [30] |

Figure 5. Hetera- and metallatriangulanes prepared from BCP during the last decade.

While BCP did not react with a photolytically generated diethynylcarbene,[31] it furnished ethyl 7-nitrodispiro[2.0.2.1]heptane-7-carboxylate with the rhodium carbenoid generated from ethyl nitrodiazoacetate, yet in only 13% yield.[32] The yield of the successful cyclopropanation with diazoacetonitrile has not been reported.[33]

IV. 1,3-Dipolar Cycloadditions of Nitrones

The Brandi-Guarna reaction, i. e. the stereo- and regioselective thermal rearrangement of spirocyclopropanated isoxazolidines and isoxazolines obtained by 1,3-dipolar cycloadditions of nitrones and nitrile oxides to alkylidenecyclopropanes into tetrahydro- or dihydropyridones (Figure 6),[34] in its extension to BCP yielded a number of spirocyclopropanated heterocyclic ketones.[3b,c] Some of these products have

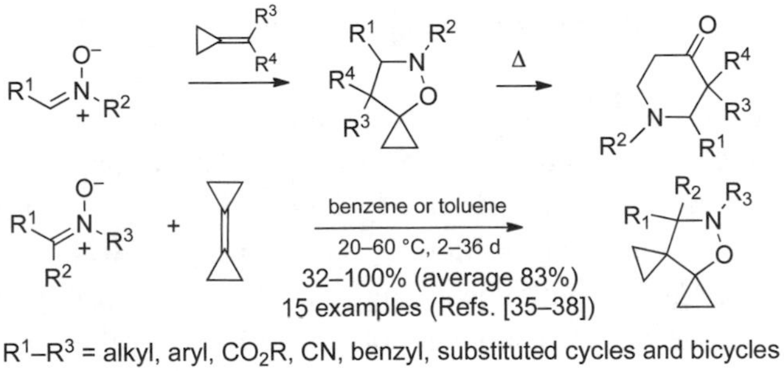

R^1–R^3 = alkyl, aryl, CO$_2$R, CN, benzyl, substituted cycles and bicycles

Figure 6. 1,3-Dipolar cycloadditions of nitrones to BCP and thermal rearrangement of the products.[35–39]

turned out to cleave supercoiled DNA in analogy to certain cytotoxic natural products like the illudines and ptaquiloside. At least 15 new functionally substituted bisspirocyclopropanated isoxazolidines have been prepared from BCP in good overall yields (average 83%) (Figure 6)[35–38] and employed in e. g. thermal rearrangements to afford spirocyclopropane-annelated

338

tetrahydropyridones.[36,37,39] In the case of *N*-arylisoxazolidines, spirocyclopropanated benzoazocinones were formed in moderate yields (21–27%).[39]

Chemoselective reduction of the N–O bond of the isoxazolidine ring in bisspirocyclopropanated isoxazolidines afforded substituted 1'-(aminomethyl)bi(cyclopropan)-1-ols (70–90% yield), which underwent a Pd(II)-catalyzed cascade rearrangement into a 1:1 mixture of 3-spirocyclopropanedihydro- and -tetrahydropyrid-4-ones (Figure 7).[36] Under the influence of trifluoroacetic acid in acetonitrile at 70 °C, these bisspirocyclopropanated isoxazolidines undergo a fragmentative thermal rearrangement to furnish the corresponding 3-spirocyclopropanated β-lactams in 75–96% yields (Figure 7).[38] Some such β-lactams could also be

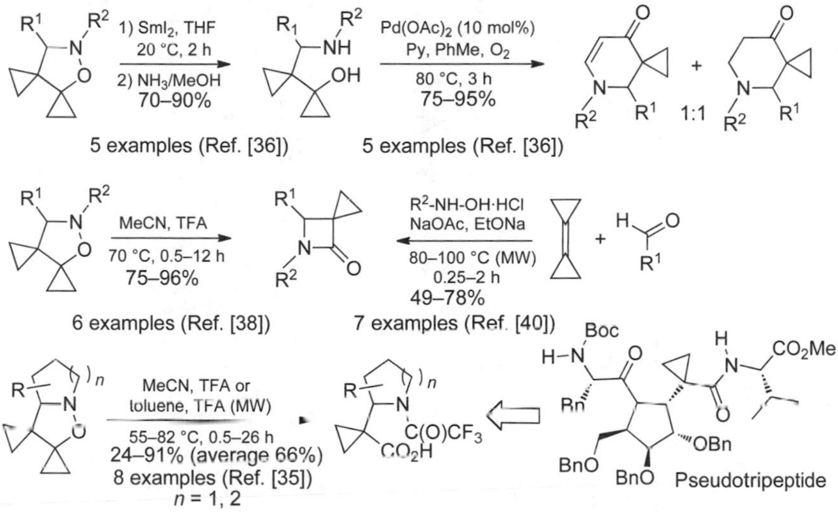

Figure 7. Transformations of bisspirocyclopropanated isoxazolidines.[35,36,38,40–42]

synthesized in a one-pot three-component reaction upon microwave heating of mixtures of BCP, certain alkylhydroxylamine hydrochlorides and formaldehyde (or an alkyl glyoxylate) in the presence of sodium acetate in ethanol for 15–120 min in 49–78% yields (Figure 7).[40]

The acid-catalyzed fragmentative thermal rearrangement of some bicyclic bisspirocyclopropanated isoxazolidines was accompanied by an opening of the β-lactam ring affording protected derivatives of α-cyclopropyl-β-homoproline (Figure 7).[35] The latter were used for the preparation of biologically active pseudotripeptides (Figure 7). At last, all the transformations discussed above have been realized in their intramolecular versions as well.[41,42]

V. Additions and Cycloadditions in the Presence of Transition Metals

Metal-mediated or -catalyzed reactions of methylenecyclopropanes including BCP have been studied extensively and found to proceed almost exclusively with opening of a cyclopropane ring,[43] whereas cyclopropane-retained compounds were observed as side products at best. In contrast to this, conservation of both rings was observed in the first ruthenium-catalyzed intermolecular hydroarylation of BCP via C–H bond functionalization, a remarkable reaction mode for a transformation proceeding through a (cyclopropylcarbinyl)metal intermediate (Figure 8).[44]

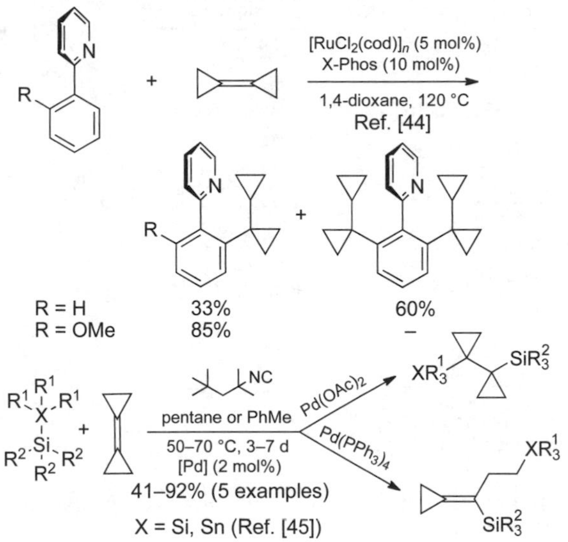

Figure 8. Metal-catalyzed additions onto the double bond in BCP.[44,45]

On the other hand, inter- and intramolecular palladium-catalyzed additions of disilanes, silylboranes, silylstannanes, and silyl cyanides across the double bond of BCP can proceed in two modes – with and without opening of one cyclopropane ring – yielding either bicyclopropyl or cyclopropylidenepropane derivatives depending on the nature of the reagent and the palladium catalyst (Figure 8).[45]

Org. Synth, **2012**. *89*, 334-349

The carbonyl ylide 1,3-dipoles, generated by dirhodium tetraacetate-catalyzed decomposition of 1-diazo-5-phenylpentane-2,5-dione and 1-(1-acetylcyclopropyl)-2-diazoethanone, cycloadd to BCP to give bridgehead-substituted di- and trispirocyclopropanated 8-oxabicyclo[3.2.1]octan-2-one and 7-oxabicyclo[2.2.1]heptan-2-one, respectively, yet only in 57 and 16% yield (Figure 9).[46] Intramolecular Pauson-Khand reactions of BCP derivatives[47] and cobalt(I)-mediated intramolecular [2+2+2] cocyclizations of (bicyclopropylidenyl)diynes[48] also proceeded with retention of both

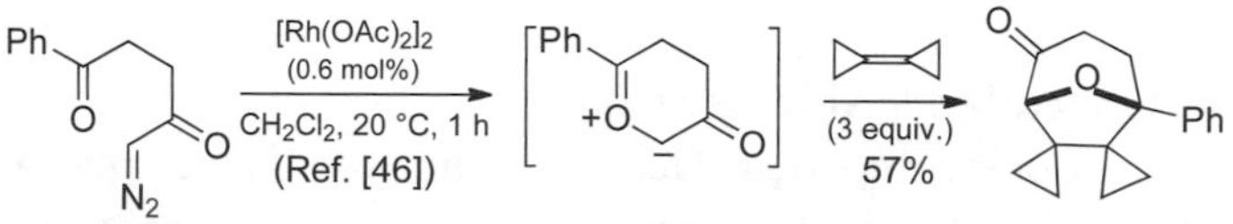

Figure 9. Cocyclizations of BCPs in the presence of transition metals with conservation of both cyclopropane rings.[46]

cyclopropane rings and afforded cyclopropane-fused and spirocyclopropanated bicyclo[3.3.0]alkenones[47] or cobalt-complexed tricyclo-[7.3.0.0^{2,6}]dodeca-1,6-dienes,[48] respectively, in moderate to good yields.

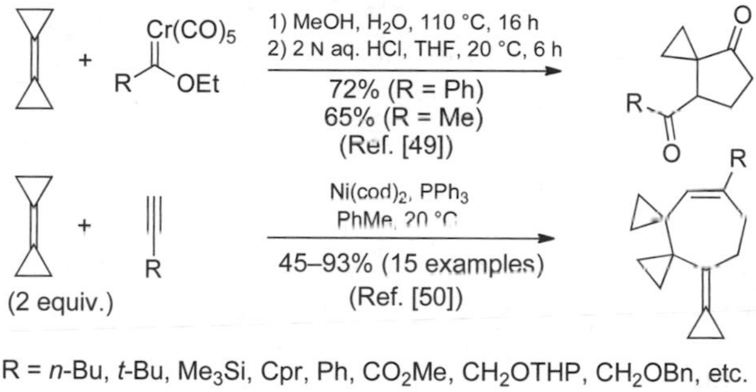

R = *n*-Bu, *t*-Bu, Me$_3$Si, Cpr, Ph, CO$_2$Me, CH$_2$OTHP, CH$_2$OBn, etc.

Figure 10. [4+1] Cocyclization of BCP with Fischer carbenechromium complexes and nickel-catalyzed intermolecular [3+2+2] cocyclizations of BCP and alkynes.[49,50]

On the other hand, Fischer carbenechromium complexes react with BCP in an unprecedented manner: All four carbon atoms of one of the methylenecyclopropane moieties of BCP along with carbon monoxide are incorporated with the formation of three new C–C-bonds to give substituted cyclopentenone derivatives in good yields (65–72%) (Figure 10).[49] Under the catalysis of bis(1,5-cyclooctadiene)nickel(0), in the presence of triphenylphosphine, two molecules of BCP and one molecule of a terminal alkyne at ambient temperature, undergo an intermolecular [3+2+2] cocyclization with opening of one cyclopropane ring to furnish bis-spirocyclopropanated cyclopropylidenecycloheptene derivatives in 45–93% yields (Figure 10).[50] For comparison, methylenecyclopropanes under these conditions experience an intermolecular [3+2+2] cocyclization of one molecule of methylenecyclopropane and two molecules of a terminal alkyne.[51]

VI. Domino Heck-Diels-Alder Reactions

The previously reported Heck coupling of haloarenes and BCP afforded 1-arylallylidenecyclopropanes, which can be trapped in a Diels-Alder reaction with an appropriate dienophile to give 8-arylspiro[2.5]octenes (Figure 11).[52,53] In the presence of trisfurylphosphine instead of triphenylphosphine the 1-arylallylidenecyclopropanes undergo readditions of the hydridopalladium halide to yield σ-allylpalladium intermediates, which are efficiently trapped with various nitrogen and carbon nucleophiles

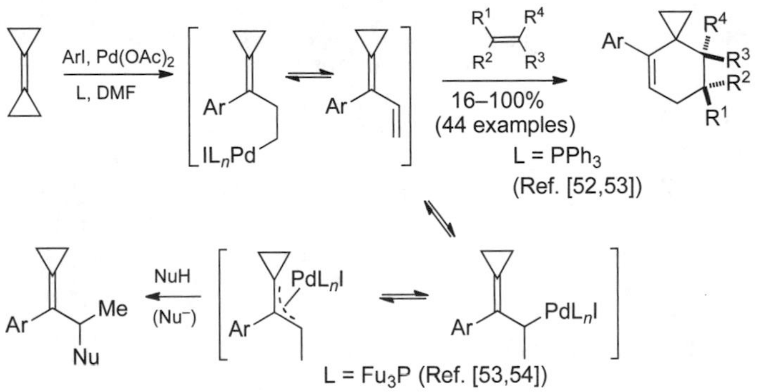

Figure 11. Two reaction modes of BCP under Heck conditions and domino Heck-Diels-Alder reactions.[52–56]

to yield functionally substituted methylenecyclopropane derivatives (Figure 11).[53,54] Both these reaction modes were further developed during the last decade. Thus, a new large family of substituted spiro[2.5]oct-4-ene derivatives has been synthesized from BCP employing this highly efficient three-component domino reaction, in most cases in good to very high yields

(49–100%) (Figure 11).[55] Moreover, the coupling of oligoiodobenzenes with BCP and subsequent cycloaddition was extended to this multicomponent reaction. For example, 1,2,4,5-tetraiodobenzene gave the fourfold domino Heck-Diels-Alder product in 47% isolated yield, in a single operation, in which 12 new carbon-carbon bonds were formed.[55]

This three-component domino reaction has also been extended to monosubstituted BCPs.[56] The second reaction mode of BCP with an intermediate σ-allylpalladium species in the presence of a secondary amine and trisfurylphosphine was further developed into a one-pot, four-component two-step queuing cascade with iodoethenes and secondary amines to provide 8-(10-aminoethyl)-substituted spiro[2.5]oct-7-ene derivatives (Figure 12).[57] The same one-pot, two-step queuing cascade can be carried out with cyclic dienophiles such as N-arylmaleinamides and N-phenyltriazolinedione to furnish highly substituted spiro[2.5]oct-4-enes and spirocyclopropanated heterobicycles.

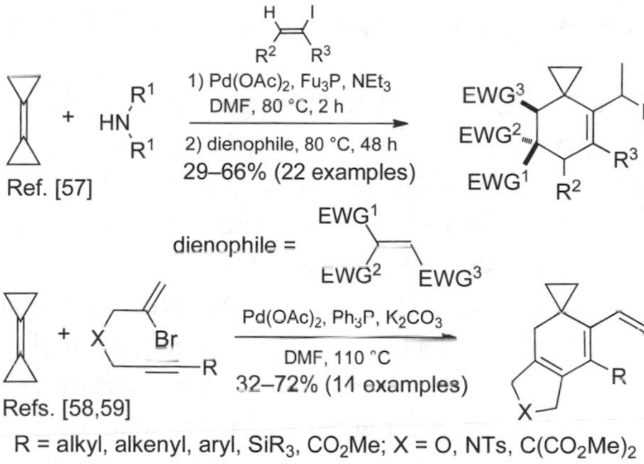

Figure 12. Four-component two-step queuing cascade reactions of BCP and palladium-catalyzed cross-coupling/electrocyclization cascade reactions.[57–59]

Finally, further cascade reactions involving Heck-type coupling of BCP with 2-bromo-1,6-enynes afforded cross-conjugated tetraenes which, at elevated temperatures, underwent 6π-electrocyclization to give spiro[cyclopropane-1,4'-bicylo[4.3.0]-1(6),2-dienes] (Figure 12).[58,59]

1. Institut für Organische und Biomolekulare Chemie der Georg-August-Universität Göttingen, Tammannstrasse 2, 37077 Göttingen, Germany.
2. de Meijere, A.; Kozhushkov, S. I.; Spaeth, T.; Zefirov, N. S. *J. Org. Chem.* **1993**, *58*, 502–505.
3. For reviews, see: (a) de Meijere, A.; Kozhushkov, S. I.; Khlebnikov, A. F. *Russ. J. Org. Chem.* **1996**, *32*, 1555–1575. (b) de Meijere, A.; Kozhushkov, S. I. *Eur. J. Org. Chem.* **2000**, 3809–3822. (c) de Meijere, A.; Kozhushkov, S. I.; Khlebnikov, A. F. *Top. Curr. Chem.* **2000**, *207*, 89–147. (d) de Meijere, A.; Kozhushkov, S. I.; Späth, T.; von Seebach, M.; Löhr, S.; Nüske, H.; Pohlmann, T.; Es-Sayed, M.; Bräse, S. *Pure Appl. Chem.* **2000**, *72*, 1745–1756. (e) In part: de Meijere, A.; Schill, H.; Kozhushkov, S. I.; Walsh, R.; Müller, E. M.; Grubmüller, H. *Russ. Chem. Bull.* **2004**, *53*, 947–959. (f) In part: de Meijere, A.; Kozhushkov, S. I.; Schill, H. *Chem. Rev.* **2006**, *106*, 4926–4996.
4. Kuznetsova, T. S.; Averina, E. B.; Kokoreva, O. V.; Zefirov, A. N.; Grishin, Y. K;. Zefirov, N. S. *Russ. J. Org. Chem.* **2000**, *36*, 205–210.
5. de Meijere, A.; Kozhushkov, S. I.; Zefirov, N. S. *Synthesis* **1993**, 681–683.
6. Löhr, S.; de Meijere, A. *Synlett* **2001**, 489–492.
7. de Meijere, A.; Bagutski, V.; Zeuner, F.; Fischer, U. K.; Rheinberger, V.; Moszner, N. *Eur. J. Org. Chem.* **2004**, 3669–3678.
8. de Meijere, A.; Kozhushkov, S. I.; Faber, D;. Bagutskii, V.; Boese, R.; Haumann, T.; Walsh, R. *Eur. J. Org. Chem.* **2001**, 3607–3614.
9. (a) von Huwyler, R.; Al-Dulayymi, A.; Neuenschwander, M. *Helv. Chim. Acta* **1999**, *82*, 2336–2347, and references cited therein. (b) von Huwyler, R.; Li, X.; Bönzli, P.; Neuenschwander, M. *Helv. Chim. Acta* **1999**, *82*, 1242–1249, and references cited therein.
10. Beckhaus, H.-D.; Rüchardt, C; Kozhushkov, S. I.; Belov, V. N.; Verevkin, S. P.; de Meijere, A. *J. Am. Chem. Soc.* **1995**, *117*, 11854–11860, and references cited therein.
11. Schill, H.; Kozhushkov, S. I.; Walsh, R.; de Meijere, A. *Eur. J. Org. Chem.* **2007**, 1510–1516.
12. (a) Norberg, D.; Larsson, P. E.; Dong, X. C.; Salhi-Benachenhou, N.; Lunell, S. *Int. J. Quant. Chem.* **2004**, *98*, 473–483. (b) Ikeda, H.; Akiyama, K.; Takahashi, Y.; Nakamura, T.; Ishizaki, S.; Shiratori, Y.; Ohaku, H.; Goodman, J. L.; Houmam, A.; Wayner, D. D. M.; Tero-Kubota, S.; Miyashi, S. *J. Am. Chem. Soc.* **2003**, *125*, 9147–9157. (c) Maier, G.; Senger, S. *Eur. J. Org. Chem.* **1999**, 1291–1294.

13. Kozhushkov, S.; Späth, T.; Fiebig, T.; Galland, B.; Ruasse, M.-F.; Xavier, P.; Apeloig, Y.; de Meijere, A. *J. Org. Chem.* **2002**, *67*, 4100–4114.

14. Löhr, S.; Jacobi, C.; Johann, A.; Gottschalk, G.; de Meijere, A. *Eur. J. Org. Chem.* **2000**, 2979–2989.

15. von Seebach, M.; Kozhushkov, S. I.; Schill, H.; Frank, D.; Boese, R.; Benet-Buchholz, J.; Yufit, D. S.; de Meijere, A. *Chem. Eur. J.* **2007**, *13*, 167–177.

16. Matteson, D. S.; Majumdar, D. *J. Am. Chem. Soc.* **1980**, *102*, 7588–7590.

17. (a) Kurahashi, T.; Kozhushkov, S. I.; Schill, H.; Meindl, K.; Rühl, S.; de Meijere, A. *Angew. Chem. Int. Ed.* **2007**, *46*, 6545–6548. (b) Bojase, G.; Nguyen, T. V.; Payne, A. D.; Willis, A. C.; Sherburn, M. S. *Chem. Sci.* **2011**, *2*, 229–232. (c) Pichierri. F. *Chem. Phys. Lett.* **2011**, *511*, 277–282.

18. For reviews, see: de Meijere, A.; Kozhushkov, S. *From spiropentane to linear and angular oligo- and polytriangulanes.* In *Advances in Strain in Organic Chemistry, Vol. 4*; Ed.: Halton, B.; JAI Press Ltd.: London, **1995**, pp. 225–282. (b) de Meijere, A.; Kozhushkov, S. I. *Chem. Rev.* **2000**, *100*, 93–142.

19. (a) Sedenkova, K. N.; Averina, E. B.; Grishin, Y. K.; Kuznetzova, T. S.; Zefirov, N. S. *Tetrahedron* **2010**, *66*, 8089–8094. (b) Averina, E. B.; Sedenkova, K. N.; Borisov, I. S.; Grishin, Y. K.; Kuznetsova, T. S.; Zefirov, N. S. *Tetrahedron* **2009**, *65*, 5693–5701. (c) Sedenkova, K. N.; Averina, E. B.; Grishin, Y. K.; Kuznetsova, T. S.; Zefirov, N. S. *Russ. J. Org. Chem.* **2008**, *44*, 950–957. (d) Averina, E. B.; Sedenkova, K. N.; Grishin, Y. K.; Kuznetzova, T. S.; Zefirov, N. S. *ARKIVOC* **2008**, *(IV)*, 71–79. (e) Averina, E. B.; Karimov, R. R.; Sedenkova, K. N.; Grishin, Y. K.; Kuznetzova, T. S.; Zefirov, N. S. *Tetrahedron* **2006**, *62*, 8814–8821.

20. Kozhushkov, S. I.; Späth, T.; Kosa, M.; Apeloig, Y.; Yufit, D. S.; de Meijere, A. *Eur. J. Org. Chem.* **2003**, 4234–4242.

21. (a) Zöllner, S. *Dissertation*, Universität Hamburg, **1991**. (b) Lukin, K. A.; Zefirov, N. S.; Yufit, D. S.; Struchkov, Y. T. *Tetrahedron* **1992**, *48*, 9977–9984. For rearrangement into the corresponding allene (so-called Doering-Skattebøl-Moore reaction) see ref. [22].

22. (a) Backes, J.; Brinker, U. *Methoden der Organischen Chemie (Houben-Weyl), Vol. E 19b*; Ed.: Regitz, M.; Thieme: Stuttgart, **1989**;

pp. 391–541. (b) Lee-Ruff, E. *Methods of Organic Chemistry (Houben-Weyl), Vol. E 17c*; Ed.: de Meijere, A.; Thieme: Stuttgart, **1997**; pp. 2388–2418. (c) Kostikov, R. R.; Molchanov, A. P.; Hopf, H. *Top. Curr. Chem.* **1990**, *155*, 41–80.

23. (a) von Seebach, M.; Kozhushkov, S. I.; Boese, R.; Benet-Buchholz, J.; Yufit, D. S.; Howard, J. A. K.; de Meijere, A. *Angew. Chem. Int. Ed.* **2000**, *39*, 2495–2498. (b) de Meijere, A.; von Seebach, M.; Zöllner, S.; Kozhushkov, S. I.; Belov, V. N.; Boese, R.; Haumann, T.; Benet-Buchholz, J.; Yufit, D. S.; Howard, J. A. K. *Chem. Eur. J.* **2001**, *7*, 4021–4034.

24. (a) de Meijere, A.; Khlebnikov, A. F.; Kozhushkov, S. I.; Yufit, D. S.; Chetina, O. V.; Howard, J. A. K.; Kurahashi, T.; Miyazawa, K.; Frank, D.; Schreiner, P. R.; Rinderspacher, B. C.; Fujisawa, M.; Yamamoto, C.; Okamoto, Y. *Chem. Eur. J.* **2006**, *12*, 5697–5721. (b) de Meijere, A.; Khlebnikov, A. F.; Kozhushkov, S. I.; Miyazawa, K.; Frank, D.; Schreiner, P. R.; Rinderspacher, B. C.; Yufit, D. S.; Howard, J. A. K. *Angew. Chem. Int. Ed.* **2004**, *43*, 6553–6557.

25. (a) de Meijere, A.; Khlebnikov, A. F.; Kozhushkov, S. I.; Kostikov, R. R.; Schreiner, P. R.; Wittkopp, A.; Rinderspacher, C.; Menzel, H.; Yufit, D. S.; Howard, J. A. K. *Chem. Eur. J.* **2002**, *8*, 828–842. (b) de Meijere, A.; Khlebnikov, A. F.; Kostikov, . R.; Kozhushkov, S. I.; Schreiner, P. R.; Wittkopp, A.; Yufit, D. S. *Angew. Chem. Int. Ed.* **1999**, *38*, 3474–3477.

26. (a) Frank, D.; Kozhushkov, S. I.; Labahn, T.; de Meijere, A. *Tetrahedron* **2002**, *58*, 7001–7007. (b) Adam, W.; Saha-Moeller, C. R.; Zhao, C.-G. *Dioxirane epoxidation of alkenes*. In: *Org. Reactions, Vol. 61*; Eds.: Overman, L. E., Adams, R.; Wiley: New York, **2001**; pp. 219 ff.

27. Adam, W.; Bargon, R. M.; Schenk, W. A. *J. Am. Chem. Soc.* **2003**, *125*, 3871–3876.

28. Kira, M.; Ishida, S.; Iwamoto, T.; de Meijere, A.; Fukitsuka, M.; Ito, O. *Angew. Chem. Int. Ed.* **2004**, *43*, 4510–4512.

29. (a) Slootweg, J. C.; de Kanter, F. J. J.; Schakel, M.; Lutz, M.; Spek, A. L.; Kozhushkov, S. I.; de Meijere, A.; Lammertsma, K. *Chem. Eur. J.* **2005**, *11*, 6982–6993. (b) Slootweg, J. C.; Schakel, M.; de Kanter, F. J. J.; Ehlers, A. W.; Kozhushkov, S. I.; de Meijere, A.; Lutz, M.; Spek, A. L.; Lammertsma, K. *J. Am. Chem. Soc.* **2004**, *126*, 3050–3051.

30. Kozhushkov, S. I.; Foerstner, J.; Kakoschke, A.; Stellfeldt, D.; Yong, L.; Warchow, R.; de Meijere, A.; Butenschön, H. *Chem. Eur. J.* **2006**, *12*, 5642–5647.

31. de Meijere, A.; Kozhushkov, S. I.; Boese, R.; Haumann, T.; Yufit, D. S.; Howard, J. A. K.; Khaikin, L. S.; Traetteberg, M. *Eur. J. Org. Chem.* **2002**, 485–492.

32. Ivanova, O. A.; Yashin, N. V.; Averina, E. B.; Grishin, Y. K.; Kuznetsova, T. S.; Zefirov, N. S. *Russ. Chem. Bull.* **2001**, *50*, 2101–2105.

33. Moss, R. A.; Zheng, F.; Krogh-Jespersen, K. *Org. Lett.* **2001**, *3*, 1439–1442.

34. For a review, see: Brandi, A.; Cicchi, S.; Cordero, F. M.; Goti, A. *Chem. Rev.*, **2003**, *103*, 1213–1269.

35. Cordero, F. M.; Salvati, M.; Vurchio, C.; de Meijere, A.; Brandi, A. *J. Org. Chem.* **2009**, *74*, 4225–4231.

36. Revuelta, J.; Cicchi, S.; de Meijere, A.; Brandi, A. *Eur. J. Org. Chem.* **2008**, 1085–1091.

37. Gensini, M.; Kozhushkov, S. I.; Frank, D.; Vidović, D.; Brandi, A.; de Meijere, A. *Eur. J. Org. Chem.* **2003**, 2001–2006.

38. Zanobini, A.; Gensini, M.; Magull, J.; Vidović, D.; Kozhushkov, S. I.; Brandi, A.; de Meijere, A. *Eur. J. Org. Chem.* **2004**, 4158–4166.

39. Cordero, F. M.; Barile, I.; Brandi, A.; Kozhushkov, S. I.; de Meijere, A. *Synlett* **2000**, 1034–1036.

40. Zanobini, A.; Brandi, A.; de Meijere, A. *Eur. J. Org. Chem.* **2006**, 1251–1255.

41. Marradi, M.; Brandi, A.; Magull, J.; Schill, H.; de Meijere, A. *Eur. J. Org. Chem.* **2006**, 5485–5494.

42. Marradi, M.; Brandi, A.; de Meijere, A. *Synlett* **2006**, 1125–1127.

43. For reviews on metal-promoted reactions of methylenecyclopropanes, see: (a) Binger, P.; Büch, H. M. *Top. Curr. Chem.* **1987**, *135*, 77–151. (b) Lautens, M.; Klute, W.; Tam, W. *Chem. Rev.* **1996**, *96*, 49–92. (c) Binger, P.; Schmidt, T. *Methods of Organic Chemistry (Houben-Weyl), Vol. E 17c*; Ed.: de Meijere, A.; Thieme: Stuttgart, **1997**; pp. 2217–2294. (d) Nakamura, I.; Yamamoto, Y. *Adv. Synth. Catal.* **2002**, *344*, 111–129. (e) Rubin, M.; Rubina, M.; Gevorgyan, V. *Chem. Rev.* **2007**, *107*, 3117–3179.

44. Kozhushkov, S. I., Yufit, D. S., Ackermann, L. *Org. Lett.* **2008**, *10*, 3409–3412.

45. Pohlmann, T.; de Meijere, A. *Org. Lett.* **2000**, *2*, 3877–3879.
46. Molchanov, A. P.; Diev, V. V.; Magull, J.; Vidović, D.; Kozhushkov, S. I.; de Meijere, A.; Kostikov, R. R. *Eur. J. Org. Chem.* **2005**, 593–599.
47. de Meijere, A.; Becker, H.; Stolle, A.; Kozhushkov, S. I.; Bes, M. T.; Salaün, J.; Noltemeyer, M. *Chem. Eur. J.* **2005**, *11*, 2471–2482.
48. Schelper, M.; Buisine, O.; Kozhushkov, S.; Aubert, C.; de Meijere, A.; Malacria, M. *Eur. J. Org. Chem.* **2005**, 3000–3007.
49. Kurahashi, T.; Wu, Y.-T.; Meindl, K.; Rühl, S.; de Meijere, A. *Synlett* **2005**, 805–808.
50. Zhao, L.; de Meijere, A. *Adv. Synth. Cat.* **2006**, *348*, 2484–2492.
51. (a) Komagawa, S.; Takeuchi, K.; Sotome, I.; Azumaya, I.; Masu, H.; Yamasaki, R.; Saito, S. *J. Org. Chem.* **2009**, *74*, 3323–3329. (b) Yamasaki, R.; Sotome, I.; Komagawa, S.; Azumaya, I.; Masu, H.; Saito, S. *Tetrahedron Lett.* **2009**, *50*, 1143–1145.
52. Bräse, S.; de Meijere, A. *Angew. Chem. Int. Ed. Engl.* **1995**, *34*, 2545–2547.
53. de Meijere, A.; Nüske, H.; Es-Sayed, M.; T. Labahn, T.; M. Schroen, S. Bräse, *Angew. Chem. Int. Ed.* **1999**, *38*, 3669–3672.
54. Nüske, H.; Noltemeyer, M.; de Meijere, A. *Angew. Chem. Int. Ed.* **2001**, *40*, 3411–3413.
55. Nüske, H.; Bräse, S.; Kozhushkov, S. I.; Noltemeyer, M.; Es-Sayed, M.; de Meijere, A. *Chem. Eur. J.* **2002**, *8*, 2350–2369.
56. Yücel, B.; Noltemeyer, M.; de Meijere, A. *Eur. J. Org. Chem.* **2008**, 1072–1078.
57. Yücel, B.; Arve, L.; de Meijere, A. *Tetrahedron* **2005**, *61*, 11355–11373.
58. de Meijere, A.; Schelper, M.; Knoke, M.; Yücel, B.; Sünnemann, H. W.; Scheurich, R. P.; Arve, L. *J. Organomet. Chem.* **2003**, *687*, 249–255.
59. Schelper, M.; de Meijere, A. *Eur. J. Org. Chem.* **2005**, 582–592.

Armin de Meijere, born 1939, studied chemistry in Freiburg and Göttingen, receiving a doctoral degree (Dr. rer. nat.) in 1966 in Göttingen, completing postdoctoral training at Yale University in 1967-1969, and receiving Habilitation in 1971 in Göttingen. He was appointed full professor of organic chemistry in Hamburg, 1977-1989 and ever since has been in Göttingen. He was visiting professor at 16 research institutions around the world including the University of Wisconsin, the Technion in Haifa, Israel, and Princeton University. His awards and honors include member of the Norwegian Academy of Sciences, Alexander von Humboldt/Gay Lussac prize, Honorary Professor of St. Petersburg State University in Russia, Adolf von Baeyer Medal of the German Chemical Society, Dr. honoris causa of the Russian Academy of Sciences. He has been or still is Editor or member of the editorial board of a number of scientific publications including Houben-Weyl, Chemical Reviews, Science of Synthesis, and Chemistry - A European Journal. His scientific achievements have been published in over 700 original publications, review articles, and book chapters as well as 25 patents.

Sergei I. Kozhushkov was born in 1956 in Kharkov, USSR. He studied chemistry at Lomonosov Moscow State University, where he obtained his doctoral degree in 1983 under the supervision of Professor N. S. Zefirov and performed his "Habilitation" in 1998. From 1983 to 1991, he worked at Moscow State University and then at Zelinsky Institute of Organic Chemistry. In 1991, he joined the research group of Professor A. de Meijere (Georg-August-Universität Göttingen, Germany) as an Alexander von Humboldt Research Fellow; since 1993 he has worked as a Research Associate, since 1996 he has held a position of a Scientific Assistant, and since 2001 a permanent position as a Senior Scientist at the University of Göttingen. His current research interests focus on the chemistry of highly strained small ring compounds. The results of his scientific activity have been published in over 200 original publications, review articles, book chapters and patents. From 2003–2011, he won eight awards from InnoCentive, Inc.

Discussion Addendum for:
Synthesis of 1,2:4,5-Di-*o*-isopropylidene–D-erythro-2,3-hexodiulo-2,6-pyranose. A Highly Enantioselective Ketone Catalyst for Epoxidation/Asymmetric Epoxidation of *trans*-β-Methylstyrene and 1-Phenylcyclohexene using a D-Fructose-derived Ketone: (*R,R*)-*trans*-β–Methylstyrene Oxide and (*R,R*)-1-Phenylcyclohexene Oxide

Prepared by Thomas A. Ramirez, O. Andrea Wong, and Yian Shi.*[1]
Original articles: Tu, Y.; Frohn, M.; Wang, Z.-X.; Shi, Y. *Org. Synth.* **2003,** *80*, 1. and Wang, Z.-X.; Shu, L.; Frohn, M.; Tu, Y.; Shi, Y. *Org. Synth.* **2003,** *80*, 9.

Prior to and subsequent to the original reports in *Organic Syntheses*, we have reported a number of carbohydrate-derived ketone epoxidation

Org. Synth. **2012,** *89*, 350-373
Published on the Web 2/16/2012
© 2012 Organic Syntheses, Inc.

catalysts[2] including **1-4** (Figure 1).   Epoxidation of olefins catalyzed by these ketones has become a valuable tool for the synthetic organic chemist.[2]

Figure 1. Four Generations of Ketone Catalysts

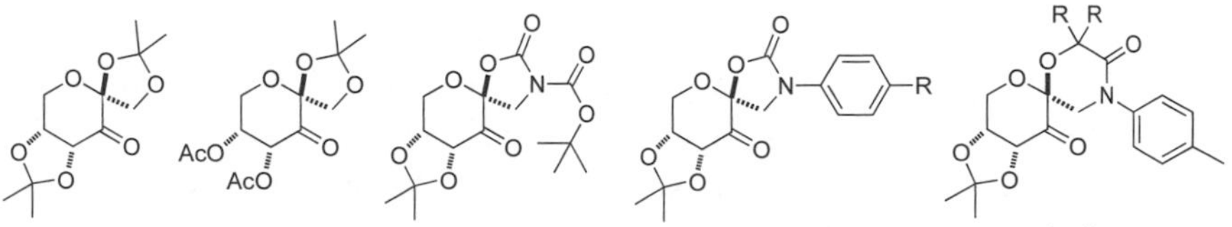

| **1** | **2** | **3a** | **3b-3f** | **4a-4b** |

3b, R = SO$_2$Me
3c, R = Me
3d, R = Et
3e, R = Ph
3f, R = *n*-Bu

4a, R = H
4b, R = Me

Generations of Catalysts:

Ketone 1:

The synthesis and use of ketone **1** was reported in the original *Organic Syntheses* articles.   Ketone **1** is an effective catalyst for the epoxidation of a variety of *trans-* and tri-substituted olefins,[3] vinylsilanes,[4] hydroxy alkenes,[5] dienes,[6] enynes,[7] enol ethers, and enol esters.[8,9]   A procedure for the preparation of *ent*-ketone **1** from L-sorbose has also been developed.[3b,10]   Ketone **1** and *ent*-ketone **1** have been successfully used in various synthetic applications.[2h-k]

Ketone 2:

Ketone **2** has proven to be a more reactive analogue of ketone **1**, particularly in the reaction of electron-deficient olefins.   It is derived from ketone **1** through a selective deketalization followed by acetylation.[11]

Ketones 3:

Ketones **3** mediate the highly stereoselective epoxidation of conjugated *cis-*, terminal, tri-substituted, and some tetra-substituted double bonds.[12-21]   While ketone **3a** is highly effective, **3c-3f** were developed due to their streamlined syntheses.[13]   Ketones **3a** and **3f** have also found efficacy in the epoxidation of non-conjugated, amphiphilic *cis*-olefins.[22]

Ketones 4:

Ketone **4a** has extended the capability of the epoxidation system to geminally-disubstituted terminal olefins and also works well in the epoxidation of some *cis-* and tri-substituted olefins.[23]   Ketone **4b** has proven effective in the reaction of *trans-* and tri-substituted double bonds.[23]

Ketone **1**, *ent*-ketone **1**, **3c**, and **3d** are now commercially available. Our group website now features "Guidelines for Asymmetric Epoxidation"[24] which includes instructive illustrations and tips for running the epoxidation reactions, and the document addresses frequently asked questions.

Synthetic Applications:

Previously published reviews provide a detailed discussion of the development of the catalysts and their respective substrate scopes.[2b-e, 2g-k] This update will focus primarily on the wealth of synthetic applications of ketone **1** and *ent*-ketone **1**.

Ketone 1:

The epoxidation with ketone **1** is practical and has been used in undergraduate teaching laboratories,[25] academic research laboratories,[2b-e,g-k] and in industry.[26] Along with other proven asymmetric epoxidation systems, ketone **1** has validated the use of stereoselective epoxidation in endgame approaches as well as those invoking the epoxide's manipulability. In addition, the ketone-catalyzed epoxidation is a process that can be used for discriminant or indiscriminant (poly-) epoxidation which can be further utilized in the rapid construction of complex molecules via cascade cyclization reactions.

Figure 2. Examples of Target Molecules Synthesized with Epoxidation Catalyzed by Ketone **1** or *ent*-Ketone **1**

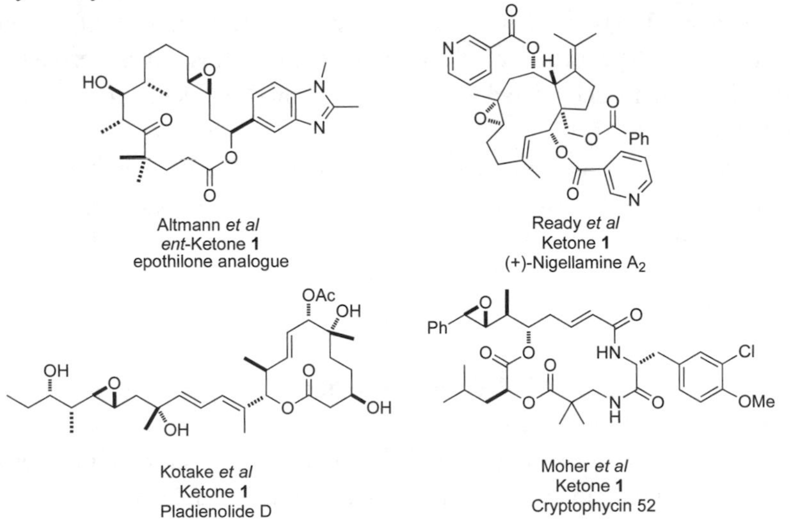

Altmann *et al*
ent-Ketone **1**
epothilone analogue

Ready *et al*
Ketone **1**
(+)-Nigellamine A$_2$

Kotake *et al*
Ketone **1**
Pladienolide D

Moher *et al*
Ketone **1**
Cryptophycin 52

352

Some examples of the use of ketone **1** in the selective installation of the epoxide functional group in biologically active molecules are shown in Figure 2. Altmann and coworkers used *ent*-ketone **1** for the epoxidization of an unsaturated macrolide to afford an epothilone analogue as a single isomer.[27] Ready and coworkers utilized ketone **1** for a highly discriminant epoxidation in their synthesis of (+)-nigellamine A_2.[28] Moher,[29] Kotake,[30] and others[31,32] have used ketone **1** to stereoselectively install epoxides in the side chains of target molecules.

The ketone-catalyzed epoxidation has found use in industry.[26] A team at DSM Pharma Chemical obtained lactone **7** in 63% overall yield and 88% ee when crude γ,δ-unsaturated carboxylate was subjected to the reaction conditions (Scheme 1).[26] The intermediate epoxide was opened by an intramolecular attack by the carboxylate anion.[22,26] After the product was crystallized from heptane, 59 lbs of lactone were obtained.

Scheme 1. Industrial Scale Synthesis of Lactone **7** via Epoxidation/Epoxide Opening

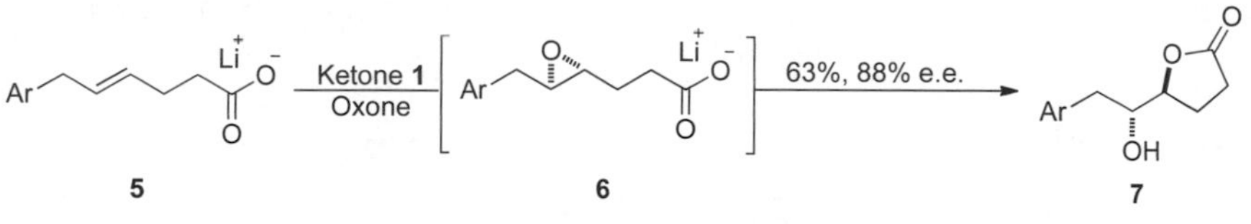

Sudalai and coworkers have used a similar process in their synthesis of (+)-L-733,060 (**10**), an antitumor agent with activity against retinoblastoma (Scheme 2).[33] The γ,δ-unsaturated carboxylate **8** was stereoselectively epoxidized and then underwent *in situ* lactonization.

Scheme 2. Synthesis of (+)-L-733,060

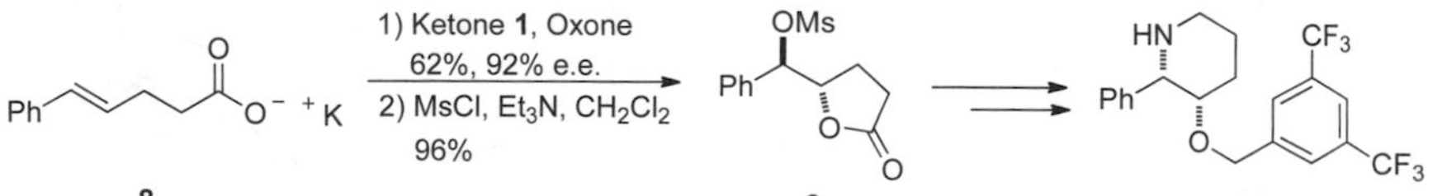

Myers and coworkers developed a method to synthesize *trans*-1,2-diol derivatives **12** using ketone **1**-catalyzed epoxidation of cyclic *E* and acyclic *E* and *Z* silyl enol ethers followed by reduction mediated by an internal hydride delivery (Scheme 3).[34,35]

Scheme 3. Epoxidation/Reduction For 1,2-Diols

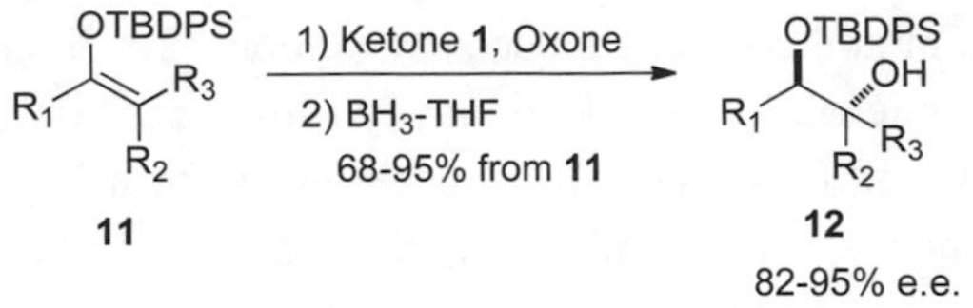

$$11 \xrightarrow[\substack{\text{2) BH}_3\text{-THF} \\ \text{68-95\% from 11}}]{\text{1) Ketone 1, Oxone}} 12$$

82-95% e.e.

McDonald and coworkers used ketone **1**-catalyzed epoxidation to access potentially antineoplastic 1-deoxy-5-hydroxysphingosine analogue **16** (Scheme 4).[36] Lewis acid catalyzed cyclization led to cyclic carbonate **15**. The resulting alcohol was then activated and displaced by an azide to yield the 2-amino-3,5-diol motif. The C-5 epimer of **16** was also synthesized when *ent*-**13** was used as the starting template.

Scheme 4. Synthesis of 1-Deoxy-5-Hydroxysphingosine Analogue

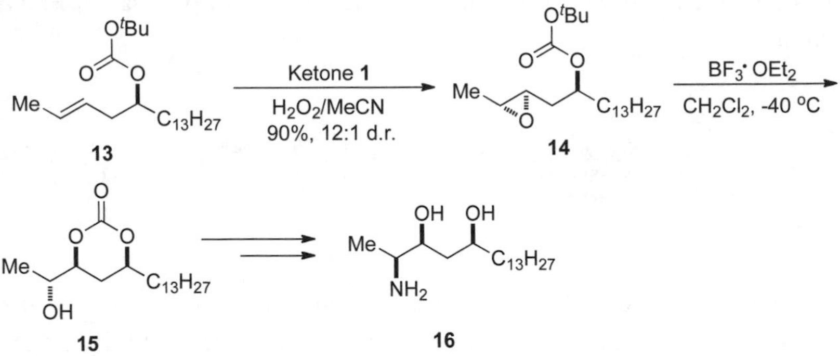

Scheme 5. Synthesis of (-)-8-Deoxyserratinine

(-)-8-deoxyserratinine (**19**)

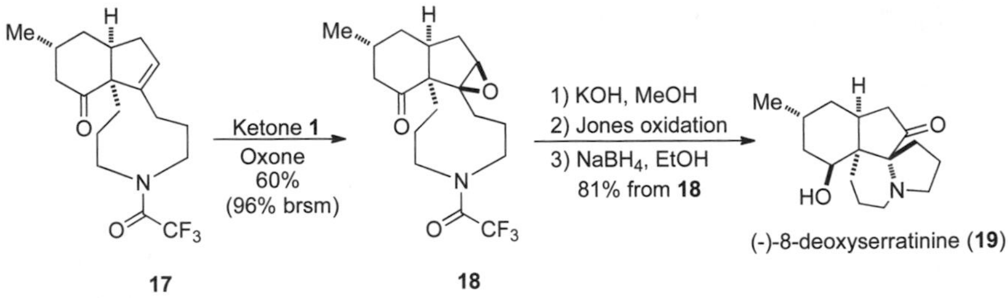

Yang and coworkers recently reported the total synthesis of (−)-8-deoxyserratinine (**19**) (Scheme 5).[37] Ketone **1** allowed them to overcome the

substrate-induced facial bias in the epoxidation of **17** which afforded **18** as the sole isomer.

Huo and coworkers recently synthesized (+)-neroplofurol (**23**) in two operations and 60% overall yield from the naturally (and commercially) available (+)-nerolidol (Scheme 6).[38] Upon epoxidation of **21**, cyclization took place *in situ*.

Scheme 6. Synthesis of (+)-Neroplofurol (**23**)

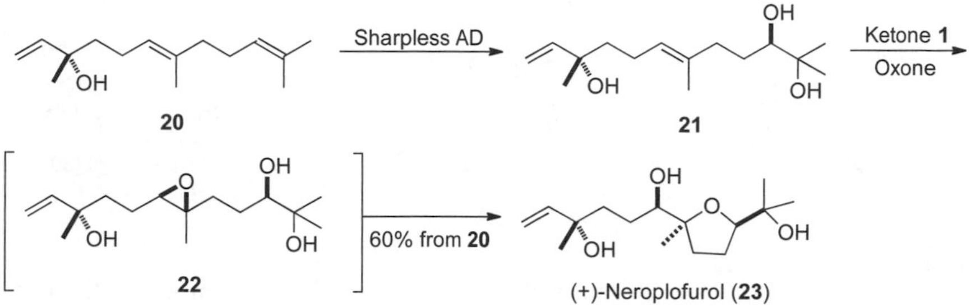

Wiemer and coworkers have also investigated the use of cascade cyclizations in the syntheses of schweinfurthins.[39] A ketone **1**-catalyzed highly site- and stereo-selective epoxidation followed by a Lewis acid catalyzed cyclization allowed them to synthesize schweinfurthin B (**27**)[39] (Scheme 7) and a diverse array of derivatives.

Scheme 7. Synthesis of Schweinfurthin B (**27**) via Epoxidation/Cascade Cyclization

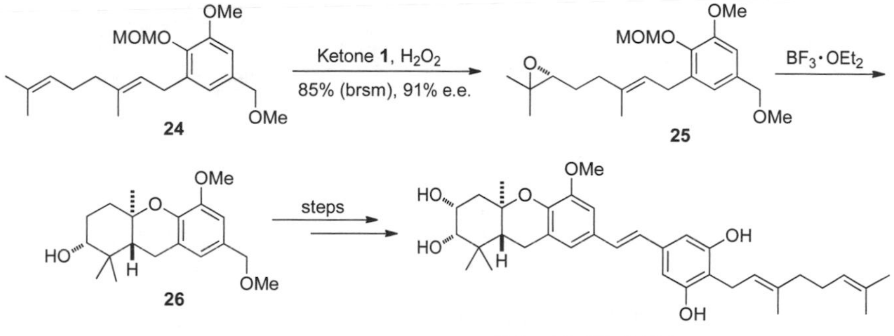

Schweinfurthin B (**27**)

In an effort to confirm the absolute stereochemistry of glabrescol (**30**), Corey and coworkers obtained the poly-tetrahydrofuran via the tetra-epoxidation of tetra-ene **28** followed by a CSA-mediated cyclization (Scheme 8).[40,41]

Scheme 8. Synthesis of Glabrescol (**30**)

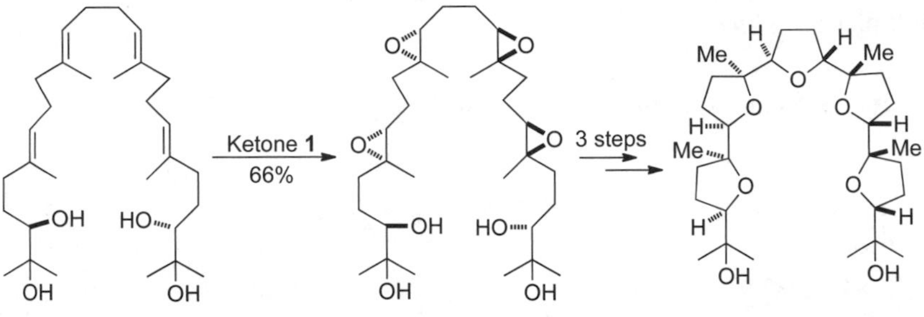

28   Ketone 1, 66% →   29   3 steps →   Glabrescol (**30**)

In 2010, (+)-omaezakianol was obtained by Xiong, Busch, and Corey via penta-epoxidation of **31** followed by cascade cyclization and dehalogenation (Scheme 9).[42]

Scheme 9. Synthesis of (+)-Omaezakianol (**34**)

31   Ketone 1, Oxone →

32   CSA, acetone, rt, 21% from **31** →

33   Na, ether, 76% →

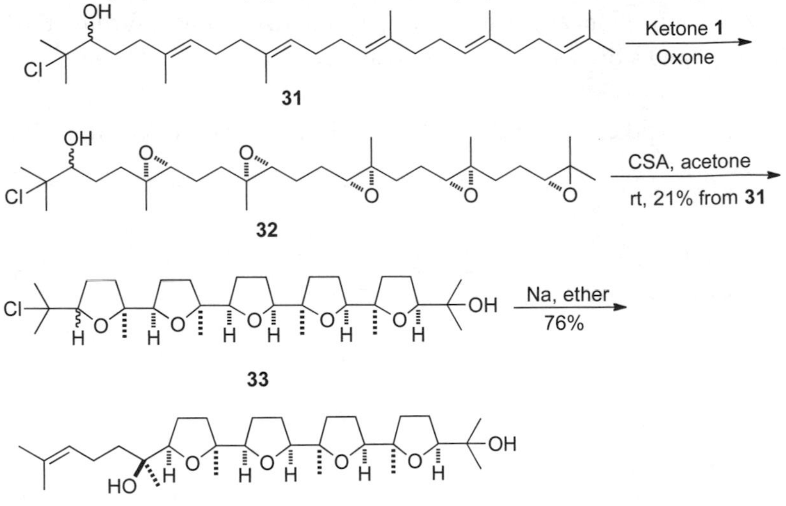

(+)-Omaezakianol (**34**)

In 2009, Morimoto and coworkers accomplished a synthesis of (+)-omaezakianol (**34**) using a convergent approach (Scheme 10).[43] Ketone **1** promoted a *bis*-epoxidation of **35** in high yield and diastereoselectivity.

356

Compound **37** was assembled, in part, by an *ent*-ketone **1** catalyzed epoxidation. Epoxide **36** and **37** were cross-coupled and a cascade cyclization along with some additional steps afforded the target product.

Scheme 10. Synthesis of (+)-Omaezakianol (**34**)

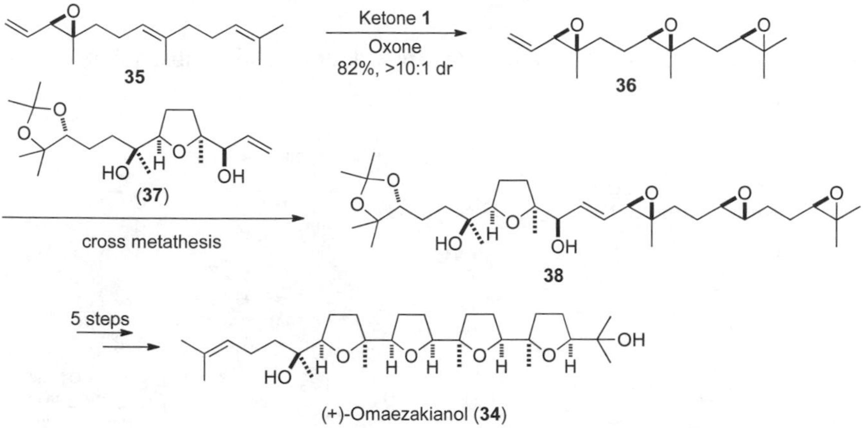

Scheme 11. Polyepoxide Cyclization

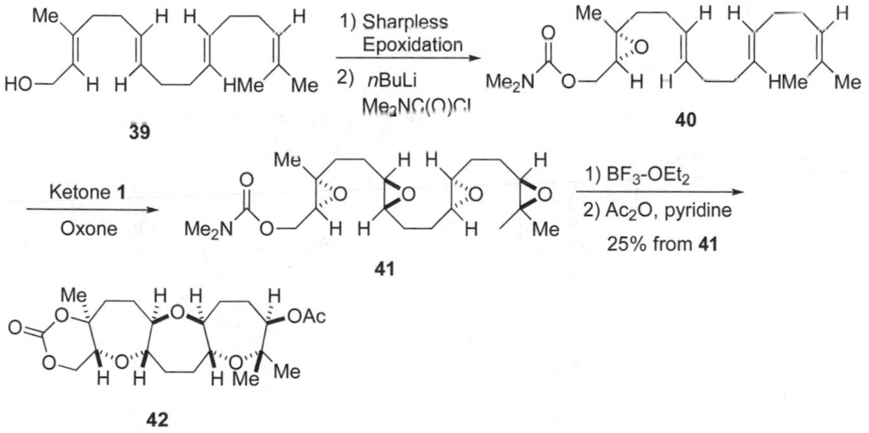

McDonald and coworkers have reported on a series of biomimetic cascade cyclizations of poly-epoxides to access complex polycyclic ethers. An example is shown in Scheme 11.[44a] The Sharpless methodology was used to epoxidize the allylic alcohol and ketone **1** facilitated the tris-epoxidation of **40**. The carbamate was used to trigger the cascade cyclization under Lewis-acidic conditions to afford fused tetracycle **42**.[44]

Jamison and coworkers have shown that poly-vinyl silane **46** can be poly-epoxidized to afford **47**, which was converted to a ladder polyether under basic conditions in methanol. The trimethylsilyl groups directed the 6-*endo* attack and then "disappeared" (Scheme 12).[45]

Scheme 12. Stereoselective *Tris*-Epoxidation and Subsequent TMS-Directed Cyclization

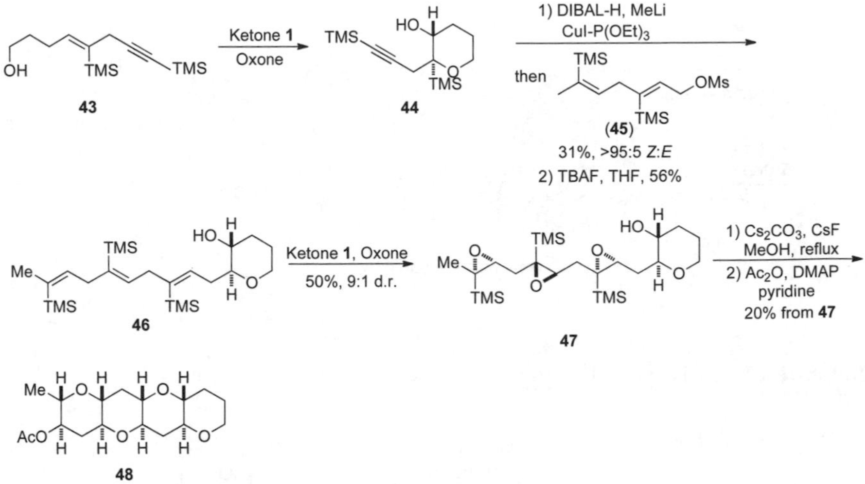

Scheme 13. Stereoselective *Tris*-Epoxidation and Subsequent Water-Promoted *Endo*-Cyclization

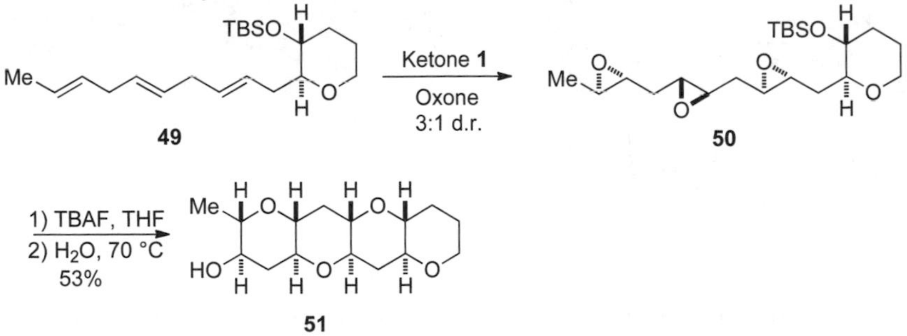

In their subsequent studies, Jamison and coworkers have reported that water can promote the highly regioselective 6-*endo* cyclization of tethered di-and tri-epoxides thus bypassing the need for the TMS-directing groups

(Scheme 13).[46] Additionally, the water allows the 6-*endo* cyclization to override regioselective bias induced by methyl groups on the epoxides.[46c] They have also shown that the 1,3-dioxane group can be a suitable template for *endo*-selective cyclizations of epoxides.[46f,47-49]

Marshall and coworkers assembled the C17-C32 fragment of the antibiotic ionomycin (Scheme 14).[50] Sharpless epoxidation installed the peripheral epoxides in **53** and ketone **1** catalyzed the epoxidation of the internal double bond, resulting in **54**.[50-52] The zinc debrominates **55** and facilitates the cascade cyclization.

Scheme 14. Synthesis of C17-C32 Fragment of Ionomycin

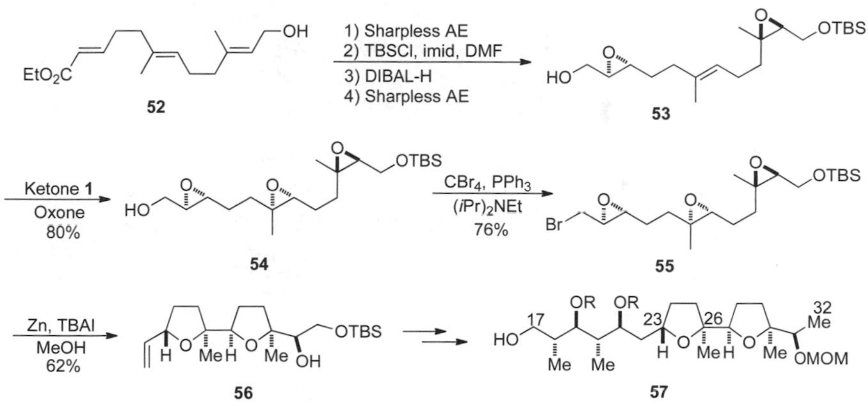

Scheme 15. Synthesis of (+)-Aurilol (**63**)

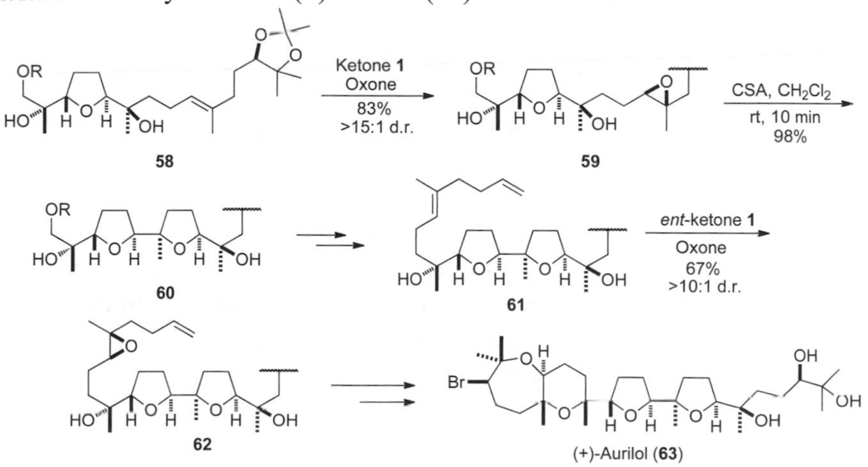

In 2005, Morimoto and coworkers reported the first total synthesis of (+)-aurilol (**63**) (Scheme 15).[53] A series of epoxidations catalyzed by ketone **1** and *ent*-ketone **1** allowed them to construct the cyclic ethers in a regio- and stereo-controlled fashion.

McDonald and coworkers synthesized *ent*-abudinol B (**69**) utilizing epoxidation/cascade cyclization (Scheme 16).[54] *Bis*-epoxidation of **64** with ketone **1** resulted in high diastereoselectivity in the formation of **65**. A second iteration of *bis*-epoxidation resulted in **68** which formed **69** upon treatment with TMSOTf. They had reported an earlier synthesis of the same molecule which also employed ketone **1**-catalyzed epoxidation.[54d]

Scheme 16. Synthesis of *ent*-Abudinol B (**69**)

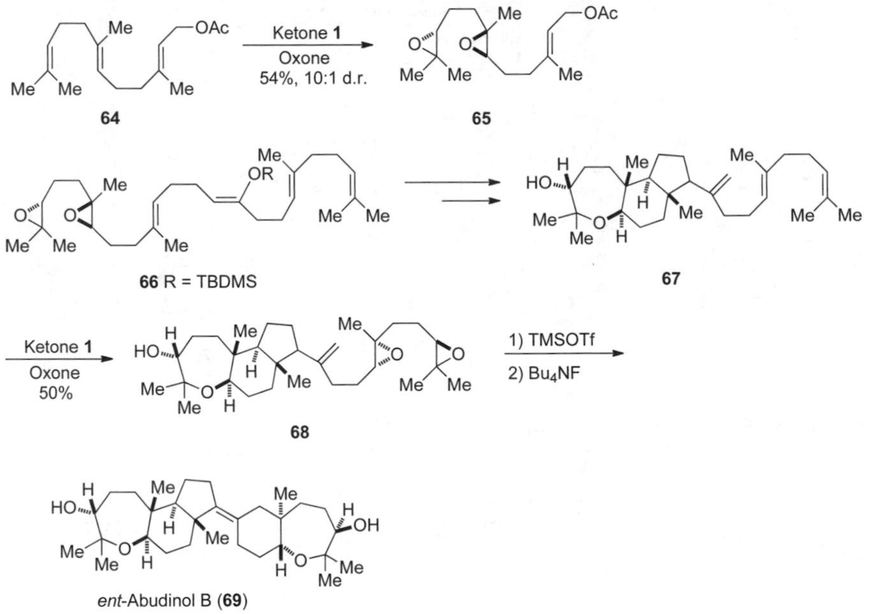

Uenishi and coworkers stereoselectively constructed linked tetrahydrofurans with a Pd(II)-catalyzed cascade cyclization (Scheme 17).[55] The metal serves to activate the epoxide while controlling the regioselectivity and also to catalyze and transfer the chirality of the displaced allylic alcohol in the S_N2'-type reaction.

Org. Synth. **2012**, *89*, 350-373

Scheme 17. Pd(II)-Catalyzed Formation of Linked Tetrahydrofurans

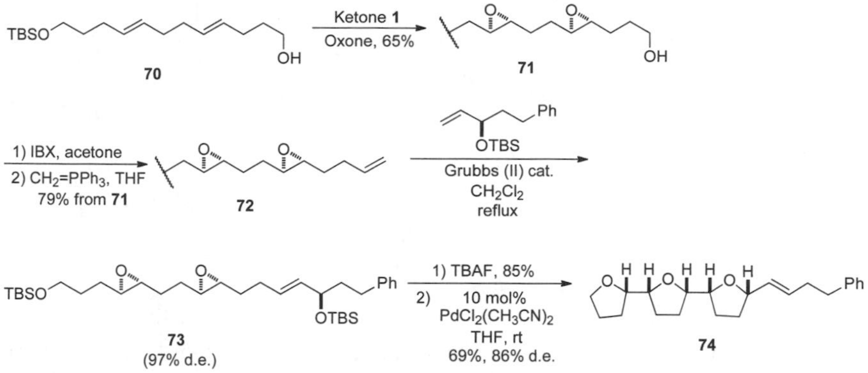

Micalizio and coworkers synthesized a tetracyclic fragment **79** of pectenotoxin-2 (Scheme 18).[56] *Ent*-Ketone **1**/H$_2$O$_2$ epoxidized the double bond in **75** with excellent diastereoselectivity and subsequent deprotection of the TES-protected alcohol led to formation of THF **77**. Deprotection and oxidation of the TBS-protected alcohol led to a domino condensation/cyclization to form **78**.

Scheme 18. Synthetic Work Toward Pectenotoxin-2

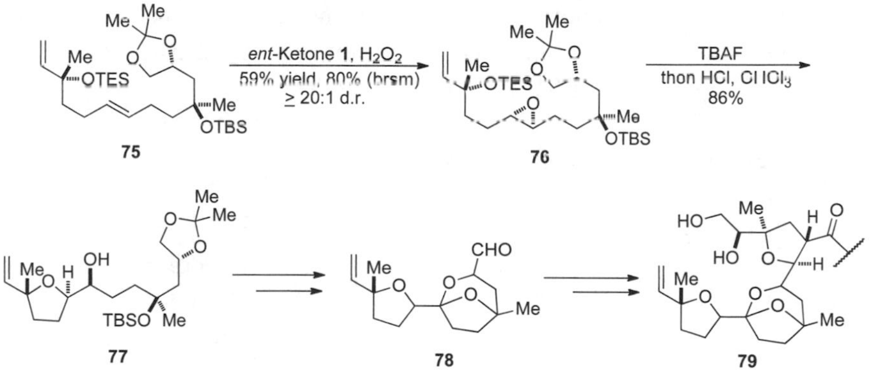

Lee, Gagné, and coworkers have shown that allenes can also initiate these cascade reactions.[57] In the example shown in Scheme 19, a tandem of Sharpless and ketone **1**-catalyzed epoxidations set the stereochemistry in **81**, and upon treatment with Au(I), the allene reacts with the proximal epoxide to set off the cascade. The product was obtained in good yield and high d.r.

Scheme 19. Allene-Initiated Cascade Cyclization

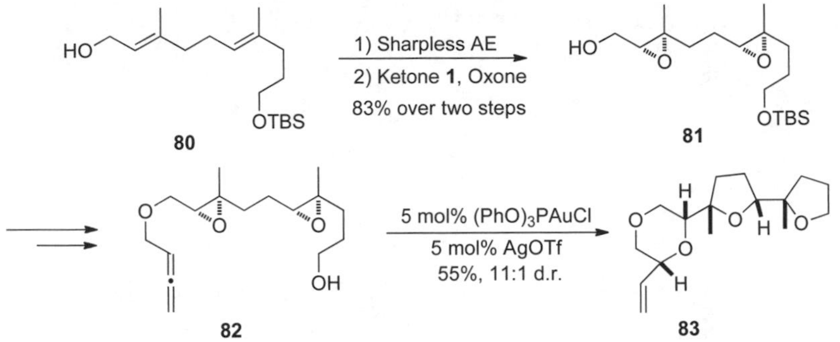

The previous section has given a snapshot of the synthetic utility of ketone **1** and *ent*-ketone **1**. The use of this methodology in synthesis[58] and dynamic method developments[58-61] are anticipated to undergo continued growth.

Ketone 2:

Ketone **2** has recently begun to receive more attention in organic synthesis. Suzuki and coworkers completed the first synthesis of *ent*-seragakinone A (**87**) (Scheme 20) and confirmed its absolute stereochemistry.[62] Substrate-controlled epoxidation of **84** resulted in low diastereoselectivity, however *ent*-ketone **2** induced good diastereoselectivity (8.2:1 d.r.). Acid-promoted hydration of the epoxide and subsequent conjugate addition provide the THF in compound **86**. This provided the tetracyclic core which was further elaborated to **87**.

Scheme 20. Synthesis of *ent*-Seragakinone A (**87**)

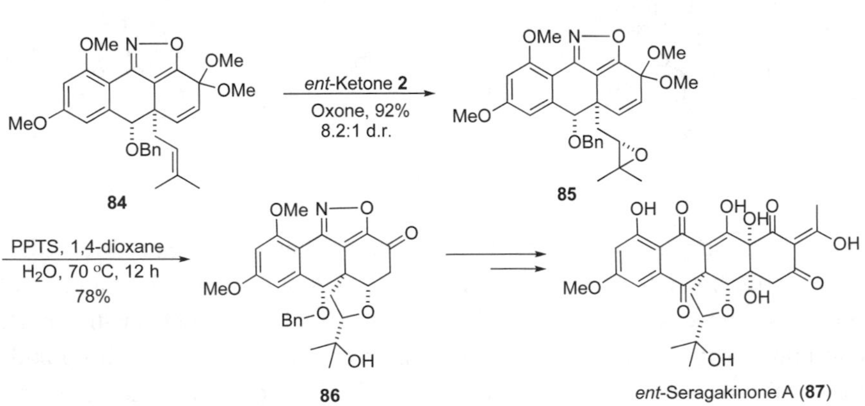

Hamada and coworkers have used ketone **2** to synthesize precursors to valuable compounds such as angelmarin, (+)-marmesin, (−)-(3'R)-decursinol, and (+)-lomatin.[63] Synthetic work toward angelmarin, a chemical that displays activity against PANC-1 cancer cells, is shown in Scheme 21.[63] Epoxidation with ketone **2** led to the formation of **89** in 88% yield and 97% e.e. Employing a two-step process for the removal of the TBS group and cyclization at low temperatures allows for highly selective (and complementary) dihydrobenzopyran formation.[63] Kan and coworkers have used ketone **2**-catalyzed epoxidation to synthesize epimeric epigallocatechin gallates.[64]

Scheme 21. Synthesis of Angelmarin (**90**) using Ketone **2**

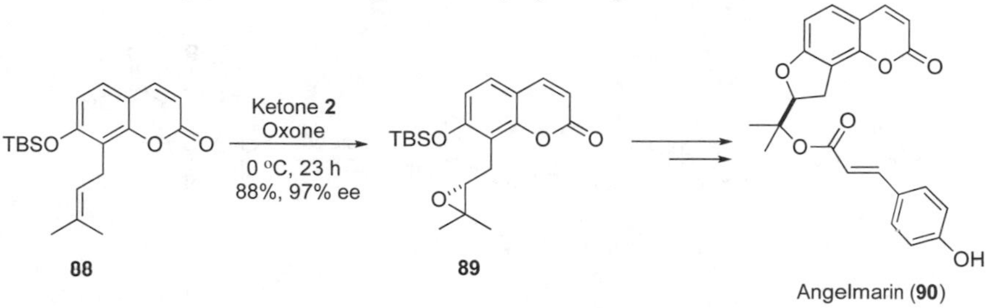

Yadav and coworkers have achieved an expedient formal synthesis of diastereomeric Hagen's gland lactones using ketone **2**.[65] Bruner and coworkers epoxidized *trans*-ethyl cinnamates and saponified the esters to study the products' ability to inhibit *Sg*TAM, an MIO-based aminomutase.[66]

Scheme 22. Synthesis of Lactodehydrothyrsiferol (**94**)

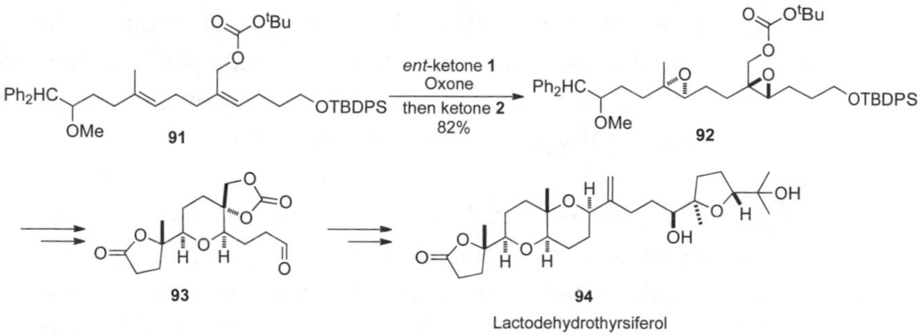

Lactodehydrothyrsiferol

Floreancig and coworkers recently accomplished the *bis*-epoxidation of **91** in high stereoselectivity using *ent*-ketone **1** and ketone **2** (Scheme 22).

Ent-ketone **1** catalyzed the epoxidation of the more reactive double bond, and upon completion, ketone **2** was added to the pot to facilitate epoxidation of the less reactive site. They then further elaborated **92** to lactodehydrothyrsiferol (**94**).[67]

Shi and coworkers recently obtained 120 g of virtually enantiopure (+)-ambrisentan (**97**) without the need for column chromatography (Scheme 23).[68] (+)-Ambrisentan, an endothelin-1 receptor antagonist, is currently used to treat hypertension. Ketone **2**-catalyzed epoxidation afforded **96** in 90% conversion and 85% ee. Compound **97** was further enriched via precipitation and filtration of the racemate.

Scheme 23. Synthesis of (*S*)-Ambrisentan

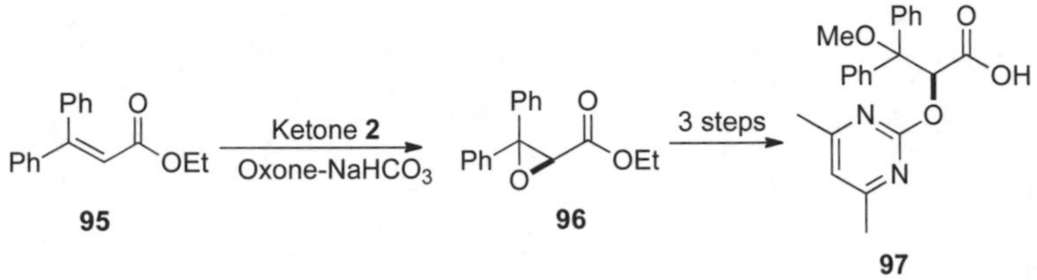

97
(+)-(*S*)-Ambrisentan

53% yield, >99% e.e.,
120.1 g obtained

Ketones 3:

Ketones **3** have been used in the asymmetric epoxidation/rearrangement of benzylidene cyclopropanes to synthetically valuable cyclobutanones and lactones and in the transformation of benzylidene cyclobutanes to cyclopentanones.[18-20] In addition, they have been used for the synthesis of oxindoles from indoles[69] and for the desymmetrization of *meso*-hydrobenzoins and the kinetic resolution of racemic hydrobenzoins.[70] They can also catalyze the stereoselective epoxidation of amphiphilic olefins; in the case of carboxylic acids, the carboxyl group can provide a directing effect as well as serve as a trigger for *in situ* lactonization.[22]

1. Department of Chemistry, Colorado State University, Fort Collins, CO 80523. We are grateful for the generous financial support provided by the General Medical Sciences of the National Institutes of Health (GM59705).
2. For leading reviews on asymmetric epoxidation mediated by chiral ketones, see: (a) Denmark, S. E.; Wu, Z. *Synlett* **1999**, 847–859. (b)

 Org. Synth. **2012**, *89*, 350-373

Frohn, M.; Shi, Y. *Synthesis* **2000**, 1979–2000. (c) Shi, Y. *J. Syn. Org. Chem., Japan* **2002**, *60*, 342–349. (d) Shi, Y. In *Modern Oxidation Methods*; Bäckvall, J-E. Ed.; Wiley-VCH: Weinheim, **2004**; Chapter 3. (e) Shi, Y. *Acc. Chem. Res.* **2004**, *37*, 488–496. (f) Yang, D. *Acc. Chem. Res.* **2004**, *37*, 497–505. (g) Shi, Y. In *Handbook of Chiral Chemicals*; Ager, D. Ed.; CRC Press, Taylor & Francis Group: Boca Raton, **2006**; Chapter 10. (h) Wong, O. A.; Shi, Y. *Chem. Rev.* **2008**, *108*, 3958–3987. (i) Goeddel, D.; Shi Y. In *Science of Synthesis*, Vol. 37, Forsyth, C.J. Ed., Georg Thieme Verlag KG, Stuttgart, Germany, **2008**, 277–320. (j) Wong, O. A.; Shi, Y. In *Top. Curr. Chem.*; List, B. Ed.; Springer: New York, **2010**; Vol 291, 201–232. (k) Shi, Y. In *Modern Oxidation Methods, 2nd edition*; Bäckvall, J-E. Ed.; Wiley-VCH: Weinheim, **2010**; Chapter 3.

3. (a) Tu, Y.; Wang, Z-X.; Shi, Y. *J. Am. Chem. Soc.* **1996**, *118*, 9806–9807. (b) Wang, Z-X.; Tu, Y.; Frohn, M.; Zhang, J-R.; Shi, Y. *J. Am. Chem. Soc.* **1997**, *119*, 11224–11235.

4. Warren, J.D.; Shi, Y. *J. Org. Chem.* **1999**, *64*, 7675–7677.

5. Wang, Z-X.; Shi, Y. *J. Org. Chem.* **1998**, *63*, 3099–3104.

6. Frohn, M.; Dalkiewicz, M.; Tu, Y.; Wang, Z-X.; Shi, Y. *J. Org. Chem.* **1998**, *63*, 2948–2953.

7. (a) Cao, G-A.; Wang, Z-X.; Tu, Y.; Shi, Y. *Tetrahedron Lett.* **1998**, *39*, 4425–4428. (b) Wang, Z-X.; Cao, G-A; Shi, Y. *J. Org. Chem.* **1999**, *64*, 7646–7650.

8. (a) Zhu, Y.; Tu, Y.; Yu, H.; Shi, Y. *Tetrahedron Lett.* **1998**, *39*, 7819–7822. (b) Zhu, Y.; Manske, K.J.; Shi, Y. *J. Am. Chem. Soc.* **1999**, *121*, 4080–4081. (c) Feng, X.; Shu, L.; Shi, Y. *J. Am. Chem. Soc.* **1999**, *121*, 11002–11003. (d) Zhu, Y.; Shu, L.; Tu, Y.; Shi, Y. *J. Org. Chem.* **2001**, *66*, 1818–1826.

9. Adam, W.; Fell, R. T.; Saha-Möller, C. R.; Zhao, C-G. *Tetrahedron: Asymmetry* **1998**, *9*, 397–401.

10. Zhao, M-X.; Shi, Y. *J. Org. Chem.* **2006**, *71*, 5377–5379.

11. (a) Wu, X.-Y.; She, X.; Shi, Y. *J. Am. Chem. Soc.* **2002**, *124*, 8792–8793. (b) Wang, B.; Wu, X.-Y.; Wong, O.A.; Nettles, B.; Zhao, M.-X.; Chen, D.; Shi, Y. *J. Org. Chem.* **2009**, *74*, 3986–3989.

12. (a) Tian, H.; She, X.; Shu, L.; Yu, H.; Shi, Y. *J. Am. Chem. Soc.* **2000**, *122*, 11551–11552. (b) Tian, H.; She, X.; Xu, J.; Shi, Y. *Org. Lett.* **2001**, *3*, 1929–1931. (c) Tian, H., She, X.; Yu, H.; Shu, L.; Shi, Y. *J.*

Org. Chem. **2002**, *67*, 2435–2446. (d) Shu, L.; Shen, Y-M.; Burke, C.; Goeddel, D.; Shi, Y. *J. Org. Chem.* **2003**, *68*, 4963–4965.

13. (a) Shu, L.; Wang, P.; Gan, Y.; Shi, Y. *Org. Lett.* **2003**, *5*, 293–296. (b) Goeddel, D.; Shu, L.; Yuan, Y.; Wong, O. A.; Wang, B.; Shi, Y. *J. Org. Chem.* **2006**, *71*, 1715–1717. (c) Zhao, M-X.; Goeddel, D.; Li, K. and Shi, Y. *Tetrahedron* **2006**, *62*, 8064–8068.

14. Shu, L.; Shi, Y. *Tetrahedron Lett.* **2004**, *45*, 8115–8117.

15. Wong, O.A.; Shi, Y. *J. Org. Chem.* **2006**, *71*, 3973–3976.

16. Burke, C. P.; Shi, Y. *Angew. Chem., Int. Ed.* **2006**, *45*, 4475–4478.

17. Burke, C. P.; Shi, Y. *J. Org. Chem.* **2007**, *72*, 4093–4097.

18. Shen, Y.-M.; Wang, B.; Shi, Y. *Angew. Chem., Int. Ed.* **2006**, *45*, 1429–1432.

19. Shen, Y.-M.; Wang, B.; Shi, Y. *Tetrahedron Lett.* **2006**, *47*, 5455–5458.

20. Wang, B.; Shen, Y-M.; Shi, Y. *J. Org. Chem.* **2006**, *71*, 9519–9521.

21. Burke, C. P.; Shu, L.; Shi, Y. *J. Org. Chem.* **2007**, *72*, 6320–6323.

22. Burke, C. P.; Shi, Y. *Org. Lett.* **2009**, *11*, 5150–5153.

23. (a) Wang, B.; Wong, O. A.; Zhao, M.-X.; Shi, Y. *J. Org. Chem.* **2008**, *73*, 9539–9543. (b) Wong, O. A.; Wang, B.; Zhao, M.-X.; Shi, Y. *J. Org. Chem.* **2009**, *74*, 6335–6338.

24. Wong, O.A., Shi, Y. "Guidelines for Asymmetric Epoxidation" <<http://www.chm.colostate.edu/shi/Epoxidation%20Guidelines/Epoxi dation%20Guidelines.html>>

25. Burke, A.; Dillon, P.; Martin, K.; Hanks, T. W. *J. Chem. Ed.* **2000**, *77*, 271–272.

26. Ager, D. J.; Anderson, K.; Oblinger, E.; Shi, Y.; VanderRoest, *J. Org. Process Res. Dev.* **2007**, *11*, 44–51.

27. (a) Altmann, K.-H.; Bold, G.; Caravatti, G.; Denni, D.; Flörsheimer, A.; Schmidt, A.; Rihs, G.; Wartmann, M. *Helv. Chim. Acta* **2002**, *85*, 4086–4110. (b) Cachoux, F.; Isarno, T.; Wartmann, M.; Altmann, K.-H. *Angew. Chem., Int. Ed.* **2005**, *44*, 7469–7473. (c) Cachoux, F.; Isarno, T.; Wartmann, M; Altmann, K.-H. *ChemBioChem* **2006**, *7*, 54–57. (d) Feyen, F.; Cachoux, F.; Gertsch, J.; Wartmann, M.; Altmann, K.-H. *Acc. Chem. Res* **2008**, *41*, 21–31.

28. Bian, J.; Van Wingerden, M.; Ready, J. M. *J. Am. Chem. Soc.* **2006**, *128*, 7428–7429.

29. Hoard, D. W.; Moher, E. D.; Martinelli, M. J.; Norman, B. H. *Org. Lett.* **2002**, *4*, 1813–1815.

30. Kanada, R. M.; Itoh, D.; Nagai, M.; Niijima, J.; Asai, N.; Mizui, Y.; Abe, S.; Kotake, Y. *Angew. Chem., Int. Ed.* **2007**, *46*, 4350–4355.

31. Mandel, A. L.; Jones, B. D.; La Clair, J. J.; Burkart, M. D. *Bioorg. Med. Chem. Lett.* **2007**, *17*, 5159–5164.

32. (a) Knöll, J.; Knölker, H.-J. *Tetrahedron Lett.* **2006**, *47*, 6079–6082. (b) Sammet, B.; Radzey, H.; Neumann, B.; Stammler, H.-G.; Sewald, N. *Synlett* **2009**, 417–420. (c) Gundluru, M. K.; Pourpak, A.; Cui, X.; Morris, S. W.; Webb, T. R. *Med. Chem. Commun.* **2011**, *2*, 904–908.

33. Emmanuvel, L.; Sudalai, A. *Tetrahedron Lett.* **2008**, *49*, 5736–5738.

34. Lim, S. M.; Hill, N.; Myers, A. G. *J. Am. Chem. Soc.* **2009**, *131*, 5763–5765.

35. For use in a related *syn*-fluorination, see: Cresswell, A. J.; Davies, S. G.; Lee, J. A.; Roberts, P. M.; Russell, A. J.; Thomson, J. E.; Tyte, M. J. *Org. Lett.* **2010**, *12*, 2936–2939.

36. Wiseman, J. M.; McDonald, F. E.; Liotta, D. C. *Org. Lett.* **2005**, *7*, 3155–3157.

37. (a) Yang, Y.-R.; Lai, Z.-W.; Shen, L.; Huang, J.-Z.; Wu, X.-D.; Yin, J.-L.; Wei, K. *Org. Lett.* **2010**, *12*, 3430–3433. (b) Yang, Y.-R.; Shen, L.; Huang, J.-Z.; Xu, T.; Wei, K. *J. Org. Chem.* **2011**, 76, 3684–3690.

38. Huo, X.; Pan, X.; Huang, G.; She, X. *Synlett* **2011**, 1149–1150.

39. (a) Mente, N. R.; Neighbors, J. D.; Wiemer, D. F. *J. Org. Chem.* **2008**, *73*, 7963–7970. (b) Neighbors, J. D.; Mente, N. R.; Boss, K. D.; Zehnder, II, D. W.; Wiemer, D. F. *Tetrahedron Lett.* **2008**, *49*, 516–519. (c) Neighbors, J. D.; Topczewski, J. J.; Swenson, D. C.; Wiemer, D. F. *Tetrahedron Lett.* **2009**, *50*, 3881–3884. (d) Topczewski, J.J.; Neighbors, J. D.; Wiemer, D. F. *J. Org. Chem.* **2009**, *74*, 6965–6972. (e) Topczewski, J. J.; Wiemer, D. F. *Tetrahedron Lett.* **2011**, *52*, 1628–1630. (f) Topczewski, J. J.; Kodet, J. G.; Wiemer, D. F. *J. Org. Chem.* **2011**, *76*, 909–919.

40. (a) Xiong, Z.; Corey, E. J. *J. Am. Chem. Soc.* **2000**, *122*, 4831–4832. (b) Xiong, Z.; Corey, E. J. *J. Am. Chem. Soc.* **2000**, *122*, 9328–9329.

41. For another synthesis of Glabrescol, see: (a) Morimoto, Y.; Iwai, T.; Kinoshita, T. *J. Am. Chem. Soc.* **2000**, *122*, 7124–7125. (b) Morimoto, Y. *Org. Biomol. Chem.* **2008**, *6*, 1709–1719.

42. Xiong, Z.; Busch, R.; Corey, E. J. *Org. Lett.* **2010**, *12*, 1512–1514.

43. Morimoto, Y.; Okita, T.; Kambara, H. *Angew. Chem., Int. Ed.* **2009**, *48*, 2538–2541.

44. (a) Valentine, J. C.; McDonald, F. E.; Neiwert, W. A.; Hardcastle, K. I. *J. Am. Chem. Soc.* **2005**, *127*, 4586–4587. For early related work, see: (b) McDonald, F. E.; Wang, X.; Do, B.; Hardcastle, K. I. *Org. Lett.* **2000**, *2*, 2917–2919. (c) McDonald, F. E.; Bravo, F.; Wang, X.; Wei, X.; Toganoh, M.; Ramón Rodríguez, J.; Do, B.; Neiwert, W. A.; Hardcastle, K. I. *J. Org. Chem.* **2002**, *67*, 2515–2523. (d) McDonald, F. E.; Wei, X. *Org. Lett.* **2002**, *4*, 593–595. (e) Bravo, F.; McDonald, F.E.; Neiwert, W. A.; Do, B.; Hardcastle, K. I. *Org. Lett.* **2003**, *5*, 2123–2126. (f) Tong, R.; McDonald, F. E.; Fang, X.; Hardcastle, K.I. *Synthesis* **2007**, 2337–2342.

45. (a) Heffron, T. P.; Jamison, T. F. *Org. Lett.* **2003**, *5*, 2339–2342. (b) Simpson, G. L.; Heffron, T. P.; Merino, E.; Jamison, T. F. *J. Am. Chem. Soc.* **2006**, *128*, 1056–1057.

46. (a) Vilotijevic, I.; Jamison, T. F. *Science* **2007**, *317*, 1189–1192. (b) Byers, J. A.; Jamison, T. F. *J. Am. Chem. Soc.* **2009**, *131*, 6383–6385. (c) Morten, C. J.; Jamison, T. F. *J. Am. Chem. Soc.* **2009**, *131*, 6678–6679. (d) Tanuwidjaja, J.; Ng, S.-S.; Jamison, T. F. *J. Am. Chem. Soc.* **2009**, *131*, 12084–12085. (e) Morten, C. J.; Jamison, T. F. *Tetrahedron* **2009**, *65*, 6648–6655. (f) Van Dyke, A.R.; Jamison, T.F. *Angew. Chem., Int. Ed.* **2009**, *48*, 4430–4432. (g) Heffron, T. P.; Simpson, G. L.; Merino, E.; Jamison, T. F. *J. Org. Chem.* **2010**, *75*, 2681–2701. (h) Morten, C. J.; Byers, J. A.; Jamison, T. F. *J. Am. Chem. Soc.* **2011**, *133*, 1902–1908.

47. For methoxy group directed 6-*endo-bis*- and *tris*-cyclizations of epoxides, see: Tokiwano, T.; Fujiwara, K.; Murai, A. *Synlett* **2000**, 335–338.

48. For electron-transfer initiated *endo-bis* cyclizations of epoxides, see: (a) Wan, S.; Gunaydin, H.; Houk, K. N.; Floreancig, P. E. *J. Am. Chem. Soc.* **2007**, *129*, 7915–7923. For earlier studies on electron-transfer initiated cyclizations of epoxides, see: (b) Kumar, V. S.; Aubele, D. L.; Floreancig, P. E. *Org. Lett.* **2002**, *4*, 2489–2492. (c) Kumar, V. S.; Wan, S.; Aubele, D. L.; Floreancig, P. E. *Tetrahedron: Asymmetry* **2005**, *16*, 3570–3578.

49. For reviews on epoxide-based cascade cyclizations, see: (a) Morimoto, Y.; Kinoshita, T.; Iwai, T. *Chirality* **2002**, *14*, 578–586. (b) Fujiwara, K.; Murai, A. *Bull. Chem. Soc., Jpn.* **2004**, *77*, 2129–2146. (c) Valentine, J. C.; McDonald, F. E. *Synlett* **2006**, 1816–1828. (d) Brimble, M. A.; Halim, R. *Pure Appl. Chem.* **2007**, *79*, 153–162. (e)

Morris, J. C.; Phillips, A.J. *Nat. Prod. Rep.* **2008**, *25*, 95–117. (f) Morimoto, Y. *Org. Biomol. Chem.* **2008**, *6*, 1709–1719. (g) Morris, J. C.; Phillips, A. J. *Nat. Prod. Rep.* **2009**, *26*, 245–265. (h) Vilotijevic, I.; Jamison, T. F. *Angew. Chem., Int. Ed.* **2009**, *48*, 5250–5281. (i) Morten, C. J.; Byers, J. A.; Van Dyke, A. R.; Vilotijevic, I.; Jamison, T. F. *Chem. Soc. Rev.* **2009**, *38*, 3175–3192. (j) Vilotijevic, I.; Jamison, T. F. *Mar. Drugs* **2010**, *8*, 763–809. (k) Nicolaou, K. C.; Aversa, R. J. *Isr. J. Chem.* **2011**, *51*, 359–377. (l) Andersen, E. A. *Org. Biomol. Chem.* **2011**, *9*, 3997–4006.

50. (a) Marshall, J. A.; Chobanian, H. R. *Org. Lett.* **2003**, *5*, 1931–1933. (b) Marshall, J. A.; Mikowski, A. M. *Org. Lett.* **2006**, *8*, 4375–4378.

51. Marshall, J. A.; Hann, R. K. *J. Org. Chem.* **2008**, *73*, 6753–6757.

52. For the use of ketone **1** and *ent*-ketone **1** in library synthesis, see: (a) Zhang, Q.; Lu, J.; Richard, C.; Curran, D.P. *J. Am. Chem. Soc.* **2004**, *126*, 36–37. (b) Das, S.; Li, L.-S.; Abraham, S.; Chen, Z.; Sinha, S.C. *J. Org. Chem.* **2005**, *70*, 5922–5931. (c) Curran, D. P.; Zhang, Q.; Richard, C.; Lu, H.; Gudipati, V.; Wilcox, C. S. *J. Am. Chem. Soc.* **2006**, *128*, 9561–9573. (d) Yoshida, T.; Yamauchi, S.; Tago, R.; Maruyama, M.; Akiyama, K.; Sugahara, T.; Kishida, T.; Koba, Y. *Biosci. Biotechnol. Biochem.* **2008**, *72*, 2342–2352. (e) Shichijo, Y.; Migita, A.; Oguri, H.; Watanabe, M.; Tokiwano, T.; Watanabe, K.; Oikawa, H. *J. Am. Chem. Soc.* **2008**, *130*, 12230–12231. (f) Matsuura, Y.; Shichijo, Y.; Minami, A.; Migita, A.; Oguri, H.; Watanabe, M.; Tokiwano, T.; Watanabe, K.; Oikawa, H. *Org. Lett.* **2010**, *12*, 2226–2229. (g) Sato, K.; Minami, A.; Ose, T.; Oguri, H.; Oikawa, H. *Tetrahedron Lett.* **2011**, *52*, 5277 5280.

53. Morimoto, Y.; Nishikawa, Y.; Takaishi, M. *J. Am. Chem. Soc.* **2005**, *127*, 5806–5807.

54. (a) Tong, R.; McDonald, F. E. *Angew. Chem., Int. Ed.* **2008**, *47*, 4377–4379. (b) Tong, R.; Boone, M. A.; McDonald, F. E. *J. Org. Chem.* **2009**, *74*, 8407–8409. (c) Boone, M. A.; Tong, R.; McDonald, F. E.; Lense, S.; Cao, R.; Hardcastle, K. I. *J. Am. Chem. Soc.* **2010**, *132*, 5300–5308. (d) Tong, R.; Valentine, J. C.; McDonald, F. E.; Cao, R.; Fang, X.; Hardcastle, K. I. *J. Am. Chem. Soc.* **2007**, *129*, 1050–1051.

55. Uenishi, J.; Fujikura, Y.; Kawai, N. *Org. Lett.* **2011**, *13*, 2350–2353.

56. Canterbury, D. P.; Micalizio, G. C. *Org. Lett.* **2011**, *13*, 2384–2387.

57. Tarselli, M. A.; Zuccarello, J. L.; Lee, S. J.; Gagné, M. R. *Org. Lett.* **2009**, *11*, 3490–3492.

58. For use of ketone **1** and *ent*-ketone **1** in the synthesis of natural products and/or biologically active molecules and/or fragments thereof, see: (a) Bluet, G.; Campagne, J.-M. *Synlett* **2000**, 221–222 (Octalactin A). (b) Guz, N. R.; Lorenz, P.; Stermitz, F. R. *Tetrahedron Lett.* **2001**, *42*, 6491–6494 [(+)-Trachypleuranin-A]. (c) Morimoto, Y.; Iwai, T.; Kinoshita, T. *Tetrahedron Lett.* **2001**, *42*, 6307–6309 [(−)-Longilene Peroxide]. (d) Smith, III, A. B.; Fox, R. J. *Org. Lett.* **2004**, *6*, 1477–1480 [(+)-Sorangicin A]. (e) Adams, C. M.; Ghosh, I.; Kishi, Y. *Org. Lett.* **2004**, *6*, 4723–4726 (Glisoprenin A). (f) Lecornué, F.; Paugam, R.; Ollivier, J. *Eur. J. Org. Chem.* **2005**, 2589–2598 (Heliannuols). (g) Taber, D. F.; He, Y. *J. Org. Chem.* **2005**, *70*, 7711–7714 [(−)-Mesembrine]. (h) Smith, III, A. B.; Walsh, S. P.; Frohn, M.; Duffey, M. O. *Org. Lett.* **2005**, *7*, 139–142 (Lituarines). (i) Halim, R.; Brimble, M. A.; Merten, J. *Org. Lett.* **2005**, *7*, 2659–2662 (Pectenotoxins). (j) Halim, R.; Brimble, M. A.; Merten, J. *Org. Biomol. Chem.* **2006**, *4*, 1387–1399 (Pectenotoxins). (k) Morimoto, Y.; Takaishi, M.; Adachi, N.; Okita, T.; Yata, T. *Org. Biomol. Chem.* **2006**, *4*, 3220-3222 [(+)-Intricatetraol]. (l) Iwasaki, J.; Ito, H.; Nakamura, M.; Iguchi, K. *Tetrahedron Lett.* **2006**, *47*, 1483–1486 (Pachyclavulide B). (m) Uchida, K.; Ishigami, K.; Watanabe, H.; Kitahara, T. *Tetrahedron* **2007**, *63*, 1281–1287 (an insecticidal tetrahydroisocoumarin). (n) Chapelat, J.; Buss, A.; Chougnet, A.; Woggon, W.-D. *Org. Lett.* **2008**, *10*, 5123–5126 (α-Tocopherol). (o) Yu, M.; Snider, B. B. *Org. Lett.* **2009**, *11*, 1031–1032 [(+) and (−) Monanchorin]. (p) Magolan, J.; Coster, M. J. *J. Org. Chem.* **2009**, *74*, 5083-5086 [(+)-Angelmarin]. (q) Kimachi, T.; Torii, E.; Ishimoto, R.; Sakue, A.; Ju-ichi, M. *Tetrahedron: Asymmetry* **2009**, *20*, 1683–1689 (Rhinacanthin A). (r) Matsushima, Y.; Kino, J. *Eur. J. Org. Chem.* **2010**, 2206–2211 (*N*-Bz-D-Ristosamine). (s) Dansey, M. V.; Di Chenna, P. H.; Veleiro, A. S.; Krištofíková, Z.; Chodounska, H.; Kasal, A.; Burton, G. *Eur. J. Med. Chem.* **2010**, *45*, 3063–3069 (A-Homo-5-Pregnenes). (t) Cheng, X.; Harzdorf, N. L.; Shaw, T.; Siegel, D. *Org. Lett.* **2010**, *12*, 1304–1307 (Caryolanemagnolol). (u) Joyasawal, S.; Lotesta, S. D.; Akhmedov, N. G.; Williams, L. J. *Org. Lett.* **2010**, *12*, 988–991 (Pectenotoxins). (v) Marshall, J. A.; Griot, C. A.; Chobanian, H. R.; Myers, W. H. *Org. Lett.* **2010**, *12*, 4328–4331 (Uprolide D). (w) Yaragorla, S.; Muthyala, R. *Tetrahedron Lett.* **2010**, *51*, 467–470 [(−)-Balanol]. (x) Kimachi, T.; Torii, E.; Kobayashi, Y.; Doe, M.; Ju-ichi, M. *Chem. Pharm. Bull.*

2011, *59*, 753–756 [(−)-Dehydroiso-β-Lapachone]. (y) Oikawa, M.; Hashimoto, R.; Sasaki, M. *Eur. J. Org. Chem.* **2011**, 538–546 (Terpendole E). (z) Nakamura, Y.; Ogawa, Y.; Suzuki, C.; Fujimoto, T.; Miyazaki, S.; Tamaki, K.; Nishi, T.; Suemune, H. *Heterocycles* **2011**, *83*, 1587–1602 (5-amino-6-dialkylamino-4-hydroxypentanamide derivatives). (aa) Harrison, T. J.; Ho, S.; Leighton, J. L. *J. Am. Chem. Soc.* **2011**, *133*, 7308–7311 (Zincophorin Methyl Ester). (ab) Butler, J. R.; Wang, C.; Bian, J.; Ready, J. M. *J. Am. Chem. Soc.* **2011**, *133*, 9956–9959 [(−)Kibdelone C]. (ac) Iwata, A.; Inuki, S.; Oishi, S.; Fujii, N.; Ohno, H. *J. Org. Chem.* **2011**, *76*, 5506–5512 [(+)-Lysergic Acid]. (ad) Aumann, K. E.; Hungerford, N. L.; Coster, M. J. *Tetrahedron Lett.* **2011**, *52*, 6988–6990 (Methyl-(+)-7-Methoxyanodendroate). (ae) Smith, III, A. B.; Dong, S.; Fox, R. J.; Brenneman, J. R.; Vanecko, J. A.; Maegawa, T. *Tetrahedron* **2011**, *67*, 9808–9828 [(+)-Sorangicin A].

59. For other applications of ketone **1**-catalyzed epoxidation, see: (a) Shen, K.-H.; Lush, S.-F.; Chen, T.-L.; Liu, R.-S. *J. Org. Chem.* **2001**, *66*, 8106–8111. (b) Olofsson, B.; Somfai, P. *J. Org. Chem.* **2002**, *67*, 8574–8583. (c) Madhushaw, R. J.; Li, C.-L.; Su, H.-L.; Hu, C.-C.; Lush, S.-F.; Liu, R.-S. *J. Org. Chem.* **2003**, *68*, 1872–1877. (d) Olofsson, B.; Somfai, P. *J. Org. Chem.* **2003**, *68*, 2514–2517. (e) Xie, X.; Yue, G.; Tang, S.; Huo, X.; Liang, Q.; She, X.; Pan, X. *Org. Lett.* **2005**, *7*, 4057–4059. (f) Ahmed, M. M.; Mortensen, M. S.; O'Doherty, G. A. *J. Org. Chem.* **2006**, *71*, 7741–7746. (g) Lovchik, M. A.; Fráter, G.; Goeke, A.; Hug, W. *Chem. & Biodiversity* **2008**, *5*, 126–138. (h) Fristrup, P.; Lassen, P. R.; Tanner, D.; Jalkanen, K. J. *Theor. Chem. Acc.* **2008**, *119*, 133–142. (i) Rousseau, G.; Robert, F.; Landais, Y. *Chem.—Eur. J.* **2009**, *15*, 11160–11173. (j) Pinacho Crisóstomo, F. R.; Lledó, A.; Shenoy, S. R.; Iwasawa, T.; Rebek, Jr., J. *J. Am. Chem. Soc.* **2009**, *131*, 7402–7410. (k) Bisol, T. B.; Bortoluzzi, A. J.; Sá, M. M. *J. Org. Chem.* **2011**, *76*, 948–962. (l) Nicolaou, K.C.; Seo, J.H.; Nakamura, T.; Aversa, R.J. *J. Am. Chem. Soc.* **2011**, *133*, 214-219. (m) Hrdina, R.; Müller, C.E.; Wende, R.C.; Lippert, K. M.; Benassi, M.; Spengler, B.; Schreiner, P. R. *J. Am. Chem. Soc.* **2011**, *133*, 7624–7627.

60. For use of ketone **1** in desymmetrizing C-H activation reactions, see: (a) Adam, W.; Saha-Möller, C. R.; Zhao, C.-G. *Tetrahedron: Asymmetry* **1998**, *9*, 4117–4122. (b) Adam, W.; Saha-Möller, C. R.; Zhao, C.-G. *J. Org. Chem.* **1999**, *64*, 7492–7497.

61. For use of ketone **1** in asymmetric sulfur oxidation, see: (a) Colonna, S.; Pironti, V.; Drabowicz, J.; Brebion, F.; Fensterbank, L.; Malacria, M. *Eur. J. Org. Chem.* **2005**, 1727–1730. (b) Dieva, S.A.; Eliseenkova, R. M.; Efremov, Y. Y.; Sharafutdinova, D. R.; Bredikhin, A. A. *Russ. J. Org. Chem.* **2006**, *42*, 12–16. (c) Khiar, N.; Mallouk, S.; Valdivia, V.; Bougrin, K.; Soufiaoui, M.; Fernández, I. *Org. Lett.* **2007**, *9*, 1255–1258.

62. Takada, A.; Hashimoto, Y.; Takikawa, H.; Hikita, K.; Suzuki, K. *Angew. Chem., Int. Ed.* **2011**, *50*, 2297–2301.

63. Jiang, H.; Sugiyama, T.; Hamajima, A.; Hamada, Y. *Adv. Synth. Catal.* **2011**, *353*, 155–162.

64. Hirooka, Y.; Nitta, M.; Furuta, T.; Kan, T. *Synlett* **2008**, 3234–3238.

65. Agrawal, D.; Sriramurthy, V.; Yadav, V. K. *Tetrahedron Lett.* **2006**, *47*, 7615–7618.

66. Montavon, T. J.; Christianson, C. V.; Festin, G. M.; Shen, B.; Bruner, S. D. *Bioorg. Med. Chem. Lett.* **2008**, *18*, 3099–3102.

67. Clausen, D. J.; Wan, S.; Floreancig, P. E. *Angew. Chem., Int. Ed.* **2011**, *50*, 5178–5181.

68. Peng, X.; Li, P.; Shi, Y. *J. Org. Chem.* **2012**, *77*, 701–703.

69. Pettersson, M.; Knueppel, D.; Martin, S. F. *Org. Lett.* **2007**, *9*, 4623–4626.

70. Jakka, K.; Zhao, C.-G. *Org. Lett.* **2006**, *8*, 3013–3015.

Yian Shi was born in Jiangsu, China in 1963. He obtained his B.Sc. degree from Nanjing University in 1983, M.Sc. degree from University of Toronto with Professor Ian W.J. Still in 1987, and Ph.D. degree from Stanford University with Professor Barry M. Trost in 1992. After a postdoctoral study at Harvard Medical School with Professor Christopher Walsh, he joined Colorado State University as assistant professor in 1995, and he was promoted to associate professor in 2000 and professor in 2003. His current research interests include the development of new synthetic methods, asymmetric catalysis, and synthesis of natural products.

Thomas A. Ramirez was born in Corpus Christi, Texas in 1980. He received his B.S. and M.S. degrees from Texas A&M University-Kingsville in 2004 and 2006, respectively, working under the guidance of Professor Apurba Bhattacharya. He is currently a graduate student in the research group of Professor Yian Shi at Colorado State University work on metal-catalyzed vicinal diamination/C-H amination methodologies. Thomas is a Ronald E. McNair Fellow and a BMS Minority Chemist Fellow (2009-2011).

On Lo Andrea Wong was born in Hong Kong in 1981. She moved to Honolulu, Hawai'i in 1994 and received her B.Sc. degree in chemistry from the University of Hawai'i at Mānoa in 2003. She completed her doctoral dissertation in 2009 with Professor Yian Shi at Colorado State University; her research was focused on carbohydrate-derived asymmetric epoxidation. She is currently a postdoctoral fellow in the laboratory of Professor Christopher Ackerson at Colorado State University working on the synthesis of monodispersed thiolate-protected gold nanoparticles.

Discussion Addendum for:
Trifluoromethylation at the α-Position of α,β-Unsaturated Ketones: 4-Phenyl-3-Trifluoromethyl-2-Butanone

Prepared by Kazuyuki Sato, Atsushi Tarui, Masaaki Omote, Itsumaro Kumadaki and Akira Ando.*[1]
Original article: Sato, K.; Omote, M.; Ando, A.; Kumadaki, I. *Org. Synth.* **2006**, *83*, 177.

Trifluoromethylated (CF_3) compounds are one of the most important classes of fluorine compounds, but the reactions for introducing a CF_3 group at the α-position of carbonyl compounds were rarely reported until recently. The reductive α-trifluoromethylation of α,β-unsaturated ketones using the combination of $RhCl(PPh_3)_3$-Et_2Zn, based on the Reformatsky-Honda reaction of non-fluorinated α-bromoacetate, became one of the breakthrough reactions in this field. The birth and advances of this α-trifluoromethylation and related reactions will be summarized here.

α-Fluoroalkylation Using Reformatsky-Honda Reaction Condition

Formerly, we applied the so-called Reformatsky-Honda reaction, using $RhCl(PPh_3)_3$-Et_2Zn,[2] to the reaction of 2-cyclohexen-1-one with ethyl bromodifluoroacetate ($BrCF_2COOEt$), and found that the reaction afforded either the α-(ethoxycarbonyl)difluoromethyl ketone or the usual Reformatsky-type product selectively only by changing the solvent (Figure 1).[3] The former was a new type of product; thus, we tried to expand this reaction to the synthesis of α-fluoroalkyl compounds, especially α-CF_3 compounds. After many efforts, we succeeded in the synthesis of α-CF_3 and other fluoroalkyl (R_f) ketones from α,β-unsaturated ketones,[4,5] and showed through the use of Et_2Zn-d_{10} that a Rh-H complex played an important role (Figure 2).[5] Furthermore, based on this mechanism, we developed a new

374

methodology for synthesis of α-R$_f$ ketones from silyl enol ethers of ketones in the absence of Et$_2$Zn, although this reaction required high temperature.[6] The addition of Et$_2$Zn allowed the reaction to proceed at lower temperature, improved the yield[7], and extended its application to various fluoroalkyl halides (Figure 3).[8] Formation of the Rh(I)Et complex appears to promte the oxidative addition of CF$_3$I at lower temperature.

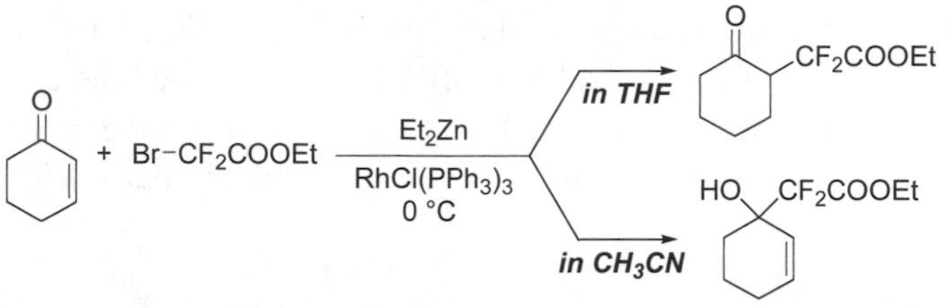

Figure 1. Solvent effect of Rh-catalyzed reaction of BrCF$_2$COOEt with 2-cyclohexen-1-one.

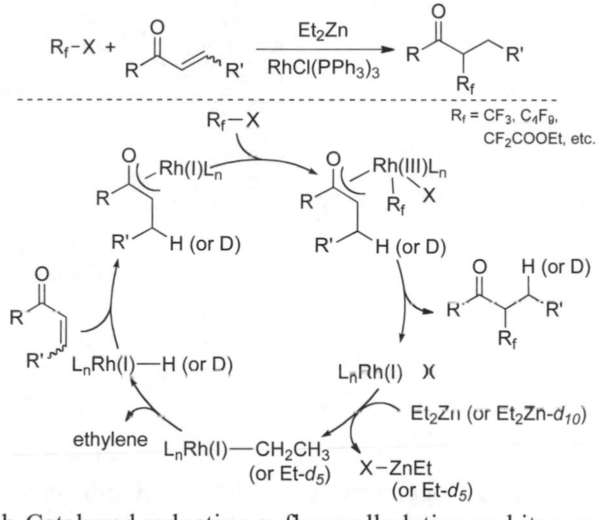

Figure 2. Rh-Catalyzed reductive α-fluoroalkylation and its mechanism.

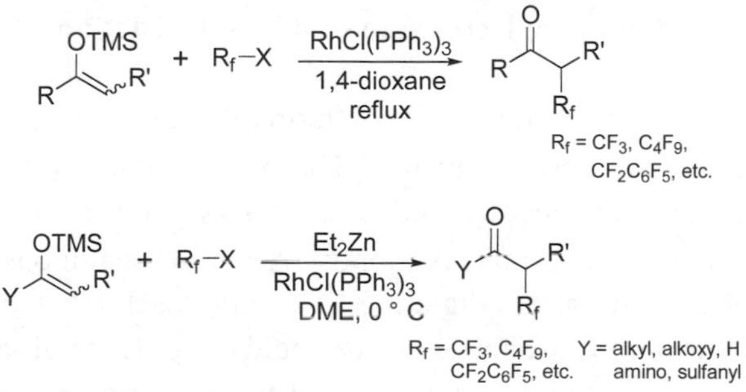

Figure 3. Synthesis of various α-fluoroalkylated carbonyl compounds.

The Reformatsky-Honda Reaction of BrCF₂COOEt with Imines

This Reformatsky-Honda reaction of $BrCF_2COOEt$ has been applied to imines. Interestingly, we succeeded in selective syntheses of difluoro-β-lactams and 3-amino-2,2-difluorocarboxylic esters (Figure 4), depending upon the presence or absence of $MgSO_4 \cdot 7H_2O$.[9] $MgSO_4 \cdot 7H_2O$ serves as a convenient source of H_2O. Addition of H_2O also provided the ester, but the yield was lower. Furthermore, highly stereoselective syntheses of difluoro-β-lactams have also been accomplished using chiral auxiliaries.[10]

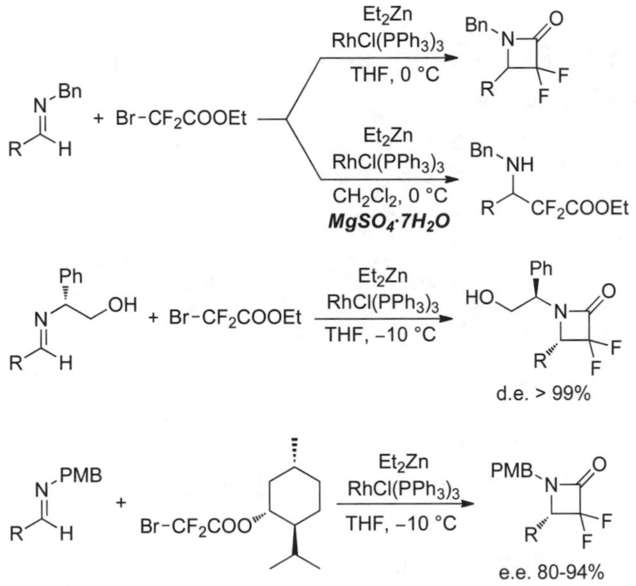

Figure 4. Reformatsky-Honda reaction of $BrCF_2COOEt$ with imines.

Reductive Reformatsky-Honda Reaction

The mechanistic studies of α-fluoroalkylation of α,β-unsaturated ketone led to another new discovery. The most recent advance has been a synthesis of 1,3-diketones by reductive α-acylation of α,β-unsaturated ketones under the Reformatsky-Honda reaction conditions that easily generate a Rh-H complex (Figure 5).[11] α,β-Unsaturated esters can be utilized with various electrophiles besides acid chlorides to give reductive α-substitution products such as aldol-type and Mannich-type products.[12]

376

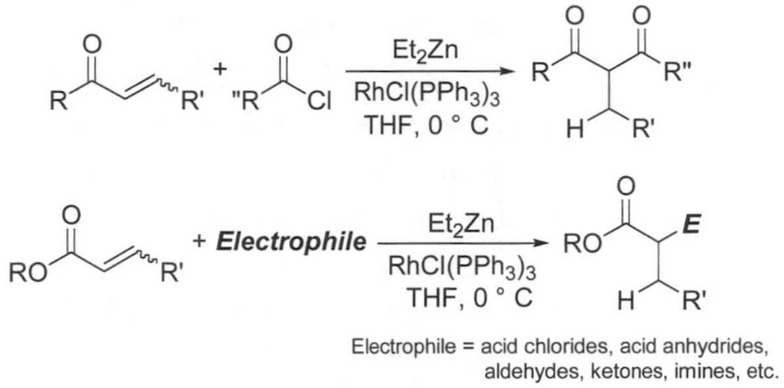

Electrophile = acid chlorides, acid anhydrides,
aldehydes, ketones, imines, etc.

Figure 5. Reductive Reformatsky-Honda reaction with acid chlorides and various electrophiles.

1. Faculty of Pharmaceutical Sciences, Setsunan University, 45-1, Nagaotoge-cho, Hirakata, Osaka 573-0101, Japan. E-mail: aando@pharm.setsunan.ac.jp

2. (a) Kanai, K.; Wakabayashi, H.; Honda, T. *Org. Lett.* **2000**, *2*, 2549–2551. (b) Honda, T.; Wakabayashi, H.; Kanai, K. *Chem. Pharm. Bull.* **2002**, *50*, 307–308. (c) Kanai, K.; Wakabayashi, H.; Honda, T. *Heterocycles* **2002**, *58*, 47–51.

3. Sato, K.; Tarui, A.; Kita, T.; Ishida, Y.; Tamura, H.; Omote, M.; Ando, A.; Kumadaki, I. *Tetrahedron Lett.* **2004**, *45*, 5735–5737.

4. Sato, K.; Omote, M.; Ando, A.; Kumadaki, I. *Org. Lett.* **2004**, *6*, 4359–4361.

5. Sato, K.; Ishida, Y.; Murata, E.; Oida, Y.; Mori, Y.; Okawa, M.; Iwase, K.; Tarui, A.; Omote, M.; Kumadaki, I.; Ando, A. *Tetrahedron* **2007**, *63*, 12735–12739.

6. Sato, K.; Higashinagata, M.; Yuki, T.; Tarui, A.; Omote, M.; Kumadaki, I.; Ando, A. *J. Fluorine Chem.* **2008**, *129*, 51–55.

7. (a) Sato, K.; Yuki, T.; Tarui, A.; Omote, M.; Kumadaki, I.; Ando, A. *Tetrahedron Lett.* **2008**, *49*, 3558–3561. (b) Sato, K.; Yuki, T.; Yamaguchi, R.; Hamano, T.; Tarui, A.; Omote, M.; Kumadaki, I.; Ando, A. *J. Org. Chem.* **2009**, *74*, 3815–3819.

8. Sato, K.; Yamazoe, S.; Akashi, Y.; Hamano, T.; Miyamoto, A.; Sugiyama, S.; Tarui, A.; Omote, M.; Kumadaki, I.; Ando, A. *J. Fluorine Chem.* **2010**, *131*, 86–90.

9. Sato, K.; Tarui, A.; Matsuda, S.; Omote, M.; Ando, A.; Kumadaki, I. *Tetrahedron Lett.* **2005**, *46*, 7679–7681.

10. (a) Tarui, A.; Kondo, K.; Taira, H.; Sato, K.; Omote, M.; Kumadaki, I.; Ando, A. *Heterocycles* **2007**, *73*, 203–208. (b) Tarui, A.; Ozaki, D.; Nakajima, N.; Yokota, Y.; Sokeirik, Y. S.; Sato, K.; Omote, M.; Kumadaki, I.; Ando, A. *Tetrahedron Lett.* **2008**, *49*, 3839–3843.

11. Sato, K.; Yamazoe, S.; Yamamoto, R.; Ohata, S.; Tarui, A.; Omote, M.; Kumadaki, I.; Ando, A. *Org. Lett.* **2008**, *10*, 2405–2408.

12. Sato, K.; Isoda, M.; Ohata, S.; Morita, S.; Tarui, A.; Omote, M.; Kumadaki, I.; Ando, A. *Adv. Synth. Catal.* **2011**, 510–514.

Kazuyuki Sato studied organofluorine chemistry at Faculty of Pharmaceutical Sciences, Setsunan University. He received Ph.D. degree from Setsunan University on the study of synthesis of fluorine compounds using copper-mediated reactions of ethyl bromodifluoroacetate under the guidance of Prof. Itsumaro Kumadaki in 2003. Since 1999, he is working as a research associate at Setsunan University. His main interest is rhodium-catalyzed synthesis of fluorine compounds. He received two awards from the Pharmaceutical Society of Japan and the Society of Synthetic Organic Chemistry, Japan.

Atsushi Tarui was born in 1981 in Wakayama (Japan). He received Ph.D. degree from Setsunan University on the research of synthesis of α,α-difluoro-β-lactam using rhodium-catalyzed reaction of bromodifluoroacetate with imines and related reactions under supervision of Prof. Akira Ando in 2010. Since 2006, he is working there as a research associate under the guidance of Prof. A. Ando. His research interest is synthesis of fluorinated β-lactams and β-amino acids.

Masaaki Omote received his Ph.D. from Faculty of Pharmaceutical Sciences, Setsunan University under the supervision of Prof. Itsumaro Kumadaki in 1997, and he became research associate of Setsunan University. In 2004, he spent a year as a postdoctoral research associate in State University of New York (Albany, New York) under the supervision of Prof. John T. Welch. In 2010 he promoted to associate professor of Setsunan University. His research interest is syntheses of fluorous and partially fluorinated chiral ligands useful for enantioselective synthesis of non-fluorinated compounds.

Itsumaro Kumadaki received Ph.D. degree from Kyushu University on the study of reaction of heteroaromatic amine *N*-oxides under the supervision of Prof. Masatomo Hamana in 1968. He joined Prof. Yoshiro Kobayashi of Tokyo College of Pharmacy in 1967 and started the studies on organic fluorine chemistry. He moved to Setsunan University in 1983 as a professor of Faculty of Pharmaceutical Sciences, and retired in 2006. His interests are still new reactions for synthesis of organofluorine compounds, new reactions of organofluorine compounds, and chemical and biological effects of fluorine substituents on organic compounds.

Akira Ando was born in 1951 in Tokyo (Japan), and obtained Ph.D. degree from Tokyo College of Pharmacy on the study of the synthesis and reactions of valence-bond isomers of tetrakis(trifluoromethyl)pyrroles under supervision of Prof. Yoshiro Kobayashi in 1980. After working as a research assistant at Tokyo College of Pharmaceutical Sciences and a research fellow at Sagami Chemical Research Center, he joined Prof. Kumadaki's group of fluorine chemistry at Setsunan University in 1985, and since 2004 he is a full professor there. His interest is to develop new methodologies for synthesis of organofluorine compounds.

Preparation of H,^{4} PyrrolidineQuin-BAM (PBAM)

A.

B.

C.

Submitted by Tyler A. Davis, Mark C. Dobish, Kenneth E. Schwieter, Aspen C. Chun, and Jeffrey N. Johnston.[1]

Checked by Alexander F. G. Goldberg and Brian M. Stoltz.

1. Procedure

> *Caution! Part A of this procedure must be carried out in a well-ventilated hood and the apparatus must be equipped with a hydrogen chloride (HCl) trap to avoid exposure to HCl gas.*

A. 2,4-Dichloroquinoline (**1**). A 250-mL, 3-necked, round-bottomed flask was equipped with a teflon-coated, oval-shaped stir bar (41 x 20 mm) and charged with malonic acid (26.0 g, 250 mmol, 1 equiv, Note 1).[2] A rubber septum was connected to one of the side necks and a condenser is attached to the middle neck. That condenser was equipped with a line that runs to an HCl trap (Note 2). The flask was lowered into an ice water bath and POCl$_3$ (100 mL) was poured into the open neck. The open neck was sealed with a rubber septum. A slight flow of nitrogen gas was introduced

380

Org. Synth. **2012**, *89*, 380-393
Published on the Web 2/22/2012
© 2012 Organic Syntheses, Inc.

into the system through a needle (16 gauge) inserted into one of the side neck septa (Note 3). Stirring was initiated at 300 rpm. Aniline (22.8 mL, 250 mmol, 1 equiv) was added slowly to the stirring mixture by syringe through a side neck septum over a period of 20 min (Note 4). The flask was then removed from the ice water bath, allowed to stir for 10 min, placed into an oil bath (~110 °C) and the mixture was stirred for 6 h (Note 5). The flask was removed from the oil bath and allowed to cool for 5 min without stirring (Note 6). The mixture was slowly and carefully poured onto crushed ice (~ 1 L) in a 2 L beaker while intermittently stirring with a spatula. The residual material in the 3-neck round-bottomed flask was removed by carefully adding water (100 mL) to the flask and scraping out with a spatula. The aqueous mixture was triturated by vigorous stirring with the spatula (Note 7). The 2 L quench beaker was placed into a bucket of ice. Aqueous NaOH (6 M, 475 mL) was added slowly (over a period of 25 min, maintaining the temperature ≤ 50 °C) while intermittently stirring (Note 8). The beaker was removed from the ice bath and the suspension was allowed to warm to ambient temperature and age with stirring (77 x 13 mm stir bar, ~100 rpm) for 12 h (Note 9). The pH of the supernatant was ~7. The suspension was then vacuum filtered through a Büchner funnel (12.5 cm diameter). Water (150 mL) was used to rinse the residual material from the beaker. The filtered orange-red solid was rinsed with water (150 mL) on the filter. After thoroughly air-drying on the filter (> 1 h), the solid was transferred into 2 cellulose thimbles (43 mm x 123 mm, int. diam. x ext. length). The contents of each thimble were subjected to a Soxhlet extraction with hexanes (300 mL, Note 10) for 16 h. The combined extracts were concentrated (Note 10) by rotary evaporation (30 °C, 13 mmHg) and further dried under vacuum (1–2 mmHg) to leave a yellow solid (13.9–14.6 g, 28–29%, Notes 11 and 12) that was used in step B without further purification (Note 13).

   B. *H,[4]ClQuin-BAM* (2). A 100 mL, round-bottomed flask (24/40 joint) equipped with a teflon-coated oval stir bar (32 x 15 mm), rubber septum and an inlet needle connected to an argon/vacuum manifold (Note 16) was first charged with (*R,R*)-1,2-diaminocyclohexane (1.63 g, 14.3 mmol, 1 equiv, Notes 17 and 18). Then, Pd(dba)₂ (164 mg, 285 µmol, 0.02 equiv), *rac*-BINAP (356 mg, 572 µmol, 0.04 equiv), 2,4-dichloroquinoline (1) (5.66 g, 28.6 mmol, 2 equiv) and sodium *tert*-butoxide (4.12 g, 42.9 mmol, 3 equiv, Note 19) were added. The reaction vessel was evacuated and backfilled with argon three times, then left under a positive pressure of argon.[3] Toluene

(35 mL) was dispensed into the flask, and the resulting red-brown solution was placed into an oil bath heated to 80 °C with stirring (Notes 20 and 21, *Caution: possible exotherm!*). The reaction was monitored by TLC (Note 22); after 1.5 h, nearly complete conversion was observed. The reaction was cooled to 25 °C, and saturated NH$_4$Cl (10 mL) and water (10 mL) were added to the flask. The suspension was stirred for 5 min and cooled to 0 °C for 10 min. The reaction mixture was filtered through a Büchner funnel (9 cm diameter) and washed with water (50 mL) and hexanes (150 mL) to afford a light yellow solid. This crude solid was transferred to a pre-weighed 100 mL round-bottomed flask with a 24/40 joint and dried under vacuum (1–2 mmHg) for at least 10 h to leave 4.57–4.80 g of a light beige-yellow solid (73–77%) (Note 23), which was used in step C without further purification.

C. H,^{4}PyrrolidineQuin-BAM (PBAM) (**3**). A 10–20 mL microwave vial equipped with an oval stir bar (12 x 8 mm) was charged with H,^{4}ClQuin-BAM (**2**) (4.30 g, 9.83 mmol, 1 equiv, Note 24), pyrrolidine (3.23 mL, 39.3 mmol, 4 equiv), and trifluorotoluene (3.3 mL) (Note 25). The vial was sealed, and this suspension was heated with stirring in the microwave at 170 °C for 3 min, then at 200 °C for 30 min (Notes 26 and 27, *Caution: exothermic!*). The reaction mixture was diluted and transferred with 20 mL of dichloromethane to a 500-mL round-bottomed flask for evaporation (Note 28). The reaction was then concentrated by rotary evaporation (80 °C, 13 mmHg), redissolved in dichloromethane (50 mL), transferred to a 125 mL separatory funnel, and washed with aqueous NaOH (3 M, 50 mL) and water (4 x 50 mL). The resulting organic layer was dispensed into a 125 mL Erlenmeyer flask and dried (MgSO$_4$, 2 g). The solution was filtered (flask and filter were rinsed with an extra 25 mL of dichloromethane) into a 250 mL round-bottomed flask and concentrated by rotary evaporation (30 °C, 13 mmHg) and high vacuum (0.2 mmHg, 1 h) to provide a light brown powder (Note 29).

As much of the material as possible was transferred as a solid from the round-bottomed flask to a pre-weighed 250 mL Erlenmeyer flask. The remaining solid was dissolved and transferred into a 25-mL pear-shaped flask with 10 mL of dichloromethane and concentrated by rotary evaporation (30 °C, 13 mmHg) and high vacuum (0.2 mmHg, 1 h). This material was then added to the 250 mL Erlenmeyer flask. The material was dissolved in a 90:10 mixture of acetone and toluene (20 mL/g crude product) with heating in a water bath (70 °C) (Note 30). The solution was removed from heat and

Org. Synth. **2012**, *89*, 380-393

placed on the bench for 1 hour to cool to room temperature. The flask was placed in an ice bath for 3 h or until no further crystallization was observed. The mother liquor was carefully decanted, leaving crystals that were washed with ice-cold acetone (2 x 10 mL). Residual solvent was removed from the crystals by placing the flask under vacuum (1–2 mmHg) for a period of 5–10 min (Note 31). The crystalline material was transferred to a pre-weighed 20 mL glass scintillation vial and dried for ~14 h in a drying pistol (0.2 mmHg) over refluxing toluene. The vial was temporarily removed from the drying pistol, and the material was thoroughly pulverized with a spatula and returned to the drying pistol for an additional 6 h. The dried material was a cream colored powder (3.08–3.54 g, 62–71% yield) (Notes 32 and 33).

2. Notes

1.   Malonic acid (99%), phosphorous oxychloride (99%) and hexanes (mixture of isomers, ACS Grade, >98.5%) were purchased from Sigma-Aldrich and used as received. Aniline (99.9%) and toluene (ACS grade) were purchased from Fisher Scientific Company and used as received.

2.   The reflux condenser is fitted with an outlet that is connected with Tygon tubing to two 250 mL bubblers, the first empty, and the second filled with water to serve as the HCl trap.

3.   The gas flow should be monitored by constantly checking the bubbler during the aniline addition to prevent the development of negative pressure in the system.

4.   A creamy, yellow heterogeneous mixture is observed after completion of the aniline addition. Some solids may aggregate near the surface and will slowly dissolve in the next stage of the reaction.

5.   Within the first 10 minutes of heating, vigorous gas evolution is observed as the mixture becomes homogeneous. Within 20 min, the color of the solution transitions from yellow to red to dark brown/black. Over approximately 90 min, the gas evolution subsides, and a gentle reflux is observed.

6.   If the solution is cooled for longer than 5 minutes, the reaction mixture may thicken to an extent that renders it difficult to transfer the entire contents of the flask to the quench beaker.

7.   This served to break up the largest fragments of material and to increase the surface area of the solid for efficient quenching in the following step.

8. This workup procedure should fully hydrolyze all P-Cl bonds leaving sodium phosphate as the byproduct of POCl$_3$ quenching.[4]

9. This served to fully quench the suspension and to produce a finer crude solid with fewer chunks. This facilitated transfer into the Soxhlet thimble and allowed for a more efficient extraction due to increased surface area of the solid.

10. It may be necessary to filter the hexanes extract if red solid precipitates from the solution: The combined extracts were filtered through Celite® (21 g) in a coarse sintered glass funnel (46 mm diameter). The filter cake was washed with hexanes (2 x 50 mL)

11. The crude 2,4-dichloroquinoline (1) exhibited the following: mp 62.0–64.2 (uncorrected); R$_f$ = 0.42 (10% EtOAc/hexanes); IR (film) 3062, 1572, 1553 cm^{-1}; ^{1}H NMR (500 MHz, CDCl$_3$) δ: 7.53 (s, 1 H), 7.67 (ddd, J = 8.3, 6.9, 1.2 Hz, 1 H), 7.81 (ddd, J = 8.5, 6.9, 1.4 Hz, 1 H), 8.05 (d, J = 8.5 Hz, 1 H), 8.21 (dd, J = 8.4, 1.4 Hz, 1 H); ^{13}C NMR (125 MHz, CDCl$_3$) δ: 122.2, 124.4, 125.4, 128.1, 129.2, 131.8, 144.6, 148.3, 150.0; HRMS (APCI-ESI) calcd for C$_9$H$_6$Cl$_2$N [M+H]$^+$ 197.9872; found 197.9870. This data matched the values given in the literature.[2] The checkers observed the development of a brown discoloration during prolonged storage at ambient temperature in air. This does not affect its use in subsequent reactions.

12. The submitters obtained yields between 33–36% using a smaller Soxhlet apparatus, which may have contributed to a greater extraction efficiency, with thimble dimensions: 33 mm x 94 mm (int. diam. x ext. length).

13. The crude material can be further purified by recrystallization. The crude solid (5.0 g) was dispensed into a 50 mL Erlenmeyer flask and dissolved in 20 mL of a 3:1 mixture of hexanes and toluene with heating in a ~70 °C water bath. The solution was allowed to slowly cool to room temperature (Note 14) before it was placed in an ice-water bath. After 2 hours, the mother liquor was decanted. The recrystallization flask was kept in the ice-water bath while the crystals were rinsed with ice-cold hexanes (5 mL). The hexanes were decanted and residual solvent was removed by vacuum (0.2 mmHg) to leave a yellow crystalline solid (2.99–3.17 g, 60–63% recovery, Note 15).

14. The submitters note that if a precipitate forms when the solution is cooled to room temperature, then the solution should be filtered before cooling in an ice-water bath.

15. The melting point range of the recrystallized 2,4-dichloroquinoline (1) was 64.1–65.0 °C (uncorrected). The submitters note that material that was purified by column chromatography (0–4% EtOAc in hexanes) exhibited a melting point of 64.0–64.5 °C (uncorrected).

16. The apparatus was flame-dried under reduced pressure (1–2 mmHg) and then maintained under a positive pressure of argon during the course of reaction.

17. (±)-2,2'-Bis(diphenylphosphino)-1,1'-binaphthalene (97%) was purchased from Sigma-Aldrich and was used as received. Sodium *tert*-butoxide (97%) was purchased from Sigma-Aldrich, stored in a glovebox under a nitrogen atmosphere, and otherwise used as received. Toluene (ACS grade) was purchased from Fisher and was dried by passage through a column of activated alumina as described by Grubbs.[5] The checkers used bis(dibenzylideneacetone)palladium as received from Sigma Aldrich and (1R,2R)-(−)-1,2-diaminocyclohexane (99%) as received from Strem Chemicals. The submitters prepared bis(dibenzylideneacetone)palladium according to the procedure of Rettig and Maitlis.[6] The submitters used (1R,2R)-(−)-1,2-diaminocyclohexane resolved from a *trans/cis* mixture (60/40) using the protocol of Jacobsen[7] with L-(+)-tartaric acid; the salt was then dissolved in 10 M NaOH and continuously extracted using benzene in a liquid/liquid extractor for 5 days. This diamine is stored in a freezer.

18. (1R,2R)-(−)-1,2-Diaminocyclohexane is a white solid but becomes a viscous oil as it warms to room temperature. To avoid difficulties in transfer, the diamine is weighed directly into the reaction vessel without allowing the diamine to warm to room temperature.

19. Sodium *tert*-butoxide was dispensed into an oven-dried vial in a glovebox under a nitrogen atmosphere. This vial was sealed and removed from the glovebox and its contents were then quickly transferred into the round-bottomed flask.

20. The reaction should be monitored very carefully for the first 10 minutes; the checkers consistently observed an exotherm within the first 5 minutes of heating. When probed, the internal temperature gradually increased to 80 °C over 4 minutes of heating, then quickly rose to its boiling point within 30 seconds. When the solution began to boil, the flask was removed from the oil bath and allowed to cool at ambient temperature until a precipitate formed (See Note 21).

21. After 5–10 minutes a solid formed and was manually dispersed after removal of the flask from the oil bath by vigorously swirling the flask by hand. The flask was then returned to the oil bath.

22. Thin layer chromatography (TLC) was performed using E. Merck silica gel 60 F254 precoated glass plates (0.25 mm). UV light was used to visualize the product. With 20% ethyl acetate in hexanes, the starting material has R_f = 0.40 and the product has R_f = 0.07. Upon reaction completion, there may still be a relatively faint spot visible of the same R_f as the starting material.

23. This material is of sufficient purity for use as a reactant to make compound **3**. The submitters note that it can be further purified by column chromatography (10% ethyl acetate in hexanes); the melting point of material purified by column chromatography is 236.0–237.0 °C. The submitters note that the crude material exhibited some variation from batch to batch in terms of melting point; a range of 235–245 °C was observed, but individual samples exhibited a routinely narrow (≤ 2 °C) melting point range. H,[4]ClQuin-BAM (**2**) exhibited the following: mp 240.7–243.0; $[\alpha]_D^{25}$ +568 (c 1.02, DMSO); R_f = 0.31 (20% EtOAc in hexanes); IR (KBr) 3250, 2942, 1614 cm^{-1}; ^{1}H NMR (500 MHz, DMSO-d$_6$) δ: 1.29–1.45 (m, 4 H), 1.71–1.79 (m, 2 H), 2.18–2.26 (m, 2 H), 4.05 (br s, 2 H), 6.84 (s, 2 H), 7.22–7.29 (m, 4 H), 7.53–7.62 (m, 4 H), 7.82 (d, J = 8.0 Hz, 2 H); ^{13}C NMR (125 MHz, DMSO-d$_6$) δ: 24.3, 31.8, 53.6, 112.6, 120.3, 122.1, 123.4, 126.0, 130.4, 140.2, 148.5, 156.5; HRMS (APCI-ESI) calcd for $C_{24}H_{23}Cl_2N_4$ [M+H]$^+$ 437.1300; found 437.1295. The submitters have found H,[4]ClQuin-BAM to be a bench stable solid: no decomposition of this compound has been observed after years of storage on the benchtop in a screw cap vial.

24. Pyrrolidine (99%) was purchased from Alfa Aesar. α,α,α-trifluorotoluene (99+%) was purchased from Acros. Dichloromethane (ACS grade) and acetone (ACS grade) were purchased from Sigma-Aldrich. All reagents were used as received.

25. Care should be taken to wash solids on the vial wall below the solvent level to prevent localized heating.

26. The reaction was performed in a Biotage Initiator EXP US 355302 microwave reactor in a sealed vial, at the "Normal" absorption level. This microwave system monitors the pressure and heat inside the sealed vial and shuts off for safety purposes when 20 bar or 250 °C is reached. A two-stage heating protocol was found by the checkers to prevent an exotherm which exceeds the safety parameters of the microwave system; this was sometimes

observed by the submitters when the reaction was heated immediately to 200 °C. Using the described protocol, the checkers observed a temperature peak at approximately 185 °C during the initial heating stage. During the second stage, the temperature was observed to peak at approximately 210 °C and the pressure at 11 bar; the temperature then stabilized at 200 °C, and the pressure at approximately 9.5 bar. Different microwave reactors may have different safety parameters and vial headspaces: *caution is strongly recommended if different microwave instrumentation is employed.*

27. Confirmation of consumption of starting material can be made by TLC with UV visualization (10% methanol/0.5% acetic acid/89.5% dichloromethane, starting material $R_f = 0.49$ (checkers) ($R_f = 0.79$ reported by submitters); product $R_f = 0.35$).

28. The contents of the microwave vial will congeal upon cooling, but the use of 20 mL of dichloromethane will sufficiently dissolve and transfer the mixture. It might be necessary to stir the contents of the vial with the added dichloromethane for several minutes to get this congealed mixture to dissolve for transfer. The use of an abnormally large flask (500 mL for less than 30 mL of solution) is necessary because this viscous mixture foams upon concentration by rotavap. The extra headspace is needed to prevent the crude product from foaming out of the flask and into the rotavap bump trap.

29. The submitters note that the material at this stage can be dried by subjecting the material to a drying pistol heated by refluxing toluene for 24–48 h and have found this dried material without further purification to give the same performance (stereoselection) as a catalyst as material further purified (column chromatography or recrystallization).

30. The crude material might fully dissolve and crystallize in a matter of seconds upon addition of the solvent mixture at room temperature. These crystals may not fully redissolve upon heating and swirling, but the recrystallization will still work when the solution is cooled.

31. Vacuum was applied with a needle through a septum secured to the top of the flask. The goal of this manipulation is merely to obtain a free-flowing solid.

32. The submitters note that the mother liquor and crystal washings can be concentrated and subjected to the same recrystallization procedure to obtain an additional 800 mg of product. The submitters note that during the determination of the melting point, the material gradually turns from white to slightly brown between 200 °C and the melting point. Despite this appearance change, the melting point observed by the submitters is

consistent at 243–245 °C. The submitters have noticed some variation in the melting point range (241–246 °C), but individual samples exhibit a routinely narrow ($\leq$ 2 °C) melting point. PBAM (**3**) exhibited the following: $[\alpha]_D^{25}$ +398 (*c* 1.02, CHCl$_3$); R_f = 0.35 (10% MeOH/CH$_2$Cl$_2$ with 0.5% acetic acid); IR (film) 3241, 3047, 2933, 2848, 1592, 1525 cm^{-1}; ^{1}H NMR (500 MHz, CDCl$_3$) δ: 1.36–1.55 (m, 4 H), 1.75–1.89 (m, 10 H), 2.26–2.34 (m, 2 H), 3.05–3.16 (m, 4 H), 3.24–3.35 (m, 4 H), 4.10 (br m, 2 H), 5.28 (s, 2 H), 5.68 (br s, 2 H), 7.01 (dd, *J* = 8.3, 6.8, 1.3 Hz, 2 H), 7.40 (ddd, *J* = 8.2, 6.8, 1.4 Hz, 2 H), 7.65 (d, *J* = 7.9 Hz, 2 H), 7.87 (dd, *J* = 8.4, 1.3 Hz, 2 H); ^{13}C NMR (125 MHz, CDCl$_3$) δ: 25.3, 25.7, 33.5, 51.7, 56.4, 93.0, 118.8, 119.3, 124.8, 126.5, 128.5, 150.0, 153.3, 158.5; HRMS (APCI-ESI) calc'd for C$_{32}$H$_{39}$N$_6$ [M+H]$^+$ 507.3236; found 507.3245; Anal. calcd for C$_{32}$H$_{38}$N$_6$: C, 75.85; H, 7.56; N, 16.59. Found: C, 75.93; H, 7.44; N, 16.65. The submitters note that PBAM purified by column chromatography has been stable upon storage in a screw cap vial on the benchtop for several years without sign of decomposition.

33. The checkers observed a ~3% acetone impurity in the ^{1}H NMR, along with a depressed melting point (135–163 °C); the submitters used a vacuum pump which reached 0.1 mmHg during the drying pistol stage. The checkers employed this material in an enantioselective aza-Henry reaction of the Boc-protected imine of *o*-tolualdehyde with nitroethane, as previously reported by the submitters, and obtained the same level of enantioselectivity as reported.[8]

Safety and Waste Disposal Information

All hazardous materials should be handled and disposed of in accordance with "Prudent Practices in the Laboratory"; National Academies Press; Washington, DC, 2011.

3. Discussion

BisAMidine ('BAM') ligands were first described in 2004 and discovered to be effective catalysts – as their protic acid complexes (e.g. **3**·HOTf, eq 1) – for the diastereo- and enantioselective aza-Henry reaction (eq 1). The nitroalkane adducts are readily reduced, delivering *vic*-diamines in a two step sequence and differentially protected at the amine nitrogens. This class of catalysts, often referred to as chiral proton catalysts to

388

emphasize the importance of the polar ionic hydrogen bond responsible for activation and orientation of the substrate, has since emerged as an increasingly general solution to reagent control of aza-Henry reactions. The title compound (H,[4]PyrrolidineQuin-BAM, 'PBAM')[8] was recently reported as a substantially more reactive and equally selective catalyst for the aza-Henry addition of nitroalkanes to aryl aldimines (eq 1).[9] This reagent, again as its triflic acid salt, expands the scope of the nitroalkane component, facilitating the stereocontrolled addition of a broad range of primary nitroalkanes and secondary nitroalkanes (e.g. nitropropane). This catalyst was discovered to be effective as a chiral Brønsted base in the alkylation of nitroalkanes using indolenine electrophiles (eq 2).[10] More recently, the addition of aryl nitromethane pronucleophiles were investigated, as these are notoriously difficult donors in the aza-Henry reaction.[11,12] In an interesting twist, the protonation state of the BAM ligand no longer affected the stereoselection in many cases.[13] Regardless, further modification of the fundamental PBAM backbone by installation of methoxy substituents (4) furnished the addition products with high diastereo- and enantioselection (eq 3). This development enabled the first enantioselective synthesis of (–)-Nutlin-3[13] (eq 3), the first potent small molecule inhibitor of p53/MDM2 discovered by Hoffmann-La Roche.[14]

The synthesis of PBAM described here is a significant improvement upon our previously reported procedure.[8] The initial procedure relied upon similar reaction conditions but flash column chromatography was used to purify after each step. The chromatographic purification for the final compound was particularly troublesome as large volumes of solvent (dichloromethane, methanol, and acetic acid) were required for the compound to fully elute from the column. The products of these two reactions are now purified by straightforward washings, filtrations, and a recrystallization after the last step. The scale of this procedure is nearly an order of magnitude greater than previously reported.

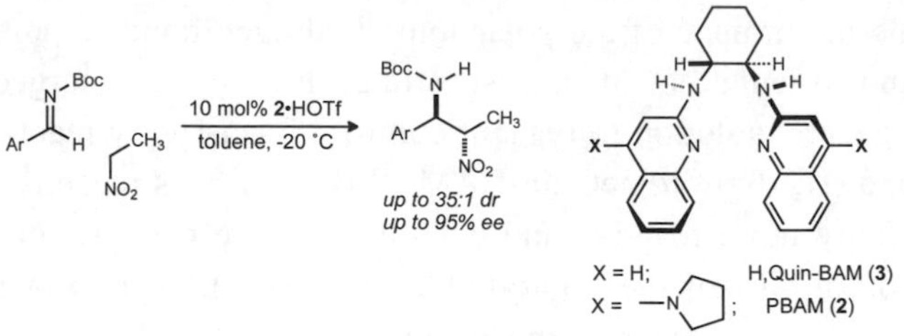

(1)

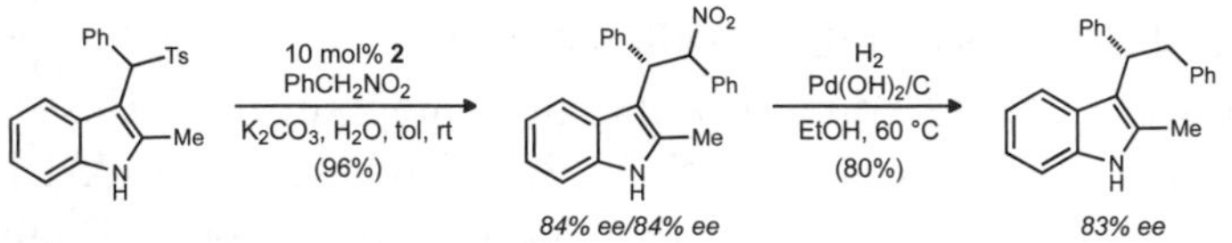

X = H;   H,Quin-BAM (**3**)

X = ⟨pyrrolidine⟩;   PBAM (**2**)

84% ee/84% ee   83% ee

(2)

13:1 dr
91% ee

(−)-Nutlin-3

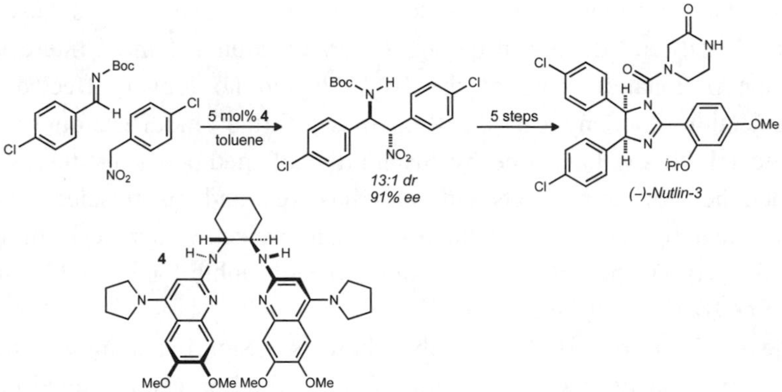

(3)

Appendix
Chemical Abstracts Nomenclature; (Registry Number)

2,4-Dichloroquinoline; (703-61-7)

Malonic acid; (141-82-2)

Phosphorus oxychloride; (10025-87-3)

Aniline; (62-53-3)

Bis(dibenzylideneacetone)palladium; (32005-36-0)

2,2′-Bis(diphenylphosphino)-1,1′-binaphthalene; (98327-87-8)

Sodium-*tert*-butoxide; (865-48-5)

(1R,2R)-(−)-1,2-Diaminocyclohexane; (20439-47-8)

Pyrrolidine; (123-75-1)

Trifluoromethyl benzene; (98-08-8)

1. Department of Chemistry & Vanderbilt Institute of Chemical Biology, Vanderbilt University, Nashville, TN 37235-1822. E-mail: jeffrey.n.johnston@vanderbilt.edu. We are grateful to the NIH (GM-084333) for financial support of this work, and the NSF-REU (KES) and VUSRP (ACC) for fellowship support.
2. A modification of the procedure reported by Jones: Jones, K.; Roset, X.; Rossiter, S.; Whitfield, P. *Org. Biomol. Chem.* **2003**, *1*, 4380-4383.
3. Adapted from Wagaw, S.; Rennels, R. A.; Buchwald, S. L. *J. Am. Chem. Soc.* **1997**, *119*, 8451–8458.
4. Achmatowicz, M. M.; Thiel, O. R.; Colyer, J. T.; Hu, J.; Elipe, M. V. S.; Tomaskevitch, J.; Tedrow, J. S.; Larsen, R. D. *Org. Process Res. Dev.* **2010**, *14*, 1490–1500.
5. Pangborn, A. B.; Giardello, M. A.; Grubbs, R. H.; Rosen, R. K.; Timmers, F. J. *Organometallics* **1996**, *15*, 1518–1520.
6. Rettig, M. F.; Maitlis, P. M. *Inorg. Synth.* **1977**, *17*, 134–137.
7. Larrow, J. F.; Jacobsen, E. N.; Gao, Y.; Hong, Y.; Nie, X.; Zepp, C. M. *J. Org. Chem.* **1994**, *59*, 1939–1942.
8. Davis, T. A.; Wilt, J. C.; Johnston, J. N. *J. Am. Chem. Soc.* **2010**, *132*, 2880–2882.
9. (a) Nugent, B. M.; Yoder, R. A.; Johnston, J. N. *J. Am. Chem. Soc.* **2004**, *126*, 3418–3419. (b) Singh, A.; Yoder, R. A.; Shen, B.; Johnston, J. N. *J. Am. Chem. Soc.* **2007**, *129*, 3466–3467. (c) Singh, A.; Johnston, J. N. *J. Am. Chem. Soc.* **2008**, *130*, 5866–5867. (d) Shen, B.; Johnston, J. N. *Org. Lett.* **2008**, *10*, 4397–4400. (e) Wilt, J. C.; Pink, M.; Johnston, J. N. *Chem. Commun* **2008**, 4177 4179. (f) Shen, B.; Makley, D. M.; Johnston, J. N. *Nature* **2010**, *465*, 1027–1032. (g) Shackleford, J. P.; Shen, B.; Johnston, J. N. *Proc. Natl. Acad. Sci. U.S.A.* **2012**, *109*, 44-46.
10. Dobish, M. C.; Johnston, J. N. *Org. Lett.* **2010**, *12*, 5744–5747.
11. Rueping, M.; Antonchick, A. P. *Org. Lett.* **2008**, *10*, 1731–1734.
12. Nishiwaki, N.; Knudsen, K. R.; Gothelf, K. V.; Jørgensen, K. A. *Angew. Chem. Int. Ed.* **2001**, *40*, 2992–2995.
13. Davis, T. A.; Johnston, J. N. *Chem. Sci.* **2011**, *2*, 1076–1079.
14. Vassilev, L. T.; Vu, B. T.; Graves, B.; Carvajal, D.; Podlaski, F.; Filipovic, Z.; Kong, N.; Kammlott, U.; Lukacs, C.; Klein, C.; Fotouhi, N.; Liu, E. A. *Science* **2004**, *303*, 844–848.

Jeffrey N. Johnston completed his B.S. Chemistry degree at Xavier University in 1992, and a Ph.D. in organic chemistry at The Ohio State University in 1997 with Prof. Leo A. Paquette. He then worked as an NIH postdoctoral fellow with Prof. David A. Evans at Harvard University. He began his independent career at Indiana University, where he was promoted to Professor of Chemistry in 2005. In 2006, his research program moved to Vanderbilt University. His group has developed a range of new reactions and reagents that are used to streamline the total synthesis of complex alkaloid natural products. The development of new Brønsted acid-catalyzed reactions, as well as chiral proton catalysts, are ongoing investigational themes.

Tyler A. Davis was born in 1983 in Nashville, TN. In the summer of 2004, he conducted research in the lab of Prof. S. Bruce King at Wake Forest University. He graduated *magna cum laude* from Lipscomb University with his B.S. degree in biochemistry in 2005. He completed his Ph. D. in the laboratory of Professor Jeffrey Johnston at Vanderbilt in 2011. His studies focused on the synthesis and application of chiral BisAMidine (BAM) organocatalysts with increased Brønsted basicity. Tyler is currently a postdoctoral scholar in the laboratory of Prof. Tomislav Rovis at Colorado State University.

Mark C. Dobish, a native of New Wilmington, PA, received his B.S. degree in chemistry from Allegheny College in 2007, where he investigated boron-mediated transfers under the tutelage of Prof. P. J. Persichini. He then began graduate studies with Prof. Johnston at Vanderbilt University as a Vanderbilt Institute of Chemical Biology Fellow. Mark's graduate research focuses on enantioselective transformations using the BisAMidine (BAM) library of catalysts as it applies to new reactions and small molecule drug targets.

Kenneth E. Schwieter was born and raised in Cincinnati, OH. He received his B.S. degree in chemistry from Xavier University in 2011 under the direction of Prof. Rick Mullins. He was a 2010 NSF-REU student in the Johnston lab, and matriculated at Vanderbilt in 2011 to continue his work.

Aspen Chun is a native of California (Los Angeles) and received her B.S. Chemistry degree from Vanderbilt in 2011. She was an undergraduate researcher in the Johnston laboratory, and is currently a staff scientist in the Vanderbilt Program in Drug Discovery.

Alex Goldberg was born in Longbranch, New Jersey in 1986 and was raised in Toronto, Canada. He received his B.ScH. degree from Queen's University in Kingston in 2008 under the supervision of Prof. Cathleen M. Crudden. During this time, he also worked in the research groups of Prof. Mark Lautens at the University of Toronto, and Prof. Ilan Marek at the Technion – Israel Institute of Technology. He began his doctoral studies at the California Institute of Technology in the fall of 2008 under the direction of Prof. Brian M. Stoltz as an NSERC Postgraduate scholar. His graduate research has focused on the total synthesis of Scandine and the Melodinus Alkaloids.

Discussion Addendum For:
Efficient Asymmetric Synthesis of *N-tert*-Butoxycarbonyl α-Amino Acids using 4-*tert*-Butoxycarbonyl-5,6-Diphenylmorpholin-2-one:
(*R*)-(*N-tert*-Butoxycarbonyl)allylglycine (4-Pentenoic acid, 2-[[(1,1-Dimethylethoxy)carbonyl]amino]-, (2*R*)-)

Prepared by Ryan E. Looper[1] and Robert M. Williams.*[2]
Original article: Williams, R. M.; Sinclair, P. J.; Demong, D. E. *Org. Synth.* **2003**, *80*, 31–37.

The use of the diphenylmorpholinone template for the synthesis of optically enriched amino acid derivatives remains a highly selective, practical and predictable method.[3] While the chiral auxiliary is destroyed (by its eventual conversion to dibenzyl) and may not be ideally suited for synthetic applications on an industrial scale, it is a versatile tool for the preparation of research quantities of unusual or rare amino acid derivatives. The sense of stereo-induction with this template is generally very high and very predictable with new substituents being added to the α-carbon anti- to the phenyl groups. Further the reactivity of this template is highly flexible.[4] The α-carbon can react not only as an enolate, as described in the original *Organic Syntheses* preparation, but can also serve as an electrophile (via the iminium ion), radical, 1,3-dipole (azomethine ylide)[5,6,7] or can be converted to and used as the α-phosphonate. It should also be noted that both antipodes of the template are commercially available, which permits the synthesis of both enantiomers of a given target with equal efficiency.

Since the original report in *Organic Syntheses* on the preparation of (*R*)-(*N-tert*-butoxycarbonyl)allylglycine a number of groups have used this

Org. Synth. **2012**, *89*, 394-403
Published on the Web 3/9/2012

allylglycine derivative for a variety of applications. Most often it is incorporated as a metathesis precursor. For example, Nolen and co-workers investigated the preparation of a more stable *C*-glycosyl serine mimic (**5**), to be used in investigations of the *O*-glycosylation of serine residues which are a ubiquitous post-translational modification for controlling biological interactions (Fig. 1).[8] The allylglycine derivative **1**, derived from the *Organic Syntheses* preparation, served as a cross-metathesis partner with gluco-heptenitol. Thus treatment of **1** and **2** with Grubb's 2nd generation Ru-catalyst gave the cyclization precursor **3** in 70% yield. Oxymercuration with the pendent alcohol generated the C-glycoside **4** in high yield and as a single diastereomer after reduction of the alkyl-mercurial species with Et_3B and $NaBH_4$. The acetoxy derivative **5**, poised for elaboration as a peptide fragment, can be obtained after hydrogenolysis and acetylation.

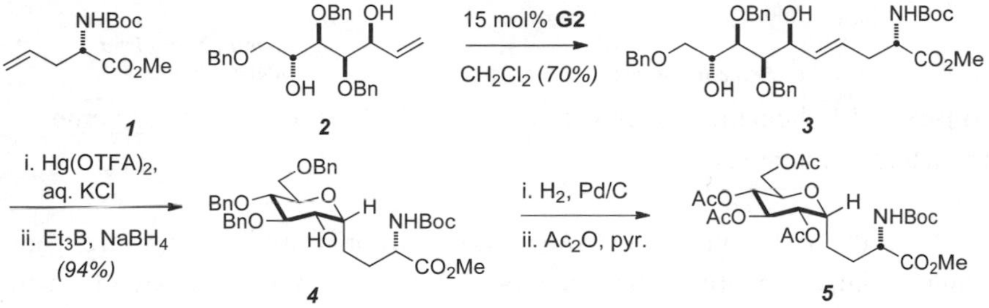

Figure 1. Application of cross methathesis to allylglycine derivatives.

Cross-metathesis between an allylglycine derivative and tetradecene, followed by hydrogenation, has also been used to create (*R*)-aminoheptadecanoic acid (Fig. 2A). This long chain amino acid has found use in the generation of truncated muraymicin analogues with increased cell permeability.[9] Ring closing metathesis (RCM) with two allylglycine units has also been reported. This technology permitted the synthesis of 14-membered macrocyclic peptides which were shown to be selective antagonists of the d and m opioid receptors (Fig. 2B).[10] Medium size rings can also be generated through RCM with allylglycine derivatives. This has found use in the development of Smac (second mitochondria-derived activator of caspases) mimetics (Fig. 2C)[11] and in the development of matrix metalloproteinase (MMP) inhibitors inspired by the natural product cobactin T (Fig. 2D).[12]

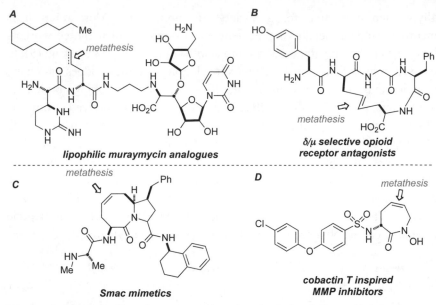

Figure 2. Medicinally relevant small molecule derived from allylglycine methathesis reactions.

Both (*R*) and (*S*)-(*N-tert*-butoxycarbonyl)allylglycine are now commercially available, albeit generally 2-3 times more expensive than the oxazinone template itself. As such, the oxazinone's inherent applicability might ultimately lie in the preparation of novel allylic amino acid derivatives. Below we report on some recent examples in natural product synthesis, medicinal chemistry, and the generation of unnatural amino acid mimetics.

Funk and Greshock described the utility of their bromomethyl vinyl ketone equivalent (6-bromomethyl-4*H*-1,3-dioxin, **6**) for the synthesis of the naturally occurring (2*S*,4*R*)-4-hydroxypipecolic acid (**11**) (Figure 3).[13] Alkylation of the oxazinone with **6** gave **7** as a single diastereomer. Thermolysis of this intermediate triggered a retrocycloaddition to reveal the vinyl ketone **8**, which underwent intramolecular Michael addition after removal of the *N-tert*-butoxycarbonyl group with TMSOTf. Reduction of the ketone group in **9** with borane gave **10** as a single diastereomer. Hydrogenolysis of the auxiliary provided **11** in good yield and high optical purity.

Org. Synth. **2012**, *89*, 394-403

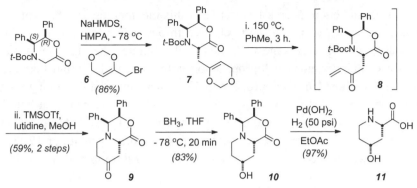

Figure 3. Allylic alkylation *en route* to (*2S,4R*)-4-hydroxypipecolic acid.

In 2004 we reported the use of the crotylglycine derivative **13** in the synthesis of cylindrospermopsin,[14] 7-epicylindrospermopsin,[15] and 7-deoxycylindrospermopsin[16] (Figure 4). Alkylation of the oxazinone with crotyl iodide gave **12** in 92% yield and as a single diastereomer. Removal of the auxiliary then gave (*R*)-(*N-tert*-butoxycarbonyl)crotylglycine in reasonable yield and with high optical purity (> 99 : 1 e.r.). This was further processed to the oxazinone-*N*-oxide **14**, which underwent an intramolecular [3+2]-cycloaddition to give the tricyclic isoxazolidine **15,** thereby setting the three contiguous stereocenters in the A ring of these natural products.[17] The key tricyclic intermediate **15** was then deployed in total syntheses of all three naturally occurring cylindrospermopsin alkaloids, through initial reductive cleavage of the N-O bond, differentiation of the two one-carbon arms and elaboration.

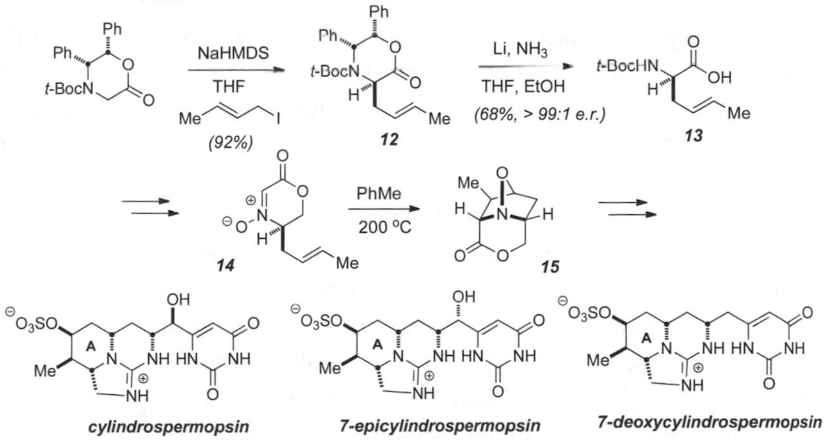

Figure 4. Allylic alkyation to access the cylindrospermopsin alkaloids.

Researchers at Merck have utilized this methodology in the synthesis of a novel series of potent and selective thrombin inhibitors (Figure 5).[18] Alkylation of the oxazinone with methallyl bromide gave the intermediate **16** in acceptable yield. Cyclopropanation with diazomethane and Pd(OAc)$_2$ then gave the pendent cyclopropane in **17**, which survived dissolving metal reduction of the auxiliary to give (R)-(N-tert-butoxycarbonyl)-(methylcyclopropyl)propanoic acid (**18**). Incorporation of this fragment into their common fluoro-proline scaffold gave the thrombin inhibitor **19** (K$_i$ = 0.37 nM), which was nearly 1,000 fold selective for thrombin over trypsin.

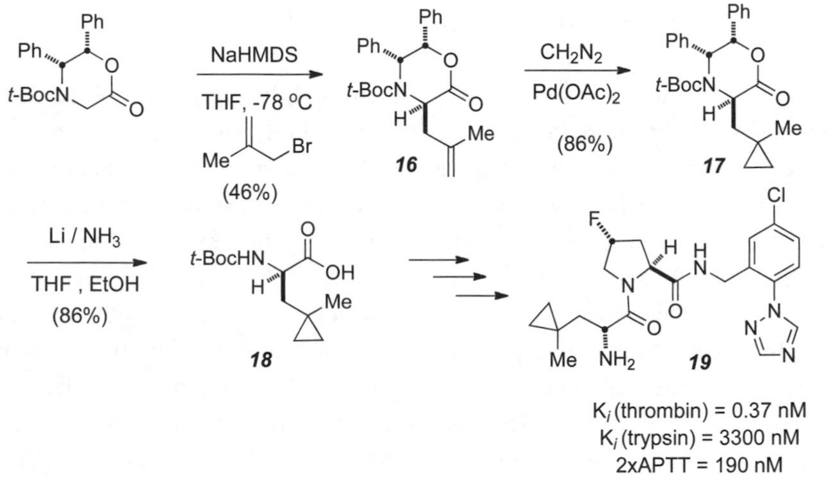

K$_i$(thrombin) = 0.37 nM
K$_i$(trypsin) = 3300 nM
2xAPTT = 190 nM

Figure 5. Allylic alkylation en route to thrombin inhibitors.

These types of reactions have been used in the preparation of other medicinally relevant compounds as shown in Figure 6. Alkylation of the oxazinone with an allylic phthalimide bearing a vinyl fluoride gave **20**, which has found use in the development of arginine mimetics as nitric oxide synthase inhibitors (e.g. **21**) (Fig. 6A).[19] It should also be noted that the allylated oxazinone originally described in the *Organic Syntheses* preparation can undergo a second round of alkylation to generate a quaternary α-carbon with good selectivity, again with the second electrophile approaching the enolate from the face opposite to the phenyl rings. For example, alkylation with a pinacolatoborane reagent give the intermediate **22**, which has found application in the synthesis of arginase inhibitors of the type **23** through oxidative cleavage of the allyl group and reductive amination (Fig. 6B).[20] A similar sequence can be carried out to

398

alkylate the allylated oxazinone with BOMCl giving **24** (Fig. 6C). Removal of the auxiliary and benzyl ether provides (*S*)-(*N-tert*-butoxycarbonyl)-(hydroxymethyl)-pentenoic acid (**25**). This intermediate has found utility in the preparation of pyrrolothiazoles as inhibitors of dipeptidyl peptidase IV (DPP-IV).[21]

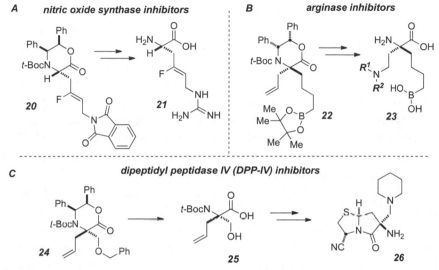

Figure 6. Allylic alkylations to generate medicinally important small molecules.

Another important use for allylglycine and its derivatives has been in the generation of conformationally constrained α-helical peptides. An early report from our lab demonstrated that the tethered *bis*-allylglycine substrate (**27**) underwent olefin metathesis to give an E/Z-mixture of macrocyclic olefin products that upon hydrogenation provided the differentially protected 2,7-diaminosuberic acid derivative **28** (Figure 7).[22]

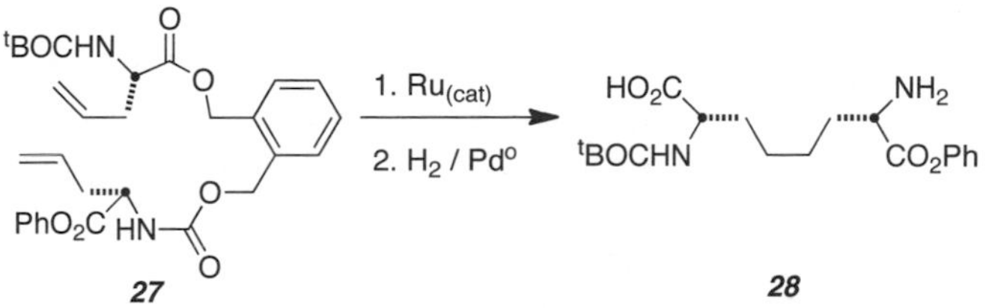

Figure 7. Synthesis of a differentially protected 2,7-diaminosuberic acid derivative.

The conceptually related intramolecular metathesis of allylglycine-containing peptides, has been pioneered by Verdine to make conformationally-constrained α-helical peptides which have been named "stapled peptides".[23] As shown in Figure 8A, the oxazinone template was used to synthesize both enantiomeric series of α-methyl-α-alkenyl amino acids. The α,α-disubstitution increases helicity, while a variation in tether length allowed the examination of chemical yield for the metathesis reaction. These amino acids were incorporated into model peptides at the i and i+4 positions. Metathesis between the two alkene termini successfully "stapled" the peptides together. The shortest metathesis reaction tolerated was that between the allylglycine derivative (S-1$_1$) and hexenyl-substituted amino acid (S-1$_4$) as shown in Figure 8B. The opposite **R,R** enantiomeric series was also successful while the **R,S** series failed to metathesize. An increase in tether length increased metathesis efficiency leading to reactions with >98% conversion. Depending on the length of the tether, these peptides possessed greater helicity and resistance to proteolysis.

The original article detailing the preparation of (R)-(N-tert-butoxycarbonyl)allylglycine provides a useful route to prepare this commonly used and now commercially available amino acid for the synthesis of more complex unnatural amino acids and elaborated peptides. Perhaps more impressive is the utility of the oxazinone to prepare more ornate allylic-amino acid derivatives for use in natural products and medicinal chemistry. Finally, the ease of preparation of the oxazinones[24] (also commercially available) allows for the synthesis of stable isotopomers of allylglycine, which can be introduced into the oxazinone template[25] via bromoacetate or via the allyl iodide fragment. The incorporation of deuterium, ^{13}C, and ^{15}N into this system can thus be accomplished in several reasonably economical ways and can be of use in a variety of applications.

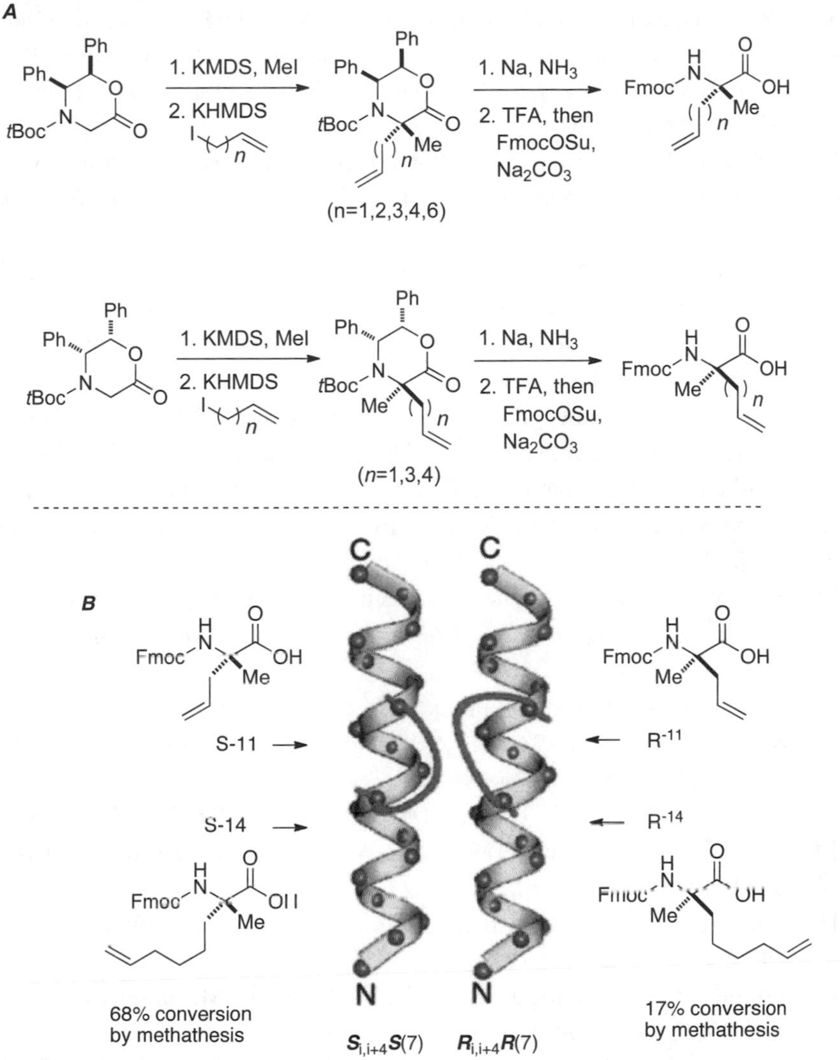

Figure 8. Synthesis of α,α-disubstituted amino acids and their incorporation into stapled peptides. Adapted with permission from *J. Am. Chem. Soc.* **2000**, *122*, 5891–5892, Copyright (2000) American Chemical Society.

1. Department Of Chemistry, University of Utah, Salt Lake City, UT 84112
2. Department of Chemistry, Colorado State University, Fort Collins, CO, 80523

3. Williams, R. M.; Im, M. N. *J. Am. Chem. Soc.* **1991**, *113*, 9276–9286.

4. Reviews: *Asymmetric Synthesis of a-Amino Acids.* Williams, R.M., *Advances in Asymmetric Synthesis*, JAI Press **1995**, Volume 1 (pp 45–94) A. Hassner, Ed.

5. Sebahar, P.; Williams, R. M. *J. Am. Chem. Soc.* **2000**, *122*, 5666–5667.

6. Sebahar, P. R.; Osada, H.; Usui T.; Williams, R. M. *Tetrahedron* **2002** *58*, 6311–6322.

7. Lo, M. M.-C.; Neumann, C. S.; Nagayama, S.; Perlstein, E. O.; Schreiber, S. L. *J. Am. Chem. Soc.* **2004**, *126*, 16077–16086.

8. Nolen, E. G.; Kurish, A. J.; Potter, J. M.; Donahue, L. A.; Orlando, M. D. *Org. Lett.* **2005**, *7*, 3383–3386.

9. Tanino, T.; Al-Dabbagh, B.; Mengin-Lecreulx, D.; Bouhss, A.; Oyama, H.; Ichikawa, S.; Matsuda, A. *J. Med. Chem.* **2011**, *54*, 8421–8439.

10. Mollica, A.; Guardiani, G.; Davis, P.; Ma, S.-W.; Porreca, F.; Lai, J.; Mannina, L.; Sobolev, A. P.; Hruby, V. J. *J. Med. Chem.* **2007**, *50*, 3138–3142.

11. Sun, W.; Nikolovska-Coleska, Z.; Qin, D.; Sun, H.; Yang, C.-Y.; Bai, L.; Qiu, S.; Wang, Y.; Ma, D.; Wang, S. *J. Med. Chem.* **2009**, *52*, 593–596.

12. .Wilson, L. J.; Wang, B.; Yang, S.-M.; Scannevin, R. H.; Burke, S. L.; Karnachi, P.; Rhodes, K. J.; Murray, W. V. *Bioorg. Med. Chem. Lett.* **2011**, *21*, 6485–6490.

13. Greshock, T. J.; Funk, R. L. *J. Am. Chem. Soc.* **2002**, *124*, 754–755.

14. Looper, R. E.; Runnegar, M. T. C.; Williams, R. M. *Tetrahedron* **2006**, *62*, 4549–4562.

15. Looper, R. E.; Williams, R. M. *Angew. Chem. Int. Ed.* **2004**, *43*, 2930–2933.

16. Looper, R. E.; Runnegar, M. T. C.; Williams, R. M. *Angew. Chem. Int. Ed.* **2005**, *44*, 3879–3881.

17. Looper, R. E.; Williams, R. M. *Tetrahedron Lett.* **2001**, *42*, 769–771.

18. Staas, D. D.; Savage, K. L.; Sherman, V. L.; Shimp, H. L.; Lyle, T. A.; Tran, L. O.; Wiscount, C. M.; McMasters, D. R.; Sanderson, P. E. J.; Williams, P. D.; Lucas, B. J.; Krueger, J. A.; Lewis, S. D.; White, R. B.; Yu, S.; Wong, B. K.; Kochansky, C. J.; Anari, M. R.; Yand, Y.; Vacca, J. P. *Bioorg. Med. Chem.* **2006**, *14,* 6900–6916.

19. Hansen, D. W., Jr.; Webber, R. K.; Awasthi, A. K.; Manning, P. T.; Sikorski, J. A. PCTInt. Appl. (**2001**), WO 2001079156 A1 20011025.

20. Van Zandt, M.; Golebiowski, A.; Ji, M. K.; Whitehouse, D.; Ryder, T.; Beckett, P. PCT Int. Appl. (**2011**), WO 2011133653 A1 20111027.
21. Wagner, H.; Schoenafinger, K.; Jaehne, G.; Gaul, H.; Buning, C.; Tschank, G.; Werner, U. PCT Int. Appl. (**2005**), WO 2005012312 A1 20050210.
22. Williams, R. M.; Liu, J. *J. Org. Chem.* **1998**, *63*, 2130–2132.
23. Schafmeister, C. E.; Po, J.; Verdine, G. L. *J. Am. Chem. Soc.* **2000**, *122*, 5891–5892.
24. Williams, R. M.; Sinclair, P. J.; DeMong, D.; Chen, D.; Zhai, D., *Org. Synth.* **2003**, *80*, 18–30.
25. Aoyagi, Y.; Iijima, A.; Williams, R. M., *J. Org. Chem.* **2001**, *66*, 8010–8014.

Robert M. Williams was born in New York in 1953 and attended Syracuse University where he received the B.A. degree in Chemistry in 1975. He obtained the Ph.D. degree in 1979 at MIT (W. H. Rastetter) and was a post-doctoral fellow at Harvard (1979-1980; R. B. Woodward/Yoshito Kishi). He joined Colorado State University in 1980 and was named a University Distinguished Professor in 2002. His interdisciplinary research program (over 280 publications) at the chemistry-biology interface is focused on the total synthesis of biomedically significant natural products, biosynthesis of secondary metabolites, studies on antitumor drug-DNA interactions, HDAC inhibitors, amino acids and peptides.

Ryan E. Looper was born in Banbury, England in 1976. He came to the U.S. to attend Western Washington University where he earned a B.S. degree in Chemistry in 1998 and an M.S. degree in 1999 under the guidance of J. R. Vyvyan. He obtained his Ph.D. degree in 2004 at Colorado State University in the laboratories of Prof. R.M. Williams. After a NIH post-doctoral fellowship at Harvard University with Prof. S.L. Schreiber, he began his independent career at the University of Utah in 2007.

A Practical Synthesis of Isocyanates from Isonitriles:
Ethyl 2-Isocyanatoacetate

Submitted by Hoang V. Le and Bruce Ganem.[1]
Checked by Kenichi Kobayashi and Viresh H. Rawal

1. Procedure

Caution: Dimethyl sulfide, a by-product of this reaction, is a highly flammable liquid and an irritant with an extremely unpleasant odor. The experimentalist should wear gloves, and the reaction and its workup should always be performed in a well-ventilated hood.

A 250-mL round-bottomed flask with a 24/40 joint equipped a Teflon-coated oval-shaped magnetic stir bar (3.2 cm x 1.5 cm) and topped with a rubber septum is flushed with nitrogen (Note 1). Dichloromethane (60 mL) is added *via* syringe (Note 2). Ethyl 2-isocyanoacetate (7.0 mL, 60.8 mmol) is added *via* syringe to afford a brown solution (Note 3). Dimethyl sulfoxide (5.4 mL, 76 mmol) is added *via* syringe (Note 4). The resulting solution is cooled to –78 °C in a dry ice/acetone bath, then trifluoroacetic anhydride (0.86 mL, 6.2 mmol) is added dropwise *via* syringe over 5 min (Note 5). The resulting solution is stirred for 5 min at –78 °C, then the dry ice/acetone bath is removed, and the dark brown solution is allowed to warm to rt over 1 h and stirred for an additional 5 min (Note 6). The magnetic stir bar is removed using a Teflon-coated magnet retriever and rinsed with dichloromethane (0.5 mL). The reaction mixture is concentrated in the hood by rotary evaporation at 20 mmHg. The resulting brown concentrate is dissolved in ice-cold dichloromethane (20 mL) and washed with three 10-mL portions of ice-cold deionized water. The organic layer is dried over anhydrous magnesium sulfate powder (1g) and filtered through a 30 mL sintered glass funnel using 3 x 5 mL portions of fresh solvent for rinsing

404

Org. Synth. **2012**, *89*, 404-408
Published on the Web 4/5/2012
© 2012 Organic Syntheses, Inc.

(Note 7). The filtrate is concentrated by rotary evaporation (25 °C, 20 mmHg) (Note 8) and then under high vacuum (0.3 mmHg) for 30 min to afford 7.58–7.66 g (96–97%) of **2** as a dark brown oil (Notes 9 and 10).

2. Notes

1.　The round-bottomed flask and the magnetic stir bar were dried in an oven, cooled to room temperature under high vacuum, and subsequently maintained under a positive pressure of nitrogen.

2.　Dichloromethane (Optima® grade) was purchased from Fisher Scientific and dried over activated alumina solvent purification system (Innovative Technology Inc. Pure Solv™).

3.　Ethyl 2-isocyanoacetate (95%) was purchased from Aldrich Chemical Company Inc. and the dark brown oil was used without further purification.

4.　Dimethyl sulfoxide (DMSO, anhydrous, ≥99.9%) was purchased from Sigma-Aldrich and used without further purification.

5.　Trifluoroacetic anhydride (≥99%) was purchased from Sigma-Aldrich and used without further purification.

6.　After stirring at room temperature for 5 min, a few drops of the reaction solution were removed *via* Pasteur pipet and submitted to IR analysis to confirm that the reaction was complete (complete disappearance of isonitrile stretch at 2164 cm^{-1}; appearance of isocyanate stretch at 2255 cm^{-1}.

7.　Magnesium sulfate anhydrous powder was purchased from Fisher Scientific and used without further purification.

8.　Removing the bulk of the solvent by rotary evaporation was judged complete when the flask no longer felt cold to the touch after the water bath was lowered from the evaporator.

9.　The product was a dark brown oil which appeared pure by ^{1}H and ^{13}C NMR spectroscopy. When stored at room temperature, the neat product decomposed with a half-life of ca. 12 h into unidentifiable solid. However, decomposition could be suppressed by wrapping the sample in aluminum foil and storing it neat at −10 °C. After 4 days, only traces of solid decomposition product were observed.

10.　The product isocyanate displayed the following spectroscopic characteristics, which matched published[2-4] spectra: ^{1}H NMR (500 MHz, CDCl$_3$) δ: 1.32 (t, 3 H, J= 7.2 Hz), 3.94 (s, 2 H), 4.29 (q, 2 H, J= 7.2 Hz);

^{13}C NMR (125 MHz, CDCl$_3$) δ: 13.9, 44.4, 62.4, 127.3 (weak), 169.2; IR (neat, cm^{-1}): 3363 (m), 2987 (m), 2943 (m), 2255 (s), 1749 (s).

Safety and Waste Disposal Information

All hazardous materials should be handled and disposed of in accordance with "Prudent Practices in the Laboratory"; National Academies Press; Washington, DC, 2011.

3. Discussion

Isocyanates, which embody the N=C=O functional group, undergo several synthetically useful reactions, including nucleophilic addition of alcohols and amines to produce urethanes and ureas, respectively. Isocyanates are also intermediates in several well-known organic reactions, including the Hofmann rearrangement of primary amides, the Curtius rearrangement of acyl azides, the Schmidt reaction of carboxylic acids with hydrazoic acid, and the Lossen rearrangement of hydroxamic acids. Diisocyanates are widely used in chemical industry to produce polyurethanes and other polymers.

The present method of synthesizing isocyanates from isonitriles is simple and easy to perform.[3] Both reagents (DMSO as oxidant; trifluoroacetic anhydride as catalyst) are inexpensive and readily available. The reaction proceeds swiftly and efficiently at low temperature and produces volatile byproducts (dimethyl sulfide, residual anhydride) that are easily removed upon evaporative workup. Furthermore, washing with cold water completely eliminates any small quantities of residual DMSO from the desired product.

Ethyl 2-isocyanatoacctate has previously been prepared by the reaction of ethanol with 2-isocyanatoacetyl chloride.[5] It has also been synthesized by the reaction of ethyl hydrogen malonate using diphenylphosphoryl azide via a Curtius rearrangement.[6] Ethyl 2-isocyanatoacetate can also be prepared from various glycine derivatives, including the reaction of N-tosylglycine ethyl ester[7] or N-trimethylsilylglycine ethyl ester[4] with phosgene, or the reaction of glycine ethyl ester hydrochloride with bis(trichloromethyl)carbonate.[8]

406

1. Department of Chemistry and Chemical Biology, Baker Laboratory, Cornell University Ithaca, NY 14853-1301, USA; bg18@cornell.edu; Funding was provided by the NIH (GM 008500 Training Grant Support for HVL). Support of the Cornell NMR Facility has been provided by NSF (CHE 7904825; PGM 8018643) and NIH (RR02002).

2. Barany, M. J.; Hammer, R. P.; Merrifield, R. B.; Barany, G. *J. Am. Chem. Soc.* **2005**, *127*, 508.

3. Le, H. V.; Ganem, B. *Org. Lett.* **2011**, *13*, 2584.

4. Lebedev, A. V.; Lebedeva, A. B.; Sheludyakov, V. D.; Ovcharuk, S. N.; Kovaleva, E. A.; Ustinova, O. L. *Russ. J. Gen. Chem.* **2006**, *76*, 1069.

5. Iwakura, Y.; Uno, K.; Kang, S. *J. Org. Chem.* **1965**, *30*, 1158.

6. Aoyama, Y.; Uenaka, M.; Konoike, T.; Hayasaki-Kajiwara, Y.; Naya, N.; Nakajima, M. *Bioorg. Med. Chem. Lett.* **2001**, *11*, 1691.

7. Schuemacher, A. C.; Hoffmann, R. W. *Synthesis* **2001**, 243.

8. Huang, J-M.; Chen, H.; Chen, R-Y. *Synth. Commun.* **2002**, *32*, 1357.

Appendix
Chemical Abstracts Nomenclature; (Registry Number)

Ethyl 2-isocyanoacetate; (2999-46-4)
Ethyl 2-isocyanatoacetate; (2949-22-6)

Bruce Ganem was born in Boston, MA. He received his B.A. degree in chemistry from Harvard College and his Ph.D in organic chemistry from Columbia University. After completing postdoctoral work at Stanford, he joined the Cornell faculty in 1974, where he is the Franz and Elisabeth Roessler Professor of Chemistry and the J. Thomas Clark Professor of Entrepreneurship. A synthetic organic chemist by training, Ganem has broad scientific research interests ranging from organic and analytical chemistry to biochemistry. His current research focuses on the development of new synthetic reactions with applications to drug discovery and development.

Hoang V. Le was born and raised in Vietnam. He graduated from the University of California, Berkeley in 2008 with a B.S. degree in Chemistry. While at Berkeley, he worked in the lab of Professor Kenneth Raymond synthesizing dicationic bridges and investigating their stability in supramolecular systems. He is currently a fourth-year graduate student in the lab of Professor Bruce Ganem at Cornell University. His research aims to expand the scope of multicomponent reaction (MCR) chemistry by improving known MCRs and developing new MCRs.

Kenichi Kobayashi received his B.S. and M.S. degrees in chemistry from Rikkyo University where he conducted research with Professor Chiaki Kuroda. After having worked at KYORIN Pharmaceutical Co., Ltd. for six years, he joined the lab of Professor Hideo Kigoshi at University of Tsukuba and received his Ph.D. in 2011. He then moved to the University of Chicago, where he is currently a postdoctoral researcher in the lab of Professor Viresh H. Rawal.

408

Discussion Addendum for:
N-Hydroxy-4-(p-chlorophenyl)thiazole-2(3H)-thione

Prepared by Christine Schur, Irina Kempter, and Jens Hartung.*[1]
Original article: Hartung, J.; Schwarz, M. *Org. Synth.* **2002**, *79*, 228–235.

O-Alkylation of N-hydroxythiazole-2(3H)-thiones (TTOHs) provides N-alkoxy derivatives, which are valuable alkoxyl radical precursors for photobiological, mechanistic, and synthetic purposes.[2,3,4] Chemical reactivity of the N-alkoxythiazole-2(3H)-thiones (TTORs) is controlled by the thiohydroxamic acid functional group that is embedded in a cross-conjugated heterocyclic π-electron system.[5,6,7] Breaking of the thiohydroxamate nitrogen-oxygen bond in TTORs allows the heterocyclic fragment to gain aromatic stabilization, which is the reason for the molecules to liberate oxygen-centered radicals under comparatively mild conditions.

Systematic exploration of O-alkyl thiohydroxamate-based O-radical precursors started with the N-alkoxypyridine-2(1H)-thiones.[8,9,10] Experimental difficulties associated with synthesis and storage of the reagents stimulated development of the N-alkoxy-4-(p-chlorophenyl)thiazole-2(3H)-thiones (CPTTORs). The latter compounds have shelf-lives from months to years,[11,12] but selectively liberate oxygen radicals if photochemically or thermally excited.[4] Problems remaining unsolved in CPTTOR-chemistry, such as synthesis of tertiary O-alkyl derivatives, were successfully addressed by introducing *N*-hydroxy-5-(*p*-methoxyphenyl)-4-methylthiazole-2(3*H*)-thione (MAnTTOH) as next generation alkoxyl radical progenitor.[3,13,14]

The Chemistry of *N*-Alkoxy-4-(*p*-chlorophenyl)thiazole-2(3*H*)-thiones

CPTTOH is an acid of similar strength to acetic acid and forms primary and secondary *O*-esters in 50–70% yield, if converted into the derived tetraalkylammonium salt and treated with a hard alkylating reagent. The largest body of chemical reactivity data compiled for the CPTTORs relate to photochemically or thermally initiated reactions with typical mediators for conducting radical chain reactions, such as tributylstannane, tris(trimethylsilyl)silane, *O*-ethyl cysteine, or bromotrichloromethane.[5,15,16,17] In a supplementary project, the propensity of *O*-acyl- and *O*-alkyl-derivatives of CPTTOH to crystallize was used to study the until then unexplored solid state chemistry of the thiohydroxamates. The results from this study show that the molecules tolerate a significant degree of steric congestion at the thiohydroxamate oxygen, because strain effects are absorbed by structural changes across the whole thiohydroxamte group.[18]

In addition to radical reactions and solid state chemistry, new ground state reactivity of CPTTORs was discovered, such as a shift of the methyl group from oxygen to sulfur, fragmentation of secondary benzylic esters leading a thiolactam and a ketone, and 4-pentenoxy rearrangements occuring specifically in the solid state (Figure 1).[19,20]

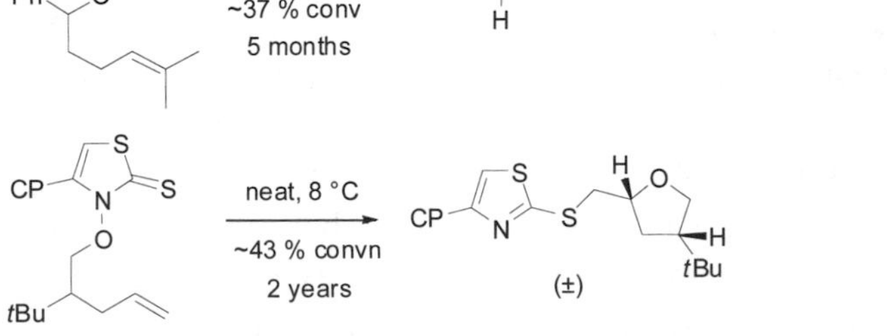

Figure 1. Background reactivity of neat CPTTORs in the solid state.

5-Substituted *N*-Hydroxy-4-methylthiazole-2(3*H*)-thiones

From a structure-reactivity study it became apparent that a substituent attached in position 5 shifts electronic absorptions of the thiazole-2(3*H*)-thione nucleus stronger than the same group does in position 4. Synthesis of 5-substituted *N*-hydroxy-4-methylthiazole-2(3*H*)-thiones (Figure 2) starts from constitutionally dissymmetric ketones. Arylpropanones that are not commercially available can be prepared from underlying arylcarbaldehydes and nitroethane via intermediate β-nitrostyrenes. Hydrolysis of nitrostyrenes in acidic aqueous solutions containing iron-powder furnishes the required ketones, which are chlorinated by sulfuryl chloride at the higher substituted α-carbon. Methods of *O*-ethyl xanthogenate- and oxime-synthesis follow the protocol of the original CPTTOH-synthesis (Figure 2). For the final step, potassium hydroxide in an equimolar volume of dichloromethane and water is the more efficient reagent to mediate the thiazolethione ring closure compared to anhydrous zinc chloride in diethyl ether, which is the recommended reagent for the final step of CPTTOH-synthesis. [3]

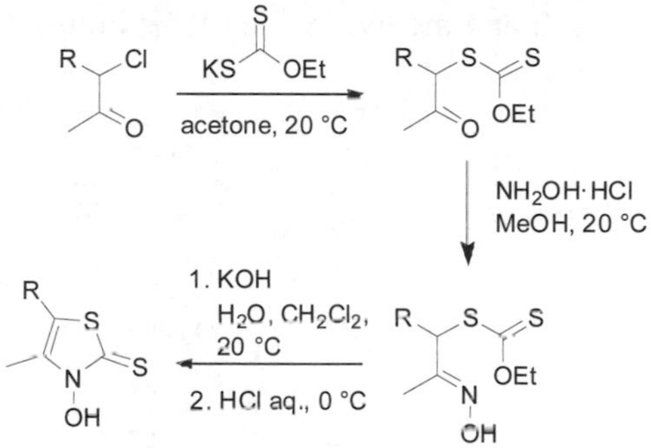

R	total yield / %	λ_{max} / nm
CH_3	74	317
C_6H_5	45	333
4-ClC_6H_4 (CP)	21	334
4-$(AcHN)C_6H_4$	59	335
4-$(H_3CO)C_6H_4$ (An)	66	334
2,4-$(H_3CO)_2C_6H_3$	54	336

16 %

λ_{max} = 376 nm

Figure 2. Synthesis of 5-substituted *N*-hydroxy-4-methylthiazole-2(3*H*)-thiones.[3]

N-Alkoxy-4-methylthiazole-2(3H)-thiones

N-Hydroxy-5-(p-methoxyphenyl)-4-methylthiazole-2(3H)-thione (MAnTTOH) was developed as successor of CPTTOR, to improve photochemical properties of alkoxyl radical progenitors, raise yields of O-alkylation products, and prepare tert-O-alkyl thiohydroxamates which until then were not available in yields higher than 5%. MAnTTORs show a longest wavelength of absorption at around 334 nm that gives some of the compounds a yellow color and allows photochemical excitation with visible light or 350 nm-light from a Rayonet®-photoreactor. MAnTTOH-derived thiohydroxamate salts show an unusual propensity to form products of O-alkylation (MAnTTORs), if treated with alkyl tosylates, alkyl chlorides, or even alkyl iodides (top graphic of Figure 3).[20] The majority of MAnTTORs are solids that crystallize by adding an excess of methanol to crude reaction mixtures. Alkylation at sulfur, which consumes the major fraction of alkylation reagent in N-alkoxypyridine-2(1H)-thione-synthesis, is restricted to comparably few instances, such the reaction between the MAnTTOH-tetraethylammonium salt and methyl iodide (46%; bottom graphic of Figure 3).

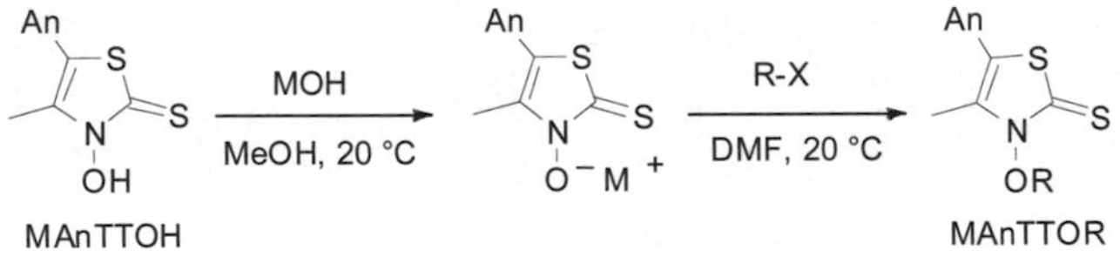

M	R-X	yield (MAnTTOR) / %
NEt$_4$	⌒OTs	87
Na	⌒OTs	57
NEt$_4$	⌒I	64
Na	⌒I	48
NEt$_4$	c-C$_5$H$_9$I	60
NEt$_4$	⌒⌒I	47
NEt$_4$	Ph⌒Cl	77

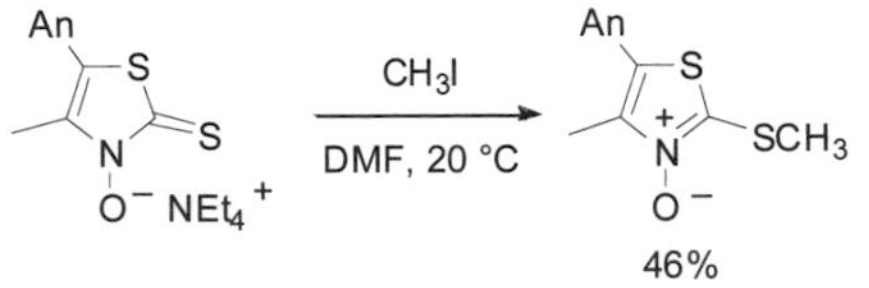

46%

Figure 3. Preparation of MAnTTORs (top and table; An = p-anisyl) and selective S-alkylation of a MAnTTOH-derived tetraethyl-ammonium salt (bottom).[20]

Org. Synth. **2012**, *89*, 409-419

Alternative methods for thiohydroxamate *O*-alkylation are the Mitsunobu-reaction, cyclic sulfate-opening, and *O*-alkyl isourea-chemistry (Figure 4). The former two methods proceed via an inversion of configuration at the attacked carbon,[21] and the latter is the only method available so far to prepare tertiary *O*-alkyl thiohydroxamates.[22]

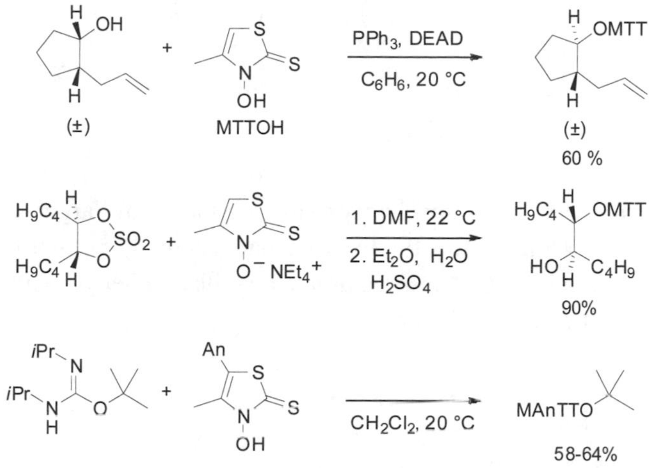

Figure 4. Synthesis of *O*-alkyl thiohydroxamates from an alcohol, cyclic sulfate, or an isourea.[21,22]

Scope of the Thiohydroxamate Method

The advantage of *N*-alkoxythiazole-2(3*H*)-thiones compared to alternative alkoxyl radical precursors is their ability to liberate alkoxyl radicals in a propagating step of a chain reaction. Nitrogen-oxygen bond homolysis thereby occurs after addition of a chain propagating radical, such as a carbon-, silicon-, or tin-radical, to the thione-sulfur (Figure 5).[14] Alkoxyl radicals are highly reactive intermediates, their selectivities are, however, predictable in a straightforward manner on the basis of kinetic and thermochemical data. Additions and hydrogen-atom abstractions, two of the main alkoxyl radical reactions, are irreversible and therefore controlled by kinetic effects (Figure 6, top and center). Selectivity of β-carbon-carbon fragmentation for synthesis of carbonyl compounds (Figure 6, bottom), and alkoxyl radical rearrangements are guided by radical stabilizing effects and therefore are thermodynamically controlled.[2,20,23]

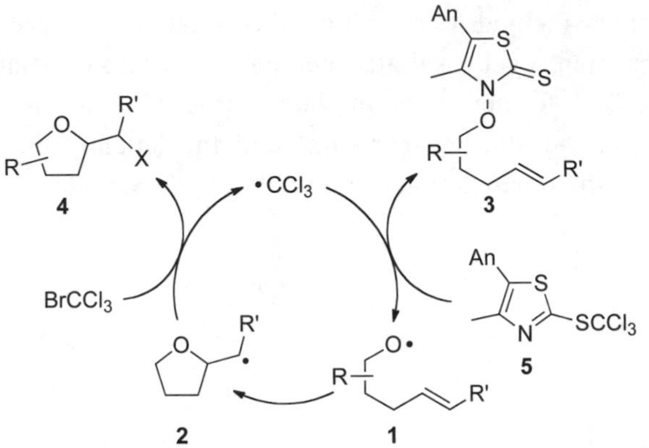

Figure 5. Radical chain mechanism for synthesis of target compounds according to the thiohydroxamate method, exemplified by the use of BrCCl₃ as mediator (R = alkyl, phenyl, acyloxy; R' = hydrogen, alkyl, phenyl).[20,22]

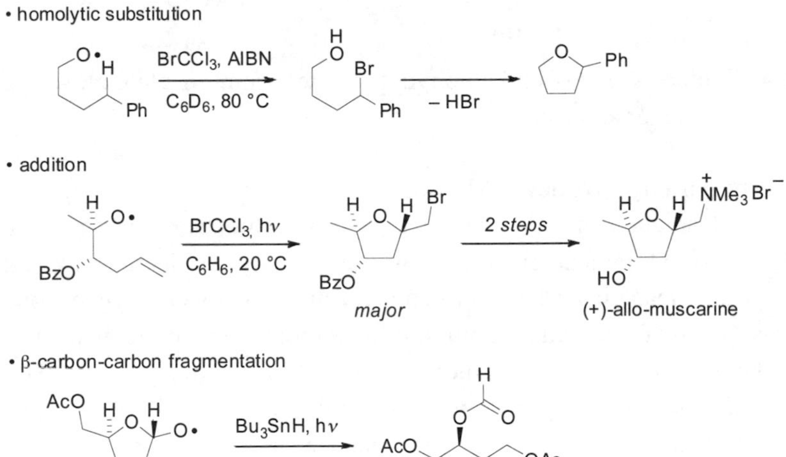

Figure 6. Application of homolytic substitution, addition, and β-carbon-carbon bond fragmentation of alkoxyl radicals in synthesis.[24]

Since the radical oxygen is electrophilic and the hydroxyl oxygen nucleophilic, additions of alkoxyl radicals and alcohols to polar carbon-carbon double bonds occur with complementary selectivity (Figure 7).[7,25,26,27]

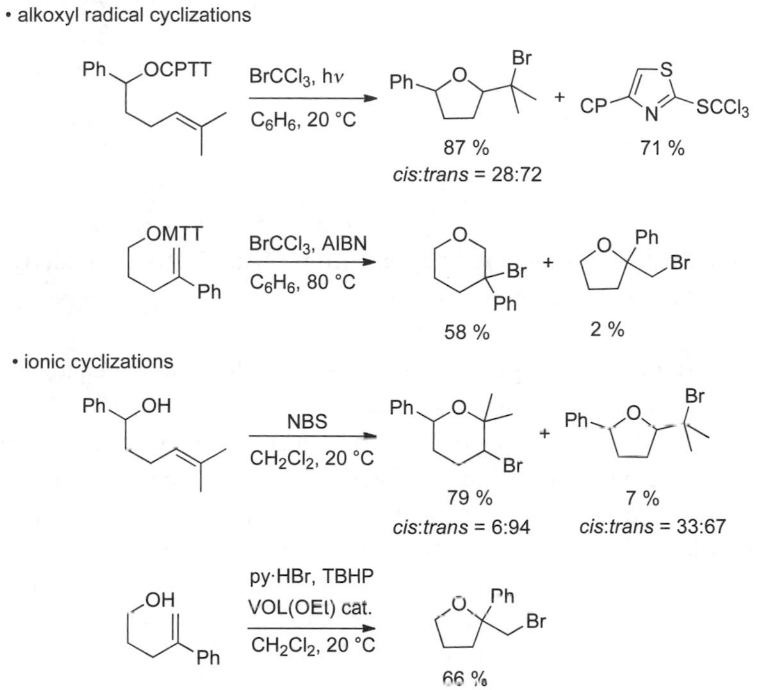

• alkoxyl radical cyclizations

87 %
cis:trans = 28:72

71 %

58 %

2 %

• ionic cyclizations

79 %
cis:trans = 6:94

7 %
cis:trans = 33:67

66 %

Figure 7. Examples showing complementary selectivity in bromocyclizations of alkenoxyl radicals (top) and alkenols (bottom: VOL(OEt) = 2-[(2-oxidophenyl)iminomethyl] (ethanolato) oxidovanadium(V); py·HBr = pyridinium hydrobromide; TBHP = tert-butylhydroperoxide}.[25,27]

Alkoxyl radical generation from N-alkoxythiazolethiones occurs under pH-neutral, non-oxidative conditions and therefore may be used for transformations that are not attainable by other alkoxyl radical-generating methods. Substituted 5-hexenyloxyl radicals, for example, cyclize onto π-bonds having two methyl groups attached at the terminal alkene carbon. The reaction stereoselectively provides tetrahydropyranylmethyl radicals, which are trapped by bromotrichlormethane to furnish bromocyclized products

(Figure 8).[28,29] Polar bromocyclization of the underlying 5-hexenols provide oxepans as major products.

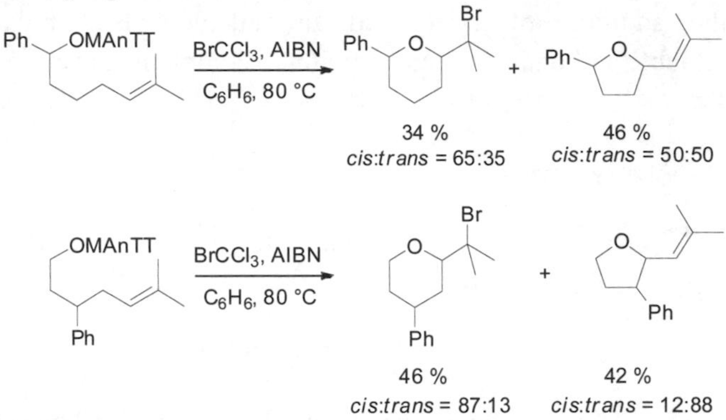

34 %
cis:trans = 65:35

46 %
cis:trans = 50:50

46 %
cis:trans = 87:13

42 %
cis:trans = 12:88

Figure 8. Synthesis of tetrahydropyrans from MAnTTORs via underlying 5-hexenoxyl radical cyclization.

The thiohydroxamate method also applies to prepare vicinal bromohydrin ethers from intermolecular alkoxyl radical addition to alkenes (Figure 9). *O*-Alkyl thiohydroxamates thus were used to synthesize norbornene-derived vicinal bromohydrin ethers, which have interesting biological properties but are difficult to prepare from ionic reactions due to complications imposed by non-classical carbocation formation.[22,30]

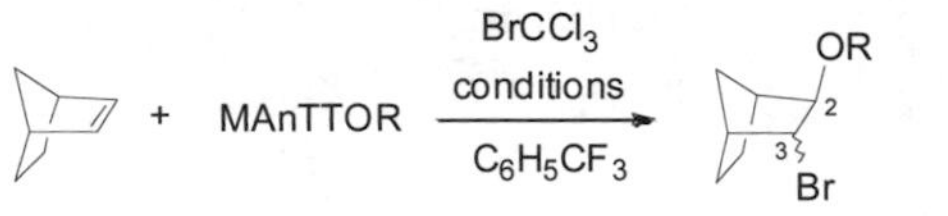

R	conditions	yield / %	3-exo : 3-endo
H	mw / 200 °C	7	13 : 87
CH$_3$	AIBN / 80 °C	64	28 : 72
CH(CH$_3$)$_2$	mw / 150 °C	33	25 : 75
C(CH$_3$)$_3$	*hv* / 20 °C	46	24 : 76

mw = laboratory microwave

Figure 9. Synthesis of β-bromoethers from intermolecular alkoxyl radical additions.

Besides adding benefits to organic synthesis, MAnTTORs are excellent tools for conducting kinetic studies on alkoxyl radical elementary reactions.[22] Kinetic data are the basis for application of alkoxyl radicals in synthesis of ethers , for complementing selectivity that for more than a century has been restricted to nucleophilic carbon-oxygen bond formation.

1. Department of Chemistry, Technische Universität Kaiserslautern, D-67663 Kaiserslautern, Germany. This work was supported by the State Rheinland-Pfalz (Scholarships for I.K. and C.S.) and the Deutsche Forschungsgemeinschaft.
2. Adam, W.; Hartung, J.; Okamoto, H.; Marquardt, S.; Nau, W.M.; Pischel, U.; Saha-Möller C.R.; Špehar, K. *J. Org. Chem.* **2002**, *67*, 6041–6049.
3. Groß, A.; Schneiders, N.; Daniel, K.; Gottwald, T.; Hartung, J. *Tetrahedron* **2008**, *64*, 10882–10889.
4. Hartung, J.; Daniel, K.; Gottwald, T.; Groß, A.; Schneiders, N. *Org. Biomol. Chem.* **2006**, *4*, 2313–2322.
5. Arnone, M.; Hartung, J.; Engels, B. *J. Phys. Chem. A*. **2005**, *109*, 5943–5950.
6. Hartung, J.; Gottwald, T.; Špehar, K. *Synthesis* **2002**, 1469–1498.
7. Greb, M.; Hartung, J. *Synlett* **2004**, 65–68.
8. Bcckwith, A.L.J, Hay, B.P. *J. Am. Chem. Soc.* **1988**, 110, 4415–4416.
9. Hartung, J.; Gallou, F. *J. Org. Chem.* **1995**, 60, 6706–6716.
10. Hartung, J.; Hiller, M.; Schwarz, M.; Svoboda, I.; Fuess, H. *Liebigs Ann. Chem.* **1996**, 2091–2097.
11. Hartung, J.; Schwarz, M.; Svoboda, I.; Fueß, H.; Duarte, M. T. *Eur. J. Org. Chem.* **1999**, 1275–1290.
12. Hartung, J.; Schwarz, M.; Paulus, E. F,; Svoboda, I.; Fuess, II. *Acta Cryst.* **2006**, *C62*, o386–o388.
13. Hartung, J.; Gottwald, T.; Špehar, K.; *Synlett* **2003**, 227–229.
14. Hartung, J.; Špehar, K.; Svoboda, I.; Fuess, H.; Arnone, M.; Engels, B. *Eur. J. Org. Chem.* **2005**, 869–881.
15. 4-(4-Chlorophenyl)-3-hydroxy-2(3*H*)thiazolethione. Hartung J. *In Reagents for Radical Chemistry – Encyclopedia of Reagents in Organic Chemistry*; Crich, D. Ed., John Wiley & Sons, New York, N.Y., 2008, 175–179.
16. Cyclization of Alkoxyl Radicals. Hartung J., *In Radicals in Organic Synthesis*; Renaud, P.; Sibi, M.P., Eds.; vol. 2, Wiley-VCH, 2001, 427–439.
17. Hartung, J. *Eur. J. Org. Chem.* **2001**, 619–632.

18. Hartung, J.; Bergsträßer, U.; Daniel, K.; Schneiders, N.; Svoboda, I.; Fuess, H. *Tetrahedron* **2009**, *65*, 2567–2573.

19. Hartung, J.; Daniel, K.; Bergsträsser, U.; Kempter, I.; Schneiders, N.; Danner, S.; Schmidt, P.; Svoboda, I.; Fuess, H. *Eur. J. Org. Chem.* **2009**, 4135–4142.

20. Hartung, J.; Schur, C.; Kempter, I.; Gottwald, T. *Tetrahedron* **2010**, *66*, 1365–1374.

21. Hartung, J.; Kempter, I.; Gottwald, T.; Kneuer, R. *Tetrahedron:Asymmetry* **2009**, *20*, 2097–2104.

22. Schur, C.; Becker, N.; Bergsträßer, U.; Gottwald, T.; Hartung, J. *Tetrahedron* **2011**, *67*, 2338–2347.

23. Hartung, J.; Kneuer, R.; Špehar, K. *Chem. Commun.* **2001**, 799–800.

24. Hartung, J.; Kneuer, R. *Tetrahedron:Asymmetry* **2003**, *14*, 3019–3031.

25. Hartung, J.; Kneuer, R.; Laug, S.; Schmidt, P.; Špehar, K.; Svoboda, I.; Fuess, H. *Eur. J. Org. Chem.* **2003**, 4033–4052.

26. Hartung, J.; Kopf, T.M.; Kneuer, R.; Schmidt, P. *C. R. Acad. Sci. Paris, Chimie/Chemistry* **2001**, 649–666.

27. Gottwald, T.; Greb, M.; Hartung, J. *Synlett* **2004**, 65–68.

28. Hartung, J.; Gottwald, T. *Tetrahedron Lett.* **2004**, *45*, 5619–5621.

29. Schneiders, N.; Gottwald, T.; Hartung, J. *Eur. J. Org. Chem.* **2009**, 799–801.

30. Hartung, J.; Schneiders, N.; Gottwald, T. *Tetrahedron Lett.* **2007**, *48*, 6327–6330.

Jens Hartung was born in Offenbach/Main, Germany in 1961. He took his diploma degree in 1987 with Klaus Hafner and his Ph.D. with Bernd Giese in 1990 at the Technische Hochschule Darmstadt, Germany. That same year, he moved to the MIT to spent a postdoctoral year with K. Barry Sharpless. In 1992 he joined the cardiovascular division of Hoechst AG (today Sanofi-Aventis). In 1994 he moved as lecturer to the Bayerische Julius-Maximilians-Universität Würzburg, where he completed his habilitation in 1998. Since 2003 he is full professor of organic chemistry at the Technische Universität Kaiserslautern. His research interests include the chemistry of reactive intermediates, in particular oxygen-centered radicals, transition metal-catalyzed oxidations, the chemistry of vanadium-dependent bromoperoxidases, static and dynamic stereochemistry, synthesis of heterocycles and marine natural products.

Christine Schur was born in Karassu, Kasachstan in 1983. She received a Diploma-degree in 2008 from the Technische Hochschule Kaiserslautern studying selectivity in inter- and intramolecular alkoxyl radical reactions, working with Jens Hartung. She continues this project as part of her Ph.D.-thesis.

Irina Kempter was born in Neustadt/Weinstraße, Germany in 1984. She received a Diploma-degree in 2008 from the Technische Hochschule Kaiserslautern studying stereoselective *O*-alkyl thiohydroxamate synthesis, working with Jens Hartung. Her research interests in a new project as part of her Ph.D.-thesis are associated with investigation of polar effects in alkoxyl radical chemistry.

Preparation of (R_a)-Methyl 3-(dimethyl(phenyl)silyl)hexa-3,4-dienoate

Submitted by Ryan A. Brawn and James S. Panek.[1]
Checked by Francois Grillet and Kay M. Brummond.

1. Procedure

A.   (±)-4-(Dimethyl(phenyl)silyl)-but-3-yn-2-ol (**2**). An oven-dried, one-necked, 500-mL, round-bottomed flask equipped with a 3-inch, egg-shaped, Teflon-coated magnetic stir bar and rubber septum is charged with flame-dried lithium chloride (6.36 g, 150 mmol, 2 equiv) (Notes 1 and 2). Tetrahydrofuran (150 mL) and 3-butyn-2-ol (**1**) (6.06 mL, 5.42 g, 75 mmol, 1 equiv) are added by syringe, and the flask is chilled to –78 °C in a dry ice/acetone bath. The rubber septum is replaced with a pressure-equalizing addition funnel and *n*-butyllithium (60 mL of a 2.5 M solution in hexanes, 150 mmol, 2 equiv) is transferred to the addition funnel using cannula techniques and then added dropwise to the cooled solution over 15 min. The solution is stirred for an additional 30 min at –78 °C (Note 3). The addition funnel is replaced with a rubber septum and chloro(dimethyl)phenylsilane (12.59 mL, 12.80 g, 75 mmol, 1 equiv) (Note 4) is added dropwise over 15 min via syringe, and the cloudy white mixture is stirred 14 h while warming

Org. Synth. **2012**, *89*, 420-431
Published on the Web 4/12/2012
© 2012 Organic Syntheses, Inc.

slowly to room temperature. After cooling to 0 °C, the septum is removed and the cloudy yellow mixture is quenched by careful addition of 0.1 M HCl (100 mL, 10 mL portions added over 5 min) and stirred at room temperature for an additional 20 min. The reaction mixture is poured into a 1-L separatory funnel and extracted with diethyl ether (3 x 50 mL), and the combined organic layers are washed with water (25 mL) and brine (25 mL). The organic layer is dried with MgSO$_4$ (15 g), filtered, and concentrated under reduced pressure. Purification by column chromatography (Note 5) (10% ethyl acetate/hexanes) yields (±)-4-(dimethyl(phenyl)silyl)-but-3-yn-2-ol (**2**) (13.73 g, 67.2 mmol, 90% yield) as a pale yellow oil (Note 6).

B. *(S)-4-(Dimethyl(phenyl)silyl)-but-3-yn-2-ol* (*S*)-**2**. A single-necked, 250-mL, round-bottomed flask equipped with a rubber septum and a 3-inch, egg-shaped, Teflon-coated magnetic stir bar is charged with racemic 4-(dimethyl(phenyl)silyl)-but-3-yn-2-ol (15.03 g, 73.5 mmol, 1 equiv) and *n*-pentane (100 mL). Lipase AK Amano (4.5 g, 0.3 weight equiv) (Note 7) is added in one portion, followed by the addition of vinyl acetate by syringe (33.9 mL, 31.66 g, 367.7 mmol, 5 equiv). The resulting brown dispersion is stirred at room temperature and the reaction is monitored by ^{1}H NMR (Note 8). Upon completion the reaction mixture is filtered through a fritted funnel to remove the lipase, the precipitate is washed with diethyl ether (2 x 100 mL), and the clear filtrate is concentrated under reduced pressure. Purification of the residue by column chromatography (Note 9) (gradient elution 5%-20% ethyl acetate/hexanes) results in *(S)*-4-(dimethyl(phenyl)silyl)-but-3-yn-2-ol (*S*)-**2** (7.04 g, 34.4 mmol, 47% yield) as a clear yellow oil (Note 10). Also isolated is the *(R)*-acetate (*R*)-**3** (7.69 g, 31.2 mmol, 43% yield) (see discussion).

C. *(R$_a$)-Methyl 3-(dimethyl(phenyl)silyl)hexa-3,4-dienoate* (*R$_a$*-**4**). A single-necked, 250-mL, round-bottomed flask equipped with a rubber septum and a 3-inch, egg-shaped, Teflon-coated magnetic stir bar is charged with (*S*)-4-(dimethyl(phenyl)silyl)-but-3-yn-2-ol (*S*-**2**) (6.61 g, 32.3 mmol, 1 equiv), trimethylorthoacetate (16.5 mL, 15.55 g, 129.3 mmol, 4 equiv) and xylenes (66 mL) (Note 11). Propionic acid (0.12 mL, 0.12 g, 1.62 mmol, 0.05 equiv) is added, and a reflux condenser is attached. The flask is placed in a pre-heated 160 °C oil bath and the solution is heated at reflux for 24 h (Note 12). Additional trimethylorthoacetate is added (8.2 mL, 7.77 g, 64.7 mmol, 2 equiv) by syringe, and the solution is refluxed for an additional 16 h. The pale yellow solution is cooled to room temperature, concentrated (Note 13) and purified by column chromatography (Note 14) (2% ethyl acetate/hexanes) to give (R)-methyl 3-(dimethyl(phenyl)silyl)hexa-3,4-dienoate (**R_a**)-**4** (6.53 g, 25.1 mmol, 78% yield) as a light yellow oil (Note 15). The product is formed in >93% ee (Note 16).

2. Notes

1. Lithium chloride was flame dried under vacuum for 10 min. All glassware is dried overnight at 95 °C prior to use. All reactions are run under 1 atmosphere of argon. All solvents and reagents are added via syringe through a rubber septum. All reagents are purchased from Sigma Aldrich and used as received unless another vendor or method of purification is specified below. The purity of 3-butyn-2-ol used was 97%. Anhydrous 99.9%, inhibitor free tetrahydrofuran was purchased from Aldrich and purified with alumina using the Sol-Tek ST-002 solvent purification system directly before use. The submitters report that reagent grade tetrahydrofuran was freshly distilled from sodium/benzophenone ketyl under an atmosphere of dry nitrogen.

2. The submitters report that the three-step reaction sequence can be performed on a 150-mmol scale with similar yields and stereocontrol.

3. Significant quantities of salts are formed during this reaction, so efficient stirring is crucial.

4. Phenyldimethylchlorosilane is purchased from Gelest and used as received.

5. The column was packed with 275 g of silica gel, then 500 mL of hexanes followed by 500 mL of 5% ethyl acetate/hexanes are used to elute unreacted silane (R_f = 0.95, 20% ethyl acetate/hexanes), followed by 4 L 10% ethyl acetate/hexanes . Fractions were collected using 50 mL test tubes, the product has an R_f = 0.40 in 20% ethyl acetate/hexanes, and stains strongly with potassium permanganate. Alcohol (**2**) exhibits the following characteristics: ^{1}H NMR (300 MHz, CDCl$_3$) δ: 0.42 (s, 6 H), 1.49 (d, J = 6.6 Hz, 3 H), 1.96 (m, 1 H), 4.56 (m, 1 H), 7.37 − 7.39 (m, 3 H), 7.60 − 7.63 (m, 2H); ^{13}C NMR (150 MHz, CDCl$_3$) δ: −1.0, 24.1, 58.6, 86.2, 109.4, 127.9, 129.4, 133.6, 136.6; IR (film) v_{max} 3330, 3076, 2978, 2181, 1429 cm^{-1}; HRMS (CI, NH$_3$) m/z calcd for C$_{12}$H$_{17}$OSi [M+H]$^+$ 205.1049, found: 205.1040; Anal. calcd. for C$_{12}$H$_{16}$OSi: C, 70.53, H, 7.89. Found: C, 70.30, H, 7.79.

6. When the reaction was performed at half-scale, 6.9 g (90%) of product was isolated.

7. Lipase AK is purchased from Amano Enzyme Inc. The lipase is washed with diethyl ether (50 mL), filtered and air dried for 30 min immediately before use.

8. Small aliquots of the solution are removed and filtered through a cotton plug to remove traces of enzyme. The filtrate is concentrated under vacuum, and a ^{1}H NMR of the residue is taken. The reaction is judged complete when the integral ratios of the resonances for the protons alpha to the alcohol (δ 4.55) and alpha to the acetate (δ 5.52) are equal, typically requiring 15–20 h.

9. The column was packed with 400 g of silica gel. Elution with 2.5 L of 5% ethyl acetate/hexanes affords acetate **3** (R_f = 0.60, 20% ethyl acetate/hexanes). The first 1.0 L of eluent is collected in 250 mL fractions, followed by 50 mL fractions. Next, elution with 4.2 L of 10% ethyl acetate/hexanes is used to collect alcohol ((S)-**2**), 50 mL fractions are collected throughout, (R_f = 0.40 in 20% ethyl acetate/hexanes and stains strongly with potassium permanganate). Enantioenriched alcohol ((S)-**2**) exhibits the following characteristics: ^{1}H NMR (600 MHz, CDCl$_3$) δ: 0.43 (s, 6 H), 1.47 (d, J = 6.6 Hz, 3 H), 1.93 (m, 1 H), 4.56 (m, 1 H), 7.26   7.40 (m, 3H), 7.62 – 7.63 (m, 2 H); ^{13}C NMR (150 MHz, CDCl$_3$) δ: -1.0, 24.1, 58.6, 86.3,   109.4, 127.9, 129.4, 133.6, 136.6; IR (film) υ_{max} 3342, 3064, 2982, 2177, 1429 cm^{-1}; HRMS (CI, NH$_3$) m/z calcd for C$_{12}$H$_{17}$OSi [M+H]$^+$ 205.1049, found: 205.1051; Anal. calcd. for C$_{12}$H$_{16}$OSi: C, 70.53, H, 7.89. Found: C, 70.25, H, 7.97; $[\alpha]_n^{20}$ –11.2 (c 5.5, CH$_2$Cl$_2$).

10. When the reaction was performed at half-scale, 3.31 g (49%) of product was isolated. The checkers established the enantiomeric purity of ((S)-**2**) by HPLC analysis (Note 17) run on a Chiralpak IB-3 column eluting in 0.5% 2-propanol/hexanes, with a 2.0 µL injection and a 1.0 mL/min flow rate. The alcohol has >98% ee. The peaks are visualized at 210 nm, with the racemic alcohol **2** exhibiting equal peaks with retention times of 11.6 and 12.6 min, and the enantioenriched alcohol ((S)-**2**) exhibiting a major peak with a retention time of 12.8 min (minor enantiomer has a retention time of 11.9 min). The submitters established the enantiomeric purity of the alcohol using a dilute solution (~0.2 mg/mL, it can be difficult to see baseline separation otherwise). HPLC is run on a ChiralCel OD column eluting in 1% 2-propanol /hexanes, with a 10 µL injection and a 1.0 mL/min flow rate. The alcohol has >95% ee. The peaks are visualized at 254 nm, with the racemic alcohol **2** exhibiting equal peaks with retention times of 18.1 and 19.5 min, and the enantioenriched alcohol ((S)-**2**) exhibiting a major peak with a

retention time of 19.2 min (minor enantiomer has a retention time of 18.4 min). Enantioenriched acetate ((**R**)-**3**) exhibits the following characteristics: ^{1}H NMR (300 MHz, CDCl$_3$) δ: 0.42 (s, 6 H), 1.51 (d, J = 6.9 Hz, 3 H), 2.08 (s, 3 H), 5.51 (m, 1 H), 7.37 − 7.39 (m, 3 H), 7.60 − 7.63 (m, 2 H); ^{13}C NMR (150 MHz, CDCl$_3$) δ: −1.2, 20.7, 21.2, 60.3, 87.1, 105.3, 127.7, 129.3, 133.4, 136.2, 169.2; IR (film) v_{max} 3068, 2991, 2954, 2178, 1744, 1434, 1364 cm^{-1}; HRMS (CI, NH$_3$) m/z calcd for C$_{14}$H$_{19}$O$_2$Si [M+H]$^+$ 245.0998, found: 245.1018; Anal. calcd. for C$_{14}$H$_{18}$O$_2$Si: C, 68.25, H, 7.36. Found: C, 68.53, H, 7.21; [α]$_D^{20}$+ 100 (7.4, CH$_2$Cl$_2$). Determination of the enantiomeric excess of the acetate required the synthesis of the racemic acetate (±)-**3** (Scheme 1).

Scheme 1. Acetylation of racemic alcohol

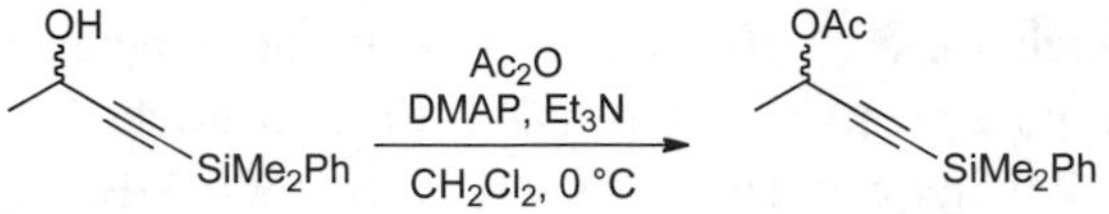

Alcohol (±)-**2** (523 mg, 2.56 mmol, 1 equiv) is dissolved in CH$_2$Cl$_2$ (10 mL) and 4-(dimethylamino)pyridine (32 mg, 0.256 mmol, 0.1 equiv), triethylamine (3.57 mL, 2.59 g, 25.6 mmol, 10 equiv) and acetic anhydride (1.2 mL, 1.31 g, 12.7 mmol, 5 equiv) are added in that order. The resulting solution is stirred at room temperature for 12 h, diluted with ammonium chloride (10 mL), filtered through a Celite plug and rinsed with CH$_2$Cl$_2$ (50 mL). The phases are separated and the aqueous layer is extracted with CH$_2$Cl$_2$ (2 x 30 mL). The combined organic layers are washed with brine (20 mL), dried over Na$_2$SO$_4$ (5 g), filtered, concentrated under reduced pressure and purified by flash chromatography (5% ethyl acetate/hexanes) to afford racemic acetate (±)-**3** (0.493 g, 78% yield) as a colorless oil. The checkers established enantiomeric purity by HPLC analysis (Note 17) run on a Chiralpak IB-3 column eluting in 0.2% 2-propanol/hexanes, with a 2.0 µL injection and a 0.3 mL/min flow rate. The acetate has >99% ee. The peaks are visualized at 210 nm, with the racemic acetate (±)-**3** exhibiting equal peaks with retention times of 10.0 and 11.6 min, and the enantioenriched alcohol (**R**)-**3** exhibiting a major peak with a retention time of 10.9 min; the minor enantiomer is not observed.

424

11. Both trimethylorthoacetate and xylenes are purchased from Aldrich and used without further purification.

12. It is critical that the bath temperature be kept at this temperature or the reaction will not go to completion.

13. Most of the xylenes is removed using a rotary evaporator with the water bath heated to 60 °C and by using dry ice/acetone in the cooling trap, the remainder is removed during the chromatography purification step.

14. The column was packed with 150 g of silica gel. Hexanes (500 mL) was used to elute the xylenes (R_f = 1.0, 20% ethyl acetate/hexanes) then elution with 1.5 L of 2% ethyl acetate/hexanes afforded the product, which has an R_f = 0.85 in 20% ethyl acetate/hexanes, and stains strongly with potassium permanganate.

15. Enantiomerically enriched allenylsilane $((R_a)$-4) exhibits the following characteristics: ^{1}H NMR (400 MHz, CDCl$_3$) δ: 0.38 (s, 6 H), 1.64 (d, J = 7.2 Hz, 3 H), 2.93 (d, J = 1.6 Hz, 2 H), 3.55 (s, 3 H), 4.94 (m, 1 H), 7.34 − 7.36 (m, 3 H), 7.53 (m, 2 H); ^{13}C NMR (100 MHz, CDCl$_3$) δ: –3.0, –2.9, 13.2, 36.0, 51.4, 81.2, 88.4, 127.6, 129.0, 133.7, 137.4, 172.0, 209.4; IR (film) υ_{max} 2954, 1945, 1740, 1434 cm^{-1}; HRMS (EI$^+$) m/z calcd for C$_{15}$H$_{20}$O$_2$Si [M]$^+$ 260.1233, found: 260.1252; $[\alpha]_D^{20}$–5.8 (c 5.7, CH$_2$Cl$_2$). When the reaction was performed at half-scale, 3.61 g (83%) of product was isolated. Purity (>98%) was established by HPLC analysis: ChiraCel OD column, eluent = i-PrOH/hexane = 0.5/99.5, flow = 1 mL/min, detection = 254 nm, injection volume : 20 µL, retention time = 4.39 mi (major isomer).

16. To determine the enantiomeric excess of allenylsilane (R_a)-4 the methyl ester was reduced to a primary alcohol using lithium borohydride (Scheme 2).

Scheme 2. Reduction of Methyl Ester

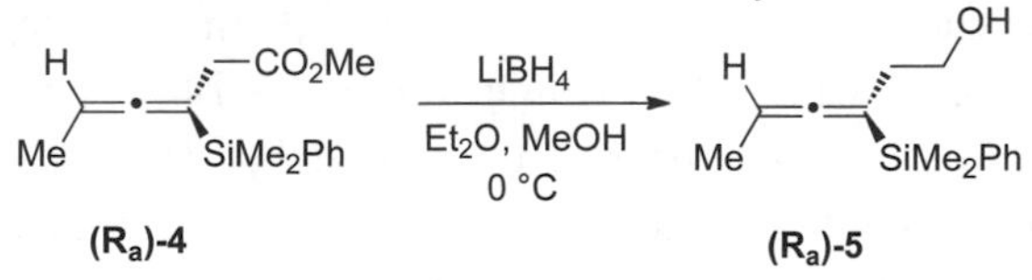

(R$_a$)-4   (R$_a$)-5

A single-necked, 25-mL round-bottomed flask equipped with a rubber septum and a 1-inch egg-shaped magnetic stir bar is charged with lithium borohydride (0.046 g, 2.12 mmol, 2 equiv), diethyl ether (3 mL) and methanol (0.2 mL). The resulting mixture is cooled to 0 °C and a solution of (R_a)-4 (0.27 g, 1.06 mmol, 1 equiv) in diethyl ether (3 mL) is added. The solution is stirred for 12 h while warming slowly to room temperature. Upon completion of the reaction, ethyl acetate (3 mL), acetone (3 mL) and water (10 mL) are added sequentially. The aqueous layer is separated and extracted with ethyl acetate (3 x 5 mL). The combined organic phases are washed with brine (10 mL), dried with sodium sulfate (200 mg), filtered and concentrated under reduced pressure. Purification over silica gel (4 g) using 100 mL of 10% ethyl acetate/hexanes results in (R)-3-(dimethyl(phenyl)silyl)hexa-3,4-dien-1-ol (**R**$_a$)-**5** (0.204 g, 0.880 mmol, 83%) as a colorless oil. The alcohol (**R**$_a$)-**5** exhibits the following characteristics: ^{1}H NMR (600 MHz, CDCl$_3$) δ: 0.37 (s, 6 H), 1.58 (m, 1 H), 1.64 (d, $J = 7.2$ Hz, 3 H), 2.16–2.18 (m, 2 H), 3.67 (m, 2 H), 4.91 (m, 1 H), 7.35 − 7.36 (m, 3 H), 7.52 (m, 2 H); ^{13}C NMR (100 MHz, CDCl$_3$) δ: –3.2, –3.1, 13.8, 32.6, 62.2, 81.1, 91.7, 127.8, 129.2, 133.7, 137.9, 172.0, 207.3; IR (film) υ$_{max}$ 3363, 3072, 2953, 2889, 1932, 1426, 1369 cm^{-1}; HRMS (EI+) m/z calcd for C$_{14}$H$_{20}$OSi [M]$^+$ 232.1252, found: 232.1255; Anal. calcd. for C$_{14}$H$_{20}$OSi: C, 72.36, H, 8.67. Found: C, 71.77, H, 8.80; [α]$_D^{20}$–18.6 (c 4.9, CH$_2$Cl$_2$); R$_f$= 0.25 in 5:1 hexanes/ethyl acetate.

The enantiomeric excess determination required the synthesis of the racemic alcohol (±)-**5**. The procedures used for the synthesis of the racemic compound were entirely analogous to the ones used to synthesize the enantiomerically enriched alcohol (**R**$_a$)-**5** (Scheme 3).

Scheme 3. Synthesis of the racemic alcohol

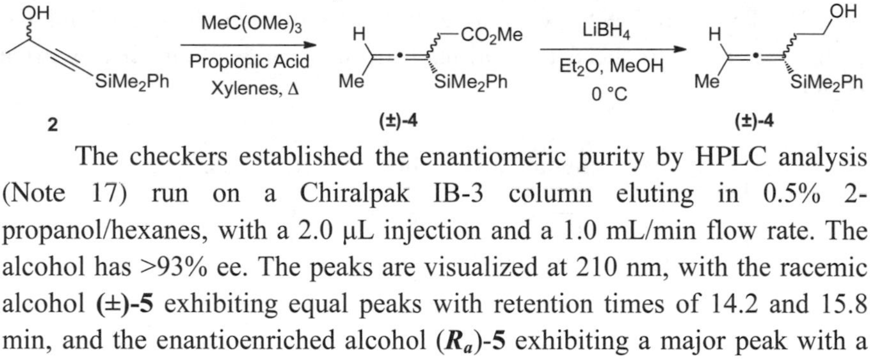

The checkers established the enantiomeric purity by HPLC analysis (Note 17) run on a Chiralpak IB-3 column eluting in 0.5% 2-propanol/hexanes, with a 2.0 μL injection and a 1.0 mL/min flow rate. The alcohol has >93% ee. The peaks are visualized at 210 nm, with the racemic alcohol (±)-**5** exhibiting equal peaks with retention times of 14.2 and 15.8 min, and the enantioenriched alcohol (**R**$_a$)-**5** exhibiting a major peak with a retention time of 12.8 min (minor enantiomer has a retention time of 11.9 min).

The submitters established enantiomeric purity of the alcohol using a ChiralCel OD column eluting in 1% 2-propanol/hexanes with a 10 μL injection and a 1.0 mL/min flow rate. The peaks are visualized at 254 nm, with the racemic alcohol (±)-5 exhibiting equal peaks with retention times of 14.3 and 15.8 min, and the enantioenriched alcohol (R_a)-5 exhibiting a major peak with a retention time of 15.7 min (minor enantiomer has a retention time of 14.6 min).

17. Enantiomeric excess determinations reported by the checkers were performed by Chiral Technologies, Inc. The checkers and Organic Syntheses gratefully acknowledge their assistance.

Safety and Waste Disposal Information

All hazardous materials should be handled and disposed of in accordance with "Prudent Practices in the Laboratory"; National Academies Press; Washington, DC, 2011.

3. Discussion

Allenylsilanes have proven to be versatile carbon nucleophiles for a variety of organic transformations.[2] In particular, chiral allenes have generated interest in the synthetic community due to the efficient transfer of the axial chirality in the allene into point chirality in the reaction products. Although allenylsilanes have proven to be effective nucleophiles for a variety of transformations, relatively few procedures exist for the multigram synthesis of highly enantioenriched reagents. Previous methods for the synthesis of chiral allenylsilane reagents typically use the S_N2' displacement strategy developed by Fleming.[3] A variety of other methods for the synthesis of enantioenriched allenylsilanes have begun to emerge in the literature, showing the increase in interest in these reagents.[4]

This submission details an efficient multi-gram synthesis of highly enantioenriched allenylsilanes. The allenes are formed in high yields with high levels of enantioselectivity by taking advantage of a C-selective silylation, followed by a lipase-catalyzed kinetic resolution, and a Johnson orthoester Claisen rearrangement. The (R) enantiomer of alcohol 2 can also be accessed by re-exposure of protected acetate (R)-3 to the lipase in aqueous buffer, followed by the orthoester Claisen rearrangement, resulting in the (S_a) enantiomer of allenylsilane 4 (Scheme 4).[5]

Scheme 4. Synthesis of allenylsilane (S_a)-4.

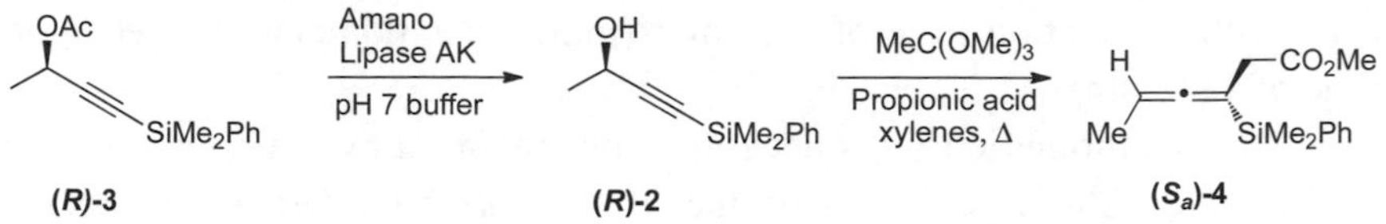

Allenylsilanes **3** have been used as carbon nuclophiles in additions to oxonium ions to form homopropargylic ethers,[5] which can be used as a template for the formation of structurally and stereochemically diverse heterocycles containing a 1,2,3-triazole functionality.[6] In addition the allenylsilanes undergo additions to iminium ions, forming dihydropyrroles, dihydrooxazines and acyclic homopropargylic sulfonamides, with the reaction product determined by the nitrogen source used in iminium ion formation.[7] Stereoselective *C*-glycosidations allow for the formation of dihydropyran products with a side chain containing an internal alkyne.[8] Current work is focused on the development of a new methodology taking advantage of the axial chirality these reagents, as well as the application of this methodology to complex molecule synthesis (Scheme 5).

Scheme 5. Reactions products formed from enantioenriched allenylsilanes.

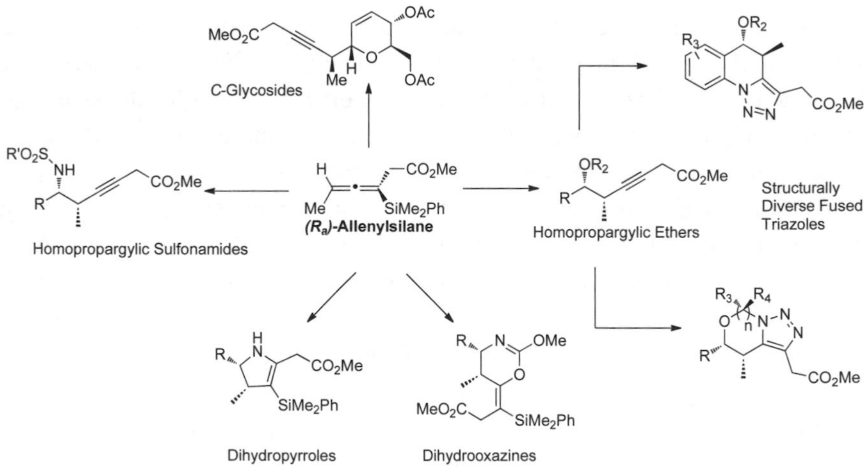

1. Department of Chemistry and Center for Chemical Methodology and Library Development, Metcalf Center for Science and Engineering, 590 Commonwealth Avenue, Boston University, Boston, Massachusetts

Org, Synth. **2012**, *89*, 420-431

02215. Email: panek@bu.edu. Funding for this work was provided by the NIH and through an AstraZeneca graduate fellowship to RAB.

2. (a) *Modern Allene Chemistry*. Krause, N.; Hashmi, A.S.K., Eds. Wiley-VCH: Weinheim, **2004**. (b) Danheiser, R. L.; Carini, D. J. *J. Org. Chem.* **1980**, *45*, 3925–3927. (c) Danheiser, R. L.; Kwasigroch, C. A.; Tsai, Y.-M. *J. Am. Chem. Soc.* **1985**, *107,* 7233–7235. (d) Danheiser, R. L.; Carini, D. J.; Kwasigroch, C. A. *J. Org. Chem.* **1986**, *51*, 3870–3878. (e) Danheiser, R. L.; Stoner, E. J. ; Koyama, H.; Yamashita, D. S.; Klade, C. A. *J. Am. Chem. Soc.* **1989**, *111,* 4407–4413. (f) Curtis-Long, M. J.; Aye, Y. *Chem. Eur. J.* **2009**, *15*, 5402–5416. (g) Lopez, F.; Mascarenas, J. L. *Chem. Eur. J.* **2011**, *17*, 418–428.

3. (a) Westmuze, H.; Vermeer, P. *Synthesis* **1979**, 390–392. (b) Fleming, I.; Terrett, N. K. *J. Organomet. Chem.* **1984**, *264*, 99–118. (c) Fleming, I.; Waterson, D. *J. Chem. Soc., Perkin Trans. 1* **1984**, 1809–1813. (d) Buckle, M. J. C.; Fleming, I. *Tetrahedron Lett.* **1993**, *34*, 2383–2386. (e) Marshall, J. A.; Maxon, K. *J. Org. Chem.* **2000**, *65*, 630–633.

4. (a) Suginome, M.; Matsumoto, A.; Ito, Y. *J. Org. Chem.* **1996,** *61,* 4884–4885. (b) Han, J. W.; Tokunaga, N.; Hayashi, T. *J. Am. Chem. Soc.* **2001**, *123,* 12915–12916. (c) Ogasawara, M.; Ito, A.; Yoshida, K.; Hayashi, T. *Organometallics* **2006**, *25,* 2715–2718. (d) Nishimura,T.; Makino, H.; Nagaosa, M.; Hayashi, T. *J. Am. Chem. Soc.* **2010**, *132*, 12865–12867. (e) Canales, E.; Gonzalez, A.; Soderquist, J. A. *Angew. Chem., Int. Ed.* **2007**, *46,* 397–399. (f) Gonzalez, A. Z.; Soderquist, J. A. *Org. Lett.* **2007**, *9*, 1081–1084. (g) Ohmiya, H.; Ito, H.; Sawamura, M. *Org. Lett.* **2009**, *11*, 5618–5621. (h) Ogasawara, M.; Okada, A.; Subbarayan, V.; Sörgel, S.; Takahashi, T. *Org. Lett.* **2010**, *12*, 5736–5739.

5. Brawn, R. A.; Panek, J. S. *Org. Lett.* **2007**, *9*, 2689–2692.

6. Brawn, R. A.; Welzel, M.; Lowe, J. T.; Panek, J. S. *Org. Lett.* **2010**, *12*, 336–339.

7. (a) Brawn, R. A.; Panek, J. S. *Org. Lett.* **2009**, *11*, 473–476. (b) Brawn, R. A.; Panek, J. S. *Org. Lett.* **2009**, *11*, 4362–4365.

8. Brawn, R. A.; Panek, J. S. *Org. Lett.* **2010**, *12*, 4624–4627.

Appendix
Chemical Abstracts Nomenclature; (Registry Number)

Lithium chloride; (7447-41-8)
3-butyn-2-ol; (2028-63-9)
n-Butyllithium; (109-72-8)
Chloro(dimethyl)phenylsilane; (768-33-2)
Vinyl Acetate; (108-05-4)
Trimethylorthoacetate; (1445-45-0)
Propionic Acid; (79-09-4)
Lithium Borohydride; (16949-15-8)
(±)-4-(Dimethyl(phenyl)silyl)-but-3-yn-2-ol; (115884-67-8)
(*S*)-4-(Dimethyl(phenyl)silyl)-but-3-yn-2-ol; (142697-17-4)
(*S*)-4-(Dimethyl(phenyl)silyl)-but-3-yn-2-yl acetate; (142611-82-3)
(R_a)-Methyl 3-(dimethyl(phenyl)silyl)hexa-3,4-dienoate; (945539-62-8)

James Panek was born in Buffalo NY in 1956. He earned a B.S. in Medicinal Chemistry, 1979, from the University at Buffalo (SUNY@Buffalo) and his Ph.D. in Medicinal Chemistry from the University of Kansas, 1984. He then pursued an NIH postdoctoral program at Yale University before joining the faculty at Boston University. Presently he serves as The Samour Family Professor of Chemistry. His research interests emphasize the field of synthetic organic chemistry, mainly in the area of acyclic stereocontrol with specific interest in the development of new reagents and new reaction methods. Ongoing studies in the area of natural product synthesis generate a complimentary research effort. Panek is also affiliated with the Center for Chemical Methodology and Library Development (CMLD) at Boston University.

Org, Synth. **2012**, *89*, 420-431

Ryan Brawn was born in Camden, ME in 1980. He earned his BA in Biochemistry from Bowdoin College in 2003. He is currently finishing his Ph.D. in Organic Chemistry at Boston University in the Panek lab. His thesis work has focused on the development of enantioenriched allenylsilane reagents, and their use as carbon nucleophiles for a variety of stereocontrolled reactions. He is currently a postdoctoral fellowship at Pfizer, Groton CT and will begin work in the CVM group in Groton, CT in 2011.

Francois Grillet was born in 1982 in Saint Remy, France. He studied chemistry at the Ecole Nationale Supérieure de Chimie de Mulhouse where he received his engineer diploma in 2005. He then moved to Grenoble to carry out his Ph.D. under the supervision of Professor Andrew E. Greene working in the area of natural products synthesis. He is currently working as a post-doctoral fellow to apply the allenic Pauson-Khand reaction to the [5-7-5] ring system of 6,12-guaianolides in the group of Professor Kay Brummond at the University of Pittsburgh.

Discussion Addendum for:
Boric Acid Catalyzed Amide Formation from Carboxylic Acids and Amines: *n*-Benzyl-4-phenylbutyramide

Prepared by Pingwah Tang.*[1]
Original article: Tang, P. W. *Org. Synth.* **2005**, *81*, 262.

In the last few years, boric acid catalyzed carboxylic amide formation directly from carboxylic acids and amines has attracted considerable attention because it has emerged as a viable alternative route to the indirect methods of preparing carboxamides. The standard methods employ carboxylic acid derivatives such as acid halides, anhydrides or activating agents[2] such as DCC/HOBt, EDC, BOP-Cl.[3] and others. These methods suffer from several disadvantages including environmental unfriendliness, poor selectivity, poor atom economy, poor step economy, and unsuitability for large-scale preparation.[4]

Scope of Boric Acid Catalyzed Amidation

Recent developments employing boric acid as a green, inexpensive and readily available catalyst have significantly expanded not only the scope of the direct formation of carboxamides from carboxylic acids and amines in general,[5,6] but also its application to the synthesis of a wide spectrum of compounds of industrial interest. Carboxamides are key components of a number of valuable intermediates for the pharmaceutical and other industries. By far, the most notable application of boric acid catalyzed amidation has been the synthesis of medicinally useful carboxamides, including carriers for drug delivery,[7] active pharmaceutical ingredients, and drug candidates.[8,9]

Boric acid catalyzed amidation was featured in the preparation of sterically hindered *N,N*-disubstituted amides, which are used as drug delivery agents, in 45–55% yield (Figure 1).[10] Mono-substituted α,β-unsaturated carboxamides such as *N*-furylmethylcinnamamides, which offer

432

Published on the Web 4/16/2012
© Organic Syntheses, Inc.

potential use as drug delivery agents and as biological models with antifungal and antiparasitic properties, were prepared by boric acid catalyzed amidation using a 2:1 cinnamic acid/amine molar ratio (Figure 1).[11]

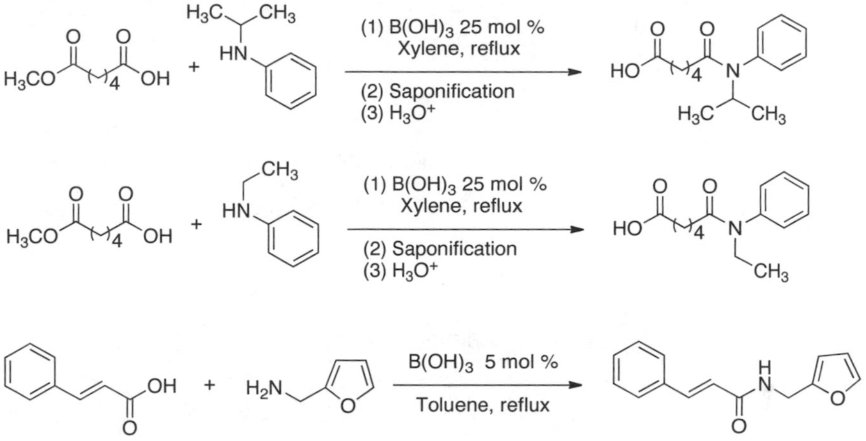

Figure 1. Boric acid catalyzed amidation

Bandichhor et al. employed this method to synthesize a number of active pharmaceutical ingredients (API).[8,9] When both primary and secondary amine functionalities are present in the same molecule, boric acid catalyzed amidation takes place chemoselectively at the primary amine functionality (Figure 2).[8]

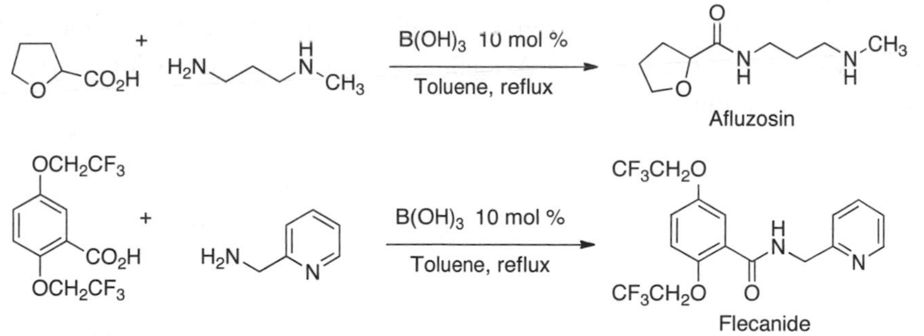

Figure 2. Boric acid catalyzed synthesis of various API intermediate.

The integrity of stereogenic centers either in the carboxylic acid and/or in the amine are preserved in the boric acid catalyzed amidation. No epimerization is observed under the studied experimental conditions (Figure 3).[4,8,12]

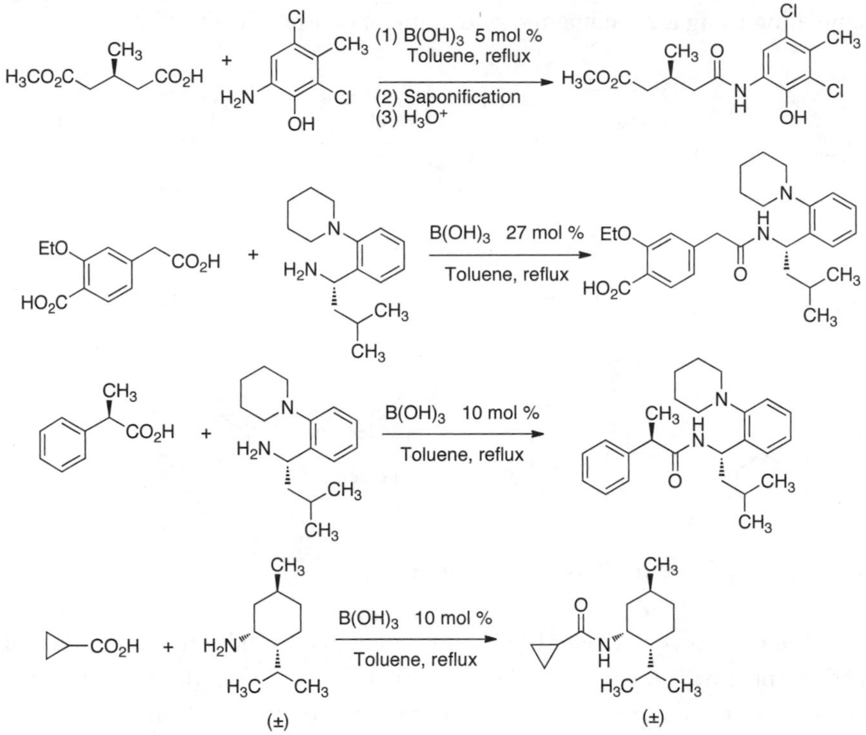

Figure 3. No racemization is observed in boric acid catalyzed amidation.

Using the same procedure in a one step reaction, the dihydroxamic acid ligands can be prepared from N-methylhydroxylamine and 1,3-phenylene diacetic acid (Figure 4).[13]

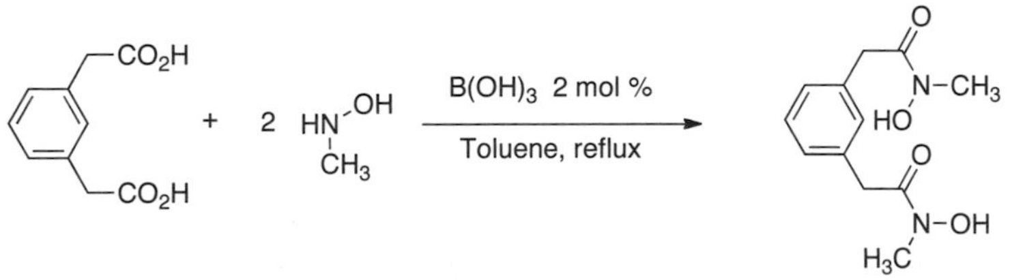

Figure 4. Boric acid catalyzed dicarboxylic amidation reactions.

Synergistic Catalytic Effect of Boric Acid and Other Compounds Containing at Least One Hydroxyl Functional Group

The ability of boric acid to form an ester or a complex with hydroxyl functionality has provided higher catalytic activity in the direct amidation reaction involving carboxylic acids and amines. This discovery allows a wider range of molecule partners to be incorporated into a cooperative catalytic system for enhancing the direct amidation. A study of the kinetics of the boric acid catalyzed amidation of 4-nitrobenzoic acid with ammonia showed that the amidation activity was enhanced by adding the co-catalyst polyethylene glycol (PEG). The latter forms a hypothetical PEG-boric ester complex with boric acid. The molecular weights of PEG have little effect on the yield of the final 4-nitrobenzamide product (Figure 5).[14] With PEG-400 as a co-catalyst, the optimal molar ratio between boric acid and PEG-400 in the amidation of 4- nitrobenzoic acid and ammonia was 1:(1.5-2.5).[14]

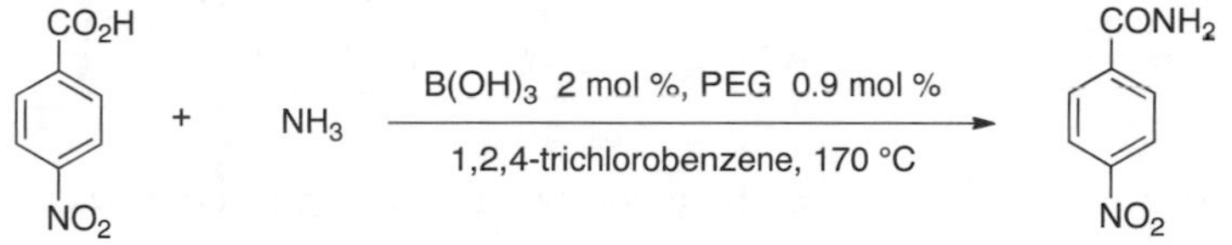

Figure 5. Synergistic catalytic effect of boric acid and polyethylene glycol (PEG).

For the preparation of amides from sterically demanding carboxylic acids and amines, such as tris(alkylsubstituted cyclohexylamide) that is used as a material for molding,[15] the catalytic activity of boric acid was enhanced when boric acid was converted *in-situ* to a boric acid ester. This conversion was achieved by means of a molecule bearing mono-hydroxyl or di-hydroxyl functional groups, such as cresol or tetrachlorocatechol prior to the addition of carboxylic acids and amines to the reaction vessel (Figure 6).[16,17]

We believe that in the near future many more applications to the synthesis of important carboxamides using this boric acid catalyzed amidation will be reported.

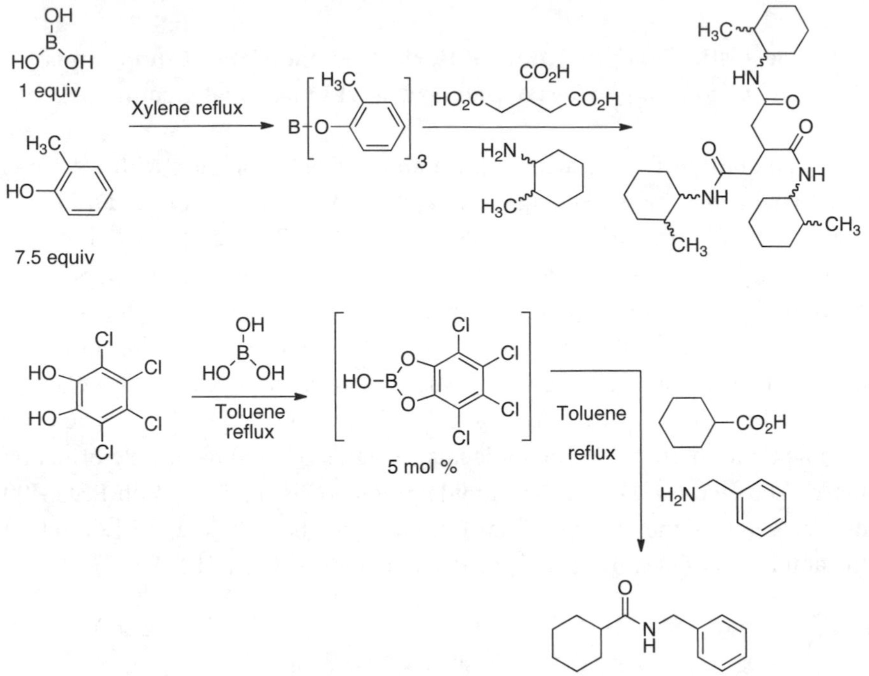

Figure 6. Amidation catalyzed by a boric acid ester prepared *in-situ* from boric acid and compounds containing hydroxyl functional groups.

1. College of Life Science and Technology, Beijing University of Chemical Technology, Box 53, Science and Technology Building, Office 304, Beisanhuan East Road, Beijing 100029, China. Email: pwtang@yahoo.com
2. Han, S. Y.; Kim, Y. A. *Tetrahedron* **2004**, *60*, 2447.
3. (a) Joullié, M. M.; Lassen, K. M. *Arkivoc* **2010**, *VIII,* 189. (b) Kim, Y.A.; Han, S.Y. Bull. Korean Chem. Soc. **2000,** *21,* 943.
4. Tang, P. W. *Org. Synth.* **2005**, *81*, 262.
5. Arnold, K.; Davies, B.; Giles, R. L.; Grosjean, C.; Smith, G. E.; Witting, A. *Adv. Synth. Catal.* **2006**, *348*, 813.
6. Charville, H.; Jackson, D.; Hodges, G.; Whiting, A. *Chem. Commun.* **2010**, *46*, 1813.
7. Tang. P. W. Preparation of Novel Delivery Agents for Delivery of Macromolecular Drugs Using Boric Acid Mediated Amidation of Carboxylic Acids and Amines. In *Advances in Controlled Drug Delivery*; Dinh, S. M.; Liu, P., Eds.; ACS Symposium, American

Chemical Society: Washington, DC, 2003; Vol. 846, Chapter 8, pp 103–112.

8. Mylavarapu, R. K.; Kondaiah, G. C. M.; Kolla, N.; Veeramalla, R.; Koikonda. P.; Bhattacharya, A.; Bandichhor, R. *Org. Proc. Res. Dev.* **2007**, *11*, 1065.

9. Bhattacharya, A.; Bandichhor, R. Green Technologies in the Generic Pharmaceutical Industry. In *Green Chemistry in the Pharmaceutical Industry,* Dunn, P.; Wells, A.; Williams, M.T., Eds.; Wiley-VCH, Weinheim, **2011**, Chapter 14, pp 289-309.

10. Gomez-Orellana, M. I.; Gschneidner, D.; Leone-Bay, A.; Moye-Sherman, D.; Pusztay, S. V.; Rath, P.; Tang, P. W.; Weidner, J. J. U.S. Patent Appl. Pub. 20080255250A1, Oct. 16, 2008.

11. Barajas, J. G. H.; Mendez, L. Y. V.; Kouznetsov, V. V.; Stashenko, E. E. *Synthesis* **2008**, *3*, 377.

12. Looft, J. l.; Vossing, T.; Ley, J.; Backes, M.; Blings, M. U.S. Patent. 7,919,133, Apr. 5, **2011**.

13. Sayed, M. A.; Smith, S. R. University of Michigan-Dearborn, Department of Natural Sciences, 15th Annual Poster Session, Chemistry Section, April 20, 2007.

14. (a) Shteinberg, L. Y. *Russ. J. Appl. Chem.* **2006**, *79*, 1282. (b) Shteinberg, L. Y. *Russ. J. Appl. Chem.* **2009**, *82*, 613. (c) Shteinberg, L. Y. *Russ. J. Appl. Chem.* **2011**, *84*, 815.

15. Kitagawa, S.; Ishikawa, M.; Ishibashi, Y.; Iwamura, T.; Kihara, Y. U.S. Patent 7,745,661, Jun 29, 2010.

16. Starkov, P.; Sheppard, T. D, *Org. Biomol. Chem.* **2011**, *9*, 1320

17. Maki, T.; Ishihara, K.; Yamamoto. H. *Org. Lett.* **2006**, *8*, 1431.

Pingwah Tang obtained his B. Sc., M. Sc., Dr. of Specialty in Organic Synthesis, and Dr. Sc. degrees from the University of Paris. After post-doctoral research at University of Montreal and Johns Hopkins University, he joined Eastman Kodak Company in Rochester, NY as a senior staff scientist (1978-1998) conducting R&D in photographic chemicals, dyes, polymers, API and novel organic synthetic methodology. He then held positions with Emisphere Technologies focusing on the research of carrier-mediated drug delivery and green chemistry. He is currently a visiting professor/research professor at Guangzhou Jinan University and Beijing University of Chemical Technology.

Preparation of 3-Alkylated Oxindoles from *N*-Benzyl Aniline via a Cu(II)-Mediated Anilide Cyclization Process

A.

B.

C.

D.

Submitted by David S. Pugh[1] and Richard J. K. Taylor.[1]
Checked by Joshua S. Alford and Huw M. L. Davies.

1. Procedure

A. *tert-Butyl 3-(benzyl(phenyl)amino)-3-oxopropanoate* (**3**). A 250-mL, 2-necked, round-bottomed flask (B24 open neck; B14 with thermometer adaptor and −10-60 °C thermometer fitted) with a stir bar (oval, 1.25 in x 0.625 in) is charged with *N*-benzylaniline (10.96 g, 60.0 mmol, 1.0 equiv) (Note 1), dichloromethane (120 mL) (Note 2) and *tert*-butyl malonate (10.56 g, 66.0 mmol, 1.1 equiv) (Note 3) to give a pale yellow solution. The flask is placed in an ice-water bath to give an internal temperature of 4 °C. 2-

Org. Synth. **2012**, *89*, 438-449
Published on the Web 4/17/2012
© 2012 Organic Syntheses, Inc.

Chloro-1-methylpyridinium iodide (Mukaiyama's reagent, 16.82 g, 66.0 mmol, 1.1 equiv) (Note 4) is added in a single portion to give a yellow suspension and an exotherm to 10 °C. The mixture is stirred for 5 min until the temperature has returned to 3 °C, before addition of triethylamine (33.3 mL, 24.29 g, 240.0 mmol, 4.0 equiv) (Note 5) *via* syringe maintaining the temperature below 10 °C (ca. 10 min) to give a yellow solution. On complete addition, the ice bath is removed and the reaction allowed to warm to room temperature. The reaction progress is monitored by TLC analysis on silica gel (Note 6). After 15 min, the yellow solution is quenched by the addition of an aqueous 3 M HCl solution (60 mL) to give a biphasic solution which is stirred for 5 min before transferring to a 500-mL separating funnel, and washing the flask with an additional 50 mL dichloromethane. The combined organic layers are separated and washed with 60 mL saturated aqueous solution of sodium bicarbonate and 120 mL saturated NaCl solution, dried over Na_2SO_4 (20 g), filtered (sinter, 60 mL, porosity 3) into a 500-mL round-bottomed flask and concentrated by rotary evaporation (38 °C water-bath, 15 to 10 mmHg) to afford 18.19 g of a yellow oil. Flash chromatography (SiO_2, 3:1 to 1:1 petroleum ether/Et_2O) (Note 7) gave 17.39 g (89%) of the product as a colorless oil (Note 8), which is utilized in the next step.

B. *tert-Butyl 3-(benzyl(phenyl)amino)-2-methyl-3-oxopropanoate* (**4**). A 250-mL, two-necked, round-bottomed flask (B24 septum and argon balloon; B14 with thermometer adaptor and −10-60 °C thermometer fitted) with a stir bar (oval, 1.25 in x 0.625 in) is charged with *tert*-butyl 3-(benzyl(phenyl)amino)-3-oxopropanoate (16.28 g, 50.0 mmol, 1.0 equiv) and tetrahydrofuran (100 mL) (Note 9) to give a pale yellow solution. The flask is placed in an ice-water bath to give an internal temperature of 3 °C. Potassium *tert*-butoxide (6.18 g, 55.0 mmol, 1.1 equiv) (Note 10) is added to give an orange suspension and an exotherm to 10 °C. The reaction is stirred until the temperature had returned to 3 °C (ca. 5 min), before addition of methyl iodide (3.4 mL, 7.61 g, 52.5 mmol, 1.05 equiv) (Note 11) *via* syringe over 30 min, maintaining the temperature below 10 °C to give a cream suspension. On complete addition, the ice bath is removed and the mixture stirred for 15 min. The reaction progress is monitored by TLC analysis on silica gel (Note 12). After this time the cream suspension is quenched by the addition of 50 mL saturated aqueous ammonium chloride solution and transferred to a 250-mL separating funnel. The washings are transferred with 20 mL of water and 100 mL Et_2O (Note 13). The organic layer is separated and washed with 100 mL saturated NaCl solution, dried over MgSO$_4$ (20 g), filtered (sinter, 60 mL, porosity 3) into a 250-mL round-bottomed flask and concentrated by rotary evaporation (32 °C water-bath, 15 to 10 mmHg) to afford 17.96 g of an orange oil. Flash chromatography (SiO$_2$, 7:3 petroleum ether/Et$_2$O) (Note 14) gave 15.61 g (92%) of the product as a colorless sticky oil (Note 15), which is utilized in the next step.

C. *tert-Butyl 1-benzyl-3-methyl-2-oxoindoline-3-carboxylate* (**5**). A 2-L 2-necked, round-bottomed flask (B24 septum and argon balloon; B19 with thermometer adaptor and 0-300 °C thermometer fitted) with a stir bar (oval, 1.25 in x 0.625 in) is charged with *tert*-butyl 3-(benzyl(phenyl)amino)-2-methyl-3-oxopropanoate (16.55 g, 48.8 mmol, 1.0 equiv) and *N,N*-dimethylformamide (800 mL) (Note 16) to give a colorless solution. Potassium *tert*-butoxide (6.02 g, 53.7 mmol, 1.1 equiv) (Note 10) is added to give a yellow solution. The mixture is stirred for 5 min before addition of copper(II) acetate monohydrate (9.78 g, 48.8 mmol, 1.00 equiv) (Note 17) in a single portion to give a green-blue suspension, which is placed in a 110 °C oil-bath. The reaction progress is monitored by TLC analysis on silica gel (Note 18). After 4 h, the brown suspension is cooled to room temperature and concentrated by rotary evaporation (78 °C water-bath, 15 to 10 mmHg) to approximately 1/10th volume. The brown suspension is poured into a 1-L separating funnel, and washed with 400 mL of water, 100 mL of 3 M HCl and 200 mL of Et$_2$O to give a brown organic and green aqueous layers. The layers are separated and the aqueous layer further extracted with of Et$_2$O (2 × 200 mL). The combined organic layers are washed with 200 mL of saturated sodium bicarbonate solution and 2 × 200 mL of a saturated NaCl solution, dried over MgSO$_4$ (20 g), filtered (sinter, 60 mL, porosity 3) into a 1-L round-bottomed flask and concentrated by rotary evaporation (32 °C water-bath, 15 to 10 mmHg) to afford 15.30 g of a brown oil. Flash chromatography (SiO$_2$, 4:1 petroleum ether/ Et$_2$O) (Note 19), followed by collection of the product on a sinter (60 mL, porosity 3), using portions of hexane (3 × 50 mL) (Note 20) gave 8.32 g (51%) of the product as a colorless solid (Note 21), which is utilized in the next step.

D. *1-Benzyl-3-methylindolin-2-one* (**6**). A 100-mL 1-necked round-bottomed flask (B14 with condenser fitted) with a stir bar (oval, 1.25 in x 0.625 in) is charged with *tert*-butyl 1-benzyl-3-methyl-2-oxoindoline-3-carboxylate (7.42 g, 22.0 mmol, 1.0 equiv) and trifluoroacetic acid (11 mL) (Note 22) to give a yellow solution which is placed in an oil-bath held at 75 °C. The reaction progress is monitored by TLC analysis on silica gel (Note

23). After 1 h, the brown solution is cooled to room temperature, transferred to a 250-mL round-bottomed flask using 3 × 20 mL of dichloromethane and concentrated by rotary evaporation (38 °C water-bath, 15 to 10 mmHg) to yield an oil. Ethanol (50 mL) (Note 24) is added to dissolve the oil and then removed by rotary evaporation (40 °C water-bath, 12 mmHg) to give a colorless solid (4.88 g). Purification by recrystallization (ethanol) (Note 25) gave 4.14 g (80%) of the product as colorless shards (Note 26).

2. Notes

1. *N*-Benzyl aniline (99%) was purchased from Aldrich Chemical Company, Inc.

2. Dichloromethane (HPLC grade) was purchased from Fisher Scientific Company and used as received.

3. *tert*-Butyl malonate (97%) was purchased from Alfa Aesar. The supplied material contained 5–10% *tert*-butyl ethyl malonate as an impurity, but was used as received.

4. 2-Chloro-1-methylpyridinium iodide (Mukaiyama's reagent; 97%) was purchased from Aldrich Chemical Company, Inc.

5. Triethylamine was purchased from Aldrich Chemical Company, Inc.

6. TLC analysis was carried out using silica gel plates (Note 28) with petroleum ether-Et$_2$O (1:1) as eluent and visualisation with UV (254 nm) and *p*-anisaldehyde. *N*-Benzyl aniline **1** has R$_f$ = 0.77 (white) and the product **3** has R$_f$ = 0.42 (purple).

7. The crude material was purified by column chromatography. A 60-mm diameter column was wet-packed with a silica gel slurry made from petroleum ether/Et$_2$O (3:1) (180 g) (Note 30) to give a 160-mm column depth. Sand (1 cm) was layered onto the silica and the crude oil suspended in the eluent (25 mL) and poured onto the sand. Washings are also loaded with further portions of the eluent (2 x 25 mL) and 400 mL of eluent was collected. Fractions (16 x 150 mm test tubes) are collected from 600 mL of petroleum ether/Et$_2$O (3:1) and 1.5 L of petroleum ether/Et$_2$O (1:1) to give 55 fractions. The product was identified by thin layer chromatography (petroleum ether/Et$_2$O (1:1); R$_f$ = 0.42) and fractions 15-53 are collected and concentrated by rotary evaporation (35 °C water-bath, 15 to 10 mmHg). The material was then dried overnight under high vacuum (0.08 mmHg) with an oval stir-bar to yield the product.

8. The checkers report an 88% yield when the reaction was performed at half-scale. Compound **3** exhibits the following physical and spectroscopic properties: ^{1}H NMR (400 MHz, CDCl$_3$) δ: 1.42 (s, 9 H), 3.14 (s, 2 H), 4.91 (s, 2 H), 7.01–7.03 (m, 2 H), 7.23–7.26 (m, 5 H), 7.29–7.31 (m, 3 H); ^{13}C NMR (100 MHz, CDCl$_3$) δ: 28.1, 43.1, 53.1, 81.6, 127.6, 128.5, 128.9, 129.8, 137.3, 142.1, 166.4, 167.0; IR (neat): 3063, 2978, 1731, 1660, 1495, 1392, 1367, 1326, 1142, 697 cm^{-1}; HRMS (ESI) *m/z* 326.1749 [326.1751 calcd for C$_{20}$H$_{24}$NO$_3$ (M+H)]. Anal. Calcd for: C, 73.82, H, 7.12, N, 4.30. Found: C, 73.59, H, 7.13, N, 4.41.

9. Tetrahydrofuran (HPLC grade) was purchased from Fisher and purified using a Innovative Technology Inc. Pure Solv™ solvent purification system.

10. Potassium *tert*-butoxide was purchased from Aldrich Chemical Company, Inc.

11. Methyl iodide was purchased from Aldrich Chemical Company, Inc.

12. TLC analysis was carried out using silica gel plates (Note 28) with petroleum ether-Et$_2$O (1:1) as eluent and visualisation with UV (254 nm) and *p*-anisaldehyde. *tert*-Butyl 3-(benzyl(phenyl)amino)-3-oxopropanoate **3** has R$_f$ = 0.38 (purple) and the product **4** has R$_f$ = 0.48 (blue).

13. Diethyl ether (Lab reagent grade) was supplied by Fisher Scientific.

14. A 60-mm diameter column was wet-packed with silica gel (220 g) (Note 30) to give a 140 mm column depth. Sand (1 cm) was layered onto the silica and the crude oil suspended in the eluent (25 mL) and poured onto the sand. Washings are also loaded in a further portion of eluent (25 mL). Fractions (45 mL) are collected from 4 L of petroleum ether/Et$_2$O (7:3). The 49 fractions were analyzed by TLC (petroleum ether/Et$_2$O :7:3) and the product (R$_f$ = 0.27) identified in fractions 11-35, which were collected and concentrated by rotary evaporation (32 °C water-bath, 15 to 10 mmHg). The residual material was dried overnight under high vacuum (0.1 mbar) with an oval stir bar to yield the product **4**.

15. The checkers report a 91% yield when the reaction was performed at half-scale. Compound **4** exhibits the following physical and spectroscopic properties: ^{1}H NMR (400 MHz, CDCl$_3$) δ: 1.28 (d, *J* = 7 Hz, 3 H), 1.41 (s, 9 H), 3.27 (q, *J* = 7.2, 1 H), 4.57 (d, *J* = 14.4 Hz, 1 H), 5.22 (d, *J* = 14.4 Hz, 1 H), 7.01–7.04 (m, 2 H), 7.22–7.27 (m, 5 H), 7.30–7.34 (m, 3 H); ^{13}C NMR (100 MHz, CDCl$_3$) δ: 14.2, 28.1, 44.6, 53.2, 81.3, 127.6, 128.5, 128.7,

442

128.9, 129.8, 137.6, 142.2, 169.9, 170.5; IR (neat): 3063, 2978, 2935, 1738, 1656, 1595, 1495, 1454, 1393, 1367, 1325, 1245, 1147, 848, 735, 697 cm^{-1}; HRMS (ESI) m/z 340.1906 [340.1907 calcd for $C_{21}H_{26}NO_3$ (M+H)]. Anal. Calcd for: C, 74.31; H, 7.42, N, 4.13. Found: C, 73.49; H, 7.29, N, 4.15.

16. *N,N*-Dimethylformamide (anhydrous, 99.8%) was purchased from Sigma-Aldrich and used as received. Submitters purchased *N,N*-dimethylformamide from Fisher Scientific and purified using a Innovative Technology Inc. Pure Solv™ solvent purification system.

17. Copper(II) acetate monohydrate was purchased from Aldrich Chemical Company, Inc.

18. TLC analysis was carried out using silica gel plates (Note 28) with petroleum ether-Et$_2$O (1:1) as eluent and visualisation with UV (254 nm) and *p*-anisaldehyde. *tert*-Butyl 3-(benzyl(phenyl)amino)-2-methyl-3-oxopropanoate **4** has R_f = 0.50 (blue) and the product **5** has R_f = 0.68 (purple).

19. A 60-mm diameter column was wet-packed with silica gel (120 g) (Note 30) to give a 110-mm column depth. Sand (1 cm) was layered onto the silica and the crude oil suspended in dichloromethane (10 mL) and poured onto the sand. Washings were also loaded in a further portion of dichloromethane (5 mL). Fractions (16 x 150 mm tubes) were collected from 1.5 L of petroleum ether/Et$_2$O (4:1). The 90 fractions were evaluated by thin layer chromatography (petroleum ether/Et$_2$O; 4:1) product R_f = 0.26) and fractions 37-81 were collected and concentrated (Rotary evaporation (32 °C water-bath, 15 to 10 mmHg)) to yield the product.

20. The colorless solid was suspended in hexane (50 mL) and filtered, collecting the solid on a sinter (60 mL, porosity 3) and washing with three additional portions of 100 mL portions of hexane (Note 27). The colorless solid obtained was transferred to a 50-mL round-bottomed flask and dried overnight under high vacuum (0.08 mmHg).

21. The checkers report a 50% yield when the reaction was performed at half-scale. Compound **5** exhibits the following physical and spectroscopic properties: mp 106–107 °C; ^{1}H NMR (400 MHz, CDCl$_3$) δ: 1.36 (s, 9 H), 4.66 (d, J = 16 Hz, 1 H), 5.22 (d, J = 16 Hz, 1 H), 6.68 (d, J = 7.6 Hz, 1 H), 7.02 (ddd, J = 1.2, 7.6, 7.6 Hz, 1 H), 7.17 (ddd, J = 1.2, 7.6, 7.6 Hz), 7.23–7.33 (m, 6 H); ^{13}C NMR (100 MHz, CDCl$_3$) δ: 20.1, 28.0, 43.9, 56.1, 82.6, 109.6, 122.8, 123.0, 127.3, 127.8, 128.9, 130.9, 135.9, 142.9, 168.9, 175.8; IR (film): 3061, 2979, 2932, 1732, 1713, 1609, 1488, 1466, 1368, 1252, 1152, 1115, 1108, 748, 697 cm^{-1}; HRMS (ESI) m/z 338.1749 [338.1751

calcd for $C_{21}H_{24}NO_3$ (M+H)]. Anal. Calcd for: C, 74.75, H, 6.87, N, 4.15. Found: C, 74.56, H, 6.94, N, 4.19.

22. Trifluoroacetic acid, 98% was purchased from Aldrich Chemical Company Inc.

23. TLC analysis was carried out using silica gel plates (Note 28) with petroleum ether-Et_2O (1:1) as eluent and visualisation with UV (254 nm) and *p*-anisaldehyde. *tert*-Butyl 1-benzyl-3-methyl-2-oxoindoline-3-carboxylate **5** has R_f = 0.68 (purple) and the product **6** has R_f = 0.46 (pink).

24. Ethanol (analytical reagent grade) was purchased from Fisher Scientific and used as received.

25. The colorless solid was dissolved in the minimum amount of boiling ethanol (ca. 15 mL) and allowed to cool to room temperature. The crystals were broken up with a spatula and collected *via* filtration in a sinter (60 mL, porosity 3) washing with three 50 mL portions of hexane (Note 27). The crystals were transferred to a 50-mL round-bottomed flask and dried overnight under high vacuum (0.08 mmHg).

26. The checkers report an 81% yield when the reaction was performed at half-scale. Compound **6** exhibits the following physical and spectroscopic properties: mp 119–120 °C; ^1H NMR (400 MHz, CDCl$_3$) δ: 1.54 (d, J = 8 Hz, 3 H), 3.54 (q, J = 7.6 Hz, 1 H), 4.92 (s, 2 H), 6.72 (d, J = 8 Hz, 1 H), 7.02 (ddd, J = 1.2, 7.6, 7.6 Hz), 7.16 (dt, J = 1.2, 2, 8 Hz, 1 H), 7.24–7.34 (m, 6 H); ^{13}C NMR (100 MHz, CDCl$_3$) δ: 15.9, 40.8, 43.9, 109.2, 122.7, 123.8, 127.5, 127.8, 128.0, 129.0, 130.9, 136.2, 143.3, 179.0 ppm; IR (film): 3057, 3031, 2930, 1704, 1613, 1487, 1466, 1454, 1348, 1202, 1169, 972, 749, 687 cm^{-1}; HRMS (ESI) *m/z* 238.1224 [238.1226 calcd. for $C_{16}H_{16}NO$ (M+H)]. Anal. Calcd. for: C, 80.98; H, 6.37, N, 5.90. Found: C, 80.70; H, 6.24, N, 5.91. The physical and spectroscopic data matches those previously reported.[2]

27. *n*-Hexane (HPLC grade) was purchased from Fisher Scientific and used as received.

28. TLC plates were supplied by Merck, aluminium backed silica gel 60 (F_{254}).

29. Petroleum ether (lab reagent grade) supplied by Fisher Scientific. Diethyl ether (anhydrous) was supplied by Fisher Scientific.

30. Silica Gel (60) was purchased from Sorbent Technologies. Submitters purchased silica gel (60) from Fluka.

Safety and Waste Disposal Information

All hazardous materials should be handled and disposed of in accordance with "Prudent Practices in the Laboratory"; National Academies Press; Washington, DC, 2011.

3. Discussion

3-Substituted oxindoles form the cornerstone of numerous natural products and bioactive lead compounds.[3] We recently reported an efficient new route to 3,3-disubstituted oxindoles such as esters **5** from anilides **4** using potassium *tert*-butoxide as base and stoichiometric Cu(OAc)$_2$.H$_2$O in DMF (Scheme 1).[4,5] This procedure is operationally straightforward, employs inexpensive reagents and does not require anhydrous conditions or the use of an inert atmosphere (other electron-withdrawing groups, such as nitrile and phosphonate, could also be employed).[4-6] Cyclization to generate the quaternary all-carbon-center occurs *via* a formal C-H, Ar-H coupling, but preliminary mechanistic studies are consistent with a sequence involving deprotonation, copper(II)-mediated radical generation and then homolytic aromatic substitution.[4,6] A related procedure was developed which utilizes catalytic Cu(OAc)$_2$•H$_2$O, but this has yet to be applied to multi-gram transformations.[7]

4, R = Alkyl, Arylalkyl, Hetarylalkyl,
R ' = Et, t-Bu
P = Me, Bn, PMB

1.1 equiv KOt-Bu
1 equiv Cu(OAc)$_2$.H$_2$O
DMF, 110 °C, air, 1 h

R' = Et, NaOH then H$^+$
or
R' = t-Bu, TFA, PhOMe

The stoichiometric Cu(OAc)$_2$.H$_2$O methodology was also extended to prepare 3-monosubstituted oxindoles **6** by subsequent decarboxyalkylation (Scheme 1). 3-Alkylated oxindoles are valuable as synthetic building blocks[8] and as drug candidates.[9] A base-mediated saponification/decarboxylation sequence using ethyl esters (**5**, R' = Et) was not widely applicable, but the use of neat TFA (with anisole as a cation trap) at room temperature on the corresponding *t*-butyl esters (**5**, R' = *t*-Bu) proved extremely efficient and versatile.[5] The scope of this process is shown in Table 1. As can be seen, this decarboxyalkylation sequence was used to prepare a range of 3-substituted oxindoles including those with saturated alkyl substituents, allyl, benzyl, phenethyl and naphthylmethyl substituents, as well as the benzyloxypropyl and 4-pyridylmethyl examples. The *N*-protecting group could be benzyl, methyl or *p*-methoxybenzyl (PMB). With *N*-PMB protection, the use of a higher temperature and a longer reaction time gave both decarboxyalkylation and *N*-deprotection.

Table 1 Scope of the copper(II)-mediated cyclisation and TFA-mediated decarboxyalkylation procedure

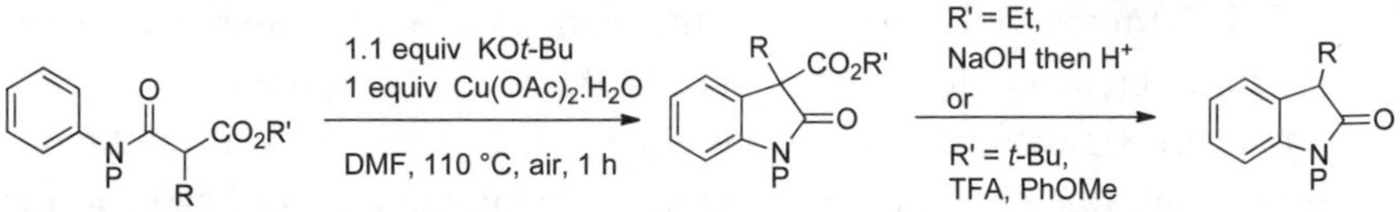

4, R = Alkyl, Arylalkyl, Hetarylalkyl,
R ' = Et, *t*-Bu
P = Me, Bn, PMB

Entry	Oxindole 5	Yield 5 (%)	Alkyoxindole 6	Yield 6 (%)
i	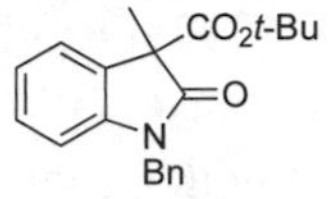	55	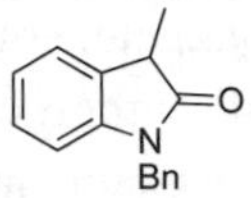	77
ii	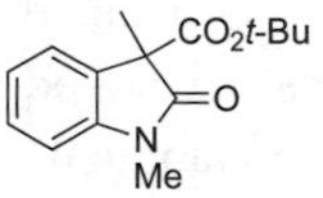	82	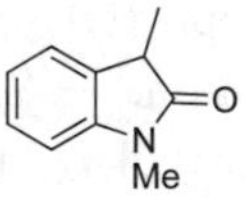	78
iii	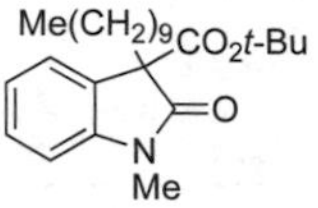	76	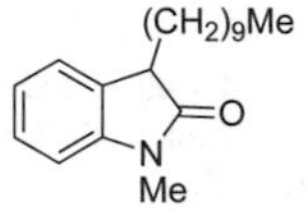	90
iv	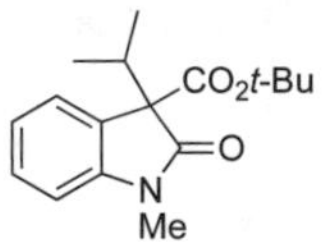	46[a]	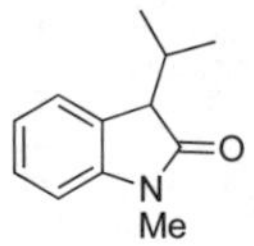	68
v	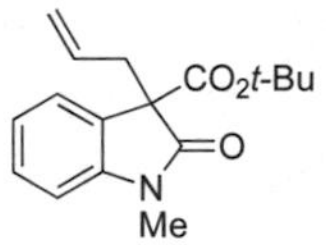	59	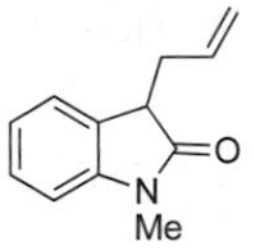	92
vi	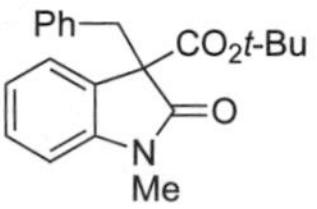	93	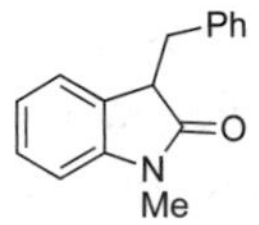	98

[a] Heat for 3 days

446

Table 1 (continued)

Entry	Oxindole 5	Yield 5 (%)	Alkyoxindole 6	Yield 6 (%)
vii		71		62
viii		73		93
ix		65		86
x		66		92
xi		69		84
xii		69		87

1. Department of Chemistry, University of York, Heslington, York, YO10 5DD, UK. E-mail: richard.taylor@york.ac.uk. We thank the EPSRC for postgraduate support (D.S.P., EP/E041302/1). We also thank Dr. Alexis Perry and Mr. Johannes Klein for helpful discussions.
2. Thomson, J. E.; Kyle, A. F.; Gallagher, K. A.; Lenden, P.; Concellón, C.; Morrill, L. C.; Miller, A. J.; Joannesse, C.; Slawin, A. M. Z.; Smith, A. D. *Synthesis* **2008**, 2805.
3. For relevant reviews see: (a) Marti, C.; Carreira, E. *Eur. J. Org. Chem.* **2003**, 2209. (b) Cerchiaro, G.; Ferreira, A. M. d. C. *J. Braz. Chem. Soc.* **2006**, *17*, 1473. (c) Galliford, C.; Scheidt, K. *Angew. Chem. Int. Ed.*

2007, *46*, 8748. (d) Trost, B. M.; Brennan, M. K. *Synthesis* **2009**, 3003. (e) Millemaggi, A.; Taylor, R. J. K. *Eur. J. Org. Chem.* **2010**, 4527.

4.   Perry, A.; Taylor, R. J. K. *Chem. Commun.* **2009**, 3249.

5.   Pugh, D. S.; Klein, J. E. M. N.; Perry, A.; Taylor, R. J. K. *Synlett* **2010**, 934.

6.   Jia and Kundig prepared 3-aryl-3-alkyl oxindoles by a related CuCl$_2$-mediated process: Jia, Y. X.; Kündig, E. *Angew. Chem. Int. Ed.* **2009**, *48*, 1636.

7.   (a) Klein, J. E. M. N.; Perry, A.; Pugh, D. S.; Taylor, R. J. K. *Org. Lett.* **2010**, *12*, 3446. (b) Moody, C. L.; Franckevičius, V.; Drouhin, P.; Klein, J. E. M. N.; Taylor, R. J. K. *Tetrahedron Lett.* **2012**, *53, 1897*.

8.   (a) Bui, T.; Borregan, M.; Barbas, C. F. *J. Org. Chem.* **2009**, *74*, 8935. (b) Qian, Z.-Q.; Zhou, F.; Du, T.-P.; Wang, B.-L.; Ding, M.; Zhao, X.-L.; Zhou, J. *Chem. Commun.* **2009**, 6753. (c) Jiang, K.; Peng, J.; Cui, H.-L.; Chen, Y.-C. *Chem. Commun.* **2009**, 3955. (d) Kato, Y.; Furutachi, M.; Chen, Z.; Mitsunuma, H.; Matsunaga, S.; Shibasaki, M. *J. Am. Chem. Soc.* **2009**, *131*, 9168. (e) Sano, D.; Nagata, K.; Itoh, T., *Org. Lett.* **2008**, *10*, 1593.

9.   (a) Shimazawa, R.; Kuriyama, M.; Shirai, R. *Bioorg. Med. Chem. Lett.* **2008**, *18*, 3350. (b) Rizzi, E.; Cassinelli, G.; Dallavalle, S.; Lanzi, C.; Cincinelli, R.; Nannei, R.; Cuccuru, G.; Zunino, F. *Bioorg. Med. Chem. Lett.* **2007**, *17*, 3962.

Appendix
Chemical Abstracts Nomenclature; (Registry Number)

tert-Butyl 3-(benzyl(phenyl)amino)-3-oxopropanoate;
N-Benzylaniline; (103-32-2)
2-Chloro-1-methylpyridinium iodide; (14338-32-0)
tert-Butyl malonate; (40052-13-9)
tert-Butyl 3-(benzyl(phenyl)amino)-2-methyl-3-oxopropanoate;
Methyl iodide; (74-88-4)
Potassium tert-butoxide; (865-47-4)
Copper(II) acetate monohydrate; (6046-93-1)
Trifluoroacetic acid; (76-05-1)
tert-Butyl 1-benzyl-3-methyl-2-oxoindoline-3-carboxylate;
1-Benzyl-3-methylindolin-2-one; (34943-91-4)

Richard Taylor obtained his B.Sc. and Ph.D. (Dr. D. Neville Jones) from the University of Sheffield. Postdoctoral periods with Dr. Ian Harrison and Professor Franz Sondheimer were followed by lectureships at the Open University and then UEA, Norwich. In 1993 he moved to the Chair of Organic Chemistry at the University of York. Taylor's research interests centre on the synthesis of bioactive natural products and the development of new synthetic methodology. His awards include the Royal Society of Chemistry's Pedler Lectureship (2007). Taylor is the current President of the International Society of Heterocyclic Chemistry, a past-President of the RSC Organic Division and an Editor of Tetrahedron.

David Pugh was born in Harrogate, North Yorkshire in 1985. He carried out his undergraduate studies and obtained an M. Chem. (Hons) at the University of York in 2007, carrying out a final year project under the supervision of Prof. Richard Taylor. He remained in the group to undertake a Ph.D. and is investigating new methodology for the synthesis of cyclopropanes and oxindoles.

Joshua S. Alford received both his B.S. and M.S. in chemistry from Missouri State University under the supervision of Dr. Chad Stearman. He later joined the Davies research group at Emory University to pursue a PhD degree. His current projects include the synthesis of pharmaceutical agents utilizing rhodium carbenoid chemistry and the development of a new type of donor/acceptor carbenoid.

Gold(I)-Catalyzed Decarboxylative Allylic Amination of Allylic N-Tosylcarbamates

Submitted by Dong Xing and Dan Yang.[1]
Checked by Kin Yang and Viresh H. Rawal.

1. Procedure

A. *(E)-Pent-2-en-1-yl tosylcarbamate.* An oven-dried, 500-mL, 24/40-necked round-bottomed flask is equipped with a Teflon-coated oval magnetic stirring bar (35 × 15 mm) (Note 1). The flask is connected to a solvent purification system, evacuated, charged with 200 mL of dichloromethane (DCM) (Note 2), refilled with nitrogen, and sealed with a rubber septum. A needle inlet with a positive pressure of nitrogen is inserted into the septum. *(E)-2-Penten-1-ol* (8.55 mL, 7.24 g, 84.0 mmol, 1.0 equiv) is added in one portion via syringe with stirring (Note 3). The resulting clear solution is then cooled to approximately 0 °C by an ice-water bath for 20 min. *p*-Toluenesulfonyl isocyanate (15.3 mL, 100.0 mmol, 1.2 equiv) is added dropwise over 15 min through the septum via syringe (Note 4). Upon complete addition, the ice-bath is removed and the reaction is allowed to warm to room temperature (23 °C). After 30 min, TLC analysis shows complete consumption of the starting material (Note 5). The reaction mixture is transferred to a 1-L separating funnel and washed with 400 mL of saturated aqueous NH_4Cl (Note 6). After the layers are separated, the aqueous phase is extracted with DCM (2 x 200 mL). The combined organic phase is stored in a 1-L Erlenmeyer flask and dried with 30 g of Na_2SO_4 for 30 min (Note 7). The solution is then filtered through a cotton plug into a 1-L recovery flask and concentrated on a rotary evaporator (20 mmHg, 25 °C)

450

Org. Synth. **2012**, *89*, 450-459
Published on the Web 4/23/2012
© 2012 Organic Syntheses, Inc.

to give the crude product as a colorless oil. The crude oil is dissolved in ~15 mL of DCM and loaded onto a chromatography column (5 x 17 cm, 120 g silica) (Note 8). After loading, the column is flushed with 400 mL of hexanes. Fraction collecting is started as 1 L of Hex:EtOAc:DCM (100:20:5) is added (flow rate: 20 mL/min, 25 mL fractions) (Note 9 and 10). Fractions containing the pure product (fractions 38–61) are combined and concentrated to yield 22.85–23.24 g (96–97%) of *(E)-pent-2-en-1-yl tosylcarbamate* as a viscous colorless oil (Note 11).

B. *N-(1-Ethyl-2-propen-1-yl)-4-methyl-benzenesulfonamide.* A 500-mL, 29/32 necked round-bottomed flask equipped with a Teflon-coated oval magnetic stirring bar (50 × 20 mm) is charged with a solution of *(E)-pent-2-en-1-yl tosylcarbamate* (3.66 g, 12.9 mmol, 1.0 equiv) in 66 mL of 1,2-dichloroethane (DCE) (Notes 12–15). Diisopropylethylamine (Hünig's base) (2.25 mL, 1.60 g, 12.9 mmol, 1.0 equiv) is then added and the clear solution is allowed to stir for 5 min. Water (66 mL) is added with vigorous stirring, resulting in a biphasic, milky solution (Notes 16 and 17). After stirring for 5 min, gold (I) chloride (AuCl) (150 mg, 0.645 mmol, 0.05 equiv) is added in portions over one minute (Note 18). Upon the addition of AuCl, the milky solution acquires a yellow tint (Note 19). Silver (I) triflate (AgOTf) (166 mg, 0.645 mmol, 0.05 equiv) is then added in one portion resulting in a much whiter solution (Note 20). A reflux condenser (30 x 3 cm) with a steady cold water flow is affixed, and the flask is then placed into a pre-heated oil bath (85 °C, bath temp.). The apparatus is then covered up with aluminum foil to minimize the exposure to light. After 1 h, the reaction progressively becomes clearer until the white color dissipates to leave a transparent/grayish solution with some black precipitate. Monitoring by ^{1}H NMR shows the reaction to be nearly complete at 3 h (Note 21). After cooling to room temperature, the reaction mixture is diluted with 75 mL of DCM, and 75 mL of brine. The organic layer is separated and the aqueous layer is extracted with DCM (2 x 75 mL). The combined organic phase is dried with 10 g of Na_2SO_4 over 30 min, then concentrated on a rotary evaporator (20 mmHg, 25 °C) to yield the crude product as a grayish/transparent oil. The oil is dissolved in ~5 mL DCM, loaded onto a chromatography column (17 x 5 cm, 80 g silica), and flushed with 200 mL of hexanes. Fractions were collected as the column is eluted with 1 L of 100:15:5 mixture of Hex:EtOAc:DCM (flow rate: 20 mL/min, 25 mL/fraction). The product is found in fractions 12–30 (Note 22), which are combined and concentrated to afford 2.57 g (83% yield) of *N-(1-ethyl-2-*

propen-1-yl)-4-methyl-benzenesulfonamide as a colorless oil that solidified upon storing neat in the refrigerator (Note 23).

2. Notes

1.   The flask was dried overnight in a 120 °C oven prior to use and cooled under vacuum.

2.   Dichloromethane (Optima Grade) was purchased from Fisher Scientific and dried by an Innovative Technologies solvent purification system.

3.   *(E)*-2-Penten-1-ol (95%) was purchased from Aldrich Chemical Company and used as received. The NMR spectrum of the Aldrich sample showed the presence of ~10% 1-pentanol, which reacts in Step A to give the corresponding saturated carbamate. The quantities used and the yields obtained are not corrected for the presence of this impurity.

4.   *p*-Toluenesulfonyl isocyanate (96%) was purchased from Aldrich Chemical Company and used as received.

5.   *(E)*-2-Penten-1-ol is not observable under UV (254 nm), but the product of step A is observable. Both of these compounds are separated using 1:3 EtOAc:Hexanes on Whatman brand silica plates (250 μm), and visualized using $KMnO_4$: R_f 0.34 (starting material); R_f 0.28 (product).

6.   Ammonium chloride (NH_4Cl) (A.R. grade) was purchased from Fisher Scientific.

7.   Sodium sulfate (Na_2SO_4), anhydrous (A.R. grade), was purchased from Fisher Scientific.

8.   Silica gel P60 (230–400 mesh) was purchased from Silicycle.

9.   Hexanes (HPLC grade) was purchased from Fischer Scientific.

10.   Ethyl acetate (G.R. grade) was purchased from Fischer Scientific.

11.   The product contains ca. 10% of the saturated carbamate, which arises from the saturated alcohol impurity in the starting material (see Note 3). The product exhibited the following characterization data: ^{1}H NMR (500 MHz, $CDCl_3$) δ: 0.96 (t, J = 7.5 Hz, 3 H), 2.04 (p, J = 6.5 Hz, 2 H), 2.44 (s, 3 H), 4.51 (dd, J = 7.0, 1.5 Hz, 2 H), 5.45 (dtt, J =15.5, 6.5, 1.5 Hz, 1 H), 5.77 (dtt, J = 15.5, 6.0, 1.0 Hz, 1 H), 7.36 (d, J = 8.5 Hz, 2 H), 7.80 (bs, 1 H), 7.93 (d, J = 8.0 Hz, 2 H); ^{13}C NMR (100 MHz, $CDCl_3$) δ: 13.1, 21.8, 25.3, 67.8, 121.7, 128.6, 129.7, 135.8, 139.5, 145.2, 150.7; IR (CH_2Cl_2): 3369, 1753, 1599 cm^{-1}; HRMS (ES+) *m/z* calcd for $C_{13}H_{17}NO_4S$

452

(M + Na): 306.0771, Found: 306.0768. Anal. Calcd. for $C_{13}H_{17}NO_4S$: C, 55.11; H, 6.05; N, 4.94. Found: C, 55.17; H, 6.08 N, 5.00.

12. It is important to use a large stir bar to promote efficient stirring of the heterogeneous reaction mixture.

13. The viscous carbamate was dissolved in DCM, transferred to a tared reaction flask, then concentrated down to get the mass before dissolving in DCE for use in the reaction. As the starting material is contaminated with 10% of the saturated carbamate, only 90% (3.294 g) is able to react in this step.

14. 1,2-Dichloroethane (anhydrous, 99.8%) was purchased from Aldrich Chemical Company.

15. The submitters used 1:9 DCE:H$_2$O solvent system, which achieved conversions of about 80%. The checkers found the reaction to go to near complete conversion in 1:1 DCE:H$_2$O.

16. Water (HPLC grade) was purchased from Aldrich Chemical Company.

17. The submitters added Hünig's base after the addition of water. N,N-Diisopropylethylamine (ReagentPlus, 99%) was purchased from Sigma-Aldrich Chemical Company and used as received.

18. Gold (I) chloride (99.9%+) was purchased from Aldrich Chemical Company and used as received.

19. A small amount of black solid is observed in the milky solution. At this point, the lights are also turned off as a precaution.

20. Silver (I) trifluoromethanesulfonate (99%) was purchased from Strem Chemical Company and used as received.

21. After 3 h, a small aliquot of the reaction mixture is taken (~0.2 mL) by syringe, extracted with DCM (1 mL), dried (Na$_2$SO$_4$), and concentrated to afford the NMR sample. The spectra reveal the reaction to have gone to ~95% conversion. Conversion is determined by the ratio of one of the internal alkene protons (5.77 or 5.45 ppm) of the starting material to the internal vinyl proton (5.54 ppm) of the product. Prolonged heating (> 10 h) does not increase conversion to the product and causes the formation of an unidentified side-product.

22. The product has an R_f = 0.23 in 100:15:5 Hex:EtOAc:DCM and is stained by KMnO$_4$.

23. The yield listed is uncorrected for the presence of 10% of the saturated alcohol starting material. When corrected for starting material purity (90%), the isolated yield for step B is 92.5%, close to the actual

conversion. For the second run on a 3.2 g (11.3 mmol) scale, the yield was 81% (2.19 g, uncorrected), or 90% if corrected for starting material purity. The submitters reported an isolated yield of 76%. The product exhibited the following characterization data: mp 32–34 °C; ^{1}H NMR (500 MHz, CDCl$_3$) δ: 0.82 (t, J = 7.5 Hz, 3 H), 1.50 (p, J = 7.0 Hz, 2 H), 2.42 (s, 3 H), 3.65–3.71 (m, 1 H), 4.77 (d, J = 8.0 Hz, 1 H), 4.95–5.01 (m, 2 H), 5.54 (ddd, J = 17.0, 10.5, 6.5 Hz, 1 H), 7.28 (d, J = 8.0 Hz, 2 H), 7.75 (d, J = 8.0 Hz, 2 H); ^{13}C NMR (100 MHz, CDCl$_3$) δ: 9.9, 21.7, 28.6, 57.9, 116.1, 127.3, 129.6, 137.7, 138.3, 143.3; IR (CH$_2$Cl$_2$): 3280, 1602, 1435 cm^{-1}; HRMS (ES+) m/z calcd for C$_{12}$H$_{17}$NO$_2$SNa (M$^+$ + Na): 262.0872 Found: 262.0870. Anal. Calcd. for C$_{12}$H$_{17}$NO$_2$S: C, 60.22; H, 7.16 ; N, 5.85. Found: C, 60.17; H, 7.11; N, 5.92.

Safety and Waste Disposal Information

All hazardous materials should be handled and disposed of in accordance with "Prudent Practices in the Laboratory"; National Academies Press; Washington, DC, 2011.

3. Discussion

Allylic amines are important and versatile synthetic intermediates in organic chemistry. Preparation of allylic amines from cheap and readily available allylic alcohols is the most powerful and convenient protocol in organic synthesis.[2] Allylic carbamates, which are easily prepared from allylic alcohols and isocyanates, have been widely applied to the low valent transition metal catalyzed allylic substitution reactions for the preparation of allylic amines.[3] In these reactions, ancillary ligands are always required in order to control the regio- and stereoselectivities. With the acidic NH group, allylic carbamates could be easily converted into its imidate form in the presence of a suitable base. Therefore, it is also a good candidate to undergo an aza-Claisen rearrangement.[4,5]

The procedures described above offer an efficient and mild gold(I)-catalyzed method for the preparation of N-tosyl allylic amines.[6] We chose allylic N-tosylcarbamates as substrates since they are highly acidic and could be easily deprotonated into imidates with simple bases. As a result of the base-induced aza-Claisen rearrangement pathway, excellent regioselectivites and high to excellent stereoselectivities can be achieved with no additional need for ancillary ligands. With this method, a series of C1 or C3-substituted and C3-disubstituted N-tosyl allylic amines can be prepared in short reaction times with moderate to excellent yields (Table 1). In addition, this reaction can be conducted by using water as the solvent, representing an environmentally benign and easily-handled protocol. However for large-scale preparation, the collection of AuCl with the large amount of insoluble and sticky carbamate substrate in water deactivated the catalytic ability of AuCl and therefore caused low conversion. To overcome this drawback, we applied a two-phase solvent system using water and 1,2-dichloroethane (DCE) in a 9:1 volume ratio for large scale synthesis. Under such reaction condition, the product could be afforded in about 80% conversion. On the other hand, the checkers found that the use of a 1:1 DCE:H$_2$O solvent system could provide the product in near complete conversion. Since this transformation undergoes a concerted aza-Claisen rearrangement pathway, enantiopure N-tosyl allylic amines can be prepared from chiral allylic alcohols through chirality transfer (Table 1, entry 12).

This transformation could be performed in one pot starting directly from allylic alcohols. This one-pot protocol was initially developed to overcome the unwanted [3,3]-oxo-rearrangement during the efforts to prepare C3-alkenyl substituted allylic N-tosylcarbamate (Scheme 1). By adding N,N-diisopropylethylamine (DIPEA) to the 1,2-dichloroethane (DCE) solution of allylic alcohols prior to the addition of tosyl isocyanates, the [3,3]-oxo-rearrangement can be inhibited since the C3-alkenyl substituted allylic N-tosylcarbamate has been converted to its imidate form in situ. AuCl and AgOTf were then directly added to the above mixture. After a short reaction time at 75 °C, a variety of synthetically important N-tosyl dienylamines, as well as C3-disubstituted N-tosyl allylic amines, could be afforded with moderate yields and good to excellent stereoselectivities (Table 2).[7]

Scheme 1. Unexpected [3,3]-Oxo-Rearrangement

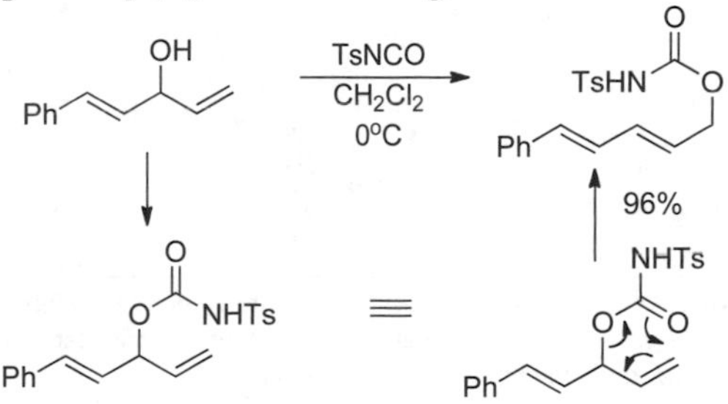

Table 1. Gold(I)-Catalyzed Decarboxylative Amination of *N*-Tosylcarbamates in Water [a]

entry	substrate	product	time (h)	yield (%)[b, c]
1			1	95
2			2	90 E:Z, 97/3
3			2	90 E:Z, 95/5
4			2	92 E:Z, 96/4
5			2	93
6			12	61 E:Z, 88/12
7			3	92
8			3	84
9[d]			24	78
10[d]			24	69
11[d]			24	42
12	ee = 94%	ee = 88%	2	94 E:Z, 90/10

[a] Unless otherwise indicated, all the reactions were carried out on a 0.3 mmol scale with 5 mol% AuCl/AgOTf and 1 equiv of DIPEA in H$_2$O for the indicated time.
[b] Isolated yield. [c] E/Z ratio determined by ^{1}H NMR. [d] At 100°C.

456

Table 2. Products Obtained via One-Pot Synthesis [a, b, c]

58%, *E:Z*, 85/15   62%, *E:Z* > 99/1   68%, *E:Z*, 96/4

58%, *E:Z*, 85/15   45%, *E:Z* > 99/1   64%, *E:Z*, 75/25

[a] Reaction conditions: 4 (0.5 mmol) and DIPEA (1.5 equiv) in DCE (5 mL) were added TsNCO (1.5 equiv), then 5 mol% AuCl/AgOTf was added and the temperature was raised to 75°C for 3 hours. [b] Isolated yield. [c] *E:Z* ratio determined by ^{1}H NMR.

1. Department of Chemistry, The University of Hong Kong, Pokfulam Road, Hong Kong, China. Email: yangdan@hku.hk. This work was supported by The University of Hong Kong and Hong Kong Research Grants Council (HKU 706109P and HKU 705807P).

2. (a) Johannsen, M.; Jørgensen, K. A. *Chem. Rev.* **1998**, *98*, 1689. (b) Cheikh, R. B.; Chaabouni, R.; Laurent, A.; Mison, P.; Nafti, A. *Synthesis* **1983**, 685. (c) Wei, Z.-Y.; Knaus, E. E. *Synthesis* **1994**, 1463.

3. (a) Mellegaard-Waetzig, S. R.; Rayabarapu, D. K.; Tunge, J. A. *Synlett* **2005**, 2759. (b) Singh, O. V.; Han, H. *J. Am. Chem. Soc.* **2007**, *129*, 774. (c) Singh, O. V.; Han, H. *Org. Lett.* **2007**, *9*, 4801.

4. For reviews and leading references on the aza-Claisen rearrangements, see: (a) Overman, L. E. *Acc. Chem. Res.* **1980**, *13*, 218. (b) Overman, L. E. *Angew. Chem., Int. Ed.* **1984**, *23*, 579. (c) Metz, P.; Mues, C.; Schoop, A. *Tetrahedron* **1992**, *48*, 1071. (d) Overman, L. E.; Carpenter, N. E. In *Organic Reactions*; Overman, L. E., Ed.; Wiley: Hoboken, NJ, 2005; Vol. 66, pp 653-760.

5. (a) Clizbe, L. A.; Overman, L. E. *Org. Syn., Coll.* Vol. 6, 507. (b) Anderson, C. E.; Overman, L. E. Watson, M. P. *Org. Syn., Coll.* Vol. 82, 134.

6. Xing, D.; Yang, D. *Org. Lett.* **2010**, *12*, 1068.

7. For the potential application of *N*-tosyl dienylamines, see: (a) Wyle, M. J.; Fowler, F. W. *J. Org. Chem.* **1984**, *49*, 4025. (b) Gaeta, F. C. A.; Lehman de Gaeta, L. S.; Kogan, T. P.; Or, Y.; Foster, C.; Czarniecki, M. *J. Med. Chem.* **1990**, *33*, 964. (c) Gulledge, B. M.; Aggen, J. B.; Eng, H.; Sweimeh, K.; Chamberlin, A. R. *Bioorg. Med. Chem. Lett.* **2003**, *13*, 2907.

Appendix
Chemical Abstracts Nomenclature; (Registry Number)

trans-2-Penten-1-ol: 2-penten-1-ol, (2E)-; (1576-96-1)
p-Toluenesulfonyl isocyanate; (4083-64-1)
N,N-Diisopropyl ethylamine; (7087-68-5)
Gold(I) chloride; (10294-29-8)
Silver trifluoromethanesulfonate: Methanesulfonic acid, trifluoro-, silver(1$^+$) salt; (2923-28-6)

Dan Yang was born in Sichuan, China. She received her BSc in Chemistry from Fudan University, P. R. China, in 1985. Through the US–China Chemistry Graduate Program, she obtained an MA in 1988 at Columbia with Professor Ronald Breslow and a PhD from Princeton under the guidance of Professor Daniel Kahne in 1991. She then spent two years as a postdoctoral fellow with Professor Stuart Schreiber at Harvard. In 1993, she joined The University of Hong Kong, where she is currently the Morningside Professor in chemical biology. Her research interests include asymmetric catalysis and total synthesis, the design and synthesis of novel foldamers and the chemical biology of natural products.

Dong Xing was born in Henan, China, in 1982. He received his BSc in Chemistry and his MSc in Organic Chemistry with Prof. Liping Yang and Prof. Zhangjie Shi (Peking University) at East China Normal University in 2003 and 2006 respectively. He obtained his PhD under the supervision of Professor Dan Yang at the University of Hong Kong in 2011. His research interests include tandem cyclization reactions and their application to total synthesis of natural products.

Kin Yang was born in 1988 in Guangdong, China. Two years later, he immigrated to the United States and settled in New York City. In 2010, he went on to obtain a B.S. degree in chemistry from Syracuse University where he worked with Prof. Don Dittmer on applications of tellurium in organic chemistry. In the same year, he enrolled in a Ph.D. program at the University of Chicago. He is currently working in the area of hydrogen-bonding catalysis under the supervision of Prof. Viresh Rawal.

Preparation of *t*-Butyl-3-Bromo-5-Formylbenzoate Through Selective Metal-Halogen Exchange Reactions

A.

B.

Submitted by Juan D. Arredondo, Hongmei Li and Jaume Balsells.[1]
Checked by Julius R. Reyes and Viresh H. Rawal.

1. Procedure

A. *t-Butyl-3,5-dibromobenzoate* (**1**). An oven-dried 500-mL 3-necked round-bottomed flask, an oven-dried Teflon-coated oval-shaped magnetic stir bar (3.8 cm x 1.6 cm), and an oven-dried 125-mL pressure-equalizing addition funnel (fitted in the leftmost neck, capped with a rubber septum) are assembled hot and fitted with rubber septa on the center and right necks. The apparatus is placed under high vacuum (< 1 mmHg) and allowed to cool to room temperature and subsequently maintained under a positive pressure of N_2 using needle inlet through the septum on the rightmost neck. Upon cooling, a thermometer is attached to the center neck using a Teflon adapter, and the apparatus is again evacuated under high vacuum. It is then placed under a positive pressure of N_2 and charged with 40.0 mL of toluene (Note 1), then cooled in a –15 °C acetone bath (Note 2). A solution of *n*-butylmagnesium chloride (16.8 mL, 33.6 mmol, 0.42 equiv., 2.0M in THF) is added (Note 3) and the resulting grey solution is cooled to an internal temperature of -10 °C (Note 2), at which point 28 mL (67.5 mmol, 0.85 equiv, 2.41 M in hexanes) of *n*-butyllithium (Note 3) is added via a disposable syringe at such a rate so as to keep the internal temperature below –5 °C (approximately 30 min addition time) (Note 4). The resulting milky

Org. Synth. **2012**, *89*, 460-470
Published on the Web 4/23/2012
© 2012 Organic Syntheses, Inc.

white slurry is stirred at –10 °C for 30 min (Note 5). During this time the addition funnel is charged with a solution of 25.00 g (79.4 mmol, 1.0 equiv) of 1,3,5-tribromobenzene in roughly 130 mL of toluene (Notes 6 and 7). Dropwise addition is initiated at such a rate that the internal temperature remains below –5 °C (approximately 45 min addition time). The resulting dark maroon slurry is stirred at –10 °C for 1 h. A solution of 22.7 g (104.0 mmol, 1.3 equiv) of di-*tert*-butyl dicarbonate (Note 8) in 30.0 mL of toluene is added slowly via cannula, keeping the internal temperature below –5 °C (approximately 40 min addition time). After addition is completed, the reaction mixture is stirred at –10 °C for 2 h (Note 5). At this point all septa are removed to facilitate ventilation of subsequently generated CO_2, and the reaction is quenched by addition of 120 mL aqueous solution of 1M citric acid, which causes the internal temperature to rise from –14 °C to 6 °C (Note 9). Upon warming to room temperature, the biphasic mixture is transferred to a 1-L separatory funnel and the organic phase is separated, then washed with an additional 120 mL of 1M aqueous citric acid. The organic phase is dried over $MgSO_4$ (40 g), filtered through a fritted glass funnel, and concentrated by rotary evaporation (40 °C, 20 mmHg). The residue is further dried under high vacuum to afford a reddish-brown oil that is purified via flash column chromatography on silica gel (675 g, 9-cm diameter column, height of silica gel is 9 inches) to yield **2** as an orange solid (18.99 g, 71%) (Notes 10, 11, and 12).

B. *t-Butyl-3-bromo-5-formylbenzoate* (**2**). A 500-mL 3-necked round-bottomed flask with a Teflon-coated oval-shaped magnetic stir bar (3.8 cm x 1.6 cm) is flame-dried and charged with 18.91 g (56.3 mmol, 1.0 equiv) of purified dibromobenzoate **1** and 135 mL THF (Note 1). The center neck is equipped with a thermometer using a Teflon adapter, and the left and right necks are fitted with rubber septa. The resulting bright orange solution is cooled to an internal temperature of 0 °C in an ice-brine bath, and treated with 44 mL (96.8 mmol, 1.7 equiv, 2.2 M in THF) of *iso*-propylmagnesium chloride (Note 13) via disposable syringe at such a rate to keep the internal temperature below 21 °C (Note 14). Upon complete addition, the solution is allowed to stir at room temperature for 1 h (Note 15). The now blood-red solution is cooled back to 0 °C, and 11.0 mL (142.7 mmol, 2.5 equiv) of *N,N*-dimethylformamide (Note 13) is added via disposable syringe at such a rate to keep the internal temperature below 25 °C (Note 16). The deep red reaction mixture is then stirred at room temperature for 30 min, then diluted with 100 mL of *tert*-butyl methyl ether (Note 13) and quenched by addition of 100 mL of saturated aqueous NH₄Cl (Note 17). The two layers are separated in a 1-L separatory funnel, and the organic phase is washed with an additional 100 mL of saturated aqueous NH₄Cl. The aqueous layers are then back-extracted once with 100 mL of ethyl acetate. The combined organic layers are dried over MgSO₄ (40 g), filtered through a fritted glass funnel, and concentrated by rotary evaporation (25 °C, 20 mmHg) to afford crude **2,** a bright orange/brown oil. Purification by flash column chromatography on silica gel (675 g, 9-cm diameter column, height of silica gel is 9 inches) affords **2** as a thick golden yellow oil that crystallizes upon standing (12.96 g, 81% yield) (Notes 18 and 19).

2. Notes

1. Toluene and tetrahydrofuran (both of Optima® grade) were purchased from Fisher Scientific and dried by passing through an activated alumina column solvent purification system (Innovative Technology Inc. Pure Solv™).

2. The bath temperature is maintained using a Neslab CC-65 immersion cooler.

3. *n*-Butylmagnesium chloride (2.0M in THF) and *n*-butyllithium (2.5M in hexanes) were purchased from Sigma-Aldrich and titrated prior to use. *n*-Butyl magnesium chloride arrived as a cloudy, grey solution, and titrated to be 2.0M; *n*-butyllithium arrived as a cloudy, yellow solution, and titrated to be 2.41M.

4. The reaction mixture changes from grey to a milky white slurry.

5. The checkers followed reaction progress by quenching an aliquot with saturated aqueous NH₄Cl and taking its ¹H NMR spectrum. The submitters monitored reaction progress by quenching an aliquot with methanol and determined progress by HPLC analysis on an Ascentis Express C18 column (10 cm x 4.6 cm; flow rate 1.8 mL/min; mobile phase acetonitrile/ 0.1% aqueous H₃PO₄, 10% to 95% acetonitrile over 6 min and hold 95% for 2 min; UV detection at 254 nm), diagnostic peaks t_R (min) at 4.6 (toluene), 5.6 (dibromobenzene), 6.4 (tribromobenzene), 6.9 (*t*-butyl-dibromobenzoate).

6. 1,3,5-Tribromobenzene (98%) was purchased from Sigma-Aldrich and used as received.

7. IMPORTANT: 1,3,5-Tribromobenzene is sparingly soluble in toluene at room temperature. Thus, prior to generation of the ate complex,

462

the checkers dissolved the tribromobenzene in toluene by heating to 90 °C and cooling the solution back to room temperature. Once dissolved, the tribromobenzene remains in solution for some time. Upon dissolution of 1,3,5-tribromobenzene in toluene, some dark solids remained. The solids were removed by filtration for the first run (half-scale), but not for the second run (full-scale). The presence of these solids had no effect on the success of the reaction.

8. Di-*tert*-butyl dicarbonate (reagent grade, 97%) was purchased from Sigma-Aldrich and used as received.

9. Citric acid monohydrate (reagent grade) was purchased from Fisher Scientific.

10. The checkers also obtained the crude product in quantitative yield, but despite numerous attempts the crude sample gave low yields when used directly in the next step. The submitters reported a quantitative yield of the crude product that was deemed to be sufficiently pure (by HPLC, see Note 5 for method conditions) for use in the next step.

The submitters purified the crude material by column chromatography using 95:5 hexanes:ethyl acetate as the eluent and obtained the purified product in 97% yield.

11. The checkers purchased silica gel from SiliCycle (SiliaFlash® P60 40-63μm). The column was wet-packed and eluted with 6:1 hexanes:dichloromethane. One large fraction (1 L) and thirty-six smaller fractions (125-150 mL) were collected. The product was found (by TLC) in fractions 9-28, which were combined and concentrated by rotary evaporation (25 °C, 20 mmHg) and then under high vacuum (< 1 mmHg) to yield **1**, with the following physical properties: mp: 58–61 °C; R_f = 0.32 (5:1 hexanes:dichloromethane, 254 nm UV lamp); ^{1}H NMR (500 MHz, CDCl$_3$) δ (ppm): 1.59 (s, 9 H), 7.81 (t, J = 2 Hz, 1 H), 8.03 (d, J = 2 Hz, 2 H); ^{13}C NMR (125 MHz, CDCl$_3$) δ (ppm): 28.1, 82.5, 122.8, 131.2, 135.2, 137.7, 163.0; IR (neat film, NaCl): 742.3, 764.7, 847.8, 1135.4, 1165.2, 1282.0, 1368.7, 1451.1, 1558.4, 1719.2, 2978.0 cm^{-1}. HRMS m/z calcd. for $(C_{11}H_{12}O_2Br_2)Na^+$: 356.9096; found (electrospray) m/z: Not detected. CHN analysis for $C_{11}H_{12}O_2Br_2$: C, 39.32; H, 3.60. Found (duplicate): C, 39.45/39.51; H, 3.65/3.74.

12. On half scale, the checkers obtained **1** in 74% yield.

13. *iso*-Propylmagnesium chloride (2.0M in THF), *N,N*-dimethylformamide (anhydrous, 99.8%) and *tert*-butyl methyl ether (reagent grade, 98%) were purchased from Sigma-Aldrich. *iso*-Propylmagnesium

chloride was titrated prior to use and found to be 2.2M. *N,N*-dimethylformamide and *tert*-butyl methyl ether were used as received.

14. The checkers found the internal temperature to rise to ~10 °C, accompanied by darkening of the reaction mixture to black/orange. The submitters noted the reaction temperature to rise to 18 °C, with darkening of the reaction mixture.

15. Progress of the reaction was monitored as described in Note 5. The submitters monitored complete consumption of starting material by quenching an aliquot of the reaction mixture with methanol and determined progress by HPLC analysis on an Ascentis Express C18 column (10 cm x 4.6 cm; flow rate 1.8 mL/min; mobile phase acetonitrile/ 0.1% aqueous H_3PO_4, 10% to 95% acetonitrile over 6 min and hold 95% for 2 min; UV detection at 254 nm), diagnostic peaks t_R (min) at 6.1 (*t*-butyl-bromobenzoate), 5.7 (aldehyde), 6.9 (*t*-butyl-dibromobenzoate).

16. A moderate exotherm is observed, with the internal temperature rising to 20–22 °C.

17. The submitters quenched the reaction with a 1M citric acid solution.

18. The sample was loaded on a wet-packed column and eluted with 9:1 hexanes:diethyl ether. One large fraction (1 L) and forty smaller fractions (100 mL) were collected. The product was found (by TLC) in fractions 16-35, which were combined and concentrated by rotary evaporation (25 °C, 20 mmHg) and then under high vacuum (< 1 mmHg) to yield **2**: mp: 61–63 °C; R_f = 0.28 (9:1 hexanes:diethyl ether, 254 nm UV lamp); ^{1}H NMR (500 MHz, CDCl$_3$) δ (ppm): 1.63 (s, 9 H), 8.16 (d, J = 2 Hz, 1 H), 8.34 (d, J = 2 Hz, 1 H), 8.38 (d, J = 2 Hz, 1 H), 10.02 (s, 1 H); ^{13}C NMR (125 MHz, CDCl$_3$) δ (ppm): 28.0, 82.6, 123.2, 129.6, 134.8, 135.0, 137.7, 137.7, 163.1, 189.9; IR (neat film, NaCl): 664.9, 719.4, 766.1, 845.2, 889.4, 1155.1, 1200.8, 1246.6, 1296.5, 1369.1, 1456.6, 1575.1, 1706.4, 2978.4 cm^{-1}. HRMS m/z calcd. for (C$_{12}$H$_{13}$O$_3$Br)Na$^+$: 306.9940; found (electrospray) m/z: 306.9940. CHN analysis for C$_{12}$H$_{13}$O$_3$Br: C, 50.55; H, 4.60. Found (duplicate): C, 50.67/50.38; H, 4.71/4.68.

19. On half-scale, the checkers obtained an 84% yield of **2**. The submitters reported obtaining **2** in 91% yield.

Waste Disposal Information

All hazardous materials should be handled and disposed of in accordance with "Prudent Practices in the Laboratory"; National Academies Press; Washington, DC, 2011.

3. Discussion

Transition metal-catalyzed carbonylation reactions have become the method of choice to prepare benzoate esters from aryl bromides.[2] These methods represent an improvement over other methodologies in terms of yields, reaction conditions and substrate scope. Alternative methods involving metallation or metal-halogen exchange, followed by trapping the resulting organometallic species with carbon dioxide, methyl chloroformate[3] or dimethyl carbonate,[4] often require cryogenic conditions and have limited compatibility to functional groups.

One of the limitations of transition metal-catalyzed carbonylations is the lack of selectivity for the monocarbonylation of substrates containing multiple halogen substituents. On these substrates, the product of the first carbonylation reaction is more reactive to carbon-halogen bond insertion than the starting aryl halide and it becomes very difficult to achieve selectivity in the reaction.[5]

Researchers have looked for solutions to this selectivity problem. Cassar and coworkers[6] showed that, under very controlled Pd-catalyzed carbonylation conditions, it was possible to prepare 4-bromobenzoic acid in 90% yield from 1,4-dibromobenzene. The key to obtaining high selectivity in this reaction was a slow addition of the substrate under phase transfer conditions. Under these reaction conditions, the product of the first carbonylation reaction was extracted as a carboxylate salt into the aqueous layer, thus being unavailable for a second palladium insertion.

Lee and coworkers[7] attempted to improve the selectivity using a metallation approach. They carried out the direct metallation of 1,4-dibromobenzene with 2.5 equivalents of magnesium under sonication conditions, followed by trapping of the resulting Grignard with diethyl dicarbonate in the presence of Lewis acids. With this method, they obtained ethyl 4-bromobenzoate in 92% yield, with only 3% of diethyl terephthalate. Lower selectivities were obtained with other dibromides studied.

In an effort to improve the performance and selectivity of this transformation, we turned our attention to a report from Iida, Mase and coworkers,[8] who showed good reactivity and very high selectivities for metal-halogen exchange reactions of dibromoarenes using lithium tri-*n*-butylmagnesium ate complex. We later discovered that di-*t*-butyl dicarbonate is an excellent electrophile to trap triarylmagnesium ate complexes affording the desired *t*-butyl esters with complete selectivity and in very high yields.

Although chemists tend to rely mostly on methyl or ethyl esters, the use of *t*-butyl esters may have advantages in some cases. The possibility of *t*-butyl ester hydrolysis under acidic conditions makes these compounds complimentary to simpler esters such as methyl or ethyl. Within the context of this chemistry, we have successfully used the *t*-butyl moiety as a protecting group for the ester functionality in subsequent Grignard chemistry. This is exemplified in the procedure above by the reaction of *t*-butyl-3,5-dibromobenzoate obtained by our method with *i*-propylmagnesium chloride followed by addition to dimethylformamide.

This methodology has been extended to other di and tri-haloarenes and heteroarenes as shown in the table below.[9] Very high selectivities for the metal-halogen exchange reaction were observed in all the substrates examined and yields for the desired ester products were uniformly high except for the pyridine substrates.

466

Table 1. n-Bu$_3$MgLi-mediated mono-carboxylation of di and tribromoarenes and heteroarenes.

	Substrate	Conditions	Product	Yield (%)
1	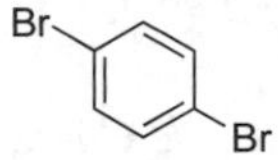 Br—C$_6$H$_4$—Br (1,4)	Toluene, -5°C	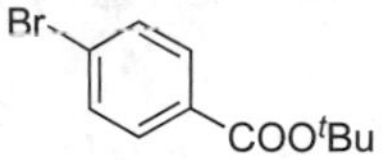 Br—C$_6$H$_4$—COO^{t}Bu	98
2	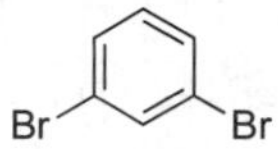 Br—C$_6$H$_4$—Br (1,3)	Toluene, -5°C	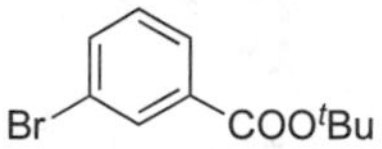 Br—C$_6$H$_4$—COO^{t}Bu	98
3	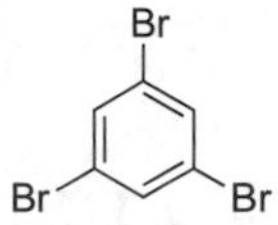 1,3,5-tribromobenzene	Toluene, -5°C	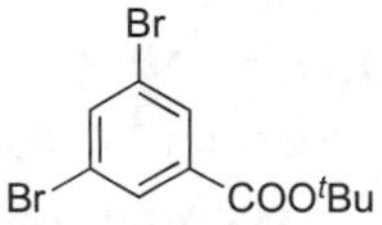 3,5-dibromo—COO^{t}Bu	97
4	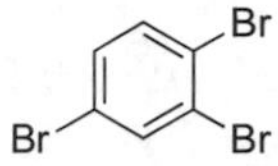 1,2,4-tribromobenzene	Toluene, -5°C	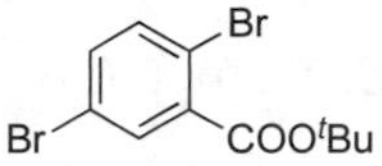 dibromo—COO^{t}Bu	71
5	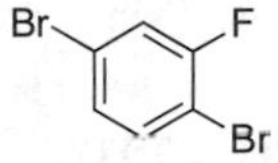 Br, F, Br benzene	Toluene, -10°C	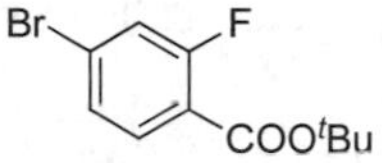 Br, F —COO^{t}Bu	92
6	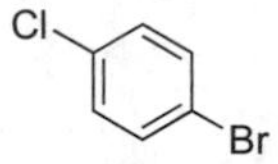 Cl—C$_6$H$_4$—Br (1,4)	Toluene, 0°C	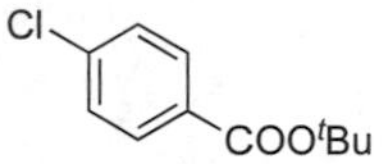 Cl—C$_6$H$_4$—COO^{t}Bu	95
7	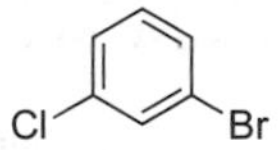 Cl—C$_6$H$_4$—Br (1,3)	Toluene, -10°C	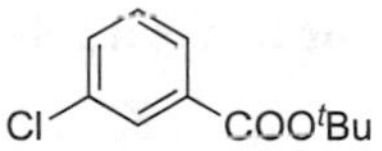 Cl—C$_6$H$_4$—COO^{t}Bu	99
8	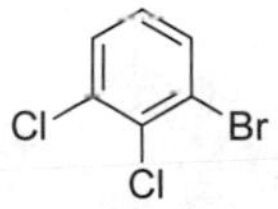 Cl, Cl, Br benzene	Toluene, -10°C	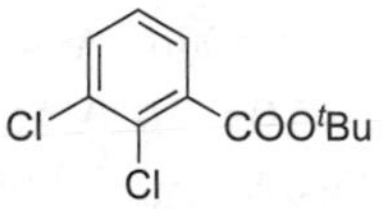 Cl, Cl —COO^{t}Bu	80
9	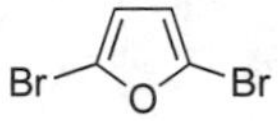 Br—furan—Br	Toluene, -10°C	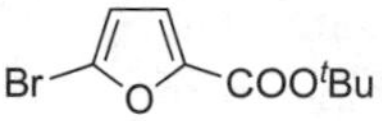 Br—furan—COO^{t}Bu	84
10	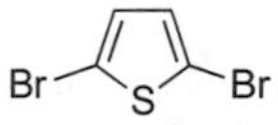 Br—thiophene—Br	Toluene, -10°C	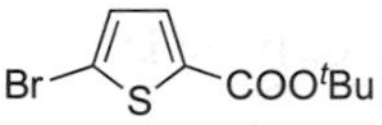 Br—thiophene—COO^{t}Bu	96
11	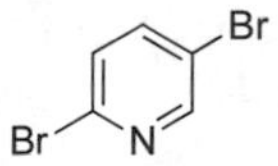 Br, Br pyridine	Toluene-THF (1:1), -10°C	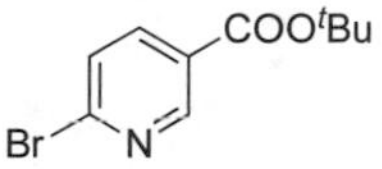 Br—pyridine—COO^{t}Bu	66

1. Department of Process Research, Merck Research Laboratories. PO Box 2000, Rahway, NJ-07065, USA. Jaume_Balsells@merck.com
2. See for example: a) Skoda-Földes, R.; Kollár, L. *Curr. Org. Chem.* **2002**, *6*, 1097–1119. b) Beller, M.; Indolese, A. F. *Chimia* **2001**, *55*, 684–687.
3. Bratton, L. D.; Huh, H.; Bartsch, R. A. *J. Heterocycl. Chem.* **2000**, *37*, 815–819.
4. Amedio, J. C. J.; Lee, G. T.; Prasad, K.; Repic, O. *Synth. Commun.* **1995**, *25*, 2599–2612.
5. Cai, C.; Rivera, N. R.; Balsells, J.; Sidler, R. R.; McWilliams, J. C.; Shultz, C. S.; Sun, Y. *Org. Lett.* **2006**, *8*, 5161–5164.
6. Cassar, L.; Foà, M.; Gardano, A. *J. Organomet. Chem.* **1976**, *121*, C55–C56.
7. Lee, A. S.-Y.; Wu, C.-C.; Lin, L./S.; Hsu, H.-F. *Synthesis* **2004**, 568–572.
8. Iida, T.; Wada, T.; Tomimoto, K.; Mase, T. *Tetrahedron Lett.* **2001**, *42*, 4841–4844.
9. Balsells, J.; Li, H. *Tetrahedron. Lett.* **2008**, *49*, 2034–2037.

Appendix
Chemical Abstracts Nomenclature; (Registry Number)

t-Butyl-3,5-dibromobenzoate; (5342-97-2)

n-Butylmagnesium chloride; (693-04-9)

n-Butyllithium; (109-72-8)

Di-*tert*-butyl dicarbonate; (24424-99-5)

1,3,5-Tribromobenzene; (626-39-1)

t-Butyl-3-bromo-5-formylbenzoate; (1018948-99-6)

iso-Propylmagnesium chloride; (1068-55-9)

N,N-Dimethylformamide; (68-12-2)

Jaume Balsells received his PhD from the University of Barcelona in 1998 under the supervision of Professor Miquel A. Pericas. After completing postdoctoral work with Professors Patrick J. Walsh and Gary A. Molander at the University of Pennsylvania, he joined the Process Research Department at Merck & Co., where he is currently an Associate Director. His Research interests include the development of practical synthetic strategies and methodology for the preparation of pharmaceutical intermediates, Heterocyclic Chemistry and Green Chemistry.

Juan D. Arredondo was born in Englewood, New Jersey. He received his undergraduate Chemistry degree in 2001 from Rutgers University under the guidance of Professor Kathryn Uhrich were he worked on the synthesis of derived salicylate-based poly(anhydride esters). He obtained his M.S. in Chemistry in 2006 from the University of Pittsburgh with Professor Peter Wipf. His research focused on zirconium-catalyzed asymmetric carboalumination. That same year he joined the Process Research Department of Merck & Co.

Hongmei Li received her B.Sc. in chemistry at the Dalian University of Technology in 1999 and her M.Sc. degree in inorganic chemistry (2002) at the Peking University with Professor Jianhua Lin. She began her Ph.D. at the University of Pennsylvania under the direction of Professor Patrick J. Walsh in 2002, during which she worked on the catalytic asymmetric addition of organozinc reagents to ketones and application of bimetallic intermediates in organic synthesis. In 2007, she joined the Department of Process Research at Merck & Co. Inc., Rahway, NJ. Her research interests focus on design and development of practical synthetic methodology and processes for the synthesis of pharmaceutical intermediates and drug candidates.

Julius R. Reyes received his B.S. in Chemistry in 2010 from the University of California, Berkeley, where he worked with Professors Richmond Sarpong and Andrew Streitwieser. The following summer he joined the labs of Prof. Viresh Rawal at the University of Chicago and is currently working on the development of novel hydrogen bond donor catalysts.

Reductive Radical Decarboxylation of Aliphatic Carboxylic Acids

Submitted by Eun Jung Ko,[1] Craig M. Williams,[1] G. Paul Savage,[2] and John Tsanaktsidis.[2]
Checked by Samantha R. Levine, Aaron Bedermann, and John L. Wood.

1. Procedure

An oven-dried (Note 1) 500-mL four-necked, round-bottomed flask (Note 2) is equipped with a 3-cm Teflon-coated oval stir bar, a 125-mL pressure-equalizing addition funnel sealed with a septum, a thermometer, a septum, and a reflux condenser fitted with a gas inlet adapter and connected to a dual manifold (Note 3). The reaction vessel is charged with chloroform (120 mL) (Note 4), 1-hydroxypyridine-2(1H)-thione, sodium salt (5.90 g, 39.6 mmol, 1.2 equiv) (Note 5) and 4-N,N-dimethylaminopyridine (0.040 g, 0.33 mmol, 0.01 equiv) to give an off-white suspension. The addition funnel is charged with palmitoyl chloride (10 mL, 9.06 g, 33.0 mmol, 1.0 equiv) followed by chloroform (60 mL), and the entire apparatus is blanketed with a slight positive pressure of nitrogen to maintain an inert atmosphere throughout the course of the reaction (Note 6).

The reaction vessel is heated to reflux over 25 min (silicon oil bath, external bath temperature 80 °C, internal temperature 57 °C) and the palmitoyl chloride solution is then added drop-wise over 85 min with concomitant irradiation from a tungsten lamp (120V, 150W) (Note 7). The reaction mixture remains a suspension, which gradually turns yellow upon the addition of palmitoyl chloride. Visible evolution of carbon dioxide is observed by 30 min. After an additional 25 min of stirring an orange suspension is observed (Note 8). Heating and irradiation is then discontinued and the resulting orange/brown suspension is allowed to cool to an internal temperature of 25 °C and transferred to a 500-mL separatory funnel containing 1M HCl (100 mL) and CH_2Cl_2 (100 mL). The aqueous phase is separated and extracted with CH_2Cl_2 (3 x 50 mL). The combined organic layers are washed with saturated NaCl solution (100 mL), dried over 5.9 g of $MgSO_4$, filtered through a 350 mL medium porosity sintered glass funnel, and concentrated by rotary evaporation (bath temperature increased from 25 to 35 °C, 250 mmHg) and then at 3 mmHg to afford a yellow-brown oil.

The neat product is charged on a plug (6 x 10 cm) of 150 g of silica gel (Note 9) and eluted with 1 L of petroleum ether 35 – 60 °C directly into a 2-L round-bottomed flask (Note 10). This solution is concentrated by rotary evaporation (bath temperature increased from 25 to 30 °C, pressure reduced from 400 to 250 mmHg) and then at 3 mmHg to afford 6.29 g (90%) (Note 11) of pentadecane as a clear, colorless oil (Note 12).

2. Notes

1.   The submitters used an oven set to 180 °C, assembled the apparatus while still hot, and allowed it to cool to ambient temperature (23 °C) under vacuum (0.9 mmHg). The checkers used an oven set to 180 °C, allowed the apparatus to cool to ambient temperature (20 °C) in a dessicator containing Drierite, assembled it, and evacuated and backfilled the system three times with nitrogen.

2.   The submitters used a 500-mL two-necked round-bottomed flask, the checkers chose to use a four-necked flask to facilitate TLC and internal reaction temperature monitoring.

3.   Depiction of the experimental set-up, including the position of the light source, is illustrated in **Figure 1**.

4.   The submitters obtained 4-*N*,*N*-dimethylaminopyridine (99%) and palmitoyl chloride (98%) from Sigma-Aldrich, Inc. which were used as received. Chloroform (99.8%) was purchased from ChemSupply Co., Inc. and distilled from P_2O_5 prior to use. The checkers obtained 4-*N*,*N*-dimethylaminopyridine (99%) from Acros Organics and palmitoyl chloride (98%) from MP Biomedicals, LLC., both of which were used as received. Chloroform (99.8%) was purchased from Mallinckrodt Chemicals, Inc., and was washed with water, dried over K_2CO_3, and distilled from Na_2SO_4 prior to use.

5.   The submitters purchased 1-hydroxypyridine-2(1*H*)-thione, sodium salt as a 40 % solution in water from Merck. The water was removed under reduced pressure (40 °C, 20 mmHg) and the resulting yellow solid

was recrystallized from ethanol to give a white powder. The checkers purchased 1-hydroxypyridine-2(1*H*)-thione, sodium salt as a 40 % solution in water from Alfa Aesar. The water was removed under reduced pressure (30 °C, 3 mmHg) and the resulting yellow solid was dissolved in ethanol and triturated with hexanes to give an off-white powder.

Figure 1. Experimental sct-up used in the reaction

6. The submitters used argon to maintain an inert atmosphere.

7. The submitters performed the addition over 40 min (1.5 mL/min) with concomitant irradiation from a tungsten lamp (240V, 500W). The reaction mixture turned bright yellow upon addition of the palmitoyl chloride, with the color fading as the evolution of carbon dioxide was observed. After 1 h the bright yellow coloration faded to an orange/brown color.

8. The progress of the reaction was monitored by TLC analysis on silica gel with 15% EtOAc-hexanes as the eluent and visualization with *p*-anisaldehyde. The acid chloride starting material has R$_f$ = 0.53 (white), the alkane product is not observable by TLC.

9. The submitters obtained silica gel (particle size 0.040 – 0.063 mm) 230-400 ASTM mesh from Advanced Molecular Technologies. The

checkers obtained silica gel (particle size 0.04 – 0.063 mm) 230-400 mesh from Silicycle.

10. The submitters diluted the brown residue with CH_2Cl_2 (10 mL), which was charged on plug (9 cm Ø) of 250 g of silica gel, eluting with petroleum ether 40 – 60 °C (1500 mL) which was obtained from Merck and purified by distillation prior to use. After removal of the solvent by rotary evaporation (40 °C, 300 mmHg) a yellow oil was obtained. This oil was further purified by bulb-to-bulb distillation using a Büchi Glass Oven B-580 Kugelrohr at 93 °C (0.9 mmHg) whose receiving bulb was cooled with dry ice to yield 5.7 g (81%) of pentadecane as a colorless oil.

11. When the reaction was carried out on a 27.9 mmol scale the checkers obtained a yield of 83%.

12. The product exhibits the following properties: v_{max}/cm^{-1} (neat) 2956, 2923, 2853; ^1H NMR (400 MHz, CDCl$_3$) δ_H 1.26 (26H, br s), 0.88 (6H, t, J = 6.8 Hz); ^{13}C NMR (100 MHz, CDCl$_3$) δ_C 32.1, 29.9, 29.8, 29.5, 22.9, 14.3; m/z GC/MS 212; Anal. calcd. for $C_{15}H_{32}$: C, 84.82; H, 15.18; found: C, 84.70; H, 14.91.

Safety and Waste Disposal Information

All hazardous materials should be handled and disposed of in accordance with "Prudent Practices in the Laboratory"; National Academies Press; Washington, DC, 2011.

3. Discussion

The Barton decarboxylation is a radical reaction in which a carboxylic acid is first converted to a thiohydroxamate ester, which upon heating (optionally in the presence of a radical initiator or light) undergoes homolytic cleavage, followed by loss of carbon dioxide (Scheme 1). The resulting aliphatic radicals can then be trapped by a variety of reagents leading to new functionality. Using this reaction it is possible to remove the carboxylic acid group from aliphatic carboxylic acids and replace it with other functional groups.[3]

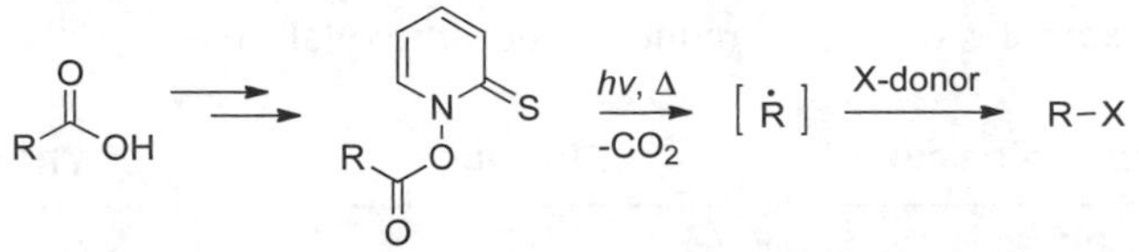

Scheme 1. Barton radical decarboxylation reaction.

The intermediate thiohydroxamate ester can be obtained by reacting an acid chloride and the sodium salt of 1-hydroxypyridine-2(1*H*)-thione (5) as in this report, or directly from the carboxylic acid using *N,N'*-dicyclohexylcarbodiimide (DCC) and similar coupling methods.[4] We found that the acid chloride method was generally more reliable. The acid chlorides can be prepared by the action of oxalyl chloride or thionyl chloride on the carboxylic acid.[5]

Reductive decarboxylation (**Scheme 1**, X = H) is an important subset of the Barton procedure, which ultimately results in replacing the carboxylic acid function with a hydrogen atom.[6] Under this protocol, reductive decarboxylation is accomplished by mild photochemical decomposition of the corresponding thiohydroxamate ester, in the presence of a suitable hydrogen donor (H-donor), originally tributyltin hydride or tert-butylthiol. We recently discovered that it is more convenient, safer, and less expensive to use chloroform as both solvent and H-donor in these reactions.[7] Aromatic carboxylic acids do not undergo this reaction. The procedure described herein is applicable to aliphatic carboxylic acids with the best results generally from primary and secondary acids (**Table 1**). The reductive decarboxylation product of especially hindered tertiary carboxylic acids using this method may sometimes be contaminated with the corresponding alkyl chloride, which arises by competing chlorine atom transfer from chloroform. In these cases, the addition of a stronger H-donor, such as tert-butyl thiol, is recommended.

Table 1. Examples of Barton reductive decarboxylations

Carboxylic acid	Product	Yield %
		66
		77
		69[a]
		82
		77
		68

[a] 2:1 mixture of adamantane:1-chloroadamantane

1. School of Chemistry and Molecular Biosciences, University of Queensland, Brisbane 4072, Queensland, Australia.
2. CSIRO Materials Science and Engineering, Clayton South, 3169 Victoria, Australia. john.tsanaktsidis@csiro.au

3. Barton, D. H. R.; Crich, D.; Motherwell, W. B. *Tetrahedron* **1985**, *41*, 3901–3924.

4. Barton, D. H. R.; Samadi, M. *Tetrahedron* **1992**, *48*, 7083–7090.
5. Allen, C. F. H.; Byers Jr., J. R.; Humphlett W. J. *Org. Synth.* **1957**, *37*, 66; Coll. Vol. 4, p.739 (**1963**).
6. Barton, D. H. R.; Crich, D.; Motherwell, W. B. *J. Chem. Soc., Chem. Commun.* **1983**, 939–941.
7. Ko, E. J.; Savage, G. P.; Williams, C. M.; Tsanaktsidis, J. *Org. Lett.* **2011**, *13*, 1944–1947.

Appendix
Chemical Abstracts Nomenclature; (Registry Number)

1-Hydroxypyridine-2(1*H*)-thione, sodium salt: 2(1*H*)-Pyridinethione, 1-hydroxy-, sodium salt; (3811-73-2)
4-*N,N*-Dimethylaminopyridine: 4-Pyridinamine, *N,N*-dimethyl-; (1122-58-3)
Palmitoyl chloride: Hexadecanoyl chloride; (112-67-4)

G. Paul Savage received his PhD under the direction of Dr R. F. Evans at the University of Queensland in 1988, and this was followed by a two-year postdoctoral position with Professor Alan R. Katritzky FRS at the University of Florida. He returned to Australia to take up a research scientist position with CSIRO, Australia's premier government research organization. He was promoted to Program Manager and completed an MBA in 2005 from the Chifley Business School, La Trobe University. His research interests include heterocyclic chemistry, dipolar cycloaddition reactions, synthesis methodology, and medicinal chemistry.

Dr Tsanaktsidis obtained his BSc (Hons) in 1983 at Flinders University and his PhD in 1988 under the supervision of Dr. Ern Della. Following postdoctoral appointments at the University of Chicago with Professor Philip Eaton, and at the Australian National University with Professor Athel Beckwith, he joined the faculty at the School of Chemistry, at the University of Melbourne in 1991. Dr Tsanaktsidis accepted an offer to join the Commonwealth Scientific and Industrial Research Organization (CSIRO) in 1995. His career achievements were recognized in 2006 through a Distinguished Alumni Award from The Flinders University of South Australia.

Eun Jung Ko was born in 1983 in Rome, Italy. She received her MSci degree in chemistry in 2006 from Imperial College London. In 2010 she was awarded her PhD from The University of Leeds, where she worked on the total synthesis of okaramine B under the supervision of Prof. Stephen P. Marsden. She is currently working as a postdoctoral fellow at the University of Queensland, under the guidance of Dr Craig M. Williams on the development of methodologies for the functionalisation of strained hydrocarbons.

Craig M. Williams received his BSc (Hons) and PhD (1997) degrees from Flinders University (Supervisor Prof. Rolf H. Prager). He worked as an Alexander von Humboldt Postdoctoral Fellow with Prof. Armin de Meijere at the Georg-August-Universität (1999) and then took up a postdoctoral fellowship at the Australian National University with Prof. Lewis N. Mander. He has held an academic position at The University of Queensland since 2000 and during this time has won a number of awards including a Thieme Chemistry Journals Award in 2007. His research focus is on constructing complex natural products and associated synthetic methodology.

Samantha R. Levine was born in Hackensack New Jersey in 1985. In 2008 she received her B. S. in chemistry from The California Institute of Technology, where she did undergraduate research in the laboratory of Professor Brian M. Stoltz. Samantha is pursuing her graduate studies in the research group of Professor John L. Wood.

Aaron Bedermann was born in Madison, Wisconsin. He received his BS in Chemistry from the University of Wisconsin, where he performed undergraduate research under the supervision of Professor Richard Hsung. He is currently pursuing graduate research at Colorado State University under the guidance of Professor John L. Wood.

Efficient Synthesis of Oxime Using
O-TBS-N-Tosylhydroxylamine:
Preparation of (2Z)-4-(Benzyloxy)but-2-enal Oxime

Submitted by Katsushi Kitahara, Tatsuya Toma, Jun Shimokawa and Tohru Fukuyama.[1]

Checked by Erica Benedetti and Kay M. Brummond.

1. Procedure

A. *O-TBS-N-tosylhydroxylamine* *(2)*. A 1-L, three-necked round-bottomed flask equipped with an internal thermometer, a 100-mL pressure-equalizing dropping funnel sealed with a rubber septum, a calcium chloride drying tube, and a 5.0-cm oval, Teflon-coated, magnetic stir bar, is charged under air with hydroxylamine hydrochloride (6.08 g, 87.5 mmol, 1.10 equiv) and *tert*-butyldimethylsilyl chloride (13.2 g, 87.5 mmol, 1.10 equiv) in 400 mL of DMF (Note 1). The clear solution is stirred (Note 2) in an ice bath for 30 min, then Et$_3$N (55.0 mL, 397.7 mmol, 5.00 equiv) is added dropwise from the dropping funnel over 20 min, such that the internal temperature does not exceed 10 °C (Note 3). A white precipitate is observed during the addition. The ice bath is removed and the suspension is vigorously stirred for 2 h (Note 4). The reaction mixture is cooled in an ice bath for 15 min and *p*-toluenesulfonyl chloride (15.1 g, 79.1 mmol, 1.00

480

Org. Synth. **2012**, *89*, 480-490
Published on the Web 5/16/2012
© 2012 Organic Syntheses, Inc.

equiv) is added in three portions (5.0 g each) at 5 min intervals, such that the internal temperature does not exceed 10 °C (Note 5). The ice bath is removed and the suspension is vigorously stirred for 2 h, at which point TLC analysis shows no signs of *p*-toluenesulfonyl chloride (Notes 6 and 7). The reaction mixture is poured into *n*-hexane (300 mL), saturated aqueous NH$_4$Cl (250 mL) and H$_2$O (250 mL) in a 2-L separatory funnel. The layers are separated, and the aqueous phase is extracted with an additional *n*-hexane (2 x 250 mL). The organic solutions are combined, washed with H$_2$O (2 x 500 mL), 10% aqueous citric acid (2 x 500 mL), brine (250 mL) and dried over magnesium sulfate (60 g). After filtration through a cotton plug and rinsing with 50 mL of *n*-hexane, the solution is concentrated on a rotary evaporator (35 °C, 150 mmHg, water bath) to about 100 mL (Note 8). White crystals precipitate on cooling to 0 °C for 15 min and are collected by Büchner funnel (diameter 90 mm) with a medium porosity fritted disk using suction filtration. The precipitate is washed with ice-cooled *n*-hexane (50 mL) and dried under vacuum (25 °C, 10 mmHg, 2 h) to give the first crop of *O*-TBS-*N*-tosylhydroxylamine (**2**) (16.4 g, 54.4 mmol, 69%). A second crop of crystals (1.24 g, 4.1 mmol, 5%) is obtained by dissolving the concentrated pale-yellow mother liquor in hot *n*-hexane (30 mL) and cooling in an ice bath for 30 min, followed by the filtration using 10 mL of ice-cooled *n*-hexane (Note 9).

B. *(Z)-N-(4-(Benzyloxy)but-2-en-1-yl)-N-((tert-butyldimethylsilyl) oxy)-4-methylbenzenesulfonamide (4)*. An oven-dried, 1-L, three-necked round-bottomed flask is equipped with a 5.0-cm oval Teflon-coated magnetic stir bar, an internal thermometer, a 50-mL pressure-equalizing dropping funnel sealed with a rubber septum, and fitted with nitrogen gas inlet adaptor. The flask is evacuated and refilled with nitrogen three times, then charged with *O*-TBS-*N*-tosylhydroxylamine (**2**) (16.1 g, 53.4 mmol, 1.00 equiv), triphenylphosphine (18.2 g, 69.3 mmol, 1.30 equiv), toluene (135 mL) and THF (45 mL) (Note 10). *cis*-4-Benzyloxy-2-buten-1-ol (**3**) (9.5 mL, 56.1 mmol, 1.05 equiv) (Note 11) is added and the solution is stirred (Note 12) in an ice bath for 10 min. Diethyl azodicarboxylate (2.2 M in toluene, 29.1 mL, 64.1 mmol, 1.20 equiv) (Note 13) is added dropwise from dropping funnel over 20 min, such that the internal temperature does not exceed 10 °C. The solution becomes cloudy during the addition (Note 14). The mixture is stirred at 0 °C for 30 min, then H$_2$O (500 μL) is added and the solution is evaporated (34 °C, 70 mmHg). *n*-Hexane (300 mL) is added at ambient temperature and the white solid that precipitates is removed in a sintered-glass Büchner funnel (diameter 80 mm) using suction filtration and washed with *n*-hexane (500 mL, ambient temperature), and the filtrate is evaporated (34 °C, 70 mmHg) to give 30.5 g of the crude product as a clear, yellow oil. This material is purified by silica gel column chromatography (elution with ethyl acetate/hexane) (Notes 15 and 16). The combined eluents are concentrated by rotary evaporation (40 °C, 27 mmHg) and then dried at 25 °C (3 mmHg) for 2 h to furnish 21.7 g (47.0 mmol, 88%) of (*Z*)-*N*-(4-(benzyloxy)but-2-en-1-yl)-*N*-((*tert*-butyldimethylsilyl)-oxy)-4- methylbenzenesulfonamide (**4**) as a clear colorless oil (Note 17).

   C. *(2Z)-4-(Benzyloxy)but-2-enal oxime (5).* An oven-dried, 500-mL, three-necked round-bottomed flask is equipped with a 3.0-cm oval Teflon-coated magnetic stir bar, a rubber septum, an internal thermometer, and fitted with nitrogen gas inlet adaptor. The flask is evacuated and refilled with nitrogen three times, then charged with (*Z*)-*N*-(4-(benzyloxy)but-2-en-1-yl)-*N*-((*tert*-butyldimethylsilyl)oxy)-4-methylbenzenesulfonamide (**4**) (21.7 g, 47.0 mmol, 1.00 equiv) and 130 mL of acetonitrile (Note 18). The septum is removed temporarily and cesium fluoride (9.29 g, 61.1 mmol, 1.30 equiv) (Note 19) is added in one portion. The resulting white suspension is stirred (Note 20) and heated in an oil bath for 2 h, so that the internal temperature is maintained at 50 °C (Note 21). The reaction mixture is cooled in an ice bath for 20 min and saturated aqueous sodium bicarbonate solution (250 mL) is added and the reaction mixture is poured into EtOAc (250 mL) in a 2-L separatory funnel. The layers are separated, and the aqueous phase is extracted with additional EtOAc (2 x 150 mL). The organic solutions are combined, washed with 150 mL of brine, and dried over sodium sulfate (30 g). Filtration through a cotton plug, rinsing with 70 mL of EtOAc and concentration on the rotary evaporator (35 °C, 42 mmHg, water bath) provides the crude oil (11.4 g). This material is purified by silica gel column chromatography (Note 22). The combined eluents are concentrated by rotary evaporation (37 °C, 34 mmHg) and then dried at 25 °C (3 mmHg) for 2 h to furnish 8.49 g (44.4 mmol, 94%) of (2*Z*)-4-(benzyloxy)but-2-enal oxime (**5**) as a clear colorless oil (Note 23).

2. Notes

   1.   The submitters purchased hydroxylamine hydrochloride (98.0%) from Kanto Chemical Co., Inc., reagent grade DMF (>99.0%) from Wako Pure Chemical Industries, Ltd. and *tert*-butyldimethylsilyl chloride (>98.0%)

from Tokyo Chemical Industry Co., Ltd. and used as received. The checkers purchased hydroxylamine hydrochloride (99.0%) from Sigma-Aldrich Chemical Company Inc., anhydrous DMF (99.8%) from Sigma-Aldrich Chemical Company Inc. and *tert*-butyldimethylsilyl chloride (98.0%) from Acros Organics Co. and used as received.

2. The submitters reported a stirring speed of 900 rpm.

3. The submitters purchased triethylamine (>99.0%) from Kanto Chemical Co., Inc. and used as received. The checkers purchased triethylamine (>99.0%) from Sigma-Aldrich Chemical Company Inc. and used as received.

4. The submitters reported it is necessary to stir the mixture vigorously for at least 50 minutes after the internal temperature reached ambient temperature for reproducible results. The checkers confirmed that the consumption of the starting material **1** could not be monitored by TLC analysis.

5. The submitters purchased *p*-toluenesulfonyl chloride (>99.0%) from Tokyo Chemical Industry Co., Ltd. and used as received. The checkers purchased *p*-toluenesulfonyl chloride (>99.0%) from Sigma-Aldrich Chemical Company Inc., and used as received.

6. The submitters reported it is necessary to stir the mixture for at least 30 minutes after the internal temperature reached ambient temperature for reproducible results.

7. The consumption of *p*-toluenesulfonyl chloride was monitored by TLC analysis on Whatman Ltd. silica gel 60 F_{254} plates (0.25 mm, aluminum-backed, visualized with 254 nm UV lamp and stained with phosphomolybdic acid) using 25% ethyl acetate in *n*-hexane as an eluant. *p*-Toluenesulfonyl chloride had $R_f = 0.67$ (UV active, black after staining) and *O*-TBS-*N*-tosylhydroxylamine (**2**) had $R_f = 0.54$ (UV active, black after staining).

8. The submitters observed the development of crystals during evaporation. The checkers did not observe crystals during evaporation.

9. The submitters observed that the appearance of the crystals depends on the workup procedure. Without washing with aqueous citric acid solution, low-density wooly crystals were obtained. Reagent **2** can be stored without decomposition over 3 months in the refrigerator (–30 °C) under argon atmosphere. A yield of 65% was obtained when the reaction was performed at half scale. The product displayed the following physicochemical properties: mp 122–123 °C; IR (neat, cm^{-1}) 3213, 2957,

2931, 2889, 2859, 1598, 1169; ^{1}H NMR (CDCl$_3$, 400 MHz) δ: 0.18 (s, 6 H), 0.87 (s, 9 H), 2.45 (s, 3 H), 6.43 (s, 1 H), 7.34 (d, J = 8.1 Hz, 2 H), 7.80 (d, J = 8.2 Hz, 2 H); ^{13}C NMR (CDCl$_3$, 100 MHz) δ: –5.3, 18.0, 21.8, 26.0, 129.0, 129.6, 133.3, 144.8; HRMS calcd for C$_{13}$H$_{23}$NO$_3$SSiNa ([M + Na]$^+$) 324.1066, found 324.1075; Anal. Calcd. for C$_{13}$H$_{23}$NO$_3$SSi: C, 51.79; H, 7.69; N, 4.65; Found: C, 51.65; H, 7.69; N, 4.66 (Submitters-in house analysis); Found C, 51.26; H, 7.29; N, 4.86 (Checkers-sent out for analysis).

10. The submitters purchased triphenylphosphine (>95.0%) from Tokyo Chemical Industry Co., Ltd., dehydrated toluene (>99.5%) and dehydrated THF (>99.5%) from Kanto Chemical Co., Inc. and used the material as received. The checkers purchased triphenylphosphine (99.0%), anhydrous toluene (99.8%) and anhydrous THF (>99.9%) from Sigma-Aldrich Chemical Company Inc. and used as received.

11. The submitters and the checkers purchased cis-4-benzyloxy-2-buten-1-ol (3) (>95%) from Sigma-Aldrich Chemical Company Inc. and used the material as received.

12. The submitters reported a stirring speed of 540 rpm.

13. The submitters purchased diethyl azodicarboxylate (2.2 M in toluene, 40%) from Tokyo Chemical Industry Co., Ltd. and used as received. The checkers purchased diethyl azodicarboxylate (2.2 M in toluene, 40%) from Chem-Impex International, Inc. and used as received.

14. The consumption of the starting material was monitored by TLC analysis on Whatman Ltd. silica gel 60 F$_{254}$ plates (0.25 mm, aluminum-backed, visualized with 254 nm UV lamp and stained with phosphomolybdic acid) using 25% ethyl acetate in n-hexane as an eluant. Ph$_3$P had R$_f$ = 0.79 (UV active, white after staining), 4 had R$_f$ = 0.68 (UV active, black after staining), 2 had R$_f$ = 0.55 (UV active, black after staining) and 3 had R$_f$ = 0.18 (UV weakly active, black after staining).

15. Silica gel (pH range: 6.5–7.5) was purchased from Sorbent Technologies, Inc. (40-63 μm).

16. The crude material is dissolved in n-hexane (10 mL) and then is charged onto a column (diameter = 6.5 cm, height = 5.5 cm) of 170 g (300 mL) of silica gel. The column was eluted with n-hexane (1500 mL) to elute Ph$_3$P, n-hexane/EtOAc (20:1) (300 mL) and n-hexane/EtOAc (10:1) (1000 mL) and 30-mL fractions were collected. Fractions 52-87 were collected.

17. A yield of 75% was obtained when the reaction was performed at half scale. The product displayed the following physicochemical properties:

Org. Synth. 2012, 89, 480-490

IR (neat, cm^{-1}) 2954, 2930, 2887, 2858, 1597, 1360, 1170; ^{1}H NMR (CDCl$_3$, 400 MHz) δ: 0.26 (s, 6 H), 0.90 (s, 9 H), 2.45 (s, 3 H), 3.66 (d, J = 6.1 Hz, 2 H), 3.98 (d, J = 6.1 Hz, 2 H), 4.45 (s, 2 H), 5.58 (dt, J = 12.3, 6.2 Hz, 1 H), 5.70 (dt, J = 12.3, 6.2 Hz, 1 H), 7.25–7.35 (m, 7 H), 7.70 (d, J = 8.2 Hz, 2 H); ^{13}C NMR (CDCl$_3$, 100 MHz) δ: 4.1, 18.2, 21.8, 26.1, 52.9, 65.8, 72.5, 126.4, 127.76, 127.80, 128.5, 129.4, 130.0, 130.20, 130.26, 138.1, 144.7; HRMS calcd for C$_{24}$H$_{35}$NO$_4$SSiNa ([M + Na]$^+$) 484.1954, found 484.1978. Anal. Calcd. for C$_{24}$H$_{35}$NO$_4$SSi: C, 62.44; H, 7.64; N, 3.03; Found: C, 62.36; H, 7.53; N, 3.13.

18. The submitters purchased dehydrated acetonitrile (99.5%) from Kanto Chemical Co., Inc. and used as received. The checkers purchased anhydrous acetonitrile (99.8%) from Sigma-Aldrich Chemical Company Inc. and used as received.

19. The submitters purchased cesium fluoride (97%) from Wako Pure Chemical Industries, Ltd. and used as received. The checkers purchased cesium fluoride (99.0%) from Sigma-Aldrich Chemical Company Inc. and used as received.

20. The submitters reported a stirring speed of 540 rpm.

21. The consumption of the starting material was monitored by TLC on Whatman Ltd. silica gel 60 F$_{254}$ plates (0.25 mm, aluminum-backed, visualized with 254 nm UV lamp and stained with cerium phosphomolybdic acid) using 25% ethyl acetate in n-hexane as an eluant. Compound **5** had R$_f$ = 0.33 and 0.23 (*E-Z* mixture, UV active, black after staining).

22. The crude material is dissolved in 10 mL of n-hexane/EtOAc (6:1) and then is charged onto a column (diameter = 6.5 cm, height = 3.7 cm) of 86 g (200 mL) of silica gel. The column was eluted with n-hexane/EtOAc (6:1) and 30-mL fractions were collected. Fractions 24-92 were collected.

23. A yield of 95% was obtained when the reaction was performed at half scale. The product was isolated as a 1.4:1 *E-Z* isomeric mixture and displayed the following physicochemical properties: IR (neat, cm^{-1}) 3286, 3063, 3032, 2859, 1454; ^{1}H NMR (CDCl$_3$, 400 MHz) δ: 4.22 (dd, J = 6.4, 1.8 Hz, 2 H *major*), 4.24 (dd, J = 6.4, 1.8 Hz, 2 H *minor*), 4.55 (s, 3 H *major*), 4.56 (s, 3 H *minor*), 5.99–6.10 (m, 1 H *minor* and 1 H *major*), 6.23 (t, J = 11.3 Hz, 1 H *major*), 6.83 (td, J = 10.0, 1.4 Hz, 1 H *minor*), 7.29–7.38 (m, 5 H *major* and 5 H *minor*), 7.46 (d, J = 10.4 Hz, 1 H *minor*), 8.11 (d, J = 10.4 Hz, 1 H *major*); ^{13}C NMR (CDCl$_3$, 100 MHz) δ: 65.98, 65.99, 72.63, 72.70, 118.1, 124.2, 128.1, 128.6, 134.5, 136.3, 137.6, 137.7, 144.1, 147.5; HRMS calcd for C$_{11}$H$_{14}$NO$_2$ ([M + H]$^+$) 192.1025, found 192.1013. Anal.

Calcd. for $C_{11}H_{13}NO_2$: C, 62.09; H, 6.85; N, 7.32; Found: C, 61.68; H, 6.39; N, 7.54.

Safety and Waste Disposal Information

All hazardous materials should be handled and disposed of in accordance with "Prudent Practices in the Laboratory"; National Academies Press; Washington, DC, 2011.

3. Discussion

The oxime functionality is most often prepared by the reactions between carbonyl compounds and hydroxylamine. This sometimes suffers from a problem if an aldehyde or a ketone is not stable enough or possesses low reactivity for the transformation. This drawback could be circumvented if the oximes are prepared without going through aldehydes or ketones. We envisaged that the elimination of a sulfonyl group attached to a hydroxylamine moiety in **6** would furnish access to an alkyl nitroso intermediate **7**, which would in turn tautomerize quickly to give oxime **8** (Scheme 1). This transformation bypasses the carbonyl intermediates and thus would not be troubled by the problems concerning the reactivity of a carbonyl intermediate. The procedure presented here is a convenient method for the synthesis of (2Z)-4-(benzyloxy)but-2-enal oxime (**5**) from the allylic alcohol **3** and *O*-TBS-*N*-tosylhydroxylamine (**2**)[3,4] minimal isomerization from *Z* to *E* within the unsaturated system.[4] The procedure utilizing a Mitsunobu reaction and subsequent treatment with cesium fluoride was found to be applicable to the preparation of various oximes shown in Table 1.

Scheme 1. Proposed mechanism.

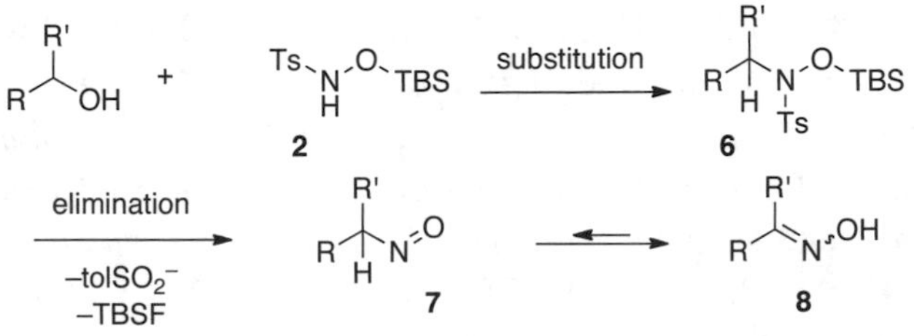

Bromides, tosylates, and mesylates can also be used as precursors in the one-pot transformation to oximes by means of an S_N2 reaction and subsequent elimination (Scheme 2).

Table 1. Substrate Scope.

entry	oxime	yield[b]	entry	oxime	yield[b]
1		99 / 99	8		99 / 94
2		91 / 92	9		100 / 92[e]
3		100 / 82	10		95 / 95[e]
4		97 / 98	11		85 / 99
5		84[c] / 90	12		99[f] / 94
6		91 / 99	13		99 / 89
7		97[d] / 91	14		93 / 98

[a] Standard conditions for the Mitsunobu reaction: 1.1 equiv of alcohol, 1.0 equiv of TsNHOTBS, 1.5 equiv of DEAD, 2.0 equiv of Ph$_3$P, toluene-THF (3:1, 0.2 M), 0 °C. Standard conditions for oxime formation: 2.0 equiv of CsF, MeCN (0.1 M), 60 °C. [b] Isolated yields for Mitsunobu reaction / oxime formation reaction. [c] Mitsunobu reaction was conducted with 2.5 equiv of DEAD and 3.0 equiv of Ph$_3$P. [d] A minimal amount (1.05 equiv) of DEAD was used because of the instability of the product with the reagent. [e] 2.0 equiv of AcOH was added for buffering the basicity. [f] Mitsunobu reaction was conducted with 1.5 equiv of alcohol and 2.0 equiv of DEAD at 40 °C.

Scheme 2. From Alkyl Bromide and Sulfonate.

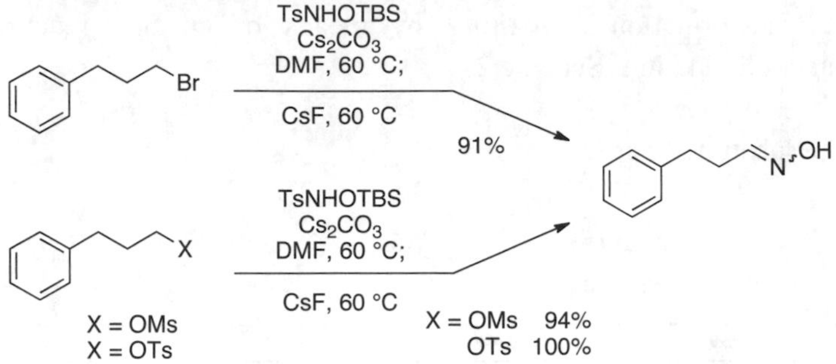

1. Graduate School of Pharmaceutical Sciences, University of Tokyo, Hongo 7-3-1, Bunkyo-ku, Tokyo 113-0033, Japan; Email: fukuyama@mol.f.u-tokyo.ac.jp. We are grateful for the Grant-in-Aid (15109001 and 18890048) from the Ministry of Education, Culture, Sports, Science, and Technology of Japan.
2. Kitahara, K.; Toma, T.; Shimokawa, J.; Fukuyama, T. *Org. Lett.* **2008**, *10*, 2259.
3. For selected applications of this method, see; (a) Rai, G.; Thomas, C. J.; Leister, W.; Maloney, D. J. *Tetrahedron Lett.* **2009**, *50*, 1710; (b) Singh, M. K.; Lakshman, M. K. *J. Org. Chem.* **2009**, *74*, 3079; (c) Su, D.-Y.; Wang, X.-Y.; Shao, C.-W.; Xu, J.-M.; Zhu, R.; Hu, Y.-F. *J. Org. Chem.* **2011**, *76*, 188.
4. Enev, V. S.; Drescher, M.; Kählig, H.; Mulzer, J. *Synlett* **2005**, 2227. According to the authors, this was the first example of a stereoselective synthesis of a *cis*-α,β-unsaturated oxime.

Appendix
Chemical Abstracts Nomenclature; (Registry Number)

Hydroxylamine hydrochloride: Hydroxylamine, hydrochloride (1:1); (5470-11-1)

tert-Butyldimethylsilyl chloride: Silane, chloro(1,1-dimethylethyl)dimethyl-; (18162-48-6)

Dimethylformamide: Formamide, *N*,*N*-dimethyl-; (68-12-2)

Triethylamine: Ethanamine, *N*,*N*-diethyl-; (121-44-8)

p-Toluenesulfonyl chloride: Benzenesulfonyl chloride, 4-mehyl-; (98-59-9)

Org. Synth. **2012**, *89*, 480-490

O-TBS-*N*-tosylhydroxylamine: Benzenesulfonamide,
   N-[[(1,1-dimethylethyl)dimethylsilyl]oxy]-4-methyl-; (1028432-04-3)
Ph₃P: Phosphine, triphenyl-; (603-35-0)
4-Benxyloxy-2-buten-1-ol: 2-Buten-1-ol, 4-(phenylmethoxy)-, (2*Z*)-;
   (81028-03-7)
DEAD: 1,2-Diazenedicarboxylic acid, 1,2-diethyl ester; (1972-28-7)
Cesium fluoride; (13400-13-0)

Katsushi Kitahara was born in 1984 in Saitama, Japan. He received his B.S. in 2007 from Tokyo University of Science, where he carried out his undergraduate research in the laboratories of Susumu Kobayashi. The same year he started his doctoral study at the University of Tokyo under the supervision of Professor Tohru Fukuyama. His current interest is enantioselective total synthesis of complex natural products.

Tatsuya Toma was born in 1984 in Saitama, Japan. He graduated in 2007 and received his M. S. degree in 2009 from the University of Tokyo. The same year he started his Ph. D. study under the supervision of Professor Tohru Fukuyama. His current interest is enantioselective total synthesis of complex natural products.

Jun Shimokawa was born in 1980 in Tokyo. He performed his Ph.D studies under the direction of Professor Tohru Fukuyama at the University of Tokyo where he conducted the research on total syntheses of complex natural products. In 2006, he was appointed as Assistant Professor in the same group. His research efforts focus on the development of novel synthetic methodology and applications to the synthesis of complex molecule.

Tohru Fukuyama received his Ph.D. in 1977 from Harvard University with Yoshito Kishi. He remained in Kishi's group as a postdoctoral fellow until 1978 when he was appointed as Assistant Professor of Chemistry at Rice University. After seventeen years on the faculty at Rice, he returned to his home country and joined the faculty of the University of Tokyo in 1995, where he is currently Professor of Pharmaceutical Sciences. He has primarily been involved in the total synthesis of complex natural products of biological and medicinal importance. He often chooses target molecules that require development of new concepts in synthetic design and/or new methodology for their total synthesis.

Erica Benedetti was born in 1984 in Como, Italy. She performed her PhD studies at the Insubria University of Como (advisors: Dr. Andrea Penoni and Pr. Giovanni Palmisano), in co-tutorship with the University Pierre et Marie Curie of Paris (advisors: Dr Jean-Philippe Goddard, Pr. Louis Fensterbank and Pr. Max Malacria), where she conducted researches on transition-metal catalyzed cycloisomerization reactions. She is now a Post-doc associate at the University of Pittsburgh, under the direction of Professor Kay M. Brummond. Her current research involves the synthesis of new fluorescent compounds.

490

Preparation of Cyclobutenone

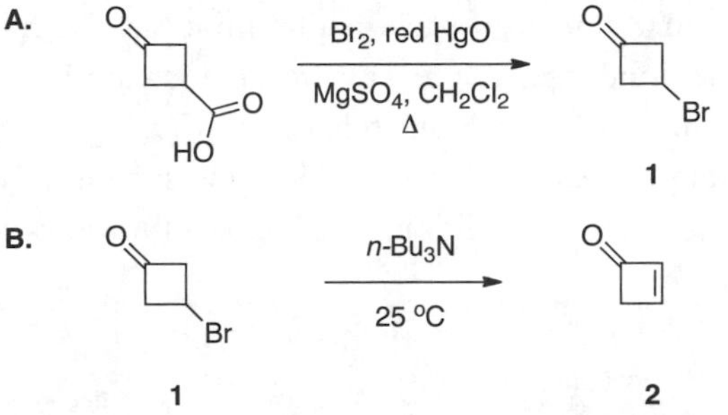

A.

B.

Submitted by Audrey G. Ross, Xiaohua Li, and Samuel J. Danishefsky.[1]
Checked by Markus Roggen and Erick M. Carreira.

> *Caution! This procedure should be carried out in an efficient fume hood. Precautions should be taken to avoid inhalation of or contact with red mercury oxide, which is highly toxic, bromine, which is corrosive and toxic, tri-n-butylamine, which is toxic especially to aquatic organisms, and cyclobutenone, which readily polymerizes and has unstudied health effects.*

1. Procedure

A.   *3-Bromocyclobutanone.*   A 500-mL, single-necked, round-bottomed flask with a 24/40 ground glass joint is equipped with a 6-cm Teflon-coated magnetic stir bar (Note 1) and charged with 3-oxocyclobutanecarboxylic acid (20 g, 175 mmol, 1 equiv) (Note 2). Dichloromethane is added (250 mL, 0.7 M) (Note 3) to dissolve the acid at ambient temperature (23 °C). While stirring gently at 150 rpm, anhydrous magnesium sulfate (21.1 g, 175 mmol, 1 equiv) is added to the reaction, followed carefully by red mercury oxide (56.9 g, 263 mmol, 1.5 equiv). A reflux condenser is assembled (Note 4) and capped with a 24/40 rubber septum, through which the reaction is maintained under argon atmosphere (Note 5). The heterogeneous mixture is then stirred at 250–300 rpm and brought to vigorous reflux (oil bath temperature 46 °C) for 5 to 10 min to ensure freely flowing solids (Note 6).

Org. Synth. **2012**, *89*, 491-500
Published on the Web 5/16/2012
© 2012 Organic Syntheses, Inc.

Separately, neat bromine (13.5 mL, 263 mmol, 1.5 equiv) is taken up in a glass-syringe with a needle approximately as long as the reflux condenser (12 inches). The needle is threaded through the rubber septum at the top of the reflux condenser such that bromine will be released near or at the bottom of the condenser (Note 7). Bromine is carefully added dropwise over about 30 min to control the release of CO_2 (g). (Figure 1) Three disposable 18-gauge needles are inserted into the septum to vent the reaction for the duration of bromine addition, and can be adjusted as needed.

Figure 1. Reaction Set-up for Step A

The reaction color gradually fades from bright orange to pale orange-yellow. After 3 h the reaction progress is checked by TLC (Note 8). As TLC

492

indicates incomplete conversion, additional bromine (7 mL, 136 mmol, 0.775 equiv) is added via syringe and the reaction is continued for a total of 6 h. After completion the reaction is cooled to ambient temperature (23 °C) and filtered through a pad of celite and silica gel (Note 9) with the aid of additional methylene chloride (250 mL). The clear, colorless filtrate is placed in a 1-L separatory funnel and washed with saturated aqueous sodium bicarbonate (3 x 200 mL). The combined milky aqueous washes are back-extracted with methylene chloride (200 mL) (Note 10). The combined organic extracts are washed with water (200 mL) and then brine (200 mL). The solution is dried over approximately 5 g of magnesium sulfate, filtered, and concentrated by rotary evaporator (25 °C, 100 to 30 mmHg) to give crude bromide 1 as a clear pale yellow liquid (23.6 g, 90% yield) (Notes 11, 12, and 13).

B. *Cyclobutenone.* A 250-mL, single-necked, round-bottomed flask equipped with a 5-cm Teflon-coated magnetic stir bar (Note 1) is charged with tri-*n*-butylamine (54.5 mL, 228 mmol, 2 equiv) (Notes 2 and 14) at ambient temperature (23 °C). Neat 3-bromocyclobutenone 1 is added to the amine dropwise over 30 min (17.0 g, 114 mmol, 1 equiv) and settles into a lower layer. The reaction mixture is stirred for 1 h at ambient temperature, and can be stirred for 2 h without adverse effect. No precautions are taken to exclude light or monitor reaction progress by TLC.

A 100-mL, single-necked, pear-bottomed flask is connected to the reaction flask via a short path distillation head with Vigreux indents (Notes 1 and 15). This receiving flask is cooled to –78 °C in a dry ice/acetone bath, and the reaction flask is placed in an oil bath held constant at 25 °C. (Figure 2) As the reaction flask is vigorously stirred at 250 rpm, pressure is carefully reduced to 2.5 mmHg and held for 20 min to ensure that cyclobutenone 2 does not foam over too rapidly (Note 16). At this time the receiving flask contains semi-frozen cyclobutenone, which rapidly melts to a clear and faint yellow liquid upon moving the receiving flask to ambient temperature (4.5–4.7 g, 58–60% yield) (Note 17). In this preparation, crude melted cyclobutenone 2 is dissolved in deuterated chloroform for use as a 1M solution (Notes 18 and 19). If desired, the cyclobutenone 2 can be briefly stored neat at –78 °C, while the product purity is quickly checked by ¹H NMR (Notes 20, 21, and 22).

Figure 2. Reaction set-up for Step B

2. Notes

1.   Glassware and stir bars were dried for at least 12 h in a 140 °C oven, cooled to ambient temperature under reduced pressure, then backfilled with argon atmosphere.

2.   3-Oxocyclobutanecarboxylic acid was purchased from Advanced ChemBlocks, Inc. and used without further purification. As a precaution the acid was stored in a refrigerator at 0 °C. Alternatively, the acid can be prepared according to literature procedure.[2] All other chemicals were used as received and stored at ambient temperature on the benchtop. Anhydrous magnesium sulfate and deuterated chloroform were purchased from Fisher Scientific International, Inc. SiliCycle SiliaFlash P60 (230-400 mesh) was

494

used.   All other chemicals were purchased from Aldrich Chemical Company, Inc.

3.   Methylene chloride (99.5%, stabilized, certified ACS) was purchased from Fisher Scientific International, Inc., and freshly distilled over calcium hydride before use.   For ease of scale, this modified Hunsdiecker reaction is notably more concentrated than those previously reported.[3]   Although some conversion of methylene chloride to bromodichloromethane (^{1}H NMR δ 7.19 in CDCl$_3$) occurs under the reaction conditions, methylene chloride is used in preference to carbon tetrachloride for reduced toxicity and easier isolation of the volatile bromide **1**.

4.   The reflux condenser was dried for at least 12 h in a 140 °C oven, then cooled to ambient temperature.   The condenser was attached to the reaction flask via a 24/40 ground glass joint lubricated with high-vacuum grease purchased from Dow Corning.

5.   The submitters recommend using a wide-gauge needle to provide pressure exchange via the argon inlet.

6.   Vigorous stirring and a large stir bar are crucial for this reaction. One can break up clumps of solids with a large spatula if needed, ideally before adding bromine.

7.   Excess reagent is required to account for volatility of bromine, and that some bromine and red mercury oxide will react with the solvent.

8.   Reaction progress can be monitored by TLC (bromide **1** R_f = 0.63 in CH$_2$Cl$_2$, visualized with KMnO$_4$ stain; 3-oxocyclobutanecarboxylic acid R_f = baseline in CH$_2$Cl$_2$, R_f = 0.41 in diethyl ether, visualized with KMnO$_4$ stain).   To reduce exposure risk, the submitters advise that the reaction as allowed to proceed for a set time of 3 h.   The submitters report full conversion after 3 h A reaction allowed to run for 1.5 h provided bromide **1** in a yield of 54%.

9.   Vacuum filtration performed on a large (>150 mL) fritted glass funnel filled with a 1-cm layer of celite and silica.   Substantial solids are collected and treated as mercury-containing hazardous waste.

10.   Spent aqueous washes are combined with and treated as mercury-containing hazardous waste.

11. Crude bromide **1** can be carried forward, since non-volatile impurities will not contaminate cyclobutenone **2**.   For characterization purposes, a sample of the volatile bromide **1** was quickly purified by silica gel chromatography to remove trace bromodichloromethane and an impurity at R_f = 0.44 (CH$_2$Cl$_2$, visualized with KMnO$_4$ stain). Bromide **1** (0.2 g) was

charged on a small glass column (1 x 10 cm) containing 1.23 g silica gel (see Note 2). Product was eluted with 10 mL of 9:1 pentane/diethyl ether, and collected in 0.6 mL fractions. Product eluted in fractions 8 – 11. Distillation of bromide **1** is reported in the literature (22–25 °C at 0.5–0.25 mmHg).[4]

12. 3-Bromocyclobutanone (**1**) characterization (purified material): $R_f =$ 0.63 (CH$_2$Cl$_2$, visualized with KMnO$_4$ stain); ^{1}H-NMR (400 MHz, CDCl$_3$) δ: 3.43–3.52 (m, 2 H), 3.72–3.81 (m, 2 H), 4.54 (tt, J = 7.9, 5.0 Hz, 1 H).^{13}C NMR (101 MHz, CDCl$_3$) δ: 29.0, 60.4, 203.2; IR (neat): 1787, 1372, 1233, 1139, 1084, 996, 967, 838, 700 cm^{-1}. MS *m/z* (relative intensity): 41.0339 (100%), 69.0335 (60.99%), 118.9489 (5.83%), 120.9484 (6.73%). Anal. Calcd for C$_4$H$_5$BrO: C, 32.25; H, 3.38.   Found: C, 32.07; H, 3.42 (The checkers found it necessary to keep purified **1** at –78 °C prior to elemental analysis.)

13.   Bromide **1** is moderately stable to neat storage in a 0 °C refrigerator. Although the material develops a yellow to orange color, NMR analysis reveals useable stability for up to two weeks. When left exposed to light on the benchtop at ambient temperatures, neat samples of bromide **1** decomposed to a black tar within two days.

14. Excess amine (2 equiv) is used in this procedure as solvent and reagent. Diethylene glycol diethyl ether (diethyl carbitol) can be used as a solvent to allow stoichiometric amine and a homogeneous solution.[3,4] To obtain more pure final product, the ether is forgone as suggested in an earlier preparation of cyclobutenone.[4]

15. The short-path distillation head was dried for at least 12 h in a 140 °C oven, then cooled to ambient temperature. All ground glass joints were lubricated with high-vacuum grease purchased from Dow Corning.

16. The boiling point of tri-*n*-butylamine is 216 °C. The boiling point range of diethylene glycol diethyl ether (diethyl carbitol) is 180–190 °C. If one prefers to use the ether in order to use stoichiometric amine, more care or a second distillation may be necessary to obtain optimally pure final product.

17. Cyclobutenone **2** characterization: R_f = 0.43 (4:1 pentane/ether, diffuse spot visualized with KMnO$_4$ stain); ^{1}H-NMR (400 MHz,) δ: 3.34 (t, J = 0.8 Hz, 2 H), 6.15 (dt, J = 2.8, 0.7 Hz, 1 H), 8.21 (dt, J = 2.9, 0.9 Hz, 1 H); ^{13}C NMR (101 MHz; CDCl$_3$) δ: 51.1, 141.4, 162.6, 189.9; [M] calcd for C$_4$H$_4$O: 68.0262. Found: 68.0252. IR (in CDCl$_3$): 2931, 1770, 1524, 1283, 1037 cm^{-1}.   GC analysis of a 1M solution of cyclobutenone in CDCl$_3$ was performed on a Shimadzu GC 17A ver. 3 instrument using a SupelCo SPB-5

column (30.0 m x 0.25 mm) with a flame ionization detector. The assay is held steady at 50 °C after injection at 220 °C. The observed retention times are 3.99 min for cyclobutenone, and 5.88 min for deuterated chloroform.

18. Solutions of cyclobutenone in deuterated chloroform are quite stable to storage, with no appreciable decomposition noted after one month in a 0 °C refrigerator. Substantial decomposition is noted by eight months under the same storage conditions. Solutions were stored in deuterated chloroform bottles and did not require argon atmosphere or seals. Other solvents, concentrations, and temperatures for storage were not examined.

19. All waste from this reaction is treated as hazardous amine waste toxic to aquatic organisms, including washes of distillation glassware.

20. Neat cyclobutenone **2** rapidly polymerizes at ambient temperature, and will polymerize at –78 °C. The submitters have kept cyclobutenone **2** neat at –78 °C for about 20 min without appreciable unwanted reaction. Cyclobutenone **2** smells like superglue and has irritating vapors.

21. If necessary, a second distillation can be performed to leave behind less volatile contaminants such as tri-*n*-butylamine. This must be done rapidly to minimize polymerization and loss of yield.

22. The submitters report a separate run which used 23.0 g of bromide **1** provided cyclobutenone **2** (6.86 g, 100.7 mmol, 65% yield) under the same reaction conditions.

Safety and Waste Disposal Information

All hazardous materials should be handled and disposed of in accordance with "Prudent Practices in the Laboratory"; National Academics Press; Washington, DC, 2011.

3. Discussion

The difficulty in preparing and handling cyclobutenone has precluded its usage for many years. In 1954, Cope and coworkers suggested its generation as a byproduct from the decomposition of 2,4,6-cyclooctatrien-1-one Diels-Alder adducts, noting a volatile carbonyl-containing compound with a sharp odor.[5] Soon after, Vogel and Hasse attempted to liberate the proposed cyclobutenone by pyrolysis of the methyl ketal of the same adduct.[6] Treatment of isolated cyclobutenone dimethyl ketal with aqueous acid gave a mixture of 3-hydroxycyclobutanone, 3-methoxycyclobutanone, and 1,3,5-triactylbenzene. Thus isolation and direct confirmation of cyclobutenone was thwarted by its reactivity.

It was not until 1971 that cyclobutenone was first conclusively prepared, when Sieja was able to access the parent compound through two independent routes based on allene and ketene.[4] Despite its curious structure and potential applications due to its inherent strain, cyclobutenone received little attention in the following decades. It is feasible that the challenging synthetic routes and inability to prevent polymerization during overnight storage at −78 °C ensured that cyclobutenone would remain of limited practical utility.

In 2004, Helal and coworkers heeded Sieja's suggestion to prepare cyclobutenone for immediate use.[7] They treated 3-acetoxycyclobutanone with 1,8-diazabicyclo[5.4.0.]-undec-7-ene (DBU) and 4-nitroimidazole, resulting in a net exchange of acetate with 4-nitroimidazole. Given the generally weak reactivity of 4-nitroimidazole in displacement reactions as noted by the authors, it is presumed that this transformation occurs by the *in-situ* generation of cyclobutenone.

To make cyclobutenone available for basic study and uses in which it cannot be conveniently generated *in-situ*, Danishefsky and Li developed an updated synthesis of cyclobutenone in 2010.[3] This procedure represents an optimized and moderately scaled-up version of said route, by which cyclobutenone can be prepared in two steps from commercial material. The first step is a mild variant of a Cristol-Firth modified Hunsdiecker reaction.[8,9] Cyclobutenone **2** is then prepared by elimination of the newly formed bromide **1**, at which time the product is distilled out for isolation.

Importantly, it was found that cyclobutenone can be kept as a stock solution in deuterated chloroform, enabling significantly longer storage, further characterization, simple measurement and handling of the elusive title compound. Thus cyclobutenone can be prepared in high purity in approximately 58% yield. The ability to prepare, isolate, and store cyclobutenone has allowed its study as a highly reactive dienophile in Diels-Alder additions, a subject of experimental and computational interest.[3,10]

1. Columbia University, Department of Chemistry, Havemeyer Hall, 3000 Broadway, New York, NY 10027; sjd15@columbia.edu. Financial support was provided by the National Institutes of Health (HL 25848).
2. Pigou, P.E.; Schiesser, C. H. *J. Org. Chem.* **1988**, *53*, 3841–3843.

3. Li, X.; Danishefsky, S. J. *J. Am. Chem. Soc.* **2010**, *132*, 11004–11005.
4. Sieja, J. B. *J. Am. Chem. Soc.*, **1971**, *93*, 2481–2483.
5. Cope, A. C.; Schaeren, S. F.; Trumbull, E. R. *J. Am. Chem. Soc.* **1954**, *76*, 1096–1100.
6. Vogel, E.; Hasse, K. *Liebigs Ann.* **1958**, *615*, 22–29.
7. Helal, C. J.; Kang, Z; Lucas, J. C.; Bohall, B. R. *Org. Lett.* **2004**, *6*, 1853–1856.
8. Cristol, S. J.; Firth, W. C., Jr. *J. Org. Chem.* **1961**, *26*, 280.
9. Cornelius, L. A. M.; Bone, R. G. A.; Hastings, R. H.; Deardorff, M. A.; Scharlach, R. A.; Hauptmann, B. E.; Stankovic, C. S.; Pinnick, H. W. *J. Org. Chem.* **1993**, *58*, 3188–3190.
10. Paton, R. S.; Kim, S.; Ross, A. G.; Danishefsky, S. J.; Houk, K. N. *Angew. Chem. Int. Ed. Engl.* **2011**, *50*, 10366–10368; *Angew. Chem.* **2011**, *123*, 10550–10552.

Appendix
Chemical Abstracts Nomenclature; (Registry Number)

3-Oxocyclobutanecarboxylic acid (23761-23-1)
Mercury (II) oxide red (21908-53-2)
Bromine (7726-95-6)
3-Bromocyclobutanone (23761-24-2)
Tri-*n*-butylamine (102-82-9)
Cyclobutenone; 2-cyclobuten-1-one (32264-87-2)
Deuterated chloroform; chloroform-d; deuterochloroform (865-49-6)

Samuel J. Danishefsky received his B.S. degree at Yeshiva University. He did graduate research under the late Professor Peter Yates, and then joined the laboratory of Professor Gilbert Stork at Columbia University as an NIH Postdoctoral Associate. His first academic position was at the University of Pittsburgh. In 1980, he moved to Yale University and became Sterling Professor in 1990. In 1993, Professor Danishefsky moved to New York as Professor of Chemistry at Columbia University and Eugene Kettering Chair and Head of the Laboratory of Bioorganic Chemistry at Memorial Sloan-Kettering Cancer Center. He shared the Wolf Prize in Chemistry with Professor Stork in 1996.

Audrey G. Ross received her B.S. in chemistry from the University of California, Los Angeles in 2007, where she did undergraduate research under the supervision of Professor Kendall N. Houk. She is currently pursuing her Ph.D. at Columbia University in the laboratory of Professor Samuel Danishefsky. In her graduate work, she has studied the total synthesis of natural products, and cyclobutenone in Diels-Alder and intramolecular Diels-Alder reactions. Audrey is a National Science Foundation Predoctoral Fellow.

Xiaohua Li received her B.S. in chemistry from Tianjin University in 1998 and her M.S. in Chemical Engineering from Beijing Institute of Technology in 2001, China. She came to the United States afterwards and pursued her Ph.D. in organic chemistry under the supervision of Professor David Mootoo at Hunter College, CUNY. Her research focused on the development and application of new spiroketalization methods in natural product synthesis. Later she worked with Professor Samuel Danishefsky at Columbia University as a postdoctoral fellow on the preparation and use of cyclobutenone as dienophile in Diels-Alder reactions. Currently she is a visiting assistant professor at University of Toledo.

Marcus Roggen received his M. Sci. in Chemistry with First Class Honors in 2008 from Imperial College in London. During his undergraduate studies he performed research at the Nanyang Technological University in Singapore under the direction of Roderick Bates and at the ETH in Zurich University under the direction of Prof. Erick Carreira. In 2008 he entered the Ph.D. program at the ETH.

Convenient Synthesis of α-Diazoacetates from α-Bromoacetates and *N,N'*-Ditosylhydrazine: Preparation of Benzyl Diazoacetate

Submitted by Eiji Ideue, Tatsuya Toma, Jun Shimokawa and Tohru Fukuyama.[1]
Checked by Pablo E. Guzmán and Huw M. L. Davies.

> *Caution! Diazoacetates are toxic and potentially explosive. They must be handled with care. This preparation should be carried out in a well-ventilated hood, and should be conducted behind a safety shield.*

1. Procedure

A. *N,N'-Ditosylhydrazine (2)*. A 3-L, three-necked round-bottomed flask equipped with an internal thermocouple temperature probe, a 20-mL pressure-equalizing dropping funnel sealed with a rubber septum, and an overhead mechanical stirrer fitted with a 8.5 x 2.0 cm rounded Teflon blade (Note 1) is charged under air with *p*-toluenesulfonyl hydrazide (**1**) (25.8 g, 134 mmol, 1.00 equiv) and *p*-toluenesulfonyl chloride (33.2 g, 174 mmol, 1.30 equiv) in 134 mL of CH_2Cl_2 (Note 2). The suspension is stirred (165 rpm) in an ice bath while pyridine (14.1 mL, 13.8 g, 174 mmol, 1.30 equiv) is added dropwise from the dropping funnel over 5 min, such that the internal temperature does not exceed 10 °C. During the addition, the reaction mixture becomes homogenous and turns yellow. White precipitate is observed within 3 min and the reaction mixture is stirred for 30 min (Note 3). *n*-Hexane (200 mL) and H_2O (300 mL) are added and stirred in an ice bath for 15 min. The white precipitates are collected in a medium-porosity sintered-glass Büchner funnel (diameter 90 mm) using suction filtration and washed with ice-cooled Et_2O (200 mL), and then dried under vacuum (25 °C, 32 mmHg, 3 h). The solid thus obtained is dissolved in boiling acetone (320 mL), and H_2O (150 mL) is slowly added (Note 4). The mixture is then cooled in an ice bath for 1 h and the white precipitates are collected by Büchner funnel (diameter 90 mm) with a medium-porosity fritted disk using suction filtration. The precipitate is washed with ice-cooled Et_2O (100 mL) and dried under vacuum over P_2O_5 (25 °C, 10 mmHg, 3 h) to give the first crop of *N,N'*-ditosylhydrazine (**2**) (35.5 g, 104 mmol, 78%). A second crop of crystals (5.05 g, 14.8 mmol, 11%) is obtained by dissolving the concentrated mother liquor in hot acetone (60 mL) and H_2O (27 mL) followed by the same cooling procedure (Note 5).

B. *Benzyl bromoacetate (5)*. A flame-dried, 1-L, three-necked round-bottomed flask is equipped with a 4.5-cm egg-shaped, Teflon-coated, magnetic stir bar, an internal thermocouple temperature probe, a 20-mL pressure-equalizing dropping funnel sealed with a rubber septum, and fitted with argon gas inlet. The flask is flushed with argon and charged with sodium hydrogen carbonate (15.9 g, 189 mmol, 3.00 equiv), benzyl alcohol (6.50 mL, 6.80 g, 63.0 mmol, 1.00 equiv) and 310 mL of acetonitrile (Note 6). The suspension is stirred in an ice bath while bromoacetyl bromide (7.70 mL, 88.2 mmol, 1.40 equiv) (Note 7) is added dropwise from a dropping funnel over 15 min. The reaction mixture is stirred for 40 min and TLC analysis shows no starting material (Note 8). Water (30 mL) is added and the reaction mixture is poured into CH_2Cl_2 (300 mL) and H_2O (300 mL) in a 2-L separatory funnel. The layers are separated, and the aqueous phase is extracted with an additional 300 mL of CH_2Cl_2. The organic solutions are combined, washed with 300 mL of brine, and dried over sodium sulfate (80 g). Filtration through a cotton plug, rinsing with 50 mL of CH_2Cl_2 and concentration on the rotary evaporator (35 °C, 50 mmHg, water bath) provides 14.6 g of the crude product as a clear amber oil. This material is purified by silica gel column chromatography (elution with ethyl acetate/ *n*-hexane) (Note 9, 10). The combined fractions are concentrated by rotary evaporation (40 °C, 27 mmHg) and then at 25 °C, 3.0 mmHg for 1 h to furnish 14.1 g (61.6 mmol, 98%) of benzyl bromoacetate (5) as a clear colorless oil. (Note 11)

C. *Benzyl diazoacetate (6)*. A flame-dried, 1-L, three-necked round-bottomed flask is equipped with a 4.5-cm egg-shaped, Teflon-coated, magnetic stir bar, an internal thermocouple temperature probe, a 50-mL pressure-equalizing dropping funnel sealed with a rubber septum, and fitted with argon gas inlet. The flask is flushed with argon and charged with benzyl bromoacetate (5) (13.8 g, 60.2 mmol, 1.00 equiv) and 300 mL of THF (Note 12). The septum is removed temporarily and *N,N'*-ditosyl hydrazine (2) (30.8 g, 90.4 mmol, 1.50 equiv) is added. The suspension is stirred in an ice bath while DBU (35.9 mL, 36.6 g, 241 mmol, 4.00 equiv) is added dropwise from dropping funnel over 5 min, such that the internal temperature does not exceed 20 °C (Note 13). During the addition, the reaction mixture becomes homogenous and turns yellow. The reaction mixture is stirred for 30 min and TLC analysis shows no starting material (Note 14). Saturated aqueous sodium hydrogen carbonate solution (30 mL) is added and the reaction mixture is poured into Et$_2$O (200 mL) and H$_2$O (300 mL) in a 2-L separatory funnel. The layers are separated, and the aqueous phase is extracted with an additional 2 x 200 mL of Et$_2$O. The organic solutions are combined, washed with 200 mL of brine, and dried over sodium sulfate (80 g). Filtration through cotton plug, rinsing with 50 mL of Et$_2$O and concentration on the rotary evaporator (30 °C, 37 mmHg, water bath) provides the crude mixture. Ice-cooled Et$_2$O (15 mL) is added and the white precipitates are removed in a fine-porosity sintered-glass Büchner funnel (diameter 90 mm) using suction filtration. The precipitates are washed with ice-cooled Et$_2$O (10 mL), and the filtrate is evaporated (30 °C, 37 mmHg) to give 11.1 g of the crude product as a clear, yellow oil. This material is purified by silica gel column chromatography (elution with ethyl acetate/*n*-hexane) (Note 15, 16). The combined fractions are concentrated by rotary evaporation (35 °C, 30 mmHg) and then at 25 °C, 4.0 mmHg for 3 h to furnish 8.56 g, (48.6 mmol, 81%) of benzyl diazoacetate (6) as a clear bright yellow oil (Note 17).

2. Notes

1. When this procedure is run on smaller scale a Teflon-coated magnetic stirring bar can be used in place of the overhead mechanical stirrer.

2. *p*-Toluenesulfonyl hydrazide (1) (97%), *p*-toluenesulfonyl chloride

(99.0%), and HPLC grade CH_2Cl_2 (>99.8%) were purchased from Sigma-Aldrich and used as received. The submitters purchased *p*-toluenesulfonyl hydrazide (**1**) (95%) and reagent grade CH_2Cl_2 (>99.0%) from Wako Pure Chemical Industries, Ltd., and the materials were used as received *p*-Toluenesulfonyl chloride (99.0%) was purchased from Tokyo Chemical Industry Co., Ltd. and the material was used as received.

3.   The consumption of the starting material **1** was monitored by TLC analysis on Merck silica gel 60 F_{254} plates (0.25 mm, glass-backed, visualized with 254 nm UV lamp and stained with PMA) using 50% ethyl acetate in *n*-hexane as an eluent. *p*-Toluenesulfonyl hydrazide (**1**) had R_f = 0.24 (weakly UV active, blue after staining).

4.   The addition of H_2O induces the formation of a white precipitate.

5.   A yield of 87% was observed when the reaction was performed on half scale.   The product displayed the following physicochemical properties: mp 226–228 °C (decomp.); ^1H NMR (DMSO-d_6, 600 MHz) δ: 2.39 (s, 3 H), 7.38 (d, J = 8.4 Hz, 2 H), 7.64 (d, J = 8.4 Hz, 2 H), 9.58 (s, 1 H); ^{13}C NMR (DMSO-d_6, 150 MHz) δ: 21.1, 127.8, 129.5, 135.5, 143.5; IR (neat, cm^{-1}) 3228, 3202, 1596, 1329, 1185, 1164, 1088, 813; HRMS Calcd. for $C_{14}H_{17}N_2O_4S_2$ (M+H$^+$) 341.0624, Found 341.0624; Anal. Calcd. for $C_{14}H_{16}N_2O_4S_2$: C, 49.40; H, 4.74; N, 8.23; Found: C, 49.26; H, 4.62; N, 8.38.

6.   Sodium hydrogen carbonate (99.7%), dehydrated MeCN (99.8%), and benzyl alcohol (>99.0%) were purchased from Sigma-Aldrich and used as received. The submitters purchased sodium hydrogen carbonate (>99.5%) and dehydrated MeCN (>99.0%) from Wako Pure Chemical Industries, Ltd., and benzyl alcohol (>99.0%) from Kanto Chemical Co., Inc. All were used as received.

7.   Bromoacetyl bromide (98.0%) was purchased from Sigma-Aldrich and used as received. The submitters purchased bromoacetyl bromide (98.0%) from Tokyo Chemical Industry Co., Ltd. and used as received.

8.   The reaction was monitored by TLC analysis on Merck silica gel 60 F_{254} plates (0.25 mm, glass-backed, visualized with 254 nm UV lamp and stained with PMA) using *n*-hexane/EtOAc (6:1) as an eluent. The product **5** had R_f = 0.53 (weakly UV active, dark brown after staining); benzyl alcohol had R_f = 0.18 (weakly UV active, black after staining).

9.   Silica gel was purchased from Silicycles Inc. ultra pure 230–400 mesh silica gel, pH 7.0, H_2O content = 6%. The submitters purchased silica gel (acidic) from Kanto Chemical Co., Inc. (40-100 μm).

10. The crude material is dissolved in eluent (5 mL) and then is charged onto a column (diameter = 7 cm, height = 5.5 cm) of 106 g (250 mL) of silica gel. The column was eluted with *n*-hexane /EtOAc (20:1) and 100-mL fractions were collected. Fractions 2-20 were combined.

11. A yield of 98% was observed when the reaction was performed on half scale. Benzyl bromoacetate (5) displayed the following physicochemical properties: IR (neat, cm^{-1}) 1732, 1273; ^{1}H NMR (CDCl$_3$, 400 MHz) δ: 3.88 (s, 2 H), 5.21 (s, 2 H), 7.35-7.39 (m, 5H); ^{13}C NMR (CDCl$_3$, 100 MHz) δ: 26.0, 68.0, 128.5, 128.7, 128.8, 135.1, 167.1; HRMS calcd for C$_9$H$_9$Br$_1$O$_2$Na$_1$ (M+Na$^+$) 250.9678, found 250.9676. Anal. Calcd. for C$_9$H$_9$O$_2$Br$_1$: C, 47.19; H, 3.96; Found: C, 47.41; H, 4.11.

12. THF (HPLC grade, H$_2$O = 0.003 %) was purchased from Fisher Scientific Company and distilled from sodium benzophenone ketyl.

13. DBU (98%) was purchased from Sigma-Aldrich and used as received. The submitters purchased DBU (98%) from Tokyo Chemical Industry Co., Ltd. and used as received.

14. The reaction was monitored by TLC analysis on Merck silica gel 60 F$_{254}$ plates (0.25 mm, glass-backed, visualized with 254 nm UV lamp and stained with PMA) using *n*-hexane/EtOAc (6:1) as an eluent. The product **6** had R$_f$ = 0.44 (UV active, black after staining).

15. Silica gel was purchased from Silicycles Inc. ultra pure 230–400 mesh silica gel, pH 7.0, H$_2$O content = 6%. The submitters purchased silica gel (neutral) from Kanto Chemical Co., Inc. (40-100 μm).

16. The crude material is dissolved in eluent (5 mL) and then is charged onto a column (diameter = 7 cm, height = 5.5 cm) of 106 g (250 mL) of neutral silica gel. The column was eluted with *n*-hexane/EtOAc (20:1) and 100-mL fractions were collected. Fractions 5-30 were combined.

17. A yield of 84% was observed when the reaction was performed on half scale. Benzyl diazoacetate (6) displayed the following physicochemical properties: IR (neat, cm^{-1}) 2105, 1683; ^{1}H NMR (CDCl$_3$, 600 MHz) δ: 4.80 (bs, 1 H), 5.20 (s, 2 H), 7.32-7.39 (m, 5 H); ^{13}C NMR (CDCl$_3$, 150 MHz) δ: 46.4, 66.5, 128.3, 128.4, 128.7, 136.0, 166.8; HRMS calcd for C$_9$H$_8$N$_2$O$_2$Na$_1$(M+Na$^+$) 199.0478, found 199.0476; Anal. Calcd. for C$_9$H$_8$N$_2$O$_2$: C, 61.36; H, 4.58; N, 15.90; Found: C, 61.43; H, 4.54; N, 15.79.

All hazardous materials should be handled and disposed of in accordance with "Prudent Practices in the Laboratory"; National Academies Press; Washington, DC, 2011.

3. Discussion

The use of diazocarbonyl compounds is becoming widespread by the advent of such important reactions as cyclopropanations and C-H insertion reactions as widely applicable and synthetically viable methods. For the preparation of diazocarbonyl compounds, diazotransfer reaction is still the most conventional approach: the α-diazo group could be introduced into an ester or ketone[2] by treatment with sulfonyl azide. Various modifications and improvements to this method have thus been developed to date towards milder reaction conditions and wider substrate scope.[2,3]

Table 1. Substrate Scope.

entry	diazoacetates	yield[a]	entry	diazoacetates	yield[a]
1		88	7		74
2		71	8		77
3		64	9		52
4		62	10		72
5		78	11		79
6		67			

a Yields for two steps from alcohol.

Nevertheless, prior introduction of a temporary activating group is necessary and sometimes become problematic, especially for the preparation of simple diazoacetates.[4] As House's procedure is one of the reliable alternatives[5] for the preparation of α-diazoacetates, this method requires the multi-step preparation of a corrosive and moisture-sensitive reagent: the *p*-toluenesulfonylhydrazone of glyoxylic acid chloride. The procedure presented here is a convenient preparative method for benzyl diazoacetate[6] from the corresponding bromoacetate and *N,N′*-ditosylhydrazine (**2**), which can be applied to the preparation of various diazoacetates. One-step synthesis of the air-stable reagent **2**, the facile removal of the water-soluble wastes, and the facile availability of various bromoacetates make this methodology reliable as a safe and general synthetic procedure for the preparation of α-diazoacetates (Table 1). α-Diazoketone could also be synthesized if the corresponding α-bromoketone is available. A typical example is the synthesis of α-diazoacetophenone, which is often synthesized from benzoyl chloride and diazomethane. The present method provides a one-step synthesis of α-diazoacetophenone, which circumvents the use of toxic and explosive diazomethane (Scheme 1). The limitation of the method is the low reactivity toward secondary bromides, to which the sulfinate ion generated in situ attacks to afford a sulfone as a sole adduct.

Scheme 1. Synthesis of α-Diazoketone.

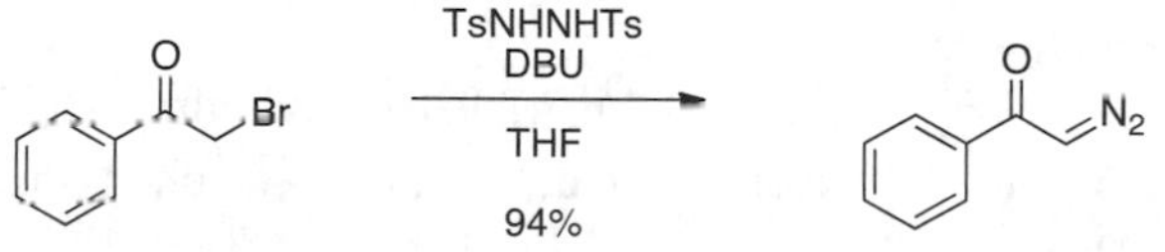

1. Graduate School of Pharmaceutical Sciences, University of Tokyo, Hongo 7-3-1, Bunkyo-ku, Tokyo 113-0033, Japan; Email: fukuyama@mol.f.u-tokyo.ac.jp. We are grateful for the Grant-in-Aid (15109001 and 18890048) from the Ministry of Education, Culture, Sports, Science, and Technology of Japan.

2. Regitz, M. *Angew. Chem. Int. Ed. Engl.* **1967**, *6*, 733.
3. For reviews, see: (a) Ye, T.; McKervey, M. A. *Chem. Rev.* **1994**, *94*, 1091. (b) Padwa, A.; Austin, D. *Angew. Chem. Int. Ed. Engl.* **1994**, *33*, 1797. (c) Doyle, M. P. *Chem. Rev.* **1986**, *86*, 919. (d) Maas, G. *Angew. Chem. Int. Ed.* **2009**, *48*, 8186. (e) Doyle, M. P.; McKervey, M. A.; Ye, T. *Modern Catalytic Methods for Organic Synthesis with Diazo*

Compounds from Cyclopropanes to Ylides; Wiley-Interscience: New York, 1998.

4. For selected modifications, see: (a) Danheiser, R. L.; Miller, R. F.; Brisbois, R. G. *Org. Synth.* **1996**, *73*, 134; (b) Taber, D. F.; Sheth, R. B.; Joshi, P. V. *J. Org. Chem.* **2005**, *70*, 2851.

5. (a) House, H. O.; Blankley, C. J. *J. Org. Chem.* **1968**, *33*, 53. (b) Blankley, C. J.; Sauter, F. J.; House, H. O. *Org. Synth.* **1972**, *49*, 22. (c) Corey and Myers reported the improved modification of this procedure; Corey, E. J.; Myers, A. G. *Tetrahedron Lett.* **1984**, *25*, 3559.

6. Toma, T.; Shimokawa, J.; Fukuyama, T. *Org. Lett.* **2007**, *9*, 3195.

Appendix
Chemical Abstracts Nomenclature; (Registry Number)

p-Toluenesulfonyl hydrazide: Benzenesulfonic acid, 4-methyl-, hydrazide; (1576-35-8)

p-Toluenesulfonyl chloride: Benzenesulfonyl chloride, 4-mehyl-; (98-59-9)

N,N'-Ditosylhydrazine: Benzenesulfonic acid, 4-methyl-, 2-[(4-Methylphenyl)sulfonyl]hydrazide; (14062-05-6)

Sodium hydrogen carbonate: Carbonic acid sodium salt (1:1); (463-79-6)

Benzyl alcohol: Benzenemethanol; (100-51-6)

Bromoacetyl bromide: Acetyl bromide, 2-bromo-; (598-21-0)

Benzyl bromoacetate: Acetic acid, 2-bromo-, phenylmethyl ester; (5437-45-6)

DBU: Pyrimido[1,2-a]azepine, 2,3,4,6,7,8,9,10-octahydro-; (6674-22-2)

Benzyl diazoacetate: Acetic acid, 2-diazo-, phenylmethyl ester; (52267-51-3)

Tohru Fukuyama received his Ph.D. in 1977 from Harvard University with Yoshito Kishi. He remained in Kishi's group as a postdoctoral fellow until 1978 when he was appointed as Assistant Professor of Chemistry at Rice University. After seventeen years on the faculty at Rice, he returned to his home country and joined the faculty of the University of Tokyo in 1995, where he is currently Professor of Pharmaceutical Sciences. He has primarily been involved in the total synthesis of complex natural products of biological and medicinal importance. He often chooses target molecules that require development of new concepts in synthetic design and/or new methodology for their total synthesis.

508

Eiji Ideue was born in Kagawa, Japan in 1986. He received his B.S. in 2009 from the University of Tokyo. In the same year, he began his graduate studies at the Graduate School of Pharmaceutical Sciences, the University of Tokyo, under the guidance of Professor Tohru Fukuyama. His research interests are in the area of the total synthesis of natural products.

Tatsuya Toma was born in 1984 in Saitama, Japan. He graduated in 2007 and received his M. S. degree in 2009 from the University of Tokyo. The same year he started his Ph. D. study under the supervision of Professor Tohru Fukuyama. His current interest is enantioselective total synthesis of complex natural products.

Jun Shimokawa was born in 1980 in Tokyo. He performed his Ph.D. studies under the direction of Professor Tohru Fukuyama at the University of Tokyo, where he conducted research on total syntheses of complex natural products. In 2006, he was appointed Assistant Professor in the same group. His research efforts focus on the development of novel synthetic methodology and applications to the synthesis of complex molecules.

Pablo E. Guzmán was born in Aguascalientes, Mexico. He earned his B.Sc. in Chemistry from Chicago State University in 2006. He is currently a doctoral student under the guidance of Professor Huw M. L. Davies. His research is focused on the development of Rh(II)-catalyzed asymmetric vinylogous [4+3] cycloaddition reactions and the exploration of cascade bis-diazo methodology directed towards complexity-generating reactions

Synthesis of Cyclobutarenes by Palladium-Catalyzed C(sp³)-H Bond Arylation: Preparation of Methyl 7-Methylbicyclo[4.2.0]Octa-1,3,5-Triene-7-Carboxylate

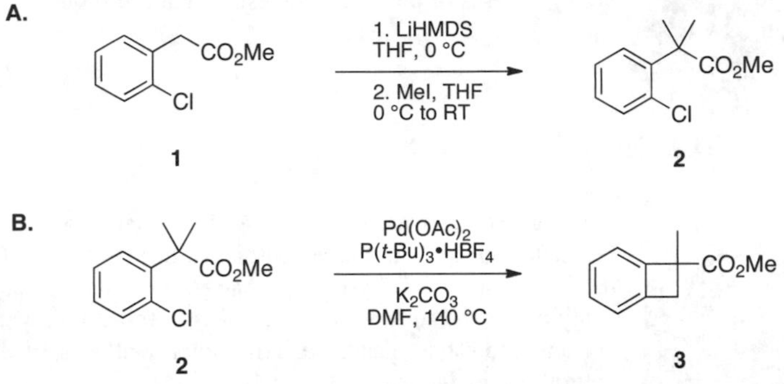

A.

B.

Submitted by Michaël Davi, Arnaud Comte, Rodolphe Jazzar, and Olivier Baudoin.[1]

Checked by Brendan T. Parr and Huw M. L. Davies.

1. Procedure

A. *Methyl 2-(2-chlorophenyl)-2-methylpropanoate* (**2**). A one-necked 500-mL round-bottomed flask (Note 1), equipped with a Teflon-coated magnetic stirring bar (Note 2), is fitted with a rubber septum and connected to a combined argon/vacuum line and evacuated/back-filled with argon twice. The reaction vessel was charged with lithium bis(trimethylsilyl)amide (135 mL, 135 mmol, 2.5 equiv) (Note 3). The flask is then cooled to 0 °C using an ice-bath. A THF (10 mL) solution (Note 4) of methyl (2-chlorophenyl)acetate (**1**) (10.2 g, 54.1 mmol, 1 equiv) (Notes 5 and 6) is added dropwise and the mixture is stirred for an additional 30 min. Finally, after dropwise addition of a THF (10 mL) (Note 4) solution of iodomethane (8.4 mL, 135.0 mmol, 2.5 equiv) (Note 7), the reaction is stirred at room temperature for 2 h (Notes 8 and 9). After drop-wise addition of saturated aqueous solution of ammonium chloride (50 mL) (Note 10), the aqueous layer is extracted twice with diethyl ether (2 x 100 mL) (Note 11). The combined organic layers are successively washed with a saturated aqueous solution of sodium thiosulfate (50 mL) (Note 12) and brine (50 mL) (Note 13), dried over anhydrous MgSO₄ (5 g) (Note 14), filtered and

Org. Synth. **2012**, *89*, 510-518
Published on the Web 5/29/2012
© 2012 Organic Syntheses, Inc.

concentrated under reduced pressure. The residue is purified by a short-path distillation under vacuum (58–61 °C/0.11 mmHg) to afford the pure product **2** (10.7–10.8 g, 93–94%) as a colorless oil (Note 15).

B. *Methyl 7-methylbicyclo[4.2.0]octa-1,3,5-triene-7-carboxylate* (**3**). A one-necked, 500-mL round-bottomed flask, equipped with a Teflon-coated magnetic stirring bar (Notes 1 and 2) is charged with palladium(II) acetate (527 mg, 2.35 mmol, 0.05 equiv) (Note 16), P(*t*-Bu)$_3$•HBF$_4$ (1.36 g, 4.69 mmol, 0.10 equiv) (Note 17) and potassium carbonate (8.44 g, 61.0 mmol, 1.30 equiv) (Note 18). The flask is fitted with a rubber septum, connected to a combined argon/vacuum line and evacuated for an hour (in order to remove traces of oxygen and water), then back-filled with argon three times (Note 19). A single-necked, 100-mL round-bottomed flask (Note 1) is charged with methyl 2-(2-chlorophenyl)-2-methylpropanoate **2** (10.00 g, 47.0 mmol, 1.0 equiv). The flask is fitted with a rubber septum, connected to a combined argon/vacuum line and evacuated for an hour, then back-filled with argon three times and charged with DMF (90 mL) (Note 20). The DMF solution of **2** is then added to the 500-mL round-bottomed flask dropwise *via* cannula. The mixture is then stirred for 15 min at room temperature and for 24 h at 140 °C (preheated oil bath) (Notes 8 and 21). After this time, the reaction mixture is cooled to ambient temperature, diluted with diethyl ether (100 mL) (Note 11) and filtered through a short pad of Celite® (10 g) (Note 22). The organic layer is washed three times with brine (3 x 100 mL) (Note 13), dried over magnesium sulfate (10 g) (Note 14) and evaporated under reduced pressure. The residue is then purified by distillation under vacuum (62 °C, 0.11 mmHg) (Note 23) to afford the title compound **3** (4.91–4.95 g, 59–60%) as a colorless oil (Note 24).

2. Notes

1. Glassware was dried in an oven at 150 °C for >2 h before being connected to a vacuum/argon line, where it was cooled to room temperature under vacuum and subsequently back-filled with argon.

2. An egg-shaped 70 mm x 15 mm magnetic stir bar was used.

3. Lithium bis(trimethylsilyl)amide (1.0 M THF solution) was purchased from Sigma-Aldrich and used as received.

4. Tetrahydrofuran (THF) was purchased from Sigma-Aldrich and was dried through a Grubbs apparatus and stored under a dry argon atmosphere prior to use.

511

5. Methyl (2-chlorophenyl)acetate (>98% pure) was purchased from TCI America and used as received.

6. Submitters prepared methyl (2-chlorophenyl)acetate in quantitative yield by reaction of 2-(2-chlorophenyl)acetic acid with catalytic H_2SO_4 in methanol at reflux for 16 h.

7. Iodomethane (99%, stabilized) was purchased from Acros Organics and used as received.

8. Submitters monitored reaction progress by analytical gas chromatography with a mass spectroscopy (GC/MS), carried out on a Shimadzu QP2010 GCMS apparatus with an electronic impact mass and equipped with a SLB-5ms column (15.0 m x 0.10 mm) containing a 0.10 μm film thickness with a flow rate of 0.58 mL/min with 474 kPa He. The profile of temperature is the following one: 1 min at 90 °C, temperature increase to 220 °C at a rate of 8 °C per minute, then temperature increase to 300 °C at a rate of 40 °C per minute, then 300 °C for 4.5 min.

9. The retention time for the starting material **1** (*m/z* 184), the monoalkylated product (*m/z* 198) and the title product **2** (*m/z* 212) were respectively 6.42, 6.90 and 7.79 min.

10. Ammonium chloride (99+%, pure) was purchased from Acros Organics and added to water until saturation.

11. Diethyl ether (absolute) was purchased from EMD Chemicals, Inc. and used as received.

12. Sodium thiosulfate (98.5% extra pure, anhydrous) was purchased from Acros Organics and added to water until saturation.

13. Sodium chloride (99.5%, for analysis) was purchased from Acros Organics and added to water until saturation.

14. Magnesium sulfate (97% pure, anhydrous) was purchased from Acros Organics and used as received.

15. Analytical data for methyl 2-(2-chlorophenyl)-2-methylpropanoate **2**: R_f 0.42 (10% EtOAc/cyclohexane); ^1H NMR (600 MHz, CDCl$_3$) δ: 1.61 (s, 6 H), 3.66 (s, 3 H), 7.20 (dt, *J* = 7.8, 1.8 Hz, 1 H), 7.27 (dt, *J* = 7.8, 1.8 Hz, 1 H), 7.34 (dd, *J* = 7.8, 1.8 Hz, 1 H), 7.41 (dd, *J* = 7.8, 1.8 Hz, 1 H). ^{13}C NMR (150 MHz, CDCl$_3$) δ: 26.0, 46.6, 52.3, 126.7, 126.9, 128.0, 130.6, 133.6, 142.2, 177.4. IR (neat) ν 2982, 2948, 1733, 1473, 1432 cm^{-1}. HRMS (APCI) calculated for $C_{11}H_{13}ClO_2$ [M+H]$^+$: 213.0677, found: 213.0673. Anal. Calcd for $C_{11}H_{13}ClO_2$: C, 62.12; H, 6.16. Found: C, 61.94; H, 6.02.

16. Palladium(II) acetate was purchased from Sigma-Aldrich and used as received.

17. Tri-*t*-butylphosphonium tetrafluoroborate (99%) was purchased from Strem Chemicals, Inc. and used as received.

18. Potassium carbonate (99+%, for analysis, anhydrous) was received from Acros Organics and dried under vacuum at 140 °C for 48 h, then stored under argon in a desiccator.

19. *Caution! The reaction is highly sensitive to the presence of traces of oxygen and moisture.*

20. *N,N*-Dimethylformamide (DMF) (absolute, over molecular sieves ($H_2O \leq 0.005\%$), $\geq 99.5\%$ (GC)) was purchased from Sigma-Aldrich and used as received.

21. Non-quantitative ratio of product (m/z 176, $t_R = 5.36$)/starting material (m/z 212, $t_R = 7.79$) is typically >10:1. The only by-product of the reaction is the protodehalogenated product (m/z 178, $t_R = 5.14$). Non-quantitative ratio of product/by-product ratio is >20:1.

22. Celite® 545, purchased from Sigma-Aldrich, was used as received.

23. A 50-mL, one-necked, round-bottomed flask equipped with a Teflon-coated magnetic stirring bar is charged with the reaction residue, then equipped with a 10-cm Vigreux column and a short-path distillation kit. The distillation vessel is gradually heated until the product slowly passes into the collection bulb (<1 drop per second).

24. Analytical data for methyl 7-methylbicyclo[4.2.0]octa-1,3,5-triene-7 carboxylate 3: R_f 0.45 (10% EtOAc/cyclohexane) visualization by UV absorption; ^{1}H NMR (400 MHz, CDCl$_3$) δ: 1.69 (s, 3 H), 3.05 (d, J = 12 Hz, 1 H), 3.69 (s, 3 H), 3.72 (d, J = 12 Hz, 1 II), 7.10 (m, 1 H), 7.15 (m, 1 H), 7.20–7.25 (m, 2 H). ^{13}C NMR (100 MHz, CDCl$_3$) δ: 22.8, 42.0, 51.9, 52.2, 121.2, 123.2, 127.1, 128.1, 142.2, 148.0, 175.0. IR (neat) ν 3069, 2951, 2929, 1729, 1457, 1433, 1277, 1148, 1140 cm^{-1}. HRMS (APCI): calculated C$_{11}$H$_{12}$O$_2$[M+H]$^+$: 177.0910, found: 177.0907. Anal. Calcd for C$_{11}$H$_{12}$O$_2$: C, 74.98; H, 6.86. Found: C, 74.72; H, 6.96.

Safety and Waste Disposal Information

All hazardous materials should be handled and disposed of in accordance with "Prudent Practices in the Laboratory"; National Academies Press; Washington, DC, 2011.

3. Discussion

The combination of palladium(II) acetate with electron-rich phosphine ligands generates versatile catalysts for the synthesis of various and synthetically useful fused carbocycles by intramolecular $C(sp^3)$-H bond arylation of aryl chlorides.[2] The $C(sp^3)$-H bond arylation of aryl bromides and iodides has been thoroughly described,[3,4] but the arylation of the corresponding aryl chlorides remains scarcely employed despite lower cost, better availability and wider structural diversity. We recently described that the combination of $Pd(OAc)_2$ and $P(t\text{-}Bu)_3$ provides an efficient catalytic system for the synthesis of cyclobutarenes by intramolecular $C(sp^3)$-H arylation of nonacidic methyl groups of aryl chlorides (Table 1).[2] This methodology allows for the synthesis of substituted cyclobutarenes containing a quaternary center in moderate to excellent yield (60–87%), some of these products being hardly accessible by other means.

The computed mechanism (DFT) of the intramolecular $C(sp^3)$-H arylation of 2-bromo-*tert*-butylbenzene indicates that the C-H activation step occurs through a base-induced concerted metalation-deprotonation (CMD) pathway.[4g,5] With *tert*-butylphosphine as the ligand, C-H activation is the rate-determining step and occurs with the carbonate coordinated to palladium *trans* to the C-H bond. An agostic interaction with a C-H bond geminal to the C-H bond to be cleaved lowers the activation barrier.

Our initial work was performed on a small scale (100 mg of aryl chloride), using 10 mol% of palladium(II) acetate for a reaction concentration of 0.25M. However, under optimized reaction conditions on a larger scale (10 g), the catalyst loading can be reduced to 5 mol% and the reaction concentration increased to 0.5M. Lower catalyst loading led to non-reproducible results. The $C(sp^3)$-H arylation reaction was shown to be effective for a broad range of aryl and heteroaryl chlorides.[2] It was demonstrated that various substituents on the benzylic carbon (Table 1, entries 1-7) as well as on the aromatic ring (Table 1, entries 8-12) are well tolerated and give rise to the corresponding cyclobutarenes in 65–87% yield. The reactions in Table 1 have been performed by using the original conditions.[2]

Table 1. Scope of the synthesis of cyclobutarenes by intramolecular C(sp^3)-H arylation[a]

Entry	Substrat	Product	Yield(%)[b]
1	(structure, CO$_2$t-Bu)	(structure, CO$_2$t-Bu)	71
2	(structure, CN)	(structure, CN)	65
3	(structure, OTIPS)	(structure, OTIPS)	66
4	(structure, CO$_2$Me)	(structure, CO$_2$Me)	70
5	(structure, MeO$_2$C, CO$_2$Me)	(structure, CO$_2$Me)	74
6	(structure, F, CO$_2$Me)	(structure, F, CO$_2$Me)	85
7[c]	(structure, Cl, CO$_2$Me)	(structure, Cl, CO$_2$Me)	72
8	(structure, F$_3$C, CO$_2$Me)	(structure, F$_3$C, CO$_2$Me)	87
9	(structure, MeO, CO$_2$Me)	(structure, MeO, CO$_2$Me)	66
10	(structure, F, CO$_2$Me)	(structure, F, CO$_2$Me)	76
11	(structure, CO$_2$Me)	(structure, CO$_2$Me)	60
12	(structure, CO$_2$Me)	(structure, CO$_2$Me)	63

(a) From ref. 2. Reaction conditions: Pd(OAc)$_2$ (10 mol%), P(t-Bu)$_3$•HBF$_4$ (20 mol%), K$_2$CO$_3$ (1.3 equiv), DMF, 140 °C. (b) Yields of isolated products after flash chromatography. (c) DMA was used instead of DMF.

515

Appendix
Chemical Abstracts Nomenclature; (Registry Number)

(2-Chlorophenyl)acetic acid; (2444-36-2)
Lithium bis(trimethylsilyl)amide; (4039-32-1)
Iodomethane; (74-88-4)
Palladium(II) acetate; (3375-31-3)
Tri-*t*-butylphosphonium tetrafluoroborate; (131274-22-1)
Potassium carbonate; (584-08-7)

1. Université Claude Bernard Lyon 1, CNRS UMR5246, Institut de Chimie et Biochimie Moléculaires et Supramoléculaires, CPE Lyon, 43 Boulevard du 11 Novembre 1918, 69622 Villeurbanne, France. Email: olivier.baudoin@univ-lyon1.fr We thank ANR (programme blanc "AlCaCHA"), Institut Universitaire de France and Bayer CropScience for financial support.
2. Rousseaux, S.; Davi, M.; Sofack-Kreutzer, J.; Pierre, C.; Kefalidis, C. E.; Clot, E.; Fagnou, K.; Baudoin, O. *J. Am. Chem. Soc.* **2010**, *132*, 10706-10716.
3. For a recent review, see: Jazzar, R.; Hitce, J.; Renaudat, A.; Sofack-Kreutzer, J.; Baudoin, O. *Chem. Eur. J.* **2010**, *16*, 2654.
4. (a) Baudoin, O.; Herrbach, A.; Guéritte, F. *Angew. Chem. Int. Ed.* **2003**, *42*, 5736. (b) Dong, C.-G.; Hu, Q.-S. *Angew. Chem. Int. Ed.* **2006**, *45*, 2289. (c) Ren, H.; Knochel, P. *Angew. Chem. Int. Ed.* **2006**, *45*, 3462. (d) Hitce, J.; Retailleau, P.; Baudoin, O. *Chem. Eur. J.* **2007**, *13*, 792. (e) Lafrance, M.; Gorelsky, S. I.; Fagnou, K. *J. Am. Chem. Soc.* **2007**, *129*, 14570. (f) Watanabe, T.; Oishi, S.; Fujii, N.; Ohno, H. *Org. Lett.* **2008**, *10*, 1759. (g) Chaumontet, M.; Piccardi, R.; Audic, N.; Hitce, J.; Peglion, J.-L.; Clot, E.; Baudoin, O. *J. Am. Chem. Soc.* **2008**, *130*, 15157.
5. Kefalidis, C. E.; Baudoin, O.; Clot, E. *Dalton Trans.* **2010**, *39*, 10528.

Olivier Baudoin studied chemistry at ENSC Paris. He completed his Ph. D. in 1998 in the group of Jean-Marie Lehn in Paris and then worked as a post-doctoral fellow with K. C. Nicolaou in the Scripps Research Institute (La Jolla, CA). He joined the Institut de Chimie des Substances Naturelles (Gif-sur-Yvette) in 1999 as a CNRS researcher. In 2006, he was appointed Associate Professor at the University of Lyon 1. His main research focuses on organopalladium catalysis and most recently on $C(sp^3)$-H functionalization. He was the recipient of the CNRS Bronze Medal in 2005 and was nominated as junior member of Institut Universitaire de France in 2009.

Michaël Davi received his M. Sc. degree in fine organic chemistry in 2001 from the University of Basse-Normandie in Caen, France. In 2001, he did a six-months internship in 3M Pharmaceuticals at Saint-Paul, USA. He completed his Ph. D. in 2009 under the supervision of Prof. H. Lebel at the University of Montréal, Canada, working on catalytic Wittig-type olefination reactions. He then joined the group of Prof. O. Baudoin at the University of Lyon 1 as a post-doctoral fellow, working on Pd-catalyzed $C(sp^3)$-H activation.

Arnaud Comte received his B. Sc. degree then his Ms. C. degree in chemistry from the University of Lyon 1 in 2010. He did his Master's degree internship under the guidance of Prof. O. Baudoin at the University of Lyon 1, where he worked on the synthesis of benzocyclobutenes.

Rodolphe Jazzar studied chemistry at the University of Poitiers (France). He completed his PhD in 2003 in the group of Mike Whittlesey in Bath (UK) on the synthesis and study of ruthenium carbene complexes. His Ph. D. was followed by post-doctoral stays with Prof. E. P. Kündig (Geneva, Switzerland), and with Prof. G. Bertrand (Riverside, CA). He then joined the group of Prof. O. Baudoin at the University of Lyon 1 in 2006 and was appointed as a CNRS researcher within the same group in 2008. His main research focuses on organopalladium catalysis towards the synthesis of allenes and C-H activation.

517

Brendan T. Parr was born in Winchester, MA in 1986. He completed his B.S. in chemistry at the University of Richmond under the supervision of Prof. William H. Myers. Currently, Brendan is a third-year graduate student in the laboratory of Prof. Huw M. L. Davies at Emory University. His research encompasses domino reactions initiated by chiral rhodium-carbenoid *in situ* activation of allyl alcohols.

An Efficient and Scalable Ritter Reaction for the Synthesis of *t*-Butyl Amides

Submitted by Jacqueline E. Milne and Jean C. Baum.[1]
Checked by Travis C. McMahon and John L. Wood.[2]

> *Caution: Flammable isobutylene gas is generated during the reaction. The use of acetic acid as solvent minimizes the amount of isobutylene gas generated by creating an equilibrium with isobutylene to regenerate tert-butyl acetate.*

1. Procedure

Methyl 4-(tert-butylcarbamoyl)benzoate. A 500-mL three-necked round-bottomed flask with 24/40 joints (Note 1) is equipped with an overhead mechanical stirrer (Teflon blade, 19 x 75 x 3 mm), a nitrogen inlet, and a temperature probe (Note 2). The nitrogen inlet is removed from the 500-mL flask. Methyl 4-cyanobenzoate (50.0 g, 310 mmol, 1 equiv) (Note 3), acetic acid (50.0 mL) (Note 4), and *tert*-butyl acetate (83.2 mL, 72.0 g, 621 mmol, 2 equiv) (Note 5) are charged to the 500-mL flask, and the nitrogen inlet is reinserted to the open neck. Stirring is initiated at a rate of 300 to 350 rpm. A temperature decrease is observed on mixing, and the slurry is allowed to warm to $\geq$ 19 °C (Note 6). Meanwhile, a 250-mL three-necked round-bottomed flask with 19/22 joints is immersed in an ambient temperature water bath (Dewar flask, 15–25 °C), 25 mm egg-shaped magnetic stir bar, temperature probe and N_2 inlet. To the 250 mL flask, acetic acid (50 mL) is charged and a rubber septum is added to the third neck. Concentrated sulfuric acid (29.8 mL, 558 mmol, 1.8 equiv) (Note 7) is charged to the 250-mL flask in 3 mL portions by syringe, maintaining the internal temperature $\leq$ 40 °C (Note 8). The solution is allowed to cool and once the internal temperature reaches $\leq$ 25 °C, it is transferred from the 250-mL flask to a 120-mL addition funnel. The N_2 line is removed from the joint of the 500-mL flask, the addition funnel is attached to the joint, and the N₂ inlet adapter is placed on top of the addition funnel. An ambient temperature water bath (Dewar flask, 15–25 °C) is fitted around the 500-mL flask. The solution is charged over 1 h, maintaining an internal temperature ≤ 30 °C (Note 9). The reaction is analyzed for completion after 2 h (Note 10). A 1000–mL three-necked round-bottomed flask with 24/40 joints is immersed in an ambient temperature water bath (Dewar flask. 15-25 °C), is equipped with an overhead mechanical stirrer (Teflon blade, 19 x 75 x 3 mm) and a thermocouple probe, and is charged with ammonium acetate (96.0 g, 124 mmol, 4 equiv) (Note 11) and water (250 mL) (Note 12). The ammonium acetate dissolves completely (Note 13). The reaction solution (slightly yellow, homogeneous, and viscous) is transferred from the 500-mL flask to a 250-mL addition funnel (Note 14). The addition funnel is attached to the 1000-mL flask and a N₂ line is attached to the top joint *via* an inlet adapter (24/40 joint). After the contents of the 1000-mL flask reaches ≥ 19 °C, the solution in the addition funnel is added over 1 h, maintaining the internal temperature ≤ 35 °C at a stir speed of 400 to 500 rpm (Note 15). White solids begin to form immediately and the turbidity increases as the addition proceeds. The slurry is stirred for 17–24 h at room temperature (Notes 15 and 16) and then filtered on a glass Büchner funnel (650 mL) (Note 17). The product is washed with water (2 × 100 mL) (Note 18) and dried on the filter for 2 h. The product is then dried under high vacuum (0.5 mmHg) in a 40 °C oil bath over two nights (Note 19) to yield 69.0 g (95%) of the amide at a purity of > 99.0 % (Note 20). No impurities were observable in the product > 0.5 % (observed at 254 nm, see Note 10).

2. Notes

1. It is not necessary to oven-dry the glassware.
2. The internal temperature was monitored using a IKA ETS-D5 theromocouple (240 mm length, 2 mm diameter, temperature range –50 to 450 °C). The probe is attached through a rubber septum. The submitters monitored the internal temperature using a J-Kem Gemini digital thermometer with a Teflon-coated T-Type thermocouple probe (12-inch length, 1/8 inch outer diameter, temperature range –200 to +250 °C). The probe is attached via a Teflon probe adapter (24/40 joint).
3. Methyl 4-cyanobenzoate (99%) is purchased from Alfa Aesar and is used as received.

520

4. Acetic acid (99.7%), purchased from JT Baker, is used as received.

5. *tert*-Butyl acetate (>99%) is purchased from Alfa Aesar and was used as received. Excess *tert*-butyl acetate is used to ensure complete reaction. Decomposition of *tert*-butyl acetate occurs from reaction with sulfuric acid to generate isobutylene gas.

6. Solids are observed that dissolve upon warming to room temperature. During a period of 1 h, the temperature rose from 12 °C to 19 °C. At 19 °C, few solids remain in a colorless solution.

7. Concentrated sulfuric acid (95%) was purchased from JT Baker and was used without further purification. Excess sulfuric acid is used to ensure reaction completion.

8. A total of 10 portions of sulfuric acid (3 mL each) were charged in 15 min. The internal temperature was 39 °C at the end of the addition. After stirring 30 min., the temperature reached 25 °C. Ice cubes may be added to the water bath to increase the efficiency of this step.

9. Temperature was maintained near 24 °C during the addition. The submitters observed the temperature to be near 30 °C during the addition. A total volume of 75 mL sulfuric acid/acetic acid solution was added.

10. The submitters monitored the reaction by HPLC under the following conditions: Luna 3 μm C18 100 Å column (100 x 3.0 mm) using 0.1% TFA in acetonitrile (A)/0.1% TFA in water (B), 90:10 A:B ramp to 10:90 A:B over 15 min with 5 min post time. Flow rate is 1 mL/min at 254 nm detection wavelength. The nitrile starting material elutes at 6.6 min and the amide product elutes at 7.7 min.

The checkers monitored the reaction by TLC (30% EtOAc, 70% hexanes, UV detection at 254 nm). R_f (methyl 4-cyanobenzoate) = 0.60; R_f (methyl 4-(*tert*-butylcarbamoyl) benzoate) = 0.46.

The submitters noted the reaction went to completion after 2 h, and the checkers also observed this on 25 g scale. However, the checkers observed that the addition of *tert*-butyl acetate (41.6 mL, 36.0 g, 310 mmol, 1 equiv) in two portions (first after 2 h, second after 3 h) was required on 50 g scale. The reaction was observed to be complete after a total of 3.5 h had elapsed from time the sulfuric acid/acetic acid addition was complete.

11. Ammonium acetate (>99%) was purchased from Mallinckrodt and used as received.

12. DI water is used.

13. No exotherm is observed.

14. The total volume is 235 mL.

15. The quench is exothermic. The temperature is maintained close to 25 °C during the addition. The submitters observed the temperature to be near 30 °C during the addition. Some off-gassing is observed initially during the quench. A fast stir speed is recommended because the slurry becomes viscous as the quench proceeds.

16. The solubility was determined by measuring the concentration of the mother liquor by HPLC. Immediately after the quench, the solubility was 5.8 mg/mL. After stirring overnight, the solubility was 3.8 mg/mL. The submitters recommend holding overnight to maximize yield. The internal temperature was 21 °C prior to filtration.

17. The glass Büchner funnel was equipped with filter paper (Whatman, Qualitative, 90 mm). The submitters used a medium-grade fritted glass funnel (650 mL) for the filtration.

18. The checkers found that the water wash is needed to wash any remaining salts from the product.

19. The submitters dried the product in a vacuum oven under house-vacuum (20–40 mmHg) at 40 °C overnight.

20. Methyl 4-(*tert*-butylcarbamoyl)benzoate has the following physical and spectroscopic data: white solid; mp: 117–120 °C (Gallenkamp melting point apparatus, uncorrected); the submitters reported 125 °C (measured by DSC); ^{1}H NMR (400 MHz, CDCl$_3$) δ: 1.48 (s, 9 H), 3.93 (s, 3 H), 5.98 (br. s., 1 H), 7.76 (d, *J*=8.23 Hz, 2 H), 8.06 (d, *J*=8.46 Hz, 2 H); ^{13}C NMR (101 MHz, CDCl$_3$) δ: 29.0, 52.1, 52.5, 126.9, 129.9, 132.5, 140.0, 166.1, 166.5; IR (Bruker Tensor 27) ν$_{max}$/cm^{-1}: 3300, 2981, 1721, 1638, 1548, 1450, 1288, 1221, 1116, 1016, 870, 825, 794, 692, 668; HRMS Calcd for C$_{13}$H$_{18}$NO$_3$ [M + H]$^+$ 236.1287 found 236.1271; Anal Calcd for C$_{13}$H$_{17}$NO$_3$: C, 66.36; H, 7.28; N, 5.95. Found: C, 66.42; H, 7.16; N, 6.12.

Safety and Waste Disposal Information

All hazardous materials should be handled and disposed of in accordance with "Prudent Practices in the Laboratory"; National Academies Press; Washington, DC, 2011.

3. Discussion

The Ritter reaction involves the conversion of a nitrile to a *N-tert*-butyl amide through reaction with a *tert*-butyl cation.[3] Typically, the *tert*-butyl cation is generated from either isobutylene gas or *tert*-butanol in the presence of an acid. However, the most commonly used reaction conditions have some inherent problems that limit their use on larger scale. For example, isobutylene is a class 4 (highly flammable) gas and there have been reports of exothermic events when used in this reaction. In addition, the use of *tert*-butanol can be problematical since it exists as a semisolid at room temperature (melting point: 26 °C). An alternative *tert*-butyl cation precursor is *tert*-butyl acetate, which is attractive due to its ease of handling (boiling point: 97–98 °C), availability as a common solvent, and low cost. Unfortunately, it was observed that when *tert*-butyl acetate is employed as both the reaction solvent and the reagent, rapid and uncontrolled gas evolution results, due to the fast decomposition to isobutylene gas.[4]

The procedure described here involves the slow addition of sulfuric acid (1.8 equiv) in acetic acid to a mixture of the nitrile substrate and *tert*-butyl acetate (2 equiv) in acetic acid at 30 °C.[5] The advantage of this reagent combination is a result of the equilibrium that exists between *tert*-butyl acetate and isobutylene in acetic acid (Figure 1). The generated carbocation either reacts along the desired reaction pathway to give *tert*-butyl amide, or is scavenged by acetic acid to regenerate the reagent. Consequently, a controlled and slow release of isobutylene occurs during the course of the reaction and the possibility of uncontrolled gas evolution is eliminated.

Figure 1. The Ritter Reaction Employing *tert*-Butyl Acetate

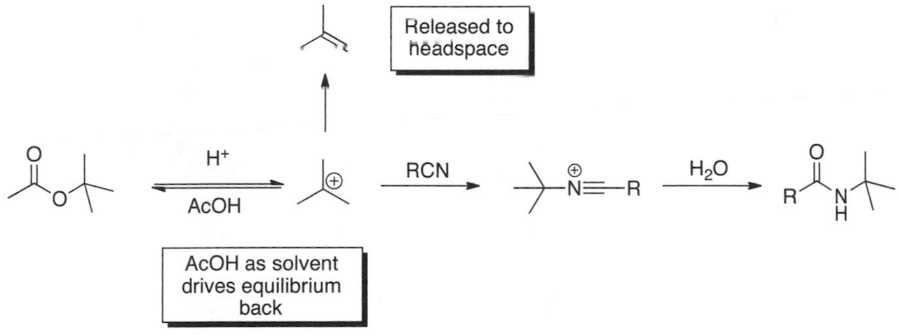

This efficient procedure has a broad substrate scope for the high yielding synthesis of both electron-donating and electron-withdrawing aromatic, alkyl, and α,β–unsaturated *tert*-butyl amides (Table 1). It is

Table 1. Substrate Scope

Entry	Substrate	Time	Yield[d]
1		3	91%
2		2	90%
3		20	85%[b]
4		2	81%
5		1	85%[a]
6		2	82%
7		5	95%
8		17	80%[c]
9		1	93%
10		2	85%
11		1	94%
12		1	88%

(a) The reaction required 2.5 equiv *t*-butyl acetate and 2 equiv sulfuric acid;
(b) Charged additional sulfuric acid (0.5 equiv) and *t*-butyl acetate (0.5 equiv)
after 17 h; (c) Corrected for purity by ^{1}H NMR; (d) Isolated product

noteworthy that in addition to the main example, methyl 4-cyanobenzoate,the reaction conditions are compatible with other acid-sensitive functional groups (Table 1, entries 6-7).

In summary, employing *tert*-butyl acetate in the presence of sulfuric acid and acetic acid provides a safe, scalable, and robust method for the conversion of nitriles to *tert*-butyl amides.

1. Chemical Process R&D, Amgen Inc., One Amgen Center Drive, Thousand Oaks, California 91320. E-mail: milne@amgen.com
2. Department of Chemistry, Colorado State University, Fort Collins, CO 80523. E-mail: john.l.wood@colostate.edu
3. For a review see: Krimen, L. I.; Cota, D. L. *Org. React.* **1969**, *17*, 213–325.
4. *tert*-Butyl acetate as *tert*-butyl cation source: (a) Fernholz, H.; Schmidt, H. J. *Angew. Chem., Int. Ed.* **1969**, *8*, 521. (b) Reddy, K. L. *Tetrahedron Lett.* **2003**, *44*, 1453–1455.
5. Baum, J. C; Milne, J. E.; Murry, J. A.; Thiel, O. R. *J. Org. Chem.* **2009**, *74*, 2207–2209.

Appendix
Chemical Abstracts Nomenclature; (Registry Number)

Methyl 4-cyanobenzoate: Benzoic acid, 4-cyano-, methyl ester; (1129-35-7)
Acetic acid; (64-19-7)
tert-Butyl acetate: Acetic acid, 1,1-dimethylethyl ester; (540-88-5)
Sulfuric acid; (7664-93-9)
Ammonium acetate; (631-61-8)
Methyl 4-(*tert*-butylcarbamoyl)benzoate: Benzoic acid, 4-[[(1,1-dimethylethyl)amino]carbonyl]-, methyl ester; (67852-98-6)

Jacqueline E. Milne received a MSci (Hons) in chemistry at the University of Glasgow in 1999 and a Ph.D. from the University of Leeds with Professor Philip Kocienski in 2003. Her graduate research explored new syntheses of α-metallated cyclic enol ethers and their application to the synthesis of D-erthro-sphingosine. In 2003 she joined Professor Stephen Buchwald's group as a postdoctoral fellow at MIT. Here she developed a new catalyst system for Negishi cross-coupling reactions. In addition, she developed an asymmetric copper-catalyzed conjugate reduction for the dynamic kinetic resolution of α,β-unsaturated lactones and applied to the synthesis of eupomatilone-3. In 2005 Jacqueline joined the Chemical Process R&D at Amgen in Thousand Oaks.

Jean Baum completed her undergraduate studies at University California, San Diego. After a six-year employment at Fluorochem, Inc. she attended the University of California, Los Angeles and obtained an M.S. degree in chemistry under the supervision of Professor Robin L. Garrell. She has been employed at Amgen since 2001 where she was recently promoted to Senior Associate Scientist within Chemical Process R&D.

Travis McMahon received his B.S. in chemistry from Western Washington University in 2008 where he did undergraduate research for Professor James R. Vyvyan. He is currently working towards his Ph.D. at Colorado State University under the supervision of Professor John L. Wood, where he is working towards the total synthesis of the tetrapetalone natural products and has also worked on the synthesis of gliotoxin, dehydrogliotoxin and analogs.

Enals via a Simple Two-Carbon Homologation of Aldehydes and Ketones

Submitted by Petr Valenta, Natalie A. Drucker, and Patrick J. Walsh.[1]
Checked by David Hughes.

1. Procedure

Diethylzinc is pyrophoric. Proper care and laboratory attire should be used when handling and quenching reaction mixtures containing diethylzinc.

(E)-5-Phenylpent-2-enal. An oven-dried, 1-L, 3-necked round-bottomed flask containing a 3-cm oval PTFE-coated magnetic stir bar is equipped with a nitrogen inlet adapter and two rubber septa, one of which is pierced with a thermocouple probe (Note 1). The flask is charged with a 50 wt % solution of ethoxyacetylene in hexanes (22 mL, 16.8 g, 0.12 mol, 1.8 equiv) (Note 2) and THF (70 mL), then cooled to 2 °C in an ice-bath. With stirring, 2M borane dimethyl sulfide complex in THF (17.5 mL, 35 mmol, 0.5 equiv) is added dropwise via a 30 mL syringe over 30 min, keeping the temperature below 7 °C. The reaction mixture is stirred for 1 h at 2 °C (Note 3), then the ice-bath is replaced with a heating mantle. The nitrogen adapter is replaced with a vacuum-distillation head connected to a 500-mL round bottom flask immersed in a dry-ice bath. The volatile materials are removed at reduced pressure (60 mmHg) by applying gentle heating to the reaction flask (Note 4). After removal of volatile materials, the reaction flask is back-filled with nitrogen and the brown oil is dissolved by addition of toluene (65 mL) (Note 5). The stirred solution is cooled to –75 °C in a dry ice/acetone bath and 1.1M diethylzinc in toluene (96 mL, 105 mmol, 1.5 equiv) (Note 6) is added by cannula over 10 min, keeping the temperature below –65 °C. After the solution is stirred for 30 min between

–70 and –75 °C, hydrocinnamaldehyde (9.1 g, 68 mmol, 1.0 equiv) (Note 7) is added dropwise via a 20 mL syringe over 2 min. The mixture is allowed to warm to 20 °C over 1 h and stirred for 3 h (Note 8). The reaction is then cooled to 5 °C in an ice-bath and *t*-butyl methyl ether (100 mL) is added. The nitrogen gas adapter is replaced with a 300-mL addition funnel. Vigorous stirring of the reaction mixture is commenced and water (100 mL) is carefully added over 5 min, keeping the temperature below 20 °C. Gas (ethane) is evolved and a white precipitate of zinc oxide is formed. The ice bath is removed and aqueous 2M HCl (200 mL) is added over 5 min via the addition funnel. The reaction mixture is stirred for 20 min until the white precipitate of zinc oxide dissolves (Note 9). The quenched reaction mixture is transferred to a 1-L separatory funnel. The aqueous layer is separated and back-extracted with MTBE (100 mL). The combined organic layer is washed with saturated aqueous sodium bicarbonate (200 mL) and brine (50 mL), then dried by filtration through 50 g sodium sulfate. The volatile materials are removed by rotary evaporation (40 °C bath temperature, 60 mmHg) to afford 12 g of a brown oil. The crude product is purified by flash column chromatography (Note 10) and concentrated by rotary evaporation (40 °C bath temperature, 60 mmHg) to afford 7.7–8.2 g (71–75 % yield) as a pale yellow oil (Notes 11 and 12).

2. Notes

1. The internal temperature was monitored using a J-Kem Gemini digital thermometer with a Teflon-coated T-Type thermocouple probe (12-inch length, 1/8 inch outer diameter, temperature range –200 to +250 °C).

2. The following reagents and solvents were used as received: ethoxyacetylene (GFC Chemicals, 50 wt % solution in hexanes), tetrahydrofuran (Sigma-Aldrich, 99.9%, inhibitor-free, Sure-Seal® bottle), 2M borane dimethyl sulfide in THF (Sigma-Aldrich), 15 wt % diethylzinc in toluene (1.1M, Sigma-Aldrich), toluene (Sigma-Aldrich, Chromasolv, 99.9%, anhydrous, Sure-Seal® bottle), hexanes (Fisher, ACS reagent, >98.5%), t-butyl methyl ether (>98.5%, Sigma-Aldrich), ethyl acetate (Fisher, ACS reagent, >99%), 2N HCl (Fisher), and silica gel (Fisher, 230-400 mesh, 60 Å). Deionized water was used throughout.

3. The conversion of ethoxyacetylene to tris(ethoxyvinyl)borane (Scheme 1) was followed by NMR by diluting a 0.05 mL reaction aliquot in 0.6 mL THF-d_8. Diagnostic signals for ^{1}H NMR were δ 4.05 (q, 2H) for

ethoxyacetylene and 3.84 (q, 6H) for tris(ethoxyvinyl)borane; ^{13}C NMR diagnostic signals were δ 74.9 and 91.5 for ethoxyacetylene and 65.3 and 163.6 for tris(ethoxyvinyl)borane. The reaction was complete within 10 min after addition of borane dimethyl sulfide. The 20% excess charge of ethoxyacetylene was reflected in its constant level in samples taken between 10 min and 1 h after addition of borane dimethyl sulfide. The tris(ethoxyvinyl)borane was about 90% pure after concentration based on NMR analysis. Tris(ethoxyvinyl)borane has the following spectroscopic properties: ^{1}H NMR (500 MHz, THF-d$_8$. ref 3.57) δ: 1.23 (t, J = 7.0 Hz, 9 H), 3.84 (q, J = 7.0 Hz, 6 H), 5.30 (d, J = 13.9 Hz, 3 H), 7.06 (d, J = 13.9 Hz, 3 H); ^{13}C NMR (125 MHz, THF-d$_8$. ref 67.57) δ: 15.2, 65.3, 104.2 (br), 163.6.

4. Contact of the reaction mixture with air was avoided at all times to avoid air oxidation of the borane intermediate. A dry ice/acetone cold trap was placed between the receiver flask and vacuum line to ensure dimethyl sulfide was fully trapped. The volatile materials were distilled under a vacuum of 60 mmHg, with most of the material distilling at a pot temperature of –2 to +2 °C. The heating mantle was removed when the pot temperature rose to 10°C, then vacuum was applied for an additional hour as the contents warmed to ambient temperature, ensuring complete removal of volatile materials.

5. Toluene was added in two portions via a 50-mL syringe from a fresh Sure-Seal® bottle.

6. The 1.1M diethylzinc in toluene solution was transferred by cannula from a 100 g bottle via an 18-gauge needle by application of slight nitrogen pressure to the bottle. About 90% of the contents were transferred; the bottle was weighed before and after addition to obtain the exact amount added (density = 0.91 g/L).

7. The checkers purchased hydrocinnamaldehyde from Acros (labeled as 95%, GC indicated 93.5%). The aldehyde (40 g) was distilled at 60 mmHg through a 10 cm Vigreaux column, collecting a center cut of 25 g (62%) boiling at 140–143 °C which was stored in the freezer. Purity by GC: 97.5% (t$_R$ = 8.4 min; conditions: Agilent DB35MS column; 30 m x 0.25 mm; initial temp 60 °C, ramp at 20 °C/min to 280 °C, hold 15 min). The submitters purchased 90% technical grade hydrocinnamaldehyde from Sigma-Aldrich, which was distilled under reduced pressure (100 °C, 13 mmHg) and stored under nitrogen in the glove box.

8. The reaction was followed by ^{1}H NMR of a reaction aliquot that was worked up as follows. A 0.1 mL sample was quenched into a vial containing 1 mL MTBE and 1 mL of 2N HCl. The sample was vigorously shaken and allowed to stand for 15 min. The MTBE layer was separated and evaporated, then dissolved in CDCl$_3$ for ^{1}H NMR analysis. The aldehyde protons were diagnostic at δ 9.51 (doublet) for the product and 9.85 (triplet) for hydrocinnamaldehyde. The reaction typically proceeds to >98% conversion. Hydrocinnamaldehyde elutes with an R$_f$ very similar to the product (0.45 product, 0.47 hydrocinnamaldehyde using 20% EtOAc/hexanes, EMD silica gel 60 F254, 2.5 x 7.5 cm plates), so following the reaction by TLC is difficult.

9. The pH of the resulting aqueous layer should be less than 3 to effect complete hydrolysis. If necessary, an additional 20 mL of 2M HCl should be added.

10. Flash column chromatography was performed with 400 g silica gel wet-packed with 5% EtOAc/hexanes and topped with 1 cm sand. The crude product was diluted with 10 mL of dichloromethane to load onto the column. Approximately 3 L of 7 % EtOAc/hexanes was used as eluent, flow 50 mL/min, collecting 50 mL fractions. Product eluting in fractions 40-55 were combined and concentrated by rotary evaporation (40 °C, 60 mmHg), then held under vacuum (60 mmHg) for 2 h at room temperature to afford product as a pale yellow oil. The chromatography was followed by TLC using PMA detection (R$_f$ = 0.45 in 20% EtOAc/hexanes). Product elution could also be followed by UV detection, but this method is not recommended since several non-UV active impurities elute just prior to product elution.

11. The submitters used a 2M solution of diethylzinc prepared by adding neat diethylzinc to toluene in a glovebox. The submitters also prepared fresh 1 M borane-dimethylsulfide complex by adding neat borane-dimethylsulfide complex to THF. The submitters reported a yield of 79 % with a GC product purity of 92 %.

12. (*E*)-5-Phenylpent-2-enal has the following physical and spectroscopic properties: ^{1}H NMR (500 MHz, CDCl$_3$) δ: 2.66–2.71 (m, 2H), 2.85 (t, J = 7.5 Hz, 2 H), 6.15 (ddt, J = 15.8, 7.8, 1.4 Hz, 1 H), 6.87 (td, J = 15.7, 6.7 Hz, 1 H), 7.20–7.25 (m, 3 H), 7.31–7.34 (m, 2 H), 9.51 (d, J = 7.8 Hz, 1 H). ^{13}C NMR (125 MHz, CDCl$_3$) δ: 34.3, 34.4, 126.6, 128.5, 128.8, 133.6, 140.4, 157.4, 194.1. IR (neat) cm^{-1}: 3064, 3031, 2930, 1685, 1490, 1120. HRMS calcd. for C$_{11}$H$_{12}$O (MH$^+$): 161.0966, found 161.0964. GC-MS (EI) *m/z*

(relative intensity), 160 (8%, M$^+$), 142 (14%), 129 (12%), 116 (75%), 92 (35%), 91 (100%), 77 (18%), 65 (60%), 51 (21%). Purity by GC: 97% (t_R = 10.5 min; conditions same as in Note 7). The material crystallizes in the freezer and has an approximate melting point of −12 to −14 °C.

Safety and Waste Disposal Information

All hazardous materials should be handled and disposed of in accordance with "Prudent Practices in the Laboratory"; National Academies Press; Washington, DC, 2011.

3. Discussion

α,β-Unsaturated aldehydes continue to attract wide-spread interest for their utility in organic synthesis,[2] organocatalysis,[3] and biochemistry.[4] Many existing approaches to synthesize enals are based on two-carbon homologation of aldehydes.[5] Reactions of aldehydes with various Wittig,[6] Horner-Emmons[7] and Peterson-type[8] reagents and with enolate of acetaldehyde[9] suffer from strongly basic reaction conditions. Such reaction conditions often result in self-condensation, destruction of base-sensitive functionalities and racemization of α-stereocenters via deprotonation. Basic conditions can be avoided by use of C-silylated imines[10] or alkoxyvinyl zirconium species,[11] which are prohibitively expensive or labor intensive reagents.

Scheme 1. Two-carbon homologation of aldehydes and ketones

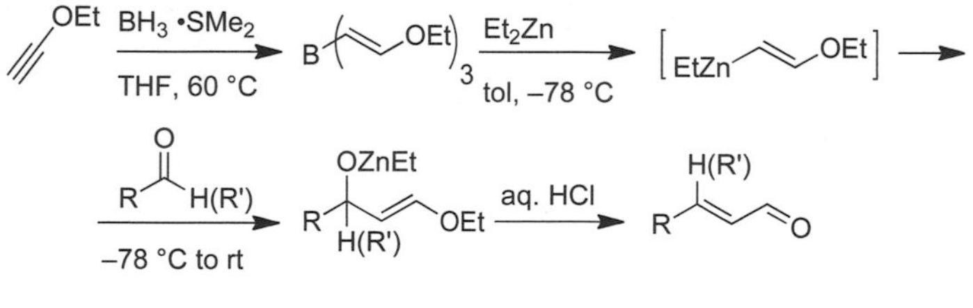

Our method utilizes only common commercially available chemicals and avoids the use of strong bases (Scheme 1).[12] The two-carbon homologation of aldehydes and ketones involves hydroboration of ethoxyacetylene with borane-dimethyl sulfide complex followed by

transmetallation of the resulting vinylboron intermediate to zinc using diethylzinc. The resulting ethoxy-vinyl zinc species adds to aldehydes and ketones to generate allylic zinc alkoxides. Protonation with aqueous acid results in elimination of water and generation of desired enal.

The broad substrate scope of the reaction is presented in Table 1. Two-carbon homologation of aliphatic (Table 1, entries 1 to 3), aromatic (entries 4 and 8), heteroaromatic (entries 5 to 7), α,β-unsaturated (entries 9 and 10), propargylic aldehydes (entry 11), and dialdehydes (entries 16 and 17) was achieved with high isolated yields. Ketones were also successfully homologated (entries 12 to 15). Reaction with unsymmetrical ketones resulted in mixtures of E/Z isomers (entry 14). When the substituents are very different in size (*t*-butyl vs. methyl) E/Z ratios up to 15:1 occurred (entry 15). *o*-Phthalaldehyde was homologated only once due to intramolecular formation of a hemiacetal, preventing the reaction with the second aldehyde (entry 17). No racemization of α-chiral aldehydes was observed under these conditions (entries 18 and 19). Isolated yields of volatile enals can be improved by use of hexanes instead of toluene as solvent, minimizing product lost during final evaporation of solvents.

Table 1. Substrate scope of two-carbon homologation of aldehydes and ketones

Entry	Product	Yield (%)[a]	Entry	Product	Yield (%)[a]
1		94	11	C_5H_{11}	87
2	Ph	94	12		86
3	Ph	89	13	Ph, Ph	54
4	Ph	96	14	Ph	83[b]
5		92	15		53[c]
6		91	16		70
7		72	17		54
8	Fe	93	18	OBn	88 (98%$_{ee}$)[d]
9		95	19	Boc	94 (98%$_{ee}$)[d]
10		95			

[a] Isolated yield. [b] E:Z = 3:1. [c] E:Z = 15:1. [d] ee detemined by HPLC

1. P. Roy and Diana T. Vagelos Laboratories, Department of Chemistry, University of Pennsylvania, 231 South 34th Street, Philadelphia, Pennsylvania 19104-6323, pwalsh@sas.upenn.edu. PJW thanks the NSF for support of this work.

2. (a) Pohjakallio, A.; Pihko, P. M.; Liu, J. *J. Org. Chem.* **2010**, *75*, 6712-6715. (b) Lifchits, O.; Reisinger, C. M.; List, B. *J. Am. Chem. Soc.* **2010**, *132*, 10227–10229.

3. (a) Fang, X.; Jiang, K.; Xing, C.; Hao, L.; Chi, Y. R. *Angew. Chem., Int. Ed.* **2011**, *50*, 1910-1913. (b) Cohen, D. T.; Cardinal-David, B.; Roberts, J. M.; Sarjeant, A. A.; Scheidt, K. A. *Org. Lett.* **2011**, *13*, 1068-1071.

4. Srivastava, S.; Ramana, K. V.; Bhatnagar, A.; Srivastava, S. K. *Methods Enzymol.* **2010**, *474*, 297-313.

5. Ebenezer, W. J.; Wight, P. *Comprehensive Organic Functional Group Transformations*; Vol. 3, page 54, Pergamon, 1995.

6. (a) Zhou, G.; Hu, Q.-Y.; Corey, E. J. *Org. Lett.* **2003**, *5*, 3979-3982. (b) Hudson, R. D. A.; Manning, A. R.; Gallagher, J. F.; Garcia, M. H.; Lopes, N.; Asselberghs, I.; Boxel, R. V.; Persoons, A.; Lough, A. J. *J. Organomet. Chem.* **2002**, *655*, 70-88. (c) Mladenova, M.; Ventelon, L.; Blanchard-Desce, M. *Tetrahedron Lett.* **1999**, *40*, 6923-6926. (d) Daubresse, N.; Francesch, C.; Rolando, C. *Tetrahedron* **1998**, *54*, 10761-10770. (e) Cristau, H.-J.; Gasc, M.-B. *Tetrahedron Lett.* **1990**, *31*, 341-344. (f) Fraser-Reid, B.; Molino, B. F.; Magdzinski, L.; Mootoo, D. R. *J. Org. Chem.* **1987**, *52*, 4505-4511. (g) Cresp, T. M.; Sargent, M. V.; Vogel, P. *J. Chem. Soc., Perkin Trans. 1* **1974**, 37-41.

7. (a) Michel, P.; Rassat, A.; Daly, J. W.; Spande, T. F. *J. Org. Chem.* **2000**, *65*, 8908-8918. (b) Paterson, I.; Cumming, J. G.; Ward, R. A.; Lamboley, S. *Tetrahedron* **1995**, *51*, 9393-9412. (c) Nicolaou, K. C.; Harter, M. W.; Gunzner, J. L.; Nadin, A. *Liebigs Ann.-Recl.* **1997**, *7*, 1283-1301. (d) Still, W. C.; Gennari, C. *Tetrahedron Lett.* **1983**, *24*, 4405-4408. (e) Ando, K. *Tetrahedron Lett.* **1995**, *36*, 4105-4108.

8. (a) Corey, E. J.; Enders, D.; Bock, M. G. *Tetrahedron Lett.* **1976**, 7-10. (b) Martin, S. F.; Dodge, J. A.; Burgess, L. E.; Limberakis, C.; Hartmann, M. *Tetrahedron* **1996**, *52*, 3229-3246.

9. Mahata, P. K.; Barun, O.; Ila, H.; Junjappa, H. *Synlett* **2000**, *9*, 1345-1347.

10. (a) Gaudemar, M.; Bellassoued, M. *Tetrahedron Lett.* **1990**, *31*, 349-352. (b) Bellassoued, M.; Majidi, A. *J. Org. Chem.* **1993**, *58*, 2517-2522.

11. Maeta, H.; Suzuki, K. *Tet. Lett.* **1992**, *33*, 5969-5972.

12. Valenta, P.; Drucker, N. A.; Bode, J. W.; Walsh, P. *J. Org. Lett.* **2009**, *11*, 2117-2119.

534

Appendix
Chemical Abstracts Nomenclature (Collective Index Number); (Registry Number)

Ethoxyacetylene: Ethyne, ethoxy-; (927-80-0)

Borane dimethyl sulfide complex: Boron, trihydro[thiobis[methane]]-; (13292-87-0)

Diethyl zinc: Zinc, diethyl-; (557-20-0)

Hydrocinnamaldehyde: Benzenepropanal; (104-53-0)

5-Phenyl-pent-2-enal: 2-Pentenal, 5-phenyl-, (2E)-; (37868-70-5)

Patrick J. Walsh received his B.A. from UC San Diego (1986) and Ph.D. in organometallic chemistry with Prof. Robert G. Bergman at UC Berkeley (1991). He was an NSF postdoctoral fellow with Prof. K. B. Sharpless at the Scripps Research Institute. Walsh became an assistant professor at San Diego State University in 1994 and from 1996-1999 was also a professor at Centro de Graduados e Investigación, Instituto Tecnológico de Tijuana (Mexico). In 1999 he moved to his current position at the University of Pennsylvania where he holds the Alan G. MacDiarmid Chair. Walsh's interests are in asymmetric catalysis, development of new methods, reaction mechanisms, and inorganic synthesis. With Prof. Marisa Kozlowski Walsh wrote *Fundamentals of Asymmetric Catalysis*.

Petr Valenta was born in Plzen, Czech Republic. He received his M.S. in Chemistry from University of Pardubice, where he performed undergraduate research with Professor Milos Sedlak. As an Alfred Bader fellow, Petr obtained his Ph.D. from University of Pennsylvania in 2010. In research group of Professor Patrick J. Walsh he studied catalytic enantioselective reactions of hetero-vinyl zinc Reagents. Petr is currently a Research Chemist at Milliken and Company in Spartanburg, SC.

Natalie A. Drucker was born in 1988 and was raised in Hollywood, Florida. She graduated with a B.A. in Chemistry from University of Pennsylvania, where she worked in the laboratory of Professor Patrick J. Walsh. She currently studies at University of Florida College of Medicine.

t-Butyl as a Pyrazole Protecting Group: Preparation and Use of 1-*tert*-Butyl-3-Methyl-1*H*-Pyrazole-5-Amine

Submitted by Patrick M. Pollock and Kevin P. Cole.[1]
Checked by Scott W. Roberts and Margaret Faul.

1. Procedure

A. *1-tert-Butyl-3-methyl-1H-pyrazol-5-amine.* A 250-mL three-necked round-bottomed flask is equipped with a large oval stir bar (3.75 cm x 0.75 cm) and placed in an oil bath. Using a large plastic funnel, the flask is charged with solid *tert*-butylhydrazine hydrochloride (25.00 g, 98%, 196.6 mmol) (Note 1) and 2 M NaOH (98.3 mL, 2 M, 196.6 mmol) (Note 2) is added. Stirring is started at ambient temperature. When a complete solution has formed (approx. 10 min), 3-aminocrotononitrile (16.82 g, 96%, 196.6 mmol) (Note 3) is added through the funnel, and the stirring is continued. The funnel is replaced with a thermocouple (Note 4), an air-cooled reflux condenser is fixed to the flask, the third joint is sealed with a glass stopper, and the top of the condenser is connected to an ambient pressure nitrogen line (Note 5). The slurry is then heated to 90 °C (internal reaction temperature) with vigorous stirring for 22 h. After 22 h, the yellow/orange biphasic mixture is cooled to 57 °C, the glass stopper removed, and a ~2 mL aliquot of the well-stirred biphasic mixture is removed with a pipette and placed in a 10-mL round-bottomed flask containing a small stir bar (1/2" x 1/8"). The 10 mL flask is then immersed in a dry ice/acetone bath for several minutes until completely frozen. The 10-mL flask is then removed from the bath and allowed to warm up slowly over a magnetic stirplate with the stirring on. As the mass melts, a slurry of seed crystals is formed (Note 6). With the bulk of the reaction being held and stirred vigorously at 57 °C, the seed slurry is introduced by pouring it through the empty joint into the 250-mL flask in a single portion. Crystallization is evidenced by the formation of large clear globules, and when this is apparent, the heating is turned off and the oil bath is removed. Vigorous stirring is continued, as the mixture cools to ambient temperature, and when the internal temperature is <30 °C, the slurry is immersed in an ice-water bath for one hour, during which time the tan solids adopt a more granular appearance. A 500-mL filter flask is fitted with a 7 cm porcelain filter funnel containing a paper filter, which is wetted with 2-3 mL of water. The slurry is filtered, and the cold filtrate (Note 7) is used to rinse the flask out and ensure maximum recovery of product (Note 8). The tan solids are allowed to suck dry on the filter for at least 5 min, and then scraped out onto a large piece of filter paper and any chunks are broken up with a spatula. The solids are then transferred into a tared 250-mL beaker, covered with a laboratory wipe, which is secured with a rubber band, and placed in vacuum oven at ambient temperature for 3 days (Note 9) to afford a light tan granular solid. The mass of the dried product is 27.0 g that was confirmed to be 97% purity by quantitative NMR. KF of the isolated solids is 3.7%, which affords a corrected yield of 87% (Notes 10, 11, 12, 13 and 14).

B. *N-(1-tert-Butyl-3-methyl-1H-pyrazol-5-yl)pyridin-2-amine.* To a 250-mL three-necked round-bottomed flask is added 2-chloropyridine (10.00 g, 99%, 87.2 mmol) (Note 15), and the flask is then equipped with an oval magnetic stir bar (3.75 cm x 0.75 cm), a water-cooled reflux condenser open to the air, an oil bath, a thermocouple (Note 4) and a plastic funnel. To the flask are then added **1** (14.46 g, estimated 97% potency, 91.5 mmol), *tetra*-butylammonium bromide (1.42 g, 99%, 4.40 mmol) (Note 16), toluene (100 mL) (Note 17) and a commercial solution of NaOH (110 mL, 2 M, 220.2 mmol) (Note 2). The funnel is replaced with a rubber septum, stirring started, and a six inch stainless steel syringe needle connected to a nitrogen source is inserted through the septum and placed below the level of the liquid in order to sparge the solution and headspace with a vigorous stream of nitrogen. Nitrogen is bubbled through the solution for 5 min, after which

538

time Pd(dba)$_2$ (0.20 g, 0.35 mmol) and Xantphos (0.31 g, 0.5 mmol) (Notes 18 and 19) are added as solids by removing the septum rapidly, pouring in the solids, and then quickly replacing the septum. Nitrogen is then bubbled through the mixture for two additional minutes, after which time the top of the condenser is attached to a static-pressure nitrogen source. The mixture is heated to vigorous reflux (internal temperature 87 °C) (Note 20) with stirring for 24 h (Notes 21, 22, and 23). After 24 h, the oil bath is turned off and removed, and the mixture is cooled to <30 °C by immersion into a cool water bath. A 500-mL filter flask is equipped with a 7 cm porcelain funnel and a filter paper. The paper is wetted with water, and the solids are isolated by filtration (Note 24). The flask is rinsed with toluene (25 mL) and this is used to rinse the solids on the filter. The solids are washed with water (25 mL) and then EtOH (25 mL) (Note 25) and allowed to suck dry for >5 min. The solids are scraped off the filter paper with a spatula onto a large piece of filter paper and any solids are broken up. The solids are placed in a tared 250-mL beaker and covered with a laboratory wipe, which is secured with a rubber band, and placed in vacuum oven at 50 °C under full house vacuum for 17 h to afford 20.0 g of light yellow/green solids (100%) (Notes 26 and 27).

  C. *N-(3-Methyl-1H-pyrazol-5-yl)pyridin-2-amine.* A 250-mL three-necked round-bottomed flask is charged with **2** (10.00 g, 43.4 mmol) and equipped with a magnetic stir bar (3.75 cm x 0.75 cm), oil bath, water-cooled reflux condenser open to the air (Note 28), thermocouple (Note 4), and plastic funnel. Deionized water (100 mL) is added, followed by trifluoroacetic acid (9.9 mL, 99%, 127.0 mmol) (Note 29). The mixture is then heated to 95 °C, and a clear light green solution is formed, with some dark solids observed floating in the solution. Once the mixture has reached 95 °C, it is held at that temperature with stirring for three hours. After three hours, the oil bath is turned off and removed, and the solution is cooled to <60 °C, at which time EtOH (50 mL) is added (Note 25). The solution is then further cooled to <30 °C using a cool water bath. A second 250-mL three-necked round-bottomed flask is equipped with two stoppers and a 4-cm coarse-fritted Büchner funnel (30 mL volume size) containing dry Celite™ (4.00 g). A few mLs of water is used to wet the Celite™ in order to form a filter pad. The Celite™ filter pad is washed until the filtrate is no longer cloudy and then the collection flask is connected to the funnel. The reaction mixture is carefully filtered through the pad into the second 250 mL flask using gentle vacuum (Note 30). The reaction flask is rinsed with water

(10 mL), which is then used to rinse the walls of the filter funnel and filter pad (Note 31). The filter and stoppers are removed, a pH probe is inserted through one of the unused side joints, and a Teflon stir paddle with a glass stir shaft assembly are introduced through the center joint and connected to an overhead stirrer (Note 32). A commercial solution of 5 N NaOH (25 mL) (Note 2) is then added via pipette rapidly at first and more slowly once pH 7 is reached, and the titration is continued until a final pH of 8-10 is reached (Notes 33, 34, 35, and 36). Upon neutralization, the solution gradually turns cloudy over a few minutes, and crystallization ensues. The slurry is placed in an ice-water bath for one hour with stirring. A 250-mL filter flask is equipped with a 5 cm porcelain funnel and filter paper. The paper is wetted with water (1-2 mL) and the product is isolated by vacuum filtration. To the flask is added water (25 mL) and any residual solids in the flask are rinsed onto the filter cake. After 2-3 minutes on vacuum, the solids are then washed with additional water (25 mL), and then sucked dry for 5 minutes. The solids are scraped onto a large piece of filter paper, broken up with a spatula, placed into a tared 250-mL beaker and covered with a laboratory wipe, which is secured with a rubber band, and placed in vacuum oven at 50 °C under full house vacuum for 15 h to afford the product (g, 6.73 g, 89%) as a pale tan powdery solid (Note 37).

2. Notes

1.   *tert*-Butylhydrazine hydrochloride (98%) was obtained from Aldrich Chemical Co. Inc. and used as received.
2.   Sodium hydroxide (2 M) was purchased from Fluka Analytical. Sodium hydroxide (5 M) was purchased from Sigma Aldrich. A 2 M solution prepared using NaOH pellets and water also works well.
3.   3-Aminocrotononitrile was obtained from Aldrich Chemical Co. Inc. (96%) and used as received.
4.   The temperature of all reactions was controlled using a programmable J-KEM™ temperature controller.
5.   If desired, the ammonia evolved from the reaction can be scrubbed by washing the outlet of the nitrogen line with diluted HCl.
6.   This method of seed crystal generation is not necessary once the reaction has been run one time, as the isolated product is suitable to induce crystallization.

540

7. It is very important to only use the cold filtrate to wash the product, as the pyrazole is highly soluble in pure water, and the recovery is greatly reduced if plain water is used.

8. If a ring of product has formed in the flask or if large clumps of product have crystallized around the stir bar or thermocouple, a spatula is used to loosen them, and they are rinsed into the funnel with cold filtrate.

9. House vacuum was used (measured to be 10-20 mbar). Temperatures higher than ambient occasionally resulted in sublimation of the product and yield reduction. It was determined that after three days, the solids were reliably dried to a constant mass (less than 100 mg loss/day).

10. The solids have been shown by single crystal X-Ray crystallography to be a 1/3 hydrate, which is in theory 3.8% water by mass; in some cases additional surface water can yield higher KF data. When the reaction was performed at one-half scale, the isolated yield was 13.0 g (86%).

11. Pyrazole 1 is commercially available in small quantities, and larger amounts can be custom made. A 1 kg batch was obtained at great cost, but its purity and physical properties were vastly inferior to material produced by this method.

12. An HPLC potency method was developed by determination of the potency of a reference standard of material by quantitative ^{1}H NMR, and these results agreed well when combined with the KF data (mass balance of >99%).

13. The reactions can be monitored by HPLC using the conditions as described. Zorbax SD C8 column (Part Number 866953-906), 4.6 x 75 mm, 3.5 μm particle size; 1.0 mL/min flow, 40 °C column temperature, detection at 215 nm; Solvent A = 1 mL/L TFA in HPLC grade water, Solvent B = 1 mL/L TFA in HPLC grade CH_3CN; Gradient Elution (min): T(0) = 5% B to 80% B at T(7), hold at 80% B until T(8), to 5% B at T(8.1), hold at 5% B until 10.5 min. HPLC samples were prepared using 5% ACN (0.1% TFA):95% water (0.1% TFA) and a 5.0 μL injection volume.

14. Properties of Pyrazole 1: mp 70–71 °C; IR (neat) cm^{-1}: 3435, 3331, 3228, 2976, 2937; ^{1}H NMR (400 MHz, DMDO-d$_6$) δ: 1.48 (s, 9 H), 1.95 (s, 3 H), 4.71 (brs, 2 H), 5.18 (s, 1 H); ^{13}C NMR (100 MHz, CDCl$_3$) δ: 14.1, 29.6, 58.2, 93.9, 145.3, 145.6; HRMS (ESI) calcd for $C_8H_{16}N_3$ (M+H$^+$): 154.1344, found: 154.13335, δ = 3.4 ppm; KF = 3.4-3.7%; weight % by quantitative ^{1}H NMR (triphenyl methane (Aldrich, 99%) used as internal

standard) = 97%; purity determined by HPLC method described by submitter (Note 13) = 100.0% at 230 nm, HPLC retention time = 3.10 min.

15. 2-Chloropyridine (99%) was purchased from the Aldrich Chemical Co. Inc and used as received.

16. *tetra*-Butylammonium bromide (99%) was purchased from Aldrich Chemical Co. Inc. and used as received.

17. Toluene (reagent grade) was purchased from Aldrich Chemical Co. Inc. and used as received.

18. The Pd(dba)$_2$ CoA for the Pd(dba)$_2$ used for this study reports carbon = 71.8%, Pd = 18.0% and Xantphos (97%) were purchased from the Aldrich Chemical Co. Inc. and used as received. They were weighed out in air, and no special precautions were taken.

19. All attempts to perform the coupling reaction *via* S$_N$Ar reaction in the absence of palladium met with failure.

20. The azeotropic boiling point of the reaction mixture has been observed to be 87–89 °C; the set point of the temperature controller was set to 95 °C to ensure a very vigorous reflux, which improves mixing and is crucial to the success and reproducibility of the reaction.

21. During the reaction, the upper walls of the flask can become coated with a thin film of black precipitate, presumably Pd(0). Furthermore, a solid precipitate of product is observed to form a suspension between the water and toluene layers.

22. The reaction may be monitored effectively by removing the oil bath to stop the reflux, and stopping stirring to allow the reaction mixture to settle. Using a long syringe needle, a small amount of the upper toluene solution is removed (no solids should be drawn up), and one drop of the solution is placed in an HPLC vial, and diluted with 1 mL of 3A EtOH (3 µL HPLC injection). After sampling, the stirring is started again and the oil bath is replaced.

23. These amination conditions are somewhat unconventional. Due to the extreme insolubility of the product these conditions were the most satisfactory that we surveyed, as the product forms an easily stirred suspension. The reaction using K$_2$CO$_3$ in *t*-amyl alcohol resulted in an extremely thick mixture, which could not be effectively stirred and also resulted in an incomplete reaction.

24. This filtration is somewhat slow and each step requires approximately 5 min ensuring that the majority of the liquid is removed from the solids before the next wash solvent is introduced.

542

25. EtOH (200 proof) was purchased from Decon Labs, Inc.

26. The solids are contaminated with some gray particles, which are presumed to be Pd(0) (Note 33). The submitters report that the gray material along with a small amount of insoluble green impurity can be removed by magnetically stirring the solids (5.00 g) with 25 mL water and 25 mL EtOH at ambient temperature while 37% HCl (5.35 mL, 3 equiv) is added over a few minutes to afford a hazy solution. Filtration of the mixture through a pad of Celite™ (2 g, wetted with water) afforded a clear and nearly colorless solution. This solution was then neutralized by the addition of 50% NaOH (3.44 mL, 3 equivalents) over five min with stirring, which afforded a thick white slurry. Filtration of the solids, washing with water (100 mL), and drying *in vacuo* resulted in pure white **2** (4.77 g, 95%, mp = 229.4–230.5 °C).

27. When the reaction was performed at one-half scale, the isolated yield was 10.2 g (100%). Properties of pyrazole **2**: mp 229–231 °C (DSC); IR (neat) cm^{-1}: 2926, 1585, 1553, 1414; ^{1}H NMR (400 MHz, AcOH-d$_4$) δ: 1.62 (s, 9 H), 2.27 (s, 3 H), 6.13 (s, 1 H), 6.79 (d, J = 9.0 Hz, 1 H), 7.06 (t, J = 6.5 Hz, 1 H), 7.96 (ddd, J = 1.6, 7.2, 7.2 Hz, 1H), 8.28 (d, J = 5.3 Hz, 1 H), 11.81 (s, 1 H); ^{13}C NMR (100 MHz, AcOH-d$_4$) δ: 12.62, 29.2, 60.3, 104.8, 110.8, 114.6, 135.3, 139.3, 144.2, 147.0, 154.6; HRMS (ESI) calcd for C$_{13}$H$_{19}$N$_4$ (M+H$^+$): 231.1610, found: 231.16062, δ = 0.9 ppm; purity determined by HPLC method (Note 13): purity = 100.0% at 230 nm, HPLC retention time = 3.88 min.

28. It was observed that the reaction was slower when it run under a nitrogen atmosphere, this may be due to the more efficient removal of iso-butylene from the system when it is open to the air. A nitrogen sweep could also be employed.

29. Trifluoroacetic acid (99%) was purchased from Sigma-Aldrich. While the deprotection can be accomplished using aqueous HCl, we chose to develop the less volatile and corrosive trifluoroacetic acid method at hand. A 5–8 °C exotherm was observed during the fast addition of TFA into the water/pyrazole slurry.

30. The clear yellow filtrate was observed to have a tendency to foam, and caution must be used in order to ensure that none of the solution is sucked into the vacuum line and lost.

31. The filtration removes the black solids as well as dark green material, which is possibly related to the Xantphos ligand from step two and affords a clear, yellow/green solution.

32. While magnetic stirring can be used during the neutralization, the mixture becomes very thick and difficulties in maintaining the stirring can be encountered when using magnetic stirring.

33. The residual palladium levels were 4200 ppm in **2** and 31 ppm in **3** as measured by ICP-ICP/MS. Further palladium rejection studies were not conducted on this system.

34. The theoretical amount of NaOH required is 26.1 mL; typically 25 mL is required to adjust the pH to the desired level as some TFA is lost presumably to evaporation.

35. Normally the color of the solution fades to nearly colorless towards the end of the NaOH addition, if the pH becomes >11, a pink color has been observed, which can be neutralized by adjustment of the pH back to 8–10 with a few drops of TFA.

36. An exotherm of 5–10 °C is normally observed during the neutralization and can be controlled by the use of a cool water bath.

37. When the reaction was performed at one-half scale, the isolated yield was 3.16 g (84%). Properties of pyrazole **3**: mp 153–154 °C; IR (neat) cm^{-1}: 3076, 2918, 1603, 1573, 1435, 1316; ^{1}H NMR (400 MHz, AcOH-d$_4$) δ: 2.31 (s, 3H), 5.95 (s, 1 H), 7.08 (t, J = 6.7 Hz, 1 H), 7.44 (d, J = 9.0 Hz, 1 H), 8.03 (dt, J = 7.6, 1.0 Hz, 1 H), 8.20 (d, J = 6.1 Hz, 1 H), 11.77 (brs, 2 H); ^{13}C NMR (100 MHz, AcOH-d$_4$) δ: 11.8, 96.5, 116.4, 116.6, 137.8, 143.4, 145.6, 149.6, 151.8; HRMS (ESI) calcd for C$_9$H$_{11}$N$_4$ (M+H$^+$): 175.0984, found: 175.09738, δ = 2.5 ppm; purity determined by HPLC method (Note 13): purity = 100.0% at 230 nm, HPLC retention time = 3.27 min.

Safety and Waste Disposal Information

All hazardous materials should be handled and disposed of in accordance with "Prudent Practices in the Laboratory"; National Academies Press; Washington, DC, 2011.

3. Discussion

5-Aminopyrazoles are ubiquitous throughout pharmaceutical and agricultural chemistry and many currently marketed products contain this residue.[2] The introduction of the unprotected aminopyrazole moiety is frequently accomplished without incident, but in some cases the use of a protected variant is required to overcome unselective reactions or to

544

facilitate product isolation and purification. We sought to develop the large-scale single-step preparation of a suitably protected aminopyrazole that could be coupled efficiently with a chloropyridine derivative and then cleanly deprotected. A previous synthesis of a protected pyrazole in our laboratories utilized three chemical steps to generate a mono-substituted hydrazine, which was then condensed with a β–keto nitrile. The unprotected 5-aminopyrazole is inexpensive and commercially available, but many protection methods either failed altogether, or suffered from poor chemo and regioselectivity. In our hands, the analogous Boc-protected 5-aminopyrazole[3] was thermally and hydrolytically unstable to the amination reaction conditions. Our efforts turned back to ring synthesis, and a brief survey of commercially available mono-substituted hydrazines quickly lead us to tert-butylhydrazine hydrochloride as an inexpensive, relatively non-toxic, easily handled solid, stable, and safe masked hydrazine. The use of 3-aminocrotononitrile is noteworthy in that our previous synthesis had relied upon in situ generation of α-cyanoacetone from methylacetate and acetonitrile; the aminonitrile is an inexpensive and convenient synthon for aminopyrazole preparation. The use of the tert-butyl protection group is also noteworthy, as the tert-butyl group on the 5-aminopyrazole system seems to exhibit unusual lability; other tert-butylpyrazoles in our laboratories have not exhibited the same relative ease of deprotection. The tert-butyl group is an atom economic protecting group, the byproduct being iso-butylene (C_4H_8).

Pyrazole 1 has previously been prepared by Giori et al. in a similar fashion in refluxing EtOH using triethylamine as base.[4] Giori's procedure employs an aqueous extraction to isolate the product, which we found difficult given the high water solubility of the product. We have developed this modified procedure, which uses no organic solvent and an inorganic base, as an environmentally friendly alternative. Furthermore, the product isolation is extremely facile, affording crystalline material directly from the reaction mixture. The byproduct from the hydrochloride neutralization-NaCl, is important in achieving high product recovery. It was found that the solubility of 1 in pure water was ca. 100 mg/mL, where as its solubility in NaCl solution (10 g NaCl diluted with 100 mL water, approximately what is generated under these reaction conditions) was ca. 30 mg/mL at ambient temperature. We believe that this salting-out effect is responsible for the ability to isolate the product in high yield as demonstrated. The isolated material contains water, NaCl, and possibly some NaOH, but is relatively free of organic impurities. Methods of further purification include vacuum sublimation, chromatography, or vacuum distillation.

The Buchwald-Hartwig amination has recently become an extremely popular method of aryl C-N bond formation.[5] We have found that the biphasic conditions as described above are very convenient in this case due to the extreme insolubility of the product. More conventional conditions using alcohol/carbonate bases afforded mixtures that could not be stirred and stalled reactions. The conditions described above use a low loading of transition metal, are simple to implement, and have proven to be robust in our hands. Both Pd(dba)$_2$ and Pd$_2$(dba)$_3$ work well for the reaction and we chose the former for stoichiometry determination convenience. We did not extensively screen phosphine ligands, but we found Xantphos to be convenient due to good reactivity, ease of use, and ready commercial availability. The addition of TBABr was introduced in order to act as a phase transfer agent and pull hydroxide base into the organic layer, as we had observed some reaction stalling, which we attributed to the organic layer being starved of base. The observation that a vigorous reflux is beneficial is likely due to additional deoxygenation of the solvent that is accomplished with the reflux, as well as improved mixing provided by the boiling action. The use of excess base in conjunction with the phase transfer catalyst and vigorous refluxing has resulted in a highly robust process.

Giori[4] demonstrated removal of the *tert*-butyl group in hot formic acid solvent. We wanted to identify an acid that could be used at lower stoichiometry and hence reduce the amount of base required for neutralization. We have found that many aqueous acids (including HCl and sulfuric) are competent at removal of the *tert*-butyl group, but TFA was convenient in that the volatility of TFA accomplishes a continuous washing of the sides of the reaction vessel during the course of the deprotection and avoids the formation of solids on the vessel walls, which has been observed with related systems.

This convenient three-step sequence highlights the efficiency with which **1** can be prepared and isolated. The utility of **1** is then demonstrated in a C-N bond formation to afford pyrazole **2**. Finally the ease with which the *tert*-butyl group can be removed is showcased with the aqueous acid deprotection and isolation of pyrazole **3**.

1. Chemical Product Research and Development, Lilly Research Laboratories, Lilly Corporate Center, Eli Lilly and Company, Indianapolis, IN 46285; k_cole@lilly.com.
2. Aggarwal, R.; Kumar, V.; Kumar, R.; Singh, S. P. *Beilstein J. Org. Chem.* **2011**, *7*, 179–197.
3. Seelen, W.; Schäfer, M.; Ernst, A. *Tetrahedron Lett.* **2003**, *44*, 4491–4493.
4. Vicentini, C. B.; Veronese, A. C.; Guarneri, M.; Manfrini, M.; Giori, P. *J. Heterocyclic Chem.* **1994**, *31*, 1477–1480.
5. Senra, J. D.; Aguiar, L. C. S.; Simas, A. B. C. *Curr. Org. Synth.* **2011**, *8*, 53–78.

Appendix
Chemical Abstracts Nomenclature; (Registry Number)

tert-Butylhydrazine hydrochloride; (7400-27-3)
3-Aminocrotononitrile; (1118-61-2)
1-*tert*-Butyl-3-methyl-1H-pyrazol-5-amine; (141459-53-2)
2-Chloropyridine; (109-09-1)
Pd(dba)$_2$: bis(dibenzylideneacetone)palladium; (32005-36-0)
4,5-Bis(diphenylphosphino)-9,9-dimethylxanthene; (161265-03-8)
Tetra-*n*-butylammonium bromide; (1643-19-2)
Trifluoroacetic acid; (76-05-1)

Kevin Cole was born in 1979 in St. Paul, Minnesota. He obtained a Ph.D. in chemistry in 2005 from the University of Minnesota Twin Cities campus, and worked in the laboratories of Professor Richard Hsung. In 2007, he completed postdoctoral studies in the laboratory of Professor K.C. Nicolaou at The Scripps Research Institute in San Diego, CA. In 2007, Kevin joined Eli Lilly and Company in Indianapolis, Indiana and works in Chemical Product Research and Development.

Patrick Pollock was born in Chicago, Illinois and grew up in San Diego, California. He attended the University of Washington and then San Diego State University working under the direction of Professor Mikael Bergdahl. He earned an M.S. in Chemistry in 2001 and then joined the Chemical Product Research and Development group at Eli Lilly and Company in Indianapolis, Indiana.

Scott W. Roberts was born in Ojai, California. He obtained a Ph.D. in Chemistry in 2007 from the University of Utah, where he worked under the direction of Professor Jon D. Rainier. In 2007, he completed postdoctoral studies under the direction of Professor Larry E. Overman. In 2009, Scott joined the department of Chemical Process Research and Development at Amgen Inc. in Thousand Oaks, California.

548

Direct Synthesis of Azaheterocycles from *N*-Aryl/Vinyl Amides. Synthesis of 4-(Methylthio)-2-phenylquinazoline and 4-(4-Methoxyphenyl)-2-phenylquinoline

A.

B.

Submitted by Omar K. Ahmad, Jonathan William Medley, Alexis Coste, and Mohammad Movassaghi.[1,2]
Checked by Hiromi Kaise and Tohru Fukuyama.

1. Procedure

A. 4-(Methylthio)-2-phenylquinazoline (3). A flame-dried, 300-mL three-necked round-bottomed flask equipped with a 3.0 cm football-shaped stir bar, rubber septum, and low temperature thermometer is charged with benzanilide (1) (Note 1) (6.02 g, 30.5 mmol, 1 equiv), sealed under an argon atmosphere, and fitted with an argon inlet. Anhydrous dichloromethane (Note 2) (60 mL) followed by 2-chloropyridine (Note 3) (5.76 mL, 6.97 g, 61.4 mmol, 2.01 equiv) is added via syringe, and the heterogeneous mixture is vigorously stirred and cooled to <–70 °C (dry-ice–acetone bath, internal temperature). After 10 min, trifluoromethanesulfonic anhydride (Note 4) (Tf$_2$O, 5.60 mL, 9.39 g, 33.3 mmol, 1.09 equiv) is added via syringe over 5 min at <–65 °C (internal temperature). After 15 min, the reaction flask is warmed to 0 °C (ice–water bath). After 5 min, the deep red solution becomes homogeneous and a solution of thiocyanic acid methyl ester (2) (Note 1) (2.52 mL, 2.67 g, 36.5 mmol, 1.20 equiv) in anhydrous dichloromethane (40 mL) is added via cannula over 5 min at 5–6 °C (Note

5). After 10 min, the cold bath is removed, and the reaction mixture is allowed to warm to 23 °C. After 2.5 h, triethylamine (Note 2) (10.0 mL, 7.26 g, 71.7 mmol, 2.35 equiv) is added via syringe over one min. The resulting mixture is concentrated with a rotary evaporator (20 mmHg, 30 °C). The remaining deep red oil is purified by flash column chromatography (Note 6) to afford quinazoline **3** (6.15 g, 80%) as an off-white solid (Note 7).

   B. *4-(4-Methoxyphenyl)-2-phenylquinoline (5)*. A flame-dried, 300-mL, three-necked round-bottomed flask equipped with a 3.0 cm football-shaped stir bar, rubber septum, and thermometer is charged with benzanilide (**1**) (Note 1) (5.29 g, 26.8 mmol, 1 equiv), sealed under an argon atmosphere, and fitted with an argon inlet. Anhydrous dichloromethane (Note 2) (45 mL) followed by 2-chloropyridine (Note 3) (5.04 mL, 6.10 g, 53.7 mmol, 2.00 equiv) is added via syringe, and the heterogeneous mixture is vigorously stirred and cooled to <–70 °C (dry ice-acetone bath, internal temperature). After 10 min, trifluoromethanesulfonic anhydride (Note 4) (Tf$_2$O, 4.96 mL, 8.33 g, 29.5 mmol, 1.10 equiv) is added via syringe over 5 min at <–65 °C (internal temperature). After 10 min, the reaction flask is warmed to 0 °C (ice-water bath). After 10 min, the yellow solution becomes homogeneous, and a solution of *p*-methoxyphenylacetylene (**4**) (Note 1) (4.18 mL, 4.26 g, 32.2 mmol, 1.20 equiv) in anhydrous dichloromethane (45 mL) is added via cannula over 5 min at <10 °C (internal temperature). After 2 h, aqueous saturated sodium bicarbonate solution (60 mL) is added carefully to the dark green reaction mixture, the cold bath is removed, and the reaction mixture is allowed to warm to 23 °C. After 10 min, the reaction mixture is transferred to a 1-L separatory funnel and diluted with aqueous saturated sodium chloride solution (200 mL) and dichloromethane (200 mL), and the layers are separated. The aqueous layer is extracted with dichloromethane (2 × 250 mL). The combined organic layers are dried over anhydrous sodium sulfate (160 g), filtered and concentrated with a rotary evaporator under reduced pressure (20 mmHg, 30 °C). To remove the 2-chloropyridine, the contents of the flask are quantitatively transferred with dichloromethane (~45 mL) into a 100-mL pear-shaped flask and concentrated with a rotary evaporator under reduced pressure (20 mmHg, 30 °C). A 3.0 cm football-shaped stir bar is added to the flask, and the residue is further concentrated with agitation under reduced pressure (<1 mmHg, 23 °C) for 12 h to afford a brown oil. The flask is placed in an oil bath while kept under reduced pressure (<1 mmHg), and the oil bath is

Org. Synth. **2012**, *89*, 549-561

warmed to 95 °C over 45 min. After 3 h, the flask is removed from the oil bath and allowed to cool to 23 °C. The residue is purified by flash column chromatography (Notes 8 and 9) to afford quinoline **5** (7.34 g, 88%) as a light brown solid (Note 10).

2. Notes

1. Benzanilide (98%) was used as purchased from Aldrich Chemical Company, Inc. Thiocyanic acid methyl ester (99+%) was used as purchased from TCI. *p*-Methoxyphenylacetylene (99%) was used as purchased from Alfa Aesar.

2. The submitters purchased dichloromethane from J. T. Baker (Cycletainer™) and triethylamine from Aldrich Chemical Company, Inc. Both were purified by the method of Grubbs et al. under positive argon pressure.[3] The checkers purchased dichloromethane from Wako Pure Chemical Industries, Ltd. (anhydrous, 99.5%) and triethylamine (>99%) was purchased from Kanto Chemical Co., Inc. and dried over KOH.

3. 2-Chloropyridine (99%) was purchased from Aldrich Chemical Company, Inc. and was purified by distillation from calcium hydride (26 mmHg, 72 °C) and stored sealed under an argon atmosphere.

4. Trifluoromethanesufonic anhydride (98%) was purchased from Oakwood Products, Inc. and used as received. Reagent quality is important for the success of these reactions. The checkers discovered that the use of trifluoromethanesulfonic anhydride purchased from TCI resulted in incomplete reactions and a significantly lower yield of **3** and **5** when the reagent was used without purification. The checkers did observe that the desired products could be obtained in good yield if freshly distilled reagent from TCI was employed for the reactions.

5. The reaction mixture became heterogeneous as precipitates formed within several minutes of thiocyanic acid methyl ester addition.

6. Flash column chromatography (8.0 cm diameter, 10 cm height) was performed using silica gel (Silica Gel 60, spherical, 40-100 mm, (Kanto, 250 g) (submitters: 60-Å pore size, 40–63 mm, standard grade, Zeochem)). The residue was loaded using 25% dichloromethane in *n*-hexane (20 mL). The eluent was 6% ethyl acetate in *n*-hexane until quinazoline **3** (TLC: R_f = 0.64, 20% ethyl acetate in *n*-hexane (submitters: TLC: R_f = 0.49, 20% ethyl acetate in hexanes), visualized by UV and $KMnO_4$ stain) completely passed through the column (fractions #9–26 collected, ~100 mL/fraction tubes).

The combined fractions were concentrated under reduced pressure (25 mmHg, 30 °C) using a rotary evaporator. Any residual 2-chloropyridine (TLC: R_f = 0.59, 20% ethyl acetate in n-hexane (submitters: R_f = 0.35, 20% ethyl acetate in hexanes), visualized by UV) can be readily removed from quinazoline **3** under reduced pressure (<1 mmHg). TLC analysis was performed on silica gel 60 F_{254} (0.25 mm, Merck) by the checkers (Submitters: EMD Silica Gel 60F plates).

7. When trifluoromethanesufonic anhydride from TCI was distilled prior to use, a yield of 74% was observed. The analytical data for quinazoline **3** is as follows: mp 89–90 °C; ^{1}H NMR (400 MHz, CDCl$_3$, 20 °C) δ: 2.85 (s, 3 H), 7.50–7.55 (m, 4 H), 7.83 (td, 1 H, J = 8.2, 1.4 Hz), 8.03 (d, 1 H, J = 9.2 Hz), 8.09 (d, 1 H, J = 8.3 Hz), 8.64 (dd, 2 H, J = 7.8, 2.3 Hz); ^{13}C NMR (100 MHz, CDCl$_3$, 20 °C) δ: 12.6, 122.5, 123.7, 126.6, 128.4, 128.5, 129.0, 130.5, 133.5, 138.1, 148.7, 158.8, 171.3; FTIR (neat) cm^{-1}: 3059, 2923, 1614, 1558, 1536, 1482, 1335, 1304; HRMS (ESI) calcd for C$_{15}$H$_{13}$N$_2$S [M+H]$^+$: 253.0799, found: 253.0805; Anal. calcd for C$_{15}$H$_{11}$N$_2$S: C, 71.40; H, 4.79; N, 11.10. Found: C, 71.47; H, 4.97; N, 11.00.

8. Flash column chromatography (7.0 cm diameter, 16 cm height (submitters: 7.0 cm diameter, 26 cm height)) was performed using silica gel (Silica Gel 60, spherical, 40–100 mm (Kanto, ca. 320 g) (submitters: 60-Å pore size, 40–63 mm, standard grade, Zeochem)). The residue was loaded using 25% dichloromethane in n-hexane (20 mL). The eluent was n-hexane for the first five fractions (fractions #1-5, ~100 mL/fraction tubes), then 5% ethyl acetate in n-hexane until fraction 42 (fractions #6-42, ~100 mL/fraction tubes), then 8.5% ethyl acetate in n-hexane until fraction 89 (fractions #43-89, ~100 mL/fraction tubes), and finally 10% ethyl acetate in n-hexane until quinoline **5** (TLC: R_f = 0.61, 20% ethyl acetate in n-hexane (submitters: TLC: R_f = 0.35, 20% ethyl acetate in hexanes), visualized by UV and KMnO$_4$ stain) completely passed through the column. The combined fractions (#27-110) were concentrated under reduced pressure (25 mmHg, 30 °C) using a rotary evaporator. Any residual 2-chloropyridine (TLC: R_f = 0.59, 20% ethyl acetate in n-hexane (submitters: R_f = 0.35, 20% ethyl acetate in hexanes), visualized by UV) can be readily removed from quinoline **5** under reduced pressure (<1 mmHg). TLC analysis was performed on EMD Silica Gel 60F plates by the submitters and on silica gel 60 F_{254} (0.25 mm, Merck) by the checkers.

9. Attempts at increasing the polarity of eluent by the checkers resulted in the contamination with 2-chloropyridine.

10. When trifluoromethanesufonic anhydride from TCI was distilled prior to use, a yield of 73% was observed. The analytical data for quinoline **5** is as follows: mp 96–100 °C; ^{1}H NMR (400 MHz, CDCl$_3$, 20 °C) δ: 3.92 (s, 3 H), 7.09 (td, 2 H, J = 8.7, 1.8 Hz), 7.44–7.55 (m, 6 H), 7.73 (tt, 1 H, J = 8.2, 1.6 Hz), 7.80 (s, 1 H), 7.95 (d, 1 H, J = 8.7 Hz), 8.19 (dd, 2 H, J = 8.7, 1.4 Hz), 8.23 (d, 1 H, J = 8.7 Hz); ^{13}C NMR (100 MHz, CDCl$_3$, 20 °C) δ: 55.4, 114.0, 119.3, 125.6, 125.9, 126.2, 127.5, 128.8, 129.2, 129.4, 130.1, 130.7, 130.8, 139.7, 148.8, 148.9, 156.9, 159.8; FTIR (neat) cm^{-1}: 3060, 2825, 1608, 1592, 1516, 1502, 1492, 1249; HRMS (ESI) calcd for C$_{22}$H$_{18}$NO [M+H]$^+$: 312.1388, found: 312.1378; Anal. calcd for C$_{22}$H$_{17}$NO: C, 84.86; H, 5.50; N, 4.50. Found: C, 84.69; H, 5.73; N, 4.35.

Safety and Waste Disposal Information

All hazardous materials should be handled and disposed of in accordance with "Prudent Practices in the Laboratory"; National Academies Press; Washington, DC, 2011.

3. Discussion

Pyridine and pyrimidine derivatives are prevalent in natural products, pharmaceuticals, and functional materials.[4,5] The majority of conventional methods for their preparation involve the condensation of ketone or carbonyl-containing fragments with amine-containing precursors. Based on key observations and guided by mechanistic insight in the electrophilic activation of amides, we developed methods for direct preparation of pyrimidines,[6] pyridines,[7] quinolines, quinazolines, and related derivatives.[8] The general methodology involves the direct condensative union of readily available N-vinyl[9] or N-aryl amides with respective nucleophiles. Our results are consistent with the formation of a uniquely electrophilic intermediate upon amide activation with the reagent combination of trifluoromethanesulfonic anhydride and 2-chloropyridine. The activated intermediate is subject to addition of various nucleophiles followed by cyclization to afford the corresponding azaheterocycle (Scheme 1).[6] More recent mechanistic studies have revealed the nature of the electrophilic intermediate is a function of the 2-halopyridine and amide substrate.[10] For example, while the use of 2-chloropyridine with N-vinyl or N-aryl amides typically results in the 2-chloropyridinium adduct, the utilization of 2-

fluoropyridine with *N*-alkyl or electron rich benzamides favors the corresponding nitrilium ion (Scheme 1).

Scheme 1 Plausible Mechanism for Condensative Azaheterocycle Synthesis

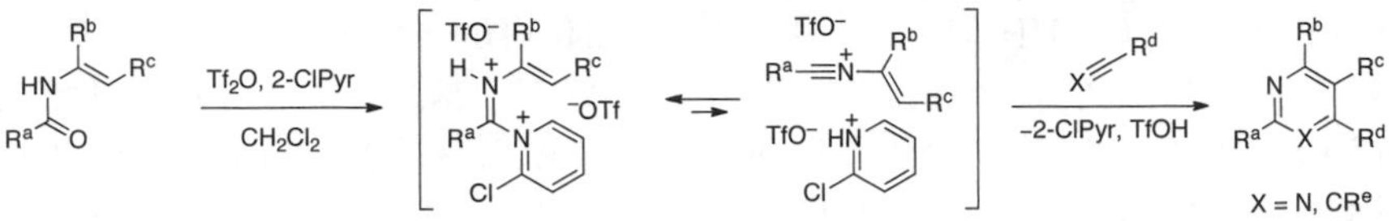

A variety of *N*-vinyl and *N*-aryl amides can be used with a range of nitriles as nucleophiles to form the desired pyrimidines and quinazolines (Table 1). As described in our original report aliphatic, aromatic, and vinyl nitriles were all competent nucleophiles in this chemistry for synthesis of the corresponding pyrimidine derivatives. Significantly, both optically active amide and nitrile substrates were used in this direct condensative pyrimidine synthesis without epimerization (Table 1, entries 2 and 3).

We recently extended this chemistry to cyanic acid–derived nucleophiles along with substrates that include carbamate, urea, or oxamate carbonyls for electrophilic activation.[8] These new substrate classes allow rapid access to densely heteroatom-substituted pyrimidine derivatives (Table 1). Furthermore, many of these rapidly accessed products are versatile precursors for preparation of other substituted azaheterocycles that were previously inaccessible under the standard amide activation conditions due to functional group incompatibility.[8] For example, the use of cyanogen bromide as the nitrile component (Table 1, entry 6) provides access to 4-bromopyrimidine derivatives. Aminolysis results in 4-amino derivatives that are not accessible using the corresponding cyanamide precursors (Scheme 2).

Scheme 2 Derivatization of 4-Bromopyrimidine Derivatives

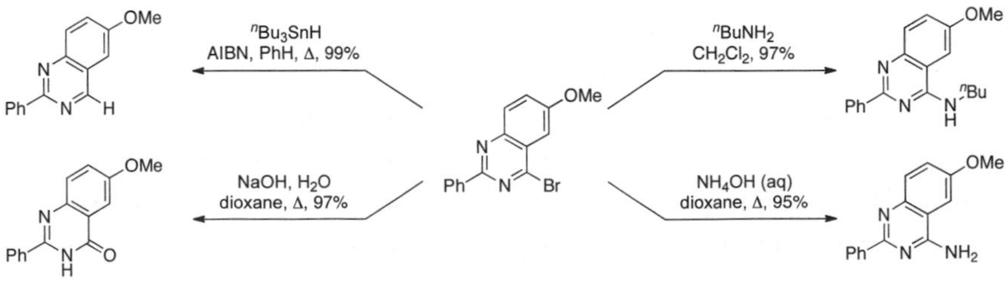

In our original communications we have reported sets of standard conditions that can be applied to a wide range of amide and nucleophile substrates. These range from highly reactive *N*-vinyl amides to less reactive electron deficient *N*-aryl benzamide substrates, and from highly reactive nitriles and ynamides to less reactive enol ethers and acetylene nucleophiles (vide infra). Most amide substrates are activated and undergo condensation with the appropriate nucleophile to the corresponding product at even subambient temperatures (Scheme 3). Typically, more forcing condensation conditions ensure further conversion to the desired product. Pyrimidine and pyridine synthesis using *N*-vinyl amides typically requires no heating and the condensation reactions are complete within minutes. As described above, even benzanilide (**1**) allows condensative azaheterocycle annulation with simple warming to room temperature. Heating in the microwave for a short time is only necessary for the most recalcitrant substrates. Thus the conditions required for the majority of substrates in this chemistry stand in contrast to a variety of other methods that require extended heating at elevated temperatures.[4,5]

Scheme 3 Condensative Azaheterocycle Synthesis at Subambient Temperatures

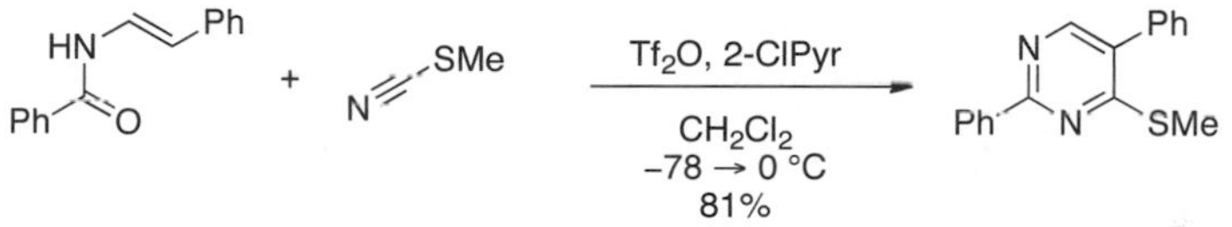

Furthermore, π-nucleophiles including alkoxy and silyloxy acetylenes, ynamines, ynamides, phenyl acetylenes, and vinyl ethers can all be used in conjunction with activated *N*-vinyl and *N*-aryl amide substrates to provide pyridines and quinolines, respectively (Table 2). Interestingly, electron-rich alkynes were found to be far more reactive than vinyl ether π-nucleophiles and provided the desired heterocycles in higher yields under milder conditions. Consistent with the observations described above in the context of pyrimidine synthesis, the condensation of an optically active amide with an ynamide provided the desired pyridine without heating (Table 2, entry 4) or loss of enantiomeric enrichment. In the case of enol ethers an excess of the nucleophile is recommended for greater efficiency in obtaining the desired product.

In addition to the synthesis of azaheterocycles, this electrophilic activation methodology can also be used for *N*-pyridinylation of amides.[10] The addition of pyridine *N*-oxides to amides activated in the presence of trifluoromethanesulfonic anhydride and 2-fluoropyridine[10] as base affords the corresponding *N*-pyridinyl amide in a single step (Scheme 4).

Scheme 4 Plausible Mechanism for *N*-Pyridylation of Amides

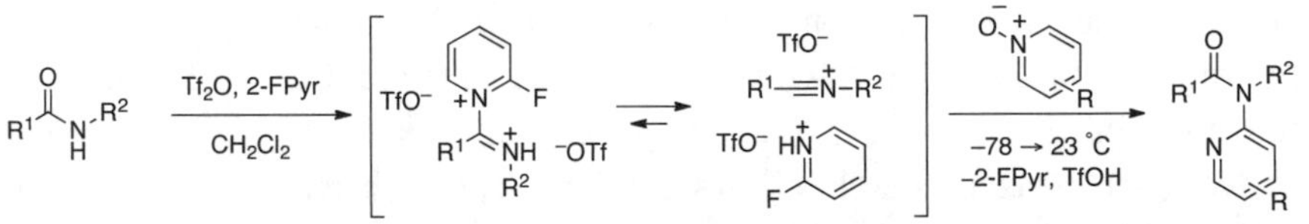

Table 1 Condensative Synthesis of Pyrimidine Derivatives

Entry	Amide	Nitrile	Conditions	Product	Yield (%)[a]
1			A		86[b]
2		96%ee[d]	B	96%ee[d]	71[c]
3	93%ee[d]		B	93%ee[d]	59[c]
4			A		99
5			C		63
6			D		68
7			A		82[b]
8			A		71
9			C		70
10			C		59

[a] Average of two experiments. Uniform conditions unless otherwise noted: Tf$_2$O (1.1 equiv), 2-ClPyr (1.2 equiv), nitrile (2.0 equiv), CH$_2$Cl$_2$, heating: A = 23 °C, 1 h; B = 45 °C, 1 h; C = microwave, 140 °C, 10 min. D = 1,2-dichloroethane, reflux, 2 h, nitrile (4.0 equiv). [b] 1.1 equiv of nitrile. [c] 3.0 equiv of nitrile. [d] E.e. determined by chiral HPLC analysis of a derivative.

Table 2 Condensative Synthesis of Pyridine Derivatives

Entry	Amide	Nucleophile	Conditions	Product	Yield (%)[a]
1	(Me, Me, cHx) enamide	≡—OEt	A	pyridine (Me, Me, cHx, OEt)	83[b,c]
2	dihydropyran enamide, Ph	nBu—≡—OTIPS	A	pyrano-pyridine (Ph, nBu, OTIPS)	61[b,c,d]
3	(Me, Me, cHx) enamide	oxazolidinone ynamide, Ph	A	pyridine (Me, Me, cHx, Ph, oxazolidinone)	80[c]
4	cyclohexenyl amide, Me, Me 94%ee[f]	azetidinone ynamide, Ph	A	pyridine (Me, Me, Ph, lactam) 94%ee[f]	67
5	(Me, Me, cHx) enamide	TMS—≡—morpholine	A	pyridine (Me, Me, cHx, morpholine, TMS)	84[b,c]
6	4-NO2-phenyl amide, cHx	Me—≡—C6H4-OMe	A	quinoline (NO2, cHx, Me, C6H4OMe)	84[c]
7	3,5-(OMe)2-phenyl amide, cHx	OTMS, Ph enol ether	B	quinoline (OMe, OMe, cHx, Ph)	61
8	4-OMe-phenyl amide, cHx	cyclopentenyl-OTBS	C	cyclopenta-quinoline (OMe, cHx)	77[e]
9	cyclohexenyl amide, Ph	vinyl-OTPS	C	tetrahydroquinoline (Ph)	69[c]
10	4-OMe-phenyl amide, cHx	indanone-OTBS	C	indeno-quinoline (OMe, cHx)	78[e]

Reagents at top: amide (1 equiv) + nucleophile (1.1 equiv) or enol ether (2.0 equiv), Tf2O, 2-ClPyr → product

[a] Average of two experiments. Uniform conditions unless otherwise noted: Tf$_2$O (1.1 equiv), 2-ClPyr (1.2 equiv), nucleophile, CH$_2$Cl$_2$, heating: A = 23 °C, 1 h; B = 45 °C, 1 h; C = microwave, 140 °C, 20 min. [b] nucleophile (2.0 equiv). [c] 2-ClPyr (2.0 equiv). [d] 10 min. [e] 1 h. [f] E.e. determined by chiral HPLC analysis of a derivative.

558

1. Department of Chemistry, Massachusetts Institute of Technology, Cambridge, MA 02139, E-mail: movassag@mit.edu.

2. The submitters acknowledge financial support from NIH-NIGMS (GM074825) and partial support from NSF (547905).

3. Pangborn, A. B.; Giardello, M. A.; Grubbs, R. H.; Rosen, R. K.; Timmers, F. J. *Organometallics* **1996**, *15*, 1518.

4. For pyrimidine reviews, see: (a) Undheim, K.; Benneche, T. In *Comprehensive Heterocyclic Chemistry II*; Katritzky, A. R.; Rees, C. W.; Scriven, E. F. V.; McKillop, A., Eds; Pergamon: Oxford, 1996; Vol. 6; p 93. (b) Lagoja, I. M. *Chem. & Biodiversity* **2005**, *2*, 1. (c) Michael, J. P. *Nat. Prod. Rep.* **2005**, *22*, 627. (d) Joule, J. A.; Mills, K. In *Heterocyclic Chemistry*, 4th ed.; Blackwell Science Ltd.: Cambridge, MA, 2000; p 194. (e) Erian, A. W. *Chem. Rev.* **1993**, *93*, 1991. (d) Joule, J. A.; Mills, K. In *Heterocyclic Chemistry*, 4th ed.; Blackwell Science Ltd.: Cambridge, MA, 2000; p 194. (e) Erian, A. W. *Chem. Rev.* **1993**, *93*, 1991.

5. For pyridine reviews, see: (a) Undheim, K.; Benneche, T. In *Comprehensive Heterocyclic Chemistry II*; Katritzky, A. R., Rees, C. W., Scriven, E. F. V., McKillop, A., Eds.; Pergamon: Oxford, 1996; Vol. 6; p 93. (b) Jones, G. In *Comprehensive Heterocyclic Chemistry II*; Katritzky, A. R.; Rees, C. W.; Scriven, E. F. V.; McKillop, A., Eds; Pergamon: Oxford, 1996; Vol. 5; p 167. (c) Joule, J. A.; Mills, K. In *Heterocyclic Chemistry*, 4th ed.; Blackwell Science Ltd.: Cambridge, MA, 2000; p 194. (d) Henry, G. D. *Tetrahedron* **2004**, *60*, 6043. (e) Michael, J. P. *Nat. Prod. Rep.* **2005**, *22*, 627. (f) Abass, M. *Heterocycles* **2005**, *65*, 901.

6. Movassaghi, M.; Hill, M. D. *J. Am. Chem. Soc.* **2006**, *128*, 14254.

7. Movassaghi, M.; Hill, M. D.; Ahmad, O. K. *J. Am. Chem. Soc.* **2007**, *129*, 10096.

8. Ahmad, O. K.; Hill, M. D.; Movassaghi, M. *J. Org. Chem.* **2009**, *74*, 8460.

9. (a) Muci, A. R.; Buchwald, S. L. *Top. Curr. Chem.* **2002**, *219*, 131. (b) Hartwig, J. F. In *Handbook of Organopalladium Chemistry for Organic Synthesis*; Negishi, E., Ed.; Wiley-Interscience: New York, 2002; p. 1051. (c) Beletskaya, I. P.; Cheprakov, A. V. *Coordin. Chem. Rev.* **2004**, *248*, 2337. (d) Dehli, J. R.; Legros, J.; Bolm, C. *Chem. Commun.* **2005**, 973.

10. Medley, J. W.; Movassaghi, M. *J. Org. Chem.* **2009**, *74*, 1341.

Appendix
Chemical Abstracts Nomenclature (Registry Number)

Benzanilide: Benzamide, *N*-phenyl-; (93-98-1)

Dichloromethane: Methane, dichloro-; (75-09-2)

2-Chloropyridine: Pyridine, 2-chloro-; (109-09-1)

Trifluoromethanesulfonic anhydride: Methanesulfonic acid, 1,1,1-trifluoro-, 1,1'-anhydride; (358-23-6)

Thiocyanic acid methyl ester: Thiocyanic acid, methyl ester; (556-64-9)

p-Methoxyphenylacetylene: Benzene, 1-ethynyl-4-methoxy-; (26797-70-6)

Mohammad Movassaghi carried out his undergraduate research with Professor Paul A. Bartlett at UC Berkeley, where he received his BS degree in chemistry in 1995. Mo then joined Professor Andrew G. Myers' group for his graduate studies and was a Roche Predoctoral Fellow at Harvard University. In 2001, Mo joined Professor Eric N. Jacobsen's group at Harvard University as a Damon Runyon–Walter Winchell Cancer Research Foundation postdoctoral fellow. In 2003, he joined the chemistry faculty at MIT where his research program has focused on the total synthesis of alkaloids in concert with the discovery and development of new reactions for organic synthesis.

Omar K. Ahmad pursued his undergraduate studies at Brown University, where he received his Sc.B. degree in chemistry in 2005. As an undergraduate he worked in the laboratories of Professor Dwight Sweigart and Professor Amit Basu. In 2005, Omar joined Professor Mohammad Movassaghi's research group at MIT for his graduate studies where his research has focused on the development of new methodologies for organic synthesis.

560

Jonathan William Medley pursued his undergraduate studies at Harvard University, where he received his A.B. degree in chemistry and physics with a minor in math in 2007. While an undergraduate, he worked in the laboratory of Professor David A. Evans. In 2007, he joined the laboratory of Professor Mohammad Movassaghi at MIT for his graduate studies, which have focused on the development of new methods for organic synthesis and their application towards the total syntheses of alkaloid natural products.

Alexis Coste carried out his graduate studies from 2007 to 2010 in the group of Professor François Couty under the supervision of Dr. Gwilherm Evano at the University of Versailles (France) as a National Cancer Institute Fellow. His work focused on the development of copper-catalyzed transformations and the total synthesis of the macrocyclic peptide TMC-95A. In 2011, he joined Professor Mohammad Movassaghi's research group at MIT for pursuing postdoctoral studies where he is currently working on the total synthesis of alkaloids.

Hiromi Kaise was born in Saitama, Japan in 1989. She received her B.S. in 2011 from the University of Tokyo. In the same year, she began her graduate studies at the Graduate School of Pharmaceutical Sciences, the University of Tokyo, under the guidance of Professor Tohru Fukuyama. Her research interests are in the area of the total synthesis of natural products.

CUMULATIVE AUTHOR INDEX FOR VOLUMES 85-89

This index comprises the names of contributors to Volumes **85**, **86**, **87**, **88**, and **89**. For authors of previous volumes, see either indices in Collective Volumes I through XI, or the single volume entitled *Organic Syntheses, Collective Volumes I-VIII, Cumulative Indices,* edited by J. P. Freeman.

Org. Synth. **2012**, 89

Bolm, C., **85**, 106
Bonjoch, J., **88**, 317, 330
Bradshaw, B., **88**, 317, 330
Braun, M., **86**, 47
Bravo-Altamirano, K., **85**, 96
Brawn, R. A., **89**, 420
Brétéché, G., **85**, 278
Brewer, M., **85**, 189
Brill, M., **89**, 55
Bringmann, G., **88**, 70
Bryan, C., **86**, 36
Bulger, P. G., **87**, 8
Bull, J. A., **87**, 170
Burke, M. D., **86**, 344
Butler, C. R., **86**, 274
Butler, J. D., **87**, 339
Buzon, R. A., **87**, 16

Cabart, F., **85**, 147
Cacchi, S., **88**, 260
Caffrey, M., **89**, 183
Caggiano, L., **86**, 121
Cai, Z., **88**, 309
Campbell, L., **85**, 15
Campeau, L. -C., **88**, 22
Carreira, E. M., **87**, 88
César, V., **85**, 34
Chalker, J. M., **87**, 288
Chaloin, O., **85**, 147
Chandrasekaran, P., **86**, 333
Chang, S., **85**, 131
Chang, Y., **87**, 245
Charette, A. B., **87**, 115, 170
Chen, Q.-Y., **87**, 126
Chen, Y., **88**, 377
Chen, Y., **89**, 294
Chiong, H. A., **86**, 105
Cho, S. H., **85**, 131
Chouai, A., **86**, 141, 151
Christie, H. S., **88**, 364
Clososki, G. C., **86**, 374
Coates, G. W., **86**, 287
Cole, K. P., **89**, 537
Comte, A., **89**, 510

564

Endo, K., **86**, 325
Erkkilä, A., **87**, 201
Etxebarria-Jardi, G., **88**, 317, 330
Evano, G., **87**, 231

Fagnou, K., **88**, 22
Fandrick, D. R., **89**, 210
Fei, X.-S., **87**, 126
Fidan, M., **86**, 47
Fillion, E., **89**, 115
Fleming, M. J., **85**, 1
Fletcher, A. M., **87**, 143
Flores-Gaspar, A., **89**, 159
Fokin, V. V., **88**, 238
Foubelo, F., **89**, 88
Franck, G., **89**, 55
Franckevičius, V., **85**, 72
Fraser, C. L., **89**, 76, 82
Fu, G. C., **87**, 299, 310, 317, 330
Fu, R., **87**, 263
Fujimoto, T., **86**, 325
Fujiwara, H., **86**, 130
Fukuyama, T., **86**, 130; **88**, 152, 388; **89**, 480, 501
Fürstner, A., **85**, 34; **86**, 298

Gálvez, E., **86**, 70, 81
Ganem, B., **89**, 404
Gaspar, B., **87**, 88
Georg, G. I., **88**, 427
Giguère-Bisson, M., **88**, 14
Gillard, J. R., **89**, 131
Gillis, E. P., **86**, 344
Glasnov, T. N., **86**, 252
Glorius, F., **85**, 267
Goldmann, P. L., **89**, 19
Gong, L., **88**, 406
González-Gómez, J. C., **89**, 88
Gooßen, L. J., **85**, 196
Goss, J. M., **86**, 236
Goudreau, S. R., **87**, 115
Graham, T. H., **88**, 42
Greszler, S., **86**, 18
Grubbs, R. H., **89**, 170
Guichard, G., **85**, 147

Guillena, G., **88**, 317, 330
Gulder, T. A. M., **88**, 70
Gulder, T., **88**, 70
Guzzo, P. R., **89**, 44

Haddadin, M. J., **87**, 339
Hahn, B. T., **85**, 267
Haney, B. P., **89**, 44
Hans, M., **87**, 77
Hansen, T. V., **89**, 220
Harada, S., **85**, 118
Harmata, M., **88**, 309
Hart, D. J., **89**, 183
Hartung, J., **89**, 409
Hartwig, J. F., **88**, 4
Hayashi, T., **89**, 126
Hein, J. E., **88**, 238
Helmchen, G., **89**, 55
Henry-Riyad, H., **88**, 121
Hierl, E., **85**, 64
Hill, M. D., **85**, 88
Hirao, T., **89**, 73
Hoang, T. T., **88**, 353
Hodgson, D. M., **85**, 1
Horning, B. D., **88**, 42
Hossain, M. M., **86**, 172
Hu, W., **88**, 406, 418
Huang, C., **88**, 309
Huang, D. S., **88**, 4
Huang, K., **87**, 26, 36
Huard, K., **86**, 59
Hughes, C. O., **88**, 364
Humphreys, P. G., **85**, 1
Hwang, S. J., **85**, 131
Hwang, S., **86**, 225

Ichikawa, J., **88**, 162
Ideue, E., **89**, 501
Ishihara, K., **89**, 105
Ishii, Y., **89**, 307
Iwasaki, M., **88**, 238

566

La Vecchia, L., **85**, 295
Landais, Y., **86**, 1
Landis, C. R., **89**, 243
Langenhan, J. M., **87**, 192
Langle, S., **85**, 231
Larock, R. C., **87**, 95; **88**, 377; **89**, 294
Lathrop, S. P., **87**, 350
Lautens, M., **85**, 172, **86**, 36
Lazareva, A., **86**, 105
Le, H. V., **89**, 404
Lebel, H., **86**, 59, 113
Lebeuf, R., **86**, 1
Lee, H., **87**, 245
Lee, H. -Y., **89**, 66
Lee, J., **89**, 66, 183
Le Gall, E., **89**, 283
Leighty, M. W., **88**, 427
Leogane, O., **86**, 113
Leonelli, F., **89**, 311
Lesser, A. B., **88**, 109
Ley, S. V., **85**, 72
Li, B., **87**, 16
Li, C. -J., **88**, 14
Li, H., **89**, 460
Li, X., **89**, 491
Liao, E-T., **89**, 115
Linder, C., **85**, 196
Lisboa, M. P., **88**, 353
List, B., **86**, 11
Liu, J. H. -C., **88**, 102
Liu, W., **88**, 406
Liu, T., **89**, 76, 82
Longbottom, D. A., **85**, 72
Looper, R. E., **89**, 394
Lou, S., **86**, 236; **87**, 299, 310, 317, 330
Lou, T., **89**, 115
Lu, C. -D., **85**, 158
Lu, K., **86**, 212
Lyons, J., **89**, 183

MacMillan, D. W. C., **88**, 42
Madanmohan, J., **89**, 183
Mani, N. S., **85**, 179
Mans, D. J., **85**, 238, 248

Marcoux, D., **87**, 115
Margarita, R., **89**, 311
Marin, J., **85**, 147
Markina, N. A., **88**, 377
Marra, A., **89**, 323
Marshall, A.-L., **87**, 192
Martens, T., **89**, 283
Martin, R., **89**, 159
Matsunaga, S., **85**, 118
Maw, G., **85**, 219
McAllister, G. D., **85**, 15
McDermott, R. E., **85**, 138
McDonald, F. E., **88**, 296
McLaughlin, E. C., **89**, 19
McLaws, M. D., **89**, 44
McNaughton, B. R., **85**, 27
Medley, J. W., **89**, 549
Meletis, P., **86**, 47
Meng, T., **87**, 137
Meyer, H., **85**, 287, 295
Miller, B. L., **85**, 27
Milne, J. E., **89**, 519
Miyaura, N., **88**, 79, 202, 207
Mohr, J. T., **86**, 181, 194
Montchamp, J.-L., **85**, 96
Moore, D. A., **85**, 10
Morán-Ramallal, R., **88**, 224
Morera, E., **88**, 260
Morken, J. P., **88**, 342
Morra, N. A., **85**, 53
Morshed, M. M., **86**, 172
Mosa, F., **85**, 219
Mousseau, J. J., **87**, 170
Movassaghi, M., **85**, 88; **89**, 230, 549
Muchalski, H., **88**, 212
Mudryk, B., **85**, 64
Müller-Hartwieg, J. C. D., **85**, 295
Murakami, K., **87**, 178

Nájera, C., **88**, 317, 330
Nakamura, E., **86**, 325
Nakamura, M., **86**, 325
Ngi, S. I., **85**, 231
Nguyen, H., **88**, 121

Nicewicz, D. A., **85**, 278
Nixon, T. D., **86**, 28
Noji, T., **88**, 388

Oelke, A. J., **85**, 72
Oh, S., **88**, 121
Ohshima, T., **85**, 118
Okano, K., **86**, 130; **88**, 152, 388
Olofsson, B., **86**, 308
Ortar, G., **88**, 260
Ortiz-Marciales, M., **87**, 26, 36
Osajima, H., **86**, 130
Oshima, K., **87**, 178

Pagenkopf, B. L., **85**, 53
Pan, S. C., **86**, 11
Partridge, B. M., **88**, 247
Pihko, P. M., **87**, 201
Pu, L., **87**, 59, 68
Punniyamurthy, T., **88**, 398

Rafferty, R. J., **86**, 262
Ragan, J. A., **85**, 138
RajanBabu, T. V., **85**, 238, 248
Reynolds, N. T., **87**, 350
Riebel, P., **85**, 64
Robert, F., **86**, 1
Roberts, P. M., **87**, 143
Rohbogner, C. J., **86**, 374
Romea, P., **86**, 70, 81
Romo, D., **88**, 121
Roush, W. R., **88**, 87, 181
Rovis, T., **87**, 350

Saha, P., **88**, 398
Schaus, S. E., **86**, 236
Schiffers, I., **85**, 106
Schramm, M. P., **87**, 253
Schwekendiek, K., **85**, 267
Scott, M. E., **85**, 172
Sebesta, R., **85**, 287
Sedelmeier, G., **85**, 72
Seebach, D., **85**, 287, 295
Sepulveda, D., **88**, 121

Shi, F., **87**, 95
Shi, Y., **86**, 263, 315
Shibasaki, M., **85**, 118
Shubinets, V., **87**, 253
Simanek, E. E., **86**, 141, 151
Singh, S. P., **87**, 275
Sirois, L. E., **88**, 109
Slatford, P. A., **86**, 28
Smith, C. R., **85**, 238, 248
Snaddon, T. N., **85**, 45
Snyder, S. A., **88**, 54
Söderberg, B. C., **88**, 291
Solano, D. M., **87**, 339
Sperotto, E., **85**, 209
Spletstoser, J. T., **88**, 427
Stambuli, J. P., **88**, 33
Stanek, K., **88**, 168
Stepanenko, V., **87**, 26
Stevens, K. L., **86**, 18
Stoltz, B. M., **86**, 161, 181, 194
Storgaard, M., **86**, 360
Struble, J. R., **87**, 362
Sugai, J., **88**, 79
Sun, H., **88**, 87, 181

Takita, R., **85**, 118
Takizawa, M., **88**, 79
Tambar, U. K., **86**, 161
Taylor, R. J. K., **85**, 15
Thibonnet, J., **85**, 231
Thirsk, C., **85**, 219
Thompson, A. L., **87**, 288
Tian, W.-S., **87**, 126
Ting, P., **87**, 137
Togni, A., **88**, 168
Tokuyama, H., **86**, 130; **88**, 152, 388
Tong, R., **88**, 296
Treitler, D. S., **86**, 287, **88**, 54
Troyer, T. L., **88**, 212
Truc, V., **85**, 64
Tseng, N. -W., **85**, 172
Turlington, M., **87**, 59, 68

Urpí, F., **86**, 70, 81

CUMULATIVE SUBJECT INDEX FOR VOLUMES 85-89

This index comprises subject matter for Volumes **85**, **86**, **87**, **88**, and **89**. For subjects in previous volumes, see either the indices in Collective Volumes I through XI or the single volume entitled *Organic Syntheses, Collective Volumes I-VIII, Cumulative Indices,* edited by J. P. Freeman.

The names of many of the compounds are isted in two forms. The first is the name used commonly in procedures. The second is the systematic name according to Chemical Abstracts nomenclature, accompanied by its registry number in parentheses. Also included are general terms for classes of compounds, types of reactions, special apparatus, and unfamiliar methods.

Most chemicals used in the procedure will appear in the index as written in the text. There generally will be entries for all starting materials, reagents, important by-products, and products, which are indicated by the use of italics.

Aldol; **86**,11, 81, 92; **88**, 121, 330, 364
Alkene Formation; **85**, 1, 15, 248
Alkenylation; **87**, 143, 170, 201, 231; **88**, 138, 291, 342, 388; **89**, 527
Alkylation; **85**, 10, 34, 45, 88, 158, 295; **86**, 1, 28, 47, 161, 194, 262, 298, 325; **87**, 36, 59, 339, 350, 362; **88**, 87, 121, 168, 181, 247; **89**, 88, 115, 159, 323, 438, 510, 519
Allenes; **88**, 138, 309; **89**, 267, 420
Allyl alcohol: 2-Propen-1-ol; (107-18-6) **86**, 194; **87**, 226
9-Allylanthracene: 9-(2-propeny-1-yl)anthracene; (23707-65-5) **89**, 230
Allyl bromide; (106-95-6) **85**, 248 ; **86**, 262
Allyl bromide: 3-Bromo-1-propene; (106-95-6) **89**, 88
(*S*)-(−)-2-Allylcyclohexanone: Cyclohexanone, 2-(2-propen-1-yl)-, (2*S*)-; (36302-35-9) **86**, 47
(+)-B-Allyldiisopinocampheylborane ((+)-(Ipc)₂B(allyl) or (ᵗIpc)₂B(allyl)); (106356-53-0) **88**, 87
Allylmagnesium bromide; (1730-25-2) **88**, 87
Allyl methyl carbonate: Carbonic acid, methyl 2-propen-1-yl ester; (35466-83-2) **86**, 47
(S)-2-(2-Allyl-2-methylcyclohexylidene)hydrazinecarboxamide: Hydrazinecarboxamide, 2-[(2S)-2-methyl-2-(2-propenyl)cyclohexylidene]-, (2E)-; (812639-25-1) **86**, 194
Allyl 1-methyl-2-oxocyclohexanecarboxylate: Cyclohexanecarboxylic acid, 1-methyl-2-oxo-, 2-propenyl ester; (7770-41-4) **86**, 194
Allylpalladium chloride dimer; (12012-95-2) **89**, 230
(3R,7aR)-7a-Allyl-3-(trichloromethyl)tetrahydropyrrolo[1,2-c]oxazol-1(3H)-one; (220200-87-3) **86**, 262
Aluminum Chloride; (7446-70-0) **89**, 210
Amide formation; **86**, 92; **87**, 1, 218, 350; **88**, 14, 42, 398, 427; **89**, 44, 66, 432
(S)-2-Amido-pyrrolidine-1-carboxylic acid benzyl ester; (34079-31-7) **85**, 72
Amination; **87**, 1, 143, 263, 362
2'-Aminoacetophenone; (551-93-9) **88**, 33
o-Aminobenzaldehyde: Benzaldehyde, 2-amino-; (529-23-7) **89**, 274
2-Aminobenzyl alcohol: Benzenemethanol, 2-amino-; (5344-90-1) **87**, 339
(Sₐ)-N-[2'-Amino-(1,1'-binaphthyl)-2-yl]-4-methylbenzene-sulfonamide; (933782-32-2) **88**, 317
3-Aminocrotononitrile; (1118-61-2) **89**, 537
Aminodiphenylmethane: Benzenemethanamine, α-phenyl-; (91-00-9) **88**, 224
(1*R*, 2*S*)-(+)-*cis*-1-Amino-2-indanol; (136030-00-7) **87**, 3502*S*)-2-Amino-3-methyl-1,1-diphenyl-butanol; (78603-95-9) **87**, 26
(S)-3-Amino-4-methylpentanoic acid; (s)-β-homovaline; (s)-β-leucine (40469-85-0) **87**, 143
2-Amino-5-methylpyridine: 2-Amino-5-picoline; (1603-41-4) **89**, 76
(*R*)-2-Amino-4-phenylbutan-1-ol: Benzenebutanol, β-amino-, (β*R*)-; (761373-40-4) **87**, 310
(*S*)-2-Amino-4-phenylbutan-1-ol: Benzenebutanol, β-amino-, (β*S*)-; (27038-09-1) **87**, 310
(E)-1-(2-Aminophenyl)ethanone oxime; (4964-49-2) **88**, 33
(2*S*)-(-)-2-Amino-3-phenyl-1-propanol; (3182-95-4) **85**, 267
Ammonium bicarbonate; (1066-33-7) **85**, 72
Androst-4-ene-3,17-dione, 4-(trifluoromethyl)-; (201664-30-4) **87**, 126
Androst-4-ene-3,17-dione, 4-bromo-; (19793-14-7) **87**, 126
Androst-4-ene-3,17-dione; (63-05-8) **87**, 126

Androst-5-en-17-one, 3-[[(1,1-dimethylethyl)dimethylsilyl]oxy]-, (3β)-; (42151-23-5) **89,** 19

Androst-5-ene-7,17-dione, 3-[[(1,1-dimethylethyl)dimethylsilyl]oxy]-, (3β)-; (157302-54-0) **89,** 19

Androst-5-ene-7,17-dione, 3-hydroxy-, (3β)-; (566-19-8) **89,** 19

Aniline: benzeneamine; (62-53-3) **88,** 212

Annulation; **88,** 330

Anthracene-9-carboxaldehyde; (642-31-9) **89,** 230

9-Anthracenemethanol; (1468-95-7) **88,** 418

1-(Anthracen-9-yl)allyl ethyl carbonate; **89,** 230

9-Anthraldehyde; (642-31-9) **88,** 418

Antimony pentachloride; (7647-18-9) **88,** 54

Aqueous HBF₄, Borate(1-), tetrafluoro-, hydrogen (1:1): (16872-11-0) **85,** 34

Arylation; **87,** 178, 209

7-Aza-bicyclo[4.1.0]heptane: Cyclohexene imine; (286-18-0) **87,** 161

Azides; **87,** 161

Aziridines; **87,** 161; **88,** 224

Benzaldehyde; (100-52-7) **85,** 118; **86,** 212; **87,** 1, 231; **88,** 14, 418

Benzaldehyde-*p*-anisidine imine; (783-08-4) **88,** 418

Benzamide, 2,2',2'',2'''-[1,2-phenylenebis[(1S,3S)-tetrahydro-5,8-dioxo-1H-
    [1,2,4]diazaphospholo[1,2-a]pyridazine-2,1,3(3H)-triyl]]tetrakis[N-[(1S)-1-
    phenylethyl]-; (851770-14-4) **89,** 243

Benzanilide: Benzamide, N-phenyl-; (93-98-1) **89,** 549

Benzenamine, 2-(2,2-dibromoethenyl)-; (167558-54-5) **86,** 36

Benzene, 1-(2,2-dibromoethenyl)-2-nitro-; (253684-24-1) **86,** 36

*(β R)-Benzenepropanoic acid, α-acetyl-β-[[[(phenylmethyl)amino]carbonyl]amino]-,
    methyl ester; (865086-76-6)* **86,** 236

*(β R)-Benzenepropanoic acid, α-acetyl-β-[[(2-propen-1-yloxy)carbonyl]amino]-, methyl
    ester; (921766-57-6)* **86,** 236

Benzenesulfinic acid sodium salt; (873-55-2) **86,** 360

Benzenesulfonamide, N-[[(1,1-dimethylethyl)dimethylsilyl]oxy]-4-methyl-; (1028432-
    04-3) **89,** 480

5-Benzo[1,3]dioxol-5-yl-3-(4-chloro-phenyl)-1-methyl-1H-pyrazole; **85,** 179

3-Benzofurancarboxylic acid, 5-chloro-, ethyl ester; (899795-65-4) **86,** 172

Benzophenone; (119-61-9) **85,** 248 ; **87,** 245

Benzophenone hydrazone; (5350-57-2) **85,** 189

Benzoyl chloride; (98-88-4) **88,** 398

Benzoyl fluoride: Benzoyl fluoride; (455-32-3) **89,** 9

N-Benzoyl pyrrolidine: Methanone, phenyl-1-pyrrolidinyl-; (3389-54-6) **87,** 1

Benzyl alcohol; (100-51-6) **86,** 28

Benzylamine; (100-46-9) **85,** 106; **89,** 432

(1R,2R)-trans-2-(N-Benzyl)amino-1-cyclohexanol; (141553-09-5) **85,** 106

(1S,2S)-trans-2-(N-Benzyl)amino-1-cyclohexanol; (322407-34-1) **85,** 106

Benzyl azide: (Azidomethyl)benzene; (622-79-7) **88,** 238

Benzyl bromide: Benzene, (bromomethyl)-; (100-39-0) **87,** 36, 137, 170

Benzyl bromoacetate: Acetic acid, 2-bromo-, phenylmethyl ester; (5437-45-6) **89,** 501

N-Benzyl-2-bromo-N-phenylbutanamide: Butanamide, 2-bromo-N-phenyl-N-
    (phenylmethyl)-; (851073-30-8) **87,** 330

Benzyl carbamate: Carbamic acid, phenylmethyl ester; (621-84-1) **85**, 287; **88**, 152

Benzyl chloroformate, carbobenzoxy chloride; (501-53-1) **88**, 42

Benzyl chloromethyl ether: Benzene, [(chloromethoxy)methyl]-; (3587-60-8) **85**, 45

1-Benzyl-3-(4-chloro-phenyl)-5-p-tolyl-1H-pyrazole; (908329-95-3) **85**, 179

(S)-N-Benzyl-7-cyano-2-ethyl-n-phenylheptanamide: Heptanamide, 7-cyano-2-ethyl-N-phenyl-N-(phenylmethyl)-, (2S)-; (851073-44-4) **87, 330**

Benzyl diazoacetate: Acetic acid, 2-diazo-, phenylmethyl ester; (52267-51-3) **89**, 501

(1R)-O-Benzyl-2,3-di-O-isopropylidene-1-(2-thiazolyl)-D-glycitol; (103795-11-5) **89**, 323

Benzylhydrazine dihydrochloride; (20570-96-1) **85**, 179

Benzyl hydroxymethyl carbamate: Carbamic acid, (hydroxymethyl)-, phenylmethyl ester; (31037-42-0) **85**, 287

(E)-N-Benzylidene-4-methylbenzenesulfonamide: Benzenesulfonamide, 4-methyl-N-(phenylmethylene)-, [N(E)]-; (51608-60-7) **86**, 212; **88**, 138

Benzyl isopropoxymethyl carbamate; **85**, 287

2-O-Benzyl-3,4-isopropylidene-D-erythrose; (103975-12-6) **89**, 323

(2R,4S) [2-Benzyl-3-(4-isopropyl-2-oxo-5,5-diphenyl-3-oxazolidinyl)-3-oxopropyl]carbamic acid benzyl ester; (218800-56-7) **85**, 295

Benzylmagnesium chloride: Magnesium, chloro(phenylmethyl)-; (6921-34-2) **88**, 54

1-Benzyl-3-methylindolin-2-one; (34943-91-4) **89**, 438

(2Z)-4-(Benzyloxy)but-2-enal oxime; **89**, 480

4-Benxyloxy-2-buten-1-ol: 2-Buten-1-ol, 4-(phenylmethoxy)-, (2Z)-; (81028-03-7) **89**, 480

*(Z)-N-(4-(Benzyloxy)but-2-en-1-yl)-N-((tert-*butyldimethylsilyl) oxy)-4-methylbenzenesulfonamide; **89**, 480

(Z)-2-Benzyloxycarbonylamino-3-(2-bromo-4,5-dimethoxyphenyl) acrylic acid methyl ester: 2-Propenoic acid, 3-(2-bromo-4,5-dimethoxyphenyl)-2-[[(phenylmethoxy)carbonyl]amino]-, methyl ester, (2Z)-; (873666-89-8) **88**, 388

(R)-2-(Benzyloxycarbonylaminomethyl)-3-phenylpropanoic acid: Benzenepropanoic acid, α-[[[(phenylmethoxy)carbonyl]amino]methyl]-, (aR) ; (132696-47-0) **85**, 295

Benzyloxymethoxy 1-hexyne: Benzene, [[(1-hexyn-1-yloxy)methoxy]methyl]-; (162552-11-6) **85**, 45

Benzyloxymethoxy-2,2,2-trifluoromethyl ether: Benzene, [[(2,2,2-trifluoroethoxy)methoxy]methyl]-; (153959-88-7) **85**, 45

*(R)-N-Benzyl-N-(α-*methylbenzyl)amine; *(R)-N-benzyl-α-*phenylethylamine; *(R)-N-*benzyl-1-phenylethylamine; (38235-77-7) **87, 143**

N-Benzyl-4-phenylbutyramide: Benzenebutanamide, N-(phenylmethyl)-; (179923-27-4) **89**, 432

Benzyl propagyl ether; (4039-82-1) **86**, 225

(S)-1-(4-Benzyl-2-thioxothiazolidin-3-yl)propan-1-one: 1-Propanone, 1-[(4S)-4-(phenylmethyl)-2-thioxo-3-thiazolidinyl]-; (263764-23-4) **88**, 364

Bicyclopropylidene (8); Cyclopropane, cyclopropylidene- (9); (27567-82-4) **89**, 334

Rac-BINAP: Phosphine, 1, 1'-[1,1'binaphthalene]-2,2'-diylbis[1,1-diphenyl-; (98327-87-8) **89**, 159

(R)-BINOL; (R)-(+)-1,1'-Bi(2-naphthol); (18531-94-7) **85**, 238; **88**, 418

(S)-BINOL: [1,1'-Binaphthalene]-2,2'-diol, (1S)-: (18531-99-2) **85**, 118

(R)-(1,1'-Binaphthalene-2,2'-dioxy)chlorophosphine: (R)-Binol-P-Cl; (155613-52-8) **85**, 238

Bis(triphenylphosphine)palladium dichloride; (13965-03-2) **89**, 294

Boc-diallylamine: Carbamic acid, di-2-propenyl-, 1,1-dimethylethyl ester; (151259-38-0) **89**, 170

Boc$_2$O: Dicarbonic acid, C,C'-bis(1,1-dimethylethyl) ester; (244424-99-5) **86**, 113, 374

N-Boc-3-pyrroline (73286-70-1) **89**, 170

9-Borabicyclo[3.3.1]nonane: 9-BBN; (280-64-8) **87**, 299; **88**, 207; **89**, 183

(5-(9-Borabicyclo[3.3.1]nonan-9-yl)pentyloxy)triethylsilane: 9-Borabicyclo[3.3.1]-nonane, 9-[5-[(triethylsilyl)oxy]pentyl]-; (157123-09-6) **87**, 299

Borane-dimethylsulfide complex: Boron, trihydro[thiobis[methane]]-(T-4)-; (13292-87-0) **88**, 102; **89**, 183, 527

Borane tetrahydrofuran complex solution, 1.0 M in tetrahydrofuran; (14044-65-6) **87**, 36

Boric acid: Boric acid (H$_3$BO$_3$); (10043-35-3) **89**, 432

(−)-Bornyl acetate: Bicyclo[2.2.1]heptan-2-ol, 1,7,7-trimethyl-, acetate, (1S,2R,4S)-; (5655-61-8) **89**, 143

Boron; **87**, 26, 299; **88**, 79, 87, 181, 202, 207, 247

Boron trifluoride etherate; (109-63-7) **86**, 81, 212

Boronic acid, phenyl-; (98-80-6) **87**, 53

Bromination; **85**, 53

Bromine; (7726-95-6) **85**, 231; **87**, 126; **89**, 491

Bromoacetyl bromide: Acetyl bromide, 2-bromo-; (598-21-0) **89**, 501

4-Bromoanisole: 1-Bromo-4-methoxybenzene; (104-92-7) **89**, 283

3'-Bromoacetophenone: 1-acetyl-3-bromobenzene; (2142-63-4) **86**, 18

2-Bromoaniline: Benzenamine, 2-bromo-; (615-36-1) **88**, 398

3-Bromoanisole: Benzene, 1-bromo-3-methoxy-; (2398-37-0) **88**, 1

4-Bromoanisole; (104-92-7) **88**, 197

4-Bromobenzaldehyde; (1122-91-4) **89**, 105

4-Bromobenzaldehyde; (1122-91-4) **88**, 224

Bromobenzene; (108-86-1) **87**, 26, 178

4-Bromobenzoic acid; (586-76-5) **89**, 105

2-Bromobenzoyl chloride: Benzoyl chloride, 2-bromo-; (7154-66-7) **86**, 181

4-Bromobenzyl alcohol: Benzenemethanol, 4-bromo-: (873-75-6) **89**, 105

N-(4-Bromobenzylidene)-1,1-diphenylmethanamine: Benzenemethanamine, N-[(4-bromophenyl)methylene]-α-phenyl-, [N(E)]-; (330455-47-5) **88**, 224

ω-Bromo-p-chloroacetophenone oxime: Ethanone, 2-bromo-1-(4-chlorophenyl)-oxime (12); (136978-96-6) **89**, 409

1-Bromo-8-chlorooctane: Octane, 1-bromo-8-chloro-; (28598-82-5) **87**, 299

3-Bromocyclobutanone (23761-24-2) **89**, 491

1-Bromo-1-cyclopropylcyclopropane: 1,1'-Bicyclopropyl, 1-bromo- (9); (60629-95-0) **89**, 334

Bromodiethylsulfonium Bromopentachloroantimonate: Sulfonium, bromodiethyl-, (OC-6-22)-bromopentachloroantimonate(1-) (1:1); (1198402-81-1) **88**, 54

6-Bromohexanenitrile: Hexanenitrile, 6-bromo-; (6621-59-6) **87**, 330

1-Bromo-2-naphthoic acid: 2-Naphthoic acid, 1-bromo-; (20717-79-7) **88**, 70

8-Bromo-1-octanol: 1-Octanol, 8-bromo-; (50816-19-8) **87**, 299

2-Bromophenol: Benzene, 2-bromo-1-hydroxy-; (95-56-7) **89**, 220

N-(2-Bromophenyl)acetamide; (614-76-6) **89**, 202

2-Bromophenylacetonitrile: Benzeneacetonitrile, 2-bromo-; (19472-74-3) **89**, 159

2-[3-(3-Bromophenyl)-2H-azirin-2-yl]-5-(trifluoromethyl)pyridine; **86**, 18

N-2-Bromophenylbenzamide; (70787-27-8) **88**, 398

4-Bromophenylboronic acid; (5467-74-3) **86**, 344

4-Bromophenylboronic MIDA ester; **86**, 344

p-Bromophenyl methyl sulfide: Benzene, 1-bromo-4-(methylthio)-; (104-95-0) **86**, 121

(S)-(–)-p-Bromophenyl methyl sulfoxide; **86**, 121

(S)-(–)-*p*-Bromophenyl methyl sulfoxide: Benzene, 1-bromo-4-[(*S*)-methylsulfinyl]-; (145266-25-0) **86**, 121

2-(2-Bromophenyl)-2-propylpentanal: Benzeneacetaldehyde, 2-bromo-α,α-dipropyl-; (1206450-98-7) **89**, 159

1-(3-Bromophenyl)-2-[5-(trifluoromethyl)-2-pyridinyl]ethanone; **86**, 18

2-(3-Bromophenyl)-6-(trifluoromethyl)pyrazolo[1,5-a]pyridine; **86**, 18

(1Z)-1-(3-Bromophenyl)-2-[5-(trifluoromethyl)-2-pyridinyl]-ethanone oxime; **86**, 18

2-Bromopropene; (557-93-7) **85**, 1, 172

2-Bromopyridine: Pyridine, 2-bromo-; (109-04-6) **88**, 79; **89**, 76

3-Bromosalicylaldehyde: Benzaldehyde, 2-hydroxy-3-bromo-; (1829-34-1) **89**, 220

N-Bromosuccimide: NBS; (128-08-5) **85**, 53, 267; **86**, 225; **87**, 16; **89**, 55

4-Bromo-5-(thiophen-2-yl)oxazole: *Oxazole, 4-bromo-5-(2-thienyl)-;* (959977-82-3) **87**, 16

(Z)-β-Bromostyrene; (103-64-0) **88**, 202

4-Bromotoluene; (106-38-7) **85**, 196 ; **88**, 22

1-Bromo-1-phenylthioethene; (80485-53-6) **88**, 207

6-Bromoveratraldehyde: Benzaldehyde, 2-bromo-4,5-dimethoxy-; (5392-10-9) **88**, 388

1-(Buta-1,2-dien-1-ylsulfonyl)-4-methylbenzene: Benzene, 1-(1,2-butadien-1-ylsulfonyl)-4-methyl-; (32140-55-9) **88**, 309

Butanal, 4-[[(1,1-dimethylethyl)dimethylsilyl]oxy]-; (87184-81-4) **89**, 243

(*S*ₛ)-*N-tert*-Butanesulfinamide: (*S*)-(–)-2-Methyl-2-propanesulfinamide; (343338-28-3) **89**, 88

(4S,S$_S$)-N-(tert-Butanesulfinyl)tridec-1-en-4-amine; (1112145-16-0) **89**, 88

2-Butanone; (78-93-3) **86**, 333

trans-Butene; (624-64-6) **88**, 181

2-tert-Butoxycarbonylamino-4-(2,2-dimethyl-4,6-dioxo-[1,3]dioxan-5-yl)-4-oxo-butyric acid tert-butyl ester; (10950-77-9) **85**, 147

(2S)-2-[(tert-Butoxycarbonyl)amino]-2-phenylethyl methanesulfonate; (110143-62-9) **85**, 219

2-[3,3'-Di-(*tert*-butoxycarbonyl)-aminodipropylamine]-4,6-dichloro-1,3,5-triazine; 12-Oxa-2,6,10-triazatetradecanoic acid, 6-(4,6-dichloro-1,3,5-triazin-2-yl)-13,13-dimethyl-11-oxo-, 1,1-dimethylethyl ester; (947602-03-1) **86**, 141, 151

N-α-*tert*-Butoxycarbonyl-L-aspartic acid α-*tert*-butyl ester (Boc-L-Asp-O*t*-Bu); (34582-32-6) **85**, 147

1-tert-Butoxycarbonyl-2,3-dihydropyrrole: 1H-Pyrrole-1-carboxylic acid, 2,3-dihydro-, 1,1-dimethylethyl ester; (73286-71-2) **85**, 64

2-(*tert*-Butoxycarbonyloxyimino)-2-phenylacetonitrile; (58632-95-4) **86**, 141

N-(*tert*-Butoxycarbonyl)-piperazine; (57260-71-6) **86**, 141

N-(*tert*-Butoxycarbonyl)-L-proline; (15761-39-4) **88**, 317

N-(*tert*-Butyloxycarbonyl)pyrrolidin-2-one; (85909-08-6) **85**, 64

tert-Butyl acetate: Acetic acid, 1,1-dimethylethyl ester; (540-88-5) **89**, 519

tert-Butyl acetoacetate; (1694-31-1) **85**, 278

N-tert-Butyl adamantan-1-yl-carbamate: Carbamic acid, tricyclo[3.3.1.13,7]dec-1-yl-, 1,1-dimethylethyl ester; (151476-40-3) **86**, 113

tert-Butanol: 2-Methyl-2-propanol; (75-65-0) **87**, 8
tert-Butylamine: 2-Propanamine, 2-methyl-; (75-64-9) **86**, 315
tert-Butyl 3-amino-4-methylpentanoate; (202072-47-7) **87**, 143
tert-Butylbenzene; (98-06-6) **86**, 308
(E)-tert-Butyl benzylidenecarbamate; **86**, 11
(E)-tert-Butyl benzylidenecarbamate: Carbamic acid, *N*-(phenylmethylene)-,1,1-
    dimethyl-ethyl ester, [*N*(*E*)]-; (177898-09-2) **86**, 11
tert-Butyl 3-[*N*-benzyl-*N*-(α-methylbenzyl)amine]-4-methylpentanoate; (38235-77-7) **87**,
    143
tert-Butyl 1-benzyl-3-methyl-2-oxoindoline-3-carboxylate; 1-Benzyl-3-methylindolin-2-
    one; (34943-91-4) **89**, 438
tert-Butyl 3-(benzyl(phenyl)amino)-2-methyl-3-oxopropanoate; **89**, 438
tert-Butyl 3-(benzyl(phenyl)amino)-3-oxopropanoate; (103-32-2) **89**, 438
(-)-2-tert-Butyl-(4S)-benzyl-(1,3)-oxazoline: 4,5-Dihydrooxazole, (4S)-benzyl, 2-tert-
    butyl; (75866-75-0) **85**, 267
tert-Butyl bromoacetate; (5292-43-3) **85**, 10
tert-Butyl-3-bromo-5-formylbenzoate; (1018948-99-6) **89**, 460
tert-Butyl-3,5-dibromobenzoate; (5342-97-2) **89**, 460
tert-Butyl *tert*-butyldimethylsilylglyoxylate; **85**, 278
(S)-tert-Butyl (4-chlorophenyl)(thiophen-2-yl)methylcarbamate; **86**, 360
tert-Butyl (1R)-2-cyano-1-phenylethylcarbamate; (126568-44-3) **85**, 219
Butyl di-1-adamantylphosphine; (321921-71-5) **86**, 105
tert-Butyl diazoacetate; (35059-50-8) **85**, 278
tert-Butyldimethylsilyl chloride: Silane, chloro(1,1-dimethylethyl)dimethyl-; (18162-48-
    6) **86**, 130
tert-Butyldimethylsilyl trifluoromethanesulfonate; (69739-34-0) **85**, 278
tert-Butyl diethylphosphonoacetate; (27784-76-5) **87**, 143
(2R,5S)-2-tert-Butyl-3,5-dimethylimidazolidin-4-one: 4-Imidazolidinone, 2-(1,1-
    dimethylethyl)-3,5-dimethyl-, hydrochloride (1:1), (2R,5S)-: (1092799-01-3) **88**,
    42
1-(tert-Butyldimethylsilyl)-1-(1-ethoxyethoxy)-1,2-propadiene; (86486-46-6) **89**, 267
(E)-1-(tert-Butyldimethylsilyl)-3-trimethylsilyl-2-propen-1-one; (83578-66-9) **89**, 267
tert Butyldimethylsilyl chloride: Silane, chloro(1,1-dimethylethyl)dimethyl-; (18162-48-
    6) **89**, 19, 267
(2R,3R)-1-(t-Butyldiphenylsilyloxy)-2-methylhex-5-en-3-ol; (112897-06-0) **88**, 87
(R)-3-(tert-Butyldiphenylsilyloxy)-2-methylpropanal: Propanal, 3-[[(1,1-
    dimethylethyl)diphenyl-silyl]oxy]-2-methyl-, (2*R*)-; (112897-04-8) **88**, 87
(1R,6S,7S)-4-(tert-Butyldimethylsilyloxy)-6-(trimethylsilyl)bicyclo[5.4.0]-undec-4-en-2-
    one; (207925-88-0) **89**, 267
tert-Butyl ethyl phthalate: *1,2-Benzenedicarboxylic acid, 1,2-bis[2-[3,5-bis(1,1-*
    dimethylethyl)-4-hydroxyphenyl]ethyl] ester; (259254-67-6) **86**, 374
3-Butyl-2-fluoro-1-tosylindole: 1H-Indole, 3-butyl-2-fluoro-1-[(4-
    methylphenyl)sulfonyl]-; (195734-36-2) **88**, 162
tert-Butylhydrazine hydrochloride; (7400-27-3) **89**, 537
tert-Butyl hydroperoxide: Hydroperoxide, 1,1-dimethylethyl; (75-91-2) **87**, 1, 88; **88**, 14;
    89, 19
tert-Butyl (1S)-2-hydroxy-1-phenylethylcarbamate; (117049-14-6) **85**, 219
tert-Butyl-hypochlorite: Hypochlorous acid, 1,1-dimethylethyl ester; (507-40-4) **86**, 315;
    88, 168

n-Butyllithium; (109-72-8) **85**, 1, 45, 53, 158, 238, 248, 295; **86**, 47, 70, 262; **87**, 137,
    143; **88**, 1, 79, 102, 212, 296, 353, 406; **89**, 82, 183, 267, 460
sec-Butyllithium; (598-30-1) **88**, 247
tert-Butyllithium: Lithium, (1,1-dimethylethyl)-; (5944-19-4) **85**, 1, 209; **88**, 296; **89**, 76
n-Butylmagnesium chloride; (693-04-9) **89**, 460
tert-Butyl malonate; (40052-13-9) **89**, 438
tert-Butyl (*E*)-4-methylbut-2-enoate; (87776-18-9) **87**, 143
*(S$_a$,S)-t-Butyl 2-[(2'-(4-methylphenylsulfonamido)-(1,1'-binaphthyl)-2-yl-
    carbamoyl]pyrrolidine-1-carboxylate; (933782-35-5)* **88**, 317
1-tert-Butyl-3-methyl-1H-pyrazol-5-amine; (141459-53-2) **89**, 537
N-(1-tert-Butyl-3-methyl-1H-pyrazol-5-yl)pyridin-2-amine; **89**, 537
3-Butyl-4-methyl-2-triisopropylsiloxy-cyclobut-2-enecarboxylic acid methyl ester;
    (731853-28-4) **87**, 253
tert-Butyl perbenzoate: Benzenecarboperoxoic acid, 1,1-dimethylethyl ester; (614-45-9)
    89, 34
tert-Butyl phenyl(phenylsulfonyl)methylcarbamate; **86**, 11
tert-Butyl phenyl(phenylsulfonyl)methylcarbamate: Carbamic acid, *N*-[phenyl(phenyl-
    sulfonyl)methyl]-1,1-dimethylether ester; (155396-71-7) **86**, 11
*tert-Butyl phenylsulfonyl(thiophen-2-yl)methylcarbamate: Carbamic acid, N-
    [(phenylsulfonyl)-2-thienylmethyl]-, 1,1-dimethylethyl ester;* (479423-34-2) **86**,
    360
*(S)-tert-ButylPHOX: Oxazole, 4-(1,1-dimethylethyl)-2-[2-(diphenylphosphino)phenyl]-
    4,5-dihydro-, (4S)-;* (148461-16-9) **86**, 181
2-Butyn-1-ol; (764-01-2) **88**, 296
3-Butyn-2-ol; (2028-63-9) **88**, 309; **89**, 420
*But-3-yn-2-yl 4-Methylbenzenesulfinate: Benzenesulfinic acid, 4-methyl-, 1-methyl-2-
    propyn-1-yl ester; (32140-54-8)* **88**, 309

ε-Caprolactam: 2*H*-Azepin-2-one, hexahydro-; (105-60-2) **89**, 19
Carbamates **86**, 11, 59, 81, 113, 141, 151, 236, 333; **88**, 152, 212, 247, 317; **89**, 170, 450
Carbamic acid, N-(2-diazoacetyl)-N-phenyl-, 1-methylethyl ester; (1198356-59-0) **88**,
    212
Carbamic acid, (hydroxymethyl)-, phenylmethyl ester; (31037-42-0) **85**, 287
Carbamic acid, *N*-[(1*S*,2*S*)-2-methyl-3-oxo-1-phenylpropyl]-1,1-dimethylethyl ester;
    (926308-17-0) **86**, 11
Carbamic acid, N-[phenyl(phenylsulfonyl)methyl]-, 2-propen-1-yl ester; (921767-12-6)
    86, 236
Carbamic acid, 2-propen-1-yl ester; (2114-11-6) **86**, 236
Carbobenzyloxy-L-proline: 1,2-Pyrrolidinedicarboxylic acid, 1-(phenylmethyl) ester,
    (2*S*)-; (1148-11-4) **85**, 72
(Carboethoxymethylene)triphenylphosphorane; (1099-45-2) **85**, 15
p-Carbomethoxybenzamide: Benzoic acid, 4-(aminocarbonyl)-, methyl ester; (6757-31-9)
    89, 66
*(2-Carbomethoxy-6-nitrobenzyl)triphenylphosphonium bromide: Phosphonium, [[2-
    (methoxycarbonyl)-6-nitrophenyl]methyl]triphenyl-, bromide; (195992-09-7)* **88**,
    291
Carbon disulfide; (75-15-0) **86**, 70; **88**, 364
Carbon monoxide; (630-08-0) **86**, 287; **88**, 291
Carbonodithioic acid *O*-ethyl ester, potassium salt; (140-89-6) **89**, 409

Carbonodithioic acid, S-[2-(4-chlorophenyl)-2-(hydroximino)ethyl], O-ethyl ester;
(195213-53-7) **89**, 409

Carbon tetrabromide: Methane, tetrabromo-; (558-13-4) **87**, 231

Carbonyl(dihydrido)tris(triphenylphosphine)ruthenium (II); (25360-32-1) **86**, 28

1,1'-Carbonyldiimidazole: Methanone, di-1*H*-imidazol-1-yl-; (530-62-1) **86**, 58

Catecholborane: 1,3,2-benzodioxaborole; (274-07-7) **88**, 202

Cesium acetate: Acetic acid, cesium salt (1:1); (3396-11-0) **88**, 388

Cesium carbonate: Carbonic acid, cesium salt (1:2); (534-17-8) **86**, 181; **87**, 231; **88**, 398;
89, 159, 183

Cesium fluoride; (13400-13-0) **86**, 161; **89**, 82, 480

Chloroacetonitrile; (107-14-2) **86**, 1

p-Chloroacetophenone: Ethanone, 1-(4-chlorophenyl); (99-91-2) **89**, 409

4-Chlorobenzaldehyde; (104-88-1) **85**, 179

Chlorobenzene; (108-90-7) **86**, 105; **87**, 362

(3-Chlorobutyl)Benzene; (4830-94-8) **87**, 88

Chloro(chloromethyl)dimethylsilane; (1719-57-9) **87**, 178

Chlorodicarbonylrhodium dimer: Di-µ-chloro-bis(dicarbonylrhodium): Tetracarbonyldi-
µ-chlorodirhodium; (14523-22-9) **88**, 109

1-Chloro-1,3-dihydro-3,3-dimethyl-1,2-benziodoxole; (69352-04-1) **88**, 168

2-Chloro-6,7-dimethoxy-1,2,3,4-tetrahydroisoquinoline; **87**, 8

(Chloromethyl)dimethylphenylsilane; (1833-51-8) **87**, 178

Chloro(dimethyl)phenylsilane; (768-33-2) **89**, 420

7-Chloro-2-methyl-1,4-dihydro-2H-isoquinolin-3-one; (877265-15-1) **89**, 44

2-Chloro-5-(3-methylphenyl)-thiophene; (1078144-58-7) **87**, 178, 184

2-Chloro-1-methylpyridinium iodide; (14338-32-0) **89**, 438

4-Chlorophenylacetic acid: *p*-Chlorophenylacetic acid, PCPA; (1878-66-6) **89**, 44

4-Chlorophenylboronic acid: Boronic acid, B-(4-chlorophenyl)-; (1679-18-1) **86**, 360

2-(4-Chlorophenyl)-N-methylacetamide; (60336-41-6) **89**, 44

2-Chloropyridine; (109-09-1) **85**, 88; **88**, 121; **89**, 537, 549

5-Chlorosalicylaldehyde: Benzaldehyde, 5-chloro-2-hydroxy-; (635-93-8) **86**, 172

2-Chlorothiophene; (96-43-5) **87**, 178, 184

(13-Chlorotridecyloxy)triethylsilane: Silane, [(13-chlorotridecyl)oxy]triethyl-; (374754-
99-1) **87**, 299

Chlorotriethylsilane: Silane, chlorotriethyl-; (994-30-9) **87**, 299

2-Chloro-5-(trifluoromethyl)pyridine; (52334-81-3) **86**, 18

m-Chloroperbenzoic acid; Peroxybenzoic acid, *m*-chloro- (8); Benzocarboperoxoic acid,
3-chloro-; (937-14-4) **86**, 308

Chlorotrimethylsilane: Silane, Chlorotrimethyl-; (75-77-4) **86**, 252; **88**, 296; **89**, 73, 82,
98

Cholesta-3,5-diene; (747-90-0) **88**, 260

Cholesta-3,5-dien-3-yl trifluoromethanesulfonate; (95667-40-6) **88**, 260

Cholest-4-en-3-one; (601-57-0) **88**, 260

Cinnamyl alcohol: 3-Phenyl-2-propen-1-ol; (104-54-1) **85**, 96

Cinnamyl-H-phosphinic acid: [(2E)-3-phenyl-2-propenyl]-Phosphinic acid; (911128-46-
6) **85**, 96

(±)-Citronellal: ((±)-3,7-dimethyl-6-octenal); (106-23-0) **87**, 201

Cobalt bromide: Cobalt(II) bromide; (7789-43-7) **89**, 283

Cobalt(II) tetrafluoroborate hexahydrate; (15684-35-2) **87**, 88

Condensation; **85**, 27, 34, 179, 248, 267; **86**, 11, 18, 92, 121, 212, 252, 262; **87**, 36, 59, 77, 88, 115, 143, 192, 201, 218, 275, 310, 339, 362; **88**, 33, 42, 138, 212, 224. 247, 338; **89**, 115, 131, 230, 255, 274, 283, 380, 409, 432, 537

Copper(II) acetate, monohydrate: Cupric acetate 1-hydrate: Acetic acid, copper(II) salt; (6046-93-1) **88**, 238; **89**, 438

Copper(I) bromide; (7787-70-4) **85**, 196

Copper-catalyzed reactions; **87**, 53, 126, 184, 231; **88**, 14, 79, 388, 398

Copper chloride: Cuprous chloride; (7758-89-6) **85**, 209 ;

Copper cyanide; (544-92-3) **85**, 131

Copper iodide; (1335-23-5) **86**, 181; **88**, 79; **89**, 294

Copper(I) iodide: Cuprous iodide; (7681-65-4) **86**, 225; **87**, 126, 178, 184, 231; **88**, 388

Copper(II) oxide nanoparticles: Copper oxide (CuO); (1317-38-0) **88**, 398

Coupling; **85**, 158, 196; **86**, 105, 225, 274; **87**, 184, 299, 317, 330; **88**, 4, 14, 22, 70,79, 102, 162, 197, 202, 207, 377, 388, 398, 406; **89**, 55, 73, 76, 183, 202

Cuprous chloride; (7758-89-6) **85**, 209

4-Cyanobenzaldehyde; (105-07-7) **89**, 131

2-(4-Cyano-phenyl)-1-[2-(3,4-dimethoxyphenyl)-ethyl]-1H-benzimidazole-5-carboxylic acid ethyl ester; **89**, 131

(S)-2-Cyano-pyrrolidine-1-carboxylic acid benzyl ester: (63808-36-6) **85**, 72

Cyanuric chloride: 2,4,6-Trichloro-1,3,5-triazine; (108-77-0) **85**, 72; **86**, 141

Cyclen: 1,4,7,10-Tetraazacyclododecane; (294-90-6) **85**, 10

Cyclization; **86**, 18, 92, 172, 181, 194, 212, 236, 252, 262, 333; **87**, 16, 77, 161, 310, 339, 350, 362; **88**, 33, 42, 54, 162, 291, 377, 388, 398; **89**, 55, 115, 131, 255, 267380, 409, 438, 537, 549

Cycloaddition; **85**, 72, 131, 138, 179; **87**, 95, 253, 263; **88**, 109, 121, 138, 224, 238; **89**, 98

Cyclobutenone; 2-cyclobuten-1-one; (32264-87-2) **89**, 491

Cycloheptane-1,3-dione; (1194-18-9) **85**, 138

Cyclohexanecarboxaldehyde; (2043-61-0) **87**, 68; **89**, 73

1,4-Cyclohexanedione mono-ethylene ketal: 1,4-Dioxaspiro[4.5]decan-8-one; (4746-97-8) **88**, 121

dl-1,2-Cyclohexylethanediol: 1,2-Ethanediol, 1,2-dicyclohexyl-; (92319-61-4) **89**, 73

Cyclohexanemethanol; (100-49-2) **86**, 58

Cyclohexanone; (108-94-1) **86**, 47

Cyclohexene; (110-83-8) **89**, 183

Cyclohexene imine: 7-Aza-bicyclo[4.1.0]heptane; (286-18-0) **87**, 161

Cyclohexene oxide; (286-20-4) **85**, 106; **87**, 161

9-[2-(3-Cyclohexenyl)ethyl]-9-BBN: 9-Borabicyclo[3.3.1]nonane, 9-[2-(3-cyclohexen-1-yl)ethyl]-; (69503-86-2) **88**, 207

4-(3-Cyclohexenyl)-2-phenylthio-1-butene; (155818-88-5) **88**, 207

(R)-(+)-α-Cyclohexyl-3-methoxy-benzenemethanol; (1036645-45-0) **87**, 68

Cyclohexylmethyl N-hydroxycarbamate; **86**, 58

Cyclohexylmethyl *N*-hydroxycarbamate: Carbamic acid, hydroxy-, cyclohexylmethyl ester; (869111-30-8) **86**, 58

Cyclohexylmethyl N-tosyloxycarbamate; **86**, 58

Cyclohexylmethyl *N*-tosyloxycarbamate: Benzenesulfonic acid, 4-methyl-, [(cyclohexylmethoxy)carbonyl]azanyl ester; (869111-41-1) **86**, 58

Cyclopropanation; **85**, 172; **87**, 115

Cyclopropanecarboxylic acid, 2-bromo-2-methyl-, ethyl ester; (89892-99-9) **85**, 172

584

Cyclopropanecarboxylic acid, 2-methylene-, ethyl ester; (18941-94-1) **85**, 172
1-Cyclopropylcyclopropanol: [1,1'-Bicyclopropyl]-1-ol; (54251-80-8) **89**, 334

DABCO: 1,4-Diazabicyclo[2.2.2]octane; (280-57-9) **87**, 104
(E)-Deca-1,3-diene: 1,3-Decadiene, (3E)-; (58396-45-5) **88**, 342
Decanal: Decyl aldehyde; (112-31-2) **89**, 88
(Z)-Dec-2-en-1-ol: 2-Decen-1-ol, (2Z)-; (4194-71-2) **88**, 342
(*R*,*R*)-deguPHOS: Pyrrolidine, 3,4-bis(diphenylphosphino)-1-(phenylmethyl)-, (3*R*,4*R*)-;
    (99135-95-2) **86**, 360
Dehydration; **85**, 34, 72
trans-Dehydroandrosterone: Androst-5-en-17-one, 3-hydroxy-, (3β)-; (53-43-0) **89**, 19
Dendrimer; **86**, 151
Deoxo-Fluor: Bis(2-methoxyethyl)aminosulfur trifluoride: Ethanamine, 2-ethoxy-N-(2-
    ethoxyethyl)-*N*-(trifluorothio)-; (202289-38-1) **87**, 245
Deoxyvasicinone: Pyrrolo[2,1-b]quinazolin-9(1H)-one, 2,3-dihydro-; (530-53-0) **89**, 274
Deprotection; **87**, 143; **88**, 317, 406; **89**, 88, 183, 267, 323, 438, 480, 537
(Diacetoxyiodo)benzene: Aldrich: Iodobenzene diacetate: Benzene, (diacetoxyiodo)-;
    Iodine, bis(aceto-*O*)phenyl-; (3240-34-4) **89**, 98
Diallyl pimelate: Pimelic acid, diallyl ester; (91906-66-0) **86**, 194
Diamination; **87**, 275
3,3'-Diaminodipropylamine; (56-18-8) **86**, 141
1,5-Diazabicyclo[4.3.0]non-5-ene (DBN): Pyrrolo[1,2-a]pyrimidine, 2,3,4,6,7,8-
    hexahydro-; (3001-72-7) **89**, 9
1,8-Diazabicyclo[5.4.0]undec-7-ene (DBU); (6674-22-2) **87**, 170; **89**, 501
Diazo compounds; **88**, 212, 418
Dibenzo[a,e]cyclooctene; (262-89-5) **89**, 55
1,2-Dibromobenzene; (583-53-9) **89**, 210
Di(μ-bromo)bis(η-allyl)nickel(II): [allylnickel bromide dimer]; (12012-90-7) **85**, 248
Di-μ-bromobis(tri-*tert*-butylphosphine)dipalladium; (185812-86-6) **88**, 1
(E)-2,3-Dibromobut-2-enoic acid: (2-Butenoic acid, 2,3-dibromo-, (2E)-; (24557-17-3)
    85, 231
*2,5-Dibromo-1,1-dimethyl-3,4-diphenylsilole: Silacyclopenta-2,4-diene, 2,5-dibromo-
    1,1 dimethyl 3,4 diphenyl-; (686290-22-2)* **85**, 53
1,2-Dibromoethane: 1,2-Dibromoethane; (106-93-4) **89**, 283
(2,2-Dibromoethenyl)benzene; (7436-90-0) **87**, 231
1-(2,2-Dibromoethenyl)-2-nitrobenzene; **86**, 36
3-(3,4-Dibromophenyl)-3-methylbutanoic acid; **89**, 210
4-(3,4-Dibromophenyl)-1,1,1-trifluoro-4-methylpentan-2-one; **89**, 210
5,11-Dibromo-5,6,11,12-tetrahydrodibenzo[a,e]-cyclooctene; (94275-22-6) **89**, 55
2-(2,2-Dibromo-vinyl)-phenylamine; **86**, 36
α,α'-Dibromo-*o*-xylene, 1,2-Bis(bromomethyl)benzene; (91-13-4) **89**, 55
Di-tert-butyldiaziridinone: 3-Diaziridinone, 1,2-bis(1,1-dimethylethyl)-; (19656-74-7) **86**,
    315; **87**, 263
Di-*tert*-butyl dicarbonate: Dicarbonic acid, C,C'-bis(1,1-dimethylethyl) ester; (24424-99-
    5) **85**, 72, 219; **86**, 113, 374; **89**, 460
Di(tert-butyl) (2S)-4,6-dioxo-1,2-piperidinedicarboxylate; (653589-10-7) **85**, 147
3,5-Di-*t*-butyl-2-hydroxybenzaldehyde: Benzaldehyde, 3,5-bis(1,1-dimethylethyl)-2-
    hydroxy-; (37942-07-7) **87**, 88
2-(3,5-Di-*tert*-butyl-2-hydroxybenzylideneamino)-2,2-diphenylacetic acid potassium salt,

(SALDIPAC); (858344-69-1) **87,** 88

Di(tert-butyl) (2S,4S)-4-hydroxy-6-oxo-1,2-piperidinedicarboxylate; (653589-16-3) **85,** 147

trans-1,3-Di-*tert*-butyl-4-phenyl-5-vinyl-imidazolidin-2-one: 2-Imidazolidinone, 1,3-bis(1,1-dimethylethyl)-4-ethenyl-5-phenyl-, (4*R*,5*R*)-rel-; (927902-91-8) **86,** 315; **87,** 263

2-(Di-*tert*-butylphosphino)biphenyl: Phosphine, [1,1'-biphenyl]-2-ylbis(1,1-dimethylethyl)-; (224311-51-7) **86,** 274

Di-tert-butylurea: Urea, N,N'-bis(1,1-dimethylethyl)-; (5336-24-3) **86,** 315

Dichloroacetyl chloride; (79-36-7) **85,** 138

1,2-Dichlorobenzene: Benzene, *o*-dichloro-; Benzene, 1,2-dichloro-; (95-50-1) **89,** 98

1,3-Dichloro-1,3-bis(dimethylamino)propenium chloride; (34057-61-9) **86,** 298

Dichloro[1,1'-bis(diphenylphosphino)ferrocene]palladium(II)• dichloromethane; (95464-05-4) **89,** 183

Dichloro[1,3-bis(diphenylphosphino)propane]palladium: Palladium, dichloro[1,1'-(1,3-propanediyl)bis[1,1-diphenylphosphine-κP]]-, (SP-4-2)-; (59831-02-6) **88,** 79

Di-μ-chloro-bis(1,5-cyclooctadiene)diiridium(I): Iridium, di-μ-chlorobis[(1,2,5,6-η)-1,5-cyclooctadiene]di-; (12112-67-3) **89,** 307

Dichlorobis(triphenylphosphine)palladium(II): Bis(triphenylphosphine)palladium(II) Dichloride; (13965-03-2) **86,** 225; **88,** 202

Dichlorodicyclopentadienylvanadium; Vanadium, dichlorobis (η^5-2,4-cyclopentadien-1-yl)-; (12083-48-6) **89,** 73

(*S*)-(−)-5,5'-Dichloro-6,6'-dimethoxy-2,2'-bis(diphenylphosphino)-1,1'-biphenyl: Phosphine, [(1*S*)-5,5'-dichloro-6,6'-dimethoxy[1,1'-biphenyl]-2,2'-diyl]bis[diphenyl-; (185913-98-8) **86,** 47

Dichlorodimethylsilane; (75-78-5) **85,** 53

1,2-Dichloroethane: Ethane, 1,2-dichloro-; (107-06-2) **89,** 19

Dichloromethylen-dimethyliminium chloride; (33842-02-3) **86,** 298

2,4-Dichloroquinoline; (703-61-7) **89,** 380

Dicyclohexylmethylamine: Cyclohexanamine, *N*-cyclohexyl-*N*-methyl-; (7560-83-0) **85,** 118

Dichloro(*N,N,N',N'*-tetramethylethylenediamine)zinc; (28308-00-1) **87,** 178

Dichlorotriphenylphosphorane: Phosphorane, dichlorotriphenyl-; (2526-64-9) **87,** 299, 317

Dicyclohexylamine: Cyclohexanamine, *N*-cyclohexyl-; (101-83-7) **88,** 1

Dicyclohexylcarbodiimide: Carbodiimide, dicyclohexyl-; Cyclohexanamine, *N,N'*-methanetetraylbis-; (538-75-0) **88,** 70; **89,** 183

2-Dicyclohexylphosphino-2',4',6'-triisopropylbiphenyl: X-Phos:; (564483-18-7) **87,** 104

Diene formation; **85,** 1; **88,** 260, 342

Diethylamine: Ethanamine, *N*-ethyl-; (109-89-7) **88,** 427

2-[(Diethylamino)methyl]benzene thiolato-copper(I); **85,** 209

Diethyl azodicarboxylate (DEAD): 1,2-Diazenedicarboxylic acid, 1,2-diethyl ester; (1972-28-7) **89,** 480

Diethyl chlorophosphate: Phosphorochloridic acid, diethyl ester; (814-49-3) **88,** 54

(±)-Diethyl (*E,E,E*)-cyclopropane-1,2-acrylate; **85,** 15

Diethyl trans-1,2-cyclopropanedicarboxylate; (3999-55-1) **85,** 15

N,N-Diethyl-3,4,5-trimethoxybenzamide; (5470-42-8) **88,** 427

*Diethyl(2-[(trimethylsilanyl)sulfanyl]benzyl)amine; * **85,** 209

Diethylzinc; (557-20-0) **87,** 68; **89,** 374, 527

4-[2-(3,4-Dimethoxyphenyl)-ethylamino]-3-nitrobenzoic acid ethyl ester (387882-07-7)
    89, 131

[(E)-3,3-Dimethoxy-2-methyl-1-propenyl]benzene: Benzene, [(1E)-3,3-dimethoxy-2-
    methyl-1-propen-1-yl]-: (137032-32-7) **86**, 81

6,7-Dimethoxy-1,2,3,4-tetrahydroisoquinoline hydrochloride: Isoquinoline, 1,2,3,4-
    tetrahydro-6,7-dimethoxy-, hydrochloride (1:1); (2328-12-3) **87**, 8

(2R,3S,4E)-N,3-Dimethoxy-N,2,4-trimethyl-5-phenyl-4-pentenamide
(3,5-Dimethoxy-1-phenyl-cyclohexa-2,5-dienyl)-acetonitrile; **86**, 1

Dimethyl acetamide; (127-19-5) **86**, 298

3,3-Dimethylacrylic acid: 2-Butenoic acid, 3-methyl-; (541-47-9) **89**, 210

Dimethylamine; (124-40-3) **86**, 298

4-(Dimethylamino)benzoic acid; (619-84-1) **87**, 201

N-(3-Dimethylaminopropyl)-N'-ethylcarbodiimide hydrochoride (EDC·HCl); (25952-53-
    8) **85**, 147

4-Dimethylaminopyridine: 4-Pyridinamine, N,N-dimethyl-; (1122-58-3) **85**, 64; **86**, 81;
    89, 143, 183, 230, 471

1,3-Dimethyl-6H-benzo[b]naphtho[1,2-d]pyran-6-one: 6H-Benzo[b]naphtho[1,2-
    d]pyran-6-one, 1,3-dimethyl-; (138435-72-0) **88**, 70

4,4'-Dimethyl-2,2'-bipyridine: 2,2'-Bipyridine, 4,4'-dimethyl-; (1134-35-6) **89**, 82

9,9-Dimethyl-4,5-bis(diphenylphosphino)xanthene: Xantphos; (161265-03-8) **85**, 96

Dimethyl-bis-phenylethynyl silane: Benzene, 1,1'-[(dimethylsilylene)di-2,1-
    ethynediyl]bis-; (2170-08-3) **85**, 53

(2R,3R)-2,3-Dimethylbutane-1,4-diol: (2R,3R) 2,3-Dimethyl-1,4-butanediol; (127253-15-
    0) **85**, 158

3,3-Dimethylbutyryl chloride: Butanoyl chloride, 3,3-dimethyl-; (7065-046-05) **88**, 138

2,2-Dimethyl-1,3-dioxane-4,6-dione (Meldrum's acid); (2033-24-1) **85**, 147; **89**, 115

(2,2-Dimethyl-1,3-dioxolan-4-yl)methyl (Z)-tetradec-7-enoate: 7-Tetradecenoic acid,
    (2,2-dimethyl-1,3-dioxolan-4-yl)methyl ester, (7Z)-; (749266-24-8) **89**, 183

4,5-Dimethyl-1,3-dithiol-2-one; (49675-88-9) **86**, 333

N,N'-Dimethylethylenediamine: 1,2-Ethanediamine, N1,N2-dimethyl-; (110-70-3) **86**,
    181; **87**, 231

(Z)-1,1-Dimethyl-1-heptenylsilanol: Silanol, (1Z)-1-heptenyldimethyl-; (261717-40-2) **88**,
    102

N,O-Dimethylhydroxylamine hydrochloride: Methanamine, N-methoxy-, hydrochloride;
    (6638-79-5) **86**, 81

Dimethyl malonate; (108-59-8) **87**, 115

(4S)-N-[(2R,3S,4E)-2,4-Dimethyl-3-methoxy-5-phenyl-4-pentenoyl]-4-isopropyl-1,3-
    thiazolidine-2-thione; **86**, 81

3,7-Dimethyl-2-methylene-6-octenal; (22418-66-2) **87**, 201

(Z)-3,7-Dimethyl-2,6-octadien-1-al (neral): (2Z)-3,7-Dimethyl-2,6-octadienal; (106-26-
    3) **89**, 311

4,4-Dimethyl-3-oxo-2-benzylpentanenitrile (875628-78-7) **86**, 28

4,4-Dimethyl-3-oxopentanenitrile; (59997-51-2) **86**, 28

2,4-Dimethyl-3-pentanol; (600-36-2) **89**, 255

3,5-Dimethylphenol: Phenol, 3,5-dimethyl-; (108-68-9) **88**, 70

3,5-Dimethylphenyl-1-bromo-2-naphthoate: 2-Naphthalenecarboxylic acid, 1-bromo-,
    3,5-dimethylphenyl ester; (138435-66-2) **88**, 70

Dimethyl 2-phenylcyclopropane-1,1-dicarboxylate (3709-20-4) **87**, 115

(±)-4-(Dimethyl(phenyl)silyl)-but-3-yn-2-ol; (115884-67-8) **89**, 420

(S)-4-(Dimethyl(phenyl)silyl)-but-3-yn-2-ol; (142697-17-4) **89**, 420

(S)-4-(Dimethyl(phenyl)silyl)-but-3-yn-2-yl acetate; (142611-82-3) **89**, 420

7,7-Dimethyl-3-phenyl-4-p-tolyl-6,7,8,9-tetrahydro-1H-pyrazolo[3,4-b]-quinolin-5(4H)-one: 5H-Pyrazolo[3,4-b]quinolin-5-one, 1,4,6,7,8,9-hexahydro-7,7-dimethyl-4-(4-methylphenyl)-3-phenyl-; (904812-68-6) **86**, 252

(S,E)-2-(2,2-Dimethylpropylidenamino)-N-methylpropanamide; (1375485-50-9) **88**, 42

N, *N*-Dimethyl-4-pyridinamine; (1122-58-3) **85**, 64

(2R,3R)-2,3-Dimethylsuccinic acid; (5866-39-7) **85**, 158

Dimethyl sulfoxide: Methyl sulfoxide; Methane, sulfinybis-; (67-68-5) **85**, 189 ; **88**, 388

(−)-(S)-1-(1,3,2-Dioxaborolan-2-yloxy)-3-methyl-1,1-diphenylbutan-2-amine; (879981-94-9) **87**, 26, 36

(1,4-Dioxaspiro[4.5]dec-7-en-8-yloxy)trimethylsilane; (144810-01-5) **88**, 121

2-(1,3-Dioxolan-2-yl)ethyl bromide: 1,3-Dioxolane, 2-(2-bromoethyl)-; (18742-02-4) **87**, 317

2-[2-(1,3-Dioxolan-2-yl)ethyl]zinc bromide: Zinc, bromo[2-(1,3-dioxolan-2-yl)ethyl]-; (864501-59-7) **87**, 317

1,3-Diphenylacetone *p*-tosylhydrazone: Benzenesulfonic acid, 4-methyl-, [2-phenyl-1-(phenylmethyl)ethylidene]hydrazide; (19816-88-7) **85**, 45; **89**, 267

Diphenyldiazomethane; (883-40-9) **85**, 189

(4-((4R,5R)-4,5-Diphenyl-1,3-dioxolan-2-yl)phenoxy)(tert-butyl)dimethylsilane; (86, 130

(1R,2R)-1,2-Diphenylethane-1,2-diol; (86, 130

α,α-Diphenylglycine: Benzeneacetic acid, α-amino-α-phenyl-; (3060-50-2) **87**, 88

Diphenylphosphine: Phosphine, diphenyl-; (829-85-6) **86**, 181

(S)-(−)-1,3-Diphenyl-2-propyn-1-ol: Benzenemethanol, α-(2-phenylethynyl)-, (αS)-; (132350-96-0) **85**, 118

Dipotassium sulfate and potassium hydrogen sulfate; (37222-66-5) **89**, 350

[8,8-Dipropylbicyclo[4.2.0]octa-1,3,5-trien-7-one]: Bicyclo[4.2.0]octa-1,3,5-trien-7-one, 8,8-dipropyl-; (1206451-50-4) **89**, 159

Dirhodium tetraacetate dihydrate; (15956-28-2) **88**, 418

Dirhodium(II) tetrakis[ε-caprolactamate]: Rhodium, tetrakis[μ-(hexahydro-2H-azepin-2-onato-κN1:κO2)]di-, (Rh-Rh); (138984-26-6) **89**, 19

N,N′-Ditosylhydrazine: Benzenesulfonic acid, 4-methyl-, 2-[(4-Methylphenyl)sulfonyl]hydrazide; (14062-05-6) **89**, 501

Eaton's reagent: Phosphorus pentoxide, 7.7 wt% solution in methanesulfonic acid; (39394-84-8) **89**, 44

Elimination; **85**, 45, 172; **86**, 11, 18, 212; **87**, 95, 231; **88**, 162, 207; **89**, 55, 98, 334, 480, 491, 501

Enamines; **88**, 388

1,2-Epoxydodecane: Oxirane, decyl-; (2855-19-8) **85**, 1

1,2-Epoxy-3-phenoxypropane: oxirane, 2-(phenoxymethyl)-; (122-60-1) **86**, 287

Esterification; **86**, 194; **88**, 70

[(Ethanol, 1-[[(1R,2S,4R)-1,7,7-trimethylbicyclo[2.2.1]hept-2-yl]oxy]-, acetate)]; (284036-61-9) **89**, 143

Ethoxyacetylene: Ethyne, ethoxy-; (927-80-0) **89**, 527

1-(1-Ethoxyethoxy)-1,2-propadiene; (20524-89-4) **89**, 267

1-(1-Ethoxyethoxy)-1,2-propadiene: 1,2-Propadiene, 1-(1-ethoxyethoxy)-; (20524-89-4) **89**, 267

3-(1-Ethoxyethoxy)-1-propyne: 1-Propyne, 3-(1-ethoxyethoxy)-; (18669-04-0) **89**, 267
Ethyl 3-amino-4-((3,4-dimethoxyphenethyl)amino)benzoate; (899374-50-6) **89**, 131
Ethyl (*R*)-2-amino-3-mercaptopropanoate hydrochloride; (868-59-7) **88**, 168
Ethyl (R)-2-amino-3-(trifluoromethylthio)propanoate hydrochloride; (931106-56-8) **88**, 168
(2R,3R)-Ethyl 1-benzhydryl-3-(4-bromophenyl)aziridine-2-carboxylate: 2-Aziridinecarboxylic acid, 3-(4-bromophenyl)-1-(diphenylmethyl)-, ethyl ester, (2 R,3R)-; (233585-43-8) **88**, 224
Ethyl benzoate: Benzoic acid, ethyl ester; (93-89-0) **86**, 374
Ethyl 1-benzyl-4-fluoropiperidine-4-carboxylate; **87,** 137
Ethyl 1-benzylpiperidine-4-carboxylate; (24228-40-8) **87,** 137
Ethyl 4-bromobenzoate; (5798-75-4) **87,** 104
Ethyl 5-(tert-butyl)-2-phenyl-1-tosyl-3-pyrroline-3-carboxylate: 1H-Pyrrole-3-carboxylic acid, 5-(1,1-dimethylethyl)-2,5-dihydro-1-[(4-methylphenyl)sulfonyl]-2-phenyl-, ethyl ester, (2R,5R)-rel-; (861668-04-4) **88**, 138
Ethyl chloroacetate; (105-39-5) **87,** 350
Ethyl chloroformate: Carbonochloridic acid, ethyl ester; (541-41-3) **88**, 317; **89**, 230
Ethyl 4-chloro-3-nitrobenzoate; (16588-16-2) **89**, 131
Ethyl (*E*)-4-chloropent-2-enoate: 2-Pentenoic acid, 4-chloro-, ethyl ester, (2*E*)-; (872455-00-0) **87,** 317
Ethyl diazoacetate: Acetic acid, 2-diazo-, ethyl ester; (623-73-4) **85**, 172; **86**, 172; **87,** 95; **88**, 224
Ethyl 5,5-dimethylhexa-2,3-dienoate: 2,3-Hexadienoic acid, 5,5-dimethyl-, ethyl ester; (35802-59-6) **88**, 138
(S,E)-Ethyl 6-(1,3-dioxolan-2-yl)-4-methylhex-2-enoate: 2-Hexenoic acid, 6-(1,3-dioxolan-2-yl)-4-methyl-, ethyl ester, (2E,4S)-; (1012041-93-8) 317
Ethylene; (74-85-1) **85**, 248
Ethylene glycol; (107-21-1) **87,** 26
Ethyl (*E*)-4-hydroxy-2-pentenoate: 2-Pentenoic acid, 4-hydroxy-, ethyl ester, (2*E*)-; (10150-92-2) **87,** 317
Ethyl 2-isocyanatoacetate; (2949-22-6) **89**, 404
Ethyl 2-isocyanoacetate; (2999-46-4) **89**, 404
Ethyl isonipecotate; (1126-09-6) **87,** 137
Ethyl 2-methylacetoacetate: Butanoic acid, 2-methyl-3-oxo-, ethyl ester; (609-14-3) **86**, 325
Ethyl 2-methylbuta-2,3-dienoate: 2,3-Butadienoic acid, 2-methyl-, ethyl ester; (5717-41-9) **86**, 212
O-Ethyl S-[oximino-2-(p-chlorophenyl)ethyl]dithiocarbonate: (195213-53-7) **89**, 409
Ethyl (*E*)-4-oxo-2-butenoate: 2-Butenoic acid, 4-oxo-, ethyl ester, (2*E*)-; (2960-66-9) **87,** 317
Ethyl 6-phenyl-1-tosyl-1,2,5,6-tetrahydropyridine-3-carboxylate: 3-Pyridinecarboxylic acid, 1,2,5,6-tetrahydro-1-[(4-methylphenyl)- sulfonyl]-6-phenyl-, ethyl ester; (528853-66-9) **86**, 212
N-(1-Ethyl-2-propen-1-yl)-4-methyl-benzenesulfonamide; (124070-39-9) **89**, 450
2-Ethyl-3-Quinolinecarboxylic acid; (888069-31-6) **85**, 27
2-Ethyl-3-Quinolinecarboxylic acid hydrochloride; (888014-11-7) **85**, 27
2-Ethyl-3-Quinolinecarboxylic acid, methyl ester; (119449-61-5) **85**, 27
Ethyl (triphenylphosphoranylidene)acetate: Acetic acid, 2-(triphenylphosphoranylidene)-, ethyl ester; (1099-45-2) **88**, 138

Ethyl 2-(triphenylphosphoranylidene)propionate: Propanoic acid, 2-
    (triphenylphosphoranylidene)-, ethyl ester; (5717-37-3) **86**, 212
Ethyl vinyl ether: Ether, ethyl vinyl; Ethene, ethoxy-; (109-92-2) **89**, 267
Ethylenediaminetetraacetic acid, disodium salt dihydrate: Acetic acid
    (ethylenedinitrilo)tetra-, disodium salt, dihydrate; Glycine, *N,N'*-1,2-
    ethanediylbis[*N*-(carboxymethyl)-, disodium salt, dihydrate; (6381-92-6) **89**, 76

Formylation; **89**, 220, 243, 460
Fluorination; **87**, 245
4-Fluorobenzaldehyde: 4-Fluorobenzaldehyde; (459-57-4) **89**, 283
N-Fluorobenzenesulfonimide; (133745-75-2) **87**, 137
(S)-2-Fluoro-1-phenylethanol; (110229-74-8) **89**, 9
1-((4-Fluorophenyl)(4-methoxyphenyl)methyl)piperidine; (96835-15-3) **89**, 283
Formaldehyde solution; (50-00-0) **85**, 287; **87**, 201
Formalin; (50-00-0) **86**, 344
Formic acid; (64-18-6) **86**, 344
4-Fluorobenzaldehyde; (459-57-4) **86**, 92
*[2-(4-Fluorophenyl-1H-indol-4-yl]-1-pyrrolidinylmethanone; **86**, 92*
2-[trans-2-(4-Fluorophenyl)vinyl]-3-nitrobenzoic acid; (917614-64-3) **86**, 92
{2-[trans-2-(4-Fluorophenyl)vinyl]-3-nitrophenyl}-1-pyrrolidinylmethanone; (917614-
    83-6) **86**, 92
Fragmentation; **88**, 121, 353

*G1-[N(CH₂CH₂CH₂NHBoc)₂]₆-Cl₃; **86**, 151*
*G1-[N(CH₂CH₂CH₂NHBoc)₂]₆-Piperidine₃; **86**, 151*
Geraniol: 2,6-Octadien-1-ol, 3,7-dimethyl-, (2*E*)-; (106-24-1) **88**, 54
Geranyl Diethyl Phosphate: Phosphoric acid, (2E)-3,7-dimethyl-2,6-octadien-1-yl diethyl
    ester; (60699-32-3) **88**, 54
D-(*R*)-Glyceraldehyde acetonide; (15186-48-8) **88**, 181
Glyoxal: Ethanedial; (107-22-2) **87**, 77
Glyoxylic acid monohydrate: Acetic acid, 2,2-dihydroxy-; (563-96-2) **88**, 152
Gold(I) chloride; (10294-29-8) **89**, 450

H,⁴ClQuin-BAM; (1229434-91-6) **89**, 380
H,⁴PyrrolidineQuin-BAM (PBAM); (1214287-91-8) **89**, 380
Halogenation; **85**, 53, 231; **87**, 8, 16, 88, 126, 137, 170; **88**, 54,102, 207, 291, 406; **89**, 9,
    55, 82, 183, 491
(Z)-1-Heptenyl-4-methoxybenzene: Benzene, 1-(1Z)-1-heptenyl-4-methoxy-; (80638-85-3)
    88, 102
1-Heptyne; (628-71-7) **88**, 102
Heterocycles; **85**, 10, 27, 34, 53, 64, 72, 88; **86**, 18, 59, 70, 81, 92, 105, 130, 141, 151,
    181, 236, 262, 315, 333; **87**, 8, 16, 36, 59, 77, 95, 137, 161, 184, 263, 310, 317,
    339, 350, 362; **88**, 22, 33, 42, 79, 121, 138, 162, 224, 238,247, 353, 364, 377,
    388, 398; **89**, 44, 76, 82, 98, 115, 131, 170, 274, 283, 294, 323, 350, 380, 409,
    438, 537, 549
Heterocyclic carbine; **85**, 34
Hexachloroethane: Ethane, hexachloro-; (67-72-1) **89**, 82
1,1,1,3,3,3-Hexafluoro-2-propanol: 2-Propanol, 1,1,1,3,3,3-hexafluoro-; (920-66-1) **89**, 9
1,2,3,3a,4,9-Hexahydropyrrolo[2,1-b]quinazoline; (495-58-9) **89**, 274

Indium(III) tris(trifluoromethanesulfonate): Methanesulfonic acid, trifluoro-, indium(3+) salt; (128008-30-0) **86**, 325

Insertion; **87**, 209; **88**, 418

Iodine; (7553-56-2) **85**, 219, 248; **86**, 1, 308; **87**, 288; **88**, 102, 406; **89**, 98, 183, 294

o-Iodoaniline: Benzenamine, 2-iodo-; (615-43-0) **88**, 162

3-Iodoanisole: Benzene, 1-iodo-3-methoxy-; (766-85-5) **87**, 68

4-Iodoanisole: Benzene, 1-iodo-4-methoxy-; (696-62-8) **88**, 102

Iodobenzene; (591-50-4) **89**, 294

Iodobenzene diacetate (IBD): Bis(acetato-κ*O*)phenyliodine; (3240-34-4) **89**, 311

(Z)-1-Iodo-1-Heptene: 1-Heptene, 1-iodo-, (1Z)-; (63318-29-6) **88**, 102

1-Iodo-1-heptyne: 1-Heptyne, 1-iodo-; (54573-13-6) **88**, 102

Iodomethane: Methane, iodo-; (74-88-4) **86**, 194; **89**, 115

1-Iodo-3-methylbenzene; (625-95-6) **87**, 178, 184

(Z)-1-Iodo-1-octene; (52356-93-1) **89**, 183

1-Iodo-1-octyne; (81438-46-2) **89**, 183

2-(2-Iodophenyl)propan-2-ol; (69352-05-2) **88**, 168

3-Iodo-4-phenylquinoline; (809274-63-3) **89**, 294

1-Iodopropane: Propane, 1-iodo-; (107-08-4) **89**, 159

Iodosobenzene diacetate; (3240-34-4) **87**, 115

(E)-(2-Iodovinyl)benzene; (42599-24-6) **87**, 170

Iron powder; (7439-896) **89**, 274

Isobutylboronic acid: Boronic acid, *B*-(2-methylpropyl)-; (84110-40-7) **88**, 247

Isobutylboronic acid pinacol ester; (67562-20-3) **88**, 247

2-Isobutylthiazole; (18640-74-9) **86**, 105

Iso-butyraldehyde; 2-methylpropionaldehyde; 2-methylpropanal (78-84-2) **87**, 143; **88**, 364

Isopropyl acetyl(phenyl)carbamate; Carbamic acid, N-acetyl-N-phenyl-, 1-methylethyl ester; (5833-25-0) **88**, 212

Isopropyl chloroformate; (108-23-6) **88**, 212

Isomerization; **89**, 267

iPrMgCl·LiCl: isopropylmagnesium chloride lithium chloride complex; (807329-97-1) **86**, 374

(4*S*)-4-Isopropyl-5,5-diphenyloxazolidin-2-one (DIOZ): 2-Oxazolidinone, 4-(1-methylethyl)-5,5-diphenyl-, (4*S*)-; (184346-45-0) **85**, 295

(4S)-4-Isopropyl-5,5-diphenyl-3-(3-phenyl-propionyl)oxazolidin-2-one: 2-Oxazolidinone, 4-(1-methylethyl)-3-(1-oxo-3-phenylpropyl)-5,5-diphenyl-, (4S)-; (213887-81-1) **85**, 295

dl-1,2-Isopropylideneglycerol: 1,3-Dioxolane-4-methanol, 2,2-dimethyl-; (100-79-8) **89**, 183

N-Isopropylidene-N'-2-nitrobenzenesulfonyl hydrazone; (6655-27-2) **89**, 230

(4*S*)-Isopropyl-2-oxazolidinone: (4*S*)-4-(1-Methylethyl)-2-oxazolidinone; (17016-83-0) **85**, 158

O-Isopropyl S-3-oxobutan-2-yl dithiocarbonate: carbonodithioic acid, O-(1-methylethyl) S-(1-methyl-2-oxopropyl) ester; (958649-73-5) **86**, 333

Isopropyl phenyl(4,4,4-trifluoro-3,3-dihydroxybutanoyl)carbamate; **88**, 212

(S)-4-Isopropyl-N-propanoyl-1,3-thiazolidine-2-thione; **86**, 70

(*S*)-4-Isopropyl-*N*-propanoyl-1,3-thiazolidine-2-thione: 2-Thiazolidinethione, 4-(1-methylethyl)-3-(1-oxopropyl)-, (4*S*)-; (102831-92-5) **86**, 70

(4S)-Isopropyl-3-propionyl-2-oxazolidinone: (4S)-4-(1-Methylethyl)-3-(1-oxopropyl)-2-oxazolidinone; (77877-19-1) **85**, 158

(S)-4-Isopropyl-1,3-thiazolidine-2-thione; **86**, 70

(*S*)-4-Isopropyl-1,3-thiazolidine-2-thione: 2-Thiazolidinethione, (4*S*)-4-(1-methylethyl)-; (76186-04-4) **86**, 70

(L)-*tert*-Leucine: L-Valine, 3-methyl-; (20859-02-3) **86**, 181

(L)-(+)-*tert*-leucinol: 1-Butanol, 2-amino-3,3-dimethyl-, (2*S*)-; (112245-13-3) **86**, 121

Lithium; (7439-93-2) **85**, 53; **86**, 1; **89**, 55

Lithium aluminum hydride; (16853-85-3) **85**, 158; **87**, 310

Lithium bis(trimethylsilyl)amide; (4039-32-1) **87**, 16; **89**, 510

Lithium borohydride; (16949-15-8) **89**, 420

Lithium *t*-butoxide; (1907-33-1) **87**, 178, 184

Lithium chloride; (7447-41-8) **86**, 47; **9**, 76

Lithium hydroxide monohydrate; (1310-66-3) **85**, 295

Lithium 2-pyridiltriolborate:Borate(1-), [2-[(hydroxy-kO)methyl]- 2-methyl-1,3-propanediolato(3-)-kO1,kO3]-2-pyridinyl-, lithium (1:1), (T-4)-; (1014717-10-2) **88**, 79

Lithium triethylborohydride; (22560-16-3) **85**, 64

(*R*)-(-)-Mandelic acid; (611-71-2) **85**, 106

(*S*)-(+)-Mandelic acid; (17199-29-0) **85**, 106

(R)-Mandelic acid salt of (1S,2S)-trans-2-(N-benzyl)amino-1-cyclohexanol; (882409-00-9) **85**, 106

(S)-Mandelic acid salt of (1R,2R)-trans-2-(N-benzyl)amino-1-cyclohexanol; (882409-01-0) **85**, 106

Manganese(IV) oxide; (1313-13-9)

Magnesium; (7439-95-4) **87**, 178; **89**, 98

Magnesium chloride; (7786-30-3) **89**, 220

Manganese(II) chloride tetrahydrate; (13446-34-9) **89**, 255

Mercury (II) oxide red; (21908-53-2) **89**, 491

Mesitylamine: Benzenamine, 2,4,6-trimethyl-; (88-05-1) **85**, 34; **87**, 77

3-(Mesitylamino)butan-2-one: 2-Butanone, 3-[(2,4,6-trimethylphenyl)-amino]-; (898552-96-0) **85**, 34

Mesitylene; (108-67-8) **85**, 196

2-Mesitylhydrazinium chloride; (76195-82-9) **87**, 362

N-Mesityl-N-(3-oxobutan-2-yl)formamide: Formamide, N-(1-methyl-2-oxopropyl)-N-(2,4,6-trimethylphenyl)-; (898553-01-0) **85**, 34

Metal complexation; **87**, 26, 104; **89**, 19, 255

Metallation, **85**, 1, 45, 209; **86**, 374; **87**, 317, 330; **88**, 79, 181; **89**, 267, 323, 420, 460

Metathesis; **89**, 170

Methanesulfonic acid, trifluoro-, 6-methyl-2-pyridinyl ester; (154447-04-8) **89**, 76

Methanesulfonyl chloride; (124-63-0) **85**, 219; **86**, 181; **88**, 33

trans-4-Methoxy-3-buten-2-one; (51731-17-0) **87**, 192

(+)-*B*-Methoxydiisopinocampheylborane ((+)-(Ipc)$_2$BOMe); (99438-28-5) **88**, 87

(4*S*)-3-[(2*R*,3*S*,4*E*)-3-Methoxy-2,4-dimethyl-1-oxo-5-phenyl-4-pentenyl]-4-(1-methylethyl)-2-thiazolidinethione; (332902-42-8) **86**, 81, 181

1-(2-Methoxyethoxy)-1-vinylcyclopropane: Cyclopropane, 1-ethenyl-1-(2-methoxyethoxy)-; (278603-80-8) **88**, 109

594

4-Methoxy-4'-nitrobiphenyl,1,1'-Biphenyl, 4-methoxy-4'-nitro-; (2143-90-0) **88**, 197

4-Methoxyphenylacetic acid; (104-01-8) **88**, 418; **89**, 255

p-Methoxyphenylacetylene: Benzene, 1-ethynyl-4-methoxy-; (26797-70-6) **89**, 549

(R)-2-Methoxy-N-(1-phenylethyl)acetamide; (162929-44-4) **89**, 255

4-(4-Methoxyphenyl)-2-phenylquinoline; (73402-86-5) **89**, 549

Methyl acetoacetate: Butanoic acid, 3-oxo-, methyl ester; (105-45-3) **86**, 161

Methyl 2-(2-acetylphenyl)acetate: Benzeneacetic acid, 2-acetyl-, methyl ester; (16535-88-9) **86**, 161

(R)-Methyl 2-allylpyrrolidine-2-carboxylate hydrochloride; (112348-46-6) **86**, 262

Methylamine; (74-89-5) **88**, 42; **89**, 44

Methyl anthranilate; (134-20-3) **87**, 226

(2S,3S)-Methyl-2-(9-anthryloxymethyl)-2-(4-methoxyphenyl)-3-(4-Methoxyphenylamino)-3-phenylpropanoate; (1034152-21-0) **88**, 418

(S)-Methyl-2-benzamido-3-methylbutanoate: L-Valine, N-benzoyl-, methyl ester; (10512-91-1) **88**, 14

(R)-α-Methylbenzylamine; *(R)*-α-phenylethylamine; *(R)*-1-phenylethylamine; (3886-69-9) **87, 143**

(±)-α-Methylbenzylamine; (618-36-0) **89**, 255

(S)-(α-Methylbenzyl)hydroxylamine oxalate salt: (α*R*)-*N*-Hydroxy-α-methyl-benzenemethanamine ethanedioate salt; (78798-33-1) **87**, 218

Methyl 2-benzyloxycarbonylamino-2-(dimethoxyphosphinyl)acetate; (88568-95-0) **88**, 152

5-Methyl-2,2'-bipyridine: 2,2'-Bipyridine, 5-methyl); (56100-20-0) **89**, 76

Methyl 4-bromobenzoate: Benzoic acid, 4-bromo-, methyl ester; *(619-42-1)* **88**, 79

Methyl 2-bromomethyl-3-nitrobenzoate: Benzoic acid, 2-bromomethyl-3-nitro-, methyl ester; (98475-07-1) **88**, 291

Methyl 4-(tert-butylcarbamoyl)benzoate: Benzoic acid, 4-[[(1,1-dimethylethyl)amino]carbonyl]-, methyl ester; (67852-98-6) **89**, 519

2-Methyl-3-butyn-2-ol; (115-19-5) **85**, 118

(S)-2-Methyl-CBS-oxazaborolidine: 1*H*, 3*H*-Pyrrolo[1,2-c][1,3,2]oxazaborole, tetrahydro-1-methyl-3,3-diphenyl-, (*S*)-; (112022-81-8) **88**, 70

Methyl 2-(2-chlorophenyl)-*methylpropanoate; (736055-19-9)* **89**, 510

N-Methyl 2-chloropyridinium trifluoromethane sulfonate: Pyridinium, 2-chloro-1-methyl-, 1,1,1-trifluoromethanesulfonate; (1:1); (84030-18-2) **88**, 121

Methyl crotonate; (623-43-8) **87**, 253

Methyl 4-cyanobenzoate: Benzoic acid, 4-cyano-, methyl ester; (1129-35-7) **89**, 66, 519

2-Methyl-1,3-cyclohexanedione: 1,3-Cyclohexanedione, 2-methyl-; (1193-55-1) **88**, 330, 353

Methyl cyclopropanecarboxylate: Cyclopropanecarboxylic acid, methyl ester; (2868-37-3) **89**, 334

Methyl 2-diazo-2-(4-methoxyphenyl) acetate: Benzeneacetic acid, α-diazo-4-methoxy-, methyl ester; (22979-36-8) **88**, 418

(R$_a$)-Methyl 3-(dimethyl(phenyl)silyl)hexa-3,4-dienoate; (945539-62-8) **89**, 420

2-Methyl-1,3-dithiane; (6007-26-7) **88**, 353

(E)-3,4-Methylenedioxy-β-nitrostyrene; (22568-48-5) **85**, 179

Methyl 2-ethenyl-3-nitrobenzoate: Benzoic acid, 2-ethenyl-3-nitro-, methyl ester; (195992-04-2) **88**, 291

Methylhydrazine; (60-34-4) **85**, 179

4-(Methylthio)-2-phenylquinazoline; (13182-63-3) **89**, 549
Methyl trifluoromethanesulfonate; (333-27-7) **88**, 121
2-Methyl-3-trifluoromethansulfonyloxy-cyclohex-2-en-1-one: Methanesulfonic acid,
    1,1,1-trifluoro-, 2-methyl-3-oxo-1-cyclohexen-1-yl ester; (150765-78-9)
5-Methyl-2-(trifluoromethanesulfonyl)oxypyridine: Methanesulfonic acid, trifluoro-, 5-
    methyl-2-pyridinyl ester; (154447-03-7) **89**, 76
3-Methyl-2-(trimethylsilyl)indole: 1H-Indole, 3-methyl-2-(trimethylsilyl)-; (135189-98-9)
    88, 377
Methyl(triphenylphosphine)gold (CH₃AuPPh₃); (23108-72-7) **89**, 126
Methyltriphenylphosphonium bromide; (1779-49-3) **85**, 248 ; 88, 342
Methyl vinyl ketone: 3-Buten-2-one; (78-94-4) **88**, 330
Microwave; **86**, 18, 252
Morpholine; (110-91-8) **87**, 59
Morpholinomethanol: 4-Morpholinemethanol; (4432-43-3) **87, 59**

Naphthalene; (91-20-3) **85**, 53
Ni(cod)₂: Bis(cyclooctadiene)nickel(0); (1295-35-8) **88**, 342
Nickel(II) chloride, dimethoxyethane adduct (NiCl₂·glyme): Nickel, dichloro[1,2-
    di(methoxy-κO)ethane]-; (29046-78-4) **87**, 317, 330
Nickel-catalyzed reactions; **87**, 317, 330; **88**, 342
Nitriles; **86**, 1,28
2-Nitrobenzaldehyde; (552-89-6) **85**, 27; **86**, 36; **87**, 339; **89**, 274
2-Nitrobenzoic acid; (552-16-9) **85**, 196
2-Nitrobenzyl bromide: Benzene, 1-(bromomethyl)-2-nitro-; (3958-60-9) **87**, 339
o-Nitrobenzenesulfonyl hydrazide; (5906-99-0) **89**, 230
Nitrone; **88**, 291
4-Nitrophenol; (100-02-7) **88**, 197
4-Nitrophenyl trifluoromethanesulfonate; (17763-80-3) **88**, 197
1,8-Nonadiyne; (2396-65-8) **89**, 126
2,8-Nonanedione; (30502-73-9) **89**, 126
trans-2-Nonenal: 2-Nonenal, (2E)-; (18829-56-6) **88**, 342
Novozym 435, Lipase acrylic resin from Candida antárctica; (9001-62-1) **89**, 255

(S)-5,5',6,6',7,7',8,8'-Octahydro-1,1'-bi-2-naphthol [(S)-H₈BINOL]: [1,1'-
    Binaphthalene]-2,2'-diol, 5,5',6,6',7,7',8,8'-octahydro-, (1S)-; (65355-00-2) **87**, 59,
    263
(S)-1-(8,9,10,11,12,13,14,15-Octahydro-3,5-dioxa-4-phosphacyclohepta[2,1-a;3,4-
    a']dinaphthalen-4-yl)-2,2,6,6-tetramethylpiperidine **87**, 263
2-Octenoic acid, 3-phenyl-, methyl ester, (2E)-; (189890-29-7) **87**, 53
n-Octyl vinyl ether; (929-62-4) **89**, 307
1-Octyne; (629-05-0) **89**, 183
2-Octynoic acid, methyl ester; (111-12-6) **87**, 53
Organocatalysis; **86**, 11; **87**, 201; **88**, 42
Oxalyl chloride: Ethanedioyl dichloride; (79-37-8) **85**, 189
Oxidation; **85**, 15, 189, 267, 278; **86**, 1, 28, 121, 308. 315; **87**, 1, 8, 253; **88**, 14, 54, 121,
    168, 247; **89**, 19, 105, 131, 274, 311, 350, 404
Oximes; **86**, 18; **87**, 36, 275; **88**, 33; **89**, 409, 480
Oxirane, 2-methyl-; (75-56-9) **87**, 126
3-Oxocyclobutanecarboxylic acid; (23761-23-1) **89**, 491

2-[(5-Oxo-1-cyclohepten-1-yl)methyl]-1H-isoindole-1,3(2H)-dione: 1H-Isoindole-
1,3(2H)-dione, 2-[(5-oxo-1-cyclohepten-1-yl)methyl]-; (278603-99-9) **88**, 109
3-(2-(2-Oxoethyl)-1,3-dioxolan-2-yl) propanoic acid; (360794-17-8) **88**, 121
Oxone® monopersulfate; (37222-66-5) **85**, 278; **89**, 105, 131

Palladium; **86**,105; **87**, 104, 143, 226, 263, 299; **88**, 4, 22, 79, 102, 207, 260, 291, 377,
406
Palladium(II) acetate; (3375-31-3) **85**, 96; **86**, 92, 105, 344; **87**, 299; **88**, 22, 70, 291, 377;
89, 159, 510
Palladium acetylacetonate; (140024-61-4) **85**, 196
Palladium, bis[(1,2,4,5-η)-1,5-diphenyl-1,4-pentadien-3-one]-: Pd(dba)₂; (32005-36-0)
88, 102; **89**, 537
Palladium hydroxide on carbon; Pearlman's catalyst; (12135-22-7) **87**, 143
Palladium tris(dibenzylideneacetone); (48243-18-1) **86**, 274
Palmitoyl chloride: Hexadecanoyl chloride; (112-67-4) **89**, 471
Paraformaldehyde; (30525-89-4) **87**, 59; **89**, 44, 220
Pentadecane; (629-62-9) **89**, 471
2-Pentafluorophenyl-6,10b-dihydro-4*H*,5a*H*-5-oxa-3,10c-diaza-2-
azoniacyclopenta[*c*]fluorene; tetrafluoroborate; (740816-14-2) **87**, 350
2-Pentafluorophenyl-6,7-dihydro-5*H*-pyrrolo[2,1-*c*][1,2,4]triazol-2-ium;
tetrafluoroborate; (862095-91-8) **87**, 350
Pentafluorophenylhydrazine; (828-73-9) **87**, 350
4-Penten-1-ol; (821-09-0) **87**, 299
(E)-Pent-2-en-1-yl tosylcarbamate; (1215019-8-8) **89**, 450
*Phenanthrene, 2-bromo-1,2,3,4,4a,9,10,10a-octahydro-1,1,4a-trimethyl-, (2R,4aR,10aS)-
rel-: (1198206-88-0)* **88**, 54
9-Phenanthreneboronic acid; (68572-87-2) **88**, 406
1,10-Phenanthroline; (66-71-7) **85**, 196; **86**, 92; **87**, 178, 184
(1,10-Phenanthroline)bis(triphenylphosphine)copper(I) nitrate; (33989-10-5) **85**, 196
(*R*)-2,2'-(9-Phenanthryl)-BINOL phosphoric acid; **88**, 418
Phenylacetylene: Benzene, ethynyl-; (536-74-3) **85**, 53, 118, 131; **86**, 325
(*S*)-Phenylalaninol; (3182-95-4) **85**, 267; **88**, 364
N-Phenylbenzenecarboxamide (benzanilide); (93-98-1) **85**, 88
2-Phenylbenzoxazole; (833-50-1) **88**, 398
4-Phenyl-1-butene: Benzene, 3-buten-1-yl-; (768-56-69) **87**, 88
4-Phenyl-3-buten-2-one; (122-57-6) **89**, 374
Phenylboronate: 1,3,2-Dioxaborinane, 5,5-dimethyl-2-phenyl-; (5123-13-7) **87**, 209
(*E*)-1-Phenyl-1,3-butadiene: Benzene, (1*E*)-1,3-butadienyl-: (16939-57-4) **86**, 315
2-Phenyl-1-butene; (2039-93-2) **85**, 248
4-Phenylbutyric acid: Benzenebutanoic acid; (1821-12-1) **89**, 432
(R,R)-1-Phenylcyclohexene oxide: 7-Oxabicyclo[4.1.0]heptane, 1-phenyl-, (1R,6R)- (9);
(17540-04-4) **89**, 350
5-Phenyl-1,3-dimethoxybenzene; (64326-17-6) **86**, 1
N-[(1S)-1-Phenylethyl]-benzeneacetamide: *2-Phenyl-N-(1-phenylethyl)acetamide;*
(17194-90-0) **87**, 218
(*S*)-Phenylglycine: Benzeneacetic acid, α-amino-, (α*S*)-; (2935-35-5) **85**, 219
5-Phenyl-2-isobutylthiazole; (600732-10-3) **86**, 105
Phenylmagnesium chloride, 2 M in THF; (100-59-4) **87**, 26
(1Z,3E)-1-Phenyl-1,3-octadiene, (Z,E)-1-Phenyl-1,3-octadiene; (39491-66-2) **88**, 202

5-Phenyl-pent-2-enal: 2-Pentenal, 5-phenyl-, (2E)-; (37868-70-5) **89**, 527
3-Phenylpropan-1-ol; (122-97-4) **88**, 247
3-Phenylpropanoyl chloride: Benzenepropanoyl chloride; (645-45-4) **85**, 295
N-(3-Phenyl-2-propynyl)aniline; (121194-51-2) **89**, 294
5-Phenyl-1H-pyrazol-3-amine: 1*H*-pyrazol-3-amine, 5-phenyl-; (1571-10-7) **86**, 252
Phenylpyruvic acid: α-Oxobenzenepropanoic acid, 2-Oxo-3-phenylpropionic acid; (156-06-9) **87, 218**
Phenylsilane: Benzene, silyl-; (694-53-1) **87,** 88
8-Phenyl-1-tetralone; (501374-10-3) **87,** 209
4-Phenyl-3-(trifluoromethyl)butan-2-one: 2-Butanone, 4,4,4-trifluoro-3- (phenylmethyl)-; (808105-43-3) **89**, 374
(Phenyl) [2-(trimethylsilyl)phenyl]iodonium triflate: Iodonium, phenyl-, 2- (trimethylsilyl)phenyl-, salt with trifluoromethanesulfonic acid (1:1); (164594-13-2) **89**, 98
N-[1-Phenyl-3-(trimethylsilyl)-2-propyn-1-ylidene]-benzeneamine; (77123-64-9) **85**, 88
Phenyl vinyl sulfide; (1822-73-7) **88**, 207
Phosphine, tris(1,1-dimethylethyl)-, tetrafluoroborate(1-) (1:1); (155026-77-0) **88**, 22
Phosphoric triamide, *N,N,N',N',N'',N''*-hexamethyl- ; (680-31-9) **87**, 126
Phosphorus oxychloride; (10025-87-3) **88**, 406; **89**, 380
Phosphorus trichloride; (7719-12-2) **85**, 238; **87**, 263; **88**, 152
Phosphorylation; **87, 263; 88,** 54, 152, 406
2-Picoline borane: Boron, trihydro(2-methylpyridine)-, (T-4)-; (3999-38-0) **87**, 339
Pimelic acid: Heptanedioic acid; (111-16-0) **86**, 194
Pinacol; (79-09-5) **88**, 247
Pinacolone: 2-Butanone, 3,3-dimethyl-; (75-97-8) **87**, 209
Piperidine; (110-89-4) **89**, 115
Piperidinium acetate; (4540-33-4) **86**, 28
Pivaldehyde, trimethylacetaldehyde; (630-19-3) **88**, 42
Potassium *tert*-butoxide: 2-Propanol, 2-methyl-, potassium salt (1:1); (865-47-4) **86**, 315; **88**, 181, 342; **89**, 267
Potassium carbonate; (584-08-7) **85**, 287; **86**, 58
Potassium *O*-ethyl xanthate; (140-89-6) **89**, 409
Potassium hydride; (7693-26-7) **89**, 230
Potassium hydroxide; (1310-58-3) **85**, 196; **87,** 115
Potassium 2-iodo-5-methylbenzenesulfonate; (1093215-92-9) **89**, 105
Potassium *O*-isopropylxanthate: Carbonodithioic acid, O-(1-methylethyl) ester, potassium salt (1:1); (140-92-1) **86**, 333
Potassium permanganate; (7722-64-7) **89**, 274
Potassium trimethylsilanolate; (10519-96-7) **86**, 274
Propanal, 3-[[(1,1-dimethylethyl)dimethylsilyl]oxy]-2-methyl-, (2R)-; (97826-89-6) **89**, 243
Propanol, 2-amino-, 3-phenyl, (*S*); (3182-95-4) **85**, 267
Propargyl alcohol: 2-Propyn-1-ol; (107-19-7) **89**, 267
Propargyl bromide; (106-96-7) **89**, 294
N-Propargylphthalimide: Phthalimide, *N*-2-propynyl-: 1*H*-Isoindole-1,3(2*H*)-dione, 2-(2-propynyl)-; (7223-50-9) **88**, 109
Prop-2-en-1-ol; (107-18-6) **89**, 243
Propionyl chloride : Propanoyl chloride; (79-03-8) **85**, 158; **86**, 70; 88. 364
(*S*)-Proline; (147-85-3) **86**, 262

2-Propenoic acid, 3,3'-(1,2-cyclopropanediyl)bis-, diethyl ester, [1α(E),2β(E)]-; (58273-88-4) **85**, 15

[(1*E*)-1-Propylbutyl-1-enyl]benzene: Benzene, [(1*E*)-1-propyl-1-buten-1-yl]-; (1151654-13-5) **89**, 159

iso-Propylmagnesium chloride; (1068-55-9) **89**, 460

N-(2-Propynyl)aniline; (14465-74-8) **89**, 294

(*R*)-*i*-Pr-Pybox: Pyridine, 2,6-bis[(4*R*)-4,5-dihydro-4-(1-methylethyl)-2- oxazolyl]-; (131864-67-0) **87**, 330

1*H*-Pyrazole, 5-(1,3-benzodioxol-5-yl)-3-(4-chlorophenyl)-1-methyl-; (908329-89-5) **85**, 179

2,6-Pyridinedicarbonitrile; (2893-33-6) **87**, 310

Pyridine *N*-oxide: Pyridine, 1-oxide; (694-59-7) **88**, 22

(*E*)-1-Pyridin-3-yl-ethanone *O*-benzyl-oxime: Ethanone, 1-(3-pyridinyl)-, *O*-(phenylmethyl)oxime, (1*E*)-; (1010079-98-7) **87**, 36

(*E*)-1-Pyridin-3-yl-ethanone oxime; (106881-77-0) **87**, 36

(*S*)-1-Pyridin-3-yl-ethylamine: 3-Pyridinemethanamine, α-methyl-, (α*S*)-; (27854-9) **87**, 36

(*S*)-1-*Pyridin-3-yl-ethylamine hydrochloride*: 3-*Pyridinemethanamine, α-methyl-, dihydrochloride, (S)-; (40154-84-5)* **87**, 36

(4S)-5-Pyrimidinecarboxylic acid, 1,2,3,4-tetrahydro-6-methyl-2-oxo-4-phenyl-1-(phenylmethyl)-, methyl ester; (865086-56-2) **86**, 236

Pyrrolidine; (123-75-1) **87**, 1, 201; **89**, 380

1-Pyrrolidinecarboxylic acid, 2-(aminocarbonyl)-, phenylmethyl ester, (2*S*)-; (34079-31-7) **85**, 72

1,2-Pyrrolidinedicarboxylic acid, 1-(phenylmethyl) ester, (2*S*)-; (1148-11-4) **85**, 72

2-Pyrrolidinone; (616-45-5) **87**, 350

2-Pyrrolidinone, 1-methyl-; (872-50-4) **87**, 126

(S)-5-Pyrrolidin-2-yl-1H-tetrazole: 2H-Tetrazole, 5-[(2S)-2-pyrrolidinyl]-; (33878-70-5) **85**, 72

Quinidine: Cinchonan-9-ol, 6'-methoxy-; (56-54-2) **88**, 121

Quinoline; (91-22-5) **85**, 196

Radical chain; **89**, 471

Rearrangements; **86**, 18, 113, 172; **88**, 309; **89**, 420, 450

Reduction; **85**, 15, 27, 53, 64, 72, 138, 147, 158, 219, 248; **86**, 1, 28, 36, 92, 181, 344; **87**, 36, 77, 143, 275, 310, 362; **88**, 22, 70, 102, 202, 260, 291, 309, 427; **89**, 115, 131, 143, 159, 230, 274, 323, 334, 471

Resolution; **85**, 106; **89**, 9, 255, 420

Rhodium (II) tetrakis(triphenylacetate);(142214-04-8) **86**, 58

Rhodium; **86**, 59; **87**, 115; **88**, 109, 418

Rhodium acetate dimer: Rhodium, tetrakis[μ-(acetato-κO:κO')]di-, (Rh-Rh); (15956-28-2) **85**, 172; **86**, 58

Rhodium, dicarbonyl(2,4-pentanedionato-κ-O2,κ-O4)-, (SP-4-2)-; (14874-82-9) **89**, 243

[RhCl(cod)]₂; (chloro(1,5-cyclooctadiene)rhodium(I) dimer); (12092-47-6) **86**, 360

Rhodium, chlorotris(triphenylphosphine)-, (SP-4-2)-; (14694-95-2) **89**, 374

Ring Expansion, **85**, 138

RuH₂(CO)(PPh₃)₃: Ruthenium, carbonyldihydrotris(triphenylphosphine); (25360-32-1) **87**, 209
Ruthenium carbonyl; (15243-33-1) **89**, 255
Ruthenium complex {{[2,3,4,5(4-OMe-C₆H₄)₄](η⁵-C₄CO)}₂H}-Ru₂(CO)₄(μ-H); (873815-22-6) **89**, **255**

Saponification; **85**, 27
Semicarbazide hydrochloride: Hydrazinecarboxamide, hydrochloride (1:1); (563-41-7) **86**, 194
Silanamine, 1,1,1-trimethyl-*N*-(trimethylsilyl)-, sodium salt (1:1): NaHMDS; (1070-89-9) **89**, 159
Silane, (1,1-dimethylethyl)dimethyl(2-propen-1-yloxy)-; (85807-85-8) **89**, 243
Silane, chloro(1,1-dimethylethyl)dimethyl-; (18162-48-6) **89**, 243
Silanes; **87**, 178, 253, 299; **88**, 102, 296, 377; **89**, 19, 34, 82, 98, 267, 323, 420, 480
Silver; **88**, 14, 309
Silver carbonate: Carbonic acid, silver(1+) salt (1:2); (534-16-7) **87**, 253, 288
Silver(I) fluoride; (7775-41-9) **86**, 225
Silver hexafluoroantimonate; (26042-64-8) **88**, 309
Silver iodate: iodic acid (HIO₃), silver(1+) salt (1:1); (7783-97-3) **88**, 14
Silver nitrate; (7761-88-8) **86**, 225
Silver trifluoromethanesulfonate: Methanesulfonic acid, trifluoro-, silver(1⁺) salt; (2923-28-6) **89**, 450
Silylation; **87**, 178, 299; **88**, 102, 121, 296; **89**, 19, 82, 98, 243, 267323, 420, 480
Sodium; (7440-23-5) **86**, 262
Sodium acetate; (127-09-3) **86**, 105, 194
Sodium acetate trihydrate; (6131-90-4) **85**, 10
Sodium amide; (7782-92-5) **86**, 298
Sodium ascorbate: L-Ascorbic acid, sodium salt: Sodium (+)-L-ascorbate; (134-03-2) **88**, 238
Sodium azide; (26628-22-8) **85**, 72, 278; **86**, 113; **87**, 161
Sodium bis(trimethylsilyl)amide; (1070-89-9) **87**, 170
Sodium borohydride: Borate(1-), tetrahydro-, sodium (1:1); (16940-66-2) **85**, 147; **86**, 181; **87**, 77; **89**, 9
Sodium carbonate; (497-19-8) **86**, 36; **88**, 377; **89**, **255**
Sodium cyanide; (143-33-9) **85**, 219
Sodium hydride; (7646-69-7) **85**, 45, 172; **86**, 18, 194; **87**, 36
Sodium hypochlorite; (7681-52-9) **87**, 8
Sodium nitrite; (7632-00-0) **87**, 362; **89**, 76
Sodium tetrafluoroborate; (13755-29-8) **85**, 248
Sodium tetrakis[(3,5-trifluoromethyl)phenyl]borate; (79060-88-1) **85**, 248
Sodium thiosulfate; (7772-98-7) **86**, 1
Sodium thiosulfate pentahydrate; (10102-17-7) **89**, 294
Sodium Triacetoxyborohydride; (56553-60-7) **89**, 115
Sodium β-trimethylsilylethanesulfonate: Ethanesulfonic acid, 2-(trimethylsilyl)-, sodium salt; (18143-40-3) **89**, 34
(–)-Sparteine: 7,14-Methano-2*H*,6*H*-dipyrido[1,2-a:1',2'-e][1,5]diazocine, dodecahydro-, (7*S*,7a*R*,14*S*,14a*S*)-; (90-39-1) **88**, 247, 364
4-Spirocyclohexyloxazolidinone; **86**, 58
4-Spirocyclohexyloxazolidinone: 3-Oxa-1-azaspiro[4.5]decan-2-one; (81467-34-7) **86**,58

*(1'S,5'R)-Spiro[1,3-dioxolane-2,3'-[6]oxabicyclo[3.2.0]heptan]-7'-one; (*360794-30-5)
88*, 121
Styrene; (100-42-5) **87,** 115
Styrene oxide: 2-Phenyloxirane; (96-09-3) **89,** 9
Substitution; **86,** 18, 141, 151, 181, 333; **87,** 104, 126, 299, 317; **88,** 33, 54, 87, 152, 168,
181, 291; **89,** 9, 44, 76, 98, 131, 210, 307, 334, 380, 409, 480, 501, 537
Sulfination; **88,** 168, 309
Sulfonation; **86,** 11; **87,** 231; **88,** 260, 317, 353; **89,** 76, 480, 501
Sulfur; (7704-34-9) **85,** 209; **87,** 231
Super-Hydride®: Lithium triethylborohydride; (22560-16-3) **85,** 64

1,4,7,10-Tetraazacyclododecane-1,4,7-triacetic acid, Tri-tert-butyl Ester Hydrobromide:
*1,4,7,10-Tetraazacyclododecane-1,4,7-tricarboxylic acid, 1,4,7-tris(1,1-
dimethylethyl) ester;* (175854-39-4) **85,** 10
Tetrabromomethane; (558-13-4) **86,** 36
Tetrabutylammonium bromide; (1643-19-2) **85,** 278
Tetra-*n*-butylammonium chloride; (1112-67-0) **88,** 377
Tetrabutylammonium fluoride; (429-41-4) **87,** 95; **88,** 162; **89,** 98
Tetrabutylammonium fluoride trihydrate: 1-Butanaminium, *N, N, N*-tributyl, fluoride,
trihydrate; (87749-50-6) **88,** 102
Tetrahydro-1,3-dimethyl-2(1H)-pyrimidinone; (7226-23-5) **87,** 178, 184
4,4a,9,9a-Tetrahydro-1-oxa-4-azafluoren-3-one: Indeno[2,1-*b*]-1,4-oxazin-3(2*H*)-one,
4,4a,9,9a-tetrahydro-, (4a*R*,9a*S*)-; (862095-79-2) **87,** 350
Tetrafluoroboric acid: Borate(1-), tetrafluoro-, hydrogen (1:1); (16872-11-0) **87,** 288
Tetrafluoroboric acid diethyl etherate; (67969-82-8) **86,** 172
5,6,11,12-Tetrahydrodibenzo[a,e]cyclooctene; (1460-59-9) **89,** 55
Tetrakis(dimethylamino)allene; (42928-64-3) **86,** 298
1,1,3,3-Tetrakis(dimethylamino)propenium tetrafluoroborate; (125254-01-5) **86,** 298
2,3,4,5-Tetrakis(4-methoxyphenyl)cyclopenta-2,4-dienone; (49764-93-4) **89,** 255
Tetrakis(triphenylphosphine)palladium; (14221-01-3) **86,** 315; **88,** 207, 406; **89,** 76, 202
α-Tetralone: 1(2*H*)-Naphthalenone, 3,4-dihydro-; (529-34-0) **87,** 209, 275
Tetramethylammonium hydroxide pentahydrate; (10424-65-4) **89,** 255
4,4,5,5-Tetramethyl-1,3,2-dioxaborolane; (25015-63-8) **88,** 342
1,1,3,3-Tetramethylguanidine: Guanidine, *N,N,N',N'*-tetramethyl-; (80-70-6) **88,** 388
2,2,6,6-Tetramethylpiperidine; (768-66-1) **86,** 374; **87,** 263
2,2,6,6-Tetramethylpiperidin-1-oxyl (TEMPO): 2,2,6,6-Tetramethyl-1-piperidinyloxy;
(2564-83-2) **89,** 311
1,3,5,7-Tetramethyl-1,3,5,7-tetravinylcyclotetrasiloxane; (2554-06-5) **86,** 274
*(S)-2-(1H-Tetrazol-5-yl)-pyrrolidin-1-carboxylic acid benzyl ester: 1-
Pyrrolidinecarboxylic acid, 2-(2H-tetrazol-5-yl)-, phenylmethyl ester, (2S)-;*
(33876-20-9) **85,** 72
Tetrolic acid: 2-Butynoic acid; (590-93-2) **85,** 231
2-Thiazolidinethione, 4-(phenylmethyl)-, (4*S*)-; (171877-39-7) **88,** 364
4,4'-Thiobis(2-*tert*-butyl-m-cresol): Phenol, 4,4'-thiobis[2-(1,1-dimethylethyl)-3-methyl-;
m-Cresol, 4,4'-thiobis[2-*tert*-butyl-; (4120-97-2) **89,** 267
Thiocyanic acid methyl ester: Thiocyanic acid, methyl ester; (556-64-9) **89,** 549
Thionyl chloride; (7719-09-7) **89,** 34, 44
2-Thiophene-carboxaldehyde; (98-03-3) **86,** 360; **87,** 16
5-(Thiophen-2-yl)oxazole: Oxazole, 5-(2-thienyl)-; (70380-70-0) **87,** 16

Thiourea; (62-56-6) **87**, 115
Tin (II) chloride; (7772-99-8) **85**, 27
Tin(II) chloride dihydrate; (10025-69-1) **87**, 362
Titanium(IV) ethoxide: Tetraethyl titanate; (3087-36-3) **89**, 88
Titanium tetrachloride; (7550-45-0) **85**, 295; **86**, 81; **88**, 34
p-Tolualdehyde: benzaldehyde, 4-Methyl-; (104-87-0) **86**, 252
p-Toluenesulfonamide: Benzenesulfonamide, 4-methyl-; (70-55-3) **86**, 212
p-Toluenesulfonic acid: Benzenesulfonic acid, 4-methyl-; (104-15-4) **85**, 287
p-Toluenesulfonic acid monohydrate: Benzenesulfonic acid, 4-methyl-, hydrate (1:1);
    (6192-52-5) **86**, 194; **89**, 267
(*S*)-3-[*N*-(*p*-Toluenesulfonyl)amino]-4-methylpentanoic acid; (936012-07-6) **87**, 143
p-Toluenesulfonyl chloride: Benzenesulfonyl chloride, 4-methyl-; (98-59-9) **86**, 58; **87**,
    88, 231; **88**, 309; **89**, 480, 501
p-Toluenesulfonyl isocyanate; (4083-64-1) **89**, 450
p-Tolylboronic acid; (5720-05-8) **86**, 344
4-(p-Tolyl)-phenylboronic acid; (393870-04-7) **86**, 344
4-(p-Tolyl)-phenylboronic acid MIDA ester; **86**, 344
N-Tosyl-(*S*_a)-binam-L-prolinamide: 2-Pyrrolidinecarboxamide, *N*-[(1S)-2'-[[(4-
    methylphenyl)sulfonyl]amino][1,1'-binaphthalen]-2-yl]-, (2*S*)-; (933782-38-8) **88**,
    330
Tosylmethylisocyanide: Benzene, [(isocyanomethyl)sulfonyl]-; (36635-63-9) **87**, 16
1,3,5-*Triacetylbenzene*; (779-90-8) **87**, 192
Tribasic potassium phosphate; (7778-53-2) **86**, 105
1,3,5-Tribromobenzene; (626-39-1) **89**, 460
Tributyl(4-methoxyphenyl)stannane; (70744-47-7) **88**, 197
Tri-*n*-butylphosphine: Phosphine, tributyl-; (998-40-3) **86**, 212; **88**, 138
Tri-*t*-butylphosphonium tetrafluoroborate; (131274-22-1) **89**, 510
Tributyltin chloride; (1461-22-9) **88**, 197
2,2,2-Trichloro-1-ethoxyethanol; (515-83-3) **86**, 262
Trichloromethane (1:1); (52522-40-4) **88**, 162
(3R,7aS)-3-(Trichloromethyl)tetrahydropyrrolo[1,2-c]oxazol-1(3H)-one; (97538-67-5)
    86, 262
Tricyclohexylphosphine: Phosphine, tricyclohexyl-, (2622-14-2) **87**, 299; **88**, 342
(4S)-1-Tridecen-4-amine (893402-88-5) **89**, 88
Triethylamine: Ethanamine, *N*,*N*-diethyl-; (121-44-8) **85**, 131, 189, 219, 295; **86**, 58, 81,
    252, 315; **87**, 231; **88**, 42, 121, 138, 212, 247, 309, 317
Triethylamine hydrochloride: Ethanamine, *N*,*N*-diethyl-, hydrochloride (1:1); (554-68-7)
    85, 72
Triethyl orthoformate: Ethane, 1,1',1''-[methylidynetris(oxy)]tris- ; (122-51-0) **87**, 77,
    350, 362
Triethyl(pent-4-enyloxy)silane: Silane, triethyl(4-penten-1-yloxy)-; (374755-00-7) **87**,
    299
Triethylphosphine; (554-70-1) **87**, 275
Trifluoroacetic acid; (76-05-1) **87**, 143; **88**, 317; **89**, 267, 283, 438
Trifluoroacetic anhydride; (407-25-0) **85**, 64; **86**, 18; **89**, 210
2,2,2-Trifluoroethanol; (75-89-8) **85**, 45
2,2,2-Trifluoroethyl *p*-toluenesulfonate: Ethanol, 2,2,2-trifluoro-, 4-
    methylbenzenesulfonate; (433-06-7) **88**, 162
2,2,2-Trifluoroethyl trifluoroacetate; (407-38-5) **88**, 212

Trifluoroiodomethane; (2314-97-8) **89**, 374

Trifluoromethanesulfonic acid; HIGHLY CORROSIVE, Methanesulfonic acid, trifluoro-
(8, 9); (1493-13-6) **86**, 308; **88**, 197; **89**, 98

Trifluoromethanesulfonic anhydride: Methanesulfonic acid, trifluoro-, anhydride; (358-
23-6) **89**, 76, 549

Trifluoromethanesulfonimide; (82113-65-3) **87**, 253

Trifluoromethylation; **87**, 126

1-(Trifluoromethyl)-1,2-benziodoxol-3(1H)-one; (887144-94-7) **88**, 168

1-Trifluoromethyl-1,3-dihydro-3,3-dimethyl-1,2-benziodoxole; (887144-97-0) **88**, 168

Trifluoro, 2-(trimethylsilyl)phenyl ester; (88284-48-4) **86**, 161

Triflic anhydride: Trifluoromethanesulfonic acid anhydride; (358-23-6) **85**, 88; **88**, 197,
353

Triisopropyl borate: Boric acid (H₃BO₃), tris(1-methylethyl)ester; (5419-55-6) **88**, 79,
181

Triisopropyl phosphite; (116-17-6) **86**, 36

(Triisopropylsilyl)acetylene: Ethynyltriisopropylsilane; (89343-06-6) **86**, 225

3,4,5-Trimethoxybenzaldehyde; (86-81-7) **88**, 427

3,4,5-Trimethoxybenzoyl chloride: Benzoyl chloride, 3,4,5-trimethoxy-; (4521-61-3) **88**,
427

Trimethylacetaldehyde: Pivaldehyde: Propanal, 2,2-dimethyl; (630-19-3) **85**, 267

Trimethylaluminum; (75-24-1) **87**, 104

Trimethylamine oxide: Methanamine, *N,N*-dimethyl-, *N*-oxide; (1184-78-7) **88**, 162

2,4,6-Trimethylaniline; (88-05-1) **87**, 362

(1*R*,2*R*,3*R*,5*S*)-2,6,6-Trimethylbicyclo[3.1.1]heptan-3-ol: (+)-isopinocampheol; (24041-
60-9) **88, 87**

Trimethyl orthoformate; (149-73-5) **86**, 81, 130

Trimethyloxonium tetrafluoroborate; (420-37-1) **87**, 350, 362

Trimethylphosphite: Phosphorous acid, trimethyl ester; (121-45-9) **88**, 152

1-(Trimethylsilyl)acetylene; (1066-54-2) **85**, 88

4-Trimethylsilyl-2-butyn-1-ol; (90933-84-9) **88**, 296

Trimethylsilyl chloride: Silane, chlorotrimethyl-; (75-77-4) **85**, 209; **88**, 121; **89**, 267

*2-Trimethylsilylethanesulfonyl chloride: Ethanesulfonyl chloride, 2-(trimethylsilyl)-;
(106018-85-3)* **89**, 34

1-Trimethylsilyloxybicyclo[3.2.0]heptan-6-one; (125302-44-5) **85**, 138

1-Trimethylsilyloxycyclopentene; (19980-43-9) **85**, 138

1-Trimethylsilyloxy-7,7-dichlorobicyclo[3.2.0]heptan-6-one; (66324-01-4) **85**, 138

2-(Trimethylsilyl)phenyl trifluoromethanesulfonate; (88284-48-4) **87**, 95

1-Trimethylsilyl-1-propyne; (6224-91-5) **88**, 377

O-Trimethylsilylquinidine; **88**, 121

2-(Trimethylsilyl)thiazole (2-TST); (7925-30-8) **89**, 323

Trimethylsilyl Trifluoromethanesulfonate; (27607-77-8) **89**, 115

Triphenylacetic acid: Benzeneacetic acid, α,α-diphenyl-; (595-91-5) **86**, 58

Triphenylarsine; (603-32-7) **89**, 183

Triphenyl borate: Boric acid (H₃BO₃), triphenyl ester; (1095-03-0) **88**, 224

Triphenylphosphine, 99%; (603-35-0) **87**, 161; **88**, 79, 121, 162, 309, 377; **89**, 230

Triphenylphospine oxide; (791-28-6) **85**, 248; **86**, 274

Triphosgene: Methanol, 1,1,1-trichloro-, 1,1'-carbonate; (32315-10-9) **86**, 315

Tripotassium phosphate, monohydrate: Phosphoric acid, tripotassium salt, monohydrate;
(27176-10-9) **87, 299**

Tripropargylamine: 2-Propyn-1-amine, *N*,*N*-di-2-propynyl: Tri-2-propynylamine; (6921-29-5) **88**, 238

Tri-*iso*-propyl borate; (5419-55-6) **87, 26**

Tri-*iso*-propyl phosphite: Phosphorous acid, tris(1-methylethyl) ester; (116-17-6) **87**, 231

Tri-*iso*-propylsilyl trifluoromethanesulfonate; (80522-42-5) **87**, 253

Tris((1-benzyl-1H-1,2,3-triazolyl)methyl)amine; (510758-28-8) **88**, 238

Tris(dibenzylideneacetone) dipalladium(0): Palladium, tris[μ-[(1,2-η:4,5-η)-(1*E*,4*E*)-1,5-diphenyl-1,4-pentadien-3-one]]di-; (51364-51-3) **86**, 194

Tris(dibenzylideneacetone)-dipalladium(0)-chloroform adduct; (52522-40-4) **86**, 47; **88**, 162

Tris(dibenzylideneacetone)dipalladium(0): Pd₂(dba)₃; (51364-51-3) **87**, 104, 263

1,1,1-Tris(hydroxymethyl)ethane: 1,3-Propanediol, 2-(hydroxymethyl)-2-methyl-; (77-85-0) **88**, 79

Tris(hydroxymethyl)phosphine: Methanol, phosphinidynetris-; (2767-80-8) **89**, 170

1,3,5-[Tris-piperazine]-triazine: 1,3,5-Triazine, 2,4,6-tri-1-piperazinyl-; (19142-26-8) **86**, 141, 151

2,4,6-Trivinylcyclotriboroxane-pyridine complex: Boron, ethenyl[(ethenylboronic acid-κ*O*) bimol. monoanhydridato(2-)] (pyridine)-; (95010-17-6) **89**, 202

L-Valine methyl ester hydrochloride; (6306-52-1) **87**, 26; **88**, 14

(*S*)-Valinol: 1-Butanol, 2-amino-3-methyl-, (2*S*)-; (2026-48-4) **86**, 70

Vanadyl acetylacetonate; (3153-26-2) **86**, 121

Vanadium; **86**, 121

(*S*)-VANOL: [2,2'-Binaphthalene]-1,1'-diol, 3,3'-diphenyl-, (2*S*)-; (147702-14-5) **88**, 224

4-Vinylbenzophenone: Methanone, (4-ethenylphenyl)phenyl-; (3139-85-3) **86**, 274

4-Vinyl-1-cyclohexene: Cyclohexene, 4-ethenyl-; (100-40-3) **88**, 207

Vinylmagnesium bromide; (1826-67-1) **89**, 230

N-(2-Vinylphenyl)acetamide: Acetamide, N-(2-ethenylphenyl)-; (29124-68-3) **89**, 202

3-Vinylquinoline: Quinoline, 3-ethenyl-; (67752-31-2) **86**, 274

Vinyltrimethylsilane: Silane, ethenyltrimethyl-; (754-05-2) **89**, 34

Wilkinson's catalyst: Rhodium, chlorotris(triphenylphosphine)-; (14694-95-2) **89**, 66

X-Phos: 2-Di-cyclohexylphosphino-2',4',6'-triisopropylbiphenyl; (564483-18-7) **87**, 104

Zinc; (7440-66-6) **87**, 330; **89**, 73, 283

Zinc-catalyzed reactions; **87**, 310

Zinc (II) chloride; (7646-85-7) **85**, 27, 53; **89**, 76

Zinc triflate; (54010-75-2) **86**, 113

Zinc trifluoromethanesulfonate: Methanesulfonic acid, 1,1,1-trifluoro-, zinc salt (2:1); (54010-75-2) **87**, 310

Zirconocene chloride hydride: Zirconium chlorobis(η⁵-2,4-cyclopentadien-1-yl)hydro-; (37342-97-5) **88**, 427